Business Mathematics

Tenth Edition

Charles D. Miller

Stanley A. Salzman
American River College

Gary Clendenen
Siena College

PEARSON
Addison
Wesley

Boston San Francisco New York London Sydney
Tokyo Singapore Madrid Mexico City Paris
Cape Town Hong Kong Montreal

Publisher: Greg Tobin
Editor in Chief: Maureen O'Connor
Project Editors: Suzanne Alley and Lauren Morse
Editorial Assistant: Marcia Emerson
Marketing Manager: Jay Jenkins
Managing Editor: Ron Hampton
Senior Production Supervisor: Jeffrey Holcomb
Cover Design: Dennis Schaefer
Senior Manufacturing Buyer: Evelyn Beaton
Senior Technical Art Specialist: Joe Vetere
Associate Media Producer: Sharon Smith
Photo Research: Beth Anderson
Project Management: Jared Sterzer, WestWords, Inc.
Composition and Proofreading: WestWords, Inc.
Copyediting: Brian Baker, Write With, Inc.
Art Illustration: Jim McLaughlin
Photography: Kathie Kelleher
Interior Design: Sandy Silva
Cover Images: © A. Schein/Masterfile, © David Zimmerman/Masterfile, © Sandra Baker/GettyImages,
© Stockfood/Rezny, © Eliot Cohen/Index Stock Imagery, © PhotoDisc/Comstock

For permission to use copyrighted material, grateful acknowledgment is made to the copyright holders
listed in the Acknowledgments on the second-to-last page of this book. These materials are used by
permission. All existing copyrights relating to these materials remain in effect and are the exclusive property
of the corporations named.

Many of the designations used by manufacturers and sellers to distinguish their products are claimed as
trademarks. Where those designations appear in this book, and Addison-Wesley was aware of a trademark
claim, the designations have been printed in initial caps or all caps.

Library of Congress Cataloging-in-Publication Data

Miller, Charles David
    Business mathematics/Charles D. Miller, Stanley A. Salzman, Gary Clendenen.—10th ed.
        p. cm.
    Includes index.
     ISBN 0-321-27782-1 (SE), ISBN 0-321-26960-8 (AIE)
     1. Business mathematics. 2. Business mathematics—Programmed instruction. I. Salzman, Stanley A. II.
    Clendenen, Gary. III. Title.

  HF5691.M465   2006
  650'.01'513—dc22

                                                      20040436185

1 2 3 4 5 6 7 8 9 10 VH 07 06 05

Contents

Preface

The 10th edition of *Business Mathematics* has been extensively revised to maximize student involvement in each chapter of the text. More than ever, real-life examples from today's business world have been incorporated; new examples from actual companies and the people who run them are woven throughout the book to serve as applications of the concepts presented. Many new photographs, news clippings, and graphs have been added to increase the relevance of chapter content to the world students know. The globalization of our society is emphasized through examples and exercises that highlight foreign countries and international topics.

The new edition reflects the extensive business and teaching experience of the authors, as well as the suggestions of many reviewers nationwide. Providing solid, practical, and up-to-date coverage of business mathematics topics, the text begins with a brief review of basic mathematics and goes on to introduce key business topics, such as bank services, payroll, business discounts and markups, simple and compound interest, stocks and bonds, consumer loans, taxes and insurance, depreciation, financial statements, and business statistics. The text is accompanied by a greatly enhanced supplements package that provides many avenues—both print and other media—for students to further practice and explore the concepts discussed in the chapters. (Please see pages x–xii of this preface for full descriptions of the student and instructor supplements available.)

■ New Content Highlights

The material in Chapter 4, "Bank Services," has been updated in keeping with the latest banking trends and practices. The statistics showing the use of banking services, today's banking charges, and currently used credit card deposit slips reflect the latest available information. The latest technologies in banking, electronic funds transfer (EFT), automated teller machines (ATM), and their usage are also presented.

In Chapter 5, "Payroll," all wages and salaries have been updated along with FICA, Medicare, and tax-withholding rates. A look at American workers and their jobs, the average annual earnings for various occupations, and the benefits offered to working moms by the top U.S. companies are shown. Average hourly pay for workers around the world, social security rates for employees and employers in major countries, today's new jobs, and the fastest growing career fields are also included.

Chapter 6, "Mathematics of Buying," introduces e-commerce and the resulting changes in business operations. A new tool to help students remember the number of days in each month of the year has been added.

Chapter 8, "Simple Interest," has been updated to reflect current interest rates. A display of historical prime interest rates shows the student how interest rates fluctuate over time. The simple interest and simple discount notes commonly used in business are shown using many examples.

Chapter 9, "Compound Interest and Annuities," shows the benefits of compounding interest over periods of time. Inflation is discussed, and its effect on spendable income is illustrated. The concepts of present value and future value are explained.

Chapter 10, "Annuities, Stocks, and Bonds," shows how regular periodic payments in an annuity such as an Individual Retirement Account benefits people or companies. All of the stock and bond data has been updated.

Greater emphasis on personal debt management in explanations and examples has been added to Chapter 11, "Business and Consumer Loans." All interest rates have been adjusted to current market rates. Many students will be able to relate easily to the section on mortgages and real-estate loans, since they may be planning to buy a home or may have bought one recently.

In Chapter 12, "Taxes and Insurance," the latest available tax forms and tables have been included in Section 12.2, "Personal Income Tax." Insurance rates for motor vehicles and life insurance have been updated to more accurately reflect today's insurance costs.

The most recent federal laws and guidelines are used in Chapter 13, "Depreciation." This coverage helps students who will be studying accounting in the future.

The company highlighted in Chapter 14, "Financial Statements and Ratios," has been changed to General Motors, which is one of the largest companies in the world. Students can learn about financial statements using this company that they know.

Many of the problems in Chapter 15, "Business Statistics," have been changed, and graphs from the business world that interest students have been added.

Appendix A, "Equations and Formulas," includes many of the basic algebra concepts needed to work business math problems. Appendix B, "The Metric System," contains the conversions needed to work with the metric system. Appendix C, "Basic Calculators," presents detailed coverage of basic calculators for professors that allow students to use calculators. Appendix D, "Financial Calculators," reviews the basic functions of financial calculators using present value and future value.

■ New Features

Chapter Openers   Many chapters introduce a new or popular business—such as Starbucks Coffee, The Home Depot, Krispy Kreme, Bank of America, General Motors, GMAC, Century 21, and Mattel—to capture student's interest. These chapter openers, identified as "Case in Point," present the owner, manager, or employee of the business, and that person is discussed throughout the chapter in the context of the company he or she represents. Exercises marked with a Case in Point icon continue to support the application of the chapter topic.

Enhanced Treatment of Real-World Applications   The 10th edition places greater emphasis on real-world applications. Application problems have been updated throughout to be as relevant as possible to today's students, and they reference well-known companies such as Starbucks, The Home Depot, Krispy Kreme, Jackson and Perkins Company, REI, FTD, Ford Motor Company, Bank of America, General Motors, GMAC, Century 21, and Mattel.

New Art Program   The art program of the 10th edition includes not only new color photographs, but also graphs and charts that utilize current data from a variety of recognized sources. Rendered to draw student attention while emphasizing the data itself, the graphs and charts, entitled Numbers in the News, help students see that the mathematics of business is inherent to the world around them.

Cumulative Reviews   The number of Cumulative Review sets has been increased, and they now follow each of these chapters: 3, 7, 9, and 11.

Here and Now   Many of the newspaper clippings, magazine articles, and other media items in the text are flagged with a Here and Now icon in the margin. By drawing students' eyes to these real media sources, the Here and Now icon helps emphasize the practical, everyday relevance of business mathematics.

Metric System   The metric system has been added to Appendix B.

■ Additional Features

Numerous Exercises   Mastering business mathematics requires working through many exercises; so we have included more than 2700 in the 10th edition. They range from simple drill problems to application exercises that require several steps to solve. All problems have been independently checked to ensure accuracy. A comprehensive Index of Applications appears at the beginning of this text.

Graded Application Exercises   All application exercises are arranged in pairs and increase progressively in difficulty. This arrangement prepares students to work the more difficult exercises as they proceed through the exercise set. Each even-numbered application exercise is the same type of problem as the previous odd-numbered exercise. This allows students to solve an odd-numbered exercise, check the answer in the answer section, and then solve the following even-numbered exercise.

Supplementary Exercises   Additional sets of supplementary exercises occur throughout the book to help students review and synthesize difficult concepts. For example, two sets in Chapter 3 require students to distinguish among the different elements of a percent problem and decide upon the correct method of solution. The set in Chapter 7 includes exercises in both markup on cost and markup on selling price. This will help the student distinguish between these two types of markup. A set in Chapter 8 gives practice in distinguishing simple interest from simple discount; a set in Chapter 10 helps students understand annuities and sinking funds; and another set in Chapter 13 combines methods of calculating depreciation.

Case in Point   Found throughout each chapter and flagged by the ◆ icon, the Case in Point application is tied to the business introduced in the chapter opener. This approach demonstrates to the student how specific topics are used by those operating the business.

Newspaper and Magazine Articles   A wide selection of current newspaper and magazine articles from various news media sources appears within each chapter to emphasize the "Here and Now" of these topics. These current, eye-catching items are tied to examples within the chapter sections and are a constant reminder to students of the relevance of the chapter content to current business trends and topics.

Calculator Solutions   Calculator solutions, identified with the calculator symbol 🖩, appear after selected examples. These solutions show students the keystrokes needed to obtain the example solution.

Cumulative Reviews   Four Cumulative Reviews, found at the end of Chapters 3, 7, 9, and 11, help students review groups of related chapter topics and reinforce understanding.

Investigative Questions   The Investigate feature now appears at the end of every chapter's Summary Exercise. The Investigate questions require higher-level thinking skills and encourage students to apply the chapter material in a practical way or go outside the classroom to seek additional knowledge. Many of these questions are ideal for collaborative assignments.

Basic Calculator Appendix   This edition includes an appendix (Appendix C) containing extensive coverage of basic calculators. A number of exercises are provided to help students develop their calculator skills.

Financial Calculators   Financial calculators are presented in Appendix D, along with exercises that may be solved by students using the financial calculator of their choice.

Cautionary Remarks   Common student difficulties and misunderstandings appear as Quick Tips. This feature is given a special graphic treatment to help students locate them.

Quick Start Solutions to Exercises   Selected exercises in the exercise sets—usually the first of each type of exercise—are denoted by a Quick Start head and include answers with solutions to help students get started. This on-the-spot reinforcement gives students both the confidence to continue working practice problems and the knowledge of which topics may require additional review.

Quick Review with Chapter Terms   The end-of-chapter Quick Review feature begins with a list of key terms from the chapter and the pages on which they first appear. The Quick Review uses a two-column format (Concepts and Examples) to help the student review all the main points presented in the chapter.

'Net Assets   In each chapter, a one-page feature emphasizes the growing importance of the World Wide Web in business by showing an example of a company's Internet home page and providing questions that relate that company's online activities to the chapter concepts. Some of the corporations highlighted operate solely over the Internet. This feature, suitable even for students without access to the Web, is ideal for self-contained assignments that will illustrate the relevance of business math to actual corporate situations.

Writing Exercises   Designed to help students better understand and relate the concepts within a section, these exercises require a written answer of a few sentences. They are flagged in the Annotated Instructor's Edition by the ✎ icon and often include references to a specific learning objective to help students formulate an answer.

Summary Exercises   Every chapter ends with a Summary Exercise that has been designed to help students apply what they have learned in the chapter. These problems require students to synthesize most or all of the topics they have covered in the chapter in order to solve one cumulative exercise. Ending with a feature labeled Investigate, these exercises offer the student an opportunity to further develop problem-solving skills beyond the classroom. The Investigate questions may be worked out as a group or individual activity, depending on the instructor's preference.

Example Titles   Each example has a title to help students understand the purpose of the example. The titles can also help students work the exercises and study for quizzes and exams.

Flexibility   After basic prerequisites have been met, the chapters in this text can be taught in any order to give instructors maximum freedom in designing courses. Chapter prerequisites are as follows:

Chapter	Prerequisite	Chapter	Prerequisite
1	None	9	Simple interest
2	None	10	Simple interest
3	Arithmetic	11	Simple interest
4	Percent	12	Percent
5	Percent	13	Percent
6	Percent	14	Percent
7	Percent	15	Percent
8	Percent		

Pretest   A business mathematics pretest is included in the introduction of the book. This pretest can help students and instructors identify individual and class strengths and weaknesses.

Chapter Tests   Each chapter ends with a chapter test that reviews all of the topics in the chapter and helps evaluate student mastery.

Equations and Formulas   A review of equations, business applications of equations, and ratios and proportions is included in Appendix A. Instructors may find it appropriate to introduce this material to lay the groundwork for an alternative approach to the mathematics of buying and selling (Chapters 6 and 7), interest (Chapters 8 and 9), annuities (Chapter 10), and consumer loans (Chapter 11).

Glossary   A glossary of key words, located at the back of the book, provides a quick reference for the main ideas of the course.

Summary of Formulas   The inside back cover of *Business Mathematics* provides a handy summary of commonly used information and business formulas from the book.

■ Supplements

INSTRUCTOR SUPPLEMENTS

ANNOTATED INSTRUCTOR'S EDITION

ISBN 0–321–26960–8

The Annotated Instructor's Edition provides immediate access directly on the page to the worked-out answers to all exercises. In addition, an answer section at the back of both the Student and Instructor's Editions gives answers to odd-numbered section exercises and answers to all chapter test exercises. Writing exercises are marked with the icon 🖉 exclusively in the AIE so that instructors may use discretion in assigning these problems.

PRINTED TEST BANK/INSTRUCTOR'S RESOURCE GUIDE

ISBN 0–321–27962–X

This extensive supplement contains teaching suggestions; two pretests—one in basic mathematics and one in business mathematics; six different test forms for each chapter (four short answer and two multiple choice); two final examinations; numerous application exercises (test items) for each chapter; answers to all test materials; suggested answers to the writing questions in the text; and a selection of tables from the text.

TESTGEN WITH QUIZMASTER

ISBN 0–321–27960–3

TestGen enables instructors to build, edit, print, and administer tests using a computerized bank of questions developed to cover all the objectives in the text. TestGen is algorithmically based, allowing instructors to create multiple, but equivalent, versions of the same question or test with the click of a button.

Instructors can also modify test bank questions or add new questions by using the built-in question editor, which allows users to create graphs, import graphics, and insert math notation, variable numbers, or text. Tests can be printed or administered online via the Internet or another network. TestGen comes packaged with Quizmaster, which allows students to take tests on a local area network. The software is available on a dual-platform Windows–Macintosh CD-ROM.

MYMATHLAB

MyMathLab is a series of text-specific, easily customized on-line courses for Addison-Wesley textbooks in mathematics and statistics. MyMathLab is powered by CourseCompass™—Pearson Education's online teaching and learning environment—and by MathXL®—our online homework, tutorial, and assessment system. MyMathLab gives you tools you need to deliver all or a portion of your course online, whether your students are in a lab setting or working from home. MyMathLab provides a rich flexible set of course materials, featuring free-response exercises that are algorithmically generated for unlimited practice and mastery. Students can also use online tools such as video lectures, animations, and a multimedia textbook, to independently improve their understanding and performance. Instructors can use MyMathLab's homework and test managers to select and assign online exercises correlated directly to the textbook, and they can import TestGen tests into MyMathLab for added flexibility. MyMathLab's online gradebook—designed specifically for mathematics and statistics—automatically tracks students' homework and test results and gives the instructor control over how to calculate final grades.

  MyMathLab is available to qualified adopters. For more information, visit our website at *www.mymathlab.com* or contact your Addison-Wesley sales representative for a product demonstration.

MATHXL® TUTORIALS ON CD

ISBN 0–321–27961-1

The InterAct Math tutorial software has been developed by professional software engineers working closely with a team of experienced math instructors. The software provides exercises that correspond to each section-level objective in the text; these exercises require the same computational and problem-solving skills as the section exercises in the book. Each InterAct Math exercise is accompanied by an example and an interactive guided solution designed to involve students in the solution process and help them identify precisely where they are having trouble. For each section of the text, the software tracks student activity and scores, which can be printed out in summary form.

MATHXL®

MathXL® is a powerful online homework, tutorial, and assessment system that accompanies your Addison-Wesley textbook in mathematics or statistics. With MathXL, instructors can create, edit, and assign online homework and tests using algorithmically generated exercises correlated at the objective level to your textbook. All student work is tracked in MathXL's on-line gradebook. Students can take chapter tests in MathXL and receive personalized study plans based on their test results. The study plan diagnoses weaknesses and links students directly to tutorial exercises for the objectives they need to study and retest. Students can also access supplemental animations and video clips directly from selected exercises. MathXL is available to qualified adopters. For more information, visit our website at *www.mathxl.com*, or contact your Addison-Wesley sales representative for a product demonstration.

STUDENT SUPPLEMENTS

MYMATHLAB

MyMathLab is a complete online course designed to help you succeed in learning and understanding mathematics. MyMathLab contains an online version of your textbook with links to multimedia resources—such as video clips, practice exercises, and animations—that are correlated to the examples and exercises in the text. MyMathLab also provides you with online homework and tests and generates a personalized study plan based on your test results. Your study plan links directly to unlimited practice exercises for the areas you need to study and retest, so you can practice until you have mastered the skills and concepts in your textbook. All of the online homework, tests, and practice work you do is tracked in your MyMathLab gradebook. For more information, visit our website at *www.mymathlab.com*.

MATHXL®

MathXL® is a powerful online homework, tutorial, and assessment system that accompanies your Addison-Wesley textbook in mathematics or statistics. With MathXL, instructors can create, edit, and assign online homework and tests using algorithmically generated exercises correlated at the objective level to your textbook. All student work is tracked in MathXL's online gradebook. Students can take chapter tests in MathXL and receive personalized study plans based on their test results. They study plan diagnoses weaknesses and links students directly to tutorial exercises for the objectives they need to study and retest. For more information, visit our website at *www.mathxl.com.*

MathXL® Tutorials on CD

This interactive tutorial CD-ROM provides algorithmically generated practice exercises that are correlated at the objective level to the exercises in the textbook. Every practice exercise is accompanied by an example and a guided solution designed to involve students in the solution process. Selected exercises may also include a video clip to help students visualize concepts. The software tracks student activity and scores can generate printed summaries of student's progress.

Videotapes

The videotapes feature an engaging team of mathematics instructors who present comprehensive coverage of each section of the text. The lecturer's presentations include examples and exercises from the text and support an approach that emphasizes visualization and problem solving. The videos include a stop-the-tape feature that encourages students to pause the videotape, work through an example, and resume play to watch the video instructor work through the solution.

Addison-Wesley Math Tutor Center

The Addison-Wesley Math Tutor Center is staffed by qualified mathematics and statistics instructors who provide students with tutoring on examples and odd-numbered exercises from the textbook. Tutoring is available via toll-free telephone, toll-free fax, e-mail, and the Internet. Interactive, Web-based technology allows tutors and students to view and work through problems together in real time over the Internet. For more information, please visit our website at *www.aw-bc.com/tutorcenter* or call us at 1-888-777-0463.

Videotape Series and Telecourse

The Southern California Community College Consortium has produced a videotape series based on *Business Mathematics* entitled *By the Numbers*, which has been aired over the Public Broadcasting System (PBS). Your school can offer a telecourse using the videotapes along with *Business Mathematics* and the teleguide for the *By the Numbers* series. The videotapes can also be purchased for use in a traditional lecture course. Contact Intelecom at 626-796-7300 or your Addison-Wesley sales consultant for further information.

■ Acknowledgments

We would like to thank the many users of the ninth edition for their insightful observations and suggestions for improving this book. We also wish to express our appreciation and thanks to the following reviewers for their contributions.

Ellen Benowitz, *Mercer County Community College*
Yvonne Block, *College of Lake County*
Donald Boyer, *Jefferson College*
Janet Ciccarelli, *Heriker County Community College*
Milton Clark, *Florence Darlington Technical College*
Bobbie Corbett, *Northern Virginia Community College*
Kathleen Crall, *Des Moines Area Community College*
Pat Cunningham, *Dawson Community College*
Dorothy Dean, *Illinois Central College*
Jecqueline Dlatt, *Collge of DuPage*
William Dorrity, *Eastern Maine Community College*
Acie Earl, *Black Hawk College*
Carolyn Fitzmorris, *Hutchinson Community College*

David Gaboardi, *Central Florida Community College*
Stephen Griffin, *Tarrant County College*
Frank Goulard, *Portland Community College*
Andrew Haaland, *Cortland Community College*
Brian Hickey, *East Central College*
Susan Hutchinson, *Florence Darlington Technical College*
Joseph Jean-Charles, *Bunker Hill Community College*
Gwen Loftis, *Rose State College*
Tristan Londre, *Blue River Community College*
Jacqualine Myers, *Alabama State University*
Dyan Pease, *Sacramento City College*
Anthony Ponder, *Sinclair Community College*
Robert Reichl, *Morton Community College*
Donald Ryktarsyk, *Schoolcraft College*
Nelda Shelton, *Tarrant County College South*
Lynn Shuster, *Central Pennsylvania College*
Natalie Smith, *Okaloosa-Walton Community College*
De Underwood, *Central Florida Community College*
Jimmie Van Alphen, *Ozarks Technical Community College*
Michael Wade, *Moraine Valley Community College*

Our appreciation goes to Steve Ouellette, John Samons, Ellen Sawyer, and Gary Williams, who checked all of the exercises and examples in the book for accuracy. We would also like to express our gratitude to our colleagues at American River College and the University of Texas at Tyler who have helped us immeasurably with their support and encouragement: Vivek Pandey, Robert Gonzalez, Meg Pollard, James Bralley, Henry Hernandez, and Rob Diamond.

The following individuals at Addison-Wesley had a large impact on this 10th edition of *Business Mathematics,* and we are grateful for their many efforts: Greg Tobin, Maureen O'Connor, Ron Hampton, Suzanne Alley, Lauren Morse, Marcia Emerson, Jeffrey Holcomb, Jay Jenkins, Sheila Spinney, Beth Anderson, Joe Vetere, Evelyn Beaton, Sharon Smith, and Dennis Schaefer. Thanks are due as well to WestWords, Inc., and Jared Sterzer in particular, for adeptly handling the production of this 10th edition.

Charles D. Miller
Stanley A. Salzman
Gary Clendenen

Introduction for Students

■ Success In Business Mathematics

With a growing need for record keeping, establishing budgets, and understanding finance, taxation, and investment opportunities, mathematics has become a greater part of our daily lives. This text applies mathematics to daily business experiences. Your success in future business courses and pursuits will be enhanced by the knowledge and skills you will gain in this course.

Studying business mathematics is different from studying subjects like English or history. The key to success is regular practice. This should not be surprising. After all, can you learn to ski or play a musical instrument without a lot of regular practice? The same is true for learning mathematics. Working out problems nearly every day is the key to becoming successful. Here are some suggestions to help you succeed in business mathematics:

1. Attend class regularly. Pay attention to what your instructor says and does in class, and take careful notes. In particular, note the problems the instructor works on the board, and copy the complete solutions. Keep these notes separate from your homework to avoid confusion when you review them later.

2. Don't hesitate to ask questions in class. It is not a sign of weakness, but of strength. There are always other students with the same question who are too shy to ask.

3. Read your text carefully. Many students read only enough to get by, usually only the examples. Reading the complete section will help you solve the homework problems. Most exercises are keyed to specific examples or objectives that will explain the procedure for working them.

4. Before you start on your homework assignment, rework the problems the teacher worked in class. This will reinforce what you have learned. Many students say, "I understand it perfectly when you do it, but I get stuck when I try to work the problem myself."

5. Do your homework assignment only after reading the text and reviewing your notes from class. Check your work against the answers in the back of the book. If you get a problem wrong and are unable to understand why, mark that problem and ask your instructor about it. Then practice working additional problems of the same type to reinforce what you have learned.

6. Work as neatly as you can using a pencil, and organize your work carefully. Write your symbols clearly, and make sure the problems are clearly separated from each other. Working neatly will help you to think clearly and also make it easier to review the homework before a test.

7. After you complete a homework assignment, look over the text again. Try to identify the main ideas that are in the lesson. Often they are clearly highlighted or boxed in the text.

8. Use the chapter test at the end of each chapter as a practice test. Work through the problems under test conditions, without referring to the text or the answers until you are finished. You may want to time yourself to see how long it takes you. When you finish, check your answers against those in the back of the book, and study the problems you missed.

9. Keep all quizzes and tests that are returned to you, and use them when you study for future tests and the final exam. These quizzes and tests indicate what concepts your instructor considers to be most important. Be sure to correct any problems on these tests that you missed so you will have the corrected work to study.

10. Don't worry if you do not understand a new topic right away. As you read more about it and work through the problems, you will gain understanding. Each time you review a topic, you will understand it a little better. Few people understand each topic completely right from the start.

Business Mathematics Pretest

This pretest will help you determine your areas of strength and weakness in the business mathematics presented in this book.

1. Round 4.38 to the nearest tenth.

2. Round $.064 to the nearest cent.

3. Round $399.49 to the nearest dollar.

4. Multiply: $\begin{array}{r} 7801 \\ \times\ 1758 \\ \hline \end{array}$

5. Divide: $35\overline{)11{,}032}$

6. Change $7\frac{3}{8}$ to an improper fraction.

7. Change $\frac{39}{28}$ to a mixed number.

8. Write $\frac{18}{21}$ in lowest terms.

9. Add: $\begin{array}{r} \frac{3}{4} \\ \frac{1}{2} \\ +\ \frac{7}{8} \\ \hline \end{array}$

10. Add: $\begin{array}{r} 2\frac{2}{3} \\ 7\frac{1}{4} \\ +\ 10\frac{1}{2} \\ \hline \end{array}$

11. Subtract: $\frac{3}{4} - \frac{14}{24}$

12. Subtract: $\begin{array}{r} 83\frac{3}{4} \\ -\ 21\frac{2}{5} \\ \hline \end{array}$

13. Multiply: $\frac{3}{8} \times \frac{3}{5}$

14. Divide: $15\frac{1}{4} \div 5\frac{1}{8}$

15. Express .875 as a common fraction.

16. Express $\frac{4}{5}$ as a decimal.

1. _____

2. _____

3. _____

4. _____

5. _____

6. _____

7. _____

8. _____

9. _____

10. _____

11. _____

12. _____

13. _____

14. _____

15. _____

16. _____

17. Subtract: 598.316
            − 79.839

18. Multiply: 30.67
          × 5.39

17. _____

18. _____

19. Divide: $1.2\overline{)309.6}$

20. Express $\frac{7}{8}$ as a percent.

19. _____

20. _____

21. Intelnet spent 5.2% of its sales on advertising. If sales amounted to $864,250, what amount was spent on advertising?

21. _____

22. What annual rate of return is needed to receive $372 in one year on an investment of $18,600?

22. _____

23. Auto Electric offers an oxygen sensor at a list price of $289 less trade discounts of 20/30. What is the net cost?

23. _____

24. A department head at Old Navy is paid $13.80 per hour with time and a half for all hours over 40 in a week. Find the employee's gross pay if she worked 43 hours in one week.

24. _____

25. How long will it take an investment of $14,500 to earn $108.75 in interest at 3% per year?

25. _____

26. An invoice from Collier Windows amounting to $20,250 is dated October 6 and offers terms of 3/10, n/30. If the invoice is paid on October 14, what amount is due?

26. _____

27. Find the percent of markup based on selling price if some home exercise equipment costing $1584 is sold for $1980.

27. _____

28. Find the single discount equivalent to a series discount of 10/20.

28. _____

29. Using the straight-line method of depreciation, find the annual depreciation on a lawn tractor that has a cost of $9375, an estimated life of six years, and a scrap value of $375.

29. _____

30. Whiting's Oak Furniture sells a big screen home-entertainment center for $731.49 after deducting 26% from the original price. Find the original price.

30. _____

Index of Applications

Office equipment, 22
Surplus-equipment auction, 127
Surveillance cameras, 471
Video equipment, 585

Construction

Airport improvement, 416
Baseboard trim, 36
Cabinet installation, 59
Ceiling fans, 279
Commercial building, 417
Commercial carpeting, 72, 582
Concrete footings, 72
Construction power tools, 576
Conveyor system, 575
Delivering concrete, 64
Double-pane windows, 280
Excavating machinery, 576
Financing construction, 345
Finish carpentry, 71
Forklift depreciation, 581
Home construction, 326
Industrial forklift, 585
Industrial tooling, 582
Kitchen island, 231
Landscape equipment, 582–583
Measuring brass trim, 64
New auditorium, 416
New roof, 334
New showroom, 416
Parking lot fencing, 64
Perimeter of fencing, 60
Remodeling, 470
Rock crusher, 345
Security fencing, 64
Steel fabrication, 72
Theater renovation, 22
Warehouse construction, 583
Warehouse shelving, 570
Weather stripping, 72
Window installation, 63
Woodworking machinery, 583
Yacht construction, 629

Domestic

Bed in a bag, 279
Ceramic dinnerware, 250
Custom made jewelry, 268
Making jewelry, 72
Producing Crafts, 71
Sewing center, 345
Tailored clothing, 64

Education

College enrollment, 105
College expenses, 127, 410
Community college enrollment, 128
Financing college expenses, 386
Paying for college, 410
Preschool manager, 21
Private school equipment, 576
Student time management, 60
University fees, 128

Employment/Employee Benefits

Aiding disabled employees, 108
Calculating gross earnings, 35
Commission with returns, 192
Communications industry layoffs, 106
Desk clerk, 184
Earnings calculation, 71
Employee health plans, 115
Employee net pay, 212
Employee population base, 105
Female lawyers, 98
Guaranteed hourly work, 191
Heating company sales, 214
Insurance office manager, 184
Layoff alternative, 116
Loan to employee, 318
Managerial earnings, 36
Office assistant, 184
Part-time work, 64
Payroll deduction, 199
Piecework with overtime, 192
Retail employment, 183
Retirement account, 107–108, 325
Retirement contribution, 402
Retirement funds, 436
Retirement income, 379
Retirement planning, 402
Retiring a manager, 410
River-rafting manager, 214
Salary plus commission, 192
Saving for retirement, 325
Self-employment, 385
Severance pay, 410
Staff-meeting cost, 30
Store manager, 184
Variable-commission payment, 192
Video rental income, 115
Workplace requirements, 98
Workforce size, 128

Entertainment/Sports

Alligator hunting, 416
Athletic socks, 295
Bowling equipment, 280
Competitive cyclist training, 21
Cruise ship travel, 21
DVD rentals, 126
English soccer equipment, 251
Exercise bicycle, 267, 286
Fly-fishing, 279
Gambling payback, 106
Gardening, 58
Golf clubs, 268
Home-workout equipment, 278
Movie projectors, 472
Parachute jumps, 14
Piano repair, 296
Ping-pong table, 280
Pricing basketball systems, 268
Rare stamps, 417
Recreation equipment rental, 15
Recreation equipment, 584
River-raft sales, 278
Scuba diving, 415
Ski boat, 463
Ski jackets, 268
Snowboard packages, 279
Sources of news, 97
Sport T-shirts, 295

Sports complex, 417
Sportswear, 280
Studio sound system, 575
Super Bowl advertising, 99
Swimmer training, 59
Swimming pool pump, 278
Telescopes, 279
Theater seating, 584
Treadmill, 286
Video games, 268
Weight-training books, 268
Weight-training equipment, 574
Youth soccer, 22

Environment

Danger of extinction, 115
Disaster relief, 410
Earthquake damage, 335
Recycling, 191
Water pollution, 22

Family

Child-care payments, 402
Divorce, 385
Engagement ring, 471
Family budget, 99, 118
Inheritance, 379
Saving for a home, 402
Wedding preferences, 97
Writing a will, 429

Food Service Industry

Beef/Turkey cost, 30
Campus vending machines, 15
Canadian food products, 250
Canned-meat sales, 98
Catering company, 244
Chicken-noodle soup, 118
Coffee consumption, 15
Commercial freezer, 581
Deep fryer, 592
Doughnut production, 14
Doughnut sales, 21
Egg production, 22
Expensive restaurants, 127
Family restaurant, 126
Fast-food restaurants, 582
Fast-food, 652
Frozen yogurt, 251
Grocery chain, 623
Grocery shopping, 25
Grocery store, 410
Health food, 317
Jell-O sales, 14
Refrigerated display case, 586
Restaurant equipment, 581
Restaurant tables, 586
Sara Lee, 126
Selling bananas, 278
Soft drink bottling, 586
Supermarket shopping, 98
Turkey and cranberry sales, 30

Foreign Affairs

Crystal from Ireland, 251
Cuban labor force, 98
Global workforce reduction, 106

CHAPTER 1

Whole Numbers and Decimals

KARA KAPPAS BEGAN WORKING PART TIME FOR KRISPY KREME when she was a community college student, and after graduation was selected to

CASE *in* POINT

attend the management class offered by the company. Upon completion of this training she was promoted to store manager. She has between 15 and 20 employees at her store, and she must continually recruit and train new people to replace the employees who go on to college or other careers. Each day, Kappas works with whole numbers and decimals as she does scheduling and payroll, computes sales and sales taxes, and orders and pays for inventory.

Often, the most difficult part of solving a problem is knowing how to set the problem up and then deciding on the procedure that will work best. The first two chapters of this book review the mathematical concepts of whole numbers, decimals, and fractions. The rest of the chapters then apply these concepts to actual business situations.

1.1 | Whole Numbers

Objectives

1. Define whole numbers.
2. Round whole numbers.
3. Add whole numbers.
4. Round numbers to estimate an answer.
5. Subtract whole numbers.
6. Multiply whole numbers.
7. Multiply by omitting zeros.
8. Divide whole numbers.

CASE *in* POINT

The employees at Krispy Kreme Doughnuts must be cross-trained so that they can perform several tasks. Food preparation, cash-register operation, and all beverage-preparation positions require basic mathematical skills.

After observing an employee give a customer too much change, Kara Kappas, the manager, began giving a short math test without the use of a calculator to all employee applicants. All employees are expected to know how to read numbers, round whole numbers, add, subtract, multiply, and divide. With this knowledge, Kappas and her employees can work more accurately and better serve the customers.

Objective 1 Define whole numbers. The standard system of numbering, the **decimal system**, uses the ten one-place **digits** 0, 1, 2, 3, 4, 5, 6, 7, 8, and 9. Combinations of these digits represent any number needed. The starting point of this system is the **decimal point (.)**. This section considers only the numbers made up of digits to the left of the decimal point—the **whole numbers**. The following diagram names the first fifteen places held by the digits to the left of the decimal point.

According to the Transportation Department, there are 203,864,307 motor vehicles in the United States. To help in reading this number, a **comma** is used at every third place, starting at the decimal point and moving left. An exception to this is that commas are frequently omitted in four-digit numbers, such as 5892 or 2318.

> **QUICK TIP**   Commas are not shown on most calculators.

The number 203,864,307 is read "two hundred three million, eight hundred sixty-four thousand, three hundred seven." Notice that the word *and* is **not** used with whole numbers. The word *and* represents the decimal point and is discussed in **Section 1.4**.

EXAMPLE 1

Expressing Whole Numbers in Words

Express the following numbers in words.

(a) 7835   **(b)** 111,356,075   **(c)** 17,000,017,000

SOLUTION

(a) seven thousand, eight hundred thirty-five
(b) one hundred eleven million, three hundred fifty-six thousand, seventy-five
(c) seventeen billion, seventeen thousand

Objective 2   **Round whole numbers.** Business applications often require **rounding** numbers. For example, money amounts are commonly rounded to the nearest cent. However, money amounts can also be rounded to the nearest dollar, hundred dollars, thousand dollars, or even hundreds of thousands of dollars and beyond.

Use the following steps for **rounding whole numbers**.

Rounding Whole Numbers

Step 1   Locate the **place** to which the number is to be rounded. Draw a line under that place.

Step 2A   If the first digit to the *right* of the underlined place is **5 or more, increase** the digit in the place to which you are rounding by one.

Step 2B   If the first digit to the right of the underlined place is **4 or less, do not change** the digit in the place to which you are rounding.

Step 3   **Change** all digits to the right of the underlined digit to zeros.

EXAMPLE 2

Rounding Whole Numbers

Round each number.

(a) 368 to the nearest ten
(b) 67,433 to the nearest thousand
(c) 5,499,059 to the nearest million

SOLUTION

(a) Step 1   Locate the place to which the number is being rounded (the tens place). Draw a line under that place.

3<u>6</u>8
     ↑————— place to which number is rounded

Step 2   The *first digit to the right* of that place is 8, which is **5 or more**, so **increase** the tens digit by 1.

Step 3   Change all digits to the right of the tens place to zero: 368 rounded to the nearest ten is 370.

(b) Step 1 Find the place to which the number is being rounded (the thousands place). Draw a line under that place.

6<u>7</u>,433

↑————— place to which number is rounded

Step 2 The *first digit to the right* of the underlined place is 4, which is **4 or less**, so **do not change** the thousands digit.

Step 3 Change all digits to the right of the thousands place to zero. 67,433 rounded to the nearest thousand is 67,000.

(c) Step 1 Find the place to which the number is being rounded (the millions place). Draw a line under that place.

<u>5</u>,499,059

↑————— place to which number is rounded

Step 2 The *first digit to the right* of the underlined place is 4, which is 4 or less, so do not change the millions digit.

Step 3 Change all digits to the right of the millions place to zero. 5,499,059 rounded to the nearest million is 5,000,000.

QUICK TIP   When rounding a number, look at the first digit to the right of the digit being rounded. Do not look beyond this digit.

The four basic **operations** that may be performed on whole numbers—**addition, subtraction, multiplication**, and **division**—are reviewed in this section.

Objective 3   Add whole numbers. In **addition**, the numbers being added are **addends**, and the answer is the **sum**, or **total**, or **amount**.

8	addend
+ 9	addend
17	sum (answer)

Add numbers by arranging them in a column with units above units, tens above tens, hundreds above hundreds, thousands above thousands, and so on. Use the decimal point as a reference for arranging the numbers. If a number does not include a decimal point, the decimal point is assumed to be at the far right.

85   no decimal point indicated; decimal point assumed to be at far right

85.   with decimal point shown

QUICK TIP   **Checking answers** is important in problem solving. The most common method of checking answers in addition is to re-add the numbers from bottom to top.

EXAMPLE 3

Adding with Checking

To find the one-day total amount of purchases at the Krispy Kreme that she manages, Kara Kappas needed to add the following amounts and check the answer.

$$
\begin{array}{r}
\$4028 \\
\$738 \\
63 \\
125 \\
2617 \\
+\ \ \ 485 \\
\hline
\$4028
\end{array}
$$

Problem (add down)

Check (add up)

By adding down and then adding up, you should arrive at the *same* answer.

Adding from the top down results in an answer of $4028. Check for accuracy by adding again—this time from the bottom up. If the answers are the same, the sum is most likely correct. If the answers are different, there is an error in either adding down or adding up, and the problem should be reworked. Both answers agree in this example, so the sum is correct.

Objective **4** **Round numbers to estimate an answer. Front-end rounding** is used to estimate an answer. With front-end rounding, each number is rounded so that all the digits are changed to zero, except the first digit, which is rounded. Only one nonzero digit remains.

EXAMPLE 4

Using Front-End Rounding to Estimate an Answer

With the information in the following graphic, use front-end rounding to estimate the total number of luxury cars sold.

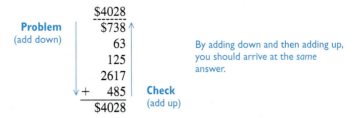

Numbers in the News

In the Driver's Seat

Luxury cars sold well during the last nine-month period. These were the seven top-selling luxury cars in the U.S.

Car	Sold
Lexus	186,009
BMW	178,463
Mercedes	159,447
Cadillac	153,680
Acura	126,227
Lincoln	119,736
Infiniti	89,856

DATA: *USA Today*

SOLUTION

$$
\begin{array}{rcl}
186,009 & \rightarrow & 200,000 \\
178,463 & \rightarrow & 200,000 \\
159,447 & \rightarrow & 200,000 \\
153,680 & \rightarrow & 200,000 \\
126,227 & \rightarrow & 100,000 \\
119,736 & \rightarrow & 100,000 \\
+\ 89,856 & \rightarrow & +\ 90,000 \\
\hline
& & 1,090,000
\end{array}
$$

all digits changed to zero except first digit, which is rounded

estimated answer

The estimated total number of luxury cars sold during the last nine months is 1,090,000.

QUICK TIP When using front-end rounding, only one nonzero digit (first digit) remains. All digits to the right are zeros.

Objective [5] **Subtract whole numbers.** A **subtraction** problem is set up much like an addition problem. The top number is the **minuend**, the number being subtracted is the **subtrahend,** and the answer is the **difference**.

$$
\begin{array}{rl}
23 & \text{minuend} \\
-7 & \text{subtrahend} \\
\hline
16 & \text{difference}
\end{array}
$$

Subtract one number from another by placing the subtrahend directly under the minuend. Be certain that units are above units, tens above tens, and so on. Then begin at the right column and subtract the subtrahend from the minuend.

When a digit in the subtrahend is **larger** than the corresponding digit in the minuend, use **borrowing**, as shown in the next example.

EXAMPLE 5

Subtracting with Borrowing

Subtract 2894 Krispy Kreme coffee cups from 3783 Krispy Kreme coffee cups in inventory. First, write the problem as follows.

$$
\begin{array}{r}
378|3 \\
-\,289|4 \\
\end{array}
$$

In the ones (units) column, subtract 4 from 3 by borrowing a 1 from the tens column in the minuend to get 1 ten + 3, or 13, in the units column with 7 now in the tens column. Then subtract 4 from 13 for a result of 9. Complete the subtraction as follows.

$$
\begin{array}{cccc}
2 & 16 & 17 & 13 \\
\cancel{3} & \cancel{7} & \cancel{8} & \cancel{3} \\
-\;2 & 8 & 9 & 4 \\
\hline
8 & 8 & \mathbf{9} & \text{coffee cups}
\end{array}
$$

In this example, the tens are borrowed from the hundreds column, and the hundreds are borrowed from the thousands column.

Check an answer to a subtraction problem by adding the answer (difference) to the subtrahend. The result should equal the minuend.

EXAMPLE 6

Subtraction with Checking

Subtract 1635 from 5383, and check the answer.

	Problem		**Check**	
Problem	5383	minuend	5383	This result should equal the minuend.
(subtract down) ↓	− 1635	subtrahend	+ 1635	
	3748	difference	3748	Check (add up)

Multiplication is actually a quick method of addition. For example, 3×4 can be found by adding 3 a total of 4 times, since 3×4 means $3 + 3 + 3 + 3 = 12$. However, it is not practical to use the addition method for large numbers. For example, 103×92 would be found by adding 103 a total of 92 times; instead, find this result with multiplication.

> **QUICK TIP** The symbol used for multiplication is "$\times$" or "$\cdot$" between two numbers. Also, two or more numbers within parentheses that are next to each other "$(\ )(\ )$" means to multiply. With a computer, the * means to multiply.

Objective 6 **Multiply whole numbers.** The number being multiplied is the **multiplicand**, the number doing the multiplying is the **multiplier**, and the answer is the **product**.

$$
\begin{array}{r}
3 \quad \text{multiplicand} \\
\times\ 4 \quad \text{multiplier} \\
\hline
12 \quad \text{product}
\end{array}
$$

When the multiplier contains more than one digit, **partial products** must be used, as in the next example, which shows the product of 25 and 34.

EXAMPLE 7

Multiplying Whole Numbers

On a recent trip, a Honda Element averaged 25 miles per gallon while using 34 gallons of gasoline. To find the total number of miles traveled, multiply 25 miles traveled per gallon of gasoline by 34 gallons of gasoline used.

$$
\begin{array}{r}
25 \quad \text{multiplicand} \\
\times\ 34 \quad \text{multiplier} \\
\hline
100 \quad \text{partial product } (4 \times 25) \\
75 \quad \text{partial product } (3 \times 25, \text{one position to the left}) \\
\hline
850 \quad \text{product}
\end{array}
$$

Find the product of 25 and 34 by first multiplying 25 by 4. (The 4 is taken from the units column of the multiplier.) The product of 25 and 4 is 100, which is a partial product. Next multiply 25 by 3 (from the tens column of the multiplier) and get 75 as a partial product. Since the 3 in the multiplier is from the tens column, write the partial product 75 one position to the left so that 5 is under the tens column. Finally, add the partial products and get the product 850.

> **QUICK TIP** If the multiplier had more digits, each partial product would be placed one additional position to the *left*.

Objective 7 **Multiply by omitting zeros.** If the multiplier or multiplicand or both end in zero, save time by first omitting any zeros at the right of the numbers and then replacing omitted zeros at the right of the final answer. This shortcut is useful even with calculators. For example, find the product of 240 and 13 as follows.

$$
\begin{array}{r}
24\cancel{0} \quad \text{Omit the zero in the calculation.} \\
\times\ 13 \\
\hline
72 \\
24 \\
\hline
3120 \quad
\end{array}
$$

Replace the omitted zero at the right of 312 for a final answer (product) of 3120.

EXAMPLE 8

Multiplying Omitting Zeros

In the following multiplication problems, omit zeros in the calculation and then replace omitted zeros to obtain the product.

(a)
$$
\begin{array}{r}
150 \qquad 15 \\
\times\ 70 \qquad \times\ 7 \\
\hline
105 \quad \textbf{+ 2 zeros}
\end{array}
$$

$$10,500 \quad \text{answer}$$

(b)
$$
\begin{array}{r}
300 \qquad 3 \\
\times\ 70 \qquad \times\ 7 \\
\hline
21 \quad \textbf{+ 3 zeros}
\end{array}
$$

$$21,000 \quad \text{answer}$$

> **QUICK TIP**   A shortcut for multiplying by 10, 100, 1000, and so on is to just attach the number of zeros to the number being multiplied. For example,
>
> $$33 \times \mathbf{10} \quad = \quad 33 \text{ and } \mathbf{1} \text{ zero } = 330$$
> $$56 \times \mathbf{100} \quad = \quad 56 \text{ and } \mathbf{2} \text{ zeros } = 5600$$
> $$732 \times \mathbf{1000} = 732 \text{ and } \mathbf{3} \text{ zeros } = 732{,}000$$

Objective 8 **Divide whole numbers.** Various symbols are used to show **division**. For example, $\div$ and $\overline{)}$ both mean "divide." Also, a — with a number above and a number below, as in a fraction, means division. In printing, or when seen on a computer screen, the bar is often written /, so that, for example, 24/6 means to divide 24 by 6.

The **dividend** is the number being divided, the **divisor** is the number doing the dividing, and the **quotient** is the answer.

Write "15 divided by 5 equals 3" in any of the following ways.

$$\underset{\text{dividend}}{15} \div \underset{\text{divisor}}{5} = \underset{\substack{\text{quotient} \\ \text{(answer)}}}{3}$$

$$\underset{\text{divisor}}{5} \overline{)\underset{\text{dividend}}{15}} \quad \overset{3}{} \quad \text{quotient (answer)}$$

$$\underset{\substack{\text{dividend} \\ \text{divisor}}}{\frac{15}{5}} = 3 \quad \text{quotient (answer)}$$

EXAMPLE 9

Dividing Whole Numbers

To divide 1095 baseball cards evenly among 73 collectors, you must divide 1095 by 73. Write the problem as follows.

$$73\overline{)1095}$$

Since 73 is larger than 1 or 10, but smaller than 109, begin by dividing 73 into 109. There is one 73 in 109, so place 1 over the digit 9 in the dividend as shown. Then multiply 1 and 73.

$$\begin{array}{r} 1 \\ 73\overline{)1095} \\ \underline{73} \qquad 1 \times 73 = 73 \\ 36 \end{array}$$

Then subtract 73 from 109 to get 36. The next step is to bring down the 5 from the dividend, placing it next to the remainder 36. This gives the number 365. The divisor, 73, is then divided into 365 with a result of 5, which is placed to the right of the 1 in the quotient. Since 73 divides into 365 exactly 5 times, the final answer (quotient) is exactly 15.

$$\begin{array}{r} 15 \\ 73\overline{)1095} \\ \underline{73} \\ 365 \\ \underline{365} \\ 0 \end{array}$$

Often, part of the quotient must be expressed as a remainder, or as a **fraction part** or **decimal part**, of the quotient. The fraction part of the quotient is discussed in the next chapter. The decimal part of the quotient, most commonly used, is discussed later in this chapter.

EXAMPLE **10**

Dividing with a Remainder in the Answer

Divide 126 by 24. Express the remainder in each of the three forms.

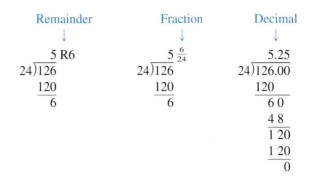

	Remainder	Fraction	Decimal

$$
\begin{array}{r}
5\ R6 \\
24\overline{)126} \\
120 \\
\hline
6
\end{array}
\qquad
\begin{array}{r}
5\frac{6}{24} \\
24\overline{)126} \\
120 \\
\hline
6
\end{array}
\qquad
\begin{array}{r}
5.25 \\
24\overline{)126.00} \\
120 \\
\hline
6\,0 \\
4\,8 \\
\hline
1\,20 \\
1\,20 \\
\hline
0
\end{array}
$$

In the first form, the answer 5 R6 is usually difficult to work with. The second form, $5\frac{6}{24}$, defines the remainder as $\frac{6}{24}$. The third form, 5.25, is also precise in its meaning. For the time being, write remainders as fractions, using the remainder as the top number (numerator) and the divisor as the bottom number (denominator). After studying decimals, express the quotient in the manner most useful in the problem being solved. After studying fractions, write fractional remainders in lowest terms, $\frac{6}{24} = \frac{1}{4}$.

If a divisor contains zeros at the far right, as in 30, 300, or 8000, first drop the zeros in the divisor. Then move the decimal point in the dividend the same number of positions to the left as there were zeros dropped from the divisor. For example, divide 108,000 by 900 by letting

$$900\overline{)108,000} \qquad \text{become} \qquad 9\overline{)1080}$$

Drop 2 zeros.   Move decimal point 2 places left.

Divide 7320 by 30 by letting

$$30\overline{)7320} \qquad \text{become} \qquad 3\overline{)732}$$

> **QUICK TIP** The shortcut of dropping zeros from the divisor and moving the decimal point the same number of places to the left in the dividend saves time and eliminates errors that may result from using larger numbers.

EXAMPLE **11**

Dropping Zeros to Divide

For each of the following, first drop zeros, and then divide.

(a) $40\overline{)11,000}$ (b) $3500\overline{)31,500}$ (c) $200\overline{)18,800}$

SOLUTION

(a)
$$
\begin{array}{r}
275 \\
4\overline{)1100} \\
8 \\
\hline
30 \\
28 \\
\hline
20 \\
20 \\
\hline
0
\end{array}
$$

(b)
$$
\begin{array}{r}
9 \\
35\overline{)315} \\
315 \\
\hline
0
\end{array}
$$

(c)
$$
\begin{array}{r}
94 \\
2\overline{)188} \\
18 \\
\hline
8 \\
8 \\
\hline
0
\end{array}
$$

EXAMPLE **12**

*Checking Division
Problems*

In a division problem, check the answer by multiplying the quotient (answer) and the divisor. Then add any remainder. The result should be the dividend. If the result is not the same as the dividend, an error exists and the problem should be reworked. Check the following division problems.

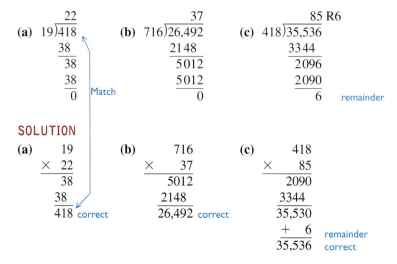

$$
\begin{array}{r}
22 \\
\textbf{(a)} \quad 19\overline{)418} \\
38 \\
\overline{38} \\
38 \\
\overline{0} \quad \text{Match}
\end{array}
$$

$$
\begin{array}{r}
37 \\
\textbf{(b)} \quad 716\overline{)26{,}492} \\
2148 \\
\overline{5012} \\
5012 \\
\overline{0}
\end{array}
$$

$$
\begin{array}{r}
85 \text{ R6} \\
\textbf{(c)} \quad 418\overline{)35{,}536} \\
3344 \\
\overline{2096} \\
2090 \\
\overline{6} \quad \text{remainder}
\end{array}
$$

SOLUTION

(a)
$$
\begin{array}{r}
19 \\
\times \ 22 \\
\hline
38 \\
38 \\
\hline
418 \ \text{correct}
\end{array}
$$

(b)
$$
\begin{array}{r}
716 \\
\times \quad 37 \\
\hline
5012 \\
2148 \\
\hline
26{,}492 \ \text{correct}
\end{array}
$$

(c)
$$
\begin{array}{r}
418 \\
\times \quad 85 \\
\hline
2090 \\
3344 \\
\hline
35{,}530 \\
+ \quad 6 \quad \text{remainder} \\
\hline
35{,}536 \quad \text{correct}
\end{array}
$$

QUICK TIP   When checking a division problem that has a remainder, be sure to add the remainder to get the check answer. Also, when an answer is rounded, know that the check answer will not be the same. The rounded answer does not allow a perfect check.

1.1 Exercises

FOR EXTRA HELP

 MyMathLab

 InterActMath.com

MathXL MathXL

 MathXL
Tutorials on CD

 Addison-Wesley
Math Tutor Center

 DVT/Videotape

The **Quick Start** *exercises in each section contain solutions to help you get started.*

Write the following numbers in words. (See Example 1.)

Quick Start

1. 7040 <u>seven thousand, forty</u>

2. 5310 <u>five thousand, three hundred ten</u>

3. 37,901 _____

4. 11,222 _____

5. 725,009 _____

6. 218,033 _____

Round each of the following numbers first to the nearest ten, then to the nearest hundred, and finally to the nearest thousand. Go back to the original number before rounding to the next position. (See Example 2.)

Quick Start

		Nearest Ten	*Nearest Hundred*	*Nearest Thousand*
7.	2065	2070	2100	2000
8.	8385	8390	8400	8000
9.	46,231	_____	_____	_____
10.	55,175	_____	_____	_____
11.	106,054	_____	_____	_____
12.	359,874	_____	_____	_____

13. Explain the three steps that you will use to round a number when the digit to the right of the place to which you are rounding is 5 or more. (See Objective 2.)

14. Explain the three steps that you will use to round a number when the digit to the right of the place to which you are rounding is 4 or less. (See Objective 2.)

C indicates an exercise that is related to the Case in Point feature within the section.

Add each of the following. Check your answers. (See Example 3.)

Quick Start

15.	16.	17.	18.
75	57	875	135
63	26	364	594
45	43	171	415
+ 27	+ 18	+ 776	+ 276
210			

19.	20.	21.	22.
750	371	311,479	803,526
91	45	77,631	759,991
8	839	+ 594,383	+ 36,024
540	3		
+ 7	+ 47		

Subtract each of the following. Check your answers. (See Examples 5 and 6.)

23.	24.	25.	26.
896	757	3715	6215
− 228	− 286	− 838	− 767

27.	28.	29.	30.
65,198	445,193	7,025,389	9,807,943
− 43,652	− 62,785	− 936,490	− 959,489

Solve the following problems. To serve as a check, the vertical and horizontal totals must be the same in the lower right-hand corner.

31. **PRODUCT PURCHASES**   The following table shows Best Buy's monthly purchases by product for each of the first six months of the year. Complete the totals by adding horizontally and vertically.

Quick Start

Product	Jan.	Feb.	Mar.	Apr.	May	June	Totals
Software	$49,802	$36,911	$47,851	$54,732	$29,852	$74,119	**$293,267**
Computers	$86,154	$72,908	$31,552	$74,944	$85,532	$36,705	
Printers	$59,854	$85,119	$87,914	$45,812	$56,314	$91,856	
Monitors	$73,951	$72,564	$39,615	$71,099	$72,918	$42,953	
Totals							

32. **DEPARTMENT SALES**   The following table shows Johnson Sporting Goods' expenses by department for the last six months of the year. Complete the totals by adding horizontally and vertically.

Department	July	Aug.	Sep.	Oct.	Nov.	Dec.	Totals
Office	$29,806	$31,712	$40,909	$32,514	$18,902	$23,514	
Production	$92,143	$86,599	$97,194	$72,815	$89,500	$63,754	
Sales	$31,802	$39,515	$58,192	$32,544	$41,920	$48,732	
Warehouse	$15,746	$12,986	$32,325	$41,983	$39,814	$20,605	
Totals							

Multiply each of the following. (See Example 7.)

Quick Start

33.
```
    218
×   43
    654
   872
   9374
```

34.
```
   672
×   56
```

35.
```
   1896
×    62
```

36.
```
   7318
×     38
```

37.
```
   6452
×    263
```

38.
```
   7143
×    295
```

39.
```
   1109
×   7311
```

40.
```
   9503
×   3411
```

Estimate answers by using front-end rounding. Then find the exact answers. (See Example 4.)

Quick Start

41. **Estimate**   **Exact**
```
   8000  ←rounds   8215
     60  ← to        56
    700  ←          729
+  4000  ←       + 3605
 12,760          12,605
```

42. **Estimate**   **Exact**
```
              ←      2685
              ←        73
              ←       592
+             ←    + 7183
```

43. **Estimate**   **Exact**
```
              ←       783
-             ←     - 238
```

44. **Estimate**   **Exact**
```
              ←       942
-             ←     - 286
```

45. **Estimate**   **Exact**
```
              ←       638
×             ←     ×  47
```

46. **Estimate**   **Exact**
```
              ←       864
×             ←     ×  74
```

Multiply, omitting zeros in the calculation and then replacing them at the right of the product to obtain the final answer. (See Example 8.)

Quick Start

47.
```
    370
×   180
     37
×    18
    666   2 zeros
         ↙
  66,600
```

48.
```
    520
×   400
```

49.
```
   3760
×   6000
```

50.
```
   7200
×   1300
```

Divide each of the following. Use fractions to express any remainders. (See Examples 9 and 10.)

Quick Start

51.
```
       1241 1/4
   4)4965
     4
     09
      8
     16
     16
     05
      4
      1
```

52. $7)\overline{13,214}$

53. $43)\overline{19,715}$

54. $93)\overline{81,452}$

55. Explain why checking the answer is an important step in solving math problems.

56. In your personal and business life, when is it most important to check your math calculations? Why?

Divide each of the following, dropping zeros from the divisor. Express any remainder as a fraction. (See Examples 10 and 11.)

Quick Start

57. 180)‾429,350

$$\begin{array}{r} 2\,385\,\tfrac{5}{18} \\ 18\overline{)42{,}935} \\ \underline{3\,6} \\ 6\,9 \\ \underline{5\,4} \\ 1\,53 \\ \underline{1\,44} \\ 95 \\ \underline{90} \\ 5 \end{array}$$

58. 320)‾360,990

59. 1300)‾75,800

60. 1600)‾253,100

Rewrite the following numbers in words. (See Example 1.)

61. DOUGHNUT PRODUCTION   Each day, Krispy Kreme Doughnuts produces 7,543,500 doughnuts. (*Source:* Krispy Kreme Doughnuts.)

62. WOMEN IN BUSINESS   There are 8,534,350 businesses owned by women in the United States. (*Source:* A. G. Edwards.)

63. PARACHUTE JUMPS   There are 3,200,000 parachute jumps in the United States each year. (*Source:* History Channel.)

64. GROSS NATIONAL PRODUCT   The annual gross national product for the United States (the sum of all goods and services produced) was $10,678,200,000,000. (*Source:* U.S. Department of Commerce.)

Rewrite the numbers from the following sentences using digits. (See Example 1.)

65. JELL-O SALES   The average number of boxes of Jell-O gelatin sold each day is eight hundred fifty-four thousand, seven hundred ninety-five. (*Source:* Kraft Foods.)

65. _____

66. CRAYON SALES   The Binney & Smith Company makes about two billion Crayola Crayons each year. (*Source:* Binney & Smith Company.)

66. _____

67. **COFFEE CONSUMPTION** Americans consume three hundred fifty-three million cups of coffee each day. (*Source: Parade* magazine.)

68. **WALT DISNEY ADVERTISING** Walt Disney Company spent one billion, seven hundred fifty-seven million, five hundred thousand dollars on advertising in one year. (*Source:* Crain Communications Inc./*Advertising Age.*)

Solve the following application problems.

69. **GYM BALL SALES** Carepanian Company, a health-care supplier, purchased 300 cartons of Thera Bond Gym Balls. If there are 10 balls in each carton, find the total number of balls purchased.

69. _____

70. **VITAMIN C INVENTORY** A medical supply house has 30 bottles of vitamin C tablets, with each bottle containing 500 tablets. Find the total number of vitamin C tablets in the supply house.

70. _____

71. **CAMPUS VENDING MACHINES** On a normal weekday, the vending machines at American River College dispense 900 sodas, 400 candy bars, 500 snack items, and 200 cups of coffee. If it takes Jim Wilson four hours to restock the vending machines, how many items does he restock each hour?

71. _____

72. **TELEMARKETING TEAMWORK** In a recent week (Monday through Friday), a telemarketing team sold 380 residential carpet-cleaning jobs, 92 commercial carpet-cleaning jobs, 208 upholstery-cleaning jobs, and 120 drapery-cleaning jobs. How many jobs were sold each day?

72. _____

RECREATION EQUIPMENT RENTAL *American River Raft Rentals lists the following daily raft rental fees. Notice that there is an additional $2 launch fee payable to the park system for each raft rented. Use this information to solve Exercises 73 and 74.*

American River Raft Rentals

Size	Rental Fee	Launch Fee
4 persons	$ 36	$2
6 persons	$ 48	$2
10 persons	$ 90	$2
12 persons	$100	$2

(*Source:* American River Raft Rentals.)

73. On a recent Tuesday, the following rafts were rented: 6 4-person rafts, 15 6-person rafts, 10 10-person rafts, and 5 12-person rafts. Find the total receipts including the $2-per-raft launch fee.

73. _____

74. On the 4th of July, the following rafts were rented: 38 4-person rafts, 73 6-person rafts, 58 10-person rafts, and 46 12-person rafts. Find the total receipts including the $2-per-raft launch fee.

74. _____

ORGANIC ACREAGE   *The following pictograph shows the states with the most organic cropland. Use this information to answer Exercises 75–78.*

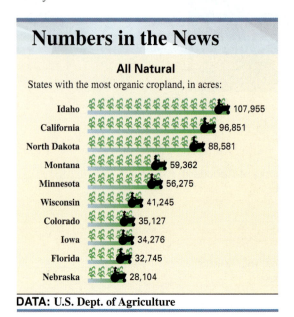

Numbers in the News

All Natural

States with the most organic cropland, in acres:

State	Acres
Idaho	107,955
California	96,851
North Dakota	88,581
Montana	59,362
Minnesota	56,275
Wisconsin	41,245
Colorado	35,127
Iowa	34,276
Florida	32,745
Nebraska	28,104

DATA: U.S. Dept. of Agriculture

75. Find the total number of organic cropland acres in Idaho, California, and North Dakota.

75. _____

76. What is the total number of organic cropland acres in Iowa, Florida, and Nebraska?

76. _____

77. How many more acres of organic cropland are there in Montana than in Colorado?

77. _____

78. How many more acres of organic cropland are there in Idaho than in Nebraska?

78. _____

RETAIL GIANTS   *The following pictograph shows the number of retail stores worldwide for the seven companies with the greatest number of outlets. Use the pictograph to answer Exercises 79–84.*

Numbers in the News

What's in Store for You?

While Wal-Mart has the greatest amount of sales, it trails the other chains in number of stores.

Dollar General
7-Eleven
Family Dollar
CVS
Walgreens
Rite-Aid
Wal-Mart

= 500 stores (rounded)

DATA: T. D. Linx

79. Find the number of Family Dollar retail stores.

79. _____

80. Approximately how many retail stores does 7-Eleven have?

80. _____

81. Which company has the greatest number of retail stores? How many stores do they have?

81. _____

82. Which companies have the least number of retail stores? How many stores do they have?

82. _____

83. How many more retail stores does Walgreens have than Wal-Mart?

83. _____

84. How many more stores does Dollar General have than Family Dollar?

84. _____

1.2 | Application Problems

Objectives

1. Find indicator words in application problems.
2. Learn the four steps for solving application problems.
3. Learn to estimate answers.

CASE in POINT

When Kara Kappas became a manager at a Krispy Kreme Doughnuts store, she had to brush up on her math skills. She remembered that certain words indicate addition, subtraction, multiplication, and division. She and her employees got together and listed some of these words.

Many business-application problems require mathematics. You must read the words carefully to decide how to solve the problem.

Objective 1 Find indicator words in application problems. Look for **indicators** in the application problem—words that indicate the necessary operations, either addition, subtraction, multiplication, or division. Some of these words appear below.

Addition	Subtraction	Multiplication	Division	Equals
plus	less	product	divided by	is
more	subtract	double	divided into	the same as
more than	subtracted from	triple	quotient	equals
added to	difference	times	goes into	equal to
increased by	less than	of	divide	yields
sum	fewer	twice	divided equally	results in
total	decreased by	twice as much	per	are
sum of	loss of			
increase of	minus			
gain of	take away			
	reduced by			

Objective 2 Learn the four steps for solving application problems.

Solving Application Problems

Step 1 Read the problem carefully, and be certain that you understand what the problem is asking. It may be necessary to read the problem several times.

Step 2 Before doing any calculations, work out a plan and try to visualize the problem. Know which facts are given and which must be found. Use word *indicators* to help determine your plan.

Step 3 Estimate a *reasonable answer* using rounding.

Step 4 *Solve* the problem by using the facts given and your plan. Does the answer make sense? If the answer is reasonable, *check* your work. If the answer is not reasonable, begin again by rereading the problem.

QUICK TIP   Be careful not to make the mistake that some students do. They begin to solve a problem before they understand what the problem is asking. Be certain that you know what the problem is asking before you try to solve it.

Objective 3 Learn to estimate answers. Each of the steps in solving an application problem is important, but special emphasis should be placed on step 3, estimating a reasonable answer. Many times an answer just *does not fit* the problem.

What is a *reasonable answer*? Read the problem and estimate the approximate size of the answer. Should the answer be part of a dollar, a few dollars, hundreds, thousands, or even millions of dollars? For example, if a problem asks for the retail price of a shirt, would an answer of $20 be reasonable? $1000? $.65? $65?

Always make an estimate of a reasonable answer. Always look at the answer and decide if it is reasonable. These steps will give greater success in problem solving.

EXAMPLE 1

Using Word Indicators to Help Solve a Problem

At a group yard sale, the total sales were $3584. The money was divided equally among the boys soccer club, the girls soccer club, the boys softball team, and the girls softball team. How much did each group receive?

SOLUTION

After reading the problem and understanding that the four groups equally divided $3584, work out a plan. The word indicators *divided equally* suggest that $3584 should be divided by 4. A reasonable answer would be slightly less than $900 each $\left(\$3600 \div 4 = \$900\right)$. Find the actual answer by dividing $3584 by 4:

$$\frac{896}{4)\overline{3584}}$$ Each group should get $896.

The answer is reasonable, so check the work:

$$\begin{array}{r} 896 \\ \times\quad 4 \\ \hline \$3584 \end{array}$$ The answer is correct.

EXAMPLE 2

Solving an Application Problem

One week, Kara Kappas decided to total her sales at Krispy Kreme. The daily sales figures were $2358 on Monday, $3056 on Tuesday, $2515 on Wednesday, $1875 on Thursday, $3978 on Friday, $3219 on Saturday, and $3008 on Sunday. Find her total sales for the week.

SOLUTION

The sales for each day are given, and the total sales are needed. The word indicators, *total sales*, tell you to add the daily sales to arrive at the weekly total. Since the sales are about $3000 each day for a week of 7 days, a reasonable estimate would be around $21,000 $\left(7 \times \$3000 = \$21,000\right)$. Find the actual answer by adding the sales for each of the 7 days.

$$\begin{array}{r} \underline{\$20,009} \\ \$2358 \\ \$3056 \\ \$2515 \\ \$1875 \\ \$3978 \\ \$3219 \\ +\ \$3008 \\ \hline \$20,009 \end{array}$$

Check ↑

$20,009 sales for the week

The answer $20,009 is reasonable.

EXAMPLE 3

Solving an Application Problem

Use the information in the following graphic to answer each question.

(a) Find the difference in annual earnings between a high school graduate and a person with an associate of arts degree.

(b) In one year, a person with a bachelor's degree will earn how much less than a person with a professional degree?

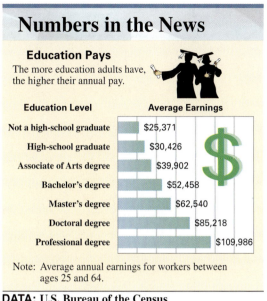

Numbers in the News

Education Pays
The more education adults have, the higher their annual pay.

Education Level	Average Earnings
Not a high-school graduate	$25,371
High-school graduate	$30,426
Associate of Arts degree	$39,902
Bachelor's degree	$52,458
Master's degree	$62,540
Doctoral degree	$85,218
Professional degree	$109,986

Note: Average annual earnings for workers between ages 25 and 64.

DATA: U.S. Bureau of the Census

SOLUTION

The word indicator in **(a)** is *difference,* and the word indicators in **(b)** are *less than.* All of these words indicate that we must subtract.

(a)
$$\begin{array}{r} \$39,902 \\ -\ 30,426 \\ \hline \$9,476 \end{array}$$

Associate of Arts degree
High school graduate
difference in annual earnings

(b)
$$\begin{array}{r} \$109,986 \\ -\ 52,458 \\ \hline \$57,528 \end{array}$$

Professional degree
Bachelor's degree
less than a person with a professional degree

EXAMPLE 4

Solving a Two-Step Problem

In May, the landlord of an apartment building received $940 from each of eight tenants. After paying $2730 in expenses, how much money did the landlord have left?

SOLUTION

Step 1   The amount of rent is given along with the number of tenants. Multiply the amount of rent by the number of tenants to arrive at the monthly income. Since the rent is about $900 and there are eight tenants, a *reasonable estimate* would be around $7200 $\left(\$900 \times 8 = \$7200\right)$.

$$\begin{array}{r} \$940 \\ \times\quad 8 \\ \hline \$7520 \end{array}$$

monthly income (this is reasonable)

Step 2   Finally, subtract the expenses from the monthly income:

$$\begin{array}{r} \$7520 \\ -\ 2730 \\ \hline \$4790 \end{array}$$

amount remaining

1.2 | Exercises

Solve the following application problems.

1. **COMPETITIVE CYCLIST TRAINING**   During a week of training, Rob Andrews rode his bike 80 miles on Monday, 75 miles on Tuesday, 135 miles on Wednesday, 40 miles on Thursday, and 52 miles on Friday. What is the total number of miles he rode in the five-day period?

    1. _____

 2. **DOUGHNUT SALES**   During a recent week, Krispy Kreme sold 2655 dozen original glazed doughnuts, 1086 dozen chocolate iced doughnuts, 978 dozen lemon filled doughnuts, and 143 dozen maple iced doughnuts. Find the total number of dozen doughnuts sold.

    2. _____

3. **CRUISE SHIP TRAVEL**   A cruise ship has 1815 passengers. When in port at Grand Cayman, 1348 passengers go ashore for the day while the others remain on the ship. How many passengers remain on the ship?

    3. _____

4. **SUV SALES**   In a recent three-month period, there were 81,465 Ford Explorers and 70,449 Jeep Grand Cherokees sold. How many more Ford Explorers were sold than Jeep Grand Cherokees? (*Source:* J. D. Power and Associates.)

    4. _____

5. **WORLD WAR II VETERANS**   World War II veterans, part of what is now called "the greatest generation," are dying at the rate of 1100 each day. How many World War II veterans are projected to die in the next year of 365 days? (*Source:* Department of Veterans Affairs.)

    5. _____

6. **TOTAL WORLD WAR II VETERANS**   There are an estimated 4,400,000 World War II veterans alive today. If only 1 in 4 are still alive, find the total number of people who were World War II veterans. (*Source:* Department of Veterans Affairs.)

    6. _____

7. **AUTOMOBILE WEIGHT**   A car weighs 2425 pounds. If its 582-pound engine is removed and replaced with a 634-pound engine, find the weight of the car after the engine change.

    7. _____

8. **PRESCHOOL MANAGER**   Gale Klein has $2324 in her preschool operating account. After spending $734 from this account, the class parents raise $568 in a rummage sale. Find the balance in the account after depositing the money from the rummage sale.

    8. _____

9. **FORD MUSTANG**   In 1964, its first year on the market, the Ford Mustang sold for $2500. In 2006, the Ford Mustang sold for $29,398. Find the increase in price. (*Source: Consumer Reports.*)

    9. _____

 indicates an exercise that is related to the Case in Point feature within the section.

10. WEIGHING FREIGHT   A truck weighs 9250 pounds when empty. After being loaded with firewood, the truck weighs 21,375 pounds. What is the weight of the firewood?

10. _____

11. LAND AREA   There are 43,560 square feet in one acre. How many square feet are there in 138 acres?

11. _____

12. WATER POLLUTION   The number of gallons of water polluted each day in an industrial area is 209,670. How many gallons of water are polluted each year? (Use a 365-day year.)

12. _____

13. HOTEL ROOM COSTS   In a recent study of hotel–casinos, the cost per night at Harrah's Reno was $79, while the cost at Harrah's Lake Tahoe was $179 per night. Find the amount saved on a five-night stay at Harrah's Reno instead of staying at Harrah's Lake Tahoe. (*Source:* Harrah's Casinos and Hotels.)

13. _____

14. LUXURY HOTELS   A hotel room at the Ritz-Carlton in San Francisco costs $495 per night, while a nearby room at a Motel 6 costs $66 per night. What amount will be saved in a four-night stay at Motel 6 instead of staying at the Ritz-Carlton? (*Source:* Ritz-Carlton; Motel 6.)

14. _____

15. PHYSICALLY IMPAIRED   The Enabling Supply House purchased 6 wheelchairs at $1256 each and 15 speech compression recorder-players at $895 each. Find the total cost.

15. _____

16. OFFICE EQUIPMENT   Find the total cost if Krispy Kreme Doughnuts buys 27 computers at $986 each and 12 printers at $280 each:

16. _____

17. YOUTH SOCCER   A youth soccer association raised $7588 through fund-raising projects. There were expenses of $838 that had to be paid first, and the remaining money was divided evenly among the 18 teams. How much did each team receive?

17. _____

18. EGG PRODUCTION   Feather Farms Ranch collects 3545 eggs in the morning and 2575 eggs in the afternoon. If the eggs are packed in flats containing 30 eggs each, find the number of flats needed for packing.

18. _____

19. THEATER RENOVATION   A theater owner is remodeling to provide enough seating for 1250 people. The main floor has 30 rows of 25 seats in each row. If the balcony has 25 rows, how many seats must be in each row of the balcony to satisfy the owner's seating requirements?

19. _____

20. PACKING AND SHIPPING   Nancy Hart makes 24 grapevine wreaths per week to sell to gift shops. She works 30 weeks a year and packages six wreaths per box. If she ships equal quantities to each of five shops, find the number of boxes each shop will receive.

20. _____

1.3 | Basics of Decimals

Objectives

1. Read and write decimal numbers.
2. Round decimal numbers.

Objective ① **Read and write decimal numbers.** A **decimal number** is any number written with a decimal point, such as 6.8, 5.375, or .000982. Decimals, like fractions, can be used to represent parts of a whole. These parts are "less than 1." Section 1.1 discussed how to read the digits to the *left* of the decimal point (whole numbers). Read the digits to the *right* of the decimal point as shown here.

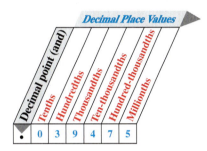

The decimal 9.7 is read as "nine and seven tenths." The word *and* represents the decimal point. Also, 11.59 is read "eleven and fifty-nine hundredths," and 72.087 is read as "seventy-two and eighty-seven thousandths."

> **QUICK TIP**   The word *and* is used only to separate a whole number and a fraction or a whole number and a decimal. Also, notice that all decimals *end* in *ths*.

EXAMPLE 1

Reading Decimal Numbers

Read the following decimals.

(a) 19.08   **(b)** .097   **(c)** 7648.9713

SOLUTION

(a) nineteen and eight hundredths
(b) ninety-seven thousandths
(c) seven thousand, six hundred forty-eight and nine thousand, seven hundred thirteen ten-thousandths.

Objective ② **Round decimal numbers.** It is important to be able to round decimals. For example, Walgreens sells two candy bars for $.79, but you want to buy only one candy bar. The price of one bar is $.79 ÷ 2, which is $.395, but you cannot pay part of a cent. So the store rounds the price up to $.40 for one bar.

Use the following steps for rounding decimals.

Rounding Decimals

Step 1   Find the **place** to which the number is to be rounded. Draw a vertical line after that place to show that you are cutting off the rest of the digits.

Step 2A   Look only at the first digit to the right of your cut-off line. If the first digit is **5 or more, increase** the digit in the place to which you are rounding by one.

Step 2B   If the first digit to the right of the line is **4 or less, do not change** the digit in the place to which you are rounding.

Step 3   **Drop** all digits to the right of the place to which you have rounded.

> **QUICK TIP**   Do not move the decimal point when rounding.

EXAMPLE 2

Rounding Decimal Numbers

Round 98.5892 to the nearest tenth.

SOLUTION

Step 1 Locate the tenths digit and draw a vertical line.

$$98.\mathbf{5}|892$$
↑ —— tenths digit

The tenths digit here is 5.

Step 2 Locate the first digit to the right of the line.

$$98.5|\mathbf{892}$$
↑ —— first digit to the right of the line

The first digit to the right of the line is 8.

Step 3 If the digit found in step 2 is 4 or less, leave the digit of step 1 alone. If the digit found in step 2 is 5 or more, increase the digit of step 1 by 1. The digit found in step 2 here is 8, so 98.5892 rounded to the nearest tenth is

$$98.\mathbf{6}$$
↑ —— increase 5 by 1

EXAMPLE 3

Rounding to the Nearest Thousandth

Round .008572 to the nearest thousandth.

SOLUTION

Locate the thousandths digit and draw a vertical line.

$$.00\mathbf{8}|572$$
↑ —— thousandths digit

Since the digit to the right of the line is 5, increase the thousandths digit by 1. The number .008572 rounded to the nearest thousandth is .009.

EXAMPLE 4

Rounding the Same Decimal to Different Places

Round 24.6483 to the nearest

(a) thousandth.   **(b)** hundredth.   **(c)** tenth.

SOLUTION

Use the method just described.

(a) 24.6483 to the nearest thousandth is 24.648.
(b) 24.6483 to the nearest hundredth is 24.65.
(c) 24.6483 to the nearest tenth is 24.6.

> **QUICK TIP**   The answer to part (c) may be surprising because of the answer in (b). However, **always round a number by going back to the *original number* instead of a number that *has already been* rounded.**

EXAMPLE 5

Rounding to the Nearest Dollar

Round each of the following to the nearest dollar.

(a) $48.69   **(b)** $594.36   **(c)** $2689.50   **(d)** $.61

SOLUTION

(a) Locate the digit representing the dollar and draw a vertical line.

$$\$48.|69$$
↑ —— dollar digit

Since the digit to the right of the line is 6, increase the dollar digit by 1. The number $48.69 rounded to the nearest dollar is $49.
(b) $594.36 rounded to the nearest dollar is $594.
(c) $2689.50 rounded to the nearest dollar is $2690.
(d) $.61 rounded to the nearest dollar is $1.

1.3 Exercises

The **Quick Start** *exercises in each section contain solutions to help you get started.*

Write the following decimals in words. (See Example 1.)

Quick Start

1. .38 thirty-eight hundredths

2. .91 ninety-one hundredths

3. 5.61 _____

4. 6.53 _____

5. 7.408 _____

6. 1.254 _____

7. 37.593 _____

8. 20.903 _____

9. 4.0062 _____

10. 9.0201 _____

11. "My answer is right, but the decimal is in the wrong place." Can this statement ever be correct? Explain. (See Objective 1.)

12. In your own words, explain the difference between *thousands* and *thousandths*.

Write the following decimals, using numbers.

Quick Start

13. four hundred thirty-eight and four tenths **438.4**

14. six hundred five and seven tenths **605.7**

15. ninety-seven and sixty-two hundredths _____

16. seventy-one and thirty-three hundredths _____

17. one and five hundred seventy-three ten-thousandths _____

18. nine and three hundred eight ten-thousandths _____

19. three and five thousand eight hundred twenty seven ten-thousandths _____

20. two thousand seventy-four ten-thousandths _____

GROCERY SHOPPING   Alan Zagorin is grocery shopping. The store will round the amount he pays for each item to the nearest cent. Write the rounded amounts. (See Examples 2–4.)

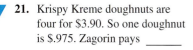 **21.** Krispy Kreme doughnuts are four for $3.90. So one doughnut is $.975. Zagorin pays _____

22. Two jars of pasta sauce cost $2.59. So one jar costs $1.295. Zagorin pays _____

 indicates an exercise that is related to the Case in Point feature within the section.

23. Muffin mix is three packages for $1.75. So one package is $.58333. Zagorin pays _____

24. Candy bars are six for $2.99. So one bar is $.4983. Zagorin pays _____

25. Barbeque sauce is three bottles for $3.50. So one bottle is $1.1666. Zagorin pays _____

26. Orange juice is two cartons for $3.89. So one carton is $1.945. Zagorin pays _____

Round each of the decimals to the nearest tenth, the nearest hundredth, and the nearest thousandth. (See Examples 2–4.)

Quick Start

	Nearest Tenth	Nearest Hundredth	Nearest Thousandth
27. 3.5218	3.5	3.52	3.522
28. 4.836	4.8	4.84	4.836
29. 2.54836	_____	_____	_____
30. 7.44652	_____	_____	_____
31. 27.32451	_____	_____	_____
32. 89.53796	_____	_____	_____
33. 36.47249	_____	_____	_____
34. 58.95651	_____	_____	_____
35. .0562	_____	_____	_____
36. .0789	_____	_____	_____

Round each of the dollar amounts to the nearest cent.

Quick Start

37. $5.056 $5.06	**38.** $16.519 $16.52	**39.** $32.493 _____
40. $375.003 _____	**41.** $382.005 _____	**42.** $12,802.965 _____
43. $42.137 _____	**44.** $.846 _____	**45.** $.0015 _____
46. $.008 _____	**47.** $1.5002 _____	**48.** $7.6009 _____
49. $1.995 _____	**50.** $28.994 _____	**51.** $752.798 _____

Round each of the dollar amounts to the nearest dollar. (Nearest whole number.)

Quick Start

52. $8.58 $9	**53.** $26.49 $26	**54.** $.57 _____
55. $.49 _____	**56.** $299.76 _____	**57.** $12,836.38 _____
58. $268.72 _____	**59.** $395.18 _____	**60.** $666.66 _____
61. $4699.62 _____	**62.** $11,285.13 _____	**63.** $378.59 _____
64. $233.86 _____	**65.** $722.38 _____	**66.** $8263.47 _____

67. Explain what happens when you round $.499 to the nearest dollar. (*See Objective 2.*)

68. Review Exercise 67. How else could you round $.499 to obtain a result that is more helpful? What kind of guideline does this suggest about rounding to the nearest dollar?

1.4 Addition and Subtraction of Decimals

Objectives

1. Add decimals.
2. Estimate answers.
3. Subtract decimals.

CASE *in* POINT

As manager of a Krispy Kreme Doughnuts, Kara Kappas is responsible for making bank deposits to the company checking account. These banking activities require the ability to accurately add and subtract decimal numbers.

Objective 1 Add decimals. Decimals are added in much the same way as whole numbers are added. The main difference with adding decimals is that the decimal points must be kept in a column.

EXAMPLE 1

Adding Decimals and Checking with Estimation

Add 9.83, 6.4, 17.592, and 3.087, or,

```
    9.83
    6.4
   17.592
+   3.087
```

by first lining up decimal points.

```
    9.83       Line up decimal points.
    6.4
   17.592
+   3.087
   36.909
```

Add by columns, just as with whole numbers. One way to keep the digits in their correct columns is to place zeros to the right of each decimal, so that each number has the same number of digits following the decimal point. Attaching zeros to this example gives

```
    9.830      All numbers now have three
    6.400      places after the decimal point.
   17.592
+   3.087
   36.909
```

QUICK TIP   Placing zeros to the right of the decimal point does not change the value of a number. For example, 4.21 = 4.210 = 4.2100, and so on.

Objective 2 Estimate answers. Check that digits were not added in the wrong columns by estimating the answer. For the numbers just added, estimate by using front-end rounding.

Problem		Estimate
9.830	⟶	10
6.400	⟶	6
17.592	⟶	20
+ 3.087	⟶	+ 3
36.909		39

<hr />

QUICK TIP   The estimate shows that the answer is reasonable and that the decimal points were lined up properly.

<hr />

EXAMPLE 2

Adding Dollars and Cents

During a recent week, Kara Kappas made the following bank deposits to the Krispy Kreme business account: $1783.38, $4341.15, $2175.94, $896.23, and $2562.53. Use front-end rounding to estimate the total deposits and then find the total deposits.

SOLUTION

Estimate		Problem
$ 2000	⟵	$ 1783.38
4000	⟵	4341.15
2000	⟵	2175.94
900	⟵	896.23
+ 3000	⟵	+ 2562.53
$11,900		$11,759.23

The total deposits for the week were $11,759.23, which is close to our estimate.

Objective 3   Subtract decimals. Subtraction is done in much the same way as addition. Line up the decimal points and place as many zeros after each decimal as needed. For example, subtract 17.432 from 21.76 as follows.

$$21.76\mathbf{0} \quad \text{Place one zero after the top decimal.}$$
$$- 17.432$$
$$\overline{4.328}$$

EXAMPLE 3

Estimating and then Subtracting Decimals

First estimate using front-end rounding and then subtract.

(a)   11.7
　　 − 4.923

(b)   39.428
　　 − 27.98

SOLUTION

Attach zeros as needed and then subtract.

(a)

Estimate		Problem
10	⟵	11.700
− 5	⟵	− 4.923
5		6.777

(b)

Estimate		Problem
40	⟵	39.428
− 30	⟵	− 27.980
10		11.448

1.4 | Exercises

The **Quick Start** exercises in each section contain solutions to help you get started.

First use front-end rounding to estimate and then add the following decimals. (See Examples 1 and 2.)

Quick Start

1. Estimate	**Problem**	**2. Estimate**	**Problem**	**3. Estimate**	**Problem**

1. Estimate **Problem**

```
   40  ←      43.36
   20  ←      15.8
 +  9  ←    +  9.3
 ─────       ─────
   69        68.46
```

2. Estimate **Problem**

```
  600  ←     623.15
  700  ←     734.29
+ 700  ←   + 686.26
─────        ──────
 2000        2043.70
```

3. Estimate **Problem**

```
             6.23
             3.6
             5.1
             7.2
    +      + 1.69
 ─────       ─────
```

4. Estimate **Problem**

```
            12.79
             2.15
            16.28
             4.39
    +      + 7.61
 ─────      ──────
```

5. Estimate **Problem**

```
          2156.38
             5.26
             2.791
    +      + 6.983
 ─────      ──────
```

6. Estimate **Problem**

```
          1889.76
            21.42
            19.35
    +      +  8.1
 ─────      ──────
```

7. Estimate **Problem**

```
          6133.78
           506.124
            18.63
    +      +  7.527
 ─────      ──────
```

8. Estimate **Problem**

```
           743.1
          3817.65
             2.908
          4123.76
    +      + 21.98
 ─────      ──────
```

9. Estimate **Problem**

```
          1798.419
            68.32
           512.807
           643.9
    +      + 428.
 ─────      ──────
```

Place each of the following numbers in a column and then add. (See Example 1.)

Quick Start

10. $45.631 + 15.8 + 7.234 + 19.63 =$ **88.295**

11. $12.15 + 6.83 + 61.75 + 19.218 + 73.325 =$ **173.273**

12. $197.4 + 83.72 + 17.43 + 25.63 + 1.4 =$

13. $27.653 + 18.7142 + 9.7496 + 3.21 =$

14. $73.618 + 19.18 + 371.82 + 355.125 =$

15. It is a good idea to estimate an answer before actually solving a problem. Why is this true? (See Objective 2.)

16. Explain why placing zeros after any digits to the right of the decimal point does not change the value of a number. (See Objective 1.)

▼C indicates an exercise that is related to the Case in Point feature within the section.

Solve the following application problems.

17. SALES COMPENSATION   Scott Salman, a sales representative, earned $2325.50 in monthly salary, $1873.26 in commissions, and $952.14 in a new product sales bonus. Find the total amount of his earnings.

17. _____

18. STAFF-MEETING COST   During the holiday season, Paul Burke has made the following purchases of Krispy Kreme doughnuts for his customers: $78.83, $125.48, $165.83, and $108.89. Find the total amount of these purchases.

18. _____

19. BEEF/TURKEY COST   The average cost of T-bone steak is $6.71 per pound, while the average cost of turkey is $.98 per pound. How much more per pound is the price of T-bone steak than turkey? (*Source:* U.S. Bureau of the Census.)

19. _____

20. TURKEY AND CRANBERRY SALES   The weight of turkeys raised in the United States in 2000 was 6.9 billion pounds, while the weight of cranberries was .639 billion pounds. How many more pounds of turkeys were raised than cranberries? (*Source:* U.S. Bureau of the Census.)

20. _____

First use front-end rounding to estimate the answer and then subtract. (See Example 3.)

Quick Start

21. Estimate	Problem	22. Estimate	Problem	23. Estimate	Problem
20 − 7 13	19.74 − 6.58 13.16	40 − 8 32	35.86 − 7.91 27.95	−	51.215 − 19.708

24. Estimate	Problem	25. Estimate	Problem	26. Estimate	Problem
−	27.613 − 18.942	−	325.053 − 85.019	−	3974.61 − 892.59

27. Estimate	Problem	28. Estimate	Problem	29. Estimate	Problem
−	7.8 − 2.952	−	27.8 − 13.582	−	5 − 1.9802

CHECKING-ACCOUNT RECORDS   Kara Kappas, manager of Krispy Kreme Doughnuts, had a bank balance of $5382.12 on March 1. During March, Kara deposited $60,375.82 received from sales, $3280.18 received as credits from suppliers, and $75.53 as a county tax refund. She paid out $27,282.75 to suppliers, $4280.83 for rent and utilities, and $12,252.23 for salaries. Find each of the following.

Quick Start

30. How much did Kappas deposit in March?
$60,375.82 (sales) + $3280.18 (credits) + $75.53 (refund) = $63,731.53

30. $63,731.53

31. How much did she pay out?

31. _____

32. What was her final balance at the end of March?

32. _____

1.5 | Multiplication and Division of Decimals

Objectives

1 Multiply decimals.
2 Divide decimals.
3 Divide a decimal by a decimal.

CASE *in* POINT

Managing a business requires the ability to multiply and divide decimal numbers. Kara Kappas, the manager at Krispy Kreme, applies these skills in many ways; some examples include payroll, purchasing, and sales.

Objective 1 **Multiply decimals.** Decimals are multiplied as if they were whole numbers. (It is not necessary to line up the decimal points.) The decimal point in the answer is then found as follows.

Positioning the Decimal Point

Step 1 Count the total number of digits to the *right* of the decimal point in each of the numbers being multiplied.

Step 2 In the answer, count from *right to left* the number of places found in step 1. It may be necessary to attach zeros to the left of the answer in order to correctly place the decimal point.

EXAMPLE 1

Multiplying Decimals

Multiply

(a) 8.34×4.2 **(b)** $.032 \times .07$

SOLUTION

(a) First multiply the given numbers as if they were whole numbers.

$$
\begin{array}{r}
8.34 \quad \longleftarrow \text{2 decimal places} \\
\times \quad 4.2 \quad \longleftarrow \text{1 decimal place} \\
\hline
1668 \\
3336 \quad\;\; \\
\hline
35.028 \quad \longleftarrow \text{3 decimal places in answer}
\end{array}
$$

There are two decimal places in 8.34 and one in 4.2. This means that there are $2 + 1 = 3$ decimal places in the final answer. Find the final answer by starting at the right and counting three places to the left:

$$35.028 \qquad \text{3 places to the left}$$

(b) Here, it is necessary to attach zeros at the left in the answer:

$$
\begin{array}{r}
.032 \quad \longleftarrow \text{3 decimal places} \\
\times \quad .07 \quad \longleftarrow \text{2 decimal places} \\
\hline
.00224 \quad \longleftarrow \text{5 decimal places in answer}
\end{array}
$$

Attach 2 zeros.

The next example uses the formula for determining the gross pay (the pay before deductions) of a worker paid by the hour.

Gross pay = Number of hours worked × Pay per hour

EXAMPLE 2

Multiplying Two Decimal Numbers

Find the gross pay of a Krispy Kreme employee working 31.5 hours at a rate of $8.65 per hour.

SOLUTION

Find gross pay by multiplying the number of hours worked by the pay per hour.

$$
\begin{array}{r}
31.5 \quad \longleftarrow \text{ I place} \\
\times\ 8.65 \quad \longleftarrow \text{ 2 places} \\
\hline
1575 \\
1890 \\
2520 \\
\hline
272.475 \quad \longleftarrow \text{ 3 places in answer}
\end{array}
$$

This worker's gross pay, rounded to the nearest cent, is $272.48.

The following graphic shows the price for 30 seconds of advertising time during the game for these Super Bowls.

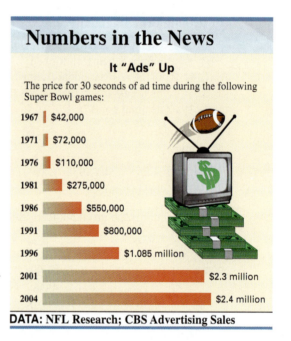

Numbers in the News

It "Ads" Up

The price for 30 seconds of ad time during the following Super Bowl games:

1967	$42,000
1971	$72,000
1976	$110,000
1981	$275,000
1986	$550,000
1991	$800,000
1996	$1.085 million
2001	$2.3 million
2004	$2.4 million

DATA: NFL Research; CBS Advertising Sales

EXAMPLE 3

Applying Decimal Multiplication

If there were 60 advertising spots of 30 seconds each during the 2004 Super Bowl, find the total amount charged for advertising during the game.

SOLUTION

Find the total amount charged for advertising during the 2004 Super Bowl by multiplying the number of 30-second advertising spots during the game by the charge for each advertisement.

$$
\begin{array}{r}
2.4 \quad \leftarrow \text{ I place} \\
\times\ \ 60 \quad \leftarrow \text{ 0 places} \\
\hline
00 \\
144 \\
\hline
144.0 \quad \leftarrow \text{ I place}
\end{array}
$$

The total amount charged for advertising during the 2004 Super Bowl was $144 million ($144,000,000).

Objective **2** **Divide decimals.** Divide the decimal 21.93 by the whole number 3 by first writing the division problem as usual.

$$3\overline{)21.93}$$

Place the decimal point in the quotient directly above the decimal point in the dividend and perform the division.

Place a decimal point directly above dividend's decimal point.

$$3\overline{)21.93}^{\,7.31}$$

Check by multiplying the divisor and the quotient. The answer should equal the dividend.

$$\begin{array}{r} 7.31 \\ \underline{3} \\ 21.93 \end{array} \quad \longleftarrow \text{ matches dividend}$$

Sometimes it is necessary to place zeros after the decimal point in the dividend. Do this if a remainder of 0 is not obtained. Attaching zeros *does not change* the value of the dividend. For example, divide 1.5 by 8 by dividing and placing zeros as needed.

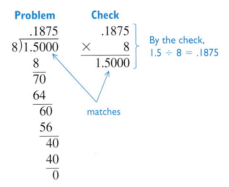

A remainder of 0 might not be obtained even though extra zeros are placed after the dividend. For example, when 4.7 is divided by 3, a remainder of 0 is never obtained. The digit 6 repeats indefinitely. In such a case, round to the nearest thousandth, although different problems might require rounding to a different number of decimal places. Rounding to the nearest thousandth gives 4.7 ÷ 3 = 1.567. Round to the nearest thousandth by carrying out the division to four places and then rounding to three places. Check by multiplying 1.567 by 3. The answer will be only approximately equal to the dividend, because of the rounding. Here is the division.

$$\begin{array}{r} 1.5\mathbf{666} \\ 3\overline{)4.7000} \\ \underline{3} \\ 17 \\ \underline{15} \\ 20 \\ \underline{18} \\ 20 \\ \underline{18} \\ 20 \\ \underline{18} \\ 2 \end{array}$$

QUICK TIP   When rounding an answer, be certain to carry the answer one place further than the position to which you are rounding.

EXAMPLE **4**

Dividing a Decimal by a Whole Number

Divide and check the following.

(a) 27.52 ÷ 32   **(b)** 153.4 ÷ 8

SOLUTION

Problem	Check	Problem	Check

(a)
```
        .86
  32)27.52
     25 6
      1 92
      1 92
         0
```
Check
```
   .86
 × 32
  172
  258
 27.52
```

(b)
```
    19.175
 8)153.400
   8
   73
   72
    1 4
      8
     60
     56
      40
      40
       0
```
Check
```
  19.175
 ×     8
 153.400
```

Objective 3 **Divide a decimal by a decimal.** To divide by a decimal, first convert the divisor to a whole number. For example, to divide 27.69 by .3, convert .3 to a whole number by moving the decimal one place to the right. Then move the decimal point in the dividend, 27.69, one place to the right so that the value of the problem does not change.

First, convert the decimal to a whole number.

$$.3.)27.6.9$$

Then, move the decimal point in the dividend the same number of places to the right as done in the divisor.

After moving the decimal point, write the original problem as 276.9 ÷ 3, and then divide as follows:

```
     92.3
 3)276.9
   27
    6
    6
    0 9
      9
      0
```

> **QUICK TIP** Sometimes it is necessary to place zeros after the dividend. As before, attach zeros and divide until the quotient has one more digit than the desired position, and then round.

EXAMPLE 5

Dividing a Decimal by a Decimal

Divide and check the following.

(a) 17.6 ÷ .25   **(b)** 5 ÷ .42

SOLUTION

(a)
```
       70.4
 .25.)17.60.0
      17 5
        100
        100
          0
```
Check
```
   70.4
 × .25
  3520
  1408
 17.600
```

(b)
```
        11.9047
 .42.)5.00.0000
      4 2
        80
        42
        38 0
        37 8
          200
          168
          320
          294
           26
```
Rounding the answer to the nearest thousandth gives 11.905.

Check
```
   11.905
 ×    .42
  23810
  47620
 5.00010
```
(The check is off a little due to rounding.)

> **QUICK TIP** In checking an answer that has been rounded, the check answer will not be exactly equal to the original dividend.

1.5 | Exercises

FOR EXTRA HELP

MyMathLab

InterActMath.com

*Math*XP MathXL

MathXL
Tutorials on CD

Tutor
Center Addison-Wesley
Math Tutor Center

DVT/Videotape

The **Quick Start** *exercises in each section contain solutions to help you get started.*

First estimate using front-end rounding and then multiply. (See Example 1.)

Quick Start

1. Estimate	Problem	2. Estimate	Problem	3. Estimate	Problem
100 ←	96.8	20 ←	16.6		34.1
× 4 ←	× 4.2	× 4 ←	× 4.2	×	× 6.8
400	406.56	80	69.72		

4. Estimate	Problem	5. Estimate	Problem	6. Estimate	Problem
	70.35		43.8		69.3
×	× 8.06	×	× 2.04	×	× 2.81

Multiply the following decimals.

Quick Start

7.	.532	8.	.259	9.	21.7
	× 3.6		× 6.2		× .431
	.53 ← 3 decimals		.259 ← 3 decimals		
	× 3.6 ← 1 decimal		× 6.2 ← 1 decimal		
	1.9152 ← 4 decimals		1.6058 ← 4 decimals		

10.	76.9	11.	.0408	12.	2481.9
	× .903		× .06		× .003

CALCULATING GROSS EARNINGS   *Find the gross pay for each Krispy Kreme employee at the given rates. Round to the nearest cent. (See Examples 2 and 3.)*

Quick Start

13. 18.5 hours at $8.25 per hour
       18.5 × $8.25 = $152.63

13. $152.63

14. 36.6 hours at $9.85 per hour

14. _____

15. 27.9 hours at $11.42 per hour, and 6.8 hours at $14.63 per hour

15. _____

16. 11.4 hours at $8.59 per hour, and 23.9 hours at $10.06 per hour

16. _____

C indicates an exercise that is related to the Case in Point feature within the section.

Divide the following, and round your answer to the nearest thousandth. (See Examples 4 and 5.)

17.
$$6\overline{)48.45}$$

18.
$$5\overline{)62.38}$$

19. $411.63 \div 15$

20.
$$2.43\overline{)9.6153}$$

21.
$$.65\overline{)37.6852}$$

22. $15.62 \div .28$

23. In your own words, write the rule for placing the decimal point in the answer of a decimal multiplication problem. (See Objective 1.)

24. Describe what must be done with the decimal point in a decimal division problem. Include the divisor, dividend, and quotient in your description. (See Objectives 2 and 3.)

Solve the following application problems:

Quick Start

25. REAL ESTATE FEES   Robert Gonzalez recently sold his home for $246,500. He paid a sales fee of .06 times the price of the house. What was the amount of the fee?
$246,500 × .06 = $14,790

25. $14,790

26. BASEBOARD TRIM   Brenda Mead bought 6.5 yards of baseboard trim to complete her kitchen remodeling. If she paid $2.68 per yard for the trim, what was her total cost?

26. _____

27. GAS MILEAGE   The Krispy Kreme delivery van used 18.5 gallons of gas and traveled 464 miles. How many miles per gallon (mpg) did the van get? Round to the nearest tenth.

27. _____

28. MANAGERIAL EARNINGS   A Krispy Kreme assistant manager earns $2528 each month for working a 48-hour week. Find **(a)** the number of hours worked each month and **(b)** the manager's hourly earnings. $\left(1 \text{ month} = 4.3 \text{ weeks}\right)$ Round to the nearest cent.

(a) _____
(b) _____

29. OLYMPIC GOLD COINS   If a five-dollar Olympic gold coin weighs 8.359 grams, find the number of coins that can be produced from 221 grams of gold. Round to the nearest whole number.

29. _____

30. MEDICINE DOSAGE   Each dosage of a medication contains 1.62 units of a certain ingredient. Find the number of dosages that can be made from 57.13 units of the ingredient. Round to the nearest whole number.

30. _____

31. U.S. PAPER MONEY   The thickness of one piece of paper money is .0043 inch.
(a) If you had a pile of 100 bills, how high would it be?

(a) _____
(b) _____

(b) How high would a pile of 1000 bills be?

32. (a) Use the information from Exercise 31 to find the number of bills in a pile that is 43 inches high.

(a) _____
(b) _____

(b) How much money would you have if the pile was all $20 bills?

Use the information from the Look Smart online catalog to answer Exercises 33–36.

43-2A   43-2B

43-3A   43-3B

Knit Shirt Ordering Information		
43–2A	Short sleeve, solid colors	$14.75 each
43–2B	Short sleeve, stripes	$16.75 each
43–3A	Long sleeve, solid colors	$18.95 each
43–3B	Long sleeve, stripes	$21.95 each
XXL size, add $2 per shirt.		
Monogram, $4.95 each. Gift box, $5 each.		

Total Price of All Items (excluding monograms and gift boxes)	Shipping, Packing, and Handling
$0–25.00	$3.50
$25.01–75.00	$5.95
$75.01–125.00	$7.95
$125.01+	$9.95
Shipping to each additional address add $4.25.	

33. Find the total cost of ordering four long-sleeve, solid-color shirts and two short-sleeve, striped shirts all of size XXL and all shipped to your home.

33. _____

34. What is the total cost of eight long-sleeve shirts, five in solid colors and three striped? Include the cost of shipping the solid shirts to your home and the striped shirts to your brother's home.

34. _____

35. (a) What is the total cost, including shipping, of sending three short-sleeve, solid-color shirts, with monograms, in a gift box to your aunt for her birthday?

(a) _____
(b) _____

(b) How much did the monogram, gift box, and shipping add to the cost of your gift?

36. (a) Suppose you order one of each type of shirt for yourself, adding a monogram on each of the solid-color shirts. At the same time, you order three long-sleeved, striped size-XXL shirts to be shipped to your father in a gift box. Find the total cost of your order.

(a) _____
(b) _____

(b) What is the difference in total cost (excluding shipping) between the shirts for yourself and the gift for your father?

Krispy Kreme

- 1937 First location in Winston-Salem, North Carolina

- Each store has a doughnut-making theater

- 2005 Stores in 42 states, Australia, Asia, Canada, Mexico, and Eastern and Western Europe.

- 504 store locations

- 5958 employees

- Daily doughnut production of 9.5 million

- 3.5 billion doughnuts produced each year

Krispy Kreme Doughnuts began in 1937 when Vernon Rudolph started selling doughnuts to local grocery stores in Winston-Salem, North Carolina. The secret yeast-raised doughnut recipe had been purchased from a French chef in New Orleans, and people soon began stopping by to ask if they could buy hot doughnuts. A hole was cut in the wall of the bakery so that hot original glazed doughnuts could be sold directly to customers. By the year 2004, Krispy Kreme had opened stores in Australia, Canada, Asia, Mexico, and both Eastern and Western Europe.

Krispy Kreme offers fund-raising programs to schools, churches and other nonprofit organizations. Throughout the nation, organizations raised $43 million to fund programs, take field trips, buy uniforms and equipment, and support other charitable causes.

1. An employee removes 32.6 pounds of flour from a sack weighing 50 pounds. Find the number of pounds of flour remaining.

2. A measuring scoop holds .0225 pounds of bakers yeast. How many measuring scoops are contained in 1 pound of bakers yeast? Round to the nearest whole number.

3. Based on your knowledge of the Krispy Kreme Doughnut company, the products that it sells, and the company business philosophy, what kind of a statement do you think the company is trying to make about itself?

4. In addition to using decimals in measuring flour, name six additional applications of decimals in a Krispy Kreme store.

Chapter 1 | Quick Review

addends [p. 4]	decimal system [p. 2]	indicator words [p. 17]	quotient [p. 7]
addition [p. 4]	difference [p. 5]	minuend [p. 5]	rounding [p. 3]
amount [p. 4]	digits [p. 2]	multiplicand [p. 6]	rounding decimals [p. 23]
borrowing [p. 5]	dividend [p. 7]	multiplication [p. 4]	rounding whole numbers [p. 3]
checking answers [p. 4]	dividing decimals [p. 33]	multiplier [p. 6]	subtraction [p. 4]
comma [p. 2]	division [p. 4	multiplying decimals [p. 31]	subtrahend [p. 5]
decimal number [p. 23]	divisor [p. 7]	operations [p. 4]	sum [p. 4]
decimal part [p. 8]	fraction part [p. 8]	partial products [p. 6]	total [p. 4]
decimal point [p. 2]	front-end rounding [p. 4]	product [p. 6]	whole numbers [p. 2]

CONCEPTS

1.1 Reading and writing whole numbers

The word *and* is not used. Commas help divide thousands, millions, and billions. A comma is not needed with a four-digit number.

1.1 Rounding whole numbers

Rules for rounding:
1. Identify the position to be rounded. Draw a line under that place.
2. If the digit to the right of the underlined place is 5 or more, increase by 1; if the digit is 4 or less, do not change.
3. Change all digits to the right of the underlined digit to zero.

1.1 Front-end rounding

Front-end rounding leaves only the first digit as a nonzero digit. All other digits are changed to zero.

1.1 Addition of whole numbers

Add from top to bottom, starting with units and working to the left. To check, add from bottom to top.

1.1 Subtraction of whole numbers

Subtract the subtrahend from the minuend to get the difference, borrowing when necessary. To check, add the difference to the subtrahend to get the minuend.

EXAMPLES

795 is written "seven hundred ninety-five." 9,768,002 is written "nine million, seven hundred sixty-eight thousand, two."

Round:

72 6 to the nearest ten — 5 or more, so add 1 to tens position
tens position

So, 726 rounds to 730

1, 498,586 to the nearest million — 4 or less, do not change
millions position

So, 1,498,586 rounds to 1,000,000

Round each of the following, using front-end rounding.

76 rounds to 80
348 rounds to 300
6512 rounds to 7000
23,751 rounds to 20,000
652,179 rounds to 700,000

Problem (add down) / Check (add up)
```
1140
 687
  26
   9
+418
1140
```

Problem
```
 621
4738
-649
4089
```
Check
```
4089
+649
4738
```

CONCEPTS	EXAMPLES

1.1 Multiplication of whole numbers

The multiplicand is multiplied by the multiplier, giving the product. When the multiplier has more than one digit, partial products must be used and then added.

$$
\begin{array}{r}
78 \\
\times\ 24 \\
\hline
312 \\
156\ \ \\
\hline
1872
\end{array}
$$

78 — multiplicand
× 24 — multiplier
312 — partial product
156 — partial product (one position left)
1872 — product

1.1 Division of whole numbers

÷ and $\overline{)}$ mean divide.
A —, as in $\frac{25}{5}$, means divide 25 by 5.
Also, the /, as in 25/5, means to divide 25 by 5.
Remainders are usually expressed as decimals.

$$
\begin{array}{r}
44 \\
2\overline{)88} \\
88 \\
\hline
0
\end{array}
$$

divisor 2)88 dividend   44 quotient

If answer is rounded, the check will not be perfect.

1.2 Application problems

Follow these steps.
1. Read the problem carefully.
2. Work out a plan, using *indicator words* before starting.
3. Estimate a reasonable answer.
4. Solve the problem. If the answer is reasonable, check; if it is not, start over.

Shauna Gallegos earns $118 on Sunday, $87 on Monday, and $63 on Tuesday. Find her total earnings for the three days. *Total* means to add.

$$
\begin{array}{r}
\$268 \\
\$118 \\
\$\ 87 \\
+\ \$\ 63 \\
\hline
\$268
\end{array}
$$

$268 Check
$118
$ 87
+ $ 63
$268 total earnings

1.3 Reading and rounding decimals

Decimal Place Values

Decimal point (and), Tenths, Hundredths, Thousandths, Ten-thousandths, Hundred-thousandths, Millionths

. 0 7 3 2 6 5

1.3 is read "one and three tenths"
Round .073265 to the nearest ten-thousandth

.0732|65

↑ ten-thousandth position

Since the digit to the right is 6, increase the ten-thousandths digit by 1 and drop all digits to the right. .073265 rounds to .0733.

1.4 Addition and subtraction of decimals

Decimal points must be in a column. Attach zeros to keep digits in their correct columns.

Add: 5.68 + 785.3 + .007 + 10.1062.
Line up the decimal points.

$$
\begin{array}{r}
5.6800 \\
785.3000 \\
.0070 \\
+\ \ 10.1062 \\
\hline
801.0932
\end{array}
$$

5.6800 ←
785.3000 ← Attach zeros.
.0070 ←
+ 10.1062
801.0932

1.5 Multiplication of decimals

Multiply as if decimals are whole numbers. Place the decimal point as follows.
1. Count digits to the right of decimal points.
2. Count from right to left the same number of places as in step 1. Zeros must be attached on the left if necessary.
.169 × .21

Multiply

$$
\begin{array}{r}
.169 \\
\times\ .21 \\
\hline
169
\end{array}
$$

.169 3 decimal places
× .21 2 decimal places
169

5 decimal places in answer
↑ Attach one zero.

CONCEPTS	EXAMPLES

1.5 Division of decimals

1. Move the decimal point in the divisor all the way to the right.
2. Move the decimal point the same number of places to the right in the dividend.
3. Place a decimal point in the answer position directly above the dividend decimal point.
4. Divide as with whole numbers.

Divide 52.8 by .75

```
          70.4          Check
  .75 )52.80.0           70.4
       52 5            ×  .75
       ───             ─────
         30            3520
         00            4928
       ───             ─────
         30 0          52.800
         30 0
       ───
           0
```

Chapter 1 | Summary Exercise

The Toll of Wedding Bells

Ladies' Home Journal recently released statistics showing the average wedding costs in the United States in 2004. In 2001, the cost of the average wedding was $20,357. This cost has increased as the average number of wedding guests has grown to over 200. The following graph gives most of the costs involved in a wedding. Use this information to answer the questions that follow.

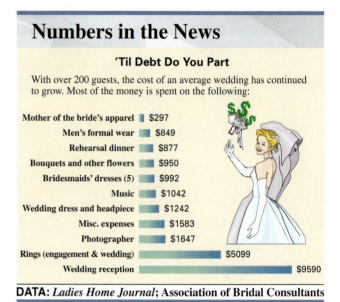

Numbers in the News

'Til Debt Do You Part

With over 200 guests, the cost of an average wedding has continued to grow. Most of the money is spent on the following:

Mother of the bride's apparel	$297
Men's formal wear	$849
Rehearsal dinner	$877
Bouquets and other flowers	$950
Bridesmaids' dresses (5)	$992
Music	$1042
Wedding dress and headpiece	$1242
Misc. expenses	$1583
Photographer	$1647
Rings (engagement & wedding)	$5099
Wedding reception	$9590

DATA: *Ladies Home Journal*; Association of Bridal Consultants

(a) What is the total of the costs shown in the graph?

(a) _____

(b) How much more expensive was a wedding in 2004 compared with 2001?

(b) _____

(c) If you decide to spend $7000 for your wedding reception, and the cost per guest is $32, how many guests can you invite? How much of your budgeted amount will be left over?

(c) _____

(d) If 150 guests are invited to the wedding and $7000 is budgeted for the reception, find the amount that can be spent per person. Round to the nearest cent.

(d) _____

(e) If you budget $4000 for the wedding reception and the cost per person is $27, how many guests can you invite and how much of your budgeted amount will be left over?

(e) _____

(f) If you budget $1000 for the wedding reception and the cost per person is $15, how many guests can you invite and how much of your budgeted amount will be left over?

(f) _____

(g) The bridal party will need five bouquets that cost $36.25 each and five boutonnieres, each costing $7.50. If a total of $863 is budgeted for flowers, find the amount that remains to be spent for other floral arrangements.

(g) _____

INVESTIGATE

List six expenses associated with a wedding that are not mentioned. List six things that could be changed to bring the cost of the wedding down. What are some costs associated with a wedding where you live? You may have to ask some friends or relatives who recently have been involved in wedding planning.

Chapter 1 | Test

To help you review, the bracketed numbers indicate the section in which the topic is discussed.

Round as indicated. **[1.1]**

1. 844 to the nearest ten

2. 21,958 to the nearest hundred

3. 671,529 to the nearest thousand

Round each of the following, using front-end rounding. **[1.1]**

4. 50,987

5. 851,004

6. One week, Suzanne Alley earned the following commissions: Monday, $124; Tuesday, $88; Wednesday, $62; Thursday, $137; Friday, $195. Find her total amount of commissions for the week. **[1.2]**

7. A rental business buys three airless sprayers at $1540 each, five rototillers at $695 each, and eight 25-foot ladders at $38 each. Find the total cost of the equipment purchased. **[1.2]**

Round as indicated. **[1.3]**

8. $21.0568 to the nearest cent

9. $364.345 to the nearest cent

10. $7246.49 to the nearest dollar

1. _____

2. _____

3. _____

4. _____

5. _____

6. _____

7. _____

8. _____

9. _____

10. _____

Solve each problem. **[1.4 and 1.5]**

11. $9.6 + 8.42 + 3.715 + 159.8 =$ _____

12.
$$\begin{array}{r} 2.715 \\ 32.78 \\ 426.3 \\ +\ \ 37 \\ \hline \end{array}$$

13.
$$\begin{array}{r} 341.4 \\ -\ 207.8 \\ \hline \end{array}$$

14. $3.8 - .0053$

15.
$$\begin{array}{r} 21.98 \\ \times\ \ \ .72 \\ \hline \end{array}$$

16.
$$\begin{array}{r} 218.6 \\ \times\ \ .037 \\ \hline \end{array}$$

17. $21.8\overline{)252.008}$

18. $57.358 \div 2.41 =$ _____

19. $79.135 \div 18.62 =$ _____

20. Find the total cost of 32.6 gallons of solvent at $13.48 per gallon and 18.5 gallons of acid at $3.56 per gallon. **[1.4 and 1.5]**

20. _____

21. Roofing material costs $84.52 per square $(10\ \text{ft} \times 10\ \text{ft})$. The roofer charges $55.75 per square for labor, plus $9.65 per square for supplies. Find the total cost for 26.3 squares of installed roof. **[1.4 and 1.5]**

21. _____

22. A federal law requires that all residential toilets sold in the United States use no more than 1.6 gallons of water per flush. Prior to this legislation, conventional toilets used 3.4 gallons of water per flush. Find the amount of water saved in one year by a family flushing the toilet 22 times each day. $(1\ \text{year} = 365\ \text{days})$ **[1.4 and 1.5]**

22. _____

23. Barry Garland bought 16.5 meters of electrical wire at $.48 per meter and three meters of brass rod at $1.05 per meter. How much change did he get from three $5 bills? **[1.4 and 1.5]**

23. _____

24. The Capital Hills Supermarket in Washington, DC, sells bananas for $1.74 per kilogram (2.2 pounds). Find the price of bananas per pound. Round to the nearest cent. (*Source:* Associated Press.) **[1.5]**

24. _____

25. A concentrated fertilizer must be applied at the rate of .058 ounce per seedling. Find the number of seedlings that can be fertilized with 14.674 ounces of fertilizer. **[1.5]**

25. _____

CHAPTER 2

Fractions

SARAH BRYN HAS BEEN EMPLOYED AT THE HOME DEPOT FOR several years. During this time, she has worked in the hardware and plumbing

 departments and now manages the cabinetry department. She has been managing this department for a year and a half and enjoys helping contractors and homeowners plan and design new and replacement cabinets for their kitchens and bathrooms. Knowing and using fractions is a key part of Bryn's job. She must be extremely accurate as she determines the exact space specifications for the building and placement of new cabinets.

The previous chapter discussed whole numbers and decimals. This chapter looks at *fractions*—numbers, like decimals, that can be used to represent parts of a whole. Fractions and decimals are two ways of representing the same quantity. Fractions are used in business and our personal lives.

2.1 | Basics of Fractions

Objectives

1. Recognize types of fractions.
2. Convert mixed numbers to improper fractions.
3. Convert improper fractions to mixed numbers.
4. Write a fraction in lowest terms.
5. Use the rules for divisibility.

A **fraction** represents part of a whole. Fractions are written in the form of one number over another, with a line between the two numbers, as in the following.

$$\frac{5}{8} \quad \frac{1}{4} \quad \frac{9}{7} \quad \frac{13}{10} \quad \longleftarrow \text{numerator}$$
$$\longleftarrow \text{denominator}$$

$\frac{2}{3}$ means 2 parts out of 3 equal parts

The number above the line is the **numerator**, and the number below the line is the **denominator**. In the fraction $\frac{2}{3}$, the numerator is 2 and the denominator is 3. The denominator is the number of equal parts into which something is divided. The numerator tells how many of these parts are needed. For example, $\frac{2}{3}$ is "2 parts out of 3 equal parts," as shown in the figure to the left.

Objective 1 Recognize types of fractions. If the numerator of a fraction is smaller than the denominator, the fraction is a **proper fraction**. Examples of proper fractions are $\frac{2}{3}, \frac{3}{4}, \frac{15}{16}$, and $\frac{1}{8}$. A fraction with a numerator greater than or equal to the denominator is an **improper fraction**. Examples of improper fractions are $\frac{17}{13}, \frac{19}{12}$, and $\frac{5}{5}$. A proper fraction has a value less than 1, while an improper fraction has a value greater than or equal to 1.

To write a whole number as a fraction, place the whole number over 1; for example, $7 = \frac{7}{1}$ and $12 = \frac{12}{1}$. The sum of a fraction and a whole number is a **mixed number**. Examples of mixed numbers include $5\frac{2}{3}$ (a short way of writing $5 + \frac{2}{3}$), $3\frac{5}{8}$, and $9\frac{5}{6}$. A mixed number can be converted to an improper fraction as shown next.

Objective 2 Convert mixed numbers to improper fractions. To convert the mixed number $4\frac{5}{8}$ to an improper fraction, first multiply the denominator of the fraction part (in this case, 8) and the whole number part (in this case, 4). This gives $4 \times 8 = 32$ ($4 = \frac{32}{8}$). Then add the product (32) to the numerator (in this case, 5). This gives $32 + 5 = 37$. This sum is the numerator of the new improper fraction. The denominator stays the same.

$$4\frac{5}{8} = \frac{37}{8} \longleftarrow (4 \times 8) + 5$$

The opening in the kitchen cabinet to the left is for a microwave oven and measures $25\frac{3}{4}$ inches wide by $16\frac{1}{2}$ inches high. These are both mixed numbers. The thickness of the oak wood used for the cabinet is $\frac{7}{8}$ inch, which is a proper fraction.

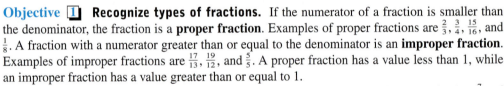

EXAMPLE 1

Converting Mixed Numbers to Improper Fractions

The width and height of the opening in the cabinet shown are both expressed as mixed numbers. Convert these mixed numbers to improper fractions.

(a) $25\frac{3}{4}$ (b) $16\frac{1}{2}$

SOLUTION

(a) First multiply 4 (the denominator) by 25 (the whole number), and then add 3 (the numerator). This gives $(4 \times 25) + 3 = 100 + 3 = 103$. The parentheses are used to show that 4 and 25 are multiplied first.

$$25\frac{3}{4} = \frac{103}{4} \longleftarrow (4 \times 25) + 3$$

(b) $16\frac{1}{2} = \frac{(2 \times 16) + 1}{2} = \frac{33}{2}$

Objective **3** **Convert improper fractions to mixed numbers.** To convert an improper fraction to a mixed number, divide the numerator of the improper fraction by the denominator. The quotient is the whole-number part of the mixed number, and the remainder is used as the numerator of the fraction part. The denominator stays the same. For example, convert $\frac{17}{5}$ to a mixed number by dividing 17 by 5.

$$\begin{array}{r} 3 \\ 5\overline{)17} \\ \underline{15} \\ 2 \end{array} \qquad \frac{17}{5} = 3\frac{2}{5}$$

The whole-number part is the quotient 3. The remainder 2 is used as the numerator of the fraction part. Keep 5 as the denominator.

$$\frac{17}{5} = 3\frac{2}{5}$$

EXAMPLE 2

Converting Improper Fractions to Mixed Numbers

Convert the following improper fractions to mixed numbers.

(a) $\frac{27}{4}$ **(b)** $\frac{29}{8}$ **(c)** $\frac{42}{7}$

SOLUTION

(a) Convert $\frac{27}{4}$ to a mixed number by dividing 27 by 4.

$$\begin{array}{r} 6 \\ 4\overline{)27} \\ \underline{24} \\ 3 \end{array} \qquad \frac{27}{4} = 6\frac{3}{4}$$

The whole-number part of the mixed number is 6. The remainder 3 is used as the numerator of the fraction. Keep 4 as the denominator.

$$\frac{27}{4} = 6\frac{3}{4}$$

(b) Divide 29 by 8 to convert $\frac{29}{8}$ to a mixed number.

$$\begin{array}{r} 3 \\ 8\overline{)29} \\ \underline{24} \\ 5 \end{array} \qquad \frac{29}{8} = 3\frac{5}{8}$$

(c) Divide 42 by 7 to convert $\frac{42}{7}$ to a mixed number.

$$\begin{array}{r} 6 \\ 7\overline{)42} \\ \underline{42} \\ 0 \end{array} \qquad \frac{42}{7} = 6$$

QUICK TIP   A proper fraction has a value that is less than 1, while an improper fraction has a value that is greater than or equal to 1.

Objective 4   Write a fraction in lowest terms. If both the numerator and denominator of a fraction cannot be divided without remainder by any number other than 1, then the fraction is in **lowest terms**. For example, 2 and 3 cannot be divided without remainder by any number other than 1, so the fraction $\frac{2}{3}$ is in lowest terms. In the same way, $\frac{1}{9}$, $\frac{4}{11}$, $\frac{12}{17}$, and $\frac{13}{15}$ are in lowest terms.

When both numerator and denominator *can* be divided without remainder by a number other than 1, the fraction is *not* in lowest terms. For example, both 15 and 25 may be divided by 5, so the fraction $\frac{15}{25}$ is not in lowest terms. Write $\frac{15}{25}$ in lowest terms by dividing both numerator and denominator by 5, as follows.

$$\frac{15}{25} = \frac{15 \div 5}{25 \div 5} = \frac{3}{5}$$

Divide by 5.

EXAMPLE 3

Writing Fractions in Lowest Terms

Write the following fractions in lowest terms.

(a) $\frac{15}{40}$ (b) $\frac{33}{39}$

SOLUTION

Look for a number that can be divided into both the numerator and denominator.

(a) Both 15 and 40 can be divided by **5**.

$$\frac{15}{40} = \frac{15 \div 5}{40 \div 5} = \frac{3}{8}$$

(b) Divide by 3.

$$\frac{33}{39} = \frac{33 \div 3}{39 \div 3} = \frac{11}{13}$$

Objective 5   Use the rules for divisibility. It is sometimes difficult to tell which numbers will divide evenly into another number. The following rules can sometimes help.

Rules for Divisibility

A number can be evenly divided by

2   if the last digit is an even number, such as 0, 2, 4, 6, or 8
3   if the sum of the digits is divisible by 3
4   if the last two digits are divisible by 4
5   if the last digit is 0 or 5
6   if the number is even and the sum of the digits is divisible by 3
8   if the last three digits are divisible by 8
9   if the sum of all the digits is divisible by 9
10   if the last digit is 0

EXAMPLE 4

Using the Divisibility Rules

Determine whether the following statements are true.

(a) 3,746,892 is divisible by 4.
(b) 15,974,802 is divisible by 9.

SOLUTION

(a) The number 3,746,892 is divisible by 4, since the last two digits form a number divisible by 4.

3,746,8**92**

92 is divisible by 4.

(b) See if 15,974,802 is divisible by 9 by adding the digits of the number.

$$1 + 5 + 9 + 7 + 4 + 8 + 0 + 2 = 36$$

36 is divisible by 9.

Since 36 is divisible by 9, the given number is divisible by 9.

QUICK TIP Testing for divisibility by adding the digits works only for 3 and 9.

The rules for divisibility only help determine whether a number is evenly divisible by another number. They cannot be used to find the result. You must carry out the division to find the quotient.

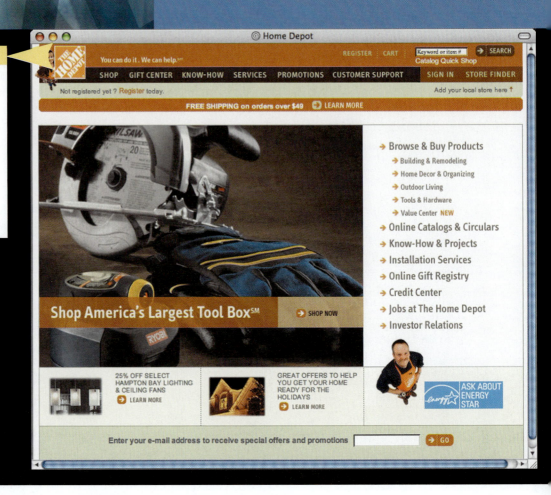

The Home Depot

- 1979: Founded in Atlanta, Georgia.

- 2002: $58.2 billion in annual sales.

- 2004: 300,000 employees worldwide.

- 2005: 2300 stores

The Home Depot is the world's largest home-improvement retailer. It is credited as being the innovator of the home-improvement industry, as well as offering a level of service unprecedented among warehouse-style retailers. Home Depot stores cater to do-it-yourselfers, as well as home-improvement, construction, and building maintenance professionals. Each store stocks approximately 35,000 building, home-improvement, and lawn-and-garden products (including variations in color and size). Newer stores are approximately 130,000 square feet or the size of four football fields.

The company's progressive corporate culture includes a philanthropic budget that directs funds back to the communities The Home Depot serves and the interests of its employees through a matching-gift program. The major focuses are affordable housing, at-risk youth, and the environment. Team Depot, an organized volunteer force, was developed in 1992 to promote volunteer activities with the local communities the stores serve.

1. A gutter downspout is 10 feet long. If a piece of gutter downspout 8 feet 8.375 inches is needed for a job, find the length of the remaining piece. (*Hint:* one foot equals 12 inches)

2. The Home Depot stock is selling for $37.09 per share. Find the number of shares that can be purchased for $10,088.48.

3. Give five specific situations in which fractions and mixed numbers would be used in a home-improvement store.

4. From your own experiences and those of family members and classmates, list eight specific activities where the ability to work with fractions would be needed.

2.1 | Exercises

FOR EXTRA HELP

 MyMathLab

 InterActMath.com

Math*XP* MathXL

 MathXL
Tutorials on CD

Tutor Center Addison-Wesley
Math Tutor Center

 DVT/Videotape

The **Quick Start** *exercises in each section contain solutions to help you get started.*

Convert the following mixed numbers to improper fractions. (See Example 1.)

Quick Start

1. $3\frac{5}{8} = \frac{29}{8}$ _____

$\frac{(8 \times 3) + 5}{8} = \frac{29}{8}$

2. $2\frac{4}{5} = \frac{14}{5}$ _____

$\frac{(5 \times 2) + 4}{5} = \frac{14}{5}$

3. $4\frac{1}{4} =$ _____

4. $3\frac{2}{3} =$ _____

5. $12\frac{2}{3} =$ _____

6. $2\frac{8}{11} =$ _____

7. $22\frac{7}{8} =$ _____

8. $17\frac{5}{8} =$ _____

9. $7\frac{6}{7} =$ _____

10. $21\frac{14}{15} =$ _____

11. $15\frac{19}{23} =$ _____

12. $7\frac{9}{16} =$ _____

Convert the following improper fractions to mixed or whole numbers and write in lowest terms.
(See Examples 2 and 3.)

Quick Start

13. $\frac{13}{4} = 3\frac{1}{4}$ _____

$\begin{array}{r} 3 \\ 4\overline{)13} \\ \underline{12} \\ 1 \end{array}$ $3\frac{1}{4}$ _____

14. $\frac{9}{5} = 1\frac{4}{5}$ _____

$\begin{array}{r} 1 \\ 5\overline{)9} \\ \underline{5} \\ 4 \end{array}$ $1\frac{4}{5}$

15. $\frac{8}{3} =$ _____

16. $\frac{23}{10} =$ _____

17. $\frac{38}{10} =$ _____

18. $\frac{56}{8} =$ _____

19. $\frac{40}{11} =$ _____

20. $\frac{78}{12} =$ _____

21. $\frac{125}{63} =$ _____

22. $\frac{195}{45} =$ _____

23. $\frac{183}{25} =$ _____

24. $\frac{720}{149} =$ _____

[C] indicates an exercise that is related to the Case in Point feature within the section.

25. Your classmate asks you how to change a mixed number to an improper fraction. Write a couple of sentences explaining how this is done. (See Objective 2.)

26. Explain in a sentence or two how to change an improper fraction to a mixed number. (See Objective 3.)

Write the following in lowest terms. (See Example 3.)

Quick Start

27. $\dfrac{8}{16} = \dfrac{1}{2}$

$\dfrac{8 \div 8}{16 \div 8} = \dfrac{1}{2}$

28. $\dfrac{15}{20} = \dfrac{3}{4}$

$\dfrac{15 \div 5}{20 \div 5} = \dfrac{3}{4}$

29. $\dfrac{16}{24} = $ ___

30. $\dfrac{25}{40} = $ ___

31. $\dfrac{36}{42} = $ ___

32. $\dfrac{27}{45} = $ ___

33. $\dfrac{60}{108} = $ ___

34. $\dfrac{30}{66} = $ ___

35. $\dfrac{112}{128} = $ ___

36. $\dfrac{165}{180} = $ ___

37. $\dfrac{12}{600} = $ ___

38. $\dfrac{96}{132} = $ ___

39. What does it mean when a fraction is expressed in lowest terms? (See Objective 4.)

40. There were eight rules of divisibility given. Write the three rules that are most useful to you. (See Objective 5.)

Put a check mark in the blank if the number at the left is evenly divisible by the number at the top.
Put an X in the blank if the number is not divisible. (See Example 4.)

Quick Start

	2	3	4	5	6	8	9	10
41. 32	✓	X	✓	X	X	✓	X	X
42. 45	X	✓	X	✓	X	X	✓	X
43. 60	—	—	—	—	—	—	—	—
44. 72	—	—	—	—	—	—	—	—
45. 90	—	—	—	—	—	—	—	—
46. 105	—	—	—	—	—	—	—	—
47. 4172	—	—	—	—	—	—	—	—
48. 5688	—	—	—	—	—	—	—	—

2.2 | Addition and Subtraction of Fractions

Objectives

1. Add and subtract like fractions.
2. Find the least common denominator.
3. Add and subtract unlike fractions.
4. Rewrite fractions with a common denominator.

CASE *in* POINT

Sarah Bryn must use fractions on a daily basis as she works with contractors and homeowners at The Home Depot. The measurements of cabinets, trim pieces, and room sizes never seem to be even numbers of inches—they always have fractions of an inch.

Objective 1 **Add and subtract like fractions.** Fractions with the same denominator are called **like fractions**. Such fractions have a **common denominator**. For example, $\frac{3}{4}$ and $\frac{5}{4}$ are *like* fractions with a common denominator of 4, while $\frac{4}{7}$ and $\frac{4}{9}$ are *not like* fractions. Add or subtract like fractions by adding or subtracting the numerators, and then place the result over the common denominator.

EXAMPLE 1

Adding and Subtracting Like Fractions

Add or subtract.

(a) $\frac{3}{4} + \frac{1}{4} + \frac{5}{4}$ (b) $\frac{11}{15} - \frac{4}{15}$

SOLUTION

The fractions in both parts of this example are like fractions. Add or subtract the numerators and place the result over the common denominator.

(a) $\frac{3}{4} + \frac{1}{4} + \frac{5}{4} = \frac{\mathbf{3 + 1 + 5}}{\mathbf{4}}$ ⟵ Add the numerators.

⟵ Write the common denominator.

$= \frac{9}{4} = 2\frac{1}{4}$ ⟵ Write the answer as a mixed number.

(b) $\frac{11}{15} - \frac{4}{15} = \frac{\mathbf{11 - 4}}{\mathbf{15}} = \frac{7}{15}$

Objective 2 **Find the least common denominator.** Fractions with different denominators, such as $\frac{3}{4}$ and $\frac{2}{3}$, are **unlike fractions**. Add or subtract unlike fractions by first writing the fractions with a common denominator. The **least common denominator (LCD)** for two or more fractions is the smallest whole number that can be divided, without remainder, by all the denominators of the fractions. For example, the LCD of the fractions $\frac{3}{4}, \frac{5}{6}$, and $\frac{1}{2}$ is 12, since 12 is the smallest number that can be divided by 4, 6, and 2.

Notice that the fractions shown in the following shelf-end base drawing are *like fractions*, $23\frac{3}{16}, 10\frac{9}{16}$, and $11\frac{3}{16}$. However, in the drawing of the shelf-end peninsula base, the fractions are *unlike fractions*, $22\frac{7}{16}, 11\frac{3}{32}$, and $11\frac{5}{8}$.

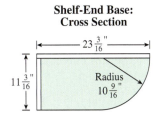

Shelf-End Base:
Cross Section

Shelf-End Peninsula Base:
Cross Section

There are two methods of finding the least common denominator.

Inspection. With small denominators, it may be possible to find the least common denominator by inspection. For example, the LCD for $\frac{1}{3}$ and $\frac{1}{5}$ is 15, the smallest number that can be divided evenly by both 3 and 5.

Method of prime numbers. If the LCD cannot be found by inspection, use the method of prime numbers, as explained in the next example.

A **prime number** is a number that can be divided without remainder by exactly two distinct numbers: itself and 1. Prime numbers are 2, 3, 5, 7, 11, 13, 17, and so on. (1 is *not* prime, because it can be divided evenly by only *one* number: the number 1.)

QUICK TIP   All prime numbers other than 2 are odd numbers. Not all odd numbers, however, are prime numbers. For example, 27 is the product of 3 and 9.

EXAMPLE 2

Finding the Least Common Denominator

Use the method of prime numbers to find the least common denominator for $\frac{5}{12}$, $\frac{7}{18}$, and $\frac{11}{20}$.

SOLUTION

First write the three denominators: 12   18   20

Begin by trying to divide the three denominators by the first prime number, 2. Write each quotient directly above the given denominator as follows.

$$
\begin{array}{r}
\quad\; 6 \quad\; 9 \quad 10 \\
2\overline{)12 \quad 18 \quad 20}
\end{array}
$$

This way of writing the division is just a handy way of writing the separate problems $2\overline{)12}$, $2\overline{)18}$, and $2\overline{)20}$. Two of the new quotients, 6 and 10, can still be divided by 2, so perform the division again. Since 9 cannot be divided evenly by 2, just bring up the 9.

$$
\begin{array}{r}
\quad\; 3 \quad\; 9 \quad\; 5 \qquad \text{Just bring 9 up.}\\
2\overline{)\; 6 \quad\; 9 \quad 10} \\
2\overline{)12 \quad 18 \quad 20}
\end{array}
$$

None of the new quotients in the top row can be divided by 2, so try the next prime number, 3. The numbers 3 and 9 can be divided by 3, and one of the new quotients can still be divided by 3, so the division is performed again.

$$
\begin{array}{r}
\quad\; 1 \quad\; 1 \quad\; 5 \\
3\overline{)\; 1 \quad\; 3 \quad\; 5} \\
3\overline{)\; 3 \quad\; 9 \quad\; 5} \\
2\overline{)\; 6 \quad\; 9 \quad 10} \\
2\overline{)12 \quad 18 \quad 20}
\end{array}
$$

Since none of the new quotients in the top row can be divided by 3, try the next prime number, 5. The number 5 can be used only once, as shown.

$$
\begin{array}{r}
\quad\; 1 \quad\; 1 \quad\; 1 \\
5\overline{)\; 1 \quad\; 1 \quad\; 5} \\
3\overline{)\; 1 \quad\; 3 \quad\; 5} \\
3\overline{)\; 3 \quad\; 9 \quad\; 5} \\
2\overline{)\; 6 \quad\; 9 \quad 10} \\
2\overline{)12 \quad 18 \quad 20}
\end{array}
$$

Now that the top row contains only 1's, find the least common denominator by multiplying the prime numbers in the left column.

The least common denominator is $2 \times 2 \times 3 \times 3 \times 5 = 180$.

> **QUICK TIP**   It is not necessary to start with the smallest prime number as shown in Example 2. In fact, no matter which prime number we start with, we will still get the same least common denominator.

EXAMPLE 3

Finding the Least Common Denominator

Find the least common denominator for $\frac{3}{8}$, $\frac{5}{12}$, and $\frac{9}{10}$.

SOLUTION

Write the denominators in a row and use the method of prime numbers.

```
                  1   1    1
              5)1   1    5
              3)1   3    5
              2)2   3    5
              2)4   6    5
Start here →  2)8  12   10
```

The least common denominator is $2 \times 2 \times 2 \times 3 \times 5 = 120$.

> **QUICK TIP**   Sometimes it is tempting to use a number that is not prime when solving for the least common denominator. This should be avoided because the result is often something different from the least common denominator.

Unlike fractions may be added or subtracted using the following steps.

Adding or Subtracting Unlike Fractions

Step 1 Find the least common denominator (LCD).

Step 2 Rewrite the unlike fractions as like fractions having the least common denominator.

Step 3 Add or subtract numerators, placing answers over the LCD and reducing to lowest terms.

Objective **3** **Add and subtract unlike fractions.** To add or subtract unlike fractions, rewrite the fractions with a common denominator. Since Example 2 shows that 180 is the least common denominator for $\frac{5}{12}$, $\frac{7}{18}$, and $\frac{11}{20}$, these three fractions can be added if each fraction is first written with a denominator of 180.

Step 1 $\qquad\qquad \frac{5}{12} = \frac{}{180} \qquad \frac{7}{18} = \frac{}{180} \qquad \frac{11}{20} = \frac{}{180}$

Objective **4** **Rewrite fractions with a common denominator.** To rewrite the preceding fractions with a common denominator, first divide each denominator from the original fractions into the common denominator.

Step 2 $\qquad\qquad 12\overline{)180}^{\,15} \qquad 18\overline{)180}^{\,10} \qquad 20\overline{)180}^{\,9}$

Next multiply each quotient by the original numerator.

$$15 \times 5 = 75 \qquad 10 \times 7 = 70 \qquad 9 \times 11 = 99$$

Now, rewrite the fractions.

$$\frac{5}{12} = \frac{75}{180} \qquad \frac{7}{18} = \frac{70}{180} \qquad \frac{11}{20} = \frac{99}{180}$$

Now add the fractions.

Step 3
$$\frac{5}{12} + \frac{7}{18} + \frac{11}{20} = \frac{75}{180} + \frac{70}{180} + \frac{99}{180} = \frac{75 + 70 + 99}{180}$$

$$= \frac{244}{180} = 1\frac{64}{180} = 1\frac{16}{45} \quad \text{Write the answer as a mixed number with the fraction in lowest terms.}$$

EXAMPLE 4

Adding and Subtracting Unlike Fractions

Add or subtract.

(a) $\frac{3}{4} + \frac{1}{2} + \frac{5}{8}$ **(b)** $\frac{9}{10} - \frac{3}{8}$

SOLUTION

(a) Inspection shows that the least common denominator is 8. Rewrite the fractions so they each have a denominator of 8. Then add.

$$\frac{3}{4} + \frac{1}{2} + \frac{5}{8} = \frac{6}{8} + \frac{4}{8} + \frac{5}{8} = \frac{6 + 4 + 5}{8} = \frac{15}{8} = 1\frac{7}{8}$$

(b) The least common denominator is 40. Rewrite the fractions so they each have a denominator of 40. Then subtract.

$$\frac{9}{10} - \frac{3}{8} = \frac{36}{40} - \frac{15}{40} = \frac{36 - 15}{40} = \frac{21}{40}$$

Fractions can also be added or subtracted vertically, as shown in the next example.

EXAMPLE 5

Adding and Subtracting Unlike Fractions

Add or subtract.

(a) $\frac{2}{9} + \frac{3}{4}$ **(b)** $\frac{11}{16} + \frac{7}{12}$ **(c)** $\frac{7}{8} - \frac{5}{12}$

SOLUTION

First rewrite the fractions with a least common denominator.

(a)
$$\begin{array}{r} \frac{2}{9} = \frac{8}{36} \\ + \frac{3}{4} = \frac{27}{36} \\ \hline \frac{35}{36} \end{array}$$

(b)
$$\begin{array}{r} \frac{11}{16} = \frac{33}{48} \\ + \frac{7}{12} = \frac{28}{48} \\ \hline \frac{61}{48} = 1\frac{13}{48} \end{array}$$

(c)
$$\begin{array}{r} \frac{7}{8} = \frac{21}{24} \\ - \frac{5}{12} = \frac{10}{24} \\ \hline = \frac{11}{24} \end{array}$$

All calculator solutions are shown using a basic calculator. The calculator solution to Example 5(b) uses the fraction key on the calculator.

$$11 \boxed{a^{b/c}} \ 16 \boxed{+} \ 7 \boxed{a^{b/c}} \ 12 \boxed{=} \ 1\tfrac{13}{48}$$

Note: Refer to Appendix C for calculator basics.

2.2 | Exercises

FOR EXTRA HELP

 MyMathLab

 InterActMath.com

MathXL MathXL

 MathXL Tutorials on CD

Tutor Center Addison-Wesley Math Tutor Center

 DVT/Videotape

The **Quick Start** exercises in each section contain solutions to help you get started.

Convert each fraction so that it has the indicated denominator. (See Objective 3.)

Quick Start

1. $\dfrac{4}{5} = \dfrac{16}{20}$

   20 ÷ 5 = 4
   4 × 4 = 16

2. $\dfrac{3}{4} = \dfrac{12}{16}$

   16 ÷ 4 = 4
   4 × 3 = 12

3. $\dfrac{9}{10} = \dfrac{}{40}$

4. $\dfrac{7}{8} = \dfrac{}{56}$

5. $\dfrac{6}{5} = \dfrac{}{40}$

6. $\dfrac{7}{8} = \dfrac{}{64}$

7. $\dfrac{6}{7} = \dfrac{}{49}$

8. $\dfrac{11}{15} = \dfrac{}{120}$

Find the least common denominator for each group of denominators using the method of prime numbers. (See Examples 2 and 3.)

Quick Start

9. 3, 8, **24**

```
      1  1
   3)3  1
   2)3  2
   2)3  4
   2)3  8
```
2 × 2 × 2 × 3 = 24

10. 18, 24, **72**

```
        1   1
   3) 3   1
   3) 9   3
   2) 9   6
   2) 9  12
   2)18  24
```
2 × 2 × 2 × 3 × 3 = 72

11. 12, 18, 20, _____

12. 18, 20, 24, _____

13. 15, 24, 32, _____

14. 6, 8, 10, 12, _____

15. 10, 35, 50, 60, _____

16. 5, 18, 25, 30, 36, _____

17. 3, 5, 8, 12, 18, _____

indicates an exercise that is related to the Case in Point feature within the section.

18. Prime numbers are used to find the least common denominator. Write the definition of a prime number in your own words. (See Objective 2.)

19. Explain how to write a fraction with an indicated denominator. Give an example changing $\frac{3}{4}$ to a fraction having 12 as a denominator. (See Objective 4.)

Add or subtract. Write answers in lowest terms. (See Examples 4 and 5.)

Quick Start

20. $\frac{2}{5} + \frac{1}{5} = \frac{3}{5}$ _____

$\frac{2+1}{5} = \frac{3}{5}$

21. $\frac{2}{9} + \frac{4}{9} = \frac{2}{3}$ _____

$\frac{2=4}{9} = \frac{6}{9} = \frac{2}{3}$

22. $\frac{5}{8} + \frac{7}{12} =$ _____

23. $\frac{11}{12} - \frac{5}{12} =$ _____

24. $\frac{5}{7} - \frac{1}{3} =$ _____

25. $\frac{5}{12} - \frac{1}{16} =$ _____

26. $\frac{2}{3} - \frac{3}{8} =$ _____

27. $\frac{3}{4} + \frac{5}{9} + \frac{1}{3} =$ _____

28. $\frac{1}{4} + \frac{1}{8} + \frac{1}{12} =$ _____

29. $\frac{3}{7} + \frac{2}{5} + \frac{1}{10} =$ _____

30. $\frac{5}{6} + \frac{3}{4} + \frac{5}{8} =$ _____

31. $\frac{7}{10} + \frac{8}{15} + \frac{5}{6} =$ _____

32. $\frac{3}{10} + \frac{2}{5} + \frac{3}{20} =$ _____

33. $\frac{3}{4}$
$\frac{2}{3}$
$+\frac{8}{9}$

34. $\frac{7}{12}$
$\frac{5}{8}$
$+\frac{7}{6}$

35. $\frac{8}{15}$
$\frac{3}{10}$
$+\frac{3}{5}$

36. $\frac{1}{6}$
$\frac{5}{9}$
$+\frac{13}{18}$

37. $\frac{7}{10}$
$-\frac{1}{4}$

38. $\frac{4}{5}$
$-\frac{2}{3}$

39. $\frac{5}{8}$
$-\frac{1}{3}$

40. $\frac{19}{24}$
$-\frac{5}{16}$

41. Where are fractions used in everyday life? Think in terms of business applications, hobbies, and personal finance. Give three examples.

42. With the exception of the number 2, all prime numbers are odd numbers. However, not all odd numbers are prime numbers. Explain why these statements are true. (See Objective 2.)

Solve the following application problems.

43. GARDENING   Carolyn Phelps is planting her flower bed and has ordered $\frac{1}{4}$ cubic yard of sand, $\frac{3}{8}$ cubic yard of mulch, and $\frac{1}{3}$ cubic yard of peat moss. Find the total cubic yards that she has ordered.

43. _____

44. AUTO REPAIR   Richard McMenamy has used his savings to repair his car. He spent $\frac{1}{4}$ of his savings for new tires, $\frac{1}{6}$ of his savings for brakes, $\frac{1}{10}$ of his savings for a tune-up, and $\frac{1}{12}$ of his savings for new belts and hoses. What fraction of his total savings has he spent?

44. _____

45. COMPUTER ASSEMBLY   When installing a printer cable to a computer, Ann Kuick must be certain that the proper type and size of mounting hardware is used. Find the total length of the bolt shown.

45. _____

46. CABINET INSTALLATION   When installing cabinets for The Home Depot, Sarah Bryn must be certain that the proper type and size of mounting screw is used. Find the total length of the screw.

46. _____

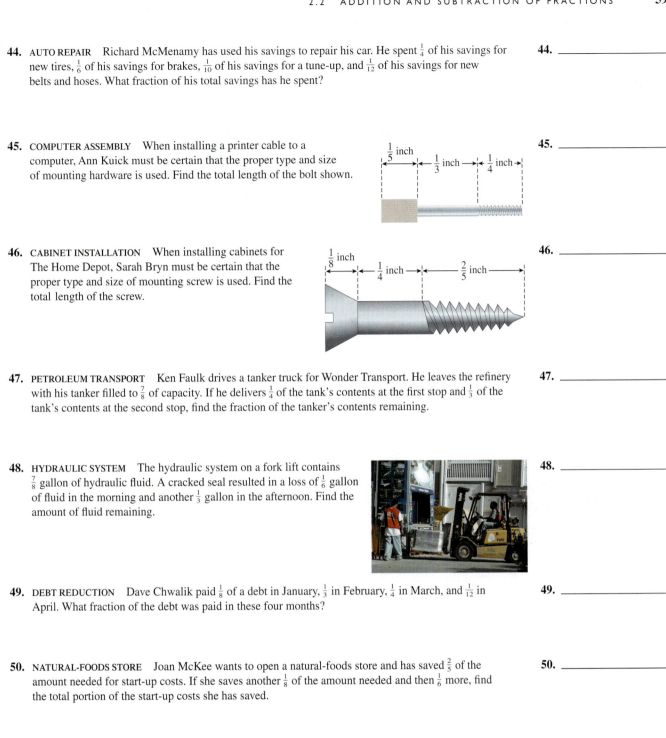

47. PETROLEUM TRANSPORT   Ken Faulk drives a tanker truck for Wonder Transport. He leaves the refinery with his tanker filled to $\frac{7}{8}$ of capacity. If he delivers $\frac{1}{4}$ of the tank's contents at the first stop and $\frac{1}{3}$ of the tank's contents at the second stop, find the fraction of the tanker's contents remaining.

47. _____

48. HYDRAULIC SYSTEM   The hydraulic system on a fork lift contains $\frac{7}{8}$ gallon of hydraulic fluid. A cracked seal resulted in a loss of $\frac{1}{6}$ gallon of fluid in the morning and another $\frac{1}{3}$ gallon in the afternoon. Find the amount of fluid remaining.

48. _____

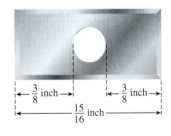

49. DEBT REDUCTION   Dave Chwalik paid $\frac{1}{8}$ of a debt in January, $\frac{1}{3}$ in February, $\frac{1}{4}$ in March, and $\frac{1}{12}$ in April. What fraction of the debt was paid in these four months?

49. _____

50. NATURAL-FOODS STORE   Joan McKee wants to open a natural-foods store and has saved $\frac{2}{5}$ of the amount needed for start-up costs. If she saves another $\frac{1}{8}$ of the amount needed and then $\frac{1}{6}$ more, find the total portion of the start-up costs she has saved.

50. _____

51. CABINET INSTALLATION   Find the diameter of the hole in the mounting bracket shown.

51. _____

52. SWIMMER TRAINING   Sheri Minkner will swim $\frac{7}{8}$ of a mile in a five-day period. If she swims $\frac{1}{8}$ of a mile on both Monday and Wednesday, $\frac{1}{6}$ of a mile on both Tuesday and Thursday, and the balance on Friday, find the distance she must swim on Friday.

52. _____

STUDENT TIME MANAGEMENT   *Refer to the circle graph to answer Exercises 53–56.*

The Day of the Student

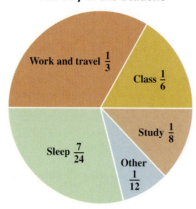

53. What fraction of the day was spent in class and study?

53. _____

54. What fraction of the day was spent in work and travel and other?

54. _____

55. In which activity was the greatest amount of time spent? How many hours did this activity take?

55. _____

56. In which activity was the least amount of time spent? How many hours did this activity take?

56. _____

57. **PERIMETER OF FENCING**   A hazardous-waste site will require $\frac{7}{8}$ mile of security fencing. The site has four sides, three of which measure $\frac{1}{4}$ mile, $\frac{1}{6}$ mile, and $\frac{3}{8}$ mile. Find the length of the fourth side.

57. _____

58. **NATIVE-AMERICAN JEWELRY**   Chakotay is fitting a turquoise stone into a bear-claw pendant. Find the diameter of the hole in the pendant. (The diameter is the distance across the center of the hole.)

58. _____

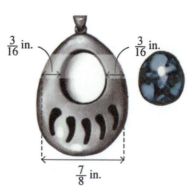

$\frac{3}{16}$ in. $\frac{3}{16}$ in. $\frac{7}{8}$ in.

2.3 | Addition and Subtraction of Mixed Numbers

Objectives

1. Add mixed numbers.
2. Add with carrying.
3. Subtract mixed numbers.
4. Subtract with borrowing.

CASE *in* **POINT**

Total customer satisfaction is important to Sarah Bryn and The Home Depot. Complete accuracy is just as important in the small jobs in hardware and trim as it is in the large jobs in cabinets and installation. To achieve this accuracy, Bryn knows that mixed numbers must be added and subtracted carefully and that all calculations must then be checked to make sure they are correct.

Objective 1 **Add mixed numbers.** To add mixed numbers, first add the fractions. Then add the whole numbers and combine the two answers. For example, add $16\frac{1}{8}$ and $5\frac{5}{8}$ as shown.

sum of fractions

$$16\frac{1}{8} + 5\frac{5}{8} = 21\frac{6}{8}$$

sum of whole numbers

Write $\frac{6}{8}$ in lowest terms as $\frac{3}{4}$, so that $16\frac{1}{8} + 5\frac{5}{8} = 21\frac{3}{4}$.

To add mixed numbers, change the mixed numbers, if necessary, so that the fraction parts have a common denominator.

EXAMPLE 1

Adding Mixed Numbers

Add $9\frac{2}{3}$ and $6\frac{1}{4}$.

SOLUTION

Inspection shows that 12 is the least common denominator. Write $9\frac{2}{3}$ as $9\frac{8}{12}$, and write $6\frac{1}{4}$ as $6\frac{3}{12}$. Then add. The work can be organized as follows.

$$9\frac{2}{3} = 9\frac{8}{12}$$
$$+ 6\frac{1}{4} = 6\frac{3}{12}$$
$$\overline{\phantom{+6\frac{1}{4} = }15\frac{11}{12}}$$

Objective 2 **Add with carrying.** If the sum of the fraction parts of mixed numbers is greater than 1, carry the excess from the fraction part to the whole number part.

EXAMPLE 2

Adding with Carrying

A rubber gasket must extend around all four edges (perimeter) of the dishwasher door panel shown below before it is installed. Find the length of gasket material needed. Add $34\frac{1}{2}$ inches, $23\frac{3}{4}$ inches, $34\frac{1}{2}$ inches, and $23\frac{3}{4}$ inches.

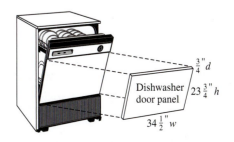

Dishwasher door panel

$\frac{3}{4}$" d

$23\frac{3}{4}$" h

$34\frac{1}{2}$" w

SOLUTION

$$34\frac{1}{2} = 34\frac{2}{4}$$

$$23\frac{3}{4} = 23\frac{3}{4}$$

$$34\frac{1}{2} = 34\frac{2}{4}$$

$$+\ 23\frac{3}{4} = 23\frac{3}{4}$$

$$\frac{10}{4} = 2\frac{2}{4}$$

$$114\frac{10}{4} = 114 + \frac{10}{4} = 114 + 2\frac{2}{4} = 116\frac{2}{4} = 116\frac{1}{2} \text{ inches} \qquad \text{length of gasket needed}$$

QUICK TIP When adding mixed numbers, first add the fraction parts. Then add the whole-number parts. Finally, combine the two answers.

Objective **3** **Subtract mixed numbers.** To subtract two mixed numbers, change the mixed numbers, if necessary, so that the fraction parts have a common denominator. Then subtract the fraction parts and the whole-number parts separately. For example, subtract $3\frac{1}{12}$ from $8\frac{5}{8}$ by first finding that the least common denominator is 24. Then rewrite the problem as shown.

$$8\frac{5}{8} - 3\frac{1}{12}$$

$$\uparrow \text{ use } \uparrow$$
$$24$$

as a common denominator

$$8\frac{15}{24} - 3\frac{2}{24}$$

$$8\frac{15}{24}$$
$$-\ 3\frac{2}{24}$$
$$\overline{\rule{0pt}{1.2em}}$$
$$5\frac{13}{24}$$

Now subtract the fraction parts and subtract the whole-number parts.

Subtract fractions.
Subtract whole numbers.

Objective **4** **Subtract with borrowing.** The following example shows how to subtract when borrowing is needed.

EXAMPLE 3

Subtracting with Borrowing

(a) Subtract $6\frac{3}{4}$ from $10\frac{1}{8}$. **(b)** Subtract $15\frac{7}{12}$ from 41.

SOLUTION

Start by rewriting each problem with a common denominator.

(a)
$$10\frac{1}{8} = 10\frac{1}{8}$$
$$-\ 6\frac{3}{4} = 6\frac{6}{8}$$

Subtracting $\frac{6}{8}$ from $\frac{1}{8}$ requires borrowing from the whole number 10.

$$10\frac{1}{8} = 9 + 1 + \frac{1}{8}$$
$$= 9 + \frac{8}{8} + \frac{1}{8} = 9\frac{9}{8} \qquad 1 = \frac{8}{8}$$

Rewrite the problem as shown. Check by adding $3\frac{3}{8}$ and $6\frac{3}{4}$. The answer should be $10\frac{1}{8}$.

$$10\frac{1}{8} = 9\frac{9}{8}$$
$$-\ 6\frac{6}{8} = 6\frac{6}{8}$$
$$\overline{\rule{0pt}{1.2em}}$$
$$3\frac{3}{8}$$

(b)
$$41$$
$$-\ 15\frac{7}{12}$$

To subtract the fraction $\frac{7}{12}$ requires borrowing 1 whole unit from 41.

$$41 = 40 + 1 = 40 + \frac{12}{12} = 40\frac{12}{12} \qquad 1 = \frac{12}{12}$$

Rewrite the problem as shown. Check by adding $25\frac{5}{12}$ and $15\frac{7}{12}$. The answer should be 41.

$$41\ \ = 40\frac{12}{12}$$
$$-\ 15\frac{7}{12} = 15\frac{7}{12}$$
$$\overline{\rule{0pt}{1.2em}}$$
$$25\frac{5}{12}$$

The calculator solution to Example 3(a) uses the fraction key.

$$10\ \boxed{a^{b/c}}\ 1\ \boxed{a^{b/c}}\ 8\ \boxed{-}\ 6\ \boxed{a^{b/c}}\ 3\ \boxed{a^{b/c}}\ 4\ \boxed{=}\ 3\frac{3}{8}$$

Note: Refer to Appendix C for calculator basics.

2.3 | Exercises

The **Quick Start** exercises in each section contain solutions to help you get started.

Add. Write each answer in lowest terms. (See Examples 1 and 2.)

Quick Start

1. $82\frac{3}{5}$ $82\frac{3}{5}$
    $+15\frac{1}{5}$ $+15\frac{1}{5}$
    $\overline{97\frac{4}{5}}$ $\overline{97\frac{4}{5}}$

2. $25\frac{2}{7}$ $25\frac{2}{7}$
    $+14\frac{3}{7}$ $+14\frac{3}{7}$
    $\overline{39\frac{5}{7}}$ $\overline{39\frac{5}{7}}$

3. $41\frac{1}{2}$
    $+39\frac{1}{4}$

4. $28\frac{1}{4}$
    $23\frac{3}{5}$
    $+19\frac{9}{10}$

5. $46\frac{3}{4}$
    $12\frac{5}{8}$
    $+37\frac{4}{5}$

6. $26\frac{5}{8}$
    $17\frac{3}{14}$
    $+32\frac{2}{7}$

7. $32\frac{3}{4}$
    $6\frac{1}{3}$
    $+14\frac{5}{8}$

8. $16\frac{7}{10}$
    $26\frac{1}{5}$
    $+8\frac{3}{8}$

9. $46\frac{5}{8}$
    $21\frac{1}{6}$
    $+38\frac{1}{10}$

Subtract. Write each answer in lowest terms. (See Example 3.)

Quick Start

10. $16\frac{3}{4}$ $16\frac{18}{24}$
    $-12\frac{3}{8}$ $-12\frac{9}{24}$
    $\overline{4\frac{3}{8}}$ $\overline{4\frac{9}{24}}=4\frac{3}{8}$

11. $25\frac{13}{24}$ $25\frac{13}{24}$
    $-18\frac{5}{12}$ $-18\frac{10}{24}$
    $\overline{7\frac{1}{8}}$ $\overline{7\frac{3}{24}}=7\frac{1}{8}$

12. $9\frac{7}{8}$
    $-6\frac{5}{12}$

13. 374
    $-211\frac{5}{6}$

14. 19
    $-12\frac{3}{4}$

15. $71\frac{3}{8}$
    $-62\frac{1}{3}$

16. $6\frac{1}{3}$
    $-2\frac{5}{12}$

17. $72\frac{3}{10}$
    $-25\frac{8}{15}$

18. $23\frac{1}{2}$
    $-18\frac{3}{4}$

19. $5\frac{1}{10}$
    $-4\frac{2}{5}$

20. $15\frac{3}{18}$
    $-12\frac{8}{9}$

21. In your own words, explain the steps you would take to add two large mixed numbers. (See Objective 1.)

22. When subtracting mixed numbers, explain when you need to borrow. Explain how to borrow using an example. (See Objective 4.)

Solve the following application problems.

Quick Start

23. **WINDOW INSTALLATION**   A contractor who installs windows for The Home Depot must attach a lead strip around all four sides of a custom-made stained glass window. If the window measures $34\frac{1}{2}$ by $23\frac{3}{4}$ inches, find the length of lead stripping needed.

    $34\frac{1}{2}+23\frac{3}{4}+34\frac{1}{2}+23\frac{3}{4}=$

    $34\frac{2}{4}+23\frac{3}{4}+34\frac{2}{4}+23\frac{3}{4}=114\frac{10}{4}=116\frac{1}{2}$ inches

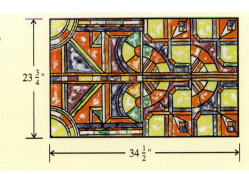

23. $116\frac{1}{2}$ inches

indicates an exercise that is related to the Case in Point feature within the section.

24. MEASURING BRASS TRIM   To complete a custom order, Sarah Bryn of The Home Depot must find the number of inches of brass trim needed to go around the four sides of the lamp base plate shown. Find the length of brass trim needed.

24. _____

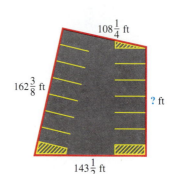

25. SECURITY FENCING   The exercise yard at the correction center has four sides and is enclosed with $527\frac{1}{24}$ feet of security fencing around it. If three sides of the yard measure $107\frac{2}{3}$ feet, $150\frac{3}{4}$ feet, and $138\frac{5}{8}$ feet, find the length of the fourth side.

25. _____

26. PARKING LOT FENCING   Three sides of a parking lot are $108\frac{1}{4}$ feet, $162\frac{3}{8}$ feet, and $143\frac{1}{2}$ feet. If the distance around the lot is $518\frac{3}{4}$ feet, find the length of the fourth side.

26. _____

$108\frac{1}{4}$ ft

$162\frac{3}{8}$ ft

? ft

$143\frac{1}{2}$ ft

27. DELIVERING CONCRETE   Chuck Stone has $8\frac{7}{8}$ cubic yards of concrete in a truck. If he unloads $2\frac{1}{2}$ cubic yards at the first stop, 3 cubic yards at the second stop, and $1\frac{3}{4}$ cubic yards at the third stop, how much concrete remains in the truck?

27. _____

28. TAILORED CLOTHING   Marv Levenson bought 15 yards of Italian silk fabric. He made two tops with $3\frac{3}{4}$ yards of the material, a suit for his wife with $4\frac{1}{8}$ yards, and a jacket with $3\frac{7}{8}$ yards. Find the number of yards of material remaining.

28. _____

29. PART-TIME WORK   Andrea Abriani, a college student, works part time at the Cyber Coffeehouse. She worked $3\frac{3}{8}$ hours on Monday, $5\frac{1}{2}$ hours on Tuesday, $4\frac{3}{4}$ hours on Wednesday, $3\frac{1}{4}$ hours on Thursday, and 6 hours on Friday. How many hours did she work altogether?

29. _____

30. AUTOMOTIVE SUPPLIES   Comet Auto Supply sold $16\frac{1}{2}$ cases of generic motor oil last week, $12\frac{1}{8}$ cases of Havoline oil, $8\frac{3}{4}$ cases of Valvoline oil, and $12\frac{5}{8}$ cases of Castrol oil. Find the total number of cases of oil that Comet Auto Supply sold during the week.

30. _____

2.4 | Multiplication and Division of Fractions

Objectives

1. Multiply proper fractions.
2. Use cancellation.
3. Multiply mixed numbers.
4. Divide fractions.
5. Divide mixed numbers.
6. Multiply or divide by whole numbers.

CASE *in* POINT

Most of the cabinets sold by The Home Depot are standard size units and modules that can be combined to satisfy varied applications and room sizes. However, all too often, Sarah Bryn finds that various components and trim pieces must be custom sized. In order to custom size items, she must multiply and divide fractions.

Objective 1 **Multiply proper fractions.** To multiply two fractions, first multiply the numerators to form a new numerator and then multiply the denominators to form a new denominator. Write the answer in lowest terms if necessary. For example, multiply $\frac{2}{3}$ and $\frac{5}{8}$ by first multiplying the numerators and then the denominators. This gives

$$\frac{2}{3} \times \frac{5}{8} = \frac{2 \times 5}{3 \times 8} = \frac{10}{24} = \frac{5}{12} \quad \text{(in lowest terms)}$$

Multiply numerators.

Multiply denominators.

Objective 2 **Use cancellation.** This problem can be simplified by **cancellation**, a modification of the method of writing fractions in lowest terms. For example, find the product of $\frac{2}{3}$ and $\frac{5}{8}$ by cancelling as follows.

$$\frac{\overset{1}{2}}{3} \times \frac{5}{\underset{4}{8}} = \frac{1 \times 5}{3 \times 4} = \frac{5}{12}$$

Divide 2 into both 2 and 8. Then multiply the numerators and, finally, multiply the denominators.

EXAMPLE 1

Multiplying Common Fractions

Multiply.

(a) $\frac{8}{15} \times \frac{5}{12}$ **(b)** $\frac{35}{12} \times \frac{32}{25}$

SOLUTION

Use cancellation in both of these problems.

(a) $\dfrac{\overset{2}{8}}{\underset{3}{15}} \times \dfrac{\overset{1}{5}}{\underset{3}{12}} = \dfrac{2 \times 1}{3 \times 3} = \dfrac{2}{9}$

Divide 4 into both 8 and 12.
Divide 5 into both 5 and 15.

(b) $\dfrac{\overset{7}{35}}{\underset{3}{12}} \times \dfrac{\overset{8}{32}}{\underset{5}{25}} = \dfrac{7 \times 8}{3 \times 5} = \dfrac{56}{15} = 3\dfrac{11}{15}$

Divide 4 into both 12 and 32.
Divide 5 into both 35 and 25.

QUICK TIP When cancelling, be certain that a numerator and a denominator are both divided by the same number.

Objective 3 Multiply mixed numbers. To multiply mixed numbers, change the mixed numbers to improper fractions, use cancellation, and then multiply them. For example, multiply $6\frac{1}{4}$ and $2\frac{2}{3}$ as follows.

$$6\frac{1}{4} \times 2\frac{2}{3} = \underbrace{\frac{25}{4} \times \frac{8}{3}}_{} = \frac{25}{\overset{}{4}} \times \frac{\overset{2}{8}}{3} = \frac{25 \times 2}{1 \times 3} = \frac{50}{3} = 16\frac{2}{3}$$

Change to improper fractions.

QUICK TIP Mixed numbers must always be changed to *improper fractions* before multiplying. When multiplying by mixed numbers, do not multiply whole numbers by whole numbers and fractions by fractions.

EXAMPLE 2

Multiplying Mixed Numbers

Multiply.

(a) $3\frac{3}{4} \times 8\frac{2}{3}$ **(b)** $1\frac{3}{5} \times 3\frac{1}{3} \times 1\frac{3}{4}$

SOLUTION

(a) $\dfrac{\overset{5}{\cancel{15}}}{\underset{2}{\cancel{4}}} \times \dfrac{\overset{13}{\cancel{26}}}{\underset{1}{\cancel{3}}} = \dfrac{5 \times 13}{2 \times 1} = \dfrac{65}{2} = 32\frac{1}{2}$

(b) $\dfrac{\overset{2}{\cancel{8}}}{\underset{1}{\cancel{5}}} \times \dfrac{\overset{2}{\cancel{10}}}{3} \times \dfrac{7}{\underset{1}{\cancel{4}}} = \dfrac{2 \times 2 \times 7}{1 \times 3 \times 1} = \dfrac{28}{3} = 9\frac{1}{3}$

The calculator solution to Example 2(b) uses the fraction key.

$$1 \boxed{a^{b/c}} 3 \boxed{a^{b/c}} 5 \boxed{\times} 3 \boxed{a^{b/c}} 1 \boxed{a^{b/c}} 3 \boxed{\times} 1 \boxed{a^{b/c}} 3 \boxed{a^{b/c}} 4 \boxed{=} 9\frac{1}{3}$$

Note: Refer to Appendix C for calculator basics.

The recipe shown next is easy to follow using proper measuring cups and spoons. Sometimes you may want to double or triple a recipe or perhaps, cooking for a small group, you need to cut the recipe in half. To double the recipe, multiply each ingredient by 2. To triple the recipe, multiply by 3. To halve the recipe you'll need to divide by 2.

Chocolate/Oat-Chip Cookies

1 cup (2 sticks) margarine or butter,
  softened
$1\frac{1}{4}$ cups firmly packed brown sugar
$\frac{1}{2}$ cup granulated sugar
2 eggs
2 tablespoons milk
2 teaspoons vanilla
$1\frac{3}{4}$ cups all-purpose flour
1 teaspoon baking soda

$\frac{1}{2}$ teaspoon salt (optional)
$2\frac{1}{2}$ cups uncooked oats
One 12-ounce package (2 cups)
  semi-sweet chocolate morsels
1 cup coarsely chopped nuts
  (optional)

Heat oven to 375°F. **Beat** margarine and sugars until creamy.

Add eggs, milk, and vanilla; beat well.

Add combined flour, baking soda, and salt; mix well. **Stir** in oats, chocolate morsels, and nuts; mix well.

Drop by rounded measuring tablespoonfuls onto ungreased cookie sheet.

Bake 9 to 10 minutes for a chewy cookie or 12 to 13 minutes for a crisp cookie.

Cool 1 minute on cookie sheet; remove to wire rack. Cool completely
  MAKES ABOUT 5 DOZEN

EXAMPLE 3

Multiplying Mixed Numbers by a Whole Number

(a) Find the amount of uncooked oats needed if the preceding recipe for chocolate/oat-chip cookies is doubled (multiplied by 2).

(b) How many cups of all-purpose flour are needed when the recipe is tripled (multiplied by 3)?

SOLUTION

(a) $2\frac{1}{2} \times 2 = \frac{5}{\overset{}{\underset{1}{2}}} \times \frac{\overset{1}{2}}{1} = \frac{5 \times 1}{1 \times 1} = \frac{5}{1} = 5$ cups

(b) $1\frac{3}{4} \times 3 = \frac{7}{4} \times \frac{3}{1} = \frac{7 \times 3}{4} = \frac{21}{4} = 5\frac{1}{4}$ cups

Objective 4 Divide fractions. To divide two fractions, invert the second fraction (divisor) and then multiply the first fraction by the inverted second fraction. (Invert a fraction by exchanging the numerator and the denominator.)

For example, divide $\frac{3}{8}$ by $\frac{7}{12}$ by *inverting* the second fraction and then *multiplying*.

$$\frac{3}{8} \div \frac{7}{12} = \frac{3}{8} \times \frac{12}{7} = \frac{3}{8} \times \frac{\overset{3}{\cancel{12}}}{7} = \frac{3 \times 3}{2 \times 7} = \frac{9}{14}$$

invert second fraction

multiply

> **QUICK TIP** Only the second fraction (divisor) is inverted when dividing by a fraction. Cancellation is done *only after inverting*.

EXAMPLE 4

Dividing Common Fractions

Divide.

(a) $\frac{7}{8} \div \frac{1}{4}$ (b) $\frac{25}{36} \div \frac{15}{18}$

SOLUTION

Invert the second fraction and then multiply.

(a) $\frac{7}{8} \div \frac{1}{4} = \frac{7}{\underset{2}{8}} \times \frac{\overset{1}{4}}{1} = \frac{7 \times 1}{2 \times 1} = \frac{7}{2} = 3\frac{1}{2}$ (b) $\frac{25}{36} \div \frac{15}{18} = \frac{\overset{5}{25}}{\underset{2}{36}} \times \frac{\overset{1}{18}}{\underset{3}{15}} = \frac{5 \times 1}{2 \times 3} = \frac{5}{6}$

Objective 5 Divide mixed numbers. To divide mixed numbers, first change all mixed numbers to improper fractions, invert the second fraction, use cancellation, and multiply.

invert

$$3\frac{5}{9} \div 2\frac{2}{5} = \frac{32}{9} \div \frac{12}{5} = \frac{\overset{8}{32}}{9} \times \frac{5}{\underset{3}{12}} = \frac{8 \times 5}{9 \times 3} = \frac{40}{27} = 1\frac{13}{27}$$

Objective 6 Multiply or divide by whole numbers. To multiply or divide a fraction by a whole number, write the whole number as a fraction over 1.

Multiply: $3\frac{3}{4} \times 16 = 3\frac{3}{4} \times \frac{16}{1} = \frac{15}{4} \times \frac{16}{1} = \frac{15}{\underset{1}{4}} \times \frac{\overset{4}{16}}{1} = 15 \times 4 = 60$

whole number over 1

Divide: $2\frac{2}{5} \div 3 = \frac{12}{5} \div \frac{3}{1} = \frac{\overset{4}{12}}{5} \times \frac{1}{\underset{1}{3}} = \frac{4 \times 1}{5 \times 1} = \frac{4}{5}$

Base-End Panel

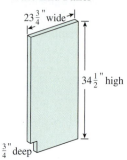

Mills Pride manufactures cabinets for kitchens and baths. The specifications for base-end panels are shown in the diagram. The lumber used is $\frac{3}{4}$ inch thick and is cut down from 24 inches to a $23\frac{3}{4}$-inch width. The panel is then cut to a height of $34\frac{1}{2}$ inches. The materials used in the manufacture of cabinets, solid oak in this case, are very expensive. Every precaution is taken to ensure a minimum of wasted material.

EXAMPLE **5**

Multiplying a Whole Number by a Mixed Number

A cabinet maker will need 80 base-end panels to complete a job. If each panel is $34\frac{1}{2}$ inches long, how many inches of oak material are needed?

SOLUTION

Multiply the number of panels needed by the length of each panel: $34\frac{1}{2}$, or $\frac{69}{2}$.

$$80 \times \frac{69}{2} = \frac{\overset{40}{\cancel{80}}}{1} \times \frac{69}{\underset{1}{\cancel{2}}} = \frac{40 \times 69}{1 \times 1} = \frac{2760}{1} = 2760 \text{ inches}$$

The length of material needed by the cabinet maker is 2760 inches.

> **QUICK TIP** It is often best to change a fraction or mixed number to a decimal number. This procedure is discussed in **Section 2.5**.

EXAMPLE **6**

Dividing a Whole Number by a Mixed Number

To complete a custom-designed cabinet, oak trim pieces must be cut exactly $2\frac{1}{4}$ inches long so that they can be used as dividers in a spice rack. Find the number of pieces that can be cut from a piece of oak that is 54 inches in length.

SOLUTION

To divide the length of the piece of oak by $2\frac{1}{4}$, or $\frac{9}{4}$, invert and then multiply.

$$54 \div 2\frac{1}{4} = 54 \div \frac{9}{4} = \frac{\overset{6}{\cancel{54}}}{1} \times \frac{4}{\underset{1}{\cancel{9}}} = \frac{6 \times 4}{1 \times 1} = \frac{24}{1} = 24$$

The number of trim pieces that can be cut from the oak stock is 24.

2.4 | Exercises

The **Quick Start** *exercises in each section contain solutions to help you get started.*

Multiply. Write each answer in lowest terms. (See Examples 1–3.)

Quick Start

1. $\frac{3}{4} \times \frac{2}{5} = \frac{3}{10}$ _____

$\frac{3}{\overset{2}{\cancel{4}}} \times \frac{\overset{1}{\cancel{2}}}{5} = \frac{3}{10}$

2. $\frac{2}{3} \times \frac{5}{8} = \frac{5}{12}$ _____

$\frac{2}{3} \times \frac{5}{\underset{4}{\cancel{8}}} = \frac{5}{12}$

3. $\frac{9}{10} \times \frac{11}{16} =$ _____

4. $\frac{2}{3} \times \frac{3}{8} =$ _____

5. $\frac{9}{22} \times \frac{11}{16} =$ _____

6. $\frac{5}{12} \times \frac{7}{10} =$ _____

7. $1\frac{1}{4} \times 3\frac{1}{2} =$ _____

8. $1\frac{2}{3} \times 2\frac{7}{10} =$ _____

9. $3\frac{1}{9} \times 3 =$ _____

10. $\frac{3}{4} \times \frac{8}{9} \times 2\frac{1}{2} =$ _____

11. $\frac{1}{4} \times 6\frac{2}{3} \times \frac{1}{5} =$ _____

12. $\frac{2}{3} \times \frac{9}{8} \times 3\frac{1}{4} =$ _____

13. $\frac{5}{9} \times 2\frac{1}{4} \times 3\frac{2}{3} =$ _____

14. $3 \times 1\frac{1}{2} \times 2\frac{2}{3} =$ _____

15. $5\frac{3}{5} \times 1\frac{5}{9} \times \frac{10}{49} =$ _____

Divide. Write each answer in lowest terms. (See Example 4.)

Quick Start

16. $\frac{1}{4} \div \frac{3}{4} = \frac{1}{3}$ _____

$\frac{1}{\underset{1}{\cancel{4}}} \times \frac{\overset{1}{\cancel{4}}}{3} = \frac{1}{3}$

17. $\frac{3}{8} \div \frac{5}{8} = \frac{3}{5}$ _____

$\frac{3}{\underset{1}{\cancel{8}}} \times \frac{\overset{1}{\cancel{8}}}{5} = \frac{3}{5}$

18. $\frac{13}{20} \div \frac{26}{30} =$ _____

19. $\frac{9}{10} \div \frac{3}{5} =$ _____

20. $\frac{7}{8} \div \frac{3}{4} =$ _____

21. $2\frac{1}{2} \div 3\frac{3}{4} =$ _____

C indicates an exercise that is related to the Case in Point feature within the section.

22. $1\frac{1}{4} \div 4\frac{1}{6} =$ _____   **23.** $5 \div 1\frac{7}{8} =$ _____   **24.** $3 \div 1\frac{1}{4} =$ _____

25. $\frac{3}{8} \div 2\frac{1}{2} =$ _____   **26.** $1\frac{7}{8} \div 6\frac{1}{4} =$ _____   **27.** $2\frac{5}{8} \div \frac{5}{16} =$ _____

28. $5\frac{2}{3} \div 6 =$ _____

29. In your own words, explain the rule for multiplying fractions. Make up an example problem of your own showing how this works.

30. A useful shortcut when multiplying fractions involves dividing a numerator and a denominator before multiplying. This is often called cancellation. Describe how this works and give an example of cancellation. (See Objective 2.)

Find the time-and-a-half pay rate for each of the following regular pay rates. (See Example 5).

Quick Start

31. $8 $12   **32.** $17 _____   **33.** $12.50 _____   **34.** $7.50 _____
$8 × 1½ = $12   (*Hint*: $12.50 = $12½)

35. Your classmate is confused on how to divide by a fraction. Write a short explanation telling how this should be done.

36. If you multiply two proper fractions, the answer is smaller than the fractions multiplied. When you divide by a proper fraction, is the answer smaller than the numbers in the problem? Show some examples to support your answer.

Solve the following application problems.

37. **ELECTRICITY RATES**   The utility company says that the cost of operating a hair dryer is $\frac{1}{5}$¢ per minute. Find the cost of operating the hair dryer for 30 minutes. (*Source*: Pacific Gas and Electric Company)

37. _____

38. ELECTRICITY RATES   The cost of electricity for brewing coffee is $\frac{2}{5}$¢ per minute. What is the cost of brewing coffee for 90 minutes? (*Source*: Pacific Gas and Electric Company.)

38. _____

39. PRODUCING CRAFTS   Matthew Genaway wants to make 16 holiday wreaths to sell at the craft fair. Each wreath needs $2\frac{1}{4}$ yards of ribbon. How many yards does he need?

39. _____

40. EARNINGS CALCULATION   Jack Horner worked $38\frac{1}{4}$ hours at $10 per hour. How much money did he make?

40. _____

41. FINISH CARPENTRY   The Home Depot estimates that a certain design for a kitchen and bathroom needs $109\frac{1}{2}$ feet of cabinet trim. How many homes can be fitted with cabinet trim if there are 1314 feet of cabinet trim available?

41. _____

42. COMMERCIAL FERTILIZER   For 1 acre of a crop, $7\frac{1}{2}$ gallons of fertilizer must be applied. How many acres can be fertilized with 1200 gallons of fertilizer?

42. _____

43. A manufacturer of floor jacks is ordering steel tubing to make the handles for this jack. How much steel tubing is needed to make 135 of these jacks? (The symbol for inch is ".")
(*Source*: Harbor Freight Tools.)

43. _____

CENTRAL HYDRAULICS
2-Ton Compact Floorjack

4000 LB CAPACITY

- $19\frac{1}{2}$" handle
- Lifts 5" to $15\frac{1}{4}$"
- Fully rolled edge for added tray strength
- 21" L × $9\frac{1}{2}$" W × 6" H
- Compact & lightweight for portability—perfect for the trunk

44. A wheelbarrow manufacturer uses handles made of hardwood. Find the amount of wood that is necessary to make 182 handles. The longest dimension shown is the handle length. (*Source*: Harbor Freight Tools.)

42. _____

6.0 CUBIC FT WHEELBARROW

- Steel construction with hardwood handles
- 14" tubeless pneumatic tire
- Fully rolled edge for added tray strength
- Overall dimensions: $61\frac{1}{2}$" L × 27" W × 24.9" H

45. STEEL FABRICATION   A fishing boat anchor requires $10\frac{3}{8}$ pounds of steel. Find the number of anchors that can be manufactured with 25,730 pounds of steel.

45. _____

46. COMMERCIAL CARPETING   The manager of the flooring department at The Home Depot determines that each apartment unit requires $62\frac{1}{2}$ square yards of carpet. Find the number of apartment units that can be carpeted with 6750 square yards of carpet.

46. _____

47. FUEL CONSUMPTION   A fishing boat uses $12\frac{3}{4}$ gallons of fuel on a full-day fishing trip and $7\frac{1}{8}$ gallons of fuel on a half-day trip. Find the total number of gallons of fuel used in 28 full-day trips and 16 half-day trips.

47. _____

48. MAKING JEWELRY   One necklace can be completed in $6\frac{1}{2}$ minutes, while a bracelet takes $3\frac{1}{8}$ minutes. Find the total time that it takes to complete 36 necklaces and 22 bracelets.

48. _____

49. DISPENSING EYEDROPS   How many $\frac{1}{8}$-ounce eyedrop dispensers can be filled with 11 ounces of eyedrops?

49. _____

50. CONCRETE FOOTINGS   Each building footing requires $\frac{5}{16}$ cubic yard of concrete. How many building footings can be constructed from 10 cubic yards of concrete?

50. _____

51. FIREWOOD SALE   Alison Romike had a small pickup truck that would carry $\frac{2}{3}$ cord of firewood. Find the number of trips needed to deliver 40 cords of wood.

51. _____

52. WEATHER STRIPPING   Bill Rhodes, an employee at The Home Depot, sells a 200-yard roll of weather stripping material. Find the number of pieces of weather stripping $\frac{5}{8}$ yard in length that may be cut from the roll.

52. _____

2.5 Converting Decimals to Fractions and Fractions to Decimals

Objectives

1. Convert decimals to fractions.
2. Convert fractions to decimals.
3. Know common decimal equivalents.

Objective 1 **Convert decimals to fractions.** A common method of converting a decimal to a fraction is by thinking of the decimal as being written in words, as in the preceding chapter. For example, think of .47 as **"forty-seven hundredths."** Then write this in fraction form as

$$.47 = \frac{47}{100}$$

In the same way, .3, read as **"three tenths,"** is written in fraction form as

$$.3 = \frac{3}{10}$$

Also, .963, read **"nine hundred sixty-three thousandths,"** is written in fraction form as

$$.963 = \frac{963}{1000}$$

Another method of converting a decimal to a fraction is by first removing the decimal point. The remaining number is the numerator of the fraction. The denominator of the fraction is 1 followed by as many zeros as there were digits to the right of the decimal point in the original number.

EXAMPLE 1

Converting Decimals to Fractions

Convert the following decimals to fractions.

(a) .3   **(b)** .98   **(c)** .654

SOLUTION

(a) There is one digit following the decimal point in .3. Make a fraction with 3 as the numerator. For the denominator, use 10, which is 1 followed by one zero.

$$.3 = \frac{3}{10}$$

    1 followed by 1 zero

This fraction is in lowest terms.

(b) There are two digits following the decimal point in .98. Make a fraction with 98 as the numerator and 100 as the denominator.

$$.98 = \frac{98}{100} = \frac{49}{50} \left(\text{lowest terms} \right)$$

    1 followed by 2 zeros

(c) There are three digits following the decimal point in .654.

$$.654 = \frac{654}{1000} = \frac{327}{500} \left(\text{lowest terms} \right)$$

    1 followed by 3 zeros

Objective 2 **Convert fractions to decimals.** Convert a fraction to a decimal by dividing the numerator of the fraction by the denominator. Place a decimal point after the numerator and attach one zero at a time to the right of the decimal point as the division is performed. Keep going until the division produces a remainder of zero or until the desired degree of accuracy is reached.

EXAMPLE 2

Converting Fractions to Decimals

Convert the following fractions to decimals.

(a) $\frac{1}{8}$ **(b)** $\frac{2}{3}$

SOLUTION

(a) Convert $\frac{1}{8}$ to a decimal by dividing 1 by 8.

$$8\overline{)1}$$

Since 8 will not divide into 1, place a 0 to the *right* of the decimal point. Now 8 goes into 10 once, with a remainder of 2.

$$\begin{array}{r} .1 \\ 8\overline{)1.0} \\ \underline{8} \\ 2 \end{array}$$ Be sure to move the decimal point up.

Continue placing zeros to the *right* of the decimal point and continue dividing. The division now gives a remainder of 0.

$$\begin{array}{r} .125 \\ 8\overline{)1.000} \\ \underline{8} \\ 20 \\ \underline{16} \\ 40 \\ \underline{40} \\ 0 \end{array}$$ Keep attaching zeros.

remainder of 0

Therefore, $\frac{1}{8} = .125$.

(b) Divide 2 by 3.

$$\begin{array}{r} 0.6666 \\ 3\overline{)2.0000} \\ \underline{1\,8} \\ 20 \\ \underline{18} \\ 20 \\ \underline{18} \\ 20 \\ \underline{18} \\ 2 \end{array}$$ Keep attaching zeros.

This division results in a repeating decimal and is often written as $.\overline{6}$, $.6\overline{6}$, or $.66\overline{6}$. Rounded to the nearest thousandth, $\frac{2}{3} = .667$.

The calculator solution to this example is

$$2 \div 3 = 0.666666667$$

Note: Refer to Appendix C for calculator basics.

Objective 3 Know common decimal equivalents. Some of the more common decimal equivalents of fractions are listed in the margin. These decimals appear from least to greatest value and are rounded to the nearest ten-thousandth. Sometimes decimals must be carried out further to give greater accuracy, while at other times they are not carried out as far and are rounded sooner.

Decimal Equivalents

$\frac{1}{16} = .0625$

$\frac{1}{10} = .1$

$\frac{1}{9} = .1111$ (rounded)

$\frac{1}{8} = .125$

$\frac{1}{7} = .1429$ (rounded)

$\frac{1}{6} = .1667$ (rounded)

$\frac{3}{16} = .1875$

$\frac{1}{5} = .2$

$\frac{1}{4} = .25$

$\frac{3}{10} = .3$

$\frac{5}{16} = .3125$

$\frac{1}{3} = .3333$ (rounded)

$\frac{3}{8} = .375$

$\frac{2}{5} = .4$

$\frac{7}{16} = .4375$

$\frac{1}{2} = .5$

$\frac{9}{16} = .5625$

$\frac{3}{5} = .6$

$\frac{5}{8} = .625$

$\frac{2}{3} = .6667$ (rounded)

$\frac{11}{16} = .6875$

$\frac{7}{10} = .7$

$\frac{3}{4} = .75$

$\frac{4}{5} = .8$

$\frac{13}{16} = .8125$

$\frac{5}{6} = .8333$ (rounded)

$\frac{7}{8} = .875$

$\frac{9}{10} = .9$

$\frac{15}{16} = .9375$

2.5 | Exercises

FOR EXTRA HELP

 MyMathLab

 InterActMath.com

 MathXL

 MathXL Tutorials on CD

 Addison-Wesley Math Tutor Center

 DVT/Videotape

The **Quick Start** exercises in each section contain solutions to help you get started.

Convert the following decimals to fractions, and write each in lowest terms. (See Example 1.)

Quick Start

1. $.75 = \frac{3}{4}$

$\frac{75}{100} = \frac{3}{4}$

2. $.55 = \frac{11}{20}$

$\frac{55}{100} = \frac{11}{20}$

3. $.24 = $ ____

4. $.64 = $ ____

5. $.73 = $ ____

6. $.33 = $ ____

7. $.85 = $ ____

8. $.68 = $ ____

9. $.34 = $ ____

10. $.288 = $ ____

11. $.444 = $ ____

12. $.125 = $ ____

13. $.625 = $ ____

14. $.875 = $ ____

15. $.805 = $ ____

16. $.791 = $ ____

17. $.096 = $ ____

18. $.012 = $ ____

19. $.0375 = $ ____

20. $.0875 = $ ____

21. $.1875 = $ ____

22. $.9845 = $ ____

23. $.0016 = $ ____

24. $.0085 = $ ____

25. A classmate of yours is confused about how to convert a decimal to a fraction. Write an explanation of this for your classmate, including changing the fraction to lowest terms. (See Objective 1.)

26. Explain how to convert a fraction to a decimal. Be sure to mention rounding in your explanation. (See Objective 2.)

Convert the following fractions to decimals. If a division does not come out evenly, round the answer to the nearest thousandth. (See Example 2.)

Quick Start

27. $\frac{1}{4} = .25$

$\begin{array}{r} .25 \\ 4\overline{)1.00} \\ \underline{8} \\ 20 \\ \underline{20} \\ 0 \end{array}$

28. $\frac{7}{8} = $ ____

29. $\frac{3}{8} = $ ____

30. $\frac{5}{8}$ = _____

31. $\frac{2}{3}$ = _____

32. $\frac{5}{6}$ = _____

33. $\frac{7}{9}$ = _____

34. $\frac{1}{9}$ = _____

35. $\frac{7}{11}$ = _____

36. $\frac{8}{25}$ = _____

37. $\frac{22}{25}$ = _____

38. $\frac{14}{25}$ = _____

39. $\frac{181}{205}$ = _____

40. $\frac{1}{99}$ = _____

41. $\frac{148}{149}$ = _____

Chapter 2 Quick Review

CONCEPTS	EXAMPLES
2.1 Types of fractions *Proper*: Numerator smaller than denominator. *Improper*: Numerator equal to or greater than denominator. *Mixed*: Whole number and proper fraction.	proper fractions $\frac{2}{3},\frac{3}{4},\frac{15}{16},\frac{1}{8}$ improper fractions $\frac{17}{8},\frac{19}{12},\frac{11}{2},\frac{5}{3},\frac{7}{7}$ mixed numbers $2\frac{2}{3},3\frac{5}{8},9\frac{5}{6}$
2.1 Converting fractions *Mixed to improper*: Multiply denominator by whole number and add numerator. *Improper to mixed*: Divide numerator by denominator and place remainder over denominator.	$7\frac{2}{3}=\frac{23}{3}\rightarrow 3\times7+2$ $\frac{17}{5}=3\frac{2}{5}\quad 5\overline{)17}\;\;\frac{15}{2}$
2.1 Writing fractions in lowest terms	$\frac{30}{42}=\frac{30\div6}{42\div6}=\frac{5}{7}$
2.2 Adding like fractions Add numerators and reduce to lowest terms.	$\frac{3}{4}+\frac{1}{4}+\frac{5}{4}=\frac{3+1+5}{4}=\frac{9}{4}=2\frac{1}{4}$
2.2 Finding a least common denominator (LCD) *Inspection method*: Look to see if the LCD can be found. *Method of prime numbers*: Use prime numbers to find the LCD.	$\frac{1}{3}+\frac{1}{4}+\frac{1}{10}$ $\begin{array}{r}5)\overline{1\;\;1\;\;5}\\3)\overline{3\;\;1\;\;5}\\2)\overline{3\;\;2\;\;5}\\\overline{3\;\;4\;\;10}\end{array}$ Multiply the prime numbers. $2\times2\times3\times5=60\text{ LCD}$
2.2 Adding unlike fractions **1.** Find the LCD. **2.** Rewrite fractions with the LCD. **3.** Add numerators, placing answers over the LCD and reduce to lowest terms.	$\frac{1}{3}+\frac{1}{4}+\frac{1}{10}\;\text{LCD}=60$ $\frac{1}{3}=\frac{20}{60},\frac{1}{4}=\frac{15}{60},\frac{1}{10}=\frac{6}{60}$ $\frac{20+15+6}{60}=\frac{41}{60}$

CONCEPTS	EXAMPLES
2.2 Subtracting fractions 1. Find the LCD. 2. Subtract numerator of subtrahend, borrowing if necessary. 3. Write the difference over the LCD and reduce to lowest terms.	$\dfrac{5}{8} - \dfrac{1}{3} = \dfrac{15}{24} - \dfrac{8}{24} = \dfrac{7}{24}$
2.3 Adding mixed numbers 1. Find the LCD, then add fractions. 2. Add whole numbers. 3. Combine the sums of whole numbers and fractions. Write the answer in simplest terms.	$9\dfrac{2}{3} = 9\dfrac{8}{12}$ $+\ 6\dfrac{3}{4} = 6\dfrac{9}{12}$ $\overline{\qquad\qquad}$ $15\dfrac{17}{12} = 16\dfrac{5}{12}$
2.3 Subtracting mixed numbers 1. Find the LCD and subtract fractions, borrowing if necessary. 2. Subtract whole numbers. 3. Combine the differences of whole numbers and fractions.	$8\dfrac{5}{8} = 8\dfrac{15}{24}$ $-\ 3\dfrac{1}{12} = 3\dfrac{2}{24}$ $\overline{\qquad\qquad}$ $5\dfrac{13}{24}$
2.4 Multiplying proper fractions 1. Multiply numerators and multiply denominators. 2. Reduce the answer to lowest terms if cancelling was not done.	$\dfrac{6}{11} \times \dfrac{7}{8} = \dfrac{\overset{3}{\cancel{6}}}{11} \times \dfrac{7}{\underset{4}{\cancel{8}}} = \dfrac{21}{44}$
2.4 Multiplying mixed numbers 1. Change mixed numbers to improper fractions. 2. Cancel if possible. 3. Multiply as proper fractions.	$1\dfrac{3}{5} \times 3\dfrac{1}{3} = \dfrac{8}{\underset{1}{\cancel{5}}} \times \dfrac{\overset{2}{\cancel{10}}}{3} = \dfrac{8}{1} \times \dfrac{2}{3}$ $= \dfrac{16}{3} = 5\dfrac{1}{3}$ Always reduce to lowest terms.
2.4 Dividing proper fractions Invert the divisor, multiply as fractions, and reduce the answer to lowest terms.	$\dfrac{25}{36} \div \dfrac{15}{18} = \dfrac{\overset{5}{\cancel{25}}}{\underset{2}{\cancel{36}}} \times \dfrac{\overset{1}{\cancel{18}}}{\underset{3}{\cancel{15}}} = \dfrac{5}{2} \times \dfrac{1}{3} = \dfrac{5}{6}$
2.4 Dividing mixed numbers Change mixed numbers to improper fractions. Invert the divisor, cancel if possible, and multiply as proper fractions.	$3\dfrac{5}{9} \div 2\dfrac{2}{5} = \dfrac{32}{9} \div \dfrac{12}{5} = \dfrac{32}{9} \times \dfrac{5}{\underset{3}{\cancel{12}}^{8}}$ $= \dfrac{40}{27} = 1\dfrac{13}{27}$
2.5 Converting decimals to fractions Think of the decimal as being written in words and write in fraction form. Reduce to lowest terms.	Convert .47 to a fraction. Think of .47 as "forty-seven hundredths." Then write as $\dfrac{47}{100}$.
2.5 Converting fractions to decimals Divide the numerator by the denominator. Round if necessary.	Convert $\tfrac{1}{8}$ to a decimal. $\begin{array}{r} .125 \\ 8\overline{)1.000} \\ \underline{8} \\ 20 \\ \underline{16} \\ 40 \\ \underline{40} \\ 0 \end{array}$ $\dfrac{1}{8} = .125$

Chapter 2 | Summary Exercise

Using Fractions with Statistics

It is often said that a picture is worth a thousand words. Visual presentation of data is often used in business in the form of graphs. A commonly used graph that shows the relationships of various data is the circle graph, also called a pie chart. The circle, which contains 360 degrees, is divided into slices, or fractional parts. The size of each slice helps to show the relationship of the various slices to each other and to the whole.

The annual operating expenses for Woodline Moldings and Trim are shown below. Use this information to answer the questions that follow.

Woodline Moldings and Trim		(Operating Expenses)	
Expense Item	Monthly Amount	Annual Amount	Fraction of Total
Salaries	$5000	_____	_____
Rent	$3000	_____	_____
Utilities	$1000	_____	_____
Insurance	$ 750	_____	_____
Advertising	$ 750	_____	_____
Miscellaneous	$1500	_____	_____
Total Expenses		_____	_____

(a) Find the total annual operating expenses for Woodline Moldings and Trim.   **(a)** _____

(b) What fraction should be used to represent each expense item as part of the total expenses?   **(b)** _____

(c) Draw a circle (pie) graph using the fractions you found in (b) to represent each expense item. Approximate the fractional part of the circle needed for each expense item. Label each segment of the circle graph with the fraction and the expense item.

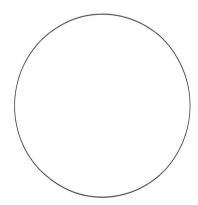

(d) Since there are 360 degrees in a circle, find the number of degrees that would be used to represent each expense item in the circle graph.   **(d)** _____

(e) In **Chapter 15**, **Statistics**, you will learn more about using graphs in business. Add all of the answers found in part (d). What is the total number of degrees? Why is this the answer?   **(e)** _____

Chapter 2   Test

To help you review, the numbers in brackets show the section in which the topic was discussed.

Write the following fractions in lowest terms.   **[2.1]**

1. $\dfrac{25}{30} =$ ___

2. $\dfrac{875}{1000} =$ ___

3. $\dfrac{84}{132} =$ ___

Convert the following improper fractions to mixed numbers, and write using lowest terms.   **[2.1]**

4. $\dfrac{65}{8} =$ ___

5. $\dfrac{56}{12} =$ ___

6. $\dfrac{120}{45} =$ ___

Convert the following mixed numbers to improper fractions.   **[2.1]**

7. $7\dfrac{3}{4} =$ ___

8. $18\dfrac{4}{5} =$ ___

9. $18\dfrac{3}{8} =$ ___

Find the LCD of each of the following groups of denominators.   **[2.2]**

10. $2, 6, 5 =$ ___

11. $6, 8, 15 =$ ___

12. $6, 9, 12, 24 =$ ___

Solve the following problems.   **[2.2–2.4]**

13.
$$\begin{array}{r} \frac{1}{5} \\ \frac{3}{10} \\ + \frac{3}{8} \end{array}$$

14.
$$\begin{array}{r} 32\frac{5}{16} \\ -17\frac{1}{4} \end{array}$$

15.
$$\begin{array}{r} 126\frac{3}{16} \\ -\ 89\frac{7}{8} \end{array}$$

16. $67\dfrac{1}{2} \times \dfrac{8}{15} =$

17. $33\dfrac{1}{3} \div \dfrac{200}{9} =$

Solve the following application problems.

18. Becky Finnerty, a pastry chef, used $23\frac{1}{2}$ pounds of powdered sugar for one recipe, $34\frac{3}{4}$ pounds of powdered sugar for another recipe, and $17\frac{5}{8}$ pounds of powdered sugar for a third recipe. If Finnerty started with two 50-pound sacks of powdered sugar, find the amount of powdered sugar remaining. **[2.3]**

18. _____

19. Rhonda Goedeker received her Social Security check of $1275. After paying $\frac{1}{3}$ of this amount for rent, she paid $\frac{2}{5}$ of the remaining amount for food, utilities, and transportation. How much money does she have left?

19. _____

20. A painting contractor arrived at a 6-unit apartment complex with $147\frac{1}{2}$ gallons of paint. If his crew sprayed $68\frac{1}{2}$ gallons on the interior walls, rolled $37\frac{3}{8}$ gallons on the masonry exterior, and brushed $5\frac{3}{4}$ gallons on the window trim, find the number of gallons of paint remaining. **[2.3]**

20. _____

21. Find the number of window-blind pull cords that can be made from $157\frac{1}{2}$ yards of cord if $4\frac{3}{8}$ yards of cord are needed for each blind.

21. _____

Convert the following decimals to fractions. **[2.5]**

22. $.625 =$ 23. $.82 =$

*Use the advertisement for this four-piece chisel set to answer Exercises 24–25. The symbol " is for inches. (**Source** Harbor Freight Tools.)* **[2.5]**

24. Convert the cutting-edge width of the smallest chisel from a fraction to a decimal.

24. _____

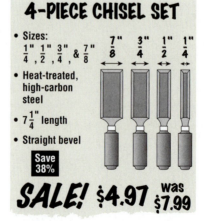

4-PIECE CHISEL SET

- Sizes: $\frac{1}{4}"$, $\frac{1}{2}"$, $\frac{3}{4}"$, & $\frac{7}{8}"$ $\frac{7}{8}"$ $\frac{3}{4}"$ $\frac{1}{2}"$ $\frac{1}{4}"$
- Heat-treated, high-carbon steel
- $7\frac{1}{4}"$ length
- Straight bevel

Save 38%

SALE! $4.97 was $7.99

25. Convert the cutting-edge width of the largest chisel from a fraction to a decimal.

25. _____

CHAPTER

3

Percent

THOMAS DUGALLY IS A REAL ESTATE BROKER WITH CENTURY 21 Realty. He says that the real estate field offers him many challenges, a great deal  of flexibility in planning his work schedule, and he loves not having to work indoors all of the time. After three years in the business, Dugally is one of the top producers in his office and continues to expand his knowledge and improve his skills.

Mathematics, especially calculations involving percent, is something that Dugally uses every day. He is constantly calculating loan charges, determining buyer loan qualifications, figuring real estate commissions, and determining and explaining property tax amounts and property insurance premiums to his clients.

Percents are widely used in business and everyday life. For example, interest rates on automobile loans, home loans, and other installment loans are almost always given as percents. Stores often advertise sale prices as being a certain percent off the regular price. In business, marketing costs, damage, and theft may be expressed as a percent of sales; profit as a percent of investment; and labor as a percent of cost of production. Current government figures about inflation, recession, and unemployment are also reported as percents.

This chapter discusses the various types of percent problems that will be used throughout the text.

3.1 | Writing Decimals and Fractions as Percents

Objectives

1. Write a decimal as a percent.
2. Write a fraction as a percent.
3. Write a percent as a decimal.
4. Write a percent as a fraction.
5. Write a fractional percent as a decimal.

Percents represent part of a whole, just as fractions or decimals do. **Percents** are **hundredths**, or parts of a hundred. "One percent" means 1 of 100 equal parts. Percents are written with a percent sign $(\%)$. For example, 25% refers to 25 out of 100 equal parts $\left(\frac{25}{100}\right)$, 50% refers to 50 out of 100 equal parts $\left(\frac{50}{100}\right)$, and 100% refers to all 100 of the 100 equal parts $\left(\frac{100}{100}\right)$. Therefore, 100% is equal to the whole item. If a percent is larger than 100% (for example, 150%), more than one item has been divided into 100 equal parts, and 150 of the parts are being considered $\left(\frac{150}{100}\right)$.

Objective **1** **Write a decimal as a percent.** To write a decimal as a percent, *move the decimal point two places to the right and attach a percent sign* $(\%)$.

For example, write .75 as a percent by moving the decimal point two places to the right and attaching a percent sign resulting in 75%.

Converting Decimals to Percents

Change a decimal to a percent by moving the decimal point two places to the right and attaching a percent sign $(\%)$.

.75	Original decimal
.75.	Move decimal 2 places to the right
75%	Attach a percent sign.

EXAMPLE 1

Changing Decimals to Percents

Change the following decimals to percents.

(a) .35   **(b)** .42   **(c)** .58

SOLUTION

Move the decimal point two places to the right and attach a percent sign.

(a) 35%   **(b)** 42%   **(c)** 58%

If there is nothing in the hundredths position, place zeros to the right of the number to hold the hundredths position. For example, the decimal .5 is expressed as 50%, and the whole number 1.2 is expressed as 120%.

$$.5 = .50.\% = 50\%$$

attach zero

$$1.2 = 1.20.\% = 120\%$$

attach zero

EXAMPLE 2

Writing Decimals as Percents

Write the following decimals as percents.

(a) .8   **(b)** 2.6   **(c)** .1   **(d)** 4

SOLUTION

It is necessary to attach zeros here.

(a) 80%   **(b)** 260%   **(c)** 10%   **(d)** 400%

   attach zero   attach zero   attach zero   attach 2 zeros

If the decimal extends past the hundredths position, the resulting percent includes decimal parts of whole percents.

EXAMPLE 3

Writing Decimals as Percents

When reading a real estate newsletter, Thomas Dugally sees the following decimals. Write these decimals as percents.

(a) .625   **(b)** .0057   **(c)** .0018

SOLUTION

(a) 62.5%   **(b)** .57%   **(c)** .18%

QUICK TIP   In Example 3, both (b) and (c) are less than 1%. They are decimal parts of one percent.

Objective ② **Write a fraction as a percent.** There are two ways to write a fraction as a percent. One way is to write the fraction first as a decimal. For example, to express the fraction $\frac{2}{5}$ as a percent, write $\frac{2}{5}$ as a decimal by dividing 2 by 5. Then write the decimal as a percent.

Fraction		Decimal		Percent
$\frac{2}{5}$	=	.4	=	40%

EXAMPLE 4

Writing Fractions as Percents

An advertising account representative is given the following data in fraction form and must change the data to percent.

(a) $\frac{1}{4}$ **(b)** $\frac{3}{8}$ **(c)** $\frac{4}{5}$

SOLUTION

First write each fraction as a decimal, and then write the decimal as a percent.

(a) $\frac{1}{4} = .25 = 25\%$ **(b)** $\frac{3}{8} = .375 = 37.5\%$ **(c)** $\frac{4}{5} = .8 = 80\%$

A second way to write a fraction as a percent is by multiplying the fraction by 100%. For example, write the fraction $\frac{4}{5}$ as a percent by multiplying $\frac{4}{5}$ by 100%.

$$\frac{4}{5} = \frac{4}{5} \times 100\% = \frac{400\%}{5} = 80\%$$

Objective ③ **Write a percent as a decimal.** To write a percent as a decimal, **move the decimal point two places to the left and drop the percent sign**. For example, 50% becomes .50 or .5, 100% becomes 1, and 352% becomes 3.52.

Converting Percents to Decimals

Change a percent to a decimal by moving the decimal point two places to the left and dropping the percent sign $(\%)$.

<table>
<tr><td>25%</td><td>Original percent</td></tr>
<tr><td>.25.%</td><td>Move decimal 2 places to the left</td></tr>
<tr><td>.25</td><td>Drop the percent sign</td></tr>
</table>

EXAMPLE 5

Writing Percents as Decimals

To calculate some insurance claims, an insurance agent must change the following percents to decimals.

(a) 35%   **(b)** 50%   **(c)** 325%   **(d)** $37\frac{1}{2}\%$ $\left(\textit{Hint: } 37\frac{1}{2}\% = 37.5\%\right)$

SOLUTION

Move the decimal point two places to the left and drop the percent sign.

(a) .35   **(b)** .5   **(c)** 3.25   **(d)** .375

QUICK TIP   In Example 5(d) change $37\frac{1}{2}\%$ to 37.5%. It is usually best to change fractional percents to the decimal percent form and then change the percent to a decimal.

Objective ☐**4** **Write a percent as a fraction.** To write a percent as a fraction, first change the percent to a decimal, then write the decimal as a fraction in lowest terms.

EXAMPLE 6

Writing Percents as Fractions

The following data are from the National Association of Realtors. The down payment amount as a percent of selling price and the percent of homebuyers paying that amount as a down payment on their first home are shown. Next to each entry below, convert each percent to a fraction. When possible, reduce each fraction to its lowest term.

Down Payment by First-Time Homebuyers

Down Payment	Percent of Homebuyers
0%–4% down	56% (a) _____
5%–9% down	31% (b) _____
10%–19% down	9% (c) _____
20% or more down	4% (d) _____

Source: National Association of Realtors.

SOLUTION

First write each percent as a decimal and then write the decimal as a fraction in lowest terms.

(a) $56\% = .56 = \dfrac{56}{100} = \dfrac{14}{25}$ **(b)** $31\% = .31 = \dfrac{31}{100}$

(c) $9\% = .09 = \dfrac{9}{100}$ **(d)** $4\% = .04 = \dfrac{4}{100} = \dfrac{1}{25}$

Objective ☐**5** **Write a fractional percent as a decimal.** A fractional percent such as $\frac{1}{2}\%$ has a value less than 1%. In fact, $\frac{1}{2}\%$ is equal to $\frac{1}{2}$ of 1%. Write a fractional percent as a decimal by first changing the fraction to a decimal, leaving the percent sign. For example, first write $\frac{1}{2}\%$ as .5%. Then write .5% as a decimal by moving the decimal point two places to the left and dropping the percent sign.

$$\frac{1}{2}\% = .5\% = .005$$

Written as a decimal with percent sign remaining.

EXAMPLE 7

Writing Fractional Percents as Decimals

The following percents appear in a newspaper article. Write each fractional percent as a decimal.

(a) $\frac{1}{5}\%$ **(b)** $\frac{3}{4}\%$ **(c)** $\frac{5}{8}\%$

SOLUTION

Begin by writing the fraction as a decimal percent.

(a) $\frac{1}{5}\% = .2\% = .002$ **(b)** $\frac{3}{4}\% = .75\% = .0075$ **(c)** $\frac{5}{8}\% = .625\% = .00625$

QUICK TIP   When writing a fraction percent as a decimal, first, change the fraction to a decimal, leaving the percent sign. Then move the decimal point two places to the left and drop the percent sign.

The following chart shows many fractions, as well as their decimal and percent equivalents. **It is helpful to memorize the more commonly used ones.**

Fraction, Decimal, and Percent Equivalents

$\frac{1}{100} = .01 = 1\%$	$\frac{9}{16} = .5625 = 56.25\%$ or $56\frac{1}{4}\%$
$\frac{1}{50} = .02 = 2\%$	$\frac{3}{5} = .6 = 60\%$
$\frac{1}{25} = .04 = 4\%$	$\frac{5}{8} = .625 = 62\frac{1}{2}\%$
$\frac{1}{20} = .05 = 5\%$	$\frac{2}{3} = .66\overline{6} = 66\frac{2}{3}\%$
$\frac{1}{16} = .0625 = 6.25\%$ or $6\frac{1}{4}\%$	$\frac{11}{16} = .6875 = 68.75\%$ or $68\frac{3}{4}\%$
$\frac{1}{12} = .083\overline{3} = 8\frac{1}{3}\%$	$\frac{7}{10} = .7 = 70\%$
$\frac{1}{10} = .1 = 10\%$	$\frac{3}{4} = .75 = 75\%$
$\frac{1}{9} = .111\overline{1} = 11\frac{1}{9}\%$	$\frac{4}{5} = .8 = 80\%$
$\frac{1}{8} = .125 = 12.5\%$ or $12\frac{1}{2}\%$	$\frac{13}{16} = .8125 = 81.25\%$ or $81\frac{1}{4}\%$
$\frac{1}{7} = .1428 = 14\frac{2}{7}\%$	$\frac{5}{6} = .833\overline{3} = 83\frac{1}{3}\%$
$\frac{1}{6} = .166\overline{6} = 16\frac{2}{3}\%$	$\frac{7}{8} = .875 = 87\frac{1}{2}\%$
$\frac{3}{16} = .1875 = 18\frac{3}{4}\%$	$\frac{9}{10} = .9 = 90\%$
$\frac{1}{5} = .2 = 20\%$	$\frac{15}{16} = .9375 = 93.75\%$ or $93\frac{3}{4}\%$
$\frac{1}{4} = .25 = 25\%$	$1 = 1.00 = 100\%$
$\frac{3}{10} = .3 = 30\%$	$1\frac{1}{10} = 1.1 = 110\%$
$\frac{5}{16} = .3125 = 31.25\%$	$1\frac{1}{4} = 1.25 = 125\%$
$\frac{1}{3} = .333\overline{3} = 33\frac{1}{3}\%$	$1\frac{1}{3} = 1.133\overline{33} = 133\frac{1}{3}\%$
$\frac{3}{8} = .375 = 37\frac{1}{2}\%$	$1\frac{1}{2} = 1.5 = 150\%$
$\frac{2}{5} = .4 = 40\%$	$1\frac{2}{3} = 1.166\overline{6} = 166\frac{2}{3}\%$
$\frac{7}{16} = .4375 = 43.75\%$ or $43\frac{3}{4}\%$	$1\frac{3}{4} = 1.75 = 175\%$
$\frac{1}{2} = .5 = 50\%$	$2 = 2.00 = 200\%$

3.1 | Exercises

The   **Quick Start**   *exercises in each section contain solutions to help you get started.*

Write the following decimals as percents. (See Examples 1–3.)

Quick Start

1. $.25 = \underline{\mathbf{25\%}}$ **2.** $.4 = \underline{\mathbf{40\%}}$ **3.** $.72 = \underline{\hspace{1cm}}$

4. $1.3 = \underline{\hspace{1cm}}$ **5.** $2.034 = \underline{\hspace{1cm}}$ **6.** $.625 = \underline{\hspace{1cm}}$

7. $3.625 = \underline{\hspace{1cm}}$ **8.** $4.6 = \underline{\hspace{1cm}}$ **9.** $.875 = \underline{\hspace{1cm}}$

10. $.005 = \underline{\hspace{1cm}}$ **11.** $.0005 = \underline{\hspace{1cm}}$ **12.** $.0012 = \underline{\hspace{1cm}}$

13. $3.45 = \underline{\hspace{1cm}}$ **14.** $.2108 = \underline{\hspace{1cm}}$ **15.** $.0308 = \underline{\hspace{1cm}}$

Write the following as decimals. (See Examples 4–6.)

Quick Start

16. $\frac{1}{5} = \underline{\mathbf{.2}}$ **17.** $\frac{5}{8} = \underline{\mathbf{.625}}$ **18.** $64\% = \underline{\hspace{1cm}}$

19. $65\% = \underline{\hspace{1cm}}$ **20.** $\frac{1}{100} = \underline{\hspace{1cm}}$ **21.** $\frac{1}{8} = \underline{\hspace{1cm}}$

22. $8\frac{1}{2}\% = \underline{\hspace{1cm}}$ **23.** $12\frac{1}{2}\% = \underline{\hspace{1cm}}$ **24.** $\frac{1}{200} = \underline{\hspace{1cm}}$

25. $\frac{1}{400} = \underline{\hspace{1cm}}$ **26.** $50\frac{3}{4}\% = \underline{\hspace{1cm}}$ **27.** $84\frac{3}{4}\% = \underline{\hspace{1cm}}$

28. $3\frac{3}{8} = \underline{\hspace{1cm}}$ **29.** $1\frac{3}{4} = \underline{\hspace{1cm}}$ **30.** $350\% = \underline{\hspace{1cm}}$

Determine the fraction, decimal, or percent equivalents for each of the following, as necessary. Write fractions in lowest terms.

Quick Start

	Fraction	Decimal	Percent
31.	$\frac{1}{2}$	$\underline{\mathbf{.5}}$	$\underline{\mathbf{50\%}}$
32.	$\frac{3}{50}$	$.06$	6%
33.	$\underline{\hspace{1.5cm}}$	$.875$	$\underline{\hspace{1.5cm}}$
34.	$\frac{4}{5}$	$\underline{\hspace{1.5cm}}$	$\underline{\hspace{1.5cm}}$
35.	$\underline{\hspace{1.5cm}}$	$\underline{\hspace{1.5cm}}$	$.8\%$
36.	$\underline{\hspace{1.5cm}}$	$.00625$	$\underline{\hspace{1.5cm}}$
37.	$10\frac{1}{2}$	$\underline{\hspace{1.5cm}}$	$\underline{\hspace{1.5cm}}$
38.	$\underline{\hspace{1.5cm}}$	$\underline{\hspace{1.5cm}}$	675%
39.	$\underline{\hspace{1.5cm}}$	$.65$	$\underline{\hspace{1.5cm}}$
40.	$4\frac{3}{8}$	$\underline{\hspace{1.5cm}}$	$\underline{\hspace{1.5cm}}$

	Fraction	*Decimal*	*Percent*
41.	_____	.005	_____
42.	_____	_____	$\frac{1}{8}\%$
43.	$\frac{1}{3}$	_____	_____
44.	_____	_____	12.5%
45.	_____	2.5	_____
46.	$\frac{7}{20}$	_____	_____
47.	_____	_____	$4\frac{1}{4}\%$
48.	_____	.7	_____
49.	$\frac{3}{200}$	_____	_____
50.	_____	5.125	_____
51.	_____	_____	1037.5%
52.	_____	_____	$\frac{3}{4}\%$
53.	_____	.0025	_____
54.	$\frac{5}{8}$	_____	_____
55.	_____	_____	$37\frac{1}{2}\%$
56.	_____	_____	$6\frac{3}{4}\%$

57. Fractions, decimals, and percents are all used to describe a part of something. The use of percent is much more common than fractions and decimals. Why do you suppose this is true?

58. List five uses of percent that are or will be part of your life. Consider the activities of working, shopping, saving, and planning for the future.

59. Select a decimal percent and write it as a fraction. Select a fraction and write it as a percent. Write an explanation of each step of your work. (See Objectives 2 and 3.)

60. The fractional percent $\frac{1}{2}\%$ is equal to .005. Explain each step as you change $\frac{1}{2}\%$ to its decimal equivalent. (See Objective 4.)

3.2 Finding Part

Objectives

1. Know the three components of a percent problem.
2. Learn the basic percent formula.
3. Solve for part.
4. Recognize the terms associated with base, rate, and part.
5. Calculate sales tax.
6. Learn the standard format of percent problems.

CASE in POINT

As a real estate agent with Century 21 Realty, Thomas Dugally is paid on a commission plan. When he produces income for the company as a result of a sale, a sale of his listing by someone else, or a rental agreement that is completed, he is paid a portion of this income. Currently, Dugally is looking for a home in the $180,000 price range for Scott and Andrea Abriani, a couple he met at an open house.

Objective 1 Know the three components of a percent problem. Problems in percent contain three main components. Usually, two of these components are given, and the third component must be found. The three key components in a percent problem are as follows.

1. **Base:** The whole or total, starting point, or that to which something is being compared.
2. **Rate:** A number followed by % or **percent**.
3. **Part:** The result of multiplying the base and the rate. The part is a *part* of the base, as sales tax is a part of the total sales, or as the number of sports cars is part of the total number of cars.

QUICK TIP *Percent* and *part* are different quantities. The stated percent in a given problem is always the rate (R). The part (P) is the product of the base (B) and the rate (R). Thus, the part is a quantity and never appears with *percent* or % following it.

Objective 2 Learn the basic percent formula. The base, rate, and part are related by the basic **percent formula**.

$$P = B \times R \qquad P = R \times B$$
Part = Base × Rate or Part = Rate × Base

Objective 3 Solve for part. If Thomas Dugally finds a $180,000 home for the Abrianis and is paid a 6% commission, use the formula $P = B \times R$ to find 6% of $180,000. Multiply the base, $180,000, by the rate, (6%). The rate must be changed to a decimal before it is multiplied.

$$P = B \times R$$
$$P = \$180,000 \times 6\%$$
$$P = \$180,000 \times .06 \quad \text{or} \quad \begin{array}{r} \$180,000 \\ \times \quad .06 \\ \hline \$10,800 \end{array}$$

Finally, 6% of $180,000 is $10,800.

EXAMPLE 1

Solving for Part

Solve for part, using $P = B \times R$.

(a) 4% of 50   **(b)** 1.2% of 180   **(c)** 140% of 225   **(d)** $\frac{1}{4}$% of 560

$\left(Hint: \frac{1}{4}\% = .25\% \right)$

SOLUTION

(a)
$$\begin{array}{r} 50 \\ \times \;\; .04 \\ \hline 2.00 \end{array}$$

(b)
$$\begin{array}{r} 180 \\ \times \;\; .012 \\ \hline 2.160 \end{array}$$

(c)
$$\begin{array}{r} 225 \\ \times \;\;\; 1.4 \\ \hline 315.0 \end{array}$$

(d)
$$\begin{array}{r} 560 \\ \times \;\; .0025 \\ \hline 1.4000 \end{array}$$

EXAMPLE 2

Finding Part

The following bar graph shows that 56% of the housing units in the United States have a dishwasher. Out of 18,000 households, how many are expected to have a dishwasher?

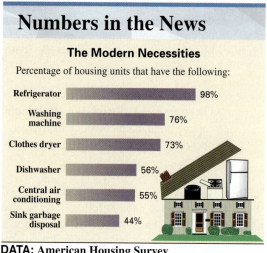

Numbers in the News

The Modern Necessities

Percentage of housing units that have the following:

Refrigerator — 98%
Washing machine — 76%
Clothes dryer — 73%
Dishwasher — 56%
Central air conditioning — 55%
Sink garbage disposal — 44%

DATA: American Housing Survey

SOLUTION

The number of households, 18,000, is the base. The rate, 56%, is the percent of the total households that have a dishwasher. Since the number of households that have a dishwasher is part of the total number of households, find the number of households that have a dishwasher by using the formula to find part.

$$P = B \times R$$
$$P = 18,000 \times 56\%$$
$$P = 18,000 \times .56$$
$$P = 10,080$$

The number of households expected to have a dishwasher is 10,080.

The calculator solution to this example is

18000 ⊠ 56 % ▣ 10080 or 18000 ⊠ .56 ▣ 10080

Note: Refer to Appendix C for calculator basics.

Objective 4   Recognize the terms associated with base, rate, and part. Percent problems have certain similarities. For example, some phrases are associated with the base in the problem. Other phrases lead to the part, while % or *percent* following a number identifies the rate. The following chart helps distinguish between the base and the part.

Words and Phrases Associated with Base and Part

Usually indicates the base (*B*)	Usually indicates the part (*P*)
Sales ————————→	Sales tax
Investment ———————→	Return
Savings ————————→	Interest
Value of bonds —————→	Dividends
Retail price ———————→	Discount
Last year's anything ———→	Increase or decrease
Value of real estate ————→	Rents
Old salary ————————→	Raise
Total sales ————————→	Commission
Value of stocks —————→	Dividends
Earnings ————————→	Expenditures
Original ————————→	Change

Objective 5 **Calculate sales tax.** Calculating **sales tax** is a good example of finding part. States, counties, and cities often collect taxes on retail sales to the consumer. The sales tax is a percent of the sale. This percent varies from as low as 3% in some states to 8% or more in other states. The formula used for finding sales tax is as follows.

$$P = B \times R$$
$$Sales\ tax = Sales \times Sales\ tax\ rate$$

EXAMPLE 3

Calculating Sales Tax

Real Estate Office Supply sold $974.50 worth of merchandise. If the sales tax rate was 6%, how much tax was paid? Find the total cost including the tax.

SOLUTION

The amount of sales, $974.50, is the starting point or base (*B*), and 6% is the rate (*R*). Since the tax is a *part* of total sales, use the formula $P = B \times R$ to find part.

$$P = B \times R$$
$$P = \$974.50 \times 6\%$$
$$P = \$974.50 \times .06 = \$58.47$$

The tax, or part, is $58.47.

To find the total amount of sales and tax, the amount of sales ($974.50) is added to the sales tax, $58.47. The total sales and tax is $1032.97 ($974.50 + $58.47).

Identify the rate, base, and part with the following hints.

Base tends to be preceded by the word *of* or *on*; tends to be the *whole*.
Rate is followed by a percent sign or the word *percent*.
Part is in the same units as the base and is usually a portion of the base.

Objective ⑥ **Learn the standard format of percent problems.** Percent problems often take the following form.

% of something is something

For example,

> 5% of the automobiles are red
> 4.2% of the workers are unemployed
> 20% of the income is income tax
> 74% of the students are full time

When expressed in this standard form, the components in the percent problem always appear in this order.

$$R \quad \times \quad B \quad = \quad P$$

Rate × **Base** = **Part**

% of **something** is **something**

QUICK TIP   Rate is identified by % (the percent sign); the word *of* means × (multiplication); the multiplicand, or number being multiplied, is the base; the word *is* means = (equals); and the product, or answer, is part of the base.

The following survey results show that 49.5% of the 822 people surveyed get most of their news from television.

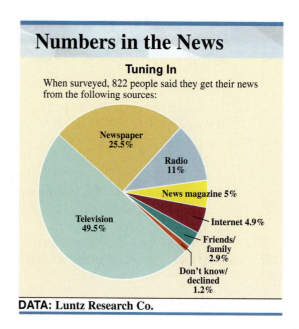

Numbers in the News

Tuning In

When surveyed, 822 people said they get their news from the following sources:

- Newspaper 25.5%
- Radio 11%
- News magazine 5%
- Internet 4.9%
- Friends/family 2.9%
- Don't know/declined 1.2%
- Television 49.5%

DATA: Luntz Research Co.

The rate is the percent selecting television, 49.5%. The base, or number of people in the survey, is 822. And the number of people selecting television as the source of "most of their news" is the part. Find part, the number of people who said television is where they get most of their news.

$$49.5\% \times 822 = 406.89 = 407 \text{ people (rounded)}$$

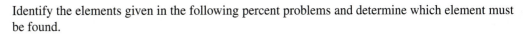

EXAMPLE **4**

*Identifying the
Elements in Percent
Problems*

Identify the elements given in the following percent problems and determine which element must be found.

(a) During a recent sale, Stockdale Marine offered a 15% discount on all new recreation equipment. Find the discount on a jet ski originally priced at $3895. First arrange this problem in standard form.

$$
\begin{array}{cccc}
\% & \text{of} & \text{something} & \text{is} & \text{something} \\
\% & \text{of} & \text{price} & \text{is} & \text{discount} \\
15\% & \text{of} & \$3895 & = & \text{discount} \\
\textbf{\textit{R}} & \times & \textbf{\textit{B}} & = & \textbf{\textit{P}}
\end{array}
$$

At this point, check that the rate and base are given, so the part must be found. Find the discount by multiplying 15%, or .15, by $3895.

$$.15 \times \$3895 = \$584.25$$

The amount of the discount is $584.25.

(b) Round Table Pizza spends an amount equal to 5.8% of its sales on advertising. If sales for the month were $28,500, find the amount spent on advertising. First arrange this problem in standard form.

$$
\begin{array}{cccc}
\% & \text{of} & \text{something} & \text{is} & \text{something} \\
\% & \text{of} & \text{sales} & \text{is} & \text{advertising} \\
5.8\% & \text{of} & \$28,500 & = & \text{advertising} \\
\textbf{\textit{R}} & \times & \textbf{\textit{B}} & = & \textbf{\textit{P}}
\end{array}
$$

Rate is given as 5.8%, base (sales) is $28,500, and part (advertising) must be found. Find the amount spent on advertising by multiplying .058 and $28,500.

$$.058 \times \$28,500 = \$1653$$

The amount spent on advertising is $1653.

3.2 | Exercises

The **Quick Start** exercises in each section contain solutions to help you get started.

Solve for part in each of the following. Round to the nearest hundredth. (See Example 1.)

Quick Start

1. 10% *of* 60 bicycles = **6 bicycles**   2. 25% of 3500 web sites = **875 web sites**

3. 75.5% of $800 = _____   4. 20.5% of $1500 = _____

5. 4% of 120 feet = _____   6. 125% of 2000 products = _____

7. 175% of 5820 miles = _____   8. 15% of 75 crates = _____

9. 17.5% of 1040 homes = _____   10. 52.5% of 1560 trucks = _____

11. 118% of 125.8 yards = _____   12. 110% of 150 apartments = _____

13. $90\frac{1}{2}$% of $5930 = _____   14. $7\frac{1}{2}$% of $150 = _____

15. Identify the three components in a percent problem. In your own words, write one sentence telling how to identify each of these three components. (See Objective 1.)

16. There are words and phrases that are usually associated with base and part. Give three examples of words that usually identify the base and the accompanying word for the part. (See Objective 4.)

Solve for part in each of the following application problems. Round to the nearest cent unless otherwise indicated. (See Examples 2–4.)

Quick Start

17. **WEDDING PREFERENCES** In a recent survey of 480 adults, 55% said that they would prefer to have their wedding at a religious site. How many said they would prefer the religious site? (*Source:* National Family Opinion Research.)

17. **264 adults**

$P = B \times R$
$P = 480 \times .55 = 264$ adults

18. **SOURCES OF NEWS** In a poll of 822 people, 25.5% said that they get their news from the newspaper. Find the number of people who said they get their news from the newspaper. Round to the nearest whole number. (*Source:* Luntz Research Co.)

18. _____

19. **SALES TAX** A real estate broker wants to purchase a Palm i705 with a built-in radio modem priced at $399. If the sales tax rate is 7.75%, find the total price including the sales tax. (*Source:* Real Estate Technology.)

19. _____

C indicates an exercise that is related to the Case in Point feature within the section.

C **20.** Thomas Dugally of Century 21 Realty is working with a mortgage company that charges borrowers $350 plus 2% of the loan amount. What is the total charge to get a home loan of $95,000?

20. _____

21. **WOMEN IN THE NAVY**   The navy guided-missile destroyer USS *Sullivans* has a 335-person crew of which 13% are female. Find the number of female crew members. Round to the nearest whole number. (*Source:* US Navy.)

21. _____

22. **SUPERMARKET SHOPPING**   The Point of Purchase Advertising Institute says that 55% of all supermarket shoppers have a written list of their needs. If there are 3680 shoppers entering the supermarket that you manage in one day, what number of shoppers would you expect to have a written shopping list?

22. _____

23. **BAR SOAP**   A bar of Ivory Soap is $99\frac{44}{100}\%$ pure. If the bar of soap weighs 9 ounces, how many ounces are pure? (Round to the nearest hundredth.)

23. _____

24. **CANNED-MEAT SALES**   According to Hormel Foods Corporation, Spam and Spam Lite together held 62.2% of the $148 million canned lunchmeat category over a 52-week period (the entire year). Find the total annual sales of these Hormel products. Round to the nearest hundredth of a million.

24. _____

25. **DRIVING DISTRACTIONS**   It is estimated that 29.5% of automobile crashes are caused by driver distractions, such as mobile communications devices. If there are 16,450 automobile crashes in a study, what number would be caused by driver distractions? Round to the nearest whole number. (*Source: National Conference of State Legislatures.*)

25. _____

26. **WORKPLACE REQUIREMENTS**   A study of office workers found that 27% would like more storage space. If there are 14 million office workers, how many would want more storage space? (*Source:* Steelcase Workplace Index.)

26. _____

27. **FEMALE LAWYERS**   There are 1,094,751 active lawyers living in the United States. If 71.4% of these lawyers are male, find (a) the percent of the lawyers who are female and (b) the number of lawyers who are female. Round to the nearest whole number. (*Source:* American Bar Association.)

(a) _____
(b) _____

28. **CUBAN LABOR FORCE**   The size of Cuba's labor force is 3.8 million people. If 61% of the labor force is male, find (a) the percent who are female and (b) the number of workers who are male.

(a) _____

(b) _____

29. DIGITAL CAMERA   A Cannon PowerShot digital camera priced at $332 is marked down 25%. Find the price of the camera after the markdown.

29. _____

30. HOSPITAL STAFFING   In a survey of 245 small- to medium-size hospitals, 45% said that they do not have enough radiologists. How many of these hospitals do have enough radiologists? Round to the nearest whole number. (*Source:* U.S. Radiology Partners Incorporated.)

30. _____

31. NEW PRODUCT FAILURE   Marketing Intelligence Service says that there were 15,401 new products introduced last year. If 86% of the products introduced last year failed to reach their business objectives, find the number of products that did reach their objectives. (Round to the nearest whole number.)

31. _____

32. FAMILY BUDGET   A family of four with a monthly income of $3800 spends 90% of its earnings and saves the balance for the down payment on a house. Find (a) the monthly savings and (b) the annual savings of this family.

(a) _____
(b) _____

33. ORANGE JUICE IN CHINA   This year the sales of Tropicana orange juice in China is $100 million. Seagram's Tropicana Beverage Group estimates that sales will increase by 35% next year. Find the amount of orange juice sales estimated for next year.

33. _____

34. SUPER BOWL ADVERTISING   The average cost of 30 seconds of advertising during the Super Bowl 5 years ago was $2.2 million. If the increase in cost over the last 5 years has been 9%, find the average cost of 30 seconds of advertising during the Super Bowl this year. Round to the nearest tenth of a million. (*Source:* NFL research.)

34. _____

35. SALES-TAX COMPUTATION   As the owner of a copy and print shop, you must collect $6\frac{1}{2}\%$ of the amount of each sale for sales tax. If sales for the month are $48,680, find the combined amount of sales and tax.

35. _____

36. TOTAL COST   A NuVac 3200 is priced at $524 with an allowed trade-in of $125 for an old unit. If sales tax of $7\frac{3}{4}\%$ is charged on the price of the new NuVac unit, find the total cost to the customer after receiving the trade-in. (*Hint:* Trade-in is subtracted last.)

36. _____

37. REAL ESTATE COMMISSIONS   Thomas Dugally of Century 21 Realty sold a home for $174,900. The commission was 6% of the sale price; however, Dugally receives only 60% of the commission, while 40% remains with his broker. Find the amount of commission received by Dugally.

37. _____

38. BUSINESS OWNERSHIP   Rick Wilson has an 82% ownership in a company called Puppets and Clowns. If the company has a value of $49,200 and Wilson receives an income of 30% of the value of his ownership, find the amount of his income.

38. _____

CONSUMER INTERNET SALES   *Country Store has a unique selection of merchandise that it sells by catalog and over the Internet. Use the shipping and insurance delivery chart below and a sales tax rate of 5% to solve Exercises 39–42. There is no sales tax on shipping and insurance.* (**Source:** Country Store catalog.)

Shipping and Insurance Delivery Chart

Up to $15.00	add $3.95
$15.01 to $25.00	add $5.95
$25.01 to $35.00	add $6.95
$35.01 to $50.00	add $7.95
$50.01 to $70.00	add $8.95
$70.01 to $99.99	add $9.95
$100.00 or more	add $10.95

39. Find the total cost of 6 Small Fry Handi-Pan electric skillets at a cost of $29.99 each.

39. _____

40. A customer ordered 5 sets of flour-sack towels at a cost of $12.99 each. What is the total cost?

40. _____

41. Find the total cost of 3 pop-up hampers at a cost of $9.99 each and 4 nonstick mini-donut pans at $10.99 each.

41. _____

42. What is the total cost of 5 coach lamp bird feeders at a cost of $19.99 each and 6 garden weather centers at $14.99 each?

42. _____

3.3 Finding Base

Objectives

1. Use the basic percent formula to solve for base.
2. Find the amount of sales when tax amount and tax rate are both known.
3. Find the amount of investment when expense and rate of expense are known.

CASE *in* **POINT**

Thomas Dugally of Century 21 Realty must help buyers select properties that they can afford. Real estate lenders have strict guidelines that determine the maximum loan that they will give a buyer. Usually, the lender will limit the borrowers' monthly house payment to no more than 28% to 36% of their monthly income.

Objective 1 **Use the basic percent formula to solve for base.** In some problems, the rate and part are given, but the base, or starting point, must be found. The formula $P = B \times R$ can be used to get the **formula for base**. The following diagram illustrates the formula $P = B \times R$. To find the formula for base, cover B. Now the letter P is left over the letter R. Think of this as meaning $\frac{P}{R}$, or part ÷ rate.

$$\text{Base} = \frac{\text{Part}}{\text{Rate}} \text{ or } B = \frac{P}{R}$$

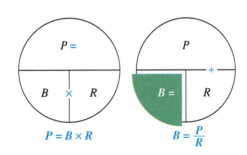

$$P = B \times R \qquad B = \frac{P}{R}$$

QUICK TIP   When using this formula, P (part) is always on top.

Suppose that Scott and Andrea Abriani told Thomas Dugally that they are able to make a monthly payment of $1155. If this is 28% of their monthly income, how can Dugally find the amount of their monthly income? The key word here, indicating that the amount of monthly income is the base, is the word *of.* To find the amount of their monthly income, insert the rate (28%) and the part ($1155) into the formula.

$$B = \frac{P}{R} \quad \text{Replace } P \text{ with 1155 and } R \text{ with .28 to get}$$

$$B = \frac{1155}{.28} = 4125. \quad \text{The amount of their monthly income is \$4125.}$$

EXAMPLE 1

Solving for Base

Solve for base, using the formula $B = \frac{P}{R}$.

(a) 8 is 4% of ____.   **(b)** 135 is 15% of ____.   **(c)** 1.25 is 25% of ____.

SOLUTION

(a) $\frac{8}{.04} = 200$ **(b)** $\frac{135}{.15} = 900$ **(c)** $\frac{1.25}{.25} = 5$

Objective ☑2 **Find the amount of sales when tax amount and tax rate are both known.**
In business problems involving sales tax, the amount of sales is always the base.

EXAMPLE 2

Finding Sales when
Sales Tax Is Given

The 5% sales tax collected by Famous Footwear was $780. What was the amount of total sales?

SOLUTION

Here, the rate of tax collection is 5%, and taxes collected are a part of total sales. The rate in this problem is 5%, the part is $780, and the base, or total sales, must be found. Arrange the problem in standard form.

$$R \quad \times \quad B \quad = \quad P$$

% of **something** is **something**

5% of **total sales** is $780 (tax)

Using the formula $B = \frac{P}{R}$, we get

$$B = \frac{780}{.05} = \$15,600 \text{ total sales.}$$

The calculator solution to this example is

$$780 \boxed{\div} .05 \boxed{=} 15600.$$

Note: Refer to Appendix C for calculator basics.

QUICK TIP   It is important to consider whether an answer is reasonable. A common error in a base problem is to confuse the base and the part. In Example 2, if the taxes, $780, had been mistakenly used as the base, the resulting answer would have been $39 ($780 × 5%). Obviously, $39 is not a reasonable amount for total sales, given $780 as sales tax.

The following newspaper clipping states that average home prices rose at a 5.56% annual pace in the third quarter. In these calculations, the average home price a year ago is the base, or starting point. It is the amount to which this year's average home is being compared.

HERE & NOW

Home prices rise faster in third quarter

Average home prices rose at a 5.56% annual pace in the third quarter as the broad economy caught up with the housing market, the Office of Federal Housing Enterprise Oversight said Monday. Prices rose 3.12% the second quarter. Third-quarter home prices rose fastest in Fresno, Calif., where the average was 16.1% above a year ago. Fresno was followed by Fort Pierce-Port St. Lucie, Fla., where price rose 14.7%.

Source: USA Today, 12/02/03.

Objective 3 **Find the amount of investment when expense and rate of expense are known.** When solving problems involving investments, the amount of the investment is the base.

EXAMPLE **3**

Finding the Amount of an Investment

The yearly utility cost of an apartment complex is $3\frac{1}{2}$% of its value. If utility cost is $73,500 per year, find the value of the apartment complex.

SOLUTION

First set up the problem:

$$R \quad \times \quad B \quad = \quad P$$

$$3\frac{1}{2}\% \quad \text{of} \quad \textbf{value} \quad \text{is} \quad \text{maintenance cost}$$

$$3\frac{1}{2}\% \quad \text{of} \quad \textbf{value} \quad \text{is} \quad \$73,500$$

Find the value of the apartment complex, the base, with the formula $B = \frac{P}{R}$.

$$B = \frac{73,500}{.035} = \$2,100,000 \text{ value of complex}$$

QUICK TIP When working with a fraction of a percent it is best to change the fraction to a decimal. In Example 3, $3\frac{1}{2}\%$ was changed to 3.5%, which equals .035.

3.3 | Exercises

FOR EXTRA HELP

MyMathLab

InterActMath.com

MathXP MathXL

MathXL
Tutorials on CD

Addison-Wesley
Math Tutor Center

DVT/Videotape

The **Quick Start** *exercises in each section contain solutions to help you get started.*

Solve for base in each of the following. Round to the nearest hundredth. (See Example 1.)

Quick Start

1. 530 firms is 25% of ___**2120**___ firms.

2. 240 letters is 80% of ___**300**___ letters.

3. 130 salads is 40% of _____ salads.

4. 32 shipments is 8% of _____ shipments.

5. 110 lab tests is 5.5% of _____ lab tests.

6. $850 is $4\frac{1}{4}$% of _____.

7. 36 employees is .75% of _____ employees.

8. 23 workers is .5% of _____ workers.

9. 66 files is .15% of _____ files.

10. 54,600 boxes is 60% of _____ boxes.

11. 50 doors is .25% of _____ doors.

12. 39 bottles is .78% of _____ bottles.

13. $33,870 is $37\frac{1}{2}$% of _____.

14. $8500 is $27\frac{1}{2}$% of _____.

15. $12\frac{1}{2}$% of _____ people is 135 people.

16. $18\frac{1}{2}$% of _____ circuits is 370 circuits.

17. 375 crates is .12% of _____ crates.

18. 3.5 quarts is .07% of _____ quarts.

19. .5% of _____ homes is 327 homes.

20. 6.5 barrels is .05% of _____ barrels.

21. 12 audits is .03% of _____ audits.

22. 8 banks is .04% of _____ banks.

23. The basic percent formula is $P = B \times R$. Show how to find the formula to solve for B (base). (See Objective 1.)

24. A problem includes amount of sales, sales tax, and a sales-tax rate. Explain how you could identify the base, rate, and part in this problem. (See Objective 2.)

Solve for base in the following application problems. (See Examples 2 and 3.)

Quick Start

25. **HOME OWNERSHIP**   The number of U.S. households owning homes last year was 72.6 million—a record setting 67.8% of all households. What is the total number of U.S. households? Round to the nearest tenth of a million. (*Source:* Habitat World.)

$B = \frac{P}{R} = \frac{72.6}{.678} = 107.07 = 107.1$ million (*rounded*)

25. __107.1 million households__

26. **EMPLOYEE POPULATION BASE**   In a large metropolitan area, 81% of the employed population is enrolled in a health maintenance organization (HMO). If 700,650 employees are enrolled, find the total number of people in the employed population.

26. _____

27. **COLLEGE ENROLLMENT**   This semester there are 1785 married students on campus. If this figure represents 23% of the total enrollment, what is the total enrollment? (Round to the nearest whole number.)

27. _____

C indicates an exercise that is related to the Case in Point feature within the section.

28. **VOTER REGISTRATION** Registered voters make up 13.8% of the county population. If there are 345,000 registered voters in the county, find the total population in the county.

28. _____

C **29.** **LOAN QUALIFICATION** Thomas Dugally found a home for Scott and Andrea Abriani that will require a monthly loan payment of $1140. If the lender insists that the buyer's monthly payment not exceed 30% of the buyer's monthly income, find the minimum monthly income required by the lender.

29. _____

30. **PERSONAL BUDGETING** Jim Lawler spends 28% of his income on housing, 15% on food, 11% on clothing, 15% on transportation, 11% on education, 7% on recreation, and saves the balance. If his savings amount to $266.50 per month, what are his monthly earnings?

30. _____

31. **DIABETES SURVEY** In a telephone survey, 749 people said that diabetes is a serious problem in the U.S. If this was 71% of the survey group, find the total number of people in the telephone survey. Round to the nearest whole number. (*Source:* Hoffman-La Roche Company.)

31. _____

32. **DRIVING TESTS** In analyzing the success of driver's license applicants, the state finds that 58.3% of those examined received a passing mark. If the records show that 8370 new driver's licenses were issued, what was the number of applicants? (Round to the nearest whole number.)

32. _____

33. **GLOBAL WORKFORCE REDUCTION** Japan's largest business daily newspaper reports that Sony will eliminate 20,000 jobs or 10% of its workforce. Find the size of Sony's workforce after the reduction. (*Source: Nihan Keizai Shimbun* newspaper.)

33. _____

34. **COMMUNICATIONS INDUSTRY LAYOFFS** Telecommunications equipment maker Nortel Networks says it will lay off 4000 workers globally. If this amounts to 4% of its total workforce, how many workers will remain after the layoffs? (*Source:* Nortel Networks.)

34. _____

35. **GAMBLING PAYBACK** An Atlantic City casino advertises that it gives a 97.4% payback on slot machines, and the balance is retained by the casino. If the amount retained by the casino is $4823, find the total amount played on the slot machines.

35. _____

36. **SMOKING OR NONSMOKING** A casino hotel in Barbados states that 45% of its rooms are for nonsmokers. If the resort allows smoking in 484 rooms, find the total number of rooms.

36. _____

Supplementary Application Exercises on Base and Part

Solve the following application problems. Read each problem carefully to determine whether base or part is being asked for.

1. **SHAMPOO INGREDIENTS** Most shampoos contain 75% to 90% water. If there are 12.5 ounces of water in a bottle of shampoo that contains 78% water, what is the size of the bottle of shampoo? (Round to the nearest whole number.)

   1. _____

2. **HOUSEHOLD LUBRICANT** The lubricant WD-40 is used in 82.3 million U.S. homes, which is 79% of all homes in the United States. Find the total number of homes in the United States. Round to the nearest tenth of a million. (*Source:* WD-40.)

   2. _____

3. **PROPERTY INSURANCE** Thomas Dugally of Century 21 sold a commercial building valued at $423,750. If the building is insured for 68% of its value, find the amount of insurance coverage.

   3. _____

4. **FLU SHOTS** In a survey of 3860 people who were 18–49 years of age, 16.3% had received an influenza vaccination (flu shot). How many of those surveyed received the vaccination? Round to the nearest whole number. (*Source:* National Health Interview Survey.)

   4. _____

5. **CAMAROS AND MUSTANGS** The Chevrolet Camaro was introduced in 1967. Camaro sales that year were 220,917, which was 46.2% of the number of Ford Mustangs sold in the same year. Find the number of Mustangs sold in 1967. (Round to the nearest whole number.)

   5. _____

6. **RETIREMENT ACCOUNT** Leah Runde has 7.5% of her earnings deposited into her retirement plan. If this amounts to $240 per month, find her annual earnings.

   6. _____

7. **DRIVER-SAFETY SURVEY** A survey at an intersection found that of 2200 drivers, 38% were wearing seat belts. How many drivers in the survey were wearing seat belts?

   7. _____

8. **BLOOD-CHOLESTEROL LEVELS** At a recent health fair, 32% of the people tested were found to have high blood-cholesterol levels. If 350 people were tested, find the number having high blood cholesterol.

   8. _____

9. **RETIREMENT ACCOUNTS**   Erin Joyce has 9.5% of her earnings deposited into a retirement account. If this amounts to $308.75 per month, find her annual earnings.

9. _____

10. **SAVINGS ACCOUNT INTEREST**   The Northridge PTA received $50.75 in annual interest on its bank account. If the bank paid $3\frac{1}{2}\%$ interest per year, how much money was in the account?

10. _____

AIDING DISABLED EMPLOYEES   The bar graph below shows how companies have accommodated their employees with disabilities. The data were collected from personnel directors, human resource directors, and executives responsible for hiring at 501 companies. Use this information to solve Exercises 11–14. Round to the nearest whole number. (**Source:** Heldrich Work Trends Survey.)

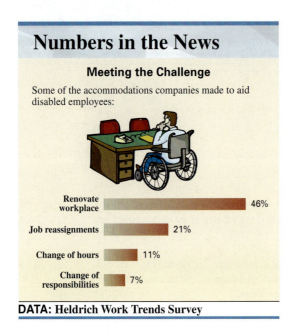

Numbers in the News

Meeting the Challenge

Some of the accommodations companies made to aid disabled employees:

Renovate workplace — 46%
Job reassignments — 21%
Change of hours — 11%
Change of responsibilities — 7%

DATA: Heldrich Work Trends Survey

11. How many companies have renovated the workplace to aid employees with disabilities.

11. _____

12. How many companies changed worker responsibilities to aid employees with disabilities.

12. _____

13. Find the number of companies who changed worker hours to aid employees with disabilities.

13. _____

14. Find the number of companies who made job reassignments to aid employes with disabilities.

14. _____

3.4 | Finding Rate

Objectives

1. Use the basic percent formula to solve for rate.
2. Find the rate of return when the amount of the return and the investment are known.
3. Solve for the percent remaining when the total amount and amount used are given.
4. Find the percent of change.

CASE *in* **POINT**

At Century 21 Realty, where Thomas Dugally is a real estate broker, all of the expenses of running the real estate office are compared to the income generated from sales and leasing activities. The most meaningful way of making these comparisons is by calculating all expense items as a percent of income. Dugally often calculates those percents as he makes decisions on how to run his business.

Objective 1 **Use the basic percent formula to solve for rate.** In the third type of percent problem, the part and base are given, and the rate must be found. The **formula for rate** is found from the formula $P = B \times R$. The diagram shows that to find the formula for rate, cover R to get $\frac{P}{B}$, or part ÷ base.

$$\text{Rate} = \frac{\text{Part}}{\text{Base}} \quad \text{or} \quad R = \frac{P}{B}$$

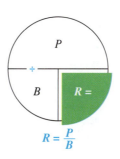

$$R = \frac{P}{B}$$

QUICK TIP   When using this formula, P (part) is always on top.

The formula $P = B \times R$ can be used to find either P, B, or R as long as the values of two of the components are known. When either B or R must be found, the known value must be divided into P, on the other side of the equal sign.

Find B: Find R:

$P = B \times R$ $P = B \times R$

$B = \dfrac{P}{R}$ $R = \dfrac{P}{B}$

See why this works by using numbers.

$$10 = \underset{\uparrow}{?} \times 2 \qquad\qquad 10 = 5 \times \underset{\uparrow}{?}$$

$$? = \frac{10}{2} \qquad\qquad ? = \frac{10}{5}$$

$$? = 5 \qquad\qquad ? = 2$$

After the division, the known numbers are on one side of the equal sign and the unknown number is on the other side of the equal sign. Refer to Appendix A, Equation and Formula Review, for additional discussion of this topic.

The rate is identified by % or *percent*. For example, what percent of 32 is 8? Using the formula $R = \frac{P}{B}$, gives

$$R = \frac{8}{32} = .25 = 25\%$$

Thus, 8 is 25% of 32, or in other words, 25% of 32 is 8.

EXAMPLE **1**

Solving for Rate

Solve for rate.

(a) 26 is ____ % of 104.   **(b)** ____ % of 300 is 60.
(c) 54 is ____ % of 12.

SOLUTION

(a) $\frac{26}{104} = .25 = 25\%$ **(b)** $\frac{60}{300} = .2 = 20\%$ **(c)** $\frac{54}{12} = 4.5 = 450\%$

> **QUICK TIP**   When finding rate, be sure to change your decimal answer to a percent.

Objective **2** **Find the rate of return when the amount of the return and the investment are known.** The rate of return in an investment problem may be found using the basic percent formula.

EXAMPLE **2**

Finding the Rate of Return

Thomas Dugally invested a total of $2010 in a cell phone, Palm Pilot, and a Compaq multimedia notebook. As a result of having this equipment, he had additional income of $1820 in the first year. Find the rate of return.

SOLUTION

First set up the problem.

$$R \quad \times \quad B \quad = \quad P$$

%   of something   is   something.

What % of investment is income?

What % of $2010   is $1820?

The amount of investment, $2010, is the base, and the return, $1820, is the part. The return is part of the total investment. Using the formula $R = \frac{P}{B}$, let $P = 1820$ and $B = 2010$.

$$R = \frac{1820}{2010} = .9054 \text{ or } 90.5\% \quad \text{rounded to the nearest tenth of a percent}$$

Objective 3 **Solve for the percent remaining when the total amount and amount used are given.** In some problems, the total amount of something and the amount used are given and the percent remaining must be found.

EXAMPLE 3

Solving for the Percent Remaining

A hot-water heater is expected to last 10 years (its total "life") before it needs replacement. If a water heater is 8 years old, what percent of the water heater's life remains?

SOLUTION

The total life of the water heater (10 years) is the base. Subtract the amount of life used, 8 years, from the total life, 10 years, to find the number of years remaining. In other words, 10 years (total life) − 8 years (life used) = 2 years (life remaining). The life remaining is part of the entire life.

$$R = \frac{P}{B} = \frac{2}{10} = .2 = 20\%$$

If the age of the water heater (8 years) had been used as part, the resulting answer, 80%, would be the percent of life used. Find the percent of remaining life by subtracting 80% from 100%. Therefore, the remaining life is 20%, which is the same answer.

QUICK TIP   Remember that base is always 100%, the whole or total.

Objective 4 **Find the percent of change.** The basic percent formula is used to find the percent of increase and the percent of decrease.

EXAMPLE 4

Finding the Percent of Increase

Due to an advertising campaign, sales of radio-controlled cars, boats, and planes at RC Country climbed from $36,600 last month to $113,460 this month. Find the percent of increase.

SOLUTION

The sales last month, $36,600, is the base. Subtract the sales last month, $36,600, from the sales this month, $113,460, to find the increase in sales volume.

$$\$113,460 - \$36,600 = \$76,860 \text{ increase in sales volume } \left(\text{part}\right)$$

Since the increase in sales volume is the part, solve for rate. Use the formula $R = \frac{P}{B}$.

$$R = \frac{\$76,860}{\$36,600} = 2.1 \text{ or } 210\%$$

The percent of increase in sales volume is 210%.

The calculator solution to this example is

(113460 − 36600) ÷ 36600 = 2.1.

Note: Refer to Appendix C for calculator basics.

QUICK TIP   Remember, to find the percent of increase, the first step is to determine the *amount of increase*. The base is *always* the original amount, such as last year's or last month's amount, and the amount of increase is the part.

The following map of the United States and the accompanying chart show that homes are becoming less affordable in many parts of the country.

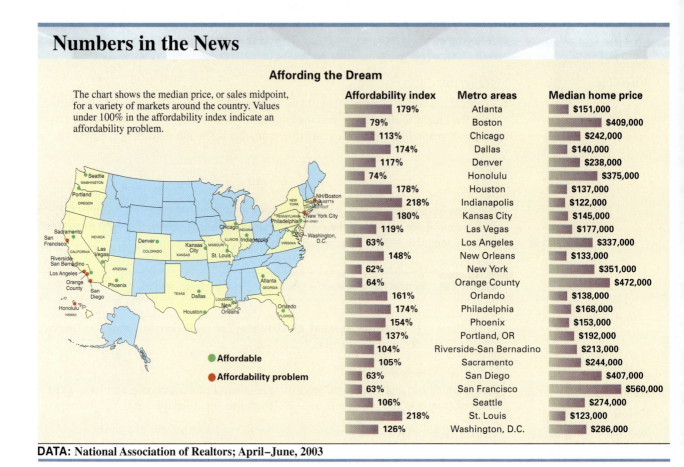

Numbers in the News

Affording the Dream

The chart shows the median price, or sales midpoint, for a variety of markets around the country. Values under 100% in the affordability index indicate an affordability problem.

Affordability index	Metro areas	Median home price
179%	Atlanta	$151,000
79%	Boston	$409,000
113%	Chicago	$242,000
174%	Dallas	$140,000
117%	Denver	$238,000
74%	Honolulu	$375,000
178%	Houston	$137,000
218%	Indianapolis	$122,000
180%	Kansas City	$145,000
119%	Las Vegas	$177,000
63%	Los Angeles	$337,000
148%	New Orleans	$133,000
62%	New York	$351,000
64%	Orange County	$472,000
161%	Orlando	$138,000
174%	Philadelphia	$168,000
154%	Phoenix	$153,000
137%	Portland, OR	$192,000
104%	Riverside-San Bernadino	$213,000
105%	Sacramento	$244,000
63%	San Diego	$407,000
63%	San Francisco	$560,000
106%	Seattle	$274,000
218%	St. Louis	$123,000
126%	Washington, D.C.	$286,000

● Affordable

● Affordability problem

DATA: National Association of Realtors; April–June, 2003

EXAMPLE 5

Finding the Percent of Decrease

As a result of lower mortgage interest rates, Scott and Andrea Abriani will be able to decrease their monthly mortgage payment from $1155 to $1050. Find the percent of decrease.

SOLUTION

The base is the previous payment, the old monthly payment of $1155. Subtract the new monthly payment of $1050, from the old payment, $1155, to find the decrease in the monthly payment.

$$\$1155 - \$1050 = \$105 \text{ decrease in monthly payment } (\text{part})$$

By the formula $R = \frac{P}{B}$:

$$R = \frac{105}{1155} = .0909 = 9.1\% \text{ (rounded)}$$

The percent of decrease is 9.1%.

The calculator solution to this example is to subtract to find the difference and then divide.

1155 ⊟ 1050 ⊜ 105 ÷ 1155 ⊜ 0.0909 = .091 (rounded)

Note: Refer to Appendix C for calculator basics.

QUICK TIP   To find the percent of decrease, the first step is to determine the *amount of decrease.* The amount of decrease is the part in the problem and the base is *always* the original amount such as last year's, last month's, or last week's amount.

3.4 | Exercises

The **Quick Start** *exercises in each section contain solutions to help you get started.*

Solve for rate in each of the following. Round to the nearest tenth of a percent. (See Example 1.)

Quick Start

1. ____10____ % of 2760 listings is 276 listings.

2. ____40____ % of 850 showings is 340 showings.

3. 35 rail cars is _____% of 70 rail cars.

4. 144 desks is _____% of 300 desks.

5. _____% of 78.57 ounces is 22.2 ounces.

6. _____% of 728 miles is 509.6 miles.

7. 114 tuxedos is _____% of 150 tuxedos.

8. $310.75 is _____% of $124.30.

9. _____% of $53.75 is $2.20.

10. _____% of 850 liters is 3.4 liters.

11. 46 shirts is _____% of 780 shirts.

12. 5.2 vats is _____% of 28.4 vats.

13. _____% of 600 acres is 7.5 acres.

14. _____% of $8 is $.06.

15. 170 cartons is _____% of 68 cartons.

16. _____% of 425 orders is 612 orders.

17. _____% of $330 is $91.74.

18. _____% of 752 employees is 470 employees.

19. The basic percent formula is $P = B \times R$. Show how to find the formula to solve for R (rate). (See Objective 1.)

20. A problem includes last year's sales, this year's sales, and asks for the percent of increase. Explain how you would identify the base, rate, and part in this problem. (See Objective 4.)

Solve for rate in the following application problems. Round to the nearest tenth of a percent. (See Examples 2–5.)

Quick Start

21. **ADVERTISING EXPENSES**   Thomas Dugally of Century 21 Realty reports that office income last month was $315,600, while advertising expenses were $19,567.20. What percent of last month's income was spent on advertising?

    21. __6.2%_____

    $R = \frac{P}{B} = \frac{19,567.20}{315,600} = .062 = 6.2\%$

22. **MUSIC STORES**   Music Land Group, Inc. will close 150 of its 1119 Sam Goody and Sun Coast Motion Picture Company Stores. What percent of the stores will close? Round to the nearest tenth of a percent. (*Source:* Associated Press.)

    22. _____

 indicates an exercise that is related to the Case in Point feature within the section.

23. WOMEN IN THE MILITARY   A recent study by Rand's National Defense Research Institute examined 48,000 military jobs, such as Army attack-helicopter pilot or Navy gunner's mate. It was found that only 960 of these jobs are filled by women. What percent of these jobs are filled by women?

23. _____

24. VOCABULARY   There are 55,000-plus words in *Webster's Dictionary*, but most educated people can identify only 20,000 of these words. What percent of the words in the dictionary can these people identify?

24. _____

25. ADVERTISING EXPENSES   Advertising expenditures for Bailey's Roofers are as follows.

Newspaper	$2250	Television	$1425
Radio	$954	Yellow Pages	$1605
Outdoor	$1950	Miscellaneous	$2775

What percent of the total advertising expenditures is spent on radio advertising?

25. _____

26. ANTIQUE SALES   Barbara's Antiquery says that of its 3800 items in inventory, 3344 are just plain junk, while the rest are antiques. What percent of the total inventory is antiques?

26. _____

27. HARLEY-DAVIDSON MOTORCYCLES   Harley-Davidson, the only major U.S.-based motorcycle maker, says that it expects to sell 317,000 motorcycles this year, up from 290,600 last year. Find the percent of increase in production. (*Source:* Harley-Davidson.)

27. _____

28. AVERAGE HOME PRICE   The average selling price of a home in the United States last year was $153,200, while the average selling price this year was $163,100. What is the percent of increase? (*Source: Realtor* magazine.)

28. _____

29. FALLING PHONE BILLS   Long-distance phone bills plummeted to an average of $24.40 a month from last year's monthly average of $30.50. What was the percent of decrease? (*Source:* J.D. Power.)

29. _____

30. SOLAR POWER   In the past five years, the cost of generating electricity from the sun has been brought down from 24 cents to 8 cents per kilowatt hour (less than the newest nuclear power plants). Find the percent of decrease.

30. _____

Supplementary Application Exercises on Base, Rate, and Part

Solve the following application problems. Read each problem carefully to determine whether base, part, or rate is being asked for. Round rates to the nearest tenth of a percent.

1. **EMPLOYEE HEALTH PLANS**   When firms with over 1000 employees were surveyed, it was found that 27% offered only one health plan to employees. If 1800 firms were surveyed, find the number offering only one health plan.

   1. _____

2. **AMERICAN CHIROPRACTIC ASSOCIATION**   There are 50,000 licensed chiropractors in the nation. If 30% of these chiropractors belong to the American Chiropractic Association (ACA), find the number of chiropractors in the ACA.

   2. _____

3. **MOTORCYCLE SAFETY**   Only 20 of the 50 states require motorcycle riders to wear helmets. What percent of the states require motorcycle riders to wear helmets? (*Source:* National Highway Traffic Safety Administration.)

   3. _____

4. **DANGER OF EXTINCTION**   Scientists tell us that there are 9600 bird species and that 1000 of these species are in danger of extinction. What percent of the bird species are in danger of extinction?

   4. _____

5. **VIDEO RENTAL INCOME**   In the United States during the last six months, there have been $.58 billion in late fees charged for late video rental returns. Since this is 12% of the total income from video rentals, find the total income. Round to the nearest tenth of a billion. (*Source: Video Store* magazine.)

   5. _____

6. **OVERWEIGHT PATIENTS**   A telephone survey of doctors found that 63% of the doctors reported an increase in the number of patients who are overweight. If the number of doctors reporting this was 189, find the number of doctors in the survey. (*Source:* Hoffman-La Roche Company.)

   6. _____

7. **COST AFTER MARKDOWN**   A fax machine priced at $398 is marked down 7% to promote the new model. If the sales tax is also 7%, find the cost of the fax machine including sales tax.

   7. _____

8. **BOOK DISCOUNT**   College students are offered a 6% discount on a dictionary that sells for $18.50. If the sales tax is 6%, find the cost of the dictionary including the sales tax.

   8. _____

9. WOMEN'S COATS A "60%-off sale" begins today. What is the sale price of women's wool coats normally priced at $335?

9. _____

10. APPLIANCES What is the sale price of a $769 Sear's Kenmore washer/dryer set with a discount of 25%?

10. _____

A weekly sales report for the top four sales people at the Family Shoe Store is shown below. Use this information to answer Exercises 11–14.

Employee	Sales	Rate of Commission	Commission
May, S.	$9480	3%	_____
Britz, C.	$10,730	3%	_____
Perdaris, A.	$8840	_____	$353.60
Terfloth, C.	$11,522	_____	$576.10

11. Find the commission for May.

11. _____

12. Find the commission for Britz.

12. _____

13. What is the rate of commission for Perdaris?

13. _____

14. What is the rate of commission for Terfloth?

14. _____

15. BIKER HELMET LAWS There were 2.48 million motorcycle riders in the country who supported biker helmet laws. If this was 62% of the total motorcycle riders in the country, what is the total number of motorcycle riders? (*Source:* National Highway Traffic Safety Administration.)

15. _____

16. AVERAGE HOME PRICE According to the National Association of Realtors, the median national sales price of a house was down 1.4%, or $2085, from last month. Find the median national sales price this month. (Round to the nearest dollar.)

16. _____

17. CREDIT-CARD DEBT According to a recent credit report, a person owes more than $30,000 total to 15 different credit-card companies. If his payments on this debt amount to $1220 this month and $298 of this is interest, what percent of his payment is interest?

17. _____

18. LAYOFF ALTERNATIVE Instead of laying off workers, a company cut all employee hours from 40 hours a week to 30 hours a week. By what percent were employee hours cut?

18. _____

NATIONWIDE HOME SALES *The number of existing single-family homes sold in four regions of the country in the same month of two separate years are shown in the figure below.*

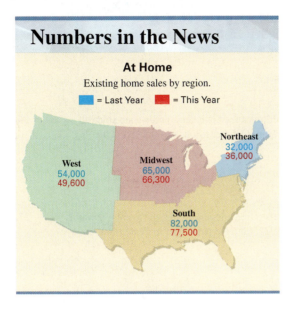

Numbers in the News

At Home
Existing home sales by region.
■ = Last Year ■ = This Year

Northeast
32,000
36,000

West
54,000
49,600

Midwest
65,000
66,300

South
82,000
77,500

Use the figure above to answer Exercises 19–22.

19. Find the percent of increase in sales in the northeastern region.

19. _____

20. Find the percent of increase in sales in the midwestern region.

20. _____

21. What is the percent of decrease in sales in the southern region?

21. _____

22. What is the percent of decrease in sales in the western region?

22. _____

23. VENDING-MACHINE SALES Of the total candy bars contained in a vending machine, 240 bars have been sold. If 25% of the bars have been sold, find the total number of candy bars that were in the machine.

23. _____

24. TOTAL SALES If the sales tax rate is $7\frac{1}{2}$% and the sales tax collected is $942.30, find the total sales.

24. _____

25. FAMILY BUDGETING Scott and Andrea Abriani established a budget allowing 25 percent of their total income for rent, 22 percent for food, 12 percent for clothing, 24 percent for travel and recreation, and the remainder for savings. Scott takes home $1950 per month and Andrea takes home $28,500 per year. How much will the couple save in one year for the down payment on a home?

25. _____

26. CHICKEN-NOODLE SOUP In one year, there were 350 million cans of chicken-noodle soup sold (all brands). If 60% of this soup is sold in the cold-and-flu season (October through March), find the number of cans sold in the cold-and-flu season.

26. _____

27. SOCIAL SECURITY BENEFITS The Social Security Administration announced that the average monthly benefit to single retirees will be increased from $903 to $922. Find the percent of increase. (*Source:* Social Security Administration.)

27. _____

28. BENEFIT INCREASE The average monthly social security benefit to couples will be increased to $1523. If the current monthly benefit is $1492, what is the percent of increase? (*Source:* Social Security Administration.)

28. _____

29. SIDE-IMPACT COLLISIONS Automobile accidents involving side-impact collision resulted in 9000 deaths last year. If automobiles were manufactured to meet a side-impact standard, it is estimated that 63.8% of these deaths would have been prevented. How many deaths would have been prevented?

29. _____

30. NEW-HOME PRICES The average price of a new home rose 4.2%. If the average price of a new home was $151,500, find the average price after the increase.

30. _____

31. U.S. PATENT RECIPIENTS Among the 50 companies receiving the greatest number of U.S. patents last year, 18 were Japanese companies. **(a)** What percent of the top 50 companies were Japanese companies? **(b)** What percent of the top 50 companies were not Japanese companies?

(a) _____
(b) _____

32. BLOOD-ALCOHOL LEVELS In the United States, 15 of the 50 states limit blood-alcohol levels for drivers to .08%. The remaining states limit these levels to .10%. **(a)** What percent of the states have a blood-alcohol limit of .08%? **(b)** What percent have a limit of .10%?

(a) _____
(b) _____

3.5 | Increase and Decrease Problems

Objectives

1 Learn to identify an increase or a decrease problem.
2 Apply the basic diagram for increase word problems.
3 Use the basic percent formula to solve for base in increase problems.
4 Apply the basic diagram for decrease word problems.
5 Use the basic percent formula to solve for base in decrease problems.

CASE *in* POINT

Thomas Dugally of Century 21 Realty knows that real estate values are always changing, usually going up, occasionally going down, but never staying the same. Dugally needs to keep track of the market and must be able to calculate increases and decreases in value on a regular basis.

Objective 1 **Learn to identify an increase or decrease problem.** Businesses commonly look at how amounts change, either up or down. For example, a manager might need to know the percent by which sales have **increased** or costs have **decreased**, while a consumer might need to know the percent by which the price of an item has changed. Identify these **increase and decrease** problems as follows.

Identifying Increase and Decrease Problems

Increase problem. The base (100%) *plus* some portion of the base gives a new value, which is the part. Phrases such as *after an increase of, more than,* or *greater than* often indicate an increase problem. The basic formula for an increase problem is

$$\text{Original} + \text{Increase} = \text{New value.}$$
$$\text{(base)} \qquad\qquad\qquad \text{(part)}$$

Decrease problem. The part equals the base (100%) *minus* some portion of the base, giving a new value. Phrases such as *after a decrease of, less than,* or *after a reduction of* often indicate a decrease problem. The basic formula for a decrease problem is

$$\text{Original} - \text{Decrease} = \text{New value.}$$
$$\text{(base)} \qquad\qquad\qquad \text{(part)}$$

QUICK TIP Base is always the original amount and both increase and decrease problems are *base* problems. Base is always 100%.

EXAMPLE **1**

*Using a Diagram to
Understand an
Increase Problem*

The value of a home sold by Thomas Dugally this year is $181,500, which is 10% more than last year's value. Find the value of the home last year.

SOLUTION

Use a diagram to help solve this problem. Remember that base is the starting point, or that to which something is compared. In this case the base is last year's value. Call base 100%, and remember that

$$\text{Original} + \text{Increase} = \text{New value.}$$

Objective 2   Apply the basic diagram for increase word problems.

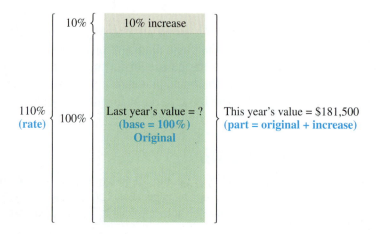

   This diagram shows that the 10% increase is based on last year's value (which is unknown) and not on this year's value of $181,500. To get this year's value, 10% of last year's value was added to the amount of last year's value.

Objective 3   Use the basic percent formula to solve for base in increase problems.

$$\begin{array}{ccccc} \text{Original} & + & \text{Increase} & = & \text{New value} \\ \mathbf{100\%} & + & \mathbf{10\%} & = & \mathbf{110\%} \end{array}$$

   This year's value is all of last year's value (100%) plus 10% of last year's value (or $100\% + 10\% = 110\%$). Since this year's value is 110% of last year's value, find last year's value, which is the base. The formula $B = \frac{P}{R}$ gives

$$B = \frac{\$181,500}{110\%} \quad \longleftarrow \begin{array}{l} \text{An amount that is all of base plus} \\ \text{10\% of base.} \\ \longleftarrow \quad 100\% + 10\% \end{array}$$

$$B = \frac{\$181,500}{1.1} \quad \longleftarrow \begin{array}{l} \text{110\% changed to 1.1} \\ \text{(change \% to decimal)} \end{array}$$

$$B = \$165,000 \quad \longleftarrow \quad \text{last year's value}$$

Now check the answer.

$$\begin{array}{rl} \$165,000 & \longleftarrow \quad \text{last year's value} \\ +\quad 16,500 & \longleftarrow \quad \text{10\% of \$110,000} \\ \hline \$181,500 & \longleftarrow \quad \text{this year's value} \end{array}$$

> **QUICK TIP**   The common error in solving an increase problem is thinking that the base is given and that the solution can be found by solving for part. Remember that the number given in Example 1, above, $181,500, is the result of having added 10% of the base to the base $\left(100\% + 10\% = 110\%\right)$. In fact, $181,500 is the part, and the base must be found.

   The following graphic shows the rate of world population growth over many decades and the length of time that it takes for the population to double at various rates of growth. Population growth rates are an example of increase problems because each year's rate of growth (increase) is based on the previous year. Similarly, last year's rate of growth was based on the year prior to that.

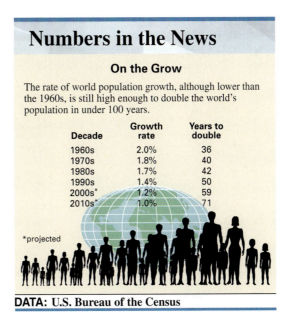

Numbers in the News

On the Grow

The rate of world population growth, although lower than the 1960s, is still high enough to double the world's population in under 100 years.

Decade	Growth rate	Years to double
1960s	2.0%	36
1970s	1.8%	40
1980s	1.7%	42
1990s	1.4%	50
2000s*	1.2%	59
2010s*	1.0%	71

*projected

DATA: U.S. Bureau of the Census

The next example shows how to handle *two* increases.

EXAMPLE 2

Finding Base after Two Increases

At Builders Doors, production last year was 20% more than the year before. This year's production is 93,600 doors, which is 20% more than last year's. Find the number of doors produced two years ago.

SOLUTION

The two 20% increases cannot be added together because the increases are from two different years, or two separate bases. The problem must be solved in two steps. First, use a diagram to find last year's production.

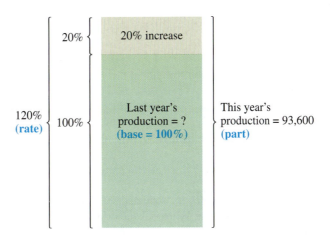

The diagram shows that last year's production plus 20% of last year's production equals this year's production. If $P = 93{,}600$ and $R = 100\% + 20\% = 120\%$, the formula $B = \frac{P}{R}$ gives

$$B = \frac{93{,}600}{120\%} = \frac{93{,}600}{1.2} = 78{,}000 \qquad \text{Last year's production}$$

Production last year was 78,000 doors. Production for the preceding year (two years ago) must now be found. Use another diagram.

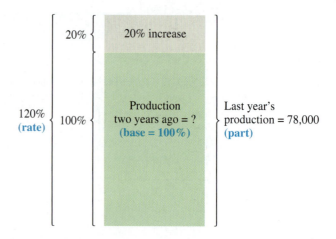

Thus, production two years ago added to 20% of production two years ago equals last year's production.

Using the formula $B = \frac{P}{R}$ with P equal to 78,000 and R equal to 120%:

$$B = \frac{78{,}000}{120\%} = \frac{78{,}000}{1.2} = 65{,}000 \text{ production 2 years ago}$$

Check the answer.

65,000	production two years ago
+ 13,000	20% increase
78,000	production last year
+ 15,600	20% increase
93,600	production this year

The calculator solution to this example is done by dividing in a series.

93600 ÷ 1.2 ÷ 1.2 = 65000

Note: Refer to Appendix C for calculator basics.

QUICK TIP It is important to realize that the two 20% increases cannot be added together to equal one increase of 40%. Each 20% increase is calculated on a different base.

EXAMPLE 3

Using a Diagram to Understand a Decrease Problem

After Fleetfeet deducted 10% from the price of a pair of competition running shoes, Tina Hight paid $135. What was the original price of the shoes?

SOLUTION

Use a diagram again, and remember that base is the starting point, which is the original price. As always, the base is 100%. Use the decrease formula.

Original	−	Decrease	=	New value
100%	−	10%	=	90%

Objective 4 Apply the basic diagram for decrease word problems.

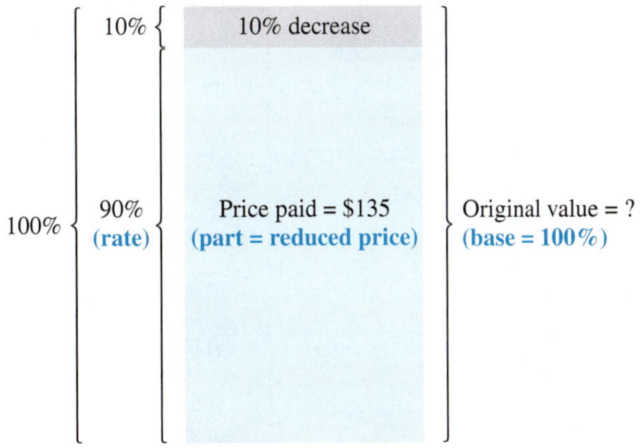

Objective 5 Use the basic percent formula to solve for the base in decrease problems. The diagram shows that 10% was deducted from the original price. The result equals the price paid, which is 90% of the original price. Since the original price is needed, and the diagram shows that the original price here is the base, use the formula $B = \frac{P}{R}$.

But what should be used as the rate? The rate 10% cannot be used because the original price is unknown (the price to which the 10% reduction was applied). The rate 90% (the difference, $100\% - 10\% = 90\%$) must be used since 90% of the original price is the $135 price paid. Now find the base.

$$B = \frac{P}{R}$$

$$B = \frac{135}{90\%} = \frac{135}{.9} = \$150 \quad \text{original price}$$

Check the answer.

$$
\begin{array}{rl}
\$150 & \text{original price} \\
-\quad 15 & \text{10\% discount} \\
\hline
\$135 & \text{price paid}
\end{array}
$$

QUICK TIP   The common mistake made in Example 3 is thinking that the reduced price, $135, is the base. The original price is the base, while the reduced price, $135, is a result after subtracting 10% from the base. The reduced price is the part or 90% *of the base* $\left(100\% - 10\% = 90\%\right)$.

'NET ASSETS Business on the Internet

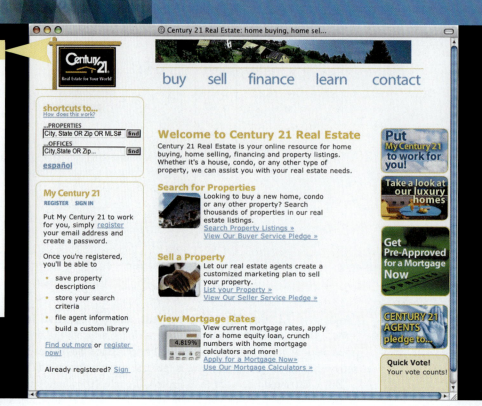

Century 21

- 2004: Headquarters in Parsippany, New Jersey

- 6800 independently owned and operated offices

- over 126,000 brokers and sales professionals worldwide

- Operations in 38 countries and territories

Century 21 Real Estate Corporation is the franchiser of the world's largest residential real estate sales organization. It provides comprehensive training, management, administrative, and marketing support for more than 6800 independently owned and operated offices in more than 38 countries and territories worldwide.

The CENTURY 21® System is the number-one consumer brand in the real estate industry. It consists of the largest broker network and has greater global coverage than any other real estate brand. They are dedicated to providing buyers and sellers of real estate with the highest-quality services possible.

1. A report by Century 21 says that home values in a certain area have increased by 8.2% since last year. Find the value of a home today that was valued at $155,000 last year.

2. Total property sales in a Century 21 office this month were $4.76 million. If total sales last month were $4.25 million, find the percent of increase.

Phone or visit a Century 21 office in your community and introduce yourself as a student before asking the following questions:

3. In what kinds of applications does the real estate industry use percent? Describe a minimum of six.

4. What are the advantages of working at Century 21? Ask for five positive characteristics of a career in real estate. Do you think that a career in real estate is for you? Why or why not?

3.5 | Exercises

FOR EXTRA HELP

 MyMathLab

 InterActMath.com

 MathXL

 MathXL Tutorials on CD

 Addison-Wesley Math Tutor Center

DVT/Videotape

The **Quick Start** *exercises in each section contain solutions to help you get started.*

Solve for base in each of the following. Round to the nearest cent.
(Hint: Original + Increase = New Value.) (See Examples 1 and 2.)

Part (after increase)	Rate of Increase	Base

Quick Start

1. $450	20%	$375
2. $800	25%	_____
3. $30.70	10%	_____
4. $10.09	5%	_____

Solve for base in each of the following. Round to the nearest cent.
(Hint: Original − Decrease = New Value.) (See Example 3.)

Part (after decrease)	Rate of Decrease	Base

Quick Start

5. $20	20%	$25
6. $1530	15%	_____
7. $598.15	30%	_____
8. $98.38	15%	_____

9. Certain words or word phrases help to identify an increase problem. Discuss how you identify an increase problem. (See Objective 1.)

10. Certain words or word phrases help to identify a decrease problem. Discuss how you identify a decrease problem. (See Objective 1.)

Solve the following application problems. Read each problem carefully to decide which are increase or decrease problems, and work accordingly. (See Examples 1–3.)

Quick Start

C 11. **HOME-VALUE APPRECIATION** Thomas Dugally of Century 21 Realty just listed a home for $178,740. If this is 8% more than what the home sold for last year, find last year's selling price.

$$B = \frac{P}{R} = \frac{178{,}740}{1.08} = \$165{,}500$$

11. $165,500

12. **DEALER'S COST** Ortiz Auto Stereos sold an auto stereo for $337.92, a loss of 12% of the dealer's original cost. Find the original cost.

$$B = \frac{P}{R} = \frac{337.92}{.88} = \$384$$

12. $384

C indicates an exercise that is related to the Case in Point feature within the section.

13. **FAMILY RESTAURANT**   Santiago Rowland owns a small restaurant and charges 8% sales tax on all orders. At the end of the day he has a total of $1026 including the sales and sales tax in his cash register. (a) What were his sales not including sales tax? (b) Find the amount that is sales tax.

(a) _____
(b) _____

14. **SALES TAX**   Tom Dugally of Century 21 purchased a Compaq iPAQH3835 personal digital assistant (PDA) for $639 including $6\frac{1}{2}$% sales tax. Find **(a)** the price of the PDA and **(b)** the amount of sales tax. (*Source:* Real Estate Technology.)

(a) _____
(b) _____
_____

15. **NOBODY DOESN'T LIKE SARA LEE**   Sara Lee, the maker of Sara Lee baked goods, Playex bras, and Hanes underwear, said it earned $230 million this quarter. If this was a 25% decrease from last quarter, find the earnings last quarter. Round to the nearest tenth of a million. (*Source:* Sara Lee.)

15. _____

16. **ANTILOCK BRAKES**   In a recent test of an automobile antilock braking system (ABS) on wet pavement, the stopping distance was 114 feet. If this was 28.75% less than the distance needed to stop the same automobile without the ABS, find the distance needed to stop without the ABS.

16. _____

17. **POPULATION GROWTH**   In 2004, the population of Rio Linda was 10% more than it was in 2003. If the population was 26,620 in 2005, which was 10% more than in 2004, find the population in 2003.

17. _____

18. **WALLPAPER SALES**   Leigh Jacka, owner of Wallpaper Plus, says that her sales have increased exactly 20% per year for the last two years. Her sales this year are $170,035.20. Find her sales two years ago.

18. _____

19. **DVD RENTALS**   Netflix, a DVD-rental company, has 1,300,000 subscribers, an increase of 74% from last year. Find the number of subscribers last year. Round to the nearest whole number. (*Source:* Netflix.)

19. _____

20. **FARMLAND PRICES**   The value of Iowa farmland increased 4.3% this year to a statewide average value of $1857 per acre. How much per acre did Iowa farmland increase this year? Round to the nearest dollar. (*Source:* Iowa State University.)

20. _____

21. **HEALTHCARE INDUSTRY**   Tenet Healthcare stock gained 3.6% in value to $15.53 per share. What was the stock value before the gain? (*Source: Wall Street Journal.*)

21. _____

22. **EXPENSIVE RESTAURANTS**   Among New York City's 20 most expensive restaurants, the average per-meal cost, excluding drinks, increased 6.5% to $69.33 in the last year. Find the price of this meal before the increase.

22. _____

23. **SURPLUS-EQUIPMENT AUCTION**   In a three-day public auction of Jackson County's surplus equipment, the first day brought $5750 in sales and the second day brought $4186 in sales, with 28% of the original equipment left to be sold on the third day. Find the value of the remaining surplus equipment.

23. _____

24. **COLLEGE EXPENSES**   After spending $3450 for tuition and $4350 for dormitory fees, Edgar Espina finds that 35% of his original savings remains. Find the amount of his savings that remains.

24. _____

25. **PAPER PRODUCTS MANUFACTURING**   International Paper Company, the world's largest paper company, reported a 16% drop in third-quarter earnings. If earnings had dropped to $122 million, find the earnings before the drop. Round to the nearest hundredth of a million. (*Source:* International Paper Company.)

25. _____

26. **NATIONAL HOME SALES**   Sales of existing homes decreased 7.4% to an annual number of 4.87 million units. Find the annual number of homes sold before the decrease. Round to the nearest hundredth of a million. (*Source:* National Association of Realtors.)

26. _____

27. **WINTER-WHEAT PLANTING**   Even though wheat prices rose during the planting season, farmers planted only 50.2 million acres of winter wheat varieties. If this is 2% fewer acres than last year, find the number of acres planted last year. (Round to the nearest tenth of a million.)

27. _____

28. STOCK VALUE   The Gateway computer company announced quarterly revenues of $883 million, a drop of 21% from last quarter. What were last quarter's revenues? Round to the nearest tenth of a million. (*Source:* Gateway.)

28. _____

29. COMMUNITY COLLEGE ENROLLMENT   In 2004, the student enrollment at American River College was 8% more than it was in 2003. If the enrollment was 29,160 students in 2005, which was 8% more than it was in 2004, find the student enrollment in 2003.

29. _____

30. UNIVERSITY FEES   Students at one state university are outraged. The annual university fees were 30% more last year than they were the year before. If the fees are $2704 per year this year, which is 30% more than they were last year, find the annual student fees two years ago.

30. _____

31. CONE ZONE DEATHS   This year there were 1181 deaths related to road construction zones in the United States. If this is an increase of 70% in the last five years, what was the number of deaths five years ago? Round to the nearest whole number. (*Source:* American Road and Transporation Builders Association.)

31. _____

32. MINORITY LOANS   A large mortgage lender made 52% more loans to minorities this year than last year. If the number of loans to minorities this year is 2660, find the number of loans made to minorities last year.

32. _____

33. NEW-HOME SALES   New-home sales this year in the Sacramento, California area were 14% fewer than last year. If the number of new homes sold this year was 5645, find the number of new homes sold last year. (Round to the nearest whole number.)

33. _____

34. WORKFORCE SIZE   Sony Corporation announced that it will reduce the size of its work force to 141,000 workers, a decrease of 12.4%. Find the size of the work force before the decrease. Round to the nearest whole number. (*Source:* Sony.)

34. _____

Chapter 3 | Quick Review

CHAPTER TERMS   *Review the following terms to test your understanding of the chapter. For each term you do not know, refer to the page number found next to that term.*

base **[p. 91]**

decrease problem **[p. 119]**

formula for base **[p. 101]**

formula for rate **[p. 109]**

hundredths **[p. 84]**

increase problem **[p. 119]**

part **[p. 91]**

percent **[p. 91]**

percent formula **[p. 91]**

percent of decrease **[p. 112]**

percent of increase **[p. 111]**

percents **[p. 84]**

rate **[p. 91]**

sales tax **[p. 102]**

CONCEPTS	EXAMPLES
3.1 Writing a decimal as a percent Move the decimal point two places to the right and attach a percent sign (%).	$.75(.75.) = 75\%$
3.1 Writing a fraction as a percent First change the fraction to a decimal. Then move the decimal point two places to the right and attach a percent sign (%).	$\dfrac{2}{5} = .4$ $.4(.40.) = 40\%$
3.1 Writing a percent as a decimal Move the decimal point two places to the left and drop the percent sign (%).	$50\% \,(.50.\%) = .5$
3.1 Writing a percent as a fraction First change the percent to a decimal. Then write the decimal as a fraction in lowest terms.	$15\% \,(.15.\%) = .15 = \dfrac{15}{100} = \dfrac{3}{20}$
3.1 Writing a fractional percent as a decimal First change the fraction to a decimal, leaving the percent sign. Then move the decimal point two places to the left and drop the percent sign (%).	$\dfrac{1}{2}\% = .5\%$ $.5\% = .005$
3.2 Solving for part, using the percent formula $$\textbf{Part} = \textbf{Base} \times \textbf{Rate}$$ $$P = B \times R$$ $$P = BR$$	A company offered a 15% discount on all sales. Find the discount on sales of \$1850. $$P = B \times R$$ $$P = \$1850 \times 15\%$$ $$P = \mathbf{\$1850 \times .15} = \$277.50 \text{ discount}$$
3.2 Using the standard format to solve percent problems Express the problem in the format $$R \times B = P$$ $$\underline{}\% \text{ of } \underline{} \text{ is } \underline{}$$ where % of something is something. Notice that *of* means $\times$ and *is* means $=$.	A shop gives a 10% discount on all repairs. Find the discount on a \$175 repair. $$R \times B = P$$ % of something is something 10% of repair is discount $\mathbf{.1 \times \$175} = \$17.50$ discount

CONCEPTS	EXAMPLES

3.3 Using the percent formula to solve for base

Use $P = B \times R$

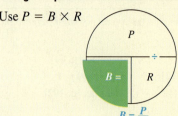

$$B = \frac{P}{R}$$

If the sales tax rate is 4%, find amount of sales when the sales tax is $18.

$$R \times B = P$$

% of something is something
4% of sales is $18 (tax).

$$B = \frac{18}{.04} = \$450 \text{ sales}$$

3.4 Using the percent formula to solve for rate

Use $P = B \times R$.

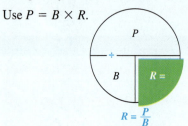

$$R = \frac{P}{B}$$

The return is $307.80 on an investment of $3420. Find the rate of return.

$$R \times B = P$$

% of something is something
What % of investment is return?
What % of $3420 is $307.80?

$$R = \frac{307.8}{3420} = .09 = 9\%$$

3.4 Finding the percent of change

Calculate the change (increase or decrease) that is the part. The base is the amount before the change.

Use $R = \frac{P}{B}$.

Production rose from 3820 units to 5157 units. Find the percent of increase.

$$5157 - 3820 = 1337 \text{ change (increase)}$$

$$R = \frac{1337}{3820} = .35 = 35\%$$

3.5 Drawing a diagram and using the percent formula to solve increase problems

Solve for the base when given the rate (110%) and the part (after increase).

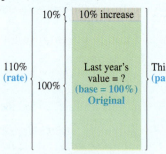

This year's sales are $121,000, which is 10% more than last year's sales. Find last year's sales.

Original + Increase = New value
100% + 10% = 110%

Using $B = \frac{P}{R}$

$$B = \frac{\$121,000}{110\%} = \frac{\$121,000}{1.1}$$
$$= 110,000 \text{ last year's sales}$$

Check:

	$110,000	last year's sales
+	11,000	(10% of $110,000)
	$121,000	this year's sales

3.5 Drawing a diagram and using an equation to solve a decrease problem

Solve for the base when given the rate (90%) and the part (after decrease).

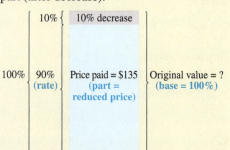

After a deduction of 10% from the price, a customer paid $135. Find the original price.

Original − Decrease = New value
100% − 10% = 90%

Using $B = \frac{P}{R}$

$$B = \frac{135}{.9} = \$150 \text{ original price}$$

Check:

	$150	original price
−	15	(10% discount)
	$135	price paid

Chapter 3 | Summary Exercise

Investment Gains and Losses

Understanding the stock market can be an important part of financial planning. Looking toward retirement, you may invest a portion of your savings by buying stock in various companies. Every day the changes in stock prices are shown in most major newspapers. Listed below are some well-known companies along with stock price information for this year and last year. Find the stock price last year, the percent of change from last year, or the stock price this year, as necessary. Round dollar amounts to the nearest cent and percents to the nearest tenth.

The Ups and Downs of Last Year				
Company Name	Stock Symbol	Stock Price Last Year	Stock Price This Year	% Change from Last Year
Amazon.com	AMZN	18.89	51.51	_____
DaimlerChrysler	DCX	30.65	_____	28.7%
Gateway Computer	GTW	_____	40.40	1186.6%
Krispy Kreme	KKD	33.77	40.02	_____
McDonald's	MCD	16.08	24.46	_____
Merck	MRK	_____	43.63	− 21.1%
Pepsi Bottling	PBG	25.71	_____	− 9.5%
R. J. Reynolds	RJR	_____	56.38	33.9%
Wal-Mart	WMT	50.51	_____	4.3%
Yahoo	YHOO	16.35	42.50	_____

(IN)VESTIGATE

Find the New York State Exchange (NYSE) listing in a newspaper. Select five stocks that you have heard of and list their closing price. Find the cost of 10 shares of each of these stocks, and round each of these costs to the nearest dollar.

Chapter 3 | Test

To help you review, the numbers in brackets show the section in which the topic was discussed.

Solve the following problems.   **[3.1–3.4]**

1. 36 home sales is 12% of what number of home sales?

    1. _____

2. What is 5% of 240 open houses?

    2. _____

3. 33 shippers is 3% of what number of shippers?

    3. _____

4. 36 accounts is what percent of 1440 accounts?

    4. _____

5. What is $\frac{1}{4}$% of $1260?

    5. _____

6. Find the fractional equivalent of 24%.

    6. _____

7. 48 purchase orders is $2\frac{1}{2}$% of how many purchase orders?

    7. _____

8. Change 87.5% to its fractional equivalent.

    8. _____

9. $141.10 is what percent of $1660?

    9. _____

10. What is the fractional equivalent of $\frac{1}{2}$%?

    10. _____

11. One share of Duke Energy (DUK) stock sells for $17.50 and pays a 6% dividend. Find the dividend per share.   **[3.2]**

    11. _____

12. A supervisor at Barrett Manufacturing finds that cabinet door hinge rejects amount to 1120 units per month. If this amounts to .5% of total monthly production, find the total monthly production of door hinges.   **[3.3]**

12. _____

13. A Ford Explorer is offered at 17% off the manufacturer's suggested retail price. Find the discount and the sale price of this Explorer, originally priced at $30,500.   **[3.2]**

13. _____

14. There are 35 million Americans 65 or older, and they make up 13 percent of the U.S. population. What is the total population of the United States? Round to the nearest tenth of a million. (*Source:* AARP.)   **[3.2]**

14. _____

15. A retail store with a monthly advertising budget of $3400 decides to set up a media budget. It plans to spend 22% for television advertising, 38% for newspaper advertising, 14% for outdoor signs, 15% for radio advertising, and the remainder for bumper stickers. **(a)** What percent of the total budget do they plan to spend on bumper stickers? **(b)** How much do they plan to spend on bumper stickers for the entire year?   **[3.2]**

(a) _____
(b) _____

16. Americans lose about 300 million golf balls each year, and about 225 million of these are recovered and resold in what has become a $200 million annual business. What percent of the lost golf balls are recovered and resold? (*Source: USA Today.*)   **[3.4]**

16. _____

17. A digital camera is marked "Reduced 25%, Now Only $337.50." Find the original price of the digital camera.   **[3.5]**

17. _____

18. Last year's backpack sales were 10% more than they were the year before. This year's sales are 1452 units, which is 10% more than last year. Find the number of backpacks sold two years ago.   **[3.5]**

18. _____

19. The local real estate board reports that the number of homes sold last month was 1658. If 1526 homes were sold in the same month last year, find the percent of increase. Round to the nearest tenth of a percent. (*Source:* Sacramento *Realtor.*)   **[3.4]**

19. _____

20. Proctor and Gamble Company, the maker of Tide detergent, Pampers diapers, and Clairol hair-care products, had quarterly earnings that rose 20% to $1.76 billion. Find the earnings in the previous quarter. Round to the nearest hundredth of a billion. (*Source:* Proctor and Gamble.)   **[3.4]**

20. _____

Chapter 3 | Cumulative Review

Chapters 1–3

To help you review, the numbers in brackets show the section in which the topic was introduced.

Round each of the following numbers as indicated. **[1.1, 1.3]**

1. 65,462 to the nearest hundred

2. 4,732,489 to the nearest thousand

3. 78.35 to the nearest tenth

4. 328.2849 to the nearest hundredth

1. _____

2. _____

3. _____

4. _____

Solve the following problems. **[1.1–1.5]**

5.
```
   351
   763
  2478
+   17
```

6.
```
   45,867
 − 37,985
```

7.
```
    634
 ×   38
```

8.
```
   2450
 ×  320
```

9. 6290 ÷ 74 = _____

10. 22,850 ÷ 102 = _____

(*Hint:* Write the answer as a mixed number.)

11. .46 + 9.2 + 8 + 17.514 = _____

12.
```
   45.36
 − 23.7
```

13.
```
   29.8
 ×  .41
```

14. 21.8$\overline{)396.76}$

Solve the following application problems.

15. Paul Altier decides to establish a budget. He will spend $650 for rent, $325 for food, $420 for child care, $182 for transportation, $250 for other expenses, and he will put the remainder in savings. If his monthly take-home pay is $2025, find his savings. **[1.1]**

15. _____

16. The Enabling Supply House purchases 6 wheelchairs at $1256 each and 15 speech-compression recorder/players at $895 each. Find the total cost. **[1.1–1.2]**

16. _____

17. Software Depot had a bank balance of $29,742.18 at the beginning of April. During the month, the firm made deposits of $14,096.18 and $6529.42. A total of $18,709.51 in checks was paid by the bank during the month. Find the firm's checking account balance at the end of April. **[1.4]**

17. _____

18. Cara Groff pays $128.11 each month to the Bank of Bolivia. How many months will it take her to pay off $4099.52? **[1.5]**

18. _____

Solve the following problems.   **[2.1–2.4]**

19. Write $\dfrac{48}{54}$ in lowest terms. _____

20. Write $8\dfrac{1}{8}$ as an improper fraction. _____

21. Write $\dfrac{107}{15}$ as a mixed number. _____

22. $1\dfrac{2}{3} + 2\dfrac{3}{4} =$ _____

23. $5\dfrac{7}{8} + 7\dfrac{2}{3} =$ _____

24. $6\dfrac{1}{3} - 4\dfrac{7}{12} =$ _____

25. $8\dfrac{1}{2} \times \dfrac{9}{17} \times \dfrac{2}{3} =$ _____

26. $3\dfrac{3}{4} \div \dfrac{27}{16} =$ _____

Solve the following application problems.

27. The area of a piece of land is $63\dfrac{3}{4}$ acres. One-third of the land is sold. What is the area of the land that is left?   **[2.4]**

27. _____

28. To prepare for the state real estate exam, Mary Rose Fink studied $5\dfrac{1}{2}$ hours on the first day, $6\dfrac{1}{4}$ hours on the second day, $3\dfrac{3}{4}$ hours on the third day, and 7 hours on the fourth day. How many hours did she study altogether?   **[2.3]**

28. _____

29. The storage area at American River Raft Rental has four sides and is enclosed with $527\dfrac{1}{24}$ feet of security fencing around it. If three sides of the yard measure $107\dfrac{2}{3}$ feet, $150\dfrac{3}{4}$ feet, and $138\dfrac{5}{8}$ feet, find the length of the fourth side.   **[2.3]**

29. _____

30. Play-It-Now Sports Center has decided to divide $\dfrac{2}{3}$ of the company's profit-sharing funds evenly among the eight store managers. What fraction of the total amount will each receive?   **[2.4]**

30. _____

Solve the following problems.   **[2.5]**

31. Change .65 to a fraction.

31. _____

32. Change $\dfrac{2}{3}$ to a decimal. Round to the nearest thousandth.

32. _____

Solve the following problems.   **[3.1–3.4]**

33. Change $\frac{7}{8}$ to a percent. _____

34. Change .25% to a decimal. _____

35. Find 35% of 6200 home loans. _____

36. Find 134% of $80. _____

37. 275 sales is what percent of 1100 sales? _____

38. 375 patients is what percent of 250 patients? _____

Solve the following application problems.

39. Analysts say that during the next three years, about 15,000 of the United States' 39,000 theater (movie) screens will be shut down. What percent of the screens will be shut down? (Round to the nearest tenth of a percent). (*Source: USA Today.*)   **[3.4]**

39. _____

40. A Canon Ultracompact Digital Camcorder normally priced at $890 is on sale for 18% off. Find the amount of discount and the sales price.   **[3.2]**

40. _____

41. Bookstore sales of the *Physicians' Desk Reference*, which contains prescription drug information, rose 13.7% this year. If sales this year were 111,150 copies, find last year's sales. (Round to the nearest whole number.)   **[3.5]**

41. _____

42. After deducting 11.8% of total sales as her commission, George-Ann Hornor, a salesperson for Marx Toy Company, deposited $35,138.88 to the company account. Find the total amount of her sales.   **[3.5]**

42. _____

43. Early investors in Dell Computer stock saw the value of the stock go up 900 times their original cost. What percent increase did they experience? Show your work explaining how you arrived at your answer. (*Source: USA Today.*)   **[3.1]**

43. _____

44. The value of a stock used to be 6 times what it is worth today. The value today is what percent of the past value? Round to the nearest tenth of a percent. Show your work, explaining how you arrived at your answer.   **[3.1]**

44. _____

WHERE'S THE BEEF? Last year the United States exported 2.3 billion pounds of beef. Use the circle graph below to solve Exercises 45–48. **[3.2]**

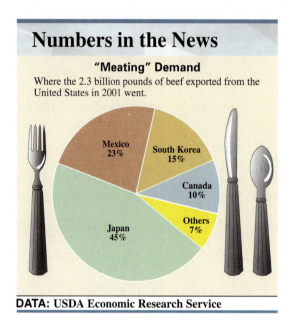

Numbers in the News

"Meating" Demand
Where the 2.3 billion pounds of beef exported from the United States in 2001 went.

Mexico 23%
South Korea 15%
Canada 10%
Others 7%
Japan 45%

DATA: USDA Economic Research Service

45. (a) What percent of the exported beef was shipped to Japan, Mexico, and South Korea combined?   **(a)** _____
(b) Find the total number of pounds exported to these three countries.   **(b)** _____

46. (a) What percent of the exported beef was shipped to countries other than Japan, Mexico, and South   **(a)** _____
Korea combined? **(b)** Find the total number of pounds exported to these three countries.   **(b)** _____

47. How much more beef was exported to Japan than to South Korea?   **47.** _____

48. How much more beef was exported to Mexico than to Canada?   **48.** _____

CHAPTER 4

Bank Services

"SAY IT WITH FLOWERS" HAS ALWAYS HAD SPECIAL MEANING FOR Jill Owens. Today she owns and operates her own flower shop, Flower Power.

 The shop sells cut flowers, floral arrangements for every occasion, and gift plants.

As her business has grown, one of the most important decisions made by Jill Owens was selecting a bank that would offer her the services to help her operate the business profitably and with the greatest efficiency. Knowing about each service that a bank offers can help anyone, regardless of whether they own or operate a business.

Modern banks and savings institutions offer many services. They are more than just places to deposit savings and take out loans. Today, many types of savings accounts and a variety of checking accounts are offered. Additional services offered are on-line banking at home and business banking, automated teller machines (ATMs), credit cards, debit cards, investment securities services, collection of notes (covered in Chapter 8), and even payroll services for the business owner.

This chapter examines checking accounts and check registers and how to use them. It also discusses business checking-account services, the depositing of credit-card transactions, and, finally, bank reconciliation (balancing a checking account).

4.1 | Checking Accounts and Check Registers

Objectives

1. Identify the parts of a check.
2. Know the types of checking accounts.
3. Calculate the monthly service charges.
4. Identify the parts of a deposit slip.
5. Identify the parts of a check stub.
6. Complete the parts of a check register.

CASE *in* **POINT**

One of the first bank services that Flower Power needed was a business checking account. Jill Owens knew that she would be receiving checks from customers and that she would be using checks to pay her suppliers and all of the other expenses of operating her business. Example 1 shows how the monthly service charge for her checking account is determined.

Objective 1 **Identify the parts of a check.** Even with the growth in **electronic commerce (EC),** where goods are purchased and sold electronically, the majority of business transactions today still involve checks.

A small business may write several hundred checks each month and take in several thousand, while large businesses can take in several million checks in a month. This heavy reliance on checks makes it important for all people in business to have a good understanding of checks and checking accounts. The various parts of a check are explained in the following diagram.

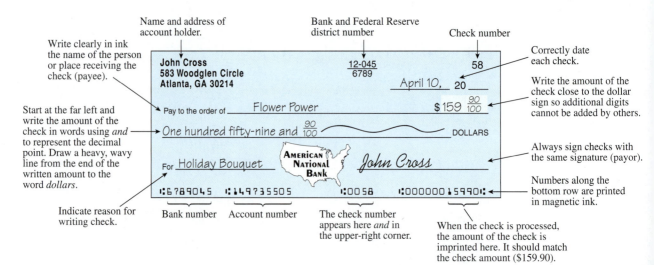

Name and address of account holder.
Bank and Federal Reserve district number
Check number

Write clearly in ink the name of the person or place receiving the check (payee).

Correctly date each check.

Write the amount of the check close to the dollar sign so additional digits cannot be added by others.

Start at the far left and write the amount of the check in words using *and* to represent the decimal point. Draw a heavy, wavy line from the end of the written amount to the word *dollars*.

Always sign checks with the same signature (payor).

Numbers along the bottom row are printed in magnetic ink.

Indicate reason for writing check.

Bank number   Account number   The check number appears here *and* in the upper-right corner.

When the check is processed, the amount of the check is imprinted here. It should match the check amount ($159.90).

John Cross
583 Woodglen Circle
Atlanta, GA 30214

12-045 / 6789   58

April 10, 20 ___

Pay to the order of ___ Flower Power ___ $159 90/100

One hundred fifty-nine and 90/100 ~~~~~~~~~~~ DOLLARS

AMERICAN NATIONAL BANK

For ___ Holiday Bouquet ___   John Cross

⑈6789045⑈ ⑈149735505⑈ ⑈0058 ⑈000000159901⑈

Objective 2 **Know the types of checking accounts.** Two main types of checking accounts are available.

Checking Facts

- Three of every four families have a checking account.
- The average adult writes over 100 checks each year.
- Checks became common after World War II.
- Today, 135 billion checks are processed each year.
- Checks represent about one-third of all consumer spending.

Personal checking accounts are used by individuals. The bank supplies printed checks (normally charging a check-printing fee) for the customer to use. Some banks offer the checking account at no charge to the customer, but most require that a minimum monthly balance remain in the checking account. If the minimum balance is not maintained during any month, a service charge is applied to the account. Today, the **flat-fee checking account** is common. For a fixed charge per month, the bank supplies the checking account, a supply of printed checks, a bank charge card, an ATM card, a debit card, and a host of other services. **Interest paid** on checking account balances is common with personal checking accounts. These accounts are offered by savings-and-loan associations, credit unions, and banks and are available to individuals, as well as to a few business customers.

Business checking accounts often receive more services and have greater activity than do personal accounts. For example, banks often arrange to receive payments on debts due to business firms. The bank automatically credits the amount to the business account.

A popular service available to personal and business customers is the **automated teller machine (ATM)**. Offered by many banks, savings and loans, and credit unions, an ATM allows the customer to perform a great number of transactions. The ATM card and **electronic banking** allow cash withdrawals and deposits, transfer of funds from one account to another, including the paying of credit-card accounts or other loans, and account-balance inquiries at *any* time. In addition, through several networking arrangements, the customer may make purchases and receive cash advances from hundreds, and in some cases thousands, of participating businesses nationally, and often worldwide.

QUICK TIP When traveling in foreign countries, you can often get cash in the local currency using your ATM card.

ATM cards are like **debit cards**, not credit cards. When you use your debit card at a **point-of-sale terminal**, the amount of your purchase is instantly subtracted from your bank account, and credit is given to the seller's bank account. When you use a credit card, you usually sign a receipt. However, when using a debit card, you enter your **personal identification number (PIN)**, your special code that authorizes the transaction. Cash can also be obtained from many ATM machines using credit cards such as Visa and MasterCard, even in other countries. The bar graph below shows additional services that adults would like to see offered at the ATMs.

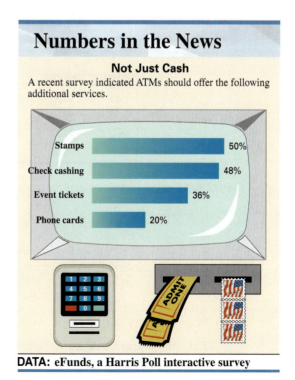

Numbers in the News

Not Just Cash

A recent survey indicated ATMs should offer the following additional services.

Service	Percent
Stamps	50%
Check cashing	48%
Event tickets	36%
Phone cards	20%

DATA: eFunds, a Harris Poll interactive survey

QUICK TIP When using your ATM card, remember to keep receipts so that the transaction can be subtracted from your own account records. Be certain to subtract any service charges made for using the ATM card.

Transaction Costs

The cost of different payment transactions.

Bank branch	$1.07
U.S. mail	$0.73
Telephone	$0.54
ATM	$0.27
Internet	$0.01

Data: Jupiter Communication *Home Banking Report*.

The use of **electronic funds transfer (EFT)** is a popular alternative to paper checks because EFT saves businesses money. The box at the side shows the cost to the company receiving payment with the use of each payment method. **Home banking (Internet banking)** is becoming very popular for its convenience and cost savings. Home banking allows a customer to pay bills, check balances, and move funds, all done over the Internet from home. The following bar graph shows the number of registered on-line users with the largest banks.

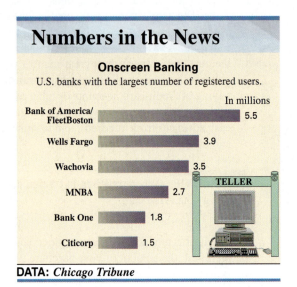

Numbers in the News

Onscreen Banking
U.S. banks with the largest number of registered users.

In millions

Bank of America/FleetBoston	5.5
Wells Fargo	3.9
Wachovia	3.5
MNBA	2.7
Bank One	1.8
Citicorp	1.5

DATA: *Chicago Tribune*

Objective ③ **Calculate the monthly service charges.** Service charges for business checking accounts are based on the average balance for the period covered by the statement. This average balance determines the **maintenance charge per month**, to which a **per-debit charge** (per-check charge) is added. The charges generally apply without regard to the amount of account activity. The following table shows some typical bank charges for a business checking account.

Average Balance	Maintenance Charge Per Month	Per-Check Charge
Less than $500	$12.00	$.20
$500–$1999	$7.50	$.20
$2000–$4999	$5.00	$.10
$5000 or more	0	0

EXAMPLE 1

Finding the Checking Account Service Charge

Find the monthly service charge for the following business accounts.

(a) Pittsburgh Glass, 38 checks written, average balance $883

According to the preceding table, an account with an average balance between $500 and $1999 has a $7.50 maintenance charge for the month. In addition, there is a per-debit (check) charge of $.20. Since 38 checks were written, find the service charge as follows:

$$\$7.50 + \mathbf{38(\$.20)} = \$7.50 + \mathbf{\$7.60} = \$15.10$$

(b) Fargo Western Auto, 87 checks written, average balance $2367

Since the average balance is between $2000 and $4999, the maintenance charge for the month is $5.00 plus $.10 per debit (check). The monthly service charge is

$$\$5.00 + \mathbf{87(\$.10)} = \$5.00 + \mathbf{\$8.70} = \$13.70.$$

The calculator solutions to this example use chain calculations with the calculator observing the order of operations.

(a) 7.5 ⊞ 38 ⊠ .2 ⊟ 15.1

(b) 5 ⊞ 87 ⊠ .1 ⊟ 13.7

Note: Refer to Appendix C for calculator basics.

Objective **4** **Identify the parts of a deposit slip.** Money, either cash or checks, is placed in a checking account with a **deposit slip**, or **deposit ticket** (see the following sample). The account number is printed at the bottom in magnetic ink. The slip contains blanks for entering any currency (bills) or coin (change), as well as any checks that are to be deposited.

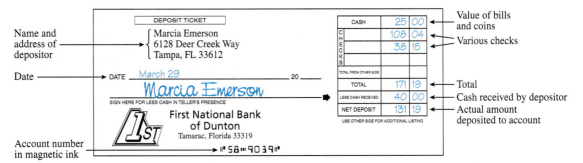

When a check is deposited, it should have "for deposit only" and either the depositor's signature or the company stamp placed on the back within 1.5 inches of the trailing edge (as seen in the following figure). In this way, if a check is lost or stolen before it is deposited, it will be worthless to anyone finding it. Such an endorsement, which limits the ability to cash a check, is called a **restricted endorsement**. An example of a restricted endorsement is shown below along with two other types of endorsements. The most common endorsement by individuals is the **blank endorsement**, where only the name of the person being paid is signed. This endorsement should be used only at the moment of cashing the check. The **special endorsement**, used to pass on the check to someone else, might be used to pay a bill on another account.

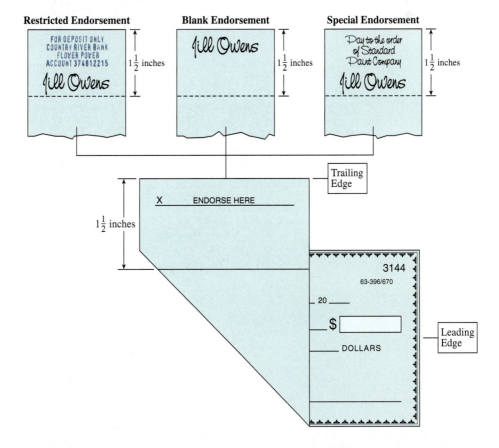

After the check is endorsed, it is normally cashed or deposited at a bank. The payee is either given cash or receives a credit in his or her account for the amount of the check. The check is then routed to a Federal Reserve Bank, which forwards the check to the payer's bank. After going through this procedure, known as **processing**, the check is then **canceled** and returned to the payer. The check will now have additional processing information on its back as follows.

The date that the bank debited (deducted) the payer's account.

The date and bank where the check was deposited are important proof against claims that a check was late or was never received.

Restricted endorsement for deposit only

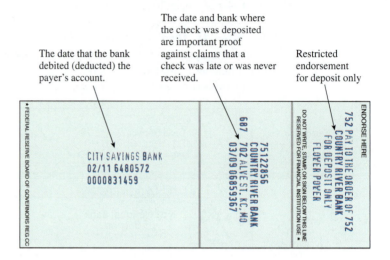

A two-sided commercial deposit slip is shown in the figure below. Notice that much more space is given for an itemized list of customers' checks that are being deposited to the business account. Many financial institutions require that the bank and Federal Reserve district numbers be shown in the description column of the deposit slip. These numbers appear near the upper right-hand corner of the check and are identified in the sample check shown on page 140.

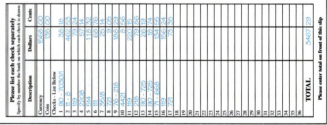

Objective **5** **Identify the parts of a check stub.** A record must be kept of all deposits made and all checks written. Business firms normally do this with a **check stub** for *each* check. These check stubs provide room to list the date, the person or firm to whom the check is paid, and the purpose of the check. Also, the check stub provides space to record the balance in the account after the last check was written (called the **balance brought forward**, abbreviated "Bal. Bro't. For'd."

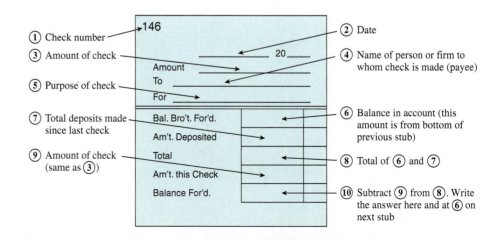

on the stub), and any sums deposited since the last check was written. The balance brought forward and amount deposited are added to provide the current balance in the checking account. The amount of the current check is then subtracted and a new balance is found. This **balance forward** from the bottom of the check stub should be written on the next check stub. A typical check stub is shown at the bottom of the previous page.

EXAMPLE 2

Completing a Check Stub

Check number 2724 was made out on June 8 to Lillburn Utilities as payment for water and power. Assume that the check was for $182.15, that the balance brought forward is $4245.36, and that deposits of $337.71 and $193.17 have been made since the last check was written. Complete the check stub.

SOLUTION

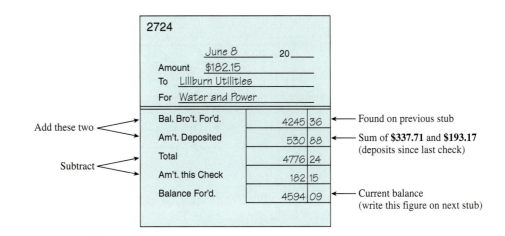

2724		
June 8 20 ____		
Amount $182.15		
To Lillburn Utilities		
For Water and Power		
Bal. Bro't. For'd.	4245	36
Am't. Deposited	530	88
Total	4776	24
Am't. this Check	182	15
Balance For'd.	4594	09

Add these two ⟶

Subtract ⟶

⟵ Found on previous stub

⟵ Sum of **$337.71** and **$193.17** (deposits since last check)

⟵ Current balance (write this figure on next stub)

Banks offer many styles of checkbooks. Notice that the following two styles shown offer two stubs and may be used for payroll. The stub next to the check can be used as the employee's record of earnings and deductions. The second style provides space on the check itself for listing a group of invoices or bills that are being paid with that same check.

Check stub   Check

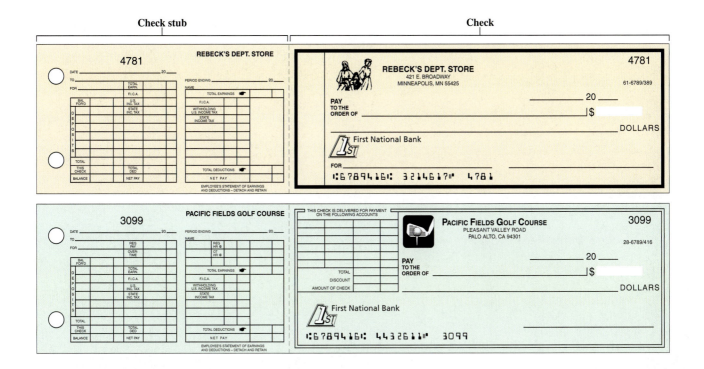

Objective 6 Complete the parts of a check register. Some depositors prefer a **check register** to check stubs, while others use both. A check register, such as the following one, shows the checks written and deposits made at a glance. The column headed with a check mark is used to record each check after it has cleared or when it is received back from the bank.

CHECK NO.	DATE	CHECK ISSUED TO	AMOUNT OF CHECK		✓	DATE OF DEP.	AMOUNT OF DEPOSIT		BALANCE	
		BALANCE BROUGHT FORWARD →							3518	72
1435	5/8	Swan Brothers	378	93					3139	79
1436	5/8	Class Acts	25	14					3114	65
1437	5/9	Mirror Lighting	519	65					2595	00
		Deposit				5/10	3821	17	6416	17
1438	5/10	Woodlake Auditorium	750	00					5666	17
		Deposit				5/12	500	00	6166	17
1439	5/12	Rick's Clowns	170	80					5995	37
1440	5/14	Y.M.C.A.	219	17					5776	20
	5/14	ATM	120	00					5656	20
		Deposit				5/15	326	15	5982	35
1441	5/16	Stage Door Playhouse	825	00					5157	35
1442	5/17	Gilbert Eckern	1785	00					3372	35
		Deposit				5/19	1580	25	4952	60

QUICK TIP   ATM transactions for cash withdrawals and purchases must be entered on check stubs or in the check register. The transaction amount and the charge for each transaction must then be subtracted to maintain an accurate balance.

4.1 | Exercises

FOR EXTRA HELP

 MyMathLab

 InterActMath.com

 MathXP MathXL

 MathXL Tutorials on CD

 Addison-Wesley Math Tutor Center

DVT/Videotape

The **Quick Start** *exercises in each section contain solutions to help you get started.*

CHECKING CHARGES Use the table on page 142 to find the monthly checking account service charge for the following accounts. (See Example 1.)

Quick Start

1. Flower Power, 92 checks, average balance $4618

$\$5.00 + (92 \times \$.10) = \$5.00 + \$9.20 = \$14.20$

1. $14.20

2. Lemon Tree Restaurant, 114 checks, average balance $3318

$\$5.00 + (114 \times \$.10) = \$5.00 + \$11.40 = \$16.40$

2. $16.40

3. Pest-X, 40 checks, average balance $491

3. _____

4. Mower Shack, Inc., 76 checks, average balance $468

4. _____

5. Mak's Smog and Tune, 48 checks, average balance $1763

5. _____

6. Direct Connection, 272 checks, average balance $8205

6. _____

7. Software and More, 72 checks, average balance $516

7. _____

8. Mart & Bottle, 74 checks, average balance $875

8. _____

 indicates an exercise that is related to the Case in Point feature within the section.

MAINTAINING BANK RECORDS   *Use the following information to complete each check stub.*
(See Example 2.)

	Date	To	For	Amount	Bal. Bro't. For'd.	Deposits
9.	Mar. 8	Nola Akala	tutoring	$380.71	$3971.28	$79.26
10.	Oct. 15	Corinn Berman	rent	$850.00	$2973.09	$1853.24
11.	Dec. 4	Paul's Pools	chemicals	$37.52	$1126.73	_____

9.

857

_____ 20 ____

Amount _____

To _____

For _____

Bal. Bro't. For'd.		
Am't. Deposited		
Total		
Am't. this Check		
Balance For'd.		

10.

1248

_____ 20 ____

Amount _____

To _____

For _____

Bal. Bro't. For'd.		
Am't. Deposited		
Total		
Am't. this Check		
Balance For'd.		

11.

735

_____ 20 ____

Amount _____

To _____

For _____

Bal. Bro't. For'd.		
Am't. Deposited		
Total		
Am't. this Check		
Balance For'd.		

12. List and explain at least six parts of a check. Draw a sketch showing where these parts appear on a check. (See Objective 1.)

13. Explain at least two advantages and two possible disadvantages of using an ATM card. (See Objective 2.)

14. Write an explanation of two types of check endorsements. Describe where these endorsements must be placed. (See Objective 4.)

15. Explain in your own words the factors that determine the service charges on a business checking account. (See Objective 3.)

COMPLETING CHECK STUBS *Using the information provided, complete the following check stubs for*
Flower Power. **The balance brought forward for check stub 5311 is $7223.69.** *(See Example 2.)*

		Checks Written				Deposits Made	
Number	Date	To	For	Amount		Date	Amount
5311	Oct. 7	Julie Davis	Seeds	$1250.80		Oct. 8	$752.18
5312	Oct. 10	County Clerk	License	$39.12		Oct. 9	$23.32
5313	Oct. 15	United Parcel	Shipping	$356.28		Oct. 13	$1025.45

16.

5311

_____ 20 ___

Amount _____
To _____
For _____

Bal. Bro't. For'd.
Am't. Deposited
Total
Am't. this Check
Balance For'd.

17.

5312

_____ 20 ___

Amount _____
To _____
For _____

Bal. Bro't. For'd.
Am't. Deposited
Total
Am't. this Check
Balance For'd.

18.

5313

_____ 20 ___

Amount _____
To _____
For _____

Bal. Bro't. For'd.
Am't. Deposited
Total
Am't. this Check
Balance For'd.

BANK BALANCES *In Exercises 19–22, complete the balance column in the following company check registers after each check or deposit transaction. (See Objective 6.)*

19. Flower Power

CHECK NO.	DATE	CHECK ISSUED TO	AMOUNT OF CHECK	✓	DATE OF DEP.	AMOUNT OF DEPOSIT	BALANCE
		BALANCE BROUGHT FORWARD →					9628 35
1221	10/4	Delta Contractors	215 71				
1222	10/5	Hand Fabricating	573 78				
1223	10/5	Photo Specialties	112 15				
		Deposit			10/6	753 28	
		Deposit			10/8	1475 69	
1224	10/9	Young Marketing	426 55				
1225	10/11	Wholesale Supply	637 93				
	10/11	ATM (fuel)	65 62				
1226	10/14	Light and Power Utilities	248 17				
		Deposit			10/16	335 85	
1227	10/16	License Board	450 50				

20. Ontime Marketing

CHECK NO.	DATE	CHECK ISSUED TO	AMOUNT OF CHECK		✓	DATE OF DEP.	AMOUNT OF DEPOSIT		BALANCE	
		BALANCE BROUGHT FORWARD →							1629	86
861	7/3	Ahwahnee Hotel	250	45						
862	7/5	Willow Creek	149	00						
863	7/5	Void								
		Deposit				7/7	117	73		
864	7/9	Del Campo High School	69	80						
		Deposit				7/10	329	86		
		Deposit				7/12	418	30		
865	7/14	Big 5 Sporting Goods	109	76						
866	7/14	Dr. Yates	614	12						
867	7/16	Greyhound	32	18						
		Deposit				7/16	520	95		

21. Stencils by Loree

CHECK NO.	DATE	CHECK ISSUED TO	AMOUNT OF CHECK		✓	DATE OF DEP.	AMOUNT OF DEPOSIT		BALANCE	
		BALANCE BROUGHT FORWARD →							832	15
1121	3/17	AirTouch Cellular	257	29						
1122	3/18	Curry Village	190	50						
		Deposit				3/19	78	29		
		Deposit				3/21	157	42		
1123	3/22	San Juan District	38	76						
1124	3/23	Macy's Gourmet	175	88						
		Deposit				3/23	379	28		
1125	3/24	Class Video	197	20						
1126	3/24	Water World	25	10						
1127	3/25	Bel Air Market	75	00						
		Deposit				3/28	722	35		

22. Roy's Handyman Services

CHECK NO.	DATE	CHECK ISSUED TO	AMOUNT OF CHECK		✓	DATE OF DEP.	AMOUNT OF DEPOSIT		BALANCE	
		BALANCE BROUGHT FORWARD →							3852	48
2308	12/6	Web Masters	143	16						
2309	12/7	Water and Power	118	40						
		Deposit				12/8	286	32		
	12/10	ATM (cash)	80	00						
2310	12/11	Ann Kuick	986	22						
2311	12/11	Account Temps	375	50						
		Deposit				12/14	1201	82		
2312	12/14	Central Chevrolet	735	68						
2313	12/15	Miller Mining	223	94						
		Deposit				12/17	498	01		
2314	12/18	Federal Parcel	78	24						

4.2 | Checking Services and Credit-Card Transactions

Objectives

1. Identify bank services available to customers.
2. Understand interest-paying checking plans.
3. Determine deposits with credit-card transactions.
4. Calculate the discount fee on credit-card deposits.

Undergrads Love Their Plastic

Undergrads with a credit card	78%
Average number of cards owned	3
Average student card debt	$1236
Students with 4 or more cards	32%
Balances of $3000 to $4000	13%
Balances over $7000	9%
Pay off card balance in full each month	50%

Source: Public Interest Research Groups and Nellie Mae (college loan provider).

Objective 1 Identify bank services available to customers. Most business checking-account charges are determined by either the average balance or the minimum balance in the account, together with specific charges for each service performed by the bank. Some of the services provided by banks, along with the *typical charges*, are listed here.

ATM cards are used as debit cards when making point-of-sale purchases. The fee for purchases varies from $.10 per transaction to $1 per month for unlimited transactions. When used at the ATM machine there is usually no fee at any branch of your bank, a fee as high as $2.50 at other banks, and an international fee as high as $5.

An **overdraft** occurs when a check is written for which there are **nonsufficient funds (NSF)** in the checking account and the customer has no overdraft protection, also referred to as *bouncing a check*. The typical charge to the writer of the "bad" check is $20 to $35 per bad check. The same charges occur when a check is returned because it was improperly completed.

Overdraft protection is given when an account balance is insufficient to cover the amount of a check and an overdraft occurs. Charges for overdraft protection vary among banks.

A **returned-deposit item** is a check that was deposited and then returned to the bank, usually because of lack of funds in the account of the person or firm writing the check. A common charge to the depositor of the check is $5. At least one bank in New York charges $35.

A **stop-payment order** is a request by a depositor that the bank not honor a check the depositor has written ($20 per request).

A **cashier's check** is a check written by the financial institution itself and is viewed as being as good as cash ($5 per check).

A **money order** is a purchased instrument that is often used in place of cash and is sometimes required instead of a personal or business check ($4 each).

A **notary service** (official certification of a signature on a document) is a service that is required on certain business documents. Occasionally this service is free to customers, but there is usually a charge ($10).

Objective 2 Understand interest-paying checking plans. Federal banking regulations allow both personal and business interest-paying checking plans. Some of the plans combine two accounts, a savings account and a checking account, while others are simply checking accounts that collect interest on the average daily balance.

While paying for purchases with checks remains popular, customer shopping habits have been changing as new payment options, such as debit cards, begin to take hold. The bar graph at the side shows how the use of credit cards has increased and projects the percent of their use in the future.

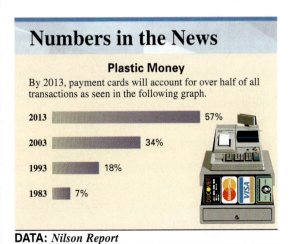

Numbers in the News

Plastic Money

By 2013, payment cards will account for over half of all transactions as seen in the following graph.

2013	57%
2003	34%
1993	18%
1983	7%

DATA: *Nilson Report*

Objective 3 Determine deposits with credit-card transactions. With the continuing growth in electronic commerce, more and more **credit-card transactions** are being completed electronically by the retailer. This results in a direct deposit to the merchants bank account and eliminates the need for a paper transaction. However, a great number of small retailers continue to process their credit-card sales mechanically. These credit-card sales are deposited into a business checking account with a **merchant batch header ticket** such as the one in Example 1. This form is used with Visa or MasterCard credit-card deposits. Notice that the form lists both sales slips and credit slips (refunds). Entries in each of these categories are totaled, and the total credits are subtracted from the total sales to give the net amount of deposit.

The merchant batch header ticket is a triplicate form. The *bank copy* along with the charge slips, credit slips, and a printed calculator tape showing the itemized deposits and credits are deposited in the business checking account.

EXAMPLE **1**

Calculating Deposits with Credit-Card Transactions

Flower Power had the following credit-card sales and refunds. Complete a merchant batch header ticket.

Sales		Refunds (Credits)
$ 82.31	$146.50	$13.83
$ 38.18	$ 78.80	$25.19
$ 65.29	$ 63.14	$78.56
$178.22	$208.67	

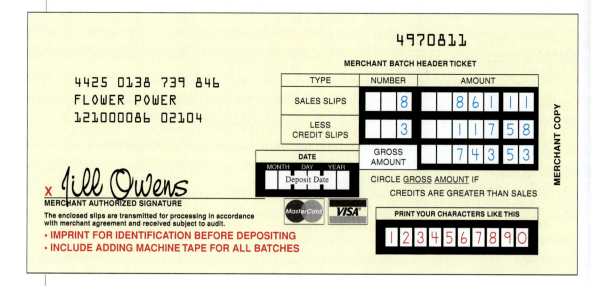

SOLUTION

All credit slips and sales slips must be totaled. The number of each of these and the totals are written at the right on the form. The sales slips total $861.11 and the credit slips total $117.58. The difference is the gross amount; here $743.53 is the gross amount of the deposit.

Objective 4 **Calculate the discount fee on credit-card deposits.** The bank collects a fee (a percent of sales) from the merchant. The bank also collects an interest charge from the card user on all accounts not paid in full at the first billing. Although credit-card transactions are deposited frequently by a business, the bank calculates the discount fee on the net amount of the credit-card deposits since the last bank statement date. The fee paid by the merchant varies from 2% to 5% of the sales slip amount and is determined by the type of processing used (electronic or manual), the dollar volume of credit-card usage by the merchant, and the average amount of the sale at the merchant's store. All credit-card deposits for the month are added, and the fee is subtracted from the total at the statement date.

EXAMPLE **2**

Finding the Discount and the Credit Given on a Credit-Card Deposit

If the deposit in Example 1 represented total credit-card deposits for the month, find the fee charged and the credit given to the merchant at the statement date if Flower Power pays a 3% fee.

SOLUTION

Since the total credit-card deposit for Flower Power is $743.53 and the fee is 3%, the discount charged is

$$\$743.53 \times .03 =$$

Out of a deposit of $743.53, the merchant will receive a credit of $743.53 − **$22.31 = $721.22**.

The calculator solution to this example is $743.53 − 3 % = 721.2241.
Note: Refer to Appendix C for calculator basics.

Jackson & Perkins

- 1872: Founded

- 1901: Marketed the first of their hybridized roses.

- 1978: Roses selected for special-issue stamp by the U.S. Postal Service.

- 2004: Developed 110 hybrid rose varieties.

- 2004: Sales of over 10 million plants.

- 2004: Cultivates over 5000 rose plants.

The Jackson & Perkins Company was founded in 1872 by Charles Perkins and A. E. Jackson. Beginning as a small company wholesaling strawberry and grape plants, they soon began selling hybrid roses as their main product. In those early days, when you bought a garden plant from Charles Perkins, he would always say, "If it doesn't grow for you, let me know." If the plant didn't grow, he would provide either a replacement or a refund. To this day, every Jackson & Perkins plant is guaranteed to grow.

Today, Jackson & Perkins sells a complete variety of roses and other plants and gardening accessories. Located in Medford, Oregon, the company distributes a full-color catalog of their products, which are sold through mail-order or by toll-free telephone number. Most recently, Jackson & Perkins has started using the Internet so customers can buy their products on-line.

1. If Jackson & Perkins pays a monthly checking-account fee of $5 plus $.10 per check, find the total checking account charge for a month when 836 checks were written.

2. The total credit-card sales for Jackson & Perkins during a certain period were $837,422, while credit-card returns for the same period were $28,225. Find **(a)** the gross amount of the credit-card deposit and **(b)** the amount of credit given to Jackson & Perkins after a fee of 2% is subtracted.

3. List and explain three advantages and three disadvantages of buying your merchandise on the Internet.

4. By asking your classmates, coworkers, and friends, find someone who has purchased items on the Internet. What was that person's experience like? Did he or she save money? Get a better selection? Get quick delivery? Would the person recommend this method of purchase to others?

4.2 | Exercises

FOR EXTRA HELP

 MyMathLab

 InterActMath.com

 MathXL

 MathXL Tutorials on CD

 Addison-Wesley Math Tutor Center

DVT/Videotape

The **Quick Start** *exercises in each section contain solutions to help you get started.*

CREDIT CARD DEPOSITS   *Mak's Tune and Smog accepts cash, checks, and credit cards from customers for auto repair and the sale of parts. The following credit-card transactions occurred during a recent period. Complete a merchant batch header ticket.*

Sales		Credits
$ 66.68	$ 18.95	$62.16
$119.63	$496.28	$106.62
$ 53.86	$ 21.85	$38.91
$178.62	$242.78	
$219.78	$176.93	

Quick Start

1. What is the total amount of the sales slips?

   $66.68 + $119.63 + $53.86 + $178.62 + $219.78 + $18.95 + $496.28 + $21.85 + $242.78 + $176.93 = $1595.36

   1. $1595.36

2. Find the total of the credit slips.

   $62.16 + $106.62 + $38.91 = $207.69

   2. $207.69

3. Find the gross deposit when the sales and credits are deposited.

   3. _____

4. If the fee paid by the business is 4%, find the amount of the charge at the statement date.

   4. _____

5. Find the amount of the credit given to Mak's Tune and Smog after the fee is subtracted.

   5. _____

 indicates an exercise that is related to the Case in Point feature within the section.

CREDIT-CARD DEPOSITS   *Flower Power does most of its business on the Internet and accepts credit cards. In a recent period, the business had the following credit-card charges and credits. (See Examples 1 and 2.)*

Sales		Credits
$ 78.56	$ 38.15	$ 29.76
$875.29	$ 18.46	$102.15
$330.82	$ 22.13	$ 71.95
$ 55.24	$707.37	
$ 47.83	$245.91	

4970812

MERCHANT BATCH HEADER TICKET

TYPE	NUMBER	AMOUNT
SALES SLIPS		
LESS CREDIT SLIPS		
	GROSS AMOUNT	

2231 0687 791
FLOWER POWER
183000215 07862

DATE
MONTH   DAY   YEAR
Deposit Date

CIRCLE <u>GROSS</u> <u>AMOUNT</u> IF
CREDITS ARE GREATER THAN SALES

X *Jill Owens*
MERCHANT AUTHORIZED SIGNATURE

The enclosed slips are transmitted for processing in accordance with merchant agreement and received subject to audit.
• **IMPRINT FOR IDENTIFICATION BEFORE DEPOSITING**
• **INCLUDE ADDING MACHINE TAPE FOR ALL BATCHES**

MasterCard   VISA

MERCHANT COPY

PRINT YOUR CHARACTERS LIKE THIS

1234567890

6. What is the total amount of the sales slips?   6. _____

7. Find the total of the credit slips.   7. _____

8. If these sales and credits are deposited, what is the gross amount of the deposit?   8. _____

9. If the bank charges Flower Power a 3% fee, find the amount of the charge at the statement date.   9. _____

10. Find the amount of the credit given to Flower Power after the fee is subtracted.   10. _____

CREDIT-CARD DEPOSITS   *Jay Jenkins owns Campus Bicycle Shop near a college campus. The shop sells new and used bicycle parts and does a major portion of its business in adjustments and repairs. The following credit-card charges and credits took place during a recent period. Complete a merchant batch header ticket.*

Sales		Credits
$ 7.84	$ 98.56	$13.86
$ 33.18	$318.72	$58.97
$ 50.76	$116.35	
$ 12.72	$ 23.78	
$ 9.36	$ 38.95	
$118.68	$235.82	

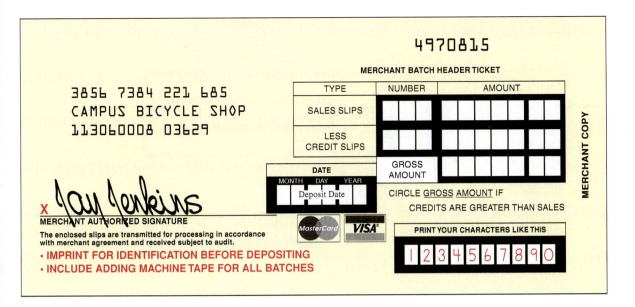

11. Find the total amount of the sales slips for the period.   11. _____

12. Find the total of the credit slips.   12. _____

13. Find the gross deposit when the sales and credits are deposited.   13. _____

14. If the fee paid by the shop is 3%, find the amount of the charge at the statement date.   14. _____

15. Find the amount of the credit given to Campus Bicycle Shop after the fee is subtracted.   15. _____

CREDIT-CARD DEPOSITS   Kaare Taylor Studios had the following credit-card transactions during a recent period. Complete a merchant batch header ticket.

Sales		Credits
$ 14.86	$ 76.15	$43.15
$ 49.70	$226.17	$17.06
$183.60	$ 63.95	
$238.75	$111.10	
$ 18.36	$ 77.86	
$ 52.08	$132.62	

4970815

MERCHANT BATCH HEADER TICKET

4425 3857 328 811
KAARE TAYLOR STUDIOS
215200081 10395

TYPE	NUMBER	AMOUNT
SALES SLIPS		
LESS CREDIT SLIPS		

DATE

MONTH   DAY   YEAR

Deposit Date

GROSS AMOUNT

CIRCLE GROSS AMOUNT IF CREDITS ARE GREATER THAN SALES

X *Kaare Taylor*
MERCHANT AUTHORIZED SIGNATURE

The enclosed slips are transmitted for processing in accordance with merchant agreement and received subject to audit.

• IMPRINT FOR IDENTIFICATION BEFORE DEPOSITING
• INCLUDE ADDING MACHINE TAPE FOR ALL BATCHES

MasterCard   VISA®

PRINT YOUR CHARACTERS LIKE THIS

1 2 3 4 5 6 7 8 9 0

MERCHANT COPY

16. Find the total amount of the sales slips for the period.

16. _____

17. Find the total of the credit slips.

17. _____

18. Find the gross deposit when the sales and credits are deposited.

18. _____

19. If the fee paid by the shop is 5%, find the amount of the charge at the statement date.

19. _____

20. Find the amount of credit given to Kaare Taylor Studios after the fee is subtracted.

20. _____

21. List and describe in your own words four services offered to business checking-account customers. (See Objective 1.)

22. The merchant accepting a credit card from a customer must pay a fee of 2% to 5% of the transaction amount. Why is the merchant willing to do this? Who really pays this fee? (See Objective 4.)

4.3 | Bank Statement Reconciliation

Objectives

1. Know the importance of reconciling a checking account.
2. Reconcile a bank statement with a checkbook.
3. List the outstanding checks.
4. Find the *adjusted bank balance* or *current balance*.

CASE *in* POINT

Jill Owens, owner of Flower Power, knows the importance of keeping accurate checking-account records. She has received customers' checks drawn on accounts with nonsufficient funds (NSF), but with great pride she says, "I have never bounced a business check." Example 1 shows how Jill Owens recently balanced her business checking account.

Objective 1 Know the importance of reconciling a checking account. Once a month, banks send their checking-account customers a **bank statement**. This bank statement shows all deposits made during the period covered by the statement, as well as all the checks paid by the bank and any automated teller machine (ATM) or debit-card transactions. Bank charges for the month covered by the statement are also listed. This is especially important with a business checking account because the bank charge normally varies from month to month. On occasion, a customer's check that was deposited has to be returned due to nonsufficient funds (NSF) in the account. This is identified as a **returned check**, and the amount of the check must be subtracted from the checkbook balance along with any other charges. The business must then resolve this matter with the writer of the bad check.

Objective 2 Reconcile a bank statement with a checkbook. Many businesses have automatic deposits made to their accounts from their customers' bank accounts. The amount of these automatic deposits must be added to the business's checkbook balance. When the bank statement is received, it is very important that its accuracy be verified. In addition, it is a good time to check the accuracy of the check register, being certain that all checks written have been listed and subtracted and that all deposits have been added to the checking-account balance. The process of checking the bank statement against the check register is called **reconciliation**. The following bank statement is for **Flower Power**.

STATEMENT							
COUNTRY RIVER BANK				3/31	STATEMENT PERIOD		4/30
					PAGE	1	

Flower Power
#6 Highway 17
Crossville, GA 38555

ACCOUNT NUMBER	PREVIOUS BALANCE	CREDITS COUNT	AMOUNT	DEBITS COUNT	AMOUNT	FEE	PRESENT BALANCE
ODA 110004565	5218.29	4	10406.71	10	8732.59		6892.41

CHECK NUMBER	CHECKS AND DEBITS		DEPOSITS	DATE	BALANCE
			Beginning Balance		5218.29
843	1836.71			4/3	3381.58
847	79.26			4/4	3302.32
841	2625.10			4/5	677.22
			4137.80	4/5	4815.02
855	1498.92			4/10	3316.10
838	478.63			4/15	2837.47
	37.75	ATM		4/19	2799.72
849	1626.63			4/19	1173.09
			3279.62	4/22	4452.71
856	142.45			4/25	4310.26
854	400.72			4/25	3909.54
			2984.51	4/28	6894.05
	6.42	SC		4/30	6887.63
			4.78 IC	4/30	6892.41

Codes Used Above	RC = Returned Check	SC = Service Charge
	IC = Interest Credit	ATM = Automated Teller Machine

Reconciliation is best done on the forms usually printed on the back of the bank statement. The codes on the bank statement indicate the following: RC means Returned Check; SC means Service Charge; IC means Interest Credit; and ATM means Automated Teller Machine. An example of the reconciliation process follows.

EXAMPLE 1

Reconciling a Checking Account

Flower Power received its bank statement. The statement shows a balance of $6892.41, after a bank service charge of $6.42 and an interest credit of $4.78. Flower Power's checkbook now shows a balance of $7576.38. Reconcile the account using the following steps.

Objective ③ **List the outstanding checks.**   Compare the list of checks on the bank statement with the list of checks written by the firm. Checks that have been written by the firm but do not yet appear on the bank statement have not been paid by the bank as of the date of the statement. These unpaid checks are called **checks outstanding**. The following table shows those checks written by the firm that are outstanding.

Number	Amount
846	$42.73
852	$598.71
853	$68.12
857	$79.80
858	$160.30

After listing the outstanding checks in the space provided on the form, total them. The total is $949.66.

Objective ④ **Find the *adjusted bank balance* or *current balance*.**   The following steps are used to reconcile the checking account of Flower Power.

Step 1   Enter the new balance from the front of the bank statement. The new balance is $6892.41. Write this number in the space provided on the reconciliation form.

Step 2   List any deposits that have not yet been recorded by the bank. These are called *deposits in transit* (DIT). Suppose that Flower Power has deposits of $892.41 and $739.58 that are not yet recorded. These numbers are written at step 2 on the form.

Step 3   Add the numbers from steps 1 and 2. At this point, the total is $8524.40.

Step 4   Write down the total of outstanding checks. The total is $949.66.

Step 5   Subtract the total in step 4 from the number in step 3. The result here is $7574.74, which is the **adjusted bank balance**, or the **current balance**. This number should represent the current checking-account balance.

Now look at the firm's own records.

Step 6   List the firm's checkbook balance. As mentioned in the problem, the checkbook balance for Flower Power is $7576.38. This number is entered on line 6.

Step 7   Enter any service charges not yet deducted. The service charge shown on the statement is $6.42. Since there are no other fees or charges, enter $6.42 on line 7.

Step 8   Subtract the charges on line 7 from the checkbook balance on line 6 to get $7569.96.

Step 9   Enter the interest credit on line 9. The interest credit here is $4.78. (This amount is interest paid on the money in the account.)

Step 10   Add the interest on line 9 to line 8 to get $7574.74, the same result as in step 5.

Since the result from step 10 is the same as the result from step 5, the account is **balanced** (reconciled). The correct current balance in the account is $7574.74.

It's Not in the Mail

Bounce a Check, and You Might Not Write Another for 5 Years

By Paul Beckett

Staff Reporter of
THE WALL STREET JOURNAL

Two years ago, Rebecca Cobos overdrew her checking account at a Bank of America branch in Los Angeles. When she couldn't immediately repay the bank, it not only closed her account, but also had her, in effect, banned for five years from opening a checking account at most other banks, too.

Bank of America did so by reporting the 23-year-old university secretary to ChexSystems, a national database to which 80% of bank branches in the country subscribe. Once lodged in ChexSystems, you automatically stay there for five years, whether your offense was bouncing a check or two or committing serious fraud. The large majority of banks using ChexSystems reject any checking-account applicant they find in the database.

Source: Wall Street Journal

Checks Outstanding		
Number	Amount	
846	$ 42	73
852	598	71
853	68	12
857	79	80
858	160	30
Total	$ 949	66

Compare the list of checks paid by the bank with your records. List and total the checks not yet paid.

(1) Enter new balance from bank statement: $ 6892.41

(2) List any deposits made by you and not yet recorded by the bank:
+ 892.41
+ 739.58
+
+

(3) Add all numbers from lines above. Total: 8524.40

(4) Write total of checks outstanding: − 949.66

(5) Subtract (4) from (3). This is adjusted bank balance: $ 7574.74

To reconcile your records:

(6) List your checkbook balance: $ 7576.38

(7) Write the total of any fees or charges deducted by the bank and not yet subtracted by you from your checkbook: − 6.42

(8) Subtract line (7) from line (6). 7569.96

(9) Enter interest credit: (Add to your checkbook) + 4.78

(10) Add line (9) to line (8). Adjusted checkbook balance. $ 7574.74

New balance of your account; this number should be same as (5).

There are several typical reasons why checking accounts do not balance.

Why Checking Accounts Do Not Balance

- Forgetting to enter a check in the check register.
- Forgetting to enter a deposit in the check register.
- Transposing numbers (writing $961.20 as $916.20, for example).
- Addition or subtraction errors.
- Forgetting to subtract one of the bank service fees, such as those charged for using your debit card or for ATM use.
- Charging the customer an amount different from the check amount.
- Check may be altered or forged.

The newspaper clipping at the side discusses what some banks are doing when customers bounce checks.

The following graphic shows how much consumers pay for monthly banking fees. The business person, as well as the consumer, must look for the best value and convenience when selecting a banking-services provider.

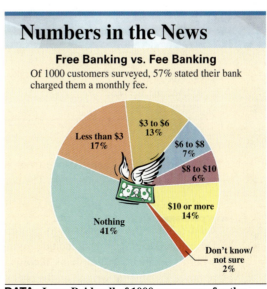

Numbers in the News

Free Banking vs. Fee Banking

Of 1000 customers surveyed, 57% stated their bank charged them a monthly fee.

- $3 to $6 — 13%
- $6 to $8 — 7%
- $8 to $10 — 6%
- $10 or more — 14%
- Less than $3 — 17%
- Nothing — 41%
- Don't know/ not sure — 2%

DATA: Ipsos Reid poll of 1000 consumers for the American Bankers Association

EXAMPLE 2

Reconciling a Checking Account

Using the information on the following check register and the bank statement, reconcile the following checking account. Compare the items appearing on the check register to the bank statement. A ✓ indicates that the check appeared on the previous month's bank statement. (Codes indicate the following: RC means Returned Check; SC means Service Charge; IC means Interest Credit; ATM means Automated Teller Machine.)

CHECK NO.	DATE	CHECK ISSUED TO	AMOUNT OF CHECK	✓	DATE OF DEP.	AMOUNT OF DEPOSIT	BALANCE	
		BALANCE BROUGHT FORWARD →					2782	95
721	7/11	Miller's Outpost	138 50	✓			2644	45
722	7/12	Barber Advertising	73 08				2571	37
723	7/18	Wayside Lumber	318 62	✓			2252	75
		Deposit			7/20	980 37	3233	12
724	7/25	I.R.S.	836 15				2396	97
725	7/26	John Lessor	450 00				1946	97
726	7/28	Sacramento Bee	67 80				1879	17
727	8/2	T.V.A.	59 25				1819	92
728	8/3	Carmichael Office	97 37				1722	55
		Deposit			8/6	875 45	2598	00
ATM	8/5	ATM Cash	80 00				2518	00

Bank Statement

CHECK NUMBER	CHECKS AND DEBITS			DEPOSITS	DATE	BALANCE
					7/20	2325.83
722	73.08				7/22	2252.75
				980.37	7/24	3233.12
724	836.15				7/28	2396.97
725	450.00	49.07	RC		7/30	1897.90
727	59.25	80.00	ATM	3.22 IC	8/4	1761.87
		7.60	SC		8/5	1754.27

Checks Outstanding	
Number	Amount
726	$ 67 80
728	97 37
Total	$ 165 17

Compare the list of checks paid by the bank with your records. List and total the checks not yet paid.

(1) Enter new balance from bank statement: $ 1754.27

(2) List any deposits made by you and not yet recorded by the bank:
   + 875.45
   + _____
   + _____
   + _____

(3) Add all numbers from lines above. Total: 2629.72

(4) Write total of checks outstanding: − 165.17

(5) Subtract (4) from (3). This is adjusted bank balance: $ 2464.55

To reconcile your records:

(6) List your checkbook balance: $ 2518.00

(7) Write the total of any fees or charges deducted by the bank and not yet subtracted by you from your checkbook: − 56.67   (Returned check and service charge)

(8) Subtract line (7) from line (6).   2461.33

(9) Enter interest credit: (Add to your checkbook)   + 3.22

(10) Add line (9) to line (8). Adjusted checkbook balance. $ 2464.55

New balance of your account; this number should be same as (5).

Since the adjusted bank balance from step 5 is the same as the new balance from step 10, the account is reconciled (balanced). The correct current balance in the account is $2464.55.

QUICK TIP   It is important to guard against identity theft and other criminal activities. Destroy all documents that include savings, checking, and credit-card account numbers, social security numbers, and all bank records. Check washing is a recent crime in which the thief removes the amount written on the check and changes it to a higher amount.

4.3 | Exercises

The **Quick Start** *exercises in each section contain solutions to help you get started.*

CURRENT CHECKING BALANCE   *Find the current balance for each of the following accounts. (See Example 1.)*

Balance from Bank Statement	Deposits Not Checks Outstanding	Current Yet Recorded	Balance

Quick Start

1. $4572.15 $225.23 $418.25 $816.14
               $ 97.68 $348.17 $571.28   **$4870.24**

   $4572.15 − $225.23 − $97.68 − $418.25 − $348.17 + $816.14 + $571.28 = $4870.24

2. $6274.76 $381.40 $681.10 $346.65
               $875.14 $ 83.15 $198.96   **$4799.58**

   $6274.76 − $381.40 − $875.14 − $681.10 − $83.15 + $346.65 + $198.96 = $4799.58

3. $7911.42 $52.38 $528.02 $492.80
               $95.42 $ 76.50 $ 38.72   _____

4. $9343.65 $840.71 $665.73 $ 971.64
               $ 78.68 $ 87.00 $3382.71   _____

5. $19,523.20 $6853.60 $340.00 $6724.93
               $ 795.77 $ 22.85 $ 78.81   _____

6. $32,489.50 $3589.70 $18,702.15 $7110.65
               $ 263.15 $ 7269.78 $2218.63   _____

7. Explain in your own words the significance of writing a bad check. What might the cost be in dollars? What are the other consequences? (See Objective 1.)

8. What are the financial costs to a business owner who receives a bad check? What would the business owner likely do regarding this customer? (See Objective 1.)

9. Briefly describe the importance of reconciling a checking account. What are the benefits derived from keeping good checking records? (See Objective 2.)

10. Suppose your checking account will not balance. Name four types of errors that you will look for in trying to correct this problem. (See Objective 4.)

C indicates an exercise that is related to the Case in Point feature within the section.

RECONCILING CHECKING ACCOUNTS *For Exercises 11 and 12, use the following table to reconcile each account and find the current balance. (See Example 1.)*

		Exercise 11.		**Exercise 12.**
Balance from bank statement		$6875.09		$14,928.42
Checks outstanding	421	$371.52	112	$84.76
(check number is given first)	424	$429.07	115	$109.38
	427	$883.69	117	$42.03
	429	$35.62	119	$1429.12
Deposits not yet recorded		$701.56		$54.21
		$421.78		$394.76
		$689.35		$1002.04
Bank charge		$8.75		$7.00
Interest credit		$10.71		$22.86
Checkbook balance		$6965.92		$14,698.28
Current balance		_____		_____

11.

Checks Outstanding	
Number	Amount
Total	

Compare the list of checks paid by the bank with your records. List and total the checks not yet paid.

(1) Enter new balance from bank statement: _____

(2) List any deposits made by you and not yet recorded by the bank:
+ _____
+ _____
+ _____
+ _____

(3) Add all numbers from lines above. Total: _____

(4) Write total of checks outstanding: − _____

(5) Subtract (4) from (3). This is adjusted bank balance: _____

To reconcile your records:

(6) List your checkbook balance: _____

(7) Write the total of any fees or charges deducted by the bank and not yet subtracted by you from your checkbook: − _____

(8) Subtract line (7) from line (6). _____

(9) Enter interest credit: (Add to your checkbook) + _____

(10) Add line (9) to line (8). Adjusted checkbook balance. _____

New balance of your account; this number should be same as (5).

12.

Checks Outstanding	
Number	Amount
Total	

Compare the list of checks paid by the bank with your records. List and total the checks not yet paid.

(1) Enter new balance from bank statement: _____

(2) List any deposits made by you and not yet recorded by the bank:
+ _____
+ _____
+ _____
+ _____

(3) Add all numbers from lines above. Total: _____

(4) Write total of checks outstanding: − _____

(5) Subtract (4) from (3). This is adjusted bank balance: _____

To reconcile your records:

(6) List your checkbook balance: _____

(7) Write the total of any fees or charges deducted by the bank and not yet subtracted by you from your checkbook: − _____

(8) Subtract line (7) from line (6). _____

(9) Enter interest credit: (Add to your checkbook) + _____

(10) Add line (9) to line (8). Adjusted checkbook balance. _____

New balance of your account; this number should be same as (5).

CHECKING-ACCOUNT RECONCILIATION   Reconcile the checking accounts for the following companies. Compare the items appearing on the bank statement with the check register. A ✓ indicates that the check appeared on the previous month's statement. (Codes indicate the following: RC means Returned Check; SC means Service Charge; IC means Interest Credit; CP means Check Printing Charge; ATM means Automated Teller Machine.) (See Example 2.)

13. Flower Power

CHECK NO.	DATE	CHECK ISSUED TO	AMOUNT OF CHECK	✓	DATE OF DEP.	AMOUNT OF DEPOSIT	BALANCE
		BALANCE BROUGHT FORWARD →					6669 34
760	2/8	Floors to Go	248 96				6420 38
762	2/9	Healthways Dist.	125 63				6294 75
		Deposit			2/11	618 34	6913 09
763	2/12	Franchise Tax	770 41	✓			6142 68
764	2/14	Foothill Repair	22 86	✓			6119 82
765	2/15	Yellow Pages	91 24				6028 58
		Deposit			2/17	826 03	6854 61
766	2/17	Morning Herald	71 59				6783 02
767	2/18	San Juan Electric	63 24				6719 78
ATM	2/22	ATM Gas	15 26				6704 52
769	2/23	West Construction	405 07				6299 45
770	2/24	Rent	525 00				5774 45
		Deposit			2/26	220 16	5994 61
771	2/28	Capital Alarm	135 76				5858 85

Bank Statement

```
***********************************************************************
CHECK          CHECKS                    DEPOSITS      DATE      BALANCE
NUMBER         AND DEBITS
***********************************************************************
                                                       2/14      5876.07
765            91.24                     618.34        2/16      6403.17
760            248.96                    826.03        2/17      6980.24
766            71.59                                   2/19      6908.65
762            125.63                                  2/21      6783.02
                          198.17  RC                   2/22      6584.85
                           15.26  ATM    8.12  IC      2/24      6577.71
769            405.07       4.85  CP                   2/26      6167.79
                            6.28  SC                   2/27      6161.51
770            525.00                                  2/28      5636.51
```

Checks Outstanding	
Number	Amount
Total	

Compare the list of checks paid by the bank with your records. List and total the checks not yet paid.

(1) Enter new balance from bank statement: _____

(2) List any deposits made by you and not yet recorded by the bank:
   + _____
   + _____
   + _____
   + _____

(3) Add all numbers from lines above. Total: _____

(4) Write total of checks outstanding: _____

(5) Subtract (4) from (3). This is adjusted bank balance: _____

To reconcile your records:

(6) List your checkbook balance: _____

(7) Write the total of any fees or charges deducted by the bank and not yet subtracted by you from your checkbook: − _____

(8) Subtract line (7) from line (6). _____

(9) Enter interest credit: (Add to your checkbook) + _____

(10) Add line (9) to line (8). Adjusted checkbook balance. _____

New balance of your account; this number should be same as (5).

14. Play It Again Sports

CHECK NO.	DATE	CHECK ISSUED TO	AMOUNT OF CHECK	✓	DATE OF DEP.	AMOUNT OF DEPOSIT	BALANCE
		BALANCE BROUGHT FORWARD →					7682 07
662	3/3	Action Packing Supplies	451 16				7230 91
663	3/3	Crown Paper	954 29	✓			6276 62
664	3/5	ATM Cash	80 00	✓			6196 62
		Deposit			3/7	913 28	7109 90
665	3/10	Fairless Water District	72 37				7037 53
666	3/12	Audia Temporary	340 88				6696 65
667	3/13	Lionel Toys	618 65				6078 00
668	3/14	Fairless Hills Power	100 50				5977 50
		Deposit			3/16	450 18	6427 68
		Deposit			3/18	163 55	6591 23
669	3/20	Hunt Roofing	238 50				6352 73
670	3/22	Standard Brands	315 62				6037 11
671	3/23	Penny-Saver Products	67 29				5969 82
		Deposit			3/24	830 75	6800 57

Bank Statement

CHECK NUMBER	CHECKS AND DEBITS			DEPOSITS	DATE	BALANCE
					3/5	6647.78
				913.28	3/7	7561.06
662	451.16				3/11	7109.90
666	340.88	82.15	RC	450.18	3/16	7137.05
665	72.37			22.48 IC	3/20	7087.16
667	618.65			163.55	3/22	6632.06
669	238.50	12.70	SC		3/26	6380.86

Checks Outstanding

Number	Amount
Total	

Compare the list of checks paid by the bank with your records. List and total the checks not yet paid.

(1) Enter new balance from bank statement: _____

(2) List any deposits made by you and not yet recorded by the bank:
+ _____
+ _____
+ _____
+ _____

(3) Add all numbers from lines above. Total: _____

(4) Write total of checks outstanding: = _____

(5) Subtract (4) from (3). This is adjusted bank balance: _____

To reconcile your records:

(6) List your checkbook balance: _____

(7) Write the total of any fees or charges deducted by the bank and not yet subtracted by you from your checkbook: − _____

(8) Subtract line (7) from line (6). _____

(9) Enter interest credit: (Add to your checkbook) + _____

(10) Add line (9) to line (8). Adjusted checkbook balance. _____

New balance of your account; this number should be same as (5).

Chapter 4 Quick Review

CONCEPTS

4.1 Checking-account service charges

A checking-account maintenance fee is usually charged and there is often a per-check charge.

4.2 Bank services offered

The checking account customer must be aware of various banking services that are offered.

4.2 Deposits with credit-card transactions

All credit-card refunds must be subtracted from total credit-card sales to find the net deposit. The discount charge is then subtracted from this total.

4.3 Reconciliation of a checking account

A checking-account customer must periodically verify checking-account records with those of the bank or financial institution. The bank statement is used for this.

EXAMPLES

Find the monthly checking-account service charge for a business with 36 checks and transactions, given a monthly maintenance charge of $7.50 and a $.20 per check charge.

$$\$7.50 + 36(\$.20) = \$7.50 + \$7.20 = \$14.70 \text{ monthly service charge}$$

Overdraft protection Offered to protect the customer from bouncing a check (NSF).
ATM card Used at an automated teller machine to get cash or used as a debit card to make purchases.
Stop-payment order Stops payment on a check written in error.
Cashier's check A check written by a financial institution, such as a bank, and that is viewed to be as good as cash.
Money order An instrument used in place of cash.
Notary service An official certification of a signature or document.

The following are credit-card charges and credits.

Charges		Credits
$28.15	$78.59	$21.86
$36.92	$63.82	$19.62

(a) Find total charges.
$28.15 + $36.92 + $78.59 + $63.82 = $207.48
(b) Find total credits.
$21.86 + $19.62 = $41.48
(c) Find gross deposit.
$207.48 − $41.48 = $166
(d) Given a 3% fee, find the amount of the charge.
$166 × .03 = $4.98
(e) Find the amount of credit given to the business.
$166 − $4.98 = $161.02

The accuracy of all checks written, deposits made, service charges incurred, and interest paid is checked and verified. The customer's checkbook balance and bank balance must be the same for the account to reconcile, or balance.

Chapter 4 | Summary Exercise

The Banking Activities of a Retailer

Keiah Fulgham owns a retail store specializing in track-and-field equipment and accessories. She sells athletic shoes, clothing, and equipment to individuals, athletic clubs, and schools. Many of her customers use credit cards for their purchases and her credit-card sales in a recent month amounted to $8752.40. During the same period, she had $573.94 in credit slips and Ms. Fulgham pays a credit-card fee of $2\frac{1}{2}\%$.

When she received her bank statement, the balance was $4228.34. The checks outstanding were found to be $758.14, $38.37, $1671.88, $120.13, $2264.75, $78.11, $3662.73, $816.25, and $400. Ms. Fulgham had both her credit-card deposit and bank deposits of $458.23, $771.18, $235.71, $1278.55, $663.52, and $1475.39 that were not recorded.

(a) Find the gross deposit when the credit-card sales and credits are deposited.

(a) _____

(b) Find the amount of the credit given to Ms. Fulgham after the fee is subtracted.

(b) _____

(c) What is the total of the checks outstanding?

(c) _____

(d) Find the total of the deposits that were not recorded.

(d) _____

(e) Find the current balance in Ms. Fulgham's checking account.

(e) _____

INVESTIGATE

Contact three banks or credit unions (other than your own) and ask about their charges for a personal checking account. Compare these fees with what you or your family members or friends pay for their checking accounts. Is there a way for you to get free checking? Name three considerations other than cost that are important to you in selecting a bank. Based on what you have learned, would it be a good idea for you to change banks?

Chapter 4 | Test

To help you review, the numbers in brackets show the section in which the topic was discussed.

Use the table on page 142 to find the monthly checking-account service charge for the following accounts. **[4.1]**

1. Tino's Italian Grocery, 62 checks, average balance $1834

   1. _____

2. Gifts Galore, 44 checks, average balance $2398

   2. _____

3. Batista Tile Works, 27 checks, average balance $418

   3. _____

Complete the following three check stubs for the Advertising Specialists. Find the balance forward at the bottom of each stub. **[4.1]**

Checks Written

Number	Date	To	For	Amount
2261	Aug. 6	WBC Broadcasting	Airtime	$6892.12
2262	Aug. 8	Lakeland Weekly	Space buy	$1258.36
2263	Aug. 14	W. Wilson	Freelance art	$416.14

Deposits made: $1572 on Aug. 7, $10,000 on Aug. 10.

4.

2261

_____ 20____
Amount _____
To _____
For _____

Bal. Bro't. For'd.	$16,409	82
Am't. Deposited		
Total		
Am't. this Check		
Balance For'd.		

5.

2262

_____ 20____
Amount _____
To _____
For _____

Bal. Bro't. For'd.		
Am't. Deposited		
Total		
Am't. this Check		
Balance For'd.		

6.

2263

_____ 20____
Amount _____
To _____
For _____

Bal. Bro't. For'd.		
Am't. Deposited		
Total		
Am't. this Check		
Balance For'd.		

Jack Wells owns E-Z Cut Mowers. The shop sells new and used lawn mowers and does repairs as well. The following credit-card transactions occurred during a recent period. **[4.2]**

Sales		Credits
$218.68	$135.82	$45.63
$37.84	$67.45	$36.36
$33.18	$461.82	
$20.76	$116.35	
$12.72	$23.78	
$8.97	$572.18	

7. Find the total amount of sales slips for the store.

7. _____

8. What is the total amount of the credit slips?

8. _____

9. Find the gross amount of the deposit.

9. _____

10. Assuming that the bank charges the retailer a $3\frac{1}{2}\%$ discount charge, find the amount of the discount charge at the statement date.

10. _____

11. Find the amount of the credit given to E-Z Cut Mowers after the fee is subtracted.

11. _____

12. Weddings by Bobbi is a regular customer of Flower Power. Use the information in the following table to reconcile her checking account on the form that follows. **[4.3]**

12. _____

Balance from bank statement		$4721.30
Checks outstanding	3221	$82.74
(check number is given first)	3229	$69.08
	3230	$124.73
	3232	$51.20
Deposits not yet recorded		$758.06
		$32.51
		$298.06
Bank charge		$2.00
Interest credit		$9.58
Checkbook balance		$5474.60
Current balance		_____

Checks Outstanding

Number	Amount
Total	

Compare the list of checks paid by the bank with your records. List and total the checks not yet paid.

(1) Enter new balance from bank statement: _____

(2) List any deposits made by you and not yet recorded by the bank:
+ _____
+ _____
+ _____
+ _____

(3) Add all numbers from lines above. Total: _____

(4) Write total of checks outstanding: = _____

(5) Subtract (4) from (3). This is adjusted bank balance: _____

To reconcile your records:

(6) List your checkbook balance: _____

(7) Write the total of any fees or charges deducted by the bank and not yet subtracted by you from your checkbook: = _____

(8) Subtract line (7) from line (6). _____

(9) Enter interest credit: (Add to your checkbook) + _____

(10) Add line (9) to line (8). Adjusted checkbook balance. _____

New balance of your account; this number should be same as (5).

CHAPTER 5

Payroll

BY THE YEAR 2004, STARBUCKS HAD OPENED STORES IN TURKEY, Chile, and Peru; opened its 1000th Asia Pacific store in Beijing, China; grown to a

total of 7225 store locations; and sold its products around the world.

Jenn Cutini worked part time for Starbucks Coffee while in college and became a full-time employee after graduation. She was recently promoted to manager of her own Starbucks Coffee location. As a store manager with over thirty-five employees, it is her responsibility to prepare the payroll. This chapter provides the essential information on calculating and working with payroll.

Preparing the payroll is one of the most important jobs in any office. Payroll records must be accurate, and the payroll must be prepared on time so that the necessary checks can be written.

5.1 | Gross Earnings: Wages and Salaries

Objectives

1. Understand the methods of calculating gross earnings for salaries and wages.
2. Find overtime earnings for over 40 hours of work per week.
3. Use the overtime premium method of calculating gross earnings.
4. Find overtime earnings for over 8 hours of work per day.
5. Understand double time, shift differentials, and split-shift premiums.
6. Find equivalent earnings for different pay periods.
7. Find overtime for salaried employees.

CASE *in* POINT

Starbucks hires most of its employees as part-time workers and pays them on an hourly basis. Jenn Cutini, the manager, will occasionally ask part-time and full-time employees to work overtime (over 8 hours in one day). When employees work more than 8 hours in one day they are paid time and a half. Having a thorough understanding of payroll helps a manager operate a business more efficiently and results in improved employee relations.

The first step in preparing the payroll is to determine the **gross earnings** (the total amount earned). There are many methods used to find gross earnings, and several of these are discussed in this chapter. A number of **deductions** may be subtracted from gross earnings to arrive at the **net pay**, the amount actually received by the employee. These various deductions also will be discussed in this chapter. Finally, the employer must keep records to maintain an efficient business and to satisfy legal requirements. Many businesses use the services of a company such as Automatic Data Processing (ADP) to professionally prepare their payroll. Other businesses use computer software such as Quick Books to help them complete this task.

Objective 1 **Understand the methods of calculating gross earnings for salaries and wages.** Several methods are used for finding an employee's pay. Two of these methods (salaries and wages) are discussed in this section; two additional methods (piecework and commission) will be discussed in the next section. Many employees live from paycheck to paycheck. Understanding how your total pay is calculated can eliminate the unwanted surprises that come with living from one paycheck to the next. The bar graph on page 173 shows how unprepared many workers are to cope with a period of unemployment.

In many businesses, the first step in preparing the payroll is to look at the **time card** maintained for each employee. The following time card shows the dates of the pay period; the employee's name and other personal information; the days, times, and hours worked; the total number of hours worked; and the signature of the employee as verification of the accuracy of the card. While the card shown is filled in by hand, many companies use a time clock that automatically stamps days, dates, and times on the card. The information on these cards is then transferred to a **payroll ledger** (a chart showing all payroll information) such as the one shown in Example 1.

Suzanne Alley, whose payroll card is shown, is a shift manager at Starbucks and is paid an **hourly wage** of $14.80. Her gross earnings can be calculated with the following formula.

$$\text{Gross earnings} = \text{Number of hours worked} \times \text{Rate per hour}$$

For example, if Alley works 7 hours at $14.80 per hour, her gross earnings would be

$$\text{Gross earnings} = 7 \times \$14.80 = \$103.60$$

Numbers in the News

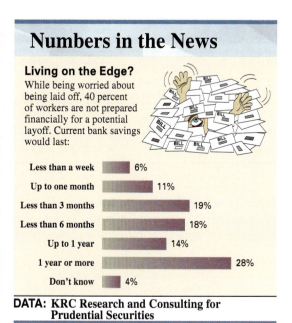

Living on the Edge?

While being worried about being laid off, 40 percent of workers are not prepared financially for a potential layoff. Current bank savings would last:

Less than a week	6%
Up to one month	11%
Less than 3 months	19%
Less than 6 months	18%
Up to 1 year	14%
1 year or more	28%
Don't know	4%

DATA: KRC Research and Consulting for Prudential Securities

EMPL. NO. 1375

PAYROLL CARD
NO TIME CLOCK REQUIRED

CARD NO. _____

FULL NAME Suzanne Alley   **AGE (IF UNDER 18)**

ADDRESS 412 Fawndale Drive   **SOCIAL SECURITY NO.** 123-45-6789

DATE EMPLOYED   **POSITION** Shift MGR   **RATE** $14.80

PAY PERIOD STARTING 7/23   **ENDING** 7/27

DATE	REGULAR TIME					OVER TIME		
	IN	**OUT**	**IN**	**OUT**	**DAILY TOTALS**	**IN**	**OUT**	**DAILY TOTALS**
7/23	8:00	11:50	12:20	4:30	8	4:30	6:30	2
7/24	7:58	12:00	12:30	4:30	8	5:00	7:30	2.5
7/25	8:00	12:00	12:30	4:32	8			
7/26	7:58	12:05	12:35	4:30	8	4:30	5:00	.5
7/27	8:01	12:00	1:00	5:00	8			

APPROVED BY PD   **TOTAL REGULAR TIME** 40   5

REGULAR DAYS WORKED	5 @	8 HRS.	@	**EARNINGS** 14.80	**$** 592.00

ADDITIONAL COMPENSATION: VALUE OF MEALS, LODGING, GIFTS, ETC.   **AMOUNT** **$** _____

COMMISSIONS, FEES, BONUSES, GOODS, ETC. OT 5 @ 22.²⁰   **AMOUNT** **$** 111.00

OTHER REMUNERATIONS (KIND)   **$** _____

DEDUCTIONS: **TOTAL EARNINGS** **$** 703.00

I CERTIFY THE FOREGOING TO BE A CORRECT ACCOUNT OF THE TIME WORKED AND WAGES RECEIVED:

SIGNATURE   **DATE PAID**

EXAMPLE 1

Completing a Payroll Ledger

Jenn Cutini is doing the payroll for two employees, Nelson and Orr. The first thing she must do is complete a payroll ledger.

Employee	Hours Worked							Total Hours	Rate	Gross Earnings
	S	**M**	**T**	**W**	**Th**	**F**	**S**			
Nelson, L.	—	2	4	8	6	3	—		$8.40	
Orr, T.	—	3.5	3	7	6.75	7	—		$7.12	

SOLUTION

First, find the total number of hours worked by each person.

Nelson: $2 + 4 + 8 + 6 + 3 =$ **23 hours**

Orr: $3.5 + 3 + 7 + 6.75 + 7 =$ **27.25 hours**

Multiply the number of hours worked and the rate per hour to find the gross earnings.

Nelson:	Orr:
23	27.25
× $8.40	× $7.12
$193.20	**$194.02**

The payroll ledger can now be completed.

Employee	Hours Worked							Total Hours	Rate	Gross Earnings
	S	M	T	W	Th	F	S			
Nelson, L.	—	2	4	8	6	3	—	23	$8.40	$193.20
Orr, T.	—	3.5	3	7	6.75	7	—	27.25	$7.12	$194.02

Workers' pay varies around the world. The bar graph at the side shows the hourly pay for manufacturing workers in the United States and other industrialized nations. A profile of American workers is shown below. The size and composition of the workforce, earnings, and various job characteristics are also shown.

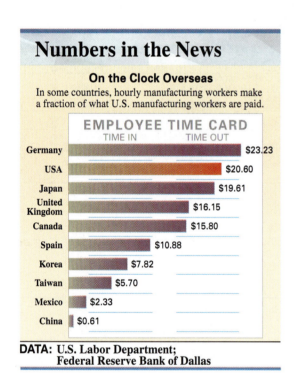

Numbers in the News

On the Clock Overseas

In some countries, hourly manufacturing workers make a fraction of what U.S. manufacturing workers are paid.

Country	Pay
Germany	$23.23
USA	$20.60
Japan	$19.61
United Kingdom	$16.15
Canada	$15.80
Spain	$10.88
Korea	$7.82
Taiwan	$5.70
Mexico	$2.33
China	$0.61

DATA: U.S. Labor Department; Federal Reserve Bank of Dallas

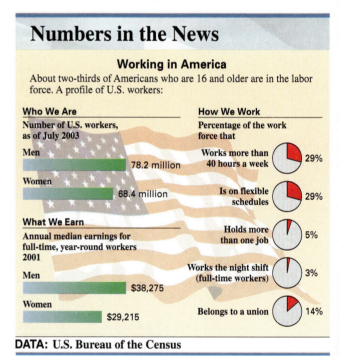

Numbers in the News

Working in America

About two-thirds of Americans who are 16 and older are in the labor force. A profile of U.S. workers:

Who We Are
Number of U.S. workers, as of July 2003

Men — 78.2 million
Women — 68.4 million

What We Earn
Annual median earnings for full-time, year-round workers 2001

Men — $38,275
Women — $29,215

How We Work
Percentage of the work force that

Works more than 40 hours a week — 29%
Is on flexible schedules — 29%
Holds more than one job — 5%
Works the night shift (full-time workers) — 3%
Belongs to a union — 14%

DATA: U.S. Bureau of the Census

Objective **2** **Find overtime earnings for over 40 hours of work per week.** The **Fair Labor Standards Act**, which covers the majority of full-time employees in the United States, establishes a workweek of 40 hours and sets the minimum hourly wage. The law states that an **overtime** wage (a higher-than-normal wage) must be paid for all hours worked over 40 hours per workweek. Also, many companies not covered by the Fair Labor Standards Act have voluntarily followed the practice of paying a **time-and-a-half rate** ($1\frac{1}{2}$ or 1.5 times the normal rate) for any work over 40 hours per week. With the time-and-a-half rate, gross earnings are found with the following formula.

Gross earnings = Earnings at regular rate + Earnings at time-and-a-half rate

EXAMPLE 2

Completing a Payroll Ledger with Overtime

Complete the following payroll ledger.

Employee	Hours Worked S	M	T	W	Th	F	S	Total Hours Reg.	O.T.	Reg. Rate	Gross Earnings Reg.	O.T.	Total
Lanier, D.	6	9	8.25	8	9	4.5	—			$8.30			
Morse, T.	—	10	6.75	9	6.25	10	4.25			$9.48			

SOLUTION

First, find the total number of hours worked.

Lanier: 6 + 9 + 8.25 + 8 + 9 + 4.5 = **44.75 hours**
Morse: 10 + 6.75 + 9 + 6.25 + 10 + 4.25 = **46.25 hours**

Both employees worked more than 40 hours. Gross earnings at the regular rate can now be found as discussed previously. Lanier earned 40 × $8.30 = $332 at the regular rate, and Morse earned 40 × $9.48 = $379.20 at the regular rate. To find overtime earnings, first find the number of overtime hours worked by each employee.

Lanier: 44.75 − **40** = **4.75 overtime hours**
Morse: 46.25 − **40** = **6.25 overtime hours**

The regular rate given for each employee can be used to find the time-and-a-half rate.

Lanier: **1.5** × $8.30 = $12.45
Morse: **1.5** × $9.48 = $14.22

Now find the overtime earnings.

Lanier: 4.75 hours × **$12.45** per hour = **$59.14** (rounded to the nearest cent)
Morse: 6.25 hours × **$14.22** per hour = **$88.88** (rounded)

The ledger can now be completed.

Employee	Hours Worked S	M	T	W	Th	F	S	Total Hours Reg.	O.T.	Reg. Rate	Gross Earnings Reg.	O.T.	Total
Lanier, D.	6	9	8.25	8	9	4.5	—	40	4.75	$8.30	$332	$59.14	$391.14
Morse, T.	—	10	6.75	9	6.25	10	4.25	40	6.25	$9.48	$379.20	$88.88	$468.08

The following newspaper clipping discusses the number of workers who depend on overtime so that they can enjoy middle-class lives.

Making Ends Meet: Overtime to the Rescue

Carmela and Alberto Gonzalez have created their version of the American dream: their own home, a daughter in college with her eye on a teaching career, and another in parochial school. But, theirs is a time-and-a-half dream, funded by the overtime both work whenever they can.

"If I didn't have overtime, it would be terrible," said Carmela, a 37-year-old mother of three. "My daughter's at university. I pay a babysitter, and I pay for school for my other daughter."

Many families are using overtime to finance middle-class lives. Although pilots, nurses, telephone operators, and para-medics are protesting forced overtime, other workers are taking second jobs or volunteering—even competing—for every hour of work they can get.

The preservation of overtime pay was a key issue that propelled Los Angeles area bus and rail operators to walk off their jobs in a strike that ran from Sept. 16 through Oct. 17.

"To maintain a standard of living, people are working longer hours. Most people don't like it but feel forced to do it," said Art Pulaski, executive secretary-treasurer for the California Labor Federation.

Source: *Los Angeles Times.*

Objective **3** **Use the overtime premium method of calculating gross earnings.** Gross earnings with overtime is sometimes calculated with the **overtime premium method**. With this method, which produces the same result as the method used in Example 2, add the total hours at the regular rate to the overtime hours at one-half the regular rate to arrive at gross earnings.

Overtime Premium Method

Straight-time earnings	← total hours worked × regular rate
+ Overtime premium	← overtime hours worked × $\frac{1}{2}$ regular rate
Gross earnings	

EXAMPLE 3

Using the Overtime Premium Method

This week, Holly Kelly worked 40 regular hours and 12 overtime hours. Her regular rate of pay is $17.40 per hour. Find her total gross pay, using the overtime premium method.

SOLUTION

Kelly's total hours are 40 + 12 = 52, and her overtime premium rate is .5 × $17.40 = $8.70.

$$
\begin{array}{rl}
52 \text{ hours} \times \$17.40 = \$904.80 & \text{regular rate earnings} \\
\underline{12 \text{ overtime hours} \times \ \ \$8.70 = \$104.40} & \text{overtime premium} \\
\$1009.20 & \text{gross earnings}
\end{array}
$$

The calculator solution uses the order of operations to first find the regular earnings, the overtime earnings, and finally adds these together.

52 ⊠ 17.4 ⊞ 12 ⊠ 17.4 ⊠ .5 ▣ 1009.2

Note: Refer to Appendix C for calculator basics.

> **QUICK TIP** Many companies prefer the overtime premium method, since it readily identifies the extra cost of overtime labor. Quite often, excessive use of overtime indicates inefficiencies in management.

Objective ④ **Find overtime earnings for over 8 hours of work per day.** Some companies pay the time-and-a-half rate for all time worked over 8 hours in any one day no matter how many hours are worked in a week. This **daily overtime** is shown in the next example.

EXAMPLE 4

Finding Overtime Each Day

Peter Harris worked 10 hours on Monday, 5 hours on Tuesday, 7 hours on Wednesday, and 12 hours on Thursday. His regular rate of pay is $10.10. Find his gross earnings for the week.

	S	M	T	W	Th	F	S	Total Hours
Reg.	—	8	5	7	8	—	—	28
O.T.	—	2	—	—	4	—	—	6

SOLUTION

Harris worked more than 8 hours on Monday and Thursday. On Monday, he had $10 - 8 = 2$ hours of overtime, with $12 - 8 = 4$ hours of overtime on Thursday. In total, he earns $2 + 4 = 6$ hours of overtime. His regular hours are 8 on Monday, 5 on Tuesday, 7 on Wednesday, and 8 on Thursday, for a total of

$$8 + 5 + 7 + 8 = 28$$

hours at the regular rate. His hourly earnings are $10.10. At the regular rate he earns

$$28 \times \textbf{\$10.10} = \$282.80.$$

If the regular rate is $10.10, the time-and-a-half rate is

$$\$10.10 \times \textbf{1.5} = \$15.15$$

He earned time and a half for 6 hours.

$$6 \times \textbf{\$15.15} = \$90.90$$

His gross earnings are

$$\underbrace{\$282.80}_{\substack{\text{total} \\ \text{regular pay}}} + \underbrace{\$90.90}_{\substack{\text{total} \\ \text{overtime}}} = \underbrace{\textbf{\$373.70}}_{\substack{\text{gross} \\ \text{earnings}}}$$

> **QUICK TIP** There are many careers that require unusual work schedules and do not pay overtime for over 40 hours worked in one week or over 8 hours worked in one day. An obvious example is the work schedule of a firefighter, who may work 24 hours and then get 48 hours off.

Objective ⑤ **Understand double time, shift differentials, and split-shift premiums.** In addition to premiums paid for overtime, other **premium payment plans** include **double time** for holidays and, in some industries, Saturdays and Sundays. A **shift differential** is often given to compensate employees for working less-desirable hours. For example, an additional amount per hour or per shift might be paid to swing shift (4 P.M. to midnight) and graveyard shift (midnight to 8:00 A.M.) employees.

Restaurant employees and telephone operators often receive a **split-shift premium**. Hours are staggered so that the employees are on the job during only the busiest times. For example, an employee may work 4 hours, be off 4 hours, and then work another 4 hours. The employee is paid a premium because of this less-desirable schedule.

Some employers offer **compensatory time**, or **comp time**, for overtime hours worked. Instead of additional money, an employee is given time off from the regular work schedule as compensation for overtime hours already worked. Quite often, the compensating time is calculated at $1\frac{1}{2}$ times the overtime hours worked. For example, 12 hours might be given as compensation for 8 hours of previously worked overtime. Occasionally, an employee is given a choice of overtime pay or comp time. Many companies reserve the use of compensatory time for their supervisory or managerial employees. Also, compensatory time is very common in government agencies.

Objective 6 **Find equivalent earnings for different pay periods.** The second common method of finding gross earnings uses a **salary**, a fixed amount given as so much per **pay period** (time between paychecks). Common pay periods are weekly, biweekly, semimonthly, and monthly.

Common Pay Periods

Monthly	12 paychecks each year
Semimonthly	Twice each month; 24 paychecks each year
Biweekly	Every 2 weeks; 26 paychecks each year
Weekly	52 paychecks each year

QUICK TIP   One person's salary might be a certain amount per month, while another person might earn a certain amount every 2 weeks. Many people receive an annual salary, divided among shorter pay periods.

EXAMPLE 5

Determining Equivalent Earnings

You are a career counselor and want to compare the earnings of four clients that you have helped to find jobs. Scott Perrine receives a weekly salary of $546, Tonya McCarley receives a biweekly salary of $1686, Julie Ward receives a semimonthly salary of $736, and Bill Leonard receives a monthly salary of $1818. For each worker, find the following: **(a)** earnings per year, **(b)** earnings per month, and **(c)** earnings per week.

SOLUTION

Scott Perrine:

(a) $546 × **52** = $28,392 per year
(b) $28,392 ÷ **12** = $2366 per month
(c) $546 per week

Tonya McCarley:

(a) $1686 × **26** = $43,836 per year
(b) $43,836 ÷ **12** = $3653 per month
(c) $1686 ÷ **2** = $843 per week

Julie Ward:

(a) $736 × **24** = $17,664 per year
(b) $736 × **2** = $1472 per month
(c) $17,664 ÷ **52** = $339.69 per week

Bill Leonard:

(a) $1818 × **12** = $21,816 per year
(b) $1818 per month
(c) $21,816 ÷ **52** = $419.54 per week

While many careers require a four-year college degree, there are still careers that do not. Notice, however, that many of these careers do require special training or an associates degree.

No Degree? Apply Here

Four years of college may be your best ticket to a high-paying career, but these solid jobs don't require an undergraduate degree:

Profession	Median Annual Earnings
Air traffic controller	$87,930
Nuclear power reactor operator	60,180
Dental hygienist*	54,700
Elevator installer/repairer	51,630
Real estate broker	51,380
Commercial pilot (non-airline)	47,410

Electrical power line installer/repairer	47,210
Locomotive engineer	46,540
Telecom equipment installer/repairer*	46,390
Funeral director*	42,010
Aircraft mechanic*	41,990
Brick mason	41,590
Police officer	40,970
Electrician	40,770
Flight attendant	40,600
Court reporter*	40,410
Real estate appraiser*	38,950

*Requires associate's degree or vocational diploma

Source: U.S. Department of Labor.

Objective 7 **Find overtime for salaried employees.** A salary is ordinarily paid for the performance of a certain job, regardless of the number of hours worked. However, the Fair Labor Standards Act requires that employees in certain salaried positions receive additional compensation for overtime. Like a wage earner, such salaried workers are paid time and a half for all hours worked over 40 hours per week.

EXAMPLE 6

Finding Overtime for Salaried Employees

Monica Wilson is paid $936 a week as an executive assistant. If her normal workweek is 40 hours, find her gross earnings for a week in which she works 45 hours.

SOLUTION

The executive assistant's salary has an hourly equivalent of

$$\frac{\$936}{\textbf{40 hours}} = \$23.40 \text{ per hour}$$

Since she must be paid overtime at the rate of $1\frac{1}{2}$ times her regular pay, she will get $1.5 \times \$23.40$, or $35.10 per hour for overtime. Her gross earnings for the week are

$936.00	salary for 40 hours
+ $175.50	overtime for 5 hours $(5 \times \$35.10)$
$1111.50	weekly gross earnings

The calculator solution to this example is

936 ➕ 936 ➗ 40 ✖ 1.5 ✖ 5 🟰 1111.5

Note: Refer to Appendix C for calculator basics.

Over the years, more and more women have become a part of the work force. The percent of women in the work force for selected countries is shown below. Many of these women are working mothers. The following article cites *Working Mothers* magazine and lists some of the best places for women to work. These best places for working mothers offer many resources not found in most U.S. companies.

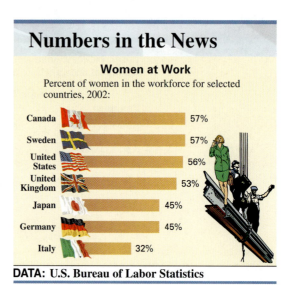

Firms rate for moms on the job

By Loretta Kalb
BEE STAFF WRITER

Northern California's working moms don't need to look far to find the best places in the nation to work, according to a list from *Working Mother* magazine released Monday.

Four California companies and several employers with large operations in the Sacramento area were among the top 100 in the magazine's 18th annual survey, including Intel Corp., Wells Fargo & Co. and USAA.

Over the years, the magazine reported in its October issue, benefits to working mothers have improved as companies strive to improve the women's lives and help them balance homes and careers.

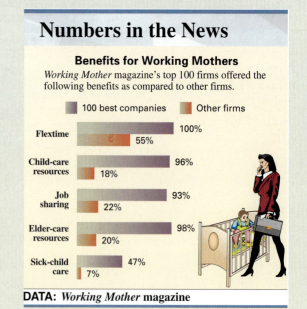

Source: Sacramento Bee.

5.1 | Exercises

FOR EXTRA HELP

 MyMathLab

 InterActMath.com

 MathXP MathXL

 MathXL
Tutorials on CD

 Tutor Center Addison-Wesley
Math Tutor Center

DVT/Videotape

The **Quick Start** *exercises in each section contain solutions to help you get started.*

*Find the number of regular hours and the overtime hours (any hours over 40) for each employee.
Then calculate the overtime rate (time and a half) for each employee. (See Examples 1 and 2.)*

Quick Start

Employee	S	M	T	W	Th	F	S	Reg. Hrs.	O.T. Hours	Reg. Rate	O.T. Rate
1. Burke, E.	—	7	4	7	10	8	4	40	0	$8.10	$12.15
2. Elbern, J.	—	6.5	9	7.5	8	9.5	7			$8.24	
3. Kling, J.	3	5	8.25	9	8.5	5	—			$10.70	
4. Scholz, K.	8.5	9	7.5	8	10	8.25	—			$9.50	
5. Tomlin, M.	—	9.5	7	9	9.25	10.5	—			$11.48	

*Using the information from Exercises 1–5, find the earnings at the regular rate, the earnings at the
overtime rate, and the gross earnings for each employee. Round to the nearest cent. (See Example 2.)*

Quick Start

Employee	Earnings at Reg. Rate	Earnings at O.T. Rate	Gross Earnings
6. Burke, E.	$324	$0	$324
7. Elbern, J.			
8. Kling, J.			
9. Scholz, K.			
10. Tomlin, M.			

*Find the overtime rate, the amount of earnings at regular pay, the amount at overtime pay, and the
total gross wages for each employee. Round to the nearest cent. (See Example 2.)*

Quick Start

	Total Hours				Gross Earnings		
Employee	Reg.	O.T.	Reg. Rate	O.T. Rate	Regular	Overtime	Total
11. Deining, M.	39.5	—	$8.80	$13.20	$347.60	$0	$347.60
12. Demaree, D.	36.25	—	$10.20				
13. Snow, P.	40	4.5	$14.40				
14. Taylor, O.	40	6.75	$12.08				
15. Weyers, C.	40	4.25	$9.18				

C indicates an exercise that is related to the Case in Point feature within the section.

Some companies use the overtime premium method to determine gross earnings. Use this method to complete the following payroll ledger. Overtime is paid at the time-and-a-half rate for all hours over 40. (See Example 3.)

Quick Start	Hours Worked							Total Hours	Reg. Rate	O.T. Hours	O.T. Premium Rate	Gross Earnings		
Employee	S	M	T	W	Th	F	S					Reg.	O.T.	Total
16. Andrews, A.	10	9	8	5	12	7	—	51	$7.40	11	$3.70	$377.40	$40.70	$418.10
17. Brownlee, K.	7.75	10	5	9.75	8	10	—		$9.50					
18. Crowley, A.	—	12	11	8	8.25	11	—		$8.60					
19. Garbin, M.	—	8.5	5.5	10	12	10.5	7		$12.50					
20. Raffaele, M.	—	10	9.75	9	11.5	10	—		$10.20					

Some companies pay overtime for all time worked over 8 hours in a given day. Use this method to complete the following payroll ledger. Overtime is paid at the time-and-a-half rate. (See Example 4.)

Quick Start	Hours Worked							Total Hours		Reg. Rate	O.T. Rate	Gross Earnings		
Employee	S	M	T	W	Th	F	S	Reg.	O.T.			Reg.	O.T.	Total
21. Bailey, M.	—	10	9	11	6	5	—	35	6	$9.40	$14.10	$329.00	$84.60	$413.60
22. Campbell, C.	—	9	8.75	7	8.5	10	—			$7.60				
23. Hales, L.	—	7.5	8	9	10.75	8	—			$10.80				
24. Jenders, P.	—	9	10	8	6	9.75	—			$17.20				
25. Jennings, A.	—	9.5	8.5	7.75	8	9.5	—			$10.20				

26. Explain in your own words what premium payment plans are. Select a premium payment plan and describe it. (See Objective 5.)

27. If you were given a choice of overtime pay or compensatory time, which would you choose? Why? (See Objective 5.)

Find the equivalent earnings for each of the following salaries as indicated. (See Example 5.)

Quick Start

	Earnings			
Weekly	**Biweekly**	**Semimonthly**	**Monthly**	**Annual**
28. $248	$496	$537.33	$1074.67	$12,896
29. $221.54	$443.08	$480	$960	$11,520
30.	$852			
31.			$2410	
32.				$26,100
33. $415				
34.				$27,600

Find the weekly gross earnings for the following people who are on salary and normally work a 40-hour week. Overtime is paid at the time-and-a-half rate. (See Example 6.)

Employee	**Weekly Salary**	**Hours Worked**	**Weekly Gross Earnings**
35. Beckenstein, J.	$360	42	
36. de Bouchel, V.	$468	45	
37. Feist-Milker, R.	$420	43	
38. Johnson, J.	$520	56	
39. Mader, C.	$640	48	

Solve the following application problems.

Quick Start

40. RETAIL EMPLOYMENT   Last week, Frank Nicolazzo worked 48 hours at Starbucks. Find his gross earnings for the week if he is paid $8.40 per hour and earns time and a half for all hours over 40 worked in a week.

40. **$436.80**

48 − 40 = 8 overtime hours
1.5 × $8.40 = $12.60 overtime rate
40 × $8.40 = $336 regular
8 × $12.60 = $100.80
$336 + $100.80 = $436.80

41. INSIDE SALES   Laura Johnson is an inside salesperson and is paid $13.60 per hour for straight time and time and a half for all hours over 40 worked in a week. Find her gross earnings for a week in which she worked 52 hours.

41. _____

42. **DESK CLERK**   Jamie Commissaris, a desk clerk, earns $8.80 per hour and is paid time and a half for all time over 8 hours worked on a given day. Find her gross earnings for a week in which she works the following hours: Monday 9.5, Tuesday 7, Wednesday 10.75, Thursday 4.5, and Friday 8.75 hours.

42. _____

43. **OFFICE ASSISTANT**   Weslie Lewis is an office assistant and worked 10 hours on Monday, 9.75 hours on Tuesday, 5.5 hours on Wednesday, 12 hours on Thursday, and 7.25 hours on Friday. Her regular rate of pay is $11.50 an hour, with time and a half paid for all hours over 8 worked in a given day. Find her gross earnings for the week.

43. _____

44. **INSURANCE OFFICE MANAGER**   Alicia Klein is paid $728 a week as an insurance office manager. Her normal workweek is 40 hours. She gets paid time and a half for overtime. Find her gross earnings for a week in which she works 46 hours.

44. _____

45. **OFFICE EMPLOYEE**   An office employee earns $630 weekly. Find the equivalent earnings if the employee is paid (**a**) biweekly, (**b**) semimonthly, (**c**) monthly, and (**d**) annually.

(**a**) _____
(**b**) _____
(**c**) _____
(**d**) _____

46. **STORE MANAGER**   Michelle Renda manages a Starbucks Coffee shop and is paid $42,900 annually. Find the equivalent earnings if this amount is paid (**a**) weekly, (**b**) biweekly, (**c**) semimonthly, and (**d**) monthly.

(**a**) _____
(**b**) _____
(**c**) _____
(**d**) _____

47. Semimonthly pay periods result in 24 paychecks per year. Biweekly pay periods result in 26 paychecks per year. Which of these pay periods results in three checks in two months of the year? Will it always be the same two months? Explain.

48. Which would you prefer: a monthly pay period or a weekly pay period? What special budgetary considerations might you consider regarding the pay period that you choose?

5.2 Gross Earnings: Piecework and Commissions

Objectives

1. Find the gross earnings for piecework.
2. Determine the gross earnings for differential piecework.
3. Find the gross pay for piecework with a guaranteed hourly wage.
4. Calculate the overtime earnings for piecework.
5. Find the gross earnings using commission rate times sales.
6. Determine a commission using the variable commission rate.
7. Find the gross earnings with a salary plus commission.

Objective 1 **Find the gross earnings for piecework.** The salaries and wages of the previous section are **time rates**, because they depend only on the actual time an employee is on the job. The methods described in this section are **incentive rates**, because they are based on production and pay an employee for actual performance on the job. The 10 help-wanted ads from the classified section of the newspaper are for jobs offering incentive rates of pay. The ad for lathers and stucco construction workers offers piecework and hourly compensation, while the ad for truck drivers lists piece rates of $.32 per mile (cpm). The ads for bill collectors and commercial roofing, health insurance, Better Business Bureau membership, automobile, swimming pool, and home security system sales positions pay on a commission plan.

A **piecework rate** pays an employee so much per item produced. Gross earnings are found with the following formula.

$$\text{Gross earnings} = \text{Pay per item} \times \text{Number of items}$$

For example, a truck driver who drives 680 miles and is paid a piecework rate of $.32 per mile would have total gross earnings as follows.

$$\text{Gross earnings} = \$.32 \times 680 = \$217.60$$

EXAMPLE 1

Finding Gross Earnings for Piecework

Stacy Arrington is paid $.73 for sewing a jacket collar, $.86 for a sleeve with cuffs, and $.94 for a lapel. One week she sewed 318 jacket collars, 112 sleeves with cuffs, and 37 lapels. Find her gross earnings.

SOLUTION

Multiply the rate per item by the number of that type of item.

Item	Rate		Number		Total
Jacket collars	$.73	×	318	=	$232.14
Sleeves with cuffs	$.86	×	112	=	$96.32
Lapels	$.94	×	37	=	$34.78

Find the gross earnings by adding the three totals from the table.

$$\$232.14 + \$96.32 + \$34.78 = \$363.24$$

Objective 2   Determine the gross earnings for differential piecework. There are many variations to the straight piecework rate just described. For example, some rates have **quotas** that must be met, with a premium for each item produced beyond the quota. These plans offer an added incentive within an incentive. A typical **differential piece rate** plan is one where the rate paid per item depends on the number of items produced.

EXAMPLE 2

Using Differential Piecework

Suppose Metro Electric pays assemblers as follows:

1–100 units	$2.10 each
101–150 units	$2.25 each
151 or more units	$2.40 each

Find the gross earnings of a worker producing **214 units**.

SOLUTION

```
   214    ← (total units)
 − 100    ← (first 100 units) ⟶     100 units at $2.10 each = $210.00
 ─────
   114
  − 50    ← (next 50 units)  ⟶       50 units at $2.25 each = $112.50
 ─────
    64    ← (number over 150) →       64 units at $2.40 each = $153.60
                                     214 total units         = $476.10
```

The gross earnings are $476.10.

> **QUICK TIP**   With differential piecework, the highest amount paid applies to only the last units produced. In Example 2, $2.10 is paid for units 1–100, $2.25 is paid for units 101–150, and $2.40 is paid on only those units beyond unit 150, which, in this case, is 64 units.

Objective 3   Find the gross pay for piecework with a guaranteed hourly wage. The piecework and differential piecework rates are frequently modified to include a guaranteed hourly pay rate. This is often necessary to satisfy federal and state laws concerning minimum wages. With this method, the employer must pay either the minimum wage or the piecework earnings, whichever is higher.

EXAMPLE 3

Finding Earnings with a Guaranteed Hourly Wage

A tire installer at the Tire Center is paid $8.40 per hour for an 8-hour day or $.95 per tire installed, whichever is higher. Find the weekly earnings for an employee having the following rate of production.

Monday	85 tires
Tuesday	70 tires
Wednesday	88 tires
Thursday	68 tires
Friday	82 tires

SOLUTION

The hourly earnings for an 8-hour day are **$67.20 (8 × $8.40)**. The larger of hourly earnings or the piecework earnings are paid each day.

Monday	85 × $.95 = $80.75 piece rate
Tuesday	~~70 × $.95~~ = **$67.20 hourly (piece rate is $66.50)**
Wednesday	88 × $.95 = $83.60 piece rate
Thursday	~~68 × $.95~~ = **$67.20 hourly (piece rate is $64.60)**
Friday	82 × $.95 = $77.90 piece rate
	$376.65 weekly earnings

QUICK TIP   The worker *can not* earn less than $8.40 per hour or $67.20 ($8.40 × 8) for the day. Since the piecework earnings on Tuesday and Thursday in Example 3 fall below the hourly minimum, the hourly rate, or $67.20 for the day, is paid on those day.

Objective 4 **Calculate the overtime earnings for piecework.** Piecework employees, like other workers, are paid time and a half for overtime. It is common for the overtime rate to be $1\frac{1}{2}$ times the regular rate per piece.

EXAMPLE 4

Calculating Earnings with Overtime Piecework

Eugene Smith is paid $.98 per child's tricycle assembled. During one week, he assembled 480 tricycles on regular time and 104 tricycles during overtime hours. Find his gross earnings for the week if time and a half per assembly is paid for overtime.

SOLUTION

$$\text{Gross earnings} = \text{Earnings at regular piece rate} + \text{Earnings at overtime piece rate}$$
$$= (480 \times \$.98) + 104 \times (1.5 \times \$.98)$$
$$= \$470.40 + \$152.88$$
$$= \$623.28$$

Have you ever wondered how people select their careers? The following bar graph shows how adults say they chose their specific careers.

Numbers in the News

Getting Started
Those surveyed chose a career path in the following way:

Saw ad, applied	25%
Studied in school	18%
Family friend, others	18%
Luck or chance	9%
Trained for the job	9%
Agency job offer	8%
Other	7%
Promoted from within	6%

DATA: Market Facts/TeleNation

A **commission rate** pays a salesperson either a fixed percent of sales or a fixed amount per item sold. Commissions are designed to produce maximum output from the salesperson, since pay is directly dependent on sales. All of the types of sales commission arrangements are discussed here.

Objective ⑤ **Find the gross earnings using commission rate times sales.** With **straight commission**, the salesperson is paid a fixed percent of sales. Gross earnings are found with the following formula.

> Gross earnings = Commission rate × Amount of sales

EXAMPLE 5

Determining Earnings Using Commission

A real-estate broker charges a 6% commission. Find the commission on a house selling for $168,500.

SOLUTION

The commission would be 6% × $168,500 = .06 × $168,500 = $10,110. The 6% is called the commission rate, or the **rate of commission**.

Before commission is calculated, any **returns** from customers, or any **allowances**, such as discounts, must be subtracted from sales.

EXAMPLE 6

Subtracting Returns when Using Commission

Amanda Roach, a food-supplements sales representative, had sales of $10,230 one month, with returns and allowances of $1120. If her commission rate is 12%, find her gross earnings.

SOLUTION

The returns and allowances must first be subtracted from gross sales. Then, multiply the difference, net sales, by the commission rate.

$$\text{Gross earnings} = \left(\$10{,}230 \text{ gross sales} - \$1120 \text{ returns and allowances}\right) \times \textbf{12\%}$$
$$= \$9110 \text{ net sales} \times .12$$
$$= \$1093.20 \text{ gross earnings}$$

QUICK TIP Before calculating the commission, all items returned are first subtracted from the amount of sales. The company will not pay a commission on sales that are not completed.

Objective ⑥ **Determine a commission using the variable commission rate.** The **sliding scale**, or **variable commission**, is a method of pay designed to retain top-producing salespeople. Under such a plan, a higher rate of commission is paid as sales get larger and larger.

EXAMPLE 7

Finding Earnings Using Variable Commission

Karen De Jamear sells food and bakery products to businesses, such as Starbucks, and is paid as follows.

Sales	Rate
Up to $10,000	6%
$10,001–$20,000	8%
$20,001 and up	9%

Find De Jamear's earnings if she has sales of $32,768 one month.

SOLUTION

$$
\begin{array}{ll}
\$32,768 & \leftarrow (\text{total sales}) \\
-\ 10,000 & \leftarrow (\text{first } \$10,000) \rightarrow \quad \$10,000 \text{ at } 6\% \quad = \$600.00 \\
\hline
\$22,768 & \\
-\ 10,000 & \leftarrow (\text{next } \$10,000) \rightarrow \quad \$10,000 \text{ at } 8\% \quad = \$800.00 \\
\hline
\$12,768 & \leftarrow (\text{over } \$20,000) \rightarrow \quad \$12,768 \text{ at } 9\% \quad = \$1149.12 \\
\end{array}
$$

$$\$32{,}768 \text{ total sales} = \$2549.12 \text{ total commissions}$$

Objective 7 **Find the gross earnings with a salary plus commission.** With a **salary plus commission**, the salesperson is paid a fixed sum per pay period, plus a commission on all sales. This method of payment is commonly used by large retail stores. Gross earnings with salary plus commission are found with the following formula.

> Gross earnings = Fixed amount per pay period + Amount earned on commission

Many salespeople like this method. It is especially attractive to beginning salespeople who lack selling experience. While providing an incentive, it offers the security of a guaranteed income to cover basic living costs. Occasionally, this income is an earnings advance or a **draw**, which is a loan against future commissions. This loan is paid back when future commissions are earned.

EXAMPLE 8

Adding Commission to a Salary

Jaime Bailey is paid $225 per week by Beverly's Creations, plus 3% on all sales over $500. During a certain week, her total sales were $972. Find her gross earnings.

SOLUTION

$$
\begin{aligned}
\text{Gross earnings} &= \text{Weekly salary} + 3\% \text{ on sales above } \$500 \\
&= \$225 + .03\,(\$972 - \$500) \\
&= \$225 + (.03 \times \$472) \\
&= \$225 + \$14.16 \\
&= \$239.16
\end{aligned}
$$

EXAMPLE 9

Subtracting a Draw to Find Earnings

Craig Johnson has office product sales of $36,850 for the month and is paid an 8% commission on all sales. He had draws of $850 for the month. Find his gross earnings after repaying the drawing account.

SOLUTION

$$
\begin{aligned}
\text{Gross earnings} &= \text{Commissions} - \text{Draw} \\
&= (.08 \times \$36{,}850) - \$850 \\
&= \$2948 - \$850 \\
&= \$2098
\end{aligned}
$$

> **QUICK TIP** Many commission-based earning plans are a strong deterrent to attracting new salespeople. It is for this reason that many companies offer the salary plus commission and draw plans to help fill sales positions.

The list at the left shows where the new jobs were in 2005. The bar graph at the right shows the fastest growing jobs over a 10-year period:

List of the Week

Where the new jobs were in 2005:

- Cashiers: 562,000 new jobs

- Janitors and cleaners: 559,000

- Retail salespeople: 532,000

- Waiters, waitresses: 479,000

- Registered nurses: 473,000

from the Labor Department via *Nations Business*

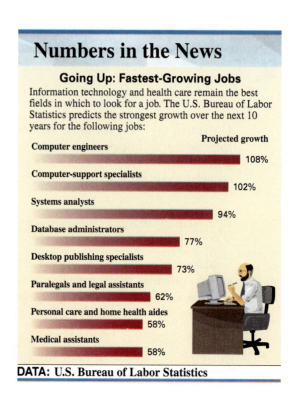

Numbers in the News

Going Up: Fastest-Growing Jobs

Information technology and health care remain the best fields in which to look for a job. The U.S. Bureau of Labor Statistics predicts the strongest growth over the next 10 years for the following jobs:

Projected growth

Computer engineers — 108%

Computer-support specialists — 102%

Systems analysts — 94%

Database administrators — 77%

Desktop publishing specialists — 73%

Paralegals and legal assistants — 62%

Personal care and home health aides — 58%

Medical assistants — 58%

DATA: U.S. Bureau of Labor Statistics

5.2 | Exercises

FOR EXTRA HELP

 MyMathLab

 InterActMath.com

 MathXL

 MathXL
Tutorials on CD

 Addison-Wesley
Math Tutor Center

DVT/Videotape

The   **Quick Start**   *exercises in each section contain solutions to help you get started.*

RECYCLING   Earth Plus pays workers $.48 per container for sorting recyclable materials. Find the gross earnings for each worker. (See Example 1.)

Quick Start

Employee	Number of Containers	Gross Earnings	Employee	Number of Containers	Gross Earnings
1. Crossett, J.	194	**$93.12**	2. Biron, C.	292	_____

1. 194 × $.48 = $93.12

3. Campbell, K.	320	_____	4. Halcomb, J.	243	_____

AGRICULTURAL WORKERS   Find the daily gross earnings for each employee. (See Example 2.)
Suppose that avocado pickers are paid as follows.

1–500 avocados	$.05 each
501–700 avocados	$.07 each
over 700 avocados	$.09 each

Quick Start

Employee	Number of Avocados	Gross Earnings	Employee	Number of Avocados	Gross Earnings
5. Hoch, R.	695	**$38.65**	6. Leonard, M.K.	907	_____

5. $(500 \times \$.05) + (195 \times \$.07) = \$38.65$

7. Matysek, J.	852	_____	8. Panunzio, K.	1108	_____

9. Wages and salaries are known as *time rates*, while commissions are called *incentive rates of pay*. Explain in your own words the difference between these payment methods. (See Objective 1.)

10. Explain in your own words the difference between a piecework rate and differential piece rate. (See Objective 2.)

GUARANTEED HOURLY WORK   Find the gross earnings for each employee by first finding the daily hourly earnings and then finding the daily piecework earnings. Each employee has an 8-hour workday and is paid $.75 for each unit of production or the hourly rate, whichever is greater. (See Example 3.)

Quick Start

Employee	Units Produced M	T	W	Th	F	Hourly Rate	Gross Earnings	
11. Coughlin, S.	66	75	58	72	68	$6.18	**$260.19**	$49.50 + $56.25 + $49.44 (hourly) + $54 + $51 = $260.19
12. Bahary, N.	62	78	79	80	81	$7.20		
13. Shea, D.	80	60	75	78	74	$6.80		
14. Ward, M.	72	70	62	88	82	$6.50		

C indicates an exercise that is related to the Case in Point feature within the section.

PIECEWORK WITH OVERTIME   Find the gross earnings for each employee. Overtime is 1.5 times the normal rate per piece. (See Example 4.)

Quick Start

	Units Produced		Rate per Unit	Gross Earnings
Employee	Reg.	O.T.		
15. Shannon, B.	430	62	$.76	**$397.48**
16. Mc Donald, M.	530	58	$.74	
17. Jennings, A.	470	70	$.82	
18. O'Brien, K.	504	36	$.80	

COMMISSION WITH RETURNS   Find the gross earnings for each of the following salespeople. (See Examples 5 and 6.)

Quick Start

Employee	Total Sales	Returns and Allowances	Rate of Commission	Gross Earnings
19. Mares, E.	$3210	$129	10%	**$308.10**

$$\Big(\$3210 - \$129\Big) \times 10\% = \$3081 \times .1 = \$308.10$$

Employee	Total Sales	Returns and Allowances	Rate of Commission	Gross Earnings
20. Peterson, J.	$5734	$415	5%	_____
21. Remington, J.	$2875	$64	15%	_____
22. Kling, J.	$3806	$108	20%	_____

VARIABLE-COMMISSION PAYMENT   Find the gross earnings for each of the following employees. (See Example 7.) Livingston's Concrete pays its salespeople the following commissions.

> 6% on first $7500 in sales
> 8% on next $7500 in sales
> 10% on any sales over $15,000

Quick Start

Employee	Total Sales	Gross Earnings		Employee	Total Sales	Gross Earnings
23. Christensen, C.	$18,550	**$1405**		**24.** Hubbard, P.	$11,225	_____

$$\Big(\$7500 \times .06\Big) + \Big(\$7500 \times .08\Big) + \Big(\$3550 \times .10\Big) = \$1405$$

Employee	Total Sales	Gross Earnings		Employee	Total Sales	Gross Earnings
25. Steed, F.	$10,480	_____		**26.** Goldstein, S.	$25,860	_____

SALARY PLUS COMMISSION   Stockdale Marine pays salespeople as follows: $452 per week plus a commission of .9% on sales above $15,000 through $25,000 with 1.1% paid on sales in excess of $25,000. Find the gross earnings for each of the following salespeople. (No commission is paid on the first $15,000 of sales.) (See Example 8.)

Quick Start

Employee	Total Sales	Gross Earnings		Employee	Total Sales	Gross Earnings
27. West, S.	$17,900	**$478.10**		**28.** Barnes, A.	$36,300	_____

$$\$452 + \$2900 \times .009 = \$478.10$$

Employee	Total Sales	Gross Earnings		Employee	Total Sales	Gross Earnings
29. Feathers, C.	$32,874	_____		**30.** Maguire, J.	$14,946	_____

5.3 Social Security, Medicare, and Other Taxes

Objectives

1. Understand FICA.
2. Find the maximum FICA tax paid by an employee in one year.
3. Understand Medicare tax.
4. Find FICA tax and Medicare tax.
5. Determine the FICA tax and the Medicare tax paid by a self-employed person.
6. Find state disability insurance deductions.

CASE in POINT

After finding the gross earnings of each employee at Starbucks, all deductions for each employee must be subtracted. Two deductions taken from most employees, no matter what they earn, are FICA (Social Security) and Medicare. As Jenn Cutini does the payroll, she must be certain that she has deducted the correct amounts from each employee's gross earnings.

Finding gross earnings is only the first step in preparing a payroll. The employer must then subtract all required deductions from gross earnings. For most employees, these deductions include Social Security tax, Medicare tax, federal income-tax withholding, and state income-tax withholding. Other deductions may include state disability insurance, union dues, retirement, vacation pay, credit union savings or loan payments, purchase of bonds, uniform expenses, group insurance plans, and charitable contributions. Subtracting these deductions from gross earnings results in **net pay**, the amount the employee receives.

Objective 1 Understand FICA. The **Federal Insurance Contributions Act (FICA)** was passed into law in the 1930s in the middle of the Great Depression. This plan, now called **Social Security**, was originally designed to give monthly benefits to retired workers and their survivors. As the number of people receiving benefits has increased along with the individual benefit amounts, people paying into Social Security have had to pay a larger amount of earnings into this fund each year. From 1937 through 1950 an employee paid 1% of income into Social Security, up to a maximum of $30 per year. This amount has increased over the years until an employee in 2005 paid 6.2% of income to FICA and 1.45% to **Medicare**, which together can total $6885 and more per year.

| Year | Social Security Tax | | Medicare Tax | |
	Social Security Tax Rate	**Employee Earnings Subject to the Tax**	**Medicare Tax Rate**	**Employee Earnings Subject to the Tax**
1994	6.2%	$59,600	1.45%	all
1995	6.2%	$61,200	1.45%	all
1996	6.2%	$62,700	1.45%	all
1997	6.2%	$65,400	1.45%	all
1998	6.2%	$68,400	1.45%	all
1999	6.2%	$72,600	1.45%	all
2000	6.2%	$76,200	1.45%	all
2001	6.2%	$80,400	1.45%	all
2002	6.2%	$84,900	1.45%	all
2003	6.2%	$87,000	1.45%	all
2004	6.2%	$87,900	1.45%	all
2005	6.2%	$90,000	1.45%	all
2006				

For many years both the Social Security tax rate and the Medicare tax rate were combined, however since 1991 these tax rates have been expressed individually. The following table shows the tax rates and the maximum earnings on which Social Security and Medicare taxes are paid by the employee. The employer pays the same rate as the employee *matching dollar for dollar* all employee contributions. Self-employed people pay double the rate paid by those who are employees since they are paying for both employee and employer.

QUICK TIP   Congress sets the tax rates and the maximum employee earnings subject to both Social Security tax and Medicare tax each year. Because these maximum employee earnings change often, **we will use 6.2% of the first $90,000 that the employee earns in a year** for Social Security tax. For Medicare tax, we will use 1.45% of everything that the employee earns in a year. **These figures are used in all examples and exercises in this chapter**.

Each employee, whether a U.S. citizen or not, must have a Social Security card. Most post offices have applications for the cards. All money set aside for an individual is credited to his or her account according to the Social Security number. Each year, the Social Security Administration sends out a Social Security statement that shows workers how Social Security fits into their future. The statements are sent three months before the employee's birthday, but only to workers who are 25 years of age and up. However, anyone may submit a **Request for Earnings and Benefit Estimate Statement** like the one shown below. Since mistakes do occur, it is important to check the statements very carefully. There is a limit of about three years, after which errors may not be corrected. To obtain one of the forms and other information about Social Security, you may phone 800-772-1213 or go on the World Wide Web to www.socialsecurity.gov.

Request for *Social Security Statement*

☐ Please check this box if you want to get your statement in Spanish instead of English.

Please print or type your answers. When you have completed the form, fold it and mail it to us. (If you prefer to send your request using the internet, contact us at http://www.ssa.gov)

1. Name shown on your Social Security card:

_____   _____
First Name   Middle Initial

_____
Last Name Only

2. Your Social Security number as shown on your card:

☐☐☐ - ☐☐ - ☐☐☐☐

3. Your date of birth (Mo.-Day-Yr.)

☐☐ - ☐☐ - ☐☐☐☐

4. Other Social Security numbers you have used:

☐☐☐ - ☐☐ - ☐☐☐☐
☐☐☐ - ☐☐ - ☐☐☐☐

5. Your Sex: ☐ Male   ☐ Female

Form SSA-7004-SM

For items 6 and 8 show only earnings covered by Social Security. Do NOT include wages from state, local, or federal government employment that are NOT covered for Social Security or that are covered ONLY by Medicare.

6. Show your actual earnings (wages and/or net self-employment income) for last year and your estimated earnings for this year.

A. Last year's actual earnings: (*Dollars Only*)

$☐☐☐,☐☐☐.**0 0**

B. This year's estimated earnings: (*Dollars Only*)

$☐☐☐,☐☐☐.**0 0**

7. Show the age at which you plan to stop working.

☐☐ (*Show only one age*)

8. Below, show the average yearly amount (not your total future lifetime earnings) that you think you will earn between now and when you plan to stop working. Include performance or scheduled pay increases or bonuses, but not cost-of-living increases.

If you expect to earn significantly more or less in the future due to promotions, job changes, part-time work, or an absence from the work force, enter the amount that most closely reflects your future average yearly earnings.

If you don't expect any significant changes, show the same amount you are earning now (the amount in 6B).

Future average yearly earnings: (*Dollars Only*)

$☐☐☐,☐☐☐.**0 0**

9. Do you want us to send the statement:
   • To you? Enter your name and mailing address.
   • To someone else (your accountant, pension plan, etc.)? Enter your name with "c/o" and the name and address of that person or organization.

_____
Name

_____

_____
Street Address (Include Apt. No., P.O. Box, or Rural Route)

_____
City   State   Zip Code

NOTICE:
I am asking for information about my own Social Security record or the record of a person I am authorized to represent. I understand that if I deliberately request information under false pretenses, I may be guilty of a federal crime and could be fined and/or imprisoned. I authorize you to use a contractor to send the statement of earnings and benefit estimates to the person named in item 9.

▶ _____
Please sign your name (Do Not Print)

_____   _____
Date   (Area Code) Daytime Telephone No.

Objective 2   Find the maximum FICA tax paid by an employee in one year.   Remember that Social Security tax is paid on only the first $90,000 (see quick tip above) of gross earnings in our examples. An employee earning $90,000 during the first 10 months of a year would pay no more Social Security tax on any additional earnings that year. The maximum Social Security tax to be paid by an employee is $90,000 × 6.2% = $90,000 × .062 = $5580.

QUICK TIP   Only 7% of all income earners are affected by the Social Security maximum.

Objective 3   **Understand Medicare tax.** Medicare tax is paid on all earnings in our examples. The total earnings are multiplied by 1.45%.

Objective 4   **Find FICA tax and Medicare tax.** When finding the amounts to be withheld for Social Security tax and Medicare tax, the employer must use the current rates and the current maximum earnings amount.

EXAMPLE **1**

Finding FICA Tax and Medicare Tax

Imagine that you are Jenn Cutini, the manager of a Starbucks. Find the Social Security tax and Medicare tax that must be withheld from the gross earnings of Kelleher and Kimbrel.

(a) Kelleher: $362.40 gross earnings   **(b)** Kimbrel: $194.02 gross earnings

SOLUTION

(a) The Social Security tax is found by multiplying gross earnings by **6.2%**.

$$\$362.40 \times \textbf{6.2\%} = \$362.40 \times .062 = \$22.47 \;(\text{rounded})$$

Medicare tax is found by multiplying gross earnings by **1.45%**.

$$\$362.40 \times \textbf{1.45\%} = \$362.40 \times .0145 = \$5.25 \;(\text{rounded})$$

(b) The Social Security tax is

$$\$194.02 \times \textbf{6.2\%} = \$194.02 \times .062 = \$12.03 \;(\text{rounded}).$$

The Medicare tax is

$$\$194.02 \times \textbf{1.45\%} = \$194.02 \times .0145 = \$2.81 \;(\text{rounded}).$$

EXAMPLE **2**

Finding FICA Tax

Shannon Woolums has earned $87,634.05 so far this year. Her gross earnings for the current pay period are $5224.03. Find her Social Security tax.

SOLUTION

Social Security tax is paid on only the first $90,000 earned in a year. Woolums has already earned $87,634.05. Subtract $87,634.05 from $90,000, to find that she has to pay Social Security tax on only $2365.95 of her earnings for the current pay period.

$$
\begin{array}{rl}
\$90,000.00 & \text{maximum earnings subject to Social Security tax} \\
- \;\$87,634.05 & \text{earnings to date} \\
\hline
\$2,365.95 & \text{earnings on which tax is due}
\end{array}
$$

The Social Security tax on $2365.95 is $146.69 ($2365.95 × 6.2%) (rounded). Therefore, Woolums pays $146.69 for the current pay period and no additional Social Security tax for the rest of the year.

The table below compares the contributions to social programs (social security and medicare) in selected countries around the world. How do U.S. workers (employee) contributions compare with those of other countries?

Social insurance contributions as a percent of total gross earnings.

Country	Employee	Employer
Italy	10.26	48.00
France	20.66	42.57
Sweden	33.00	7.00
Belgium	14.24	32.73
Mexico	22.00	3.00
Germany	21.12	21.12
Netherlands	7.04	13.05
Japan	11.00	12.00
Ireland	5.52	10.54
U.K.	7.08	10.00
Switzerland	6.55	8.37
Canada	5.00	8.00
U.S.A.	7.65	7.65

Source: Benefits Report Europe, USA, and Canada; Watson Wyatt Worldwide, and Knight Ridder Tribune.

The following bar graph shows where people 25 to 69 years of age think they will get their retirement income.

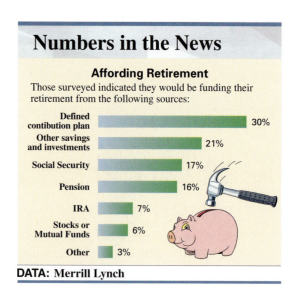

Numbers in the News

Affording Retirement

Those surveyed indicated they would be funding their retirement from the following sources:

Source	Percent
Defined contribution plan	30%
Other savings and investments	21%
Social Security	17%
Pension	16%
IRA	7%
Stocks or Mutual Funds	6%
Other	3%

DATA: Merrill Lynch

Objective **5**   **Determine the FICA tax and the Medicare tax paid by a self-employed person.** People who are self-employed pay higher Social Security tax and higher Medicare tax than people who work for others. There is no employer to match the employee contribution so the self-employed person pays a rate that is double that of an employee. In our examples the self-employed person pays 12.4% of gross earnings for Social Security tax and 2.9% of gross earnings for Medicare tax.

EXAMPLE 3

Finding FICA and Medicare Tax for the Self-Employed

Find the Social Security tax and the Medicare tax paid by Ta Shon Williams, a self-employed Web designer who earned $53,820 this year.

SOLUTION

$$\text{Social Security tax} = \$53,820 \times 12.4\% = \$53,820 \times .124 = \$6673.68$$
$$\text{Medicare tax} = \$53,820 \times 2.9\% = \$53,820 \times .029 = \$1560.78$$

QUICK TIP All employers and those who are self-employed should have the current tax rates for both Social Security and Medicare. These can always be found in **Circular E, Employer's Tax Guide**, which is available from the Internal Revenue Service.

Have you ever wondered why we haven't run out of social security numbers. Marilyn Vos Savant answers this question for a reader in her weekly column in *Parade* magazine. The reader's question and her answers are shown at the left.

Objective 6 Find state disability insurance deductions. Many states have a state disability program. Qualifying employees must pay a portion of their earnings to the program. If an employee is injured and unable to work, the program pays the employee during the period of disability. A typical program requires an **SDI deduction** of 1% of gross earnings on the first $31,800 earned each year, with no payment on earnings above this amount.

EXAMPLE 4

Finding State Disability Insurance Deductions

Find the state disability deduction for an employee at Comet Auto Parts with gross earnings of $418 this pay period. The SDI rate is 1%, and the employee has not earned $31,800 this year.

SOLUTION

The state disability deduction is $4.18 ($418 × **.01**).

EXAMPLE 5

Knowing SDI Maximum Deductions

Jenoa Perkins has earned $29,960 so far this year. Find the SDI deduction if gross earnings this pay period are $2872. Use an SDI rate of 1% on the first $31,800.

SOLUTION

The SDI deduction will be taken on $1840 of the current gross earnings.

$31,800	maximum earnings subject to SDI
− $29,960	earnings this year
$1,840	earnings subject to SDI

The SDI deduction is $18.40 ($1840 × .01).

QUICK TIP Those involved in payroll work must always be up to date on the current rates and the maximum annual earning amounts against which FICA, Medicare, and SDI payroll deductions may be taken.

5.3 | Exercises

FOR EXTRA HELP

 MyMathLab

 InterActMath.com

Math*XP* MathXL

 MathXL Tutorials on CD

Tutor Center Addison-Wesley Math Tutor Center

DVT/Videotape

The **Quick Start** *exercises in each section contain solutions to help you get started.*

Find the Social Security tax and the Medicare tax for each of the following amounts of gross earnings. Assume a 6.2% FICA rate and a 1.45% Medicare tax rate. (See Example 1.)

Quick Start

1. $324.72	**$20.13**	**$4.71**	**2.** $207.25	**$12.85**	**$3.01**	**3.** $463.24	___ ___
4. $606.35	___	___	**5.** $854.71	___	___	**6.** $683.65	___ ___

SOCIAL SECURITY TAX Find the Social Security tax for each employee for the current pay period. Assume a 6.2% FICA rate up to a maximum of $90,000. (See Example 2.)

Quick Start

Employee	Gross Earnings This Year (So Far)	Earnings Current Pay Period	Social Security Tax
7. Dandridge, T.	$86,945.32	$4218.48	**$189.39**
8. Hale, R.	$87,438.75	$3200	**$158.80**
9. Hall, T.	$85,016.22	$5260	___
10. Maurin, J.	$88,971.95	$2487.52	___
11. Saraniti, S.	$89,329.75	$1053.73	___
12. De Bouchel, V.	$88,974.08	$4160.86	___

PAYROLL DEDUCTIONS Find the regular earnings, overtime earnings, gross earnings, Social Security tax (6.2%), Medicare tax (1.45%), and state disability insurance deduction (1%) for each employee. Assume that no employee will have earned more than the FICA or SDI maximum at the end of the current pay period. Assume that time and a half is paid for any overtime in a 40-hour week. (See Examples 1–4.)

Quick Start

Employee	Hours Worked	Regular Rate	Regular Earnings	Overtime Earnings	Gross Earnings	Social Security Tax	Medicare Tax	SDI Deduction
13. Garrett, R.	45.5	$9.22	**$368.80**	**$76.07**	**$444.87**	**$27.58**	**$6.45**	**$4.45**
14. Harcos, W.	47.75	$7.52	___	___	___	___	___	___
15. Plescia, P.	45	$6.58	___	___	___	___	___	___
16. Eckern, G.	45	$5.10	___	___	___	___	___	___
17. McIntosh, R.	47	$11.68	___	___	___	___	___	___
18. Wright, R.	46.75	$6.24	___	___	___	___	___	___

Solve the following application problems. Round to the nearest cent.

Quick Start

 19. SOCIAL SECURITY AND MEDICARE Maria Ortega worked 43.5 hours last week at Starbucks. She is paid $8.58 per hour, plus time and a half for all hours over 40 per week. Find her **(a)** Social Security tax and **(b)** Medicare tax for the week.

(a) $24.07
(b) $5.63

Gross income is $\left(40 \times \$8.58\right) + \left(3.5 \times 1\frac{1}{2}\right) \times \$8.58 = \$388.25$
(a) Social Security tax is .062 × $388.25 = $24.07
(b) Medicare tax is .0145 × $388.25 = $5.63

 indicates an exercise that is related to the Case in Point feature within the section.

20. Chriscelle Merquillo receives 7% commission on all sales. Her sales on Monday of last week were $1412.20, with $1928.42 on Tuesday, $598.14 on Wednesday, $1051.12 on Thursday, and $958.72 on Friday. Find her **(a)** Social Security tax and **(b)** Medicare tax for the week.

(a) _____
(b) _____

21. STATE DISABILITY DEDUCTION   Donna Laughman is paid an 8% commission on sales. During a recent pay period, she had sales of $19,482 and returns and allowances of $193. Find the amount of **(a)** her Social Security tax, **(b)** her Medicare tax, and **(c)** her state disability for this pay period. (The FICA rate is 6.2%, the Medicare rate is 1.45%, the SDI rate is 1%, and the earnings will not exceed $31,800.)

(a) _____
(b) _____
(c) _____

22. Peter Phelps is a representative for Delta International Machinery and is paid $675 per week plus a commission of 2% on sales. His sales last week were $17,240. Find the amount of **(a)** his Social Security tax, **(b)** his Medicare tax, and **(c)** his state disability for the pay period. (The FICA rate is 6.2%, the Medicare rate is 1.45%, the SDI rate is 1%, and earnings will not exceed $31,800.)

(a) _____
(b) _____
(c) _____

SELF-EMPLOYMENT DEDUCTIONS   The following problems refer to self-employed individuals. These people pay a Social Security tax of 12.4% and Medicare tax of 2.9%. Find both of the taxes on the following annual gross incomes. (See Example 3.)

23. Tony Romano, owner of The Cutlery, earned $58,238.74

23. _____

24. Rachel Leach, an interior designer, earned $36,724.72

24. _____

25. Krystal McClellan, cosmetics consultant, earned $29,104.80

25. _____

26. Ron Morris, a Chic-Filet franchise owner, earned $48,007.14

26. _____

27. Peggy Kelleher, shop owner, earned $26,843.60

27. _____

28. Julie Heslin, senior account executive, earned $52,748.32

28. _____

29. A young person who has just received her first paycheck is puzzled by the amounts that have been deducted from gross earnings. Briefly explain both the FICA and Medicare deductions to this person. (See Objectives 1–4.)

30. Describe the difference between the FICA paid by an employee and that paid by a self-employed person. (See Objective 5.)

5.4 | Income Tax Withholding

Objectives

1. Understand the Employee's Withholding Allowance Certificate.
2. Find the federal withholding tax using the wage bracket method.
3. Find the federal withholding tax using the percentage method.
4. Find the state withholding tax using the state income tax rate.
5. Find net pay when given gross wages, taxes, and other deductions.
6. Find the quarterly amount owed to the Internal Revenue Service.
7. Understand additional employer responsibilities and employee benefits.

CASE in POINT

After completing the payroll at Starbucks, Jenn Cutini must be certain that all FICA taxes, Medicare taxes, and federal withholding taxes withheld from employees are sent to the Internal Revenue Service. Additionally, it's essential that Cutini keeps current with all of the changes in the tax codes that affect withholding.

The **personal income tax** is the largest single source of money for the federal government. The law requires that the bulk of the tax owed by an individual be paid as the income is earned. For this reason, employers must deduct money from the gross earnings of almost every employee. These deductions, called **income tax withholdings**, are sent periodically to the Internal Revenue Service. Most recently, the Internal Revenue Service has introduced EFTPS, an electronic funds transfer payment system, which allows employers to transfer these funds electronically. The amount of money withheld from employees depends on several factors.

Marital status. Generally, the withholding tax for a married person is less than the withholding tax for a single person making the same income.

Objective 1 Understand the Employee's Withholding Allowance Certificate. Each employee must file a W-4 form, as shown, with his or her employer. On this form, the employee states the number of **withholding allowances** being claimed along with additional information so that the employer can withhold the proper amount for income tax.

Form **W-4**	**Employee's Withholding Allowance Certificate**	OMB No. 1545-0010
Department of the Treasury Internal Revenue Service	◁ **Your employer must send a copy of this form to the IRS if: (a) you claim more than 10 allowances or (b) you claim "Exempt" and your wages are normally more than $200 per week.**	200

1 Type or print your first name and middle initial	Last name	2 Your social security number

Home address (number and street or rural route)	3 ☐ Single ☐ Married ☐ Married, but withhold at higher Single rate.
	Note: *If married, but legally separated, or spouse is a nonresident alien, check the "Single" box.*
City or town, state, and ZIP code	4 **If your last name differs from that shown on your social security card, check here. You must call 1-800-772-1213 for a new card** ◁ ☐

5	Total number of allowances you are claiming (from line **H** above **or** from the applicable worksheet on page 2)	**5**	
6	Additional amount, if any, you want withheld from each paycheck	**6** $
7	I claim exemption from withholding for 2004, and I certify that I meet **both** of the following conditions for exemption:	

- Last year I had a right to a refund of **all** Federal income tax withheld because I had **no** tax liability **and**
- This year I expect a refund of **all** Federal income tax withheld because I expect to have **no** tax liability.

If you meet both conditions, write "Exempt" here . . . . ◁ **7**

Under penalties of perjury, I certify that I am entitled to the number of withholding allowances claimed on this certificate, or I am entitled to claim exempt status.
Employee's signature
(Form is not valid
unless you sign it.)◁   Date◁

8 Employer's name and address (Employer: Complete lines 8 and 10 only if sending to the IRS.)	9 Office code (optional)	10 Employer identification number (EIN)

For Privacy Act and Paperwork Reduction Act Notice, see page 2.   Cat. No. 10220Q   Form **W-4**

Tables for Wage Bracket Method of Withholding

SINGLE Persons—WEEKLY Payroll Period

(For Wages Paid through December 2004)

If the wages are—		And the number of withholding allowances claimed is—										
At least	But less than	0	1	2	3	4	5	6	7	8	9	10
		The amount of income tax to be withheld is—										
125	130	8	2	0	0	0	0	0	0	0	0	0
130	135	8	2	0	0	0	0	0	0	0	0	0
135	140	9	3	0	0	0	0	0	0	0	0	0
140	145	9	3	0	0	0	0	0	0	0	0	0
145	150	10	4	0	0	0	0	0	0	0	0	0
150	155	10	4	0	0	0	0	0	0	0	0	0
155	160	11	5	0	0	0	0	0	0	0	0	0
160	165	11	5	0	0	0	0	0	0	0	0	0
165	170	12	6	0	0	0	0	0	0	0	0	0
170	175	12	6	0	0	0	0	0	0	0	0	0
175	180	13	7	1	0	0	0	0	0	0	0	0
180	185	13	7	1	0	0	0	0	0	0	0	0
185	190	14	8	2	0	0	0	0	0	0	0	0
190	195	14	8	2	0	0	0	0	0	0	0	0
195	200	15	9	3	0	0	0	0	0	0	0	0
200	210	16	9	3	0	0	0	0	0	0	0	0
210	220	18	10	4	0	0	0	0	0	0	0	0
220	230	19	11	5	0	0	0	0	0	0	0	0
230	240	21	12	6	1	0	0	0	0	0	0	0
240	250	22	13	7	2	0	0	0	0	0	0	0
250	260	24	15	8	3	0	0	0	0	0	0	0
260	270	25	16	9	4	0	0	0	0	0	0	0
270	280	27	18	10	5	0	0	0	0	0	0	0
280	290	28	19	11	6	0	0	0	0	0	0	0
290	300	30	21	12	7	1	0	0	0	0	0	0
300	310	31	22	13	8	2	0	0	0	0	0	0
310	320	33	24	15	9	3	0	0	0	0	0	0
320	330	34	25	16	10	4	0	0	0	0	0	0
330	340	36	27	18	11	5	0	0	0	0	0	0
340	350	37	28	19	12	6	0	0	0	0	0	0

MARRIED Persons—WEEKLY Payroll Period

(For Wages Paid through December 2004)

If the wages are—		And the number of withholding allowances claimed is—										
At least	But less than	0	1	2	3	4	5	6	7	8	9	10
		The amount of income tax to be withheld is—										
440	450	30	23	17	11	5	0	0	0	0	0	0
450	460	31	24	18	12	6	0	0	0	0	0	0
460	470	33	25	19	13	7	1	0	0	0	0	0
470	480	34	26	20	14	8	2	0	0	0	0	0
480	490	36	27	21	15	9	3	0	0	0	0	0
490	500	37	28	22	16	10	4	0	0	0	0	0
500	510	39	30	23	17	11	5	0	0	0	0	0
510	520	40	31	24	18	12	6	0	0	0	0	0
520	530	42	33	25	19	13	7	1	0	0	0	0
530	540	43	34	26	20	14	8	2	0	0	0	0
540	550	45	36	27	21	15	9	3	0	0	0	0
550	560	46	37	29	22	16	10	4	0	0	0	0
560	570	48	39	30	23	17	11	5	0	0	0	0
570	580	49	40	32	24	18	12	6	0	0	0	0
580	590	51	42	33	25	19	13	7	1	0	0	0
590	600	52	43	35	26	20	14	8	2	0	0	0
600	610	54	45	36	27	21	15	9	3	0	0	0
610	620	55	46	38	29	22	16	10	4	0	0	0
620	630	57	48	39	30	23	17	11	5	0	0	0
630	640	58	49	41	32	24	18	12	6	0	0	0
640	650	60	51	42	33	25	19	13	7	1	0	0
650	660	61	52	44	35	26	20	14	8	2	0	0
660	670	63	54	45	36	27	21	15	9	3	0	0
670	680	64	55	47	38	29	22	16	10	4	0	0
680	690	66	57	48	39	30	23	17	11	5	0	0
690	700	67	58	50	41	32	24	18	12	6	0	0
700	710	69	60	51	42	33	25	19	13	7	1	0
710	720	70	61	53	44	35	26	20	14	8	2	0
720	730	72	63	54	45	36	27	21	15	9	3	0
730	740	73	64	56	47	38	29	22	16	10	4	0

Tables for Wage Bracket Method of Withholding, continued
SINGLE Persons—MONTHLY Payroll Period
(For Wages Paid through December 2004)

If the wages are—		And the number of withholding allowances claimed is—										
At least	But less than	0	1	2	3	4	5	6	7	8	9	10
		The amount of income tax to be withheld is—										
840	880	67	38	12	0	0	0	0	0	0	0	0
880	920	73	43	16	0	0	0	0	0	0	0	0
920	960	79	46	20	0	0	0	0	0	0	0	0
960	1,000	85	50	24	0	0	0	0	0	0	0	0
1,000	1,040	91	54	28	2	0	0	0	0	0	0	0
1,040	1,080	97	58	32	6	0	0	0	0	0	0	0
1,080	1,120	103	64	36	10	0	0	0	0	0	0	0
1,120	1,160	109	70	40	14	0	0	0	0	0	0	0
1,160	1,200	115	76	44	18	0	0	0	0	0	0	0
1,200	1,240	121	82	48	22	0	0	0	0	0	0	0
1,240	1,280	127	88	52	26	1	0	0	0	0	0	0
1,280	1,320	133	94	56	30	5	0	0	0	0	0	0
1,320	1,360	139	100	61	34	9	0	0	0	0	0	0
1,360	1,400	145	106	67	38	13	0	0	0	0	0	0
1,400	1,440	151	112	73	42	17	0	0	0	0	0	0
1,440	1,480	157	118	79	46	21	0	0	0	0	0	0
1,480	1,520	163	124	85	50	25	0	0	0	0	0	0
1,520	1,560	169	130	91	54	29	3	0	0	0	0	0
1,560	1,600	175	136	97	58	33	7	0	0	0	0	0
1,600	1,640	181	142	103	64	37	11	0	0	0	0	0
1,640	1,680	187	148	109	70	41	15	0	0	0	0	0
1,680	1,720	193	154	115	76	45	19	0	0	0	0	0
1,720	1,760	199	160	121	82	49	23	0	0	0	0	0
1,760	1,800	205	166	127	88	53	27	1	0	0	0	0
1,800	1,840	211	172	133	94	57	31	5	0	0	0	0
1,840	1,880	217	178	139	100	62	35	9	0	0	0	0
1,880	1,920	223	184	145	106	68	39	13	0	0	0	0
1,920	1,960	229	190	151	112	74	43	17	0	0	0	0
1,960	2,000	235	196	157	118	80	47	21	0	0	0	0
2,000	2,040	241	202	163	124	86	51	25	0	0	0	0
2,040	2,080	247	208	169	130	92	55	29	3	0	0	0

MARRIED Persons—MONTHLY Payroll Period
(For Wages Paid through December 2004)

If the wages are—		And the number of withholding allowances claimed is—										
At least	But less than	0	1	2	3	4	5	6	7	8	9	10
		The amount of income tax to be withheld is—										
2,040	2,080	149	114	88	62	36	10	0	0	0	0	0
2,080	2,120	155	118	92	66	40	14	0	0	0	0	0
2,120	2,160	161	123	96	70	44	18	0	0	0	0	0
2,160	2,200	167	129	100	74	48	22	0	0	0	0	0
2,200	2,240	173	135	104	78	52	26	0	0	0	0	0
2,240	2,280	179	141	108	82	56	30	4	0	0	0	0
2,280	2,320	185	147	112	86	60	34	8	0	0	0	0
2,320	2,360	191	153	116	90	64	38	12	0	0	0	0
2,360	2,400	197	159	120	94	68	42	16	0	0	0	0
2,400	2,440	203	165	126	98	72	46	20	0	0	0	0
2,440	2,480	209	171	132	102	76	50	24	0	0	0	0
2,480	2,520	215	177	138	106	80	54	28	3	0	0	0
2,520	2,560	221	183	144	110	84	58	32	7	0	0	0
2,560	2,600	227	189	150	114	88	62	36	11	0	0	0
2,600	2,640	233	195	156	118	92	66	40	15	0	0	0
2,640	2,680	239	201	162	123	96	70	44	19	0	0	0
2,680	2,720	245	207	168	129	100	74	48	23	0	0	0
2,720	2,760	251	213	174	135	104	78	52	27	1	0	0
2,760	2,800	257	219	180	141	108	82	56	31	5	0	0
2,800	2,840	263	225	186	147	112	86	60	35	9	0	0
2,840	2,880	269	231	192	153	116	90	64	39	13	0	0
2,880	2,920	275	237	198	159	120	94	68	43	17	0	0
2,920	2,960	281	243	204	165	126	98	72	47	21	0	0
2,960	3,000	287	249	210	171	132	102	76	51	25	0	0
3,000	3,040	293	255	216	177	138	106	80	55	29	3	0
3,040	3,080	299	261	222	183	144	110	84	59	33	7	0
3,080	3,120	305	267	228	189	150	114	88	63	37	11	0
3,120	3,160	311	273	234	195	156	118	92	67	41	15	0
3,160	3,200	317	279	240	201	162	124	96	71	45	19	0
3,200	3,240	323	285	246	207	168	130	100	75	49	23	0

A W-4 form is usually completed when a person starts a new job. A married person with three children normally claims five allowances (one each for the employee and spouse and one for each child). However, if both spouses are employed, each may claim himself or herself. The number of allowances may be raised if an employee has been receiving a refund of income taxes or the number may be lowered if the employee has had a balance due in previous tax years. The W-4 form has instructions to help determine the proper number of allowances. Some people enjoy receiving a tax refund when filing their income tax return, so they claim fewer allowances, having more withheld from each check. Other individuals would rather receive more of their income each pay period, so they claim the maximum number of allowances to which they are entitled. The exact number of allowances *must* be claimed when the income tax return is filed.

Amount of gross earnings. The withholding tax is found on the basis of the gross earnings per pay period. Income tax withholding is applied to all earnings—not just earnings up to a certain amount as with Social Security. Generally, the higher a person's gross earnings, the more withholding tax paid.

There are two methods that employers use to determine the amount of federal withholding tax to deduct from paychecks: the **wage bracket method** and the **percentage method**.

Objective 2 Find the federal withholding tax using the wage bracket method. The Internal Revenue Service supplies withholding tax tables to be used with the wage bracket method. These tables are very extensive, covering weekly, biweekly, semimonthly, monthly, and daily pay periods. The preceding pages show samples of the withholding tables. Two of the tables are for people who are paid weekly, both single and married. The other two tables are for both single and married people who are paid monthly.

EXAMPLE 1

Finding Federal Withholding Using the Wage Bracket Method

Kirsten Starr is single and claims no withholding allowances. (Some employees do this to receive a refund from the government or to avoid owing taxes at the end of the year. The proper number of allowances will be used when filing her income tax return.) Use the wage bracket method (tax tables) to find her withholding tax if her weekly gross earnings are $334.88.

SOLUTION

Use the table for single persons—weekly payroll period. The given earnings are found in the row "at least $330 but less than $340." Go across this row to the column headed "0" (for no withholding allowances). From the table, the withholding is $36.

EXAMPLE 2

Using the Wage Bracket Method for Federal Withholding

Pat Rowell is married, claims three withholding allowances, and has monthly gross earnings of $3016.47. Find her withholding tax using the wage bracket method.

SOLUTION

Use the table for married persons—monthly payroll period. Look down the two left columns, and find the range that includes Rowell's gross earnings: "at least $3000 but less than $3040." Read across the table to the column headed "3" (for the three withholding allowances). The withholding tax is $177. Had Rowell claimed six withholding allowances, her withholding tax would have been only $80.

Objective 3 Find the federal withholding tax using the percentage method. Many companies today prefer to use the *percentage method* to determine federal withholding tax. The percentage method does not require the several pages of tables needed with the wage bracket method and is more easily adapted to computer applications in the processing of payrolls.

Percentage Method: Amount for One Withholding Allowance

Payroll Period	One Withholding Allowance
Weekly	$ 59.62
Biweekly	119.23
Semimonthly	129.17
Monthly	258.33
Quarterly	775.00
Semiannually	1550.00
Annually	3100.00
Daily or miscellaneous (each day of the payroll period)	11.92

Tables for Percentage Method of Withholding

TABLE 1—WEEKLY Payroll Period

(a) SINGLE person (including head of household)—

If the amount of wages (after subtracting withholding allowances) is: — The amount of income tax to withhold is:

Not over $51 . . . . . $0

Over—	But not over—		of excess over—
$51	—$187	. 10%	—$51
$187	—$592	. $13.60 plus 15%	—$187
$592	—$1,317	. $74.35 plus 25%	—$592
$1,317	—$2,860	. $255.60 plus 28%	—$1,317
$2,860	—$6,177	. $687.64 plus 33%	—$2,860
$6,177	. . . .	. $1,782.25 plus 35%	—$6,177

(b) MARRIED person—

If the amount of wages (after subtracting withholding allowances) is: — The amount of income tax to withhold is:

Not over $154 . . . . $0

Over—	But not over—		of excess over—
$154	—$429	. 10%	—$154
$429	—$1,245	. $27.50 plus 15%	—$429
$1,245	—$2,270	. $149.90 plus 25%	—$1,245
$2,270	—$3,568	. $406.15 plus 28%	—$2,270
$3,568	—$6,271	. $769.59 plus 33%	—$3,568
$6,271	. . . .	. $1,661.58 plus 35%	—$6,271

TABLE 2—BIWEEKLY Payroll Period

(a) SINGLE person (including head of household)—

If the amount of wages (after subtracting withholding allowances) is: — The amount of income tax to withhold is:

Not over $102 . . . . $0

Over—	But not over—		of excess over—
$102	—$373	. 10%	—$102
$373	—$1,185	. $27.10 plus 15%	—$373
$1,185	—$2,635	. $148.90 plus 25%	—$1,185
$2,635	—$5,719	. $511.40 plus 28%	—$2,635
$5,719	—$12,354	. $1,374.92 plus 33%	—$5,719
$12,354	. . . .	. $3,564.47 plus 35%	—$12,354

(b) MARRIED person—

If the amount of wages (after subtracting withholding allowances) is: — The amount of income tax to withhold is:

Not over $308 . . . . $0

Over—	But not over—		of excess over—
$308	—$858	. 10%	—$308
$858	—$2,490	. $55.00 plus 15%	—$858
$2,490	—$4,540	. $299.80 plus 25%	—$2,490
$4,540	—$7,137	. $812.30 plus 28%	—$4,540
$7,137	—$12,542	. $1,539.46 plus 33%	—$7,137
$12,542	. . . .	. $3,323.11 plus 35%	—$12,542

TABLE 3—SEMIMONTHLY Payroll Period

(a) SINGLE person (including head of household)—

If the amount of wages (after subtracting withholding allowances) is: — The amount of income tax to withhold is:

Not over $110 . . . . $0

Over—	But not over—		of excess over—
$110	—$404	. 10%	—$110
$404	—$1,283	. $29.40 plus 15%	—$404
$1,283	—$2,854	. $161.25 plus 25%	—$1,283
$2,854	—$6,196	. $554.00 plus 28%	—$2,854
$6,196	—$13,383	. $1,489.76 plus 33%	—$6,196
$13,383	. . . .	. $3,861.47 plus 35%	—$13,383

(b) MARRIED person—

If the amount of wages (after subtracting withholding allowances) is: — The amount of income tax to withhold is:

Not over $333 . . . . $0

Over—	But not over—		of excess over—
$333	—$929	. 10%	—$333
$929	—$2,698	. $59.60 plus 15%	—$929
$2,698	—$4,919	. $324.95 plus 25%	—$2,698
$4,919	—$7,731	. $880.20 plus 28%	—$4,919
$7,731	—$13,588	. $1,667.56 plus 33%	—$7,731
$13,588	. . . .	. $3,600.37 plus 35%	—$13,588

TABLE 4—MONTHLY Payroll Period

(a) SINGLE person (including head of household)—

If the amount of wages (after subtracting withholding allowances) is: — The amount of income tax to withhold is:

Not over $221 . . . . $0

Over—	But not over—		of excess over—
$221	—$808	. 10%	—$221
$808	—$2,567	. $58.70 plus 15%	—$808
$2,567	—$5,708	. $322.55 plus 25%	—$2,567
$5,708	—$12,392	. $1,107.80 plus 28%	—$5,708
$12,392	—$26,767	. $2,979.32 plus 33%	—$12,392
$26,767	. . . .	. $7,723.07 plus 35%	—$26,767

(b) MARRIED person—

If the amount of wages (after subtracting withholding allowances) is: — The amount of income tax to withhold is:

Not over $667 . . . . $0

Over—	But not over—		of excess over—
$667	—$1,858	. 10%	—$667
$1,858	—$5,396	. $119.10 plus 15%	—$1,858
$5,396	—$9,838	. $649.80 plus 25%	—$5,396
$9,838	—$15,463	. $1,760.30 plus 28%	—$9,838
$15,463	—$27,175	. $3,335.30 plus 33%	—$15,463
$27,175	. . . .	. $7,200.26 plus 35%	—$27,175

EXAMPLE 3

Finding Federal Withholding Using the Percentage Method

Joseph Fillipi is married, claims four withholding allowances, and has weekly gross earnings of $1575. Use the percentage method to find his withholding tax.

SOLUTION

Step 1 Find the withholding allowance for *one* on the weekly payroll period in the percentage method income tax withholding table. The amount is $59.62. Since Fillipi claims four allowances, multiply the one withholding allowance ($59.62) by his number of withholding allowances (4).

$$\$59.62 \times 4 = \$238.48$$

amount for one withholding allowance ⟶   ⟶ amount for four withholding allowances

Step 2 Subtract the amount in step 1 from gross earnings.

$$\$1575 - \$238.48 = \$1336.52$$

gross earnings ⟶   ⟶ amount used in Table 1(b)

Step 3 Find the "married person weekly" section of the percentage method withholding table. Since $1336.52 is over $1245 but not over $2270, an amount of $149.90 is added to 25% of the excess over $1245 as shown in the highlighted line in Table 1(b).

$$\$1336.52 - \$1245 = \$91.52 \quad \text{excess over } \$1245$$
$$\$91.52 \times 25\% = \$91.52 \times .25 = \$22.88$$
$$\$149.90 + \$22.88 = \$172.78 \quad \text{withholding tax}$$

The calculator solution to this example is

1575 − 4 × 59.62 − 1245 = × .25 + 149.9 = 172.78

Note: Refer to Appendix C for calculator basics.

EXAMPLE 4

Finding Federal Withholding Using the Percentage Method

Sharon MacDonald is married, claims three withholding allowances, and has monthly gross earnings of $6420. Use the percentage method to find her withholding tax.

SOLUTION

Step 1 Find the withholding allowance for *one* on the monthly payroll period in the percentage method income tax withholding table. The amount is $258.33. Since MacDonald claims three withholding allowances, multiply the one withholding allowance ($258.33) by her number of withholding allowances (3).

$$\$258.33 \times 3 = \$774.99$$

amount for one withholding allowance ⟶   ⟶ amount for three withholding allowances

Step 2 Subtract the amount in step 1 from gross earnings.

$$\$6420 - \$774.99 = \$5645.01$$

gross earnings ⟶   ⟶ amount used in Table 4(b)

Step 3 Find the "married person monthly" section of the percentage method withholding table. Since $5645.01 is over $5396, but not over $9838, an amount of $649.80 is added to 25% of the excess over $5396 as shown in the highlighted line in Table 4(b).

$$\$5645.01 - \$5396 = \$249.01 \quad \text{excess over } \$5396$$
$$\$249.01 \times 25\% = \$249.01 \times .25 = \$62.25$$
$$\$649.80 + \$62.25 = \$712.05 \quad \text{withholding tax}$$

EXAMPLE 5

Finding Federal Withholding Using the Percentage Method

Carol Dixon is single, claims two withholding allowances, has weekly gross earnings of $1538. Use the percentage method to find her withholding tax.

SOLUTION

Step 1 Find the withholding allowance for *one* on the weekly payroll period ($59.62) and multiply by the number of withholding allowances (2).

$$\$59.62 \times 2 = \$119.24$$

amount of one withholding allowance ⬏   ⬏ amount of two withholding allowances

Step 2 Subtract the amount in Step 1 from gross earnings.

$$\$1538 - \$119.24 = \$1418.76$$

gross earnings ⬏   ⬏ amount used in Table 1(a)

Step 3 Find the "single person weekly" section of the percentage method withholding table. Since $1418.76 is over $1317, but not over $2860, an amount of $255.60 is added to 28% of the excess over $1317 as shown in the highlighted line in Table 1(a).

$$\$1418.76 - \$1317 = \$101.76 \quad \text{excess over } \$1317$$
$$\$101.76 \times 28\% = \$101.76 \times .28 = \$28.49$$
$$\$255.60 + \$28.49 = \$284.09 \quad \text{withholding tax}$$

> **QUICK TIP**   The amount of withholding tax found using the wage bracket method can vary slightly from the amount of withholding tax found using the percentage method. Any differences would be eliminated when the income tax return is filed.

Objective ☐**4** **Find the state withholding tax using the state income tax rate.** Many states and cities also have an income tax collected by withholding. Income taxes vary from state to state, with no state income tax in the states of Alaska, Florida, Nevada, South Dakota, Texas, Washington, and Wyoming. A few states have a flat tax rate (percent of income) as a **state income tax**, while the majority of the states issue tax tables with taxes going as high as 9% and 10%. A few of the states' income tax rates are shown in the following figure.

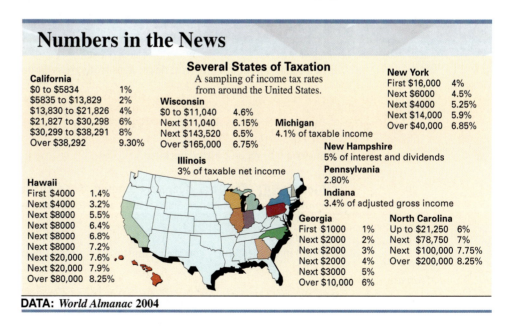

Numbers in the News

Several States of Taxation
A sampling of income tax rates from around the United States.

California
$0 to $5834	1%
$5835 to $13,829	2%
$13,830 to $21,826	4%
$21,827 to $30,298	6%
$30,299 to $38,291	8%
Over $38,292	9.30%

Wisconsin
$0 to $11,040	4.6%
Next $11,040	6.15%
Next $143,520	6.5%
Over $165,000	6.75%

Illinois
3% of taxable net income

Hawaii
First $4000	1.4%
Next $4000	3.2%
Next $8000	5.5%
Next $8000	6.4%
Next $8000	6.8%
Next $8000	7.2%
Next $20,000	7.6%
Next $20,000	7.9%
Over $80,000	8.25%

Michigan
4.1% of taxable income

New Hampshire
5% of interest and dividends

Pennsylvania
2.80%

Indiana
3.4% of adjusted gross income

New York
First $16,000	4%
Next $6000	4.5%
Next $4000	5.25%
Next $14,000	5.9%
Over $40,000	6.85%

Georgia
First $1000	1%
Next $2000	2%
Next $2000	3%
Next $2000	4%
Next $3000	5%
Over $10,000	6%

North Carolina
Up to $21,250	6%
Next $78,750	7%
Next $100,000	7.75%
Over $200,000	8.25%

DATA: *World Almanac 2004*

EXAMPLE 6

Finding State Withholding Tax

Suzanne Holba has gross earnings for the month of $4118. If her state has a 3.4% income tax rate, find the state withholding tax.

SOLUTION

State withholding tax can be found by multiplying 3.4% by the amount of earnings.

$$3.4\% \times \$4118 = .034 \times \$4118 = \$140.01 \ (\text{rounded})$$

The amount of state withholding tax is $140.01.

QUICK TIP   Several computer software programs are available to find state income tax. For example, Quickbooks, Turbo Tax, and Peachtree Accounting are often used.

Objective 5 Find net pay when given gross wages, taxes, and other deductions. It is common for employees to request additional deductions, such as union dues and credit union payments. The final amount of pay received by the employee equals gross earnings – deductions. This final amount is called net pay and is given by the following formula.

$$\text{Net pay} = \text{Gross earnings} - \text{FICA tax} \ (\text{Social Security})$$
$$- \text{ Medicare tax} - \text{Federal withholding tax}$$
$$- \text{ State withholding tax} - \text{Other deductions}$$

EXAMPLE 7

Determining Net Pay after Deductions

Kizzy Johnstone is married and claims three withholding allowances. Her weekly gross earnings are $643.35. Her state withholding is 2.5% and her union dues are $25. Find her net pay using the percentage method of withholding.

SOLUTION

First find FICA (Social Security) tax, which is $39.89, then Medicare, which is $9.33. Federal withholding tax is $32.82 and state withholding is $16.08. Total deductions are

$39.89	FICA tax (6.2%)
9.33	Medicare tax (1.45%)
32.82	federal withholding
16.08	state withholding (2.5%)
+ 25.00	union dues
$123.12	total deductions

Find net pay by subtracting total deductions from gross earnings.

$643.35	gross earnings
− 123.12	total deductions
$520.23	net pay

Johnstone will receive a paycheck for $520.23.

Objective 6 Find the quarterly amount owed to the Internal Revenue Service. An employee's contribution to Social Security and Medicare must be matched by the employer.

EXAMPLE 8

Finding the Amount of FICA and Medicare Tax Due

If the employees at Fair Oaks Automotive Repair pay a total of $789.10 in Social Security tax and $182.10 in Medicare tax, how much must the employer send to the Internal Revenue Service?

SOLUTION

The employer must match this amount and send a total of $971.20 ($789.10 + $182.10 from employees) + $971.20 (from employer) = $1942.40 to the government.

QUICK TIP   In addition to the employee's Social Security tax and a matching amount paid by the employer, the employer must also send the amount withheld for income tax to the Internal Revenue Service on a quarterly basis.

EXAMPLE 9

Finding the Employer's Amount Due the IRS

Suppose that during a certain quarter one Starbucks store has collected $2765.42 from its employees for FICA tax, $646.75 for Medicare tax, and $3572.86 in federal withholding tax. Compute the total amount due to the government from that store.

SOLUTION

$2765.42	collected from *employees* for FICA tax
2765.42	equal amount paid by *employer* for FICA tax
646.75	collected from *employees* for Medicare tax
646.75	equal amount paid by *employer* for Medicare tax
3572.86	federal withholding tax
$10,397.20	total due to government

The store must send $10,397.20 to the Internal Revenue Service.

Objective **7** **Understand additional employer responsibilities and employee benefits.** Each quarter, employers must file **Form 941, the Employer's Quarterly Federal Tax Return**. The form itemizes total employee wages and earnings, the income taxes withheld from employees, the FICA taxes withheld from employees, and the FICA taxes paid by the employer. In addition, Form 941 divides the quarter into its three months and the amounts of the tax liability (employee and employer) are entered on the line of the proper month in that quarter.

The **Federal Unemployment Tax Act (FUTA)** requires employers to pay an additional tax. This **unemployment insurance tax**, paid entirely by employers, is used to pay unemployment benefits to an individual who has become unemployed and is unable to find work. In general, all employers who paid wages of $1500 or more in a calendar quarter, or had one or more employees for some part of a day in 20 different weeks, must file an Employer's Annual Federal Unemployment (FUTA) Tax Return.

The employer must pay 6.2% of the first $7000 in earnings for that year for each employee. Since most states have unemployment taxes, the employer is given credit for these when filing the FUTA return. As soon as the employee reaches earnings of $7000, no additional unemployment tax must be paid.

The **fringe benefits** offered today are very important to employees. These are the extras being offered by the employer that go beyond the paycheck.

The following bar graph shows recent trends in employee fringe benefits (or perks). It includes the most frequently offered "work perks" and the percent of the responding companies offering them. The survey was conducted on-line and included 4800 companies ranging in size from 2 to 5000 employees.

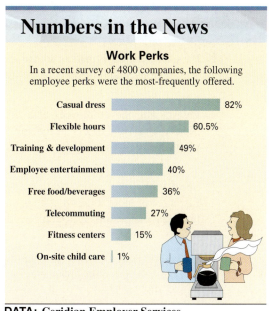

Numbers in the News

Work Perks

In a recent survey of 4800 companies, the following employee perks were the most-frequently offered.

Casual dress	82%
Flexible hours	60.5%
Training & development	49%
Employee entertainment	40%
Free food/beverages	36%
Telecommuting	27%
Fitness centers	15%
On-site child care	1%

DATA: Ceridian Employer Services

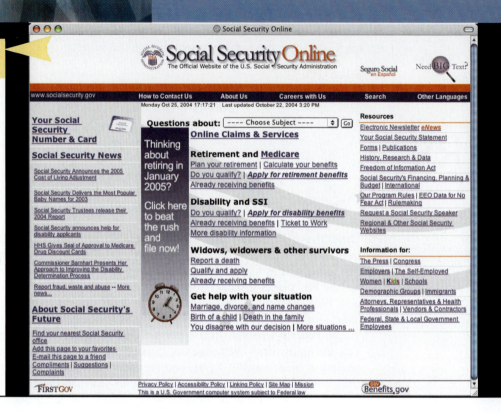

Social Security Administration

- 1935: Social Security Act passed.

- 1937: First Social Security taxes collected.

- 1965: Medicare bill signed into law.

- 2004: 27.8 million retired workers receive benefits.

- 2004: 1 in 7 Americans receive Social Security benefit checks—44 million in all.

In the 1930s, during the Great Depression, the U.S. economy continued to move away from one based primarily on agriculture toward one based on industrial production. The uncertainties associated with disability and old age, which in the past were the responsibility of family and local community, were becoming a much greater concern to many Americans. This concern resulted in the Social Security Act of 1935. Both the employee and emloyer were to contribute to a fund that would then provide benefits to retiring workers.

On January 1, 1940, the first monthly Social Security check was issued to Ida May Fuller of Ludlow, Vermont, in the amount of $22.54. Miss Fuller received the monthly checks for 35 years until her death in 1975 at the age of 100. Today, on the Internet, we can get electronic publications from the Social Security Administration in addition to information on benefits and many more on-line services.

1. If a store manager is paid an annual salary of $31,200, find her equivalent earnings for monthly, semimonthly, and biweekly pay periods.

2. Last month, the employees at a Starbucks outlet paid a total of $792 in FICA, $185 in Medicare tax, and $2217 in federal withholding tax. Find the total amount that the employer must send to the Internal Revenue Service.

3. Speak with a family member, a friend, or someone else who receives Social Security benefits. List three or more things that he or she likes about Social Security. List three or more suggestions that the person has to improve the Social Security system.

4. Do you think that participation in the Social Security system should be voluntary or mandatory? List the possible advantages of each choice.

5.4 | Exercises

FOR EXTRA HELP

 MyMathLab

 InterActMath.com

 MathXL

 MathXL Tutorials on CD

 Addison-Wesley Math Tutor Center

 DVT/Videotape

The **Quick Start** *exercises in each section contain solutions to help you get started.*

FEDERAL WITHHOLDING TAX Find the federal withholding tax for the following employees. Use the wage bracket method. (See Examples 1 and 2.)

Quick Start

Employee	Gross Earnings	Married?	Withholding Allowances	Withholding Tax
1. Stanegna, J.	$3128.51 monthly	yes	3	$195
2. De Simone, D.	$325.76 weekly	no	2	$16
3. Koehler, J.	$507.52 weekly	yes	0	_____
4. Jones, C.	$2416.88 monthly	yes	1	_____
5. Jenders, P.	$1268.29 monthly	no	2	_____
6. Hollabaugh, A.	$1953.35 monthly	no	1	_____
7. Esposito, C.	$2817.32 monthly	yes	6	_____
8. Miller, M.	$341.18 weekly	no	1	_____
9. Mehta, S.	$1622.41 monthly	no	4	_____
10. Sommerfield, K.	$465.92 weekly	yes	0	_____
11. Weisner, W.	$1834.57 monthly	no	3	_____
12. NgauTibbits, C.	$618.43 weekly	yes	2	_____

STATE WITHHOLDING TAX Use the state income tax rate given to find the state withholding tax for the following employees. Round to the nearest cent. (See Example 4.)

Quick Start

Employee	Gross Weekly Earnings	State Income Tax Rate	State Withholding
13. Azzaro, J.	$245.18	2.8%	$6.87
14. Burner, P.	$368.53	3.5%	$12.90
15. Davis, L.	$466.71	6%	_____
16. Lundborg, J.	$541.45	5%	_____
17. Fox, D.	$1607.23	4.3%	_____
18. Ticarro, C.	$2802.58	4.95%	_____

C indicates an exercise that is related to the Case in Point feature within the section.

*EMPLOYEE NET PAY   Use the percentage method of withholding to find federal withholding tax, a
6.2% FICA rate to find FICA tax, and 1.45% to find Medicare tax for the following employees.
Then find the net pay for each employee. The number of withholding allowances and the marital
status are listed after each employee's name. Assume that no employee has earned over $90,000 so
far this year. (See Examples 3 and 5.)*

Quick Start

	Employee	Gross Earnings	FICA	Medicare Tax	Federal Withholding Tax	Net Pay
19.	Hamel; 4, M	$576.28 weekly	$35.73	$8.36	$18.38	$513.81
20.	Guardino; 3, S	$2878.12 monthly	$178.44	$41.73	$252.97	$2404.98
21.	Foster; 1, S	$2512.53 monthly				
22.	Erb; 2, M	$625 weekly				
23.	Terry; 3, M	$2276.83 semimonthly				
24.	Galluccio; 1, S	$420.17 weekly				
25.	Derma; 6, M	$1971.06 semimonthly				
26.	Eddy; 2, M	$1020 weekly				
27.	Wann; 3, S	$710.56 biweekly				
28.	Zamost; 2, S	$6625.24 monthly				
29.	Reilly; 1, S	$1786.44 weekly				
30.	Gertz; 4, M	$2618.50 biweekly				

31. Write an explanation of how to determine the federal withholding tax using the wage bracket (tax tables) method. (See Objective 2.)

32. Write an explanation of how to find the federal withholding tax using the percentage method. (See Objective 3.)

33. In your present or past job, which deductions were subtracted from gross earnings to arrive at net pay? Which was the largest deduction?

34. If you were an employer, would you prefer to use the wage bracket method or the percentage method to determine federal withholding tax? Why? (See Objectives 2 and 3.)

AMOUNT OWED THE IRS *Calculate the total amount owed to the Internal Revenue Service from each of the following firms. (See Examples 6 and 7.)*

Quick Start

Firm	FICA Tax Collected from Employees	Medicare Tax Collected from Employees	Total Federal Withholding Tax	Amount Due IRS
35. Starbucks	$1483.59	$342.37	$5096.13	**$8748.05**
36. Atlasta Ranch	$265.36	$61.24	$4111.68	_____
37. Tony Balony's	$8212.18	$1895.37	$33,117.42	_____
38. Plescia Produce	$212.78	$49.10	$958.68	_____
39. Todd Consultants	$7271.39	$1678.24	$26,423.84	_____
40. Hartmann Shoes	$6538.42	$1508.87	$22,738.57	_____

Use the percentage method of withholding, a FICA rate of 6.2%, a Medicare rate of 1.45%, an SDI rate of 1%, and a state withholding tax of 3.4% in the following problems.

Quick Start

41. MARKETING REPRESENTATIVE   David Horwitz, industrial salesperson, has weekly earnings of $975. He is married and claims four withholding allowances. His deductions include FICA, Medicare, federal withholding, state disability insurance, state withholding, union dues of $15.50, and credit union savings of $100. Find his net pay for a week in February.

Net pay = $975 − $60.45 − $14.14 − $73.63 − $9.75 − $33.15 − $15.50 − $100 = $668.38

41. $668.38 _____

42. Awanata Jackson, a shift manager at Starbucks, has earnings of $742 in one week of March. She is single and claims four withholding allowances. Her deductions include FICA, Medicare, federal withholding, state disability insurance, state withholding, a United Way contribution of $15, and a savings bond of $100. Find her net pay for the week.

42. _____

43. EDUCATIONAL SALES   Karen Jordon, the top salesperson for Weber Scientific, is paid a salary of $410 per week plus 7% of all sales over $5000. She is single and claims two withholding allowances. Her deductions include FICA, Medicare, federal withholding, state disability insurance, state withholding, credit union savings of $50, a Salvation Army contribution of $10, and dues of $15 to the National Association of Professional Saleswomen. Find her net pay for a week in April during which she has sales of $11,284 with returns and allowances of $424.50.

43. _____

44. TRAVEL-AGENCY SALES   Scott Salman, a travel agent, is paid on a variable commission, is married, and claims four withholding allowances. He receives 3% of the first $20,000 in sales, 4% of the next $10,000 in sales, and 6% of all sales over $30,000. This week he has sales of $45,550 and the following deductions: FICA, Medicare, federal withholding, state disability insurance, state withholding, a retirement contribution of $45, a savings bond of $50, and charitable contributions of $20. Find his net pay after subtracting all of his deductions.

44. _____

45. HEATING-COMPANY REPRESENTATIVE   Sheri Minkner, a commission sales representative for Alternative Heating Company, is paid a monthly salary of $4200 plus a bonus of 1.5% on monthly sales. She is married and claims three withholding allowances. Her deductions include FICA, Medicare, federal withholding, state disability insurance, no state withholding, credit union savings of $150, charitable contributions of $25, and a savings bond of $50. Find her net pay for a month in which her sales were $42,618. The state in which Minkner works has no state income tax.

45. _____

46. RIVER-RAFTING MANAGER   River Raft Adventures pays its manager Kathryn Speers a monthly salary of $2880 plus a commission of .8% based on total monthly sales volume. In the month of May, River Raft Adventures has total sales of $86,280. Speers is married and claims five withholding allowances. Her deductions include FICA, Medicare, federal withholding, state disability insurance, state withholding of $159.30, credit union payment of $300, March of Dimes contributions of $20, and savings bonds of $250. Find her net pay for May.

46. _____

Chapter 5 | Quick Review

CHAPTER TERMS   *Review the following terms to test your understanding of the chapter. For each term you do not know, refer to the page number found next to that term.*

allowances [**p. 188**]

commission rate [**p. 188**]

compensatory (comp) time
   [**p. 178**]

daily overtime [**p. 177**]

deductions [**p. 172**]

differential piece rate [**p. 186**]

double time [**p. 177**]

draw [**p. 189**]

Fair Labor Standards Act
   [**p. 174**]

Federal Insurance Contributions
   Act (FICA) [**p. 193**]

Federal Unemployment Tax Act
   (FUTA) [**p. 209**]

Form 941 [**p. 209**]

fringe benefits [**p. 209**]

gross earnings [**p. 172**]

hourly wage [**p. 172**]

incentive rates [**p. 185**]

income tax withholdings
   [**p. 201**]

marital status [**p. 201**]

Medicare [**p. 193**]

net pay [**p. 172**]

overtime [**p. 174**]

overtime premium method
   [**p. 176**]

pay period [**p. 178**]

payroll ledger [**p. 172**]

percentage method [**p. 202**]

personal income tax
   [**p. 201**]

piecework rate [**p. 185**]

premium payment plans
   [**p. 177**]

quotas [**p. 186**]

rate of commission [**p. 188**]

Request for Earnings and
   Benefit Estimate Statement
   [**p. 194**]

returns [**p. 188**]

salary [**p. 178**]

salary plus commission [**p. 189**]

SDI deduction [**p. 197**]

shift differential [**p. 177**]

sliding scale [**p. 188**]

Social Security [**p. 193**]

split-shift premium [**p. 178**]

state income tax [**p. 207**]

straight commission [**p. 188**]

time card [**p. 172**]

time rates [**p. 185**]

time-and-a-half rate [**p. 174**]

unemployment insurance tax
   [**p. 209**]

variable commission [**p. 188**]

wage bracket method [**p. 204**]

withholding allowances [**p. 201**]

CONCEPTS	EXAMPLES
5.1 Gross earnings Gross earnings = Hours worked × Rate per hour	40 hours worked at $8.40 per hour. Gross earnings = **40** × $8.40 = $336
5.1 Gross earnings with overtime First, find the regular earnings. Then, determine overtime pay at overtime rate. Finally, add regular and overtime earnings. Gross earnings = Earnings at regular rate+ Earnings at time-and-half rate	40 regular hours at $8.40 per hour. 10 overtime hours at time and a half. Gross earnings = $$(40 \times \$8.40) + (\mathbf{10 \times 8.40 \times 1.5})$$ $$= \$336 + \$126 = \$462$$
5.1 Common pay periods <table><tr><th>Pay Period</th><th>Paychecks Per Year</th></tr><tr><td>Monthly</td><td>12</td></tr><tr><td>Semimonthly</td><td>24</td></tr><tr><td>Biweekly</td><td>26</td></tr><tr><td>Weekly</td><td>52</td></tr></table>	Find the earnings equivalent of $1400 per month for other pay periods. $$\text{Semimonthly} = \frac{1400}{2} = \$700$$ $$\text{Biweekly} = \frac{1400 \times 12}{26} = \$646.15$$ $$\text{Weekly} = \frac{1400 \times 12}{52} = \$323.08$$
5.1 Overtime for salaried employees First, find the hourly equivalent. Next, multiply the hourly equivalent rate by the overtime hours by 1.5. Finally, add overtime earnings to the salary.	Salary is $648 per week for 40 hours. Find the earnings for 46 hours. $$\$648 \div \mathbf{40} = \$16.20 \text{ per hour}$$ $$\$16.20 \times \mathbf{6 \times 1.5} = \$145.80 \text{ overtime}$$ $$\$648 + \$145.80 = \$793.80$$
5.2 Gross earnings for piecework Gross earnings = Pay per item × Number of items	Items produced, 175; pay per item, $.65; find the gross earnings. $$\mathbf{\$.65} \times 175 = \$113.75$$

CONCEPTS	EXAMPLES
5.2 Gross earnings for differential piecework The rate paid per item produced varies with level of production.	1–100 items $.75 each 101–150 items $.90 each 151 or more items $1.04 each Find the gross earnings for producing 214 items. $100 \times \textbf{\$.75} = \$75.00$ (first 100 units) $50 \times \textbf{\$.90} = \$45.00$ (next 50 units) $64 \times \textbf{\$1.04} = \$66.56$ (number over 150) 214 total items $186.56 total earnings
5.2 Overtime earnings on piecework Gross earnings = Earnings at regular rate + Earnings at overtime rate	Items produced on regular time, 530; items produced on overtime, 110; piece rate $.34; find the gross earnings. Gross earnings = $(530 \times \$.34) + 110(1.5 \times \$.34) = \$180.20 + \$56.10$ $= \$236.30$
5.2 Straight commission Gross earnings = Commission rate × Amount of sales	Sales of $25,800; commission rate is 5%. $.05 \times \$25,800 = \$1290$
5.2 Variable commission Commission rate varies at different sales levels.	Up to $10,000, 6%; $10,001–$20,000, 8%; $20,001 and up, 9% Find the commission on sales of $32,768. $.06 \times \$10,000 = \$600.00$ (first $10,000) $.08 \times \$10,000 = \$800.00$ (next $10,000) $.09 \times \$12,768 = \$1149.12$ (amount over $20,000) $32,768 $2549.12 total commission
5.2 Salary and commission Gross earnings = Fixed earnings + Commission	Salary, $250 per week; commission rate, 3%; find the gross earnings on sales of $6848. Gross earnings = $250 + (.03 \times \$6848) = \$250 + \$205.44$ $= \$455.44$
5.2 Commission with a drawing account Gross earnings = Commission − Draw	Sales for month, $28,560; commission rate, 7%; draw is $750 for month; find the gross earnings. Gross earnings = $(.07 \times \$28,560) - \$750 = \$1999.20 - \$750$ $= \$1249.20$
5.3 FICA; Social Security tax The gross earnings are multiplied by the tax rate. When the maximum earnings are reached, no additional FICA is withheld that year.	Gross earnings, $458; Social Security tax rate, 6.2%; find the Social Security tax. $458 \times .062 = \$28.40$
5.3 Medicare tax The gross earnings are multiplied by the Medicare tax rate. Medicare tax is paid on all earnings.	Gross earnings, $458; Medicare tax rate, 1.45%; find the Medicare tax. $458 \times .0145 = \$6.64$

CONCEPTS	EXAMPLES
5.3 State disability insurance deductions Multiply the gross earnings by the SDI tax rate. When the maximum earnings are reached, no additional taxes are paid in that year.	Gross earnings, \$2880; SDI tax rate, 1%; find SDI tax. $$\$2880 \times .01 = \$28.80$$
5.4 Federal withholding tax Tax is paid on total earnings. No maximum as with FICA.	Single employee with 3 allowances; weekly earnings of \$338; find the federal withholding tax. Using wage bracket amount "at least \$330, but less than \$340," withholding is \$11. Use the percentage method to find the withholding tax. $$\$338 - (\$59.62 \times 3) = \$159.14$$ $$\$159.14 - \$51 = \$108.14$$ $$\$108.14 \times .1 = \$10.81$$
5.4 State withholding tax Tax is paid on total earnings. No maximum as with FICA.	Married employee with weekly earnings of \$392; find the state withholding tax given a state withholding tax rate of 4.5%. $$4.5\% \times \$392 = .045 \times \$392 = \$17.64$$
5.4 Quarterly report, Form 941 Filed each quarter; FICA and federal withholding are sent to the IRS (FICA + Medicare) × 2 (employer matches) + federal withholding tax	If quarterly FICA withheld from employees is \$5269, Medicare tax is \$1581, and federal withholding tax is \$14,780, find the total owed to the IRS by the employer. $$(\$5269 + \$1581) \times 2 + \$14,780$$ $$= \$28,480$$

Chapter 5 | Summary Exercise

Payroll: Finding Your Take-Home Pay

Jenn Cutini, the manager of a Starbucks Coffee shop, receives an annual salary of $42,536, which is paid weekly. Her normal workweek is 40 hours and she is paid time and a half for all overtime. She is single and claims one withholding allowance. Her deductions include FICA, Medicare, federal withholding, state disability insurance, state withholding, credit union payments of $125, retirement deductions of $75, association dues of $12, and a Diabetes Association contribution of $25. Find each of the following for a week in which she works 52 hours.

(a) Regular weekly earnings

(a) _____

(b) Overtime earnings

(b) _____

(c) Total gross earnings

(c) _____

(d) FICA

(d) _____

(e) Medicare

(e) _____

(f) Federal withholding

(f) _____

(g) State disability

(g) _____

(h) State withholding (Assume that the state income tax rate is 4.4%)

(h) _____

(i) Net pay

(i) _____

(IN)VESTIGATE

Look at the statement that you received with your last paycheck. Be certain that your gross earnings are correct. Understand and check all of the deductions made by your employer. Subtract all deductions from your gross earnings to be certain that your net pay is accurate.

Chapter 5 | Test

To help you review, the numbers in brackets show the section in which the topic was discussed.

Complete the following payroll ledger. Find the total gross earnings for each employee. Time and a half is paid on all hours over 40 in one week.   **[5.1]**

Employee	Hours Worked	Reg. Hrs.	O.T. Hrs.	Reg. Rate	Gross Earnings
1. Bianchi	46.5	___	___	$10.80	___
2. Hanna	47.5	___	___	$8.60	___

Solve the following application problems.

3. Judy Martinez is paid $34,060 annually. Find the equivalent earnings if this amount is paid **(a)** weekly, **(b)** biweekly, **(c)** semimonthly, and **(d)** monthly.   **[5.1]**

3. (a) _____
   (b) _____
   (c) _____
   (d) _____

4. At Jalisco Electronics, assemblers are paid according to the following differential piece rate scale: 1–20 units in a week, $4.50 each; 21–30 units, $5.50 each; and $7 each for every unit over 30. Adrian Ortega assembled 28 units in one week. Find his gross pay.   **[5.2]**

4. _____

5. Rheonna Winston receives a commission of 6% for selling a $235,500 house. One-half of the commission goes to the broker and one-half of the remainder to another salesperson. Winston gets the rest. Find the amount she receives.   **[5.2]**

5. _____

An employee is paid a salary of $8050 per month. If the current FICA rate is 6.2% on the first $90,000 of earnings, and the Medicare tax rate is 1.45% of all earnings, how much should be withheld for (a) FICA tax and (b) Medicare tax during the following months?   **[5.3]**

6. March: **(a)** _____   **(b)** _____   **7.** December: **(a)** _____   **b)** _____

Find the federal withholding tax using the wage bracket method for each of the following employees.   **[5.4]**

8. Ahearn: 2 withholding allowances, single, $315.82 weekly earnings.

8. _____

9. Zanotti: 2 withholding allowances, married, $675.25 weekly earnings.

9. _____

10. Allgier: 3 withholding allowances, married, $3210.55 monthly earnings

10. _____

11. Yeoman: 4 withholding allowances, single, $1859.62 monthly earnings

11. _____

12. Benner: 6 withholding allowances, married, $2864.47 monthly earnings

12. _____

Find the net pay for each of the following employees after FICA, Medicare, federal withholding tax, state disability, and other deductions have been made. Assume that none has earned over $90,000 so far this year. Assume a FICA rate of 6.2%, Medicare rate of 1.45%, and a state disability rate of 1%. Use the percentage method of withholding.   **[5.3 and 5.4]**

13. Tran: $1852.75 monthly earnings, 1 withholding allowance, single, $37.80 in other deductions

13. _____

14. Blumka: $522.11 weekly earnings, 4 withholding allowances, married, state withholding of $15.34, credit union savings of $20, educational television contribution of $7.50.

14. _____

15. Comar: $677.92 weekly earnings, 6 withholding allowances, married, state withholding of $22.18, union dues of $14, charitable contribution of $15

15. _____

Solve the following application problems.

16. A salesperson is paid $452 per week plus a commission of 2% on all sales. The salesperson sold $712 worth of goods on Monday, $523 on Tuesday, $1002 on Wednesday, $391 on Thursday, and $609 on Friday. Returns and allowances for the week were $114. Find the employee's **(a)** Social Security tax (6.2%), **(b)** Medicare tax (1.45%), and **(c)** state disability insurance (1%) for the week   **[5.3 and 5.4]**

(a) _____
(b) _____
(c) _____

17. Neta Fitzgerald earned $88,760.28 so far this year. This week she earned $2418.65. Find her **(a)** FICA tax and **(b)** Medicare tax for this week's earnings.   **[5.3]**

(a) _____
(b) _____

For Exercises 18 and 19, find (a) the Social Security tax and (b) the Medicare tax for each of the following self-employed people. Use a FICA tax rate of 12.4% and a Medicare tax rate of 2.9%. **[5.3]**

18. Kirby: $36,714.12

(a) _____
(b) _____

19. Biondi: $42,380.62

(a) _____
(b) _____

20. The employees of Quick-Lube paid a total of $418.12 in Social Security tax last month, $96.48 in Medicare tax, and $1217.34 in federal withholding tax. Find the total amount that the employer must send to the Internal Revenue Service.

20 _____

CHAPTER

6

Mathematics of Buying

ANDREW RYAN OWNS KITCHENS GALORE, A SPECIALTY STORE THAT carries many unique and hard-to-find kitchen, houseware, and gift items.

The store purchases its inventory from a variety of suppliers at a retailer's discounted price (trade discount) and then receives an invoice from the supplier. When the invoice is paid it is common for another discount (cash discount) to be taken. The owner of the store realizes how important these discounts are in contributing to the profitability of his business.

Retail businesses make a profit by purchasing items and then selling them for more than they cost. There are several steps in this process: **manufacturers** buy raw materials and component parts and assemble them into products that can be sold to other manufacturers or **wholesalers**. The **wholesaler**, often called a "middleman," buys from manufacturers or other wholesalers and sells to the retailer. **Retailers** sell directly to the ultimate user, the **consumer**.

Documents called **invoices** help businesses keep track of sales, while various types of discounts help them buy products at lower costs so that they can increase profits. Recent technology has enabled businesses to replace much of their paper-based business processes with electronic solutions, known collectively as **electronic commerce (EC)**. Expect to see further changes in how business is conducted in the future. This chapter covers the mathematics needed for working with invoices and discounts—the mathematics of buying.

6.1 | Invoices and Trade Discounts

Objectives

1. Complete an invoice.
2. Understand common shipping terms.
3. Identify invoice abbreviations.
4. Calculate trade discounts and understand why they are given.
5. Differentiate between single and series discounts.
6. Calculate each series discount separately.
7. Use complements to calculate series discounts.
8. Use a table to find the net cost equivalent of series discounts.

CASE *in* POINT

Andrew Ryan, owner of Kitchens Galore, must have a thorough understanding of invoices, trade discounts, and cash discounts. As the owner of a small, independent store, he must carry first-quality items, buy at the best price, and take all earned discounts.

An invoice is a printed record of a purchase and sale. For the seller, it is a **sales invoice** that records a sale. For the buyer, it is a **purchase invoice** that records a purchase. The invoice identifies the seller and the buyer, describes the items purchased, states the quantity purchased, and provides the unit price of each item. In addition, the invoice shows the number of items purchased times the price per unit, applies any discounts and shipping and insurance charges, and provides the invoice total.

Objective 1 Complete an invoice. The document on the next page, serves as a sales invoice for J. B. Sherr Company and as a purchase invoice for Kitchens Galore. The numbers in the units **shipped** column multiplied by the **unit price** give the **amount**, or **extension total**, for each item. The **invoice total** is the sum of the extension totals.

Trade and cash discounts, discussed later in this chapter, *are never applied to shipping and insurance charges*. For this reason, shipping and insurance charges are often not included in the invoice total, so the purchaser must add them to the invoice total to find the total amount due. In the J. B. Sherr Company invoice, the freight (shipping) charges of $19.45 are included in the INVOICE TOTAL space.

Objective 2 Understand common shipping terms. The shipping term **FOB (free on board)**, followed by the words **shipping point** or **destination**, commonly appears on invoices. The term *FOB shipping point* means that the *buyer* pays for shipping and that ownership of the merchandise passes to the purchaser prior to shipment. The term *FOB destination* means that the *seller* pays the shipping charges and retains ownership until the goods reach the destination. This becomes very important in the event that the merchandise is lost or damaged during shipment.

```
J. B. SHERR Co.                    SHOWROOM AND WAREHOUSE
                                   1704 ROLLINS ROAD
                                   BURLINGAME, CA 94010          INVOICE
                                   TELEPHONE: (650) 697-3430
                                   TO ORDER 1-800-660-1422
```

SOLD TO	KITCHENS GALORE OAK007 10100 FAIR OAKS BLVD. FAIR OAKS CA 95628

PAGE NO. 1 OF

INVOICE DATE	INVOICE NO.
03/17	0002271-IN

SHIP TO	KITCHENS GALORE 10100 FAIR OAKS BLVD. FAIR OAKS CA 95628

TERMS   1% 15 DAYS, NET 30

SALESMAN	ENTRY NUMBER	ENTRY DATE	SHIPPING DATE	SHIPPED VIA	CUSTOMER ORDER NO./DEPT.
0008	0002280	03/17		UPS	3-13

TAG #	QTY. ORD.	SHIPPED	UNIT	STOCK NUMBER	DESCRIPTION	SUGG. RETAIL	UNIT PRICE	AMOUNT
1	12	12	EACH	736-080	IMPT NATURAL SEA SPONGE	.00	2.400	28.80
2	1	0	EA	267-6682	ACRYLIC BUTTER DISH	6.39	3.760	.00
3	1	1	EA	267-6683	ACRYLIC CREAM & SUGAR S	6.39	3.760	3.76
4	1	1	EA	267-6684	ACRYLIC NAPKIN HOLDER	5.29	3.140	3.14
5	6	6	EACH	694-322	IMPT WENOL METAL POLISH	7.79	4.650	27.90
6	6	6	EACH	694-353	IMPTRED BEAR POLISH	7.39	4.370	26.22
7	1	1	EA	274-10012	FRIENDSHIP MIXING BOWL	31.50	18.750	18.75
8	1	0	EACH	274-10014	FRIENDSHIP MIXING BOWL	44.90	26.500	.00
9	2	2	EA	589-31008	FLEXIBLE CHOPPING MATS	3.93	2.360	4.72
10	1	1	EA	589-22153	CORNER SINK SHELF W/SUC	2.79	1.600	1.60
11	6	0	EACH	281-7950	GEMCO JUICER W/GLASS JA	4.59	2.730	.00
12	2	2	EACH	54-611	ARDEN WAFFLE TOWEL–BLUE	3.19	1.900	3.80
13	2	2	EACH	54-612	ARDEN WAFFLE TOWEL–GREE	3.19	1.900	3.80
14	2	2	EACH	60-6	ASHLAND TIRE MAT 18.5 X	18.75	11.250	22.50
15	2	2	EACH	998-713	3 HALF/RD DRAGON 18 X 30	.00	5.200	10.40
16	2	2	EACH	998-143	3 WELCOME MAT 18 X 30	.00	6.910	13.82
17	3	3	EACH	998-303	3 MB PLAIN MAT 18 X 30	.00	8.640	25.92
18	2	2	EACH	998-504	4 HALF/RD MB PLAIN 20 X 3	.00	10.550	21.10
19								
20								

NON-TAX TOTAL	TAXABLE TOTAL	SALES TAX	FREIGHT	MISC.	INVOICE TOTAL
216.23	.00	.00	19.45		235.68

ALL ITEMS NOT SHIPPED ARE CANCELLED. PLEASE REORDER   PLEASE PAY FROM THIS INVOICE

ALL ORDERS SUBJECT TO CREDIT ACCEPTANCE. PRICES SUBJECT TO CHANGE WITHOUT NOTICE. SHORTAGES MUST BE REPORTED WITHIN 10 DAYS. NO RETURNS ACCEPTED WITHOUT PRIOR AUTHORIZATION. PAST DUE ACCOUNTS SUBJECT TO INTEREST (1½% PER MO.) PLUS COLLECTION CHARGES.   CUSTOMER'S COPY

The shipping term **COD** means **cash on delivery**. When goods are sent COD, the shipper delivers to the purchaser on receipt of enough cash to pay for the goods. When goods are moved over water, the shipping term **FAS**, which means **free alongside ship**, is common. Goods shipped this way are delivered to the dock with all freight charges paid to that point by the shipper.

Objective 3 **Identify invoice abbreviations.** A number of abbreviations are used on invoices to identify measurements, quantities of merchandise, shipping terms, and additional discounts. The most common ones are found in the table on this and the following page.

Invoice Abbreviations

ea.	each	drm.	drum
doz.	dozen	cs.	case
gro.	gross (144 items)	bx.	box
gr gro.	great gross (12 gross)	sk.	sack
qt.	quart	pr.	pair
gal.	gallon (4 quarts)	C	Roman numeral for 100
bbl.	barrel	M	Roman numeral for 1000

mL	milliliter	cwt.	per hundredweight
cL	centiliter	cpm.	cost per thousand
L	liter	@	at
in.	inch	lb.	pound
ft.	foot	oz.	ounce
yd.	yard	g	gram
mm	millimeter	kg	kilogram
cm	centimeter	ROG	receipt of goods
m	meter	ex. or x	extra dating
km	kilometer	FOB	free on board
ct.	crate	EOM	end of month
cart	carton	COD	cash on delivery
ctn.	carton	FAS	free alongside ship

Objective 4   Calculate trade discounts and understand why they are given. Trade discounts are often given to businesses or individuals who buy an item for resale or produce an item that will then be sold. The seller usually gives the price of an item as its **list price** (the suggested price at which the item can be sold to the public). Then the seller gives a trade discount that is subtracted from the list price. The result is the **net cost** or **net price** which is the amount paid by the buyer. Find the net cost with the following formula.

$$\text{Net cost} = \text{List price} - \text{Trade discount} \quad \text{or} \quad \begin{array}{r} \text{List price} \\ - \text{Trade discount} \\ \hline \text{Net cost} \end{array}$$

QUICK TIP   The terms *net cost* and *net price* both refer to the amount paid by the buyer. However, net cost is the preferred term since this is the cost of an item to the business.

EXAMPLE 1

Calculating a Single Trade Discount

The list price of a KitchenAid Artisan Stand Mixer is $298.80, and the trade discount is 25%. Find the net cost.

SOLUTION
First find the amount of the trade discount by finding 25% of $298.80.

$$\begin{array}{ccc} R & \times & B & = & P \\ 25\% \times \$298.80 & = & .25 \times \$298.80 & = & \$74.70 \end{array}$$

Subtract $74.70 from the list price of $298.80.

$$\begin{array}{rl} \$298.80 & \text{list price} \\ - \quad 74.70 & \textbf{trade discount} \\ \hline \$224.10 & \text{net cost} \end{array}$$

The net cost of the food processor is $224.10.

Objective 5   Differentiate between single and series discounts. In Example 1, a **single discount** of 25% was offered. Sometimes two or more discounts are combined into a **series** or **chain discount**. A series discount is written, for example, as 20/10, which means that a 20% discount is subtracted from the list price, and *from this difference* another 10% discount is subtracted. Another discount of 15% could be attached to the series discount of 20/10, giving a new series discount of 20/10/15.

Why Trade Discounts Change

Price changes may cause trade discounts to be raised or lowered.

As the *quantity purchased* increases, the discount offered may increase.

The buyer's position in *marketing channels* (manufacturer → wholesaler → retailer → consumer) may determine the amount of discount offered. For example, a wholesaler would receive a larger discount than a succeeding retailer.

Geographic location may influence the trade discount. An additional discount may be offered to increase sales in a particular area.

Seasonal fluctuations in sales may influence the trade discounts offered.

Competition from other companies may cause the raising or lowering of trade discounts

The following advertisement indicates that the retail store received some very large trade discounts so that they are able to offer customers as much as 25%–60% off the list price. These high discounts may have resulted from very large quantities of merchandise purchased, or perhaps it was the end of the season or the last of a production cycle for the manufacturer.

Objective 6 **Calculate each series discount separately.** Three methods can be used to calculate a series discount and net cost. The first of these is by **calculating discounts separately**.

EXAMPLE 2

Calculating Series Trade Discounts

Kitchens Galore is offered a series discount of 20/10 on a Meyer stainless steel cookware set with a list price of $150. Find the net cost after the series discount.

SOLUTION

First, multiply the decimal equivalent of 20% (.2) by $150. Then subtract the product ($30) from $150, getting $120. Then multiply the decimal equivalent of the second discount, 10% (.1), by $120. Finally, subtract the product ($12) from $120, getting $108. The result is the net cost.

$$\begin{array}{rl} \$150 & \text{list price} \qquad \text{Discount: 20/10} \\ -\ \ 30 & (.2 \times \$150) \longleftarrow \\ \hline \$120 & \\ -\ \ 12 & (.1 \times \$120) \longleftarrow \\ \hline \$108 & \text{net cost} \end{array}$$

After the first discount, each discount is applied to the balance remaining after the preceding discount or discounts have been subtracted. This method demonstrates how trade discounts are applied, but this is usually *not* the preferred method for finding the invoice amount.

> **QUICK TIP** **Single discounts in a series are *never* added together**. For example, a series discount of 20/10 is *not the same* as a discount of 30%.

Objective 7 **Use complements to calculate series discounts.** By this second method of finding the net cost, first find the **complement** (with respect to 1, or 100%) of each single discount. The complement is the number that must be added to a given discount to get 1. For example, the complement (with respect to 1) of 10%, or .1, is .9 since .1 + .9 = 1. The complement (with respect to 1) of 40%, or .4, is .6. Other typical complements (with respect to 1) are as follows.

Discount	Complement with Respect to 100%	Decimal Equivalent of Discount	Complement with Respect to 1
10%	90%	.1	.9
15%	85%	.15	.85
20%	80%	.2	.8
25%	75%	.25	.75
30%	70%	.3	.7
35%	65%	.35	.65
50%	50%	.5	.5

The complement of the discount is the portion actually paid. For example, 10% discount means 90% is paid, 25% discount means 75% is paid, and 50% discount means 50% is paid.

Multiply each of the complements of the single discounts to get the **net cost equivalent**. The net cost equivalent is the percent paid. Then multiply the net cost equivalent (percent paid) by the list price to obtain the net cost.

EXAMPLE 3

Using Complements to Find the Net Cost

Kitchens Galore is offered a series discount of 20/10 on a George Foreman Grilling Machine with a list price of $150. Find the net cost after the series discount.

SOLUTION

For a series discount of 20/10, the complements (with respect to 1) of 20% and 10% are .8 and .9. Multiplying the complements gives .8 × .9 = .72, the net cost equivalent. In other words, receiving a series discount of 20/10 is the same as paying 72% of the list price. Find the net cost by multiplying .72 by the list price of $150, to get $108 as the net cost. The calculation is shown as follows.

Step 1 Series discount   20 / 10

Step 2 Find complements with respect to 1   .8 .9

Step 3 Multiply complements   .8 × .9 = .72   net cost equivalent

```
    $150   list price
 ×   .72   net cost equivalent
    300
   1050
 $108.00   net cost
```

Find the amount of the discount by subtracting the net cost from the list price.

```
   $150   list price
 − 108    net cost
   $ 42   amount of discount
```

On many calculators, you can subtract the discount percents from the list price in a series calculation.

150 − 20 % − 10 % = 108

Note: Refer to Appendix C for calculator basics.

EXAMPLE 4

Using Complements to Solve Series Discounts

The list price of a Heartland 30-inch combination gas and electric stove is $3095. Find the net cost after a series discount of 20/10/10.

SOLUTION

Start by finding the complements with respect to 1 of each discount.

Series discount ⟶	20/10/10	
Find complements with respect to 1 ⟶	↓ ↓ ↓ .8 .9 .9	
Multiply complements ⟶	↓ ↓ ↓ .8 × .9 × .9 = **.648**	net cost equivalent

$$
\begin{array}{ll}
\$3095 & \text{list price} \\
\times \quad \textbf{.648} & \textbf{net cost equivalent} \\
\hline
\$2005.56 & \text{net cost}
\end{array}
$$

> **QUICK TIP** **Never round the net cost equivalent**. Doing so will often result in a net cost that is incorrect. If the net cost equivalent in Example 4 had been rounded to .65 the resulting net cost would have been $2011.75 ($.65 \times \$3095$). This error of $6.19 demonstrates the importance of not rounding the net cost equivalent.

Objective **8** **Use a table to find the net cost equivalent of series discounts.** People working with series discounts every day often use a table to find the net cost equivalents for various series discounts. For example, the following table shows that the net cost equivalent for a series discount of 20/10/10 is .648, the number located both to the right of 10/10 and below 20%.

Because changing the order in which numbers are multiplied does not change the answer, the order of the discounts in a series does not change the net cost equivalent. A 10/20 series is the same as a 20/10 series, and a 15/10/20 is identical to a 20/15/10 or to a 15/20/10.

Net Cost Equivalents of Series Discounts

	5%	10%	15%	20%	25%	30%	35%	40%
5	.9025	.855	.8075	.76	.7125	.665	.6175	.57
10	.855	.81	.765	.72	.675	.63	.585	.54
10/5	.81225	.7695	.72675	.684	.64125	.5985	.55575	.513
10/10	.7695	.729	.6885	.648	.6075	.567	.5265	.486
15	.8075	.765	.7225	.68	.6375	.595	.5525	.51
15/10	.72675	.6885	.65025	.612	.57375	.5355	.49725	.459
20	.76	.72	.68	.64	.6	.56	.52	.48
20/15	.646	.612	.578	.544	.51	.476	.442	.408
25	.7125	.675	.6375	.6	.5625	.525	.4875	.45
25/20	.57	.54	.51	.48	.45	.42	.39	.36
25/25	.534375	.50625	.478125	.45	.421875	.39375	.365625	.3375
30	.665	.63	.595	.56	.525	.49	.455	.42
40	.57	.54	.51	.48	.45	.42	.39	.36

> **QUICK TIP** Do not round net cost equivalents. Doing so can cause an error in the net cost.

EXAMPLE 5

Using a Table of Net Cost Equivalents

Using the table of net cost equivalents, find the net cost equivalent of the following series discounts.

(a) 10/10 **(b)** 20/10 **(c)** 25/25/5 **(d)** 35/20/15

SOLUTION

(a) .81 **(b)** .72 **(c)** .534375 **(d)** .442

New and improved technology is helping small businesses keep better records of purchases and discounts, sales, and expenses. Less costly software programs have helped small businesses achieve this goal. The next article shows that greater tech spending by small and medium sized businesses has attracted the attention of the software giants.

Software Giants Think Small

IBM, Microsoft move into niche markets

By Byron Acohido
USA TODAY

SEATTLE – Not long after sales at his Houston-area Krispy Kreme operation began to soar, Jason Gordon discovered that his $200 Intuit accounting software couldn't keep up.

So two years ago, the small-business owner nearly shelled out $100,000 for high-powered accounting software. At the last minute, Gordon learned that Intuit was about to roll out a juiced-up product for $2,500. He jumped on it and hasn't looked back. Intuit, the dominant supplier of bookkeeping software for very small firms, savors the win.

It had better.

That's because small and midsize businesses, or SMBs, loom as the tech industry's key to growth amid anemic spending by big companies. IBM and Microsoft, the two biggest tech companies, are squaring off to dominate the still-fragmented market. In doing so, they're forcing thousands of independent tech resellers, specialty software makers and tech consultants – the middlemen who sell to small firms – to pick sides.

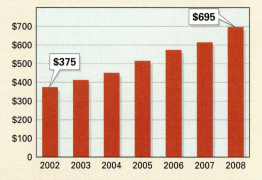

Numbers in the News

Up with Spending

Spending on technology products and services by small and medium-size businesses is projected to almost double over a seven-year period.

DATA: AMI-Partners

The bar graph below shows the number of visitors to on-line shopping sites in a one-week period. In addition to consumers, many businesses buy from Internet shopping sites such as eBay and Yahoo! Shopping.

Numbers in the News

Cyber Shopping

The top five Internet shopping sites, ranked by visitors to sites in millions, were as follows:

Site	Visitors (millions)
eBay	12.6
Amazon	10.2
Yahoo! Shopping	7.4
MSN Shopping	3.1
Wal-Mart Stores	2.8

6.1 Exercises

FOR EXTRA HELP

MyMathLab

InterActMath.com

MathXL

MathXL
Tutorials on CD

Addison-Wesley
Math Tutor Center

DVT/Videotape

The **Quick Start** *exercises in each section contain solutions to help you get started.*

USING INVOICES   Compute the extension totals and the invoice total for the following invoices.

Quick Start

HOME ACCESSORIES WHOLESALERS

Sold to: Kitchens Galore
10100 Fair Oaks Blvd.
Fair Oaks, CA 95628

Date: June 10
Order. No.: 796152
Shipped by: UPS
Terms: Net

	Quantity	Order No./Description	Unit Price	Extension Total
1.	6 doz.	pastry brush, wide	$37.80 doz.	$226.80
2.	3 gro.	napkins, cotton	$12.60 gro.	$37.80
3.	9 doz.	cherry pitters	$14.04 doz.	
4.	8	food processors (3 qt.)	$106.12 ea.	
5.	53 pr.	stainless tongs	$68.12 pr.	
6.			Invoice Total	
			Shipping and Insurance	$37.45
7.			Total Amount Due	

J & K'S MUSTANG PARTS
New and Used

Sold to: Dave's Auto Body & Paint
4443-B Auburn Blvd.
York, PA 17402

Date: July 17
Order. No.: 100603
Shipped by: Emery
Terms: Net

	Quantity	Order No./Description	Unit Price	Extension Total
8.	24	filler tube gaskets	$2.25 ea.	
9.	12 pr.	taillight lens gaskets	$4.75 pr.	
10.	6 pr.	taillight bezels to body	$10.80 pr.	
11.	2 gr.	door panel fasteners	$14.20 gr.	
12.	18	bumper bolt kits	$16.50 ea.	
13.			Invoice Total	
			Shipping and Insurance	$23.75
14.			Total Amount Due	

ABBREVIATIONS ON INVOICES   What do each of the following abbreviations represent?

Quick Start

15. ft. **foot** _____

16. sk. **sack** _____

17. pr. _____

18. gr. gro. _____

19. kg _____

20. qt. _____

 indicates an exercise that is related to the Case in Point feature within the section.

21. cs. _____   **22.** gro. _____

23. drm. _____   **24.** yd. _____

25. L _____   **26.** cpm. _____

27. gal. _____   **28.** cwt. _____

29. COD _____   **30.** FOB _____

31. Name six items that appear on an invoice. Try to do this without looking at an invoice. (See Objective 1.)

32. Explain in your own words the difference between *FOB shipping point* and *FOB destination*. In each case, who pays for shipping? When does ownership of the merchandise transfer? (See Objective 2.)

Using complements (with respect to 1) of the single discounts, find the net cost equivalents for each of the following discounts. Do not round. (See Examples 3 and 4.)

Quick Start

33. 10/20 .9 × .8 = .72   **34.** 20/20 .8 × .8 = .64

35. 10/10/10 .9 × .9 × .9 = .729   **36.** 15/20/25 .85 × .8 × .75 = .51

37. 25/5 _____   **38.** 5/15 _____

39. 40/30/20 _____   **40.** 20/20/10 _____

41. 50/10/20/5 _____   **42.** 25/10/20/10 _____

Find the net cost of each of the following list prices. Round to the nearest cent. (See Examples 1–4.)

Quick Start

43. $418 less 20/20 **$267.52**   **44.** $148 less 25/10 **$99.90**

45. $16.40 less 5/10 _____   **46.** $860 less 20/40 _____

47. $1260 less 15/25/10 _____   **48.** $8.80 less 40/10/20 _____

49. $380 less 20/10/20 _____   **50.** $2008 less 10/5/20 _____

51. $22 less 10/15 _____   **52.** $25 less 30/20 _____

53. $980 less 10/10/10 _____   **54.** $8220 less 30/5/10 _____

55. $2000 less 10/40/10 _____   **56.** $1630 less 10/5/10 _____

57. $1250 less 20/20/20 _____   **58.** $1410 less 10/20/5 _____

59. Identify and explain four reasons that might cause series trade discounts to change. (See Objective 5.)

60. Explain the difference between a single trade discount and a series or chain trade discount.

61. Explain what a complement (with respect to 1 or 100%) is. Give an example. (See Objective 7.)

62. Using complements, explain how to find the net cost equivalent of a 25/20 series discount. Explain why a 25/10/10 series discount is not the same as a 25/20 discount. (See Objective 7.)

Solve the following application problems in trade discount. Round to the nearest cent.

Quick Start

63. PURCHASING A DIGITAL CAMERA   The list price of an Olympus 6 megapixel digital camera is $399.99. If the series discount offered is 10/10/25, what is the net cost after trade discounts?
.9 × .9 × .75 = .6075
$399.99 × .6075 = $242.993 = $242.99

63. $242.99

64. NURSING-CARE PURCHASES   Roger Wheatley, a restorative nursing assistant (RNA), finds that the list price of one dozen adjustable walkers is $1680. Find the cost per walker if a series discount of 40/25 is offered.

64. _____

65. KITCHEN ISLAND   Kitchens Galore purchases a tile-topped wooden kitchen island list priced at $480. It is available at either a 10/15/10 discount or a 20/15 discount. **(a)** Which discount gives the lower price? **(b)** Find the difference in net cost.

(a) _____
(b) _____

66. LIQUID FERTILIZER   Continental Fertilizer Supply offers a series discount of 10/20/20 on major purchases. If a 58,000-gallon tank (bulk) of liquid fertilizer is list priced at $27,200, what is the net cost after trade discounts?

66. _____

67. Kitchens Galore receives a 10/5/20 series trade discount from a supplier. If they purchase three dozen 32-ounce smoothie makers list priced at $468 per dozen, find the net cost.

67. _____

68. WHOLESALE AUTO PARTS   Kimara Swenson, an automotive mechanics instructor, is offered mechanic's net prices on all purchases at Foothill Auto Supply. If mechanic's net prices mean a 10/20 discount, how much will Swenson spend on a dozen sets of metallic brake pads that are list priced at $648 per dozen?

68. _____

69. TRADE-DISCOUNT COMPARISON   The Door Store offers a series trade discount of 30/20 to its builder customers. Robert Gonzalez, a new employee in the billing department, understood the 30/20 terms to mean 50% and computed this trade discount on a list price of $5440. How much difference did this error make in the amount of the invoice?

69. _____

70. FIBER OPTICS   Pam Gondola has a choice of two suppliers of fiber optics for her business. Tyler Suppliers offers a 20/10/25 discount on a list price of $5.70 per unit. Irving Optics offers a 30/20 discount on a list price of $5.40 per unit. **(a)** Which supplier gives her the lower price? **(b)** Find the amount saved if she buys 12,500 units from the lower-priced supplier. (*Hint:* Do not round.)

(a) _____
(b) _____

6.2 Single Discount Equivalents

Objectives

1. Express a series discount as an equivalent single discount.
2. Find the net cost by multiplying the list price by the complements of the single discounts in a series.
3. Find the list price given the series discount and the net cost.

Objective 1 Express a series discount as an equivalent single discount. Series or chain discounts must often be expressed as a single discount rate. Find a **single discount equivalent to a series discount** by multiplying the complements (with respect to 1 or 100%) of the individual discounts. As in the previous section, the result is the net cost equivalent. Then subtract the net cost equivalent from 1 $(1 = 100\%)$. The result is the single discount that is the equivalent to the series discount. *The single discount equivalent is expressed as a percent.*

Finding the Single Discount Equivalent

Single discount equivalent = 1 − Net cost equivalent

EXAMPLE 1

Finding a Single Discount Equivalent

If the Optimum Energy Company offered a 20/10 discount to wholesale accounts on all heating, ventilation, and cooling systems, what would the single discount equivalent be?

SOLUTION

Series discount ⟶ 20/10

Find complements with respect to 1 ⟶ .8   .9

Multiply complements ⟶ .8 × .9 = **.72**   net cost equivalent

$$
\begin{array}{rl}
1.00 & \text{base } (100\%) \\
- \ .72 & \textbf{net cost equivalent (remains)} \\
\hline
.28 & \text{or 28\% was discounted}
\end{array}
$$

The single discount equivalent of a 20/10 series discount is 28%.

Objective 2 Find the net cost by multiplying the list price by the complements of the single discounts in a series. Net cost can be found by multiplying the list price by the complements of each of the single discounts in a series as shown in the next example.

EXAMPLE 2

Finding the Net Cost Using Complements

The list price of an oak dinette set is $495. Find the net cost if trade discounts of 20/15/5 are offered.

SOLUTION

Multiply as follows.

Net cost = List price × Complements of individual discounts
Net cost = $495 × **.8** × **.85** × **.95**

20/15/5

Net cost = $319.77

The net cost of the oak dinette set is $319.77.

The calculator solution to this example is

$$495 \;\boxed{\times}\; .8 \;\boxed{\times}\; .85 \;\boxed{\times}\; .95 \;\boxed{=}\; 319.77$$

Note: Refer to Appendix C for calculator basics.

Objective $\boxed{3}$ **Find the list price given the series discount and the net cost.** Sometimes the net cost after trade discounts is given, along with the series discount, and the list price must be found.

EXAMPLE 3

Solving for List Price

Find the list price of a Kohler kitchen sink that has a net cost of $243.20 after trade discounts of 20/20.

SOLUTION

Use a net cost equivalent. Start by finding the percent paid, using complements.

Series discount $\rightarrow$ 20/20

Complements with respect to 1 $.8 \times .8 = .64$ remains (net cost equivalent)

As the work shows, **.64** or **64%**, of the list price was paid. Find the list price with the standard percent formula.

%	of	something is something
64%	of	list price = $243.20
Rate	×	Base = Part
R	×	*B* = *P*

or

$$B = \frac{P}{R} = \frac{243.2}{.64} = \$380 \text{ list price}$$

Check the answer.

$$
\begin{array}{rl}
\$380 & \text{list price} \\
-\;\;\;\;76 & (.2 \times \$380) \\
\hline
\$304 & \\
-\;\;60.80 & (.2 \times \$304) \\
\hline
\$243.20 & \text{net cost}
\end{array}
$$

The list price of the sink is $380.

EXAMPLE 4

Solving for List Price

Find the list price of a Melita espresso/cappuccino/coffeemaker having a series discount of 10/30/20 and a net cost of $113.40.

SOLUTION

Use complements to find the percent paid.

Series discount $\longrightarrow$ 10 / 30 / 20

$.9 \times .7 \times .8 = .504$ remains (percent paid)

Complements with respect to 1

Therefore, **.504** of the list price is $113.40. Use the formula for base.

$$B = \frac{P}{R} = \frac{\$113.40}{.504} = \$225 \text{ list price}$$

The list price of the espresso/cappuccino coffeemaker is $225. Check this answer as in the previous example.

QUICK TIP   Notice that Examples 3 and 4 are decrease problems similar to those shown in **Chapter 3, Section 5**. They are still base problems but may look different because the discount is now shown as a series of two or more discounts rather than a single percent decrease as in Chapter 3. If you need help, refer to **Section 3.5**.

6.2 | Exercises

FOR EXTRA HELP

MyMathLab

InterActMath.com

*Math*XP MathXL

MathXL
Tutorials on CD

Tutor
Center Addison-Wesley
Math Tutor Center

DVT/Videotape

The **Quick Start** *exercises in each section contain solutions to help you get started.*

Find the net cost equivalent and the single discount equivalent of each of the following series discounts. Do not round net cost equivalents or single discount equivalents. (See Example 1.)

Quick Start

Series Discount	Net Cost Equivalent	Single Discount Equivalent
1. 10/20	.72	28%
	$.9 \times .8 = .72$; $1.00 - .72 = 28\%$	
2. 10/10	.81	19%
	$.9 \times .9 = .81$; $1.00 - .81 = 19\%$	
3. 20/15	_____	_____
4. 25/25	_____	_____
5. 10/30/20	_____	_____
6. 5/10/15	_____	_____
7. 20/10/10/20	_____	_____
8. 25/10/5/20	_____	_____

9. Using complements, show that the single discount equivalent of a 25/20/10 series discount is 46%. (See Objective 1.)

10. Suppose that you own a business and are offered a choice of a 10/20 trade discount or a 20/10 trade discount. Which do you prefer? Why? (See Objective 1.)

Find the list price, given the net cost and the series discount. (See Examples 3 and 4.)

Quick Start

11. Net cost $518.40; trade discount 20/10
$.8 \times .9 = .72$; $\$518.40 \div .72 = \$720$

11. $720 _____

12. Net cost $813.75; trade discount 30/25

12. _____

indicates an exercise that is related to the Case in Point feature within the section.

13. Net cost $1559.52; trade discount 5/10/20

13. _____

14. Net cost $2697.30; trade discount 10/10/10

14. _____

Solve the following application problems in trade discount. Round to the nearest cent.

Quick Start

15. COMPARING DISCOUNTS   A S'mores Maker Kit has a list price of $39.95 and is offered to wholesalers with a series discount of 20/10/10. The same appliance is offered to Kitchens Galore (a retailer) with a series discount of 20/10. **(a)** Find the wholesaler's price. **(b)** Find Kitchens Galore's price. **(c)** Find the difference between the two prices.

(a) $25.89 wholesale
(b) $28.76 retailer's price
(c) $2.87

(a) .8 × .9 × .9 = .648; $39.95 × .648 = $25.887 = $25.89 wholesale
(b) .8 × .9 = .72; $39.95 × .72 = $28.764 = $28.76 retailer's price
(c) $28.76 − $25.89 = $2.87

16. VIDEO CONFERENCING DEVICE   A Via Video conferencing device is list priced at $395. The manufacturer offers a series discount of 25/20/10 to wholesalers and a 25/20 series discount to retailers. **(a)** What is the wholesaler's price? **(b)** What is the retailer's price? **(c)** What is the difference between the prices?

(a) _____
(b) _____
(c) _____

17. COMPARING DISCOUNTS   Express Video offers a series discount of 20/20/20 while States Video offers a series discount of 40/10/5. **(a)** Which discount is higher? **(b)** Find the difference.

(a) _____
(b) _____

18. COMPARING COST   Rheonna Winston is offered an oak stair railing by The Turning Point for $1370 less 30/10. Sierra Stair Company offers the same railing for $1220 less 10/10. **(a)** Which offer is better? **(b)** How much does Winston save by taking the better offer?

(a) _____
(b) _____

19. PRICING POTTED PLANTS   The Plant Place paid a net price of $414.40 for a shipment of potted plants after a trade discount of 30/20 from the list price. Find the list price.

19. _____

20. VITAMIN SUPPLEMENTS   SJ's Nutrition Center received a shipment of vitamins, minerals, and diet supplements at a net cost of $1125. This cost was the result of a trade discount of 25/20 from the list price. Find the list price of this shipment.

20. _____

21. SINGLE TRADE DISCOUNT   An automotive battery charger with a list price of $295.95 is sold by a wholesaler at a net cost of $221.95. Find the single trade discount rate being offered. Round to the nearest tenth of a percent.

21. _____

22. SINGLE TRADE DISCOUNT   Modern Glassware offers crystal wine glasses to Kitchens Galore at a net cost of $864 per gross. If the list price of the wine glasses is $1350 per gross, find the single trade discount rate.

22. _____

6.3 | Cash Discounts: Ordinary Dating Methods

Objectives

1. Calculate net cost after discounts.
2. Use the ordinary dating method.
3. Determine whether cash discounts are earned.
4. Use postdating when calculating cash discounts.

CASE *in* POINT

At Kitchens Galore, Andrew Ryan pays close attention to all invoices received from suppliers. Besides the fact that invoices can frequently have errors on them, he wants to be certain that all of these invoices are paid early enough to receive any additional cash discounts that are offered. He prides himself that he has never missed a final due date and has said, "I'm never overdue on an account."

Objective 1   Calculate net cost after discounts. **Cash discounts** are offered by sellers to encourage prompt payment by customers. In effect, the seller is saying, "Pay me quickly, and receive a discount." Businesses often borrow money for their day-to-day operation. Immediate cash payments from customers decrease the need for borrowed money.

To find the net cost when a cash discount is offered, begin with the list price and subtract any trade discounts. From the result, subtract the cash discount. Use the following formula.

Finding the Net Cost

$$\text{Net cost} = (\text{List price} - \text{Trade discount}) - \text{Cash discount}$$

QUICK TIP   A cash discount is never allowed on shipping and insurance charges. If an invoice amount includes shipping and insurance charges, subtract these charges first, before a cash discount is taken. These charges are then added back to find net cost after the cash discount is subtracted.

The type of cash discount appears on the invoice, under TERMS, which can be found in the bottom right-hand corner of the Hershey invoice on the next page. Many companies using automated billing systems state the exact amount of the cash discount on the invoice. This eliminates all calculations on the part of the buyer. The Hershey invoice is an example of an invoice stating the exact amount of the cash discount. This exact amount is found at the bottom of the invoice. Not all businesses do this, however, so it is important to know how to determine cash discounts.

Objective 2   Use the ordinary dating method. There are many methods for finding cash discounts, but nearly all of these are based on the **ordinary dating method**. The methods discussed here and in the next section are the most common in use today. The ordinary dating method of cash discount, for example, is expressed on an invoice as

2/10, n/30   or   2/10, net 30

and is read "two ten, net thirty." The first digit is the rate of discount (2%), the second digit is the number of days allowed to take the discount (10 days), and n/30 (net 30) is the total number of days given to pay the invoice in full, if the buyer does not use the cash discount. If this invoice is paid within 10 days from the date of the invoice, a 2% discount is subtracted from the amount owed. If payment is made between the 11th and 30th days from the invoice date, the entire amount of the invoice is due. After 30 days from the date of the invoice, the invoice is considered overdue and may be subject to a late charge.

Hershey Chocolate U.S.A.
✕ Hershey Foods Corporation
Hershey, Pennsylvania 17033-0819 U.S.A.   04 98 921006 00

CLAIMS FOR LOSS, DAMAGE, SHORTAGE MUST BE PROMPTLY FILED AND FORWARDED TO:
HERSHEY CHOCOLATE U.S.A.
ATTN: CREDIT DEPARTMENT
HERSHEY, PA 17033

	INVOICE NO.	DATE
0	00003	10/06
	1956003	

CHARGE TO:
HANCOCK & OBRIEN
309 SPRING RD
TREMONT PA
17981

SHIP TO:
HANCOCK & OBRIEN
309 SPRING RD
TREMONT PA

COPY OF INVOICE TO CREDIT ON ALL
MISC/CASH SALES
INVOICE TO ART GINGRICH FOR MAILING

HERSHEY CHOC CO. HERSHEY, PA 17033

PLEASE RETURN DUPLICATE COPY WITH PAYMENT TO

REGION/DIST.	CL BILL	WHSE NO.	CT	CARRIER CODS	SALES REP NO.	ORDER DATE	CUSTOMER ORDER NUMBER	SCN.	DATE SHIPPED		PLEASE REFER TO THIS NUMBER WHEN REMITTING	ACCOUNT NUMBER
9091	04		09	HEST		10 06	3406	5	10 06		088500	0091

QTY. SHIPPED	UNIT	UPC CASH CODE	VER	CUSTOMER STOCK NO.	DESCRIPTION	PACK	RETAIL UNIT UPC	QTY. BILLED	UNIT PRICE	ALLOW	NET UNIT PRICE	AMOUNT
		34000			MANUFACTURER ID FOR THE FOLLOWING ITEMS:							
2	CS	11614			RE PCS368X12			2	6845		6845	13690
2	CS	15500			5TH AVE REG CT			2	12320		12320	24640
2	CS	48400			REESE 4-CUP24/6			2	7020		7020	14040
2	CS	22600			KITKAT KING 6/24			2	7020		7020	14040
2	CS	31800			STRWBRY SYP 22OZ			2	1603		1603	3206
2	CS	37014			SKOR 36BX12			2	12320		12320	24640
2	CS	24100			ALMOND 36CT12			2	12320		12320	24640
2	CS	06630			YORK MINT 200/8			2	8580		8580	17160
2	CS	22200			KRACKL KNG 18/12			2	10530		10530	21060
2	CS	24000			MILK 36CT12			2	12320		12320	24640

20	846	ITM HAS 243458 QTY HAS 20			2/10 N30 TERMS 2% DISCOUNT OF $36.35 IF PAID BY 10/16 NET 30 DAYS	181756
TOTAL PCS	GROSS WT.					PAY LAST AMOUNT IF DISCOUNT IS NOT EARNED

To find the due date of an invoice, use the number of days in each month, given in the following chart.

The Number of Days in Each Month

30-Day Months	31-Day Months		Exception
April	January	August	February
June	March	October	(28 days normally;
September	May	December	29 days in leap year)
November	July		

The number of days in each month of the year can also be remembered using the following rhyme or "knuckle" methods:

Rhyme Method:
30 days hath September,
April, June, and November.
All the rest have 31, except February,
which has 28 and in a leap year 29.

Knuckle Method:

QUICK TIP Leap years occur every 4 years. They are the same as Summer Olympic years and presidential-election years in the United States. If a year is evenly divisible by the number 4, it is a leap year. The years 2008, 2012, and 2016 are all leap years because they are evenly divisible by 4.

Objective **3** **Determine whether cash discounts are earned.** Find the date that an invoice is due by counting from the next day after the date of the invoice. The date of the invoice is never counted. Another way to determine due dates is to add the given number of days to the starting date. For example, to determine 10 days from April 7, add the number of days to the date $(7 + 10 = 17)$. The due date, or 10 days from April 7, is April 17.

When the discount date or net payment date falls in the next month, the number of days remaining in the current month is found by subtracting the invoice date from the number of days in the month. Then find the number of days in the next month needed to equal the discount period or net payment period. For example, determine the date that is 15 days from October 20.

$$
\begin{array}{rl}
31 & \text{days in October} \\
- 20 & \text{the beginning date is October 20} \\
\hline
11 & \text{days remaining in October} \\[4pt]
15 & \text{total number of days} \\
- 11 & \text{days remaining in October} \\
\hline
4 & \text{November (future date)}
\end{array}
$$

Therefore, November 4 is 15 days from October 20.

EXAMPLE 1

Finding Cash Discount Dates

A Hershey invoice is dated January 2 and offers terms of 2/10, net 30. Find **(a)** the last date on which the 2% discount may be taken, and **(b)** the net payment date.

SOLUTION

(a) Beginning with the invoice date, January 2, the last date for taking the discount is January 12 $(2 + 10)$.

(b) The net payment date is February 1 $(31 - 2 = 29$ days remaining in January plus 1 day in February.)

EXAMPLE 2

Finding the Amount Due on the Invoice

An invoice received by Kitchens Galore for $840 is dated July 1 and offers terms of 2/10, n/30. If the invoice is paid on July 8 and the shipping and insurance charges, which were FOB shipping point, are $18.70, find the total amount due.

SOLUTION

1. The invoice was paid 7 days after its date $(8 - 1 = 7)$; therefore, the 2% cash discount is taken.
2. The **2%** cash discount is found on $840. The discount to be taken is $840 \times .02 = $16.80.
3. The cash discount is subtracted from the invoice amount to determine the amount due.

$840 invoice amount − **$16.80 cash discount (2%)** = $823.20 amount due

4. The shipping and insurance charges are added to find the total amount due.

$823.20 amount due + **$18.70 shipping and insurance** = $841.90 total amount due.

QUICK TIP When the terms of an invoice are 2/10, a 2% discount is taken. Only **98%** $(100\% - 2\%)$ of the invoice must be paid if the invoice is paid during the first 10 days. In Example 2, the amount due may be found as follows.

$$840 \times .98 = $823.20$$

invoice amount   complement of 2%   amount due

> **QUICK TIP**   A cash discount is never taken on shipping and insurance charges. Be certain that shipping and insurance charges are excluded from the invoice amount before calculating the cash discount. Shipping and insurance charges must then be added back in to find the total amount due.

Objective 4 **Use postdating when calculating cash discounts.** In the ordinary dating method, the cash discount date and net payment date are both counted from the date of the invoice. Occasionally, an invoice is **postdated**. This may be done to give the purchaser more time to take the cash discount on the invoice or maybe to more closely fit the accounting practices of the seller. The seller places a date that is after the actual invoice date, sometimes labeling it **AS OF**. For example, the following Levi Strauss invoice is dated 07/25 AS OF 08/01. Both the cash discount period and the net payment date are counted from 08/01 (August 1). This results in giving additional time for the purchaser to pay the invoice and receive the discount.

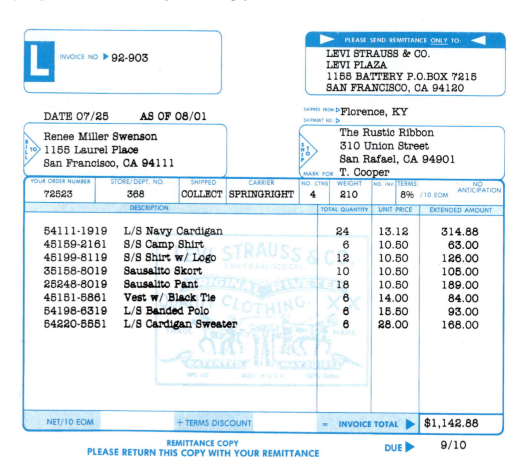

EXAMPLE 3

Using Postdating AS OF with Invoices

An invoice for a shipment of Henkels cutlery from Germany is dated October 21 AS OF November 1 with terms of 3/15, n/30. Find **(a)** the last date on which the cash discount may be taken and **(b)** the net payment date.

SOLUTION

(a) Beginning with the postdate (AS OF) of November 1, the last date for taking the discount is November 16 (1 + 15).

(b) The net payment date is December 1 (29 days remaining in November and 1 day in December).

Business customers, as well as consumers, look for the best prices and the best discounts. A greater number of discounts resulting in lower costs allows the business to operate more efficiently, which results in higher profits. The following newspaper article reminds us that even businesses must comparison shop to save money.

Businesses Turn to Smart Shopping

NEW YORK—Being a savvy business owner means being a smart shopper—deciding which retailers and service establishments provide the best deal. That means shopping like a consumer—and at times, buying in the same places a consumer does.

Whether you're looking for mundane items like office supplies or bigger investments such as computer equipment or telephone service, you'll do better if you comparison shop.

If your business is very small, chances are you won't be able to negotiate a volume discount with an office supplies wholesaler or distributor. So your first thought might be to head to a big superstore chain such as Staples, Office Depot or OfficeMax that tend to have lower prices than independent retailers.

But you might find you'll do better at a local warehouse retailer such as Sam's Wholesale Club or Costco, which have business memberships.

Source: Sacramento Bee.

QUICK TIP   Sometimes, a sliding scale of cash discounts is offered. For example, with the cash discount 3/10, 2/20, 1/30, n/60, a discount of 3% is given if payment is made within 10 days, 2% if paid from the 11th through the 20th day, and 1% if paid from the 21st through the 30th day. The entire amount (net) must be paid no later than 60 days from the date of the invoice.

EXAMPLE 4

Determining Cash Discount Due Dates for a Sliding Scale

An invoice from Cellular Products is dated May 18 and offers terms of 4/10, 3/25, 1/40, n/60. Find **(a)** the three final dates for cash discounts and **(b)** the net payment date.

SOLUTION

(a) The three final cash discount dates are

4% if paid by May 28   10 days from May 18

3% if paid by June 12   25 days from May 18

1% if paid by June 27   40 days from May 18

(b) The net payment date is July 17 (20 days beyond the cash discount period).

QUICK TIP   Never take more than one of the cash discounts. *With all methods of giving cash discounts, if the net payment period is not given, the net payment due date is assumed to be 20 days beyond the cash discount period.* After that date, the invoice is considered overdue. If either the final discount date or the net payment date is on a Sunday or holiday, the next business day is used. Many companies insist that payment is made when the payment is received. In non-retail transactions, payment is considered to be made when it is mailed.

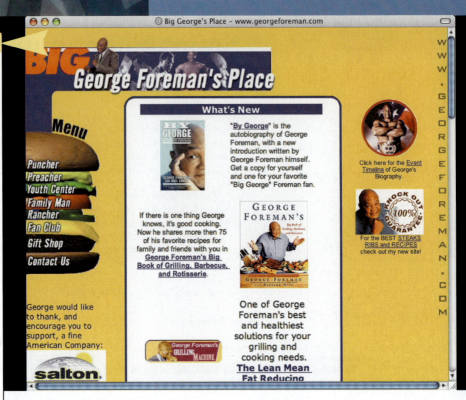

George Foreman

- 1949: Born January 10th.

- 1968: Receives Olympic gold medal for boxing.

- 1973: Defeats Joe Frazier to become world heavyweight champion.

- 1974: Loses heavyweight title to Muhammad Ali.

- 1994: Wins heavyweight title at age of 45 from Michael Moore, age 26.

- 1995: Partners with Salton to promote the George Foreman Lean Mean Grilling Machine.

- 1999: Crowned King of the Grill after selling over 10 million grills.

George E. Foreman was born on January 10, 1949, in Marshal, Texas. He grew up on the mean streets of Houston's Fifth Ward area, and, as a young kid, he was always getting into trouble. He joined the Job Corps, and his life changed. One of the counselors noticed that he was always into fights so he decided that George should put all of his energy into something positive. This started his boxing career. Since 1995, Foreman has been a partner with Salton in promoting the George Foreman cooking grills and other barbecue products. Today, "if there is one thing George Foreman knows, it's good cooking."

1. Kitchens Galore purchased George Foreman Lean Mean Grilling Machines that were list priced at $59.99. If the supplier offered a trade discount of 20/20, find the cost of one dozen grilling machines.

2. The list price of a George Foreman Party grill is $119.99. The manufacturer gives a 25/10 trade discount and offers a cash discount of 3/15, net/30. Find the cost to Kitchens Galore if both discounts are earned and taken.

3. The list price of a *George Foreman Knock-Out-the-Fat Barbecue and Grilling Cookbook* is $16.95. If the cookbook is offered at a reduced price of $13.95 when purchased on the Internet, what is the percent of markdown? Round to the nearest tenth of a percent.

4. There are many celebrities and athletes who use their popularity to promote the products of various companies. However, George Foreman has his own line of products. Name four other well-known people who don't just promote the products of others but have their own line of products.

6.3 | Exercises

FOR EXTRA HELP

MyMathLab

InterActMath.com

MathXP MathXL

MathXL
Tutorials on CD

Tutor Center Addison-Wesley
Math Tutor Center

DVT/Videotape

The **Quick Start** *exercises in each section contain solutions to help you get started.*

Find the final discount date and the net payment date for each of the following. (See Examples 1 and 3.)

Quick Start

	Invoice Date	AS OF	Terms	Final Discount Date	Net Payment Date
1.	May 4		2/10, n/30	May 14	June 3
2.	Apr. 12		3/10, net 30	April 22	May 12
3.	June 30	July 10	3/15, n/60	_____	_____
4.	Nov. 7	Nov. 18	3/10, n/40	_____	_____
5.	Sep. 11		4/20, n/30	_____	_____
6.	July 31		2/15, net 20	_____	_____

Solve for the amount of discount and the total amount due on each of the following invoices. Add shipping and insurance charges if given. (See Examples 2 and 4.)

Quick Start

	Invoice Amount	Invoice Date	Terms	Date Invoice Paid	Shipping and Insurance	Amount of Discount	Total Amount Due
7.	$85.18	Nov. 2	2/10, net 30	Nov. 11	$8.72	$1.70	$92.20
	$85.18 − $1.70 + $8.72 = $92.20						
8.	$66.10	Mar. 8	6/10, n/30	Mar. 14	$4.39	$3.97	$66.52
	$66.10 − $3.97 + $4.39 = $66.52						
9.	$78.07	May 5	net 30	June 1	$3.18	_____	_____
10.	$294	Apr. 5	net 30	May 2	$16.20	_____	_____
11.	$1080	July 8	5/10, 2/20, n/30	July 26	$62.15	_____	_____
12.	$1282	July 1	4/15, net 40	July 7	$21.40	_____	_____

13. Describe the difference between a trade discount and a cash discount. Why are cash discounts offered? (See Objective 1.)

14. Using 2/10, n/30 as an example, explain what an ordinary cash discount means. (See Objective 2.)

 indicates an exercise that is related to the Case in Point feature within the section.

Solve the following application problems. Round to the nearest cent.

Quick Start

15. **CATERING COMPANY**   Kitchens Galore offers cash discounts of 4/10, 2/20, net 30 to all catering companies. An invoice is dated June 18 amounting to $4635.40 and is paid on July 7. Find the amount needed to pay the invoice.

   June 18 to July 7 = $(30 - 18) + 7 = 19$ days
   Discount is 2%; $4635.40 $\times$ .98 = $4542.69

   15. $4542.69

16. **RUSSIAN ELECTRICAL SUPPLIES**   A shipment of electrical supplies is received from the Lyskovo Electrotechnical Works. The invoice is dated March 8, amounts to $6824.58, and has terms of 2/15, 1/20 as of March 20. Find the amount needed to pay the invoice on April 2.

   16. _____

17. **PETROCHEMICAL PRODUCTS**   Century Petrochemical Products offers customers a trade discount of 10/20/5 on all products, with terms of net 30. Find the customer's price for products with a total list price of $2630 if the invoice was paid within 30 days.

   17. _____

18. **GEORGE FOREMAN GRILL**   A George Foreman Jumbo Grill plus a free Toastmaster Ultravection oven is list priced at $140 with a trade discount of 20/5/10 and terms of 4/10, n/30. Find the cost to Kitchens Galore assuming that both discounts are earned.

   18. _____

19. **FINDING DISCOUNT DATES**   An invoice is dated January 18 and offers terms of 6/10, 4/20, 1/30, n/50. Find **(a)** the three final discount dates and **(b)** the net payment date.

   (a) _____
   (b) _____

20. An invoice with terms 4/15, 3/20, 1/30, n/60 is dated September 4. Find **(a)** the three final discount dates and **(b)** the net payment date.

   (a) _____
   (b) _____

21. An invoice is dated March 28 AS OF April 5 with terms of 4/20, n/30. Find **(a)** the final discount date and **(b)** the net payment date.

   (a) _____
   (b) _____

22. An invoice is dated May 20 AS OF June 5 with terms of 2/10, n/30. Find **(a)** the final discount date and **(b)** the net payment date.

   (a) _____
   (b) _____

23. How do you remember the number of days in each month of the year? List the months and the number of days in each. (See Objective 2.)

24. Explain in your own words how AS OF dating (postdating) works. Why is it used? (See Objective 4.)

6.4 | Cash Discounts: Other Dating Methods

Objectives

[1] Solve cash discount problems with end-of-month dating.
[2] Use receipt-of-goods dating to solve cash discount problems.
[3] Use extra dating to solve cash discount problems.
[4] Determine credit given for partial payment of an invoice.

> **CASE** *in* **POINT**
>
> In addition to the ordinary dating method of cash discounts, there are several other cash discount methods that are in common use. Andrew Ryan of Kitchens Galore must be able to understand and use each of these.

Objective [1] **Solve cash discount problems with end-of-month dating.** This section discusses several other methods of finding cash discounts. **End-of-month** and **proximo** dating, abbreviated **EOM** and **prox.**, are treated the same. For example, both

3/10 EOM   and   **3/10 prox.**

mean that 3% may be taken as a cash discount if payment is made by the 10th of the month that follows the sale. The 10 days are counted from the *end of the month* in which the invoice is dated. For example, an invoice dated July 14 with terms of 3/10 EOM would have a discount date 10 days from the end of the month, or the 10th of August (August 10).

Since this is a method of increasing the length of time during which a discount may be taken, it has become common business practice to add an extra month when the date of an invoice is the 26th of the month or later. For example, if an invoice is dated March 25 and the discount offered is 3/10 EOM, the last date on which the discount may be taken is April 10. *However, if the invoice is dated March 26 (or any later date in March), and the cash discount offered is 3/10 EOM, then the last date on which the discount may be taken is May 10.*

SALE 59.99
OSTER TOASTER OVEN/BROILER
6-slice capacity, four heat settings and continuous-clean interior, #6232. Reg. 69.99

Muffins not included.

> **QUICK TIP** The practice of adding an extra month when the invoice is dated the 26th of a month or after is used *only* with the end-of-month (proximo) dating cash discount. It does *not* apply to any of the other cash discount methods.

EXAMPLE 1

Using End-of-Month Dating

If an invoice from Oster is dated June 10 with terms of 3/20 EOM, find **(a)** the final date on which the cash discount may be taken and **(b)** the net payment date.

SOLUTION

(a) The discount date is July 20 (20 days after the end of June).
(b) When no net payment due date is given, common business practice is to allow 20 days after the last discount date. The net payment date is August 9, which is **20 days** after the last discount date (July 20), since no net payment date is given.

EXAMPLE 2

Using Proximo Dating

Find the amount due on an invoice of $782 for some Braun hand blenders dated August 3, if terms are 1/10 prox. and the invoice is paid on September 4.

SOLUTION

The last date on which the discount may be taken is September 10 (**10 days** after the end of August). September 4 is within the discount period, so the discount is earned. The 1% cash discount is computed on $782, the amount of the invoice. Subtract the discount ($782 × .01 = $7.82) from the invoice amount to find the amount due.

$782.00   invoice amount
−   7.82   **cash discount (1%)**
$774.18   amount due

> **QUICK TIP**   With all methods of cash discounts, if the net payment period is not given, the net payment due date is assumed to be *20 days beyond the cash discount date*.

Objective 2 Use receipt-of-goods dating to solve cash discount problems. Receipt-of-goods dating, abbreviated **ROG**, offers cash discounts determined from the date on which goods are actually received. This method is often used when shipping time is long. The invoice might arrive overnight by mail, but the goods may take several weeks. Under the ROG method of cash discount, the buyer is given the time to receive and inspect the merchandise and then is allowed to benefit from a cash discount. For example, the discount

3/15 ROG

allows a 3% cash discount if the invoice is paid within 15 days from receipt of goods. The date that goods are received is determined by the delivery date. If the invoice was dated March 5 and goods were received on April 7, the last date to take the 3% cash discount would be April 22 (April 7 plus 15 days). The net payment date, since it is not stated, is 20 days after the last discount date, or May 12 (April 22 plus 20 days).

EXAMPLE 3

Using Receipt-of-Goods Dating

Kitchens Galore received an invoice dated December 12, with terms of 2/10 ROG. The goods were received on January 2. Find **(a)** the final date on which the cash discount may be taken and **(b)** the net payment date.

SOLUTION

(a) The discount date is January 12 (**10 days** after receipt of goods, January 2 plus **10 days**).
(b) The net payment date is February 1 (**20 days** after the last discount date).

EXAMPLE 4

Working with ROG Dating

Find the amount due to Sir Speedy on an invoice of $285 for some printing services, with terms of 3/10 ROG, if the invoice is dated June 8, the goods are received June 18, and the invoice is paid June 30.

SOLUTION

The last date to take the 3% cash discount is June 28, 10 days after June 18. Since the invoice is paid on June 30, 2 days after the last discount date, **no cash discount may be taken**. The entire amount of the invoice must be paid.

$285   invoice amount
−   0   **no cash discount**
$285   amount due

Objective 3 Use extra dating to solve cash discount problems. Extra dating (extra, ex., or x) gives the buyer additional time to take advantage of a cash discount. For example, the discount

2/10−50 extra   or   2/10−50 ex.   or   2/10−50 x

allows a 2% cash discount if the invoice is paid within 10 + 50 = 60 days from the date of the invoice. The discount is expressed 2/10–50 ex. (rather than 2/60) to show that the 50 days are *extra*, or in addition to the normal 10 days offered.

There are several reasons for using extra dating. A supplier might extend the discount period during a slow sales season to generate more sales or to gain a competitive advantage. For example, the seller might offer Christmas merchandise with extra dating to allow the buyer to take the cash discount after the holiday selling period.

EXAMPLE 5

Using Extra Dating

An invoice for Corning cookware sets is dated November 23 with terms 2/10–50 ex. Find **(a)** the final date on which the cash discount may be taken and **(b)** the net payment date.

SOLUTION

(a) The discount date is January 22 (**7 days** remaining in November + 31 days in December = 38; thus, **22 more days** are needed in January to total 60).

(b) The net payment date is February 11 (**20 days** after the last discount date).

EXAMPLE 6

Understanding Extra Dating

An invoice from KitchenAid is dated August 5, amounts to $8180, offers terms of 3/10–30 x, and is paid on September 12. Find the net payment.

SOLUTION

1. The last day to take the 3% cash discount is September 14 (**August 5 + 40 days = September 14**). Since the invoice is paid on September 12, the **3%** discount may be taken.

2. The **3%** cash discount is computed on $8180, the amount of the invoice. The discount to be taken is $245.40.

3. Subtract the cash discount from the invoice amount to determine the amount of payment.

$$\begin{array}{rl} \$8180.00 & \text{invoice amount} \\ -\ \$245.40 & \textbf{3\% cash discount} \\ \hline \$7934.60 & \text{amount of payment} \end{array}$$

> **QUICK TIP** In Example 6, the amount of payment may be found by multiplying the invoice amount by $(100\% - 3\%)$ 97%. Here, $8180 × .97 = $7934.60 is the amount of payment.

Objective 4 Determine credit given for partial payment of an invoice. Occasionally, a customer may pay only a portion of the total amount due on an invoice. If this **partial payment** is made within a discount period, the customer is entitled to a discount on the portion of the invoice that is paid.

If the terms of an invoice are 3%, 10 days, then only 97% $(100\% - 3\%)$ of the invoice amount must be paid during the first 10 days. So, for each $.97 paid, the customer is entitled to $1.00 of credit. When a partial payment is made, the credit given for the partial payment (base) is found by dividing the partial payment by the complement of the cash discount percent. Then, to find the balance due, subtract the credit given from the invoice amount. The cash discount is found by subtracting the partial payment from the credit given.

EXAMPLE 7

Finding Credit for Partial Payment

Dave's Body and Paint receives an invoice for $1140 dated March 8 that offers terms of 2/10 prox. A partial payment of $450 is made on April 5. Find **(a)** the amount credited for the partial payment, **(b)** the balance due on the invoice, and **(c)** the cash discount earned.

SOLUTION

(a) The cash discount is earned on the $450 partial payment made on April 5 (April 10 was the last discount date). The amount paid ($450) is part of the base (amount for which credit is given).

$$\textbf{100\% - 2\% = 98\%}$$

The rate 98% is used to solve for base using the formula Base = $\dfrac{\text{Part}}{\text{Rate}}$.

$$B = \frac{P}{R}$$

$$B = \frac{450}{98\%} = \frac{450}{.98} = \$459.18$$

The amount credited for partial payment is $459.18.

(b) Balance due = Invoice amount − Credit given
Balance due = $1140 − **$459.18**
Balance due = $680.82

(c) Cash discount = Credit given − Partial payment
Cash discount = **$459.18** − $450
Cash discount = $9.18

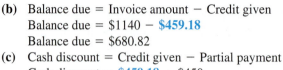 A calculator solution to this example will include these three steps.

Step 1 First, find the amount of credit given.

450 ÷ .98 = 459.18 (rounded)

Step 2 Now, store the amount of credit and subtract this amount from the invoice amount to find the balance due.

[STO] 1140 [−] [RCL] [=] 680.82 (rounded)

Step 3 Finally, subtract the partial payment from the amount of credit given to find the cash discount.

[RCL] [−] 450 [=] 9.18 (rounded)

Note: Refer to Appendix C for calculator basics.

> **QUICK TIP**   Cash discounts are important, and a business should make the effort to pay invoices early to earn the cash discounts. In many cases, the money saved through cash discounts has a great effect on the profitability of a business. Often, companies will borrow money to enable them to take advantage of cash discounts. The mathematics of this type of loan is discussed in **Section 8.1**, Simple Interest.

The cash discounts discussed here are normally not used when selling to foreign customers or purchasing from foreign suppliers. Instead, other types of discounts may be offered to reduce the price of goods sold to foreign buyers. These discounts may be given as allowances for tariffs paid (import duties) by the customer, reimbursement for shipping and insurance paid by the customer, or in the form of an advertising allowance.

6.4 Exercises

FOR EXTRA HELP

MyMathLab

InterActMath.com

MathXL

MathXL
Tutorials on CD

Addison-Wesley
Math Tutor Center

DVT/Videotape

The **Quick Start** *exercises in each section contain solutions to help you get started.*

Find the discount date and net payment date for each of the following. (The net payment date is 20 days after the final discount date; See Examples 1, 3, and 5.)

Quick Start

	Invoice Date	Terms	Date Goods Received	Final Discount Date	Net Payment Date
1.	Feb. 8	3/10 EOM		Mar. 10	Mar. 30
2.	July 14	2/15 ROG	Sept. 3	Sept. 18	Oct. 8
3.	Nov. 22	1/10–20 x		_____	_____
4.	July 6	2/10 EOM		_____	_____
5.	Apr. 12	3/15–50 ex.		_____	_____
6.	Jan. 15	3/15 ROG	Feb. 5	_____	_____

Solve for the amount of discount and the amount due on each of the following invoices. (See Examples 2, 4, and 6.)

Quick Start

	Invoice Amount	Invoice Date	Terms	Date Goods Received	Date Invoice Paid	Amount of Discount	Amount Due
7.	$682.28	June 4	3/20 ROG	July 25	Aug. 10	$20.47	$661.81
	$20.47 discount; $682.28 − $20.47 = $661.81						
8.	$356.20	May 17	3/15 prox.		June 12	$10.69	$345.51
	10.686 = $10.69 discount; $356.20 − $10.69 = $345.51 due						
9.	$785.64	Sept. 8	2/10–20 x		Oct. 14	_____	_____
10.	$12.38	Mar. 29	2/15 ROG	Apr. 15	Apr. 30	_____	_____
11.	$11,480	Apr. 6	2/15 prox.		Apr. 30	_____	_____
12.	$1380	May 28	1/15 EOM		June 10	_____	_____
13.	$23.95	Aug. 2	3/10–20 extra		Sept. 1	_____	_____
14.	$3250.60	Oct. 17	3/15–20 ex.		Oct. 20	_____	_____

C indicates an exercise that is related to the Case in Point feature within the section.

15. Quite often there is no mention of a net payment date on an invoice. Explain the common business practice when no net payment date is given. (See Objective 1.)

16. Describe why ROG dating is offered to customers. Use an example in your description. (See Objective 2.)

Solve the following application problems. Round to the nearest cent.

Quick Start

17. CERAMIC DINNERWARE   Fiesta, a dinnerware manufacturer, offers terms of 2/10–30 ex. to stimulate slow sales in the winter months. Kitchens Galore purchased $2382.58 worth of dinnerware and was offered the above terms. If the invoice was dated November 3, find **(a)** the final date on which the cash discount may be taken and **(b)** the amount paid if the discount was earned.

(a) Dec. 13

(b) $2334.93

   (a)   27 days remain in November
     + 13 days in December
      40 days is December 13

   (b) $2382.58 × .02 = 47.651 = $47.65 discount
      $2382.58 − $47.65 = $2334.93 paid

18. CANADIAN FOOD PRODUCTS   Vanitha Evans, a wholesaler of Canadian food products, offers terms of 4/15–40 ex. to encourage the sales of her products. In a recent order, a retailer purchased $9864.18 worth of Canadian foods and was offered the above terms. If the invoice was dated March 10, find **(a)** the final date on which the cash discount may be taken and **(b)** the amount paid if the discount was earned.

(a) _____

(b) _____

19. ELECTRIC SCOOTER   A recent invoice for 10 Scoot-'N-Go electric scooters amounting to $4358.50, was dated February 20, and offered terms of 2/20 ROG. If the equipment was received on March 20 and the invoice was paid on April 8, find the amount due.

19. _____

20. CUSTOM WHEELS   Scott Ryder purchased some 19-inch and 21-inch custom wheels for his performance auto parts store and was offered a cash discount of 2/10 EOM. The invoice amounted to $7218.80 and was dated June 2. The wheels were received 7 days later, and the invoice was paid on July 7. Find the amount necessary to pay the invoice in full.

20. _____

21. ENGLISH SOCCER EQUIPMENT   An invoice dated December 8 is received with a shipment of soccer equipment from England on April 18 of the following year. The list price of the equipment is $2538, with allowed series discounts of 25/10/10. If cash terms of sale are 3/15 ROG, find the amount necessary to pay in full on April 21.

21. _____

22. CRYSTAL FROM IRELAND   Kitchens Galore receives an invoice for Waterford crystal from Ireland amounting to $5382.40 and dated May 17. The terms of the invoice are 5/20–90 x and the invoice is paid on September 2. Find the amount necessary to pay the invoice in full.

22. _____

23. PARTIAL INVOICE PAYMENT   Mail Boxes Etc. receives an invoice amounting to $672.30 with terms of 8/10, net 30 and dated August 20 AS OF September 1. If a partial payment of $450 is made on September 8, find **(a)** the credit given for the partial payment and **(b)** the balance due on the invoice.

(a) _____
(b) _____

24. FROZEN YOGURT   Yogurt for You receives an invoice amounting to $263.40 with terms of 2/20 EOM and dated September 6. If a partial payment of $150 is made on October 15, find **(a)** the credit given for the partial payment and **(b)** the balance due on the invoice.

(a) _____
(b) _____

C **25.** DISCOUNT DATES   An invoice received by Kitchens Galore is dated May 12 with terms of 2/10 prox. **(a)** _____
Find **(a)** the final date on which the discount may be taken and **(b)** the net payment date. **(b)** _____

26. DISCOUNT DATES   An invoice from Dollar Distributors is dated November 11 with terms of 3/20 ROG **(a)** _____
and the goods are received on December 3. Find **(a)** the final date on which the cash discount may **(b)** _____
be taken and **(b)** the net payment date.

27. INVOICE AMOUNT DUE   Find the amount due on an invoice of $1525 with terms of 1/20 ROG. The **27.** _____
invoice is dated October 20, goods are received December 1, and the invoice is paid on December 20.

28. PAYMENT DUE   Find the payment that should be made on an invoice dated September 28, amounting **28.** _____
to $4680, offering terms of 2/10–50 x and paid on November 25.

29. PARTIAL INVOICE PAYMENT   An invoice received for Lenox crystal has terms of 3/15–30 x and is dated **(a)** _____
May 20. The amount of the invoice is $4402.58, and a partial payment of $3250 is made on July 1. Find **(b)** _____
(a) the credit given for the partial payment and **(b)** the balance due on the invoice.

30. AUTOMOTIVE   Jeepers Supply makes a partial payment of $660 **(a)** _____
on an invoice of $1491.54. If the invoice is dated April 14 with **(b)** _____
terms of 4/20 prox. and the partial payment is made on May 13,
find **(a)** the credit given for the partial payment and **(b)** the balance
due on the invoice.

31. Write a short explanation of partial payment. Why would a company accept a partial payment? Why
would a customer make a partial payment? (See Objective 4.)

32. Of all the different types of cash discounts presented in this section, which type seemed most
interesting to you? Explain your reasons.

Chapter 6 | Quick Review

amount [p. 222]	extra dating (extra, ex., x)	net price [p. 224]	sales invoice [p. 222]
AS OF [p. 240]	[p. 246]	ordinary dating method [p. 237]	series discount [p. 224]
cash discount [p. 237]	FAS (free alongside ship) [p. 223]	partial payment [p. 247]	single discount [p. 224]
chain discount [p. 224]	FOB (free on board) [p. 222]	postdated "AS OF" [p. 240]	single discount equivalent
COD (cash on delivery) [p. 223]	invoices [p. 222]	prox. [p. 245]	[p. 233]
consumer [p. 222]	invoice total [p. 222]	proximo [p. 245]	trade discounts [p. 224]
electronic commerce (EC)	list price [p. 224]	purchase invoice [p. 222]	unit price [p. 222]
[p. 222]	manufacturers [p. 222]	receipt-of-goods dating [p. 246]	units shipped [p. 222]
end of month (EOM) [p. 245]	net cost [p. 224]	retailer [p. 222]	wholesalers [p. 222]
extension total [p. 222]	net cost equivalent [p. 226]	ROG [p. 246]	

CONCEPTS	EXAMPLES
6.1 Trade discount and net cost First find the amount of the trade discount. Then use the formula for net cost. **Net cost = List price − trade discount**	List price, $28; trade discount, 25%; find the net cost. $$\$28 \times .25 = \$7$$ $$\text{Net cost} = \$28 - \$7 = \$21$$
6.1 Complements with respect to 1 (100%) The complement is the number that must be added to a given discount to get 1 or 100%.	Find the complement with respect to 1 (100%) for each of the following. **(a)** 10% $$10\% + \underline{\qquad} = 100\%$$ or $100\% - 10\% = 90\%$ **(b)** 50% complement = 50% **(c)** 5% complement = 95%
6.1 Complements and series discounts The complement of a discount is the percent paid. Multiply the complements of the series discounts to get the *net cost equivalent.*	Series discount, 10/20/10; find the net cost equivalent. 10/ 20/ 10 ↓ ↓ ↓ $$.9 \times .8 \times .9 = .648$$
6.1 Net cost equivalent (percent paid) and the net cost Multiply the net cost equivalent (percent paid) by the list price to get the net cost.	List price, $280; series discount, 10/30/20; find the net cost. 10/ 30/ 20 ↓ ↓ ↓ $$.9 \times .7 \times .8 = .504 \text{ percent paid}$$ $$.504 \times \$280 = \$141.12$$
6.2 Single discount equivalent to a series discount Often needed to compare one series discount to another, the single discount equivalent is found by multiplying the complements of the individual discounts to get the net cost equivalent, then subtracting from 1. $$1 - \frac{\text{Net cost}}{\text{equivalent}} = \frac{\text{Single discount}}{\text{equivalent}}$$	What single discount is equivalent to a 10/20/20 series discount? 10/ 20/ 20 ↓ ↓ ↓ $$.9 \times .8 \times .8 = .576$$ $$1 - .576 = .424 = 42.4\%$$

CONCEPTS	EXAMPLES

6.2 Finding net cost, using complements of individual discounts

Net cost = List price × complements of individual discounts

List price, $510; series discount, 30/10/5; find the net cost.

$$30/ \quad 10/ \quad 5$$
$$\downarrow \quad \downarrow \quad \downarrow$$
$$\$510 \times .7 \times .9 \times .95 = \$305.24 \ (\text{rounded})$$

6.2 Finding list price if given the series discount and the net cost

First, find the net cost equivalent, percent paid. Then use the standard percent formula to find the list price (base).

$$B = \frac{P}{R}$$

Net cost; $224; series discount, 20/20; find list price.

$$20/ \quad 20$$
$$\downarrow \quad \downarrow$$
$$.8 \times .8 = .64$$

$$B = \frac{P}{R} = \frac{224}{.64} = \$350 \text{ list price}$$

6.3 Determining number of days and dates

30-Day Months	31-Day Months
April	All the rest
June	except February with
September	28 days (29 days in leap year)
November	

Date, July 24; find 10 days from date.
July 31 − 24 = 7 days remaining in July

$$\begin{array}{r} 10 \ \text{total number of days} \\ -\ 7 \ \text{days remaining in July} \\ \hline \text{August 3 (future date)} \end{array}$$

6.3 Ordinary dating and cash discounts

With ordinary dating, count days from the date of the invoice. Remember:

2/ 10, n/ 30
↓ ↓ ↓ ↓
% days net days

Invoice amount $182; terms 2/10, n/30; find the cash discount and amount due.

Cash discount: $182 × .02 = $3.64
Amount due: $182 − $3.64 = $178.36

6.4 Cash discounts with end-of-month dating (EOM or proximo)

The final discount date and the net date are counted from the end of the month. If the invoice is dated the 26th or after, add the entire following month when determining the dates. If not stated, the net date is 20 days beyond the discount date.

Terms, 2/10 EOM; invoice date, Oct. 18; find the final discount date and the net payment date.

Final discount date:
November 10, which is 10 days from the end of October.

Net payment date:
November 30, which is 20 days beyond the discount date.

6.4 Receipt-of-goods dating and cash discounts (ROG)

Time is counted from the date goods are received to determine the final cash discount date and the net payment date. If not stated, the net date is 20 days beyond the discount date.

Terms 3/10 ROG; invoice date, March 8; goods received, May 10; find the final discount date and the net payment date.

Final discount date:
May 20 (May 10 + 10 days)

Net payment date:
June 9 (May 20 + 20 days)

CONCEPTS	EXAMPLES
6.4 Extra dating and cash discounts Extra dating adds extra days to the usual cash discount period, so, 3/10–20 x means 3/30. If not stated, the net date is 20 days beyond the discount date.	Terms, 3/10–20 x, invoice date, January 8; find the final discount date and the net payment date. Final discount date: February 7 $\left(23 \text{ days in January} + 7 \text{ days in February} = 30\right)$ Net payment date: February 27 $\left(\text{February } 7 + 20 \text{ days}\right)$
6.4 Partial payment credit When only a portion of an invoice amount is paid within the cash discount period, credit will be given for the partial payment. Use the standard percent formula. $$B = \frac{P}{R}$$ Where the credit given is the base, the partial payment is the part, and $\left(100\% - \text{the cash discount}\right)$ is the rate.	Invoice, \$400; terms, 2/10, n/30; invoice date, Oct. 10; partial payment of \$200 on Oct. 15; find credit given for partial payment and the balance due on the invoice. $$B = \frac{P}{R} = \frac{\$200}{100\% - 2\%} = \frac{\$200}{.98}$$ Credit = \$204.08 $\left(\text{rounded}\right)$ Balance due = \$400 − \$204.08 = \$195.92

Chapter 6 | Summary Exercise

The Retailer: Invoices, Trade Discounts, and Cash Discounts

Andrew Ryan of Kitchens Galore buys much of his merchandise from Gourmet Kitchen Wholesalers. In early September, he ordered kitchen flatware, dinnerware, and cutlery having a total list price of $9748, and appliances and cookware having a list price of $17,645. Gourmet Kitchen Wholesalers offers trade discounts of 20/10/10 on these items and charges for shipping.

The invoice for this order arrived a few days later, is dated September 4, has terms of 3/15 EOM, and shows a shipping charge of $748.38.

Kitchens Galore will need to know all of the following. Round to the nearest cent.

(a) The total amount of the invoice excluding shipping.

(a) _____

(b) The final discount date.

(b) _____

(c) The net payment date.

(c) _____

(d) The amount necessary to pay the invoice in full on October 11 including the shipping.

(d) _____

(e) Suppose that on October 11 the invoice is not paid in full, but a partial payment of $10,000 is made instead. Find the credit given for the partial payment and the balance due on the invoice including shipping.

(e) _____
_____

⟨IN⟩VESTIGATE

Talk with a business owner or store manager. Ask what kinds of trade and cash discounts are standard in their particular line of business. Do they take their earned discounts? Are these discounts important to them? Who is responsible for making sure that discounts are taken? Do they do any of their purchasing electronically? How does their electronic purchasing work?

Chapter 6 | Test

To help you review, the numbers in brackets show the section in which the topic was discussed.

Find the net cost (invoice amount) for the following. Round to the nearest cent. **[6.1]**

1. List price: $348.22 less 10/20/10

1. _____

2. List price: $1308 less 20/25

2. _____

Find (a) the net cost equivalent and (b) the single discount equivalent for the following series discounts. **[6.2]**

3. 30/10

(a) _____
(b) _____

4. 20/10/20

(a) _____
(b) _____

Find the final discount date for the following. **[6.4]**

Invoice Date	Terms	Date Goods Received	Final Discount Date
5. Feb. 10	4/15 EOM	Feb. 16	_____
6. May 8	2/10 ROG	May 20	_____
7. Dec. 8	4/15 prox.	Jan. 5	_____
8. Oct. 20	2/20–40 extra	Oct. 31	_____

9. The following invoice was paid on November 15. Find **(a)** the invoice total, **(b)** the amount that should be paid after the cash discount, and **(c)** the total amount due, including shipping and insurance. **[6.1–6.4]**

(a) _____
(b) _____
(c) _____

GOURMET KITCHEN WHOLESALER			
Terms: 2/10, 1/15, n/60			November 6
Quantity	**Description**	**Unit Price**	**Extension Total**
16	tablecloths, linen	@ 17.50 ea.	
8	rings, napkin	@ 3.25 ea.	
4	cups, ceramic	@ 12.65 ea.	
12	bowls, 1 qt. stainless	@ 3.15 ea.	
		(a) Invoice Total	
		Cash Discount	
		(b) Due after Cash Discount	
		Shipping and Insurance	$11.55
		(c) Total Amount Due	

Solve the following application problems involving cash and trade discounts. Round to the nearest cent.

10. While You Were Out Decorators made purchases at a net cost of $46,746 after a series discount of 20/20/20. Find the list price. **[6.2]**

10. _____

11. An invoice of $1056 from K-D Gadgets has cash terms of 4/20 EOM and is dated June 5. Find **(a)** the final date on which the cash discount may be taken and **(b)** the amount necessary to pay the invoice in full if the cash discount is earned. **[6.4]**

(a) _____
(b) _____

12. Subway Sandwich Shop purchased paper products list priced at $348 less series discounts of 10/20/10, with terms of 3/10–50 extra. If the retailer paid the invoice within 60 days, find the amount paid. **[6.4]**

12. _____

13. The Fireside Shop offers chimney caps for $120 less 25/10. The same chimney cap is offered by Builders Supply for $111 less 25/5. Find **(a)** the firm that offers the lower price and **(b)** the difference in price. **[6.1]**

(a) _____
(b) _____

14. The amount of an invoice from Cloverdale Creamery is $1780 with terms of 2/10, 1/15, net 30. The invoice is dated March 8. **(a)** What amount should be paid on March 20? **(b)** What amount should be paid on April 3? **[6.3]**

(a) _____
(b) _____

15. Diamond Consulting receives an invoice dated November 23 for $2514 with terms of 3/15 EOM. If the invoice is paid on December 14, find the amount necessary to pay the invoice in full. **[6.4]**

15. _____

16. Ron Hill receives an invoice amounting to $2916 with cash terms of 3/10 prox. and dated June 7. If a partial payment of $1666 is made on July 8, find **(a)** the credit given for the partial payment and **(b)** the balance due on the invoice. **[6.4]**

(a) _____
(b) _____

CHAPTER 7

Mathematics of Selling

OLYMPIC SPORTS CARRIES A FULL LINE OF SPORTS EQUIPMENT, sportswear, and athletic shoes for the entire family. The store is best known for its quality and selection of merchandise, but is also competitive in its pricing.

CASE *in* POINT   All offered business discounts are taken to keep the costs of merchandise down. The company must also adjust the markups on various merchandise to keep prices "in the ballpark," relative to competition. Olympic Sports reduces prices on merchandise on a regular basis, using these sale opportunities to clear out existing merchandise and make room for incoming orders.

The success of a business depends on many things. One of the most important is the price it charges for goods and services. The difference between the amount a business pays for an item (cost) and the price at which the item is sold is called **markup**. For example, if Olympic Sports buys Timex Ironman Watches for $16 and sells them for $19.99, the markup is $3.99. This chapter discusses (1) the two standard methods of calculating markup as a percent of cost and as a percent of selling price, (2) converting markups from one method to the other, (3) markdown, and (4) turnover and valuation of inventory.

7.1 | Markup on Cost

Objectives

1. Recognize the terms used in selling.
2. Use the basic formula for markup.
3. Calculate markup based on cost.
4. Apply percent to markup problems.

CASE *in* **POINT**

Last week, Olympic Sports received a shipment of fishing tackle boxes. Store manager Maureen O'Connor must determine the selling price of each tackle box by using a basic markup formula. She knows her regular monthly operating expenses. She also knows that she will be sending out an advertising flyer soon. She must price all merchandise with just enough markup that it is still attractive to customers, while at the same time generating enough revenue to attain the store's profit goals.

Objective 1 **Recognize the terms used in selling.** The terms used in markup are summarized here.

Cost is the amount paid to the manufacturer or supplier after trade and cash discounts have been taken. Shipping and insurance charges are included in cost.

Selling price is the price at which merchandise is offered for sale to the public.

Markup, margin, or **gross profit** is the difference between the cost and the selling price. These three terms are often used interchangeably.

Operating expenses, or **overhead,** include the expenses of operating the business, such as wages and salaries of employees, rent for buildings and equipment, utilities, insurance, and advertising. Even an expense like postage can add up. Mailing costs average from 6.2% of operating expenses for small companies to as high as 9.2% for the largest companies. Notice in the chart to the side how postal rates have changed during the last thirty years.

Net profit (net earnings) is the amount (if any) remaining for the business after operating expenses and the cost of goods have been paid. (Income tax is computed on net profit.)

Most manufacturers, many wholesalers, and some retailers calculate markup as a percent of cost (**markup on cost**). Manufacturers usually express inventories in terms of cost, a method most consistent with their operations. Retailers, on the other hand, usually compute **markup on selling price,** since retailers compare most areas of their business operations to sales revenue. Such items of expense as sales commissions, sales taxes, and advertising are expressed as a percent of sales. It is reasonable, then, for the retailer to express markup as a percent of sales. Wholesalers use either cost or selling price.

U.S. Postal Rates Since 1971	
May 16, 1971	8 cents
March 2, 1974	10 cents
Dec. 31, 1975	13 cents
May 29, 1978	15 cents
March 22, 1981	18 cents
Nov. 1, 1981	20 cents
Feb. 17, 1985	22 cents
April 3, 1988	25 cents
Feb. 3, 1991	29 cents
Jan. 1, 1995	32 cents
Jan. 1, 1999	33 cents
Jan. 1, 2001	34 cents
June 30, 2002	37 cents

Greater use of the Internet has resulted in increased online shopping by consumers. Selling online results in lower operating expenses, which may give greater profits. The newspaper clipping below shows the growth of online retail sales, while the bar graph shows the growing use of the Internet by age group.

Online Sales Soar Over Last Year

Online sales have continued to surge with Amazon.com calling this year its best holiday season ever. Overall, experts forecast online sales to soar by 46 percent over last year.

"Many retailers will be a little disappointed about ... consumer spending," said retail analyst Kurt Barnard of Barnard's Retail Trend Report in New Jersey.

"Discounters are OK. Some (upscale) retailers will be very happy. The stock market has been very ... friendly to a lot of people who are feeling a lot wealthier than they did a year ago."

Source: Sacramento Bee.

Numbers in the News

Americans Online

The increase in the number of Americans using the Internet, by percentage:

Age Group	Percent Increase
2–11	0%
12–17	7%
18–24	13%
25–34	3%
35–49	1%
50–54	1%
55–64	15%
65+	25%

DATA: Nielson; NetRatings

Objective 2   Use the basic formula for markup. Whether markup is based on cost or on selling price, the same basic **markup formula** is always used: $C + M = S$, or

Basic Markup Formula

$$
\begin{array}{ccc}
\text{Cost} & & C \\
+\ \text{Markup} & \text{or} & +\ M \\
\hline
\text{Selling price} & & S
\end{array}
$$

This markup formula is shown by the following diagram.

EXAMPLE 1

*Using the Basic
Markup Formula*

Most markup problems give two of the items in the formula and ask you to find the third. Olympic Sports received assorted nylon hooded vests. Determine the selling price, markup, and cost of the vests in the following problems.

$$
\begin{array}{lll}
& C & \$10 \\
\textbf{(a)} & + M & \$\ 5 \\
\hline & S & \$
\end{array}
\qquad
\begin{array}{lll}
& C & \$10 \\
\textbf{(b)} & + M & \$ \\
\hline & S & \$15
\end{array}
\qquad
\begin{array}{lll}
& C & \$ \\
\textbf{(c)} & + M & \$\ 5 \\
\hline & S & \$15
\end{array}
$$

SOLUTION

$$
\begin{array}{lll}
& C & \$10 \\
\textbf{(a)} & + M & \$\ 5 \\
\hline & S & \$15
\end{array}
\qquad
\begin{array}{lll}
& C & \$10 \\
\textbf{(b)} & + M & \$\ 5 \\
\hline & S & \$15
\end{array}
\qquad
\begin{array}{lll}
& C & \$10 \\
\textbf{(c)} & + M & \$\ 5 \\
\hline & S & \$15
\end{array}
$$

Objective 3 **Calculate markup based on cost.** **Markup based on cost** is expressed as a percent of cost. As shown in the discussion of percent in **Section 3.5**, the base is always 100%. Therefore, cost has a value of 100%. Markup and selling price also have percent values found by comparing their dollar values to the dollar value of the cost. Solve markup problems with the basic formula $C + M = S$, or

$$
\begin{array}{l}
C \\
+ M \\
\hline S
\end{array}
$$

Write the dollar values of cost, markup, and selling price on the right of the formula, and place the rate or percent value for each of these on the left of the formula.

   Suppose an item costs $2 and sells for $3, and that markup is **based on cost**. To find markup, percent of markup on cost, and the selling price as a percent of cost, begin as follows.

$$
\begin{array}{lll}
100\% & C & \$2 \\
\% & M & \$ \\
\hline \% & S & \$3
\end{array}
$$

The dollar amount of cost and selling price have been written in their corresponding positions to the right of the formula, and **100%** has been written to the left of cost, **since cost *is* the base**. The dollar amount of markup is the difference between cost and selling price, or

$$
\begin{array}{llll}
100\% & C & \$2 & \text{base} \\
\% & M & \$1 & \\
\hline \% & S & \$3 &
\end{array}
$$

Next, find the percent of markup based on cost. Do this by comparing the amount of markup, $1, to the cost, $2. The comparison of 1 to 2 is $\frac{1}{2}$, or **50%**.

> **QUICK TIP**   With markup on cost, the base is cost and markup is part.

$$
\begin{array}{lllll}
& 100\% & C & \$2 & \text{base} \\
\text{rate} & 50\% & M & \$1 & \text{part} \\
\hline & \% & S & \$3 &
\end{array}
$$

Finally, add 100% to 50%.

$$
\begin{array}{llll}
& 100\% & C & \$2 & \text{base} \\
+ & 50\% & M & \$1 & \text{part} \\
\hline & 150\% & S & \$3 &
\end{array}
$$

> **QUICK TIP**   **Cost plus markup always equals selling price**, both with dollar amounts and rate amounts.

Objective 4 **Apply percent to markup problems.** Knowledge of percent is used to solve markup problems.

EXAMPLE 2

Solving for Percent of Markup on Cost

The manager of the shoe department bought some hiking shoes manufactured in Mexico for $30 a pair and will sell them for $37.50 a pair. Find the percent of markup based on cost.

SOLUTION

Set up the problem using the basic markup formula.

	100%	*C*	$30.00	**base**
rate	?%	*M*	$	**part**
	%	*S*	$37.50	

The dollar amount of markup is the difference between $37.50 and $30.00, or $7.50.

100%	*C*	$30.00
?%	*M*	**$ 7.50**
%	*S*	$37.50

The cost, $30.00, is the base—it is identified by the 100%. There are two rates and two corresponding parts. Find percent of markup (a rate) by using the part corresponding to markup, $7.50. Identify the components in this example as follows.

	100%	*C*	$30.00	**base**
rate	?%	*M*	$ 7.50	**part**
rate	%	*S*	$37.50	**part**

Find the percent of markup based on cost, using the formula for rate.

$$\text{Rate} = \frac{\textbf{Part}}{\textbf{Base}} = \frac{\textbf{\$7.50}}{\textbf{\$30.00}} = .25 = 25\% \text{ markup based on cost}$$

Complete the problem by adding the rate for cost to the rate for markup and arriving at a rate for selling price.

100%	*C*	$30.00
+ **25%**	*M*	$ 7.50
125%	*S*	$37.50

 The calculator solution to this example is as follows.

$$\boxed{(}\ 37.5\ \boxed{-}\ 30\ \boxed{)}\ \boxed{\div}\ 30\ \boxed{=}\ .25$$

Note: Refer to Appendix C for calculator basics.

This method can be used for solving all problems involving markup, as shown in the next few examples.

EXAMPLE 3

Finding Cost When Cost Is Base

Olympic Sports has a markup on a 100-lb iron barbell set of $16, which is 50% based on cost. Find the cost and the selling price.

SOLUTION

Set up the problem.

100%	*C*	$?
50%	*M*	$16
%	*S*	$

Identify the components.

	100%	*C*	$?	**base**
rate	50%	*M*	$16	**part**
rate	150%	*S*	$	**part**

The rate of markup, 50%, and the corresponding part, $16, are used in the formula to find the base. Solve for base.

$$\text{Base} = \frac{\textbf{Part}}{\textbf{Rate}} = \frac{\textbf{\$16}}{.5} = \$32 \text{ cost}$$

The cost of the weight set is $32.

Now find the selling price by adding the cost and the markup.

100%	*C*	**$32**
50%	*M*	**$16**
150%	*S*	**$48**

The selling price of the weight set is $48.

EXAMPLE 4

Finding the Markup and the Selling Price

Find the markup and the selling price for a Casio solar-powered Sea Pathfinder watch if the cost is $23.60 and the markup is 25% of cost.

SOLUTION

Set up the problem.

100%	C	$23.60
25%	M	$?
%	S	$

Identify the components.

100%	C	$23.60	base
25%	M	$?	part
125%	S	$	part

(left "rate" labels: rate for M row, rate for S row)

If the rate for selling price, 125%, is used in the formula, the resulting part is the selling price. Since markup is to be found, use the rate for markup in the formula. Solve for the markup part.

$$\text{Part} = \textbf{Base} \times \textbf{Rate} = \textbf{\$23.60} \times \textbf{.25} = \$5.90 \text{ markup}$$

The markup is $5.90.

Now solve for selling price by adding cost and markup.

100%	C	$23.60
25%	M	$ 5.90
125%	S	$29.50

The selling price of the watch is $29.50.

This calculator solution uses the percent add-on feature found on many calculators.

$$23.6 \;\boxed{+}\; 25 \;\boxed{\%}\; \boxed{=}\; 29.5$$

Note: Refer to Appendix C for calculator basics.

QUICK TIP   Be certain that you use the corresponding rate and part when working with markup problems. If 125% was used as the rate in Example 4, the answer (part) would have been the selling price. This would work, but you would have to remember to subtract the cost from the selling price $\left(\$29.50 - \$23.60\right)$ to get the markup of $5.90.

EXAMPLE 5

Finding Cost When Cost Is Base

Olympic Sports is selling a Wilson baseball glove for $42, which is 140% of the cost. How much did Olympic Sports pay for the baseball glove?

SOLUTION

Set up the problem.

100%	C	$?
%	M	$
140%	S	$42

Identify the elements.

100%	C	$?	base
40%	M	$	rate
140%	S	$42	part

The rate for markup, 40%, *cannot* be used in the formula because there is no corresponding part. Solve for base using the *corresponding* rate and part.

$$\text{Base} = \frac{\text{Part}}{\text{Rate}} = \frac{\$42}{1.4} = \$30 \text{ cost}$$

The cost of the glove is $30. Check: .40 $\times$ $30 = $12 (markup); then $30 + $12 = $42.

EXAMPLE **6**

Finding the Cost and
the Markup

The retail (selling) price of a Kenmore Elite 21.6 cubic foot refrigerator is $978.75. If the markup is 35% of cost, find the cost and the markup.

SOLUTION

Set up the problem.   Identify the elements.

100%	*C*	$?
35%	*M*	$
%	*S*	$978.75

	100%	*C*	$?	**base**
rate	35%	*M*	$	**part**
rate	135%	*S*	$978.75	**part**

The rate of markup, 35%, *cannot* be used in the formula since there is no corresponding part. Instead, solve for base using the *corresponding* rate and part.

$$\text{Base} = \frac{\text{Part}}{\text{Rate}} = \frac{\$978.75}{1.35} = \$725 \text{ cost}$$

The cost of the refrigerator is $725.
   Now solve for the markup by subtracting cost from selling price.

100%	*C*	$725
35%	*M*	**$253.75**
135%	*S*	$978.75

The markup is $253.75 ($978.75 − $725).

QUICK TIP   Remember, when calculating markup on cost, *cost is always the base* and 100% always goes next to cost.

7.1 Exercises

The **Quick Start** *exercises in each section contain solutions to help you get started.*

Solve for the missing numbers. Markup is based on cost. Round dollar amounts to the nearest cent. (See Examples 1–6.)

Quick Start

1.	100%	C	$12.40	**2.**	100%	C	**$5.40**	**3.**	%	C	$	
	40%	M	$ 4.96		25%	M	$1.35		%	M	$	
	140%	S	**$17.36**		**125%**	S	**$6.75**		120%	S	$32.60	

4.	%	C	$	**5.**	%	C	$	**6.**	100%	C	$78.00	
	50%	M	$ 50.00		30%	M	$ 50.40		%	M	$17.94	
	%	S	$		%	S	$		%	S	$	

Find the missing numbers. Round rates to the nearest tenth of a percent and dollar amounts to the nearest cent. (See Examples 1–6.)

Quick Start

	Cost Price	Markup	% Markup on Cost	Selling Price
7.	$9.00	**$2.70**	30%	**$11.70**
8.	**$36.00**	$7.20	**20%**	$43.20
9.	$12.00	$7.20	___	___
10.	___	___	100%	$68.98
11.	$153.60	___	___	$215.04
12.	___	$54.38	50%	___
13.	___	$8.45	___	$42.25
14.	$7.75	___	28%	___

15. Markup may be calculated on cost or on selling price. Explain why most manufacturers use cost as base when calculating markup. (See Objective 1.)

16. Write the markup formula in vertical form. Define each term. (See Objective 2.)

Solve the following application problems, using cost as a base. Round rates to the nearest tenth of a percent and dollar amounts to the nearest cent.

Quick Start

17. **EXERCISE BICYCLE**   Olympic Sports pays $264.30 for a Schwinn Exercise Bike and the markup is 45% of cost. Find the markup.

     100%   *C* $264.30   base
     45%   *M* $?
rate   145%   *S* $ part $P = B \times R = \$264.30 \times .45 = 118.935 = \$118.94$

17. *$118.94 markup*

 indicates an exercise that is related to the Case in Point feature within the section.

C   18.   SKI JACKETS   Olympic Sports offers ski jackets, sizes S, M, and L, for $138.   18. _____
If the markup is 35% of cost, find the cost.

19.   TAIWAN TOOL PRODUCTS   The cost of some socket-wrench sets from Taiwan is $10.36 per set. Harbor   19. _____
Tool decides to use a markup of 25% on cost. Find the selling price of the socket wrench set.

20.   WEIGHT-TRAINING BOOKS   Gold's Gym sells a weight-training book for $15.60 per copy. If this   20. _____
includes a markup of 50% on cost, find the cost.

C   21.   PRICING BASKETBALL SYSTEMS   Olympic Sports purchases   21. _____
Spalding basketball systems at a cost of $180 each. If the
company's operating expenses are 16% of cost, and a net profit
of 7% of cost is desired, find the selling price of one basketball
system.

22.   PRICING MERCHANDISE   Barstools are purchased by Factory Outlet Stores at a cost of $59 each. If the   22. _____
company's operating expenses are 12% of cost, and a net profit of 9% of cost is needed, find the
selling price of six barstools.

23.   VIDEO GAMES   Action Games placed a markup of $18 on a video game sold for $58. Find **(a)** the cost,   (a) _____
(b) the markup percent on cost, and **(c)** the selling price as a percent of cost.   (b) _____
                                                                                                                 (c) _____

C   24.   GOLF CLUBS   Olympic Sports had a markup of $46.64 on golf   (a) _____
clubs sold for $222.64. Find **(a)** the cost, **(b)** the markup percent   (b) _____
on cost, and **(c)** the selling price as a percent of cost.   (c) _____

25.   TRACTOR PARTS   Bismark Tractor put a markup of 26% on cost on some parts for which they paid   (a) _____
$4.50. Find **(a)** the selling price as a percent of cost, **(b)** the selling price, and **(c)** the markup.   (b) _____
                                                                                                                 (c) _____

26.   CUSTOM-MADE JEWELRY   A jewelry dealer sold custom-made   (a) _____
necklaces at a selling price that was 250% of his cost. If the   (b) _____
markup is $135, find **(a)** the markup percent on cost, **(b)** the cost,   (c) _____
and **(c)** the selling price.

7.2 | Markup on Selling Price

Objectives

1. Understand the phrase *markup based on selling price.*
2. Solve markup problems when selling price is the base.
3. Use the markup formula to solve variations of markup problems.
4. Determine percent markup on cost and the equivalent percent markup on selling price.
5. Convert markup percent on cost to selling price.
6. Convert markup percent on selling price to cost.
7. Find the selling price for perishables.

CASE *in* POINT

Olympic Sports faces stiff competition from companies like Sportmart. Sportmart has placed a newspaper ad about their baseball equipment specifically aimed at teams, which buy in large quantities. Sportmart offers to pay customers double the difference if the customer finds any item for a lower price at another store. Olympic Sports must compete by buying their merchandise at the lowest possible price and keeping markups at a minimum.

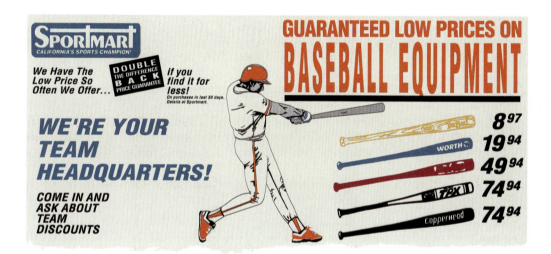

Objective 1   Understand the phrase *markup based on selling price.* As mentioned in the previous section, wholesalers sometimes calculate markup based on cost and other times calculate markup based on selling price. In retailing, it is common to calculate markup based on selling price. In each method, markup is stated as being "on cost" or "on selling price." If markup is based on selling price, then selling price is the base.

When **markup on selling price** is calculated, the same basic markup formula is used, $C + M = S$, or

$$\begin{array}{c|cc} \% & C & \$ \\ \% \; + & M & \$ \\ \hline \% & S & \$ \end{array}$$

Objective 2   Solve markup problems when selling price is the base. The dollar amounts for cost, markup, and selling price are still written to the right of the formula, and the rate amounts for each of these are still written to the left of the formula. However, the base is now the selling price. Since the selling price is the base, which is always 100%, place 100% next to the selling price on the left-hand side:

$$
\begin{array}{lll}
\% & C & \$ \\
\% & M & \$ \\
\hline
\mathbf{100\%} & S & \$ \quad \text{base}
\end{array}
$$

EXAMPLE 1

Solving for Markup on Selling Price

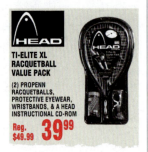

To remain competitive, Olympic Sports must sell Head TI-Elite XL racquetball value packs for $39.99. They pay $35 for each value pack and calculate markup on selling price. Find the amount of markup, the percent of markup on selling price, and the percent of cost on selling price.

SOLUTION
Set up the problem.

$$
\begin{array}{lll}
?\% & C & \$35.00 \\
?\% & M & \$? \\
\hline
\mathbf{100\%} & S & \$39.99
\end{array}
$$

Solve for markup.

$$
\begin{array}{lll}
?\% & C & \$35.00 \\
?\% & M & \$\ 4.99 \\
\hline
100\% & S & \$39.99
\end{array}
$$

Identify the components.

$$
\begin{array}{lllll}
\mathbf{rate} & ?\% & C & \$35.00 & \mathbf{part} \\
\mathbf{rate} & ?\% & M & \$\ 4.99 & \mathbf{part} \\
\hline
& 100\% & S & \$39.99 & \mathbf{base}
\end{array}
$$

Solve for either of the rates, and subtract the result from 100% to find the other. Solve for markup rate.

$$
\text{Rate} = \frac{\mathbf{Part}}{\mathbf{Base}} = \frac{4.99}{39.99} = .1247 = 12.5\% \text{ markup on selling price } (\text{rounded})
$$

The rate of markup on selling price is 12.5%, and cost as a percent of selling price is 87.5% $(100\% - 12.5\% = 87.5\%)$. The result is shown below.

$$
\begin{array}{lll}
87.5\% & C & \$35.00 \\
12.5\% & M & \$\ 4.99 \\
\hline
100\% & S & \$39.99
\end{array}
$$

QUICK TIP   Remember that the *part* and *rate* must *always correspond*. If you use the markup part in the formula, the resulting rate will be the markup rate. If you use the cost part in the formula, the resulting rate will be the cost rate.

Markups vary widely from industry to industry and from business to business. This variation is a result of different costs of merchandise, operating costs, levels of profit margin, and local competition. The next table shows average markups for different types of retail stores.

Average Markups for Retail Stores (Markup on Selling Price)

Type of Store	Markup	Type of Store	Markup
General merchandise stores	29.97%	Furniture and home furnishings	35.75%
Grocery stores	22.05%	Bars	52.49%
Other food stores	27.31%	Restaurants	56.35%
Motor vehicle dealers (new)	12.83%	Drug and proprietary stores	30.81%
Gasoline service stations	14.47%	Liquor stores	20.19%
Other automotive dealers	29.57%	Sporting goods and bicycle shops	29.72%
Apparel and accessories	37.64%	Gift, novelty, and souvenir shops	41.86%

Source: Sole-proprietorship income tax returns, U.S. Treasury Dept., Internal Revenue Service, Statistics Division.

Objective ③ **Use the markup formula to solve variations of markup problems.** As with problems with markup based on cost, this basic formula may be used for all markup problems in which selling price is the base. In each of these examples, the selling price has a percent value of 100%.

EXAMPLE 2

Finding Cost When Selling Price Is Base

An Olympic Sports employee knows that the professional-grade leather weight gloves in stock have a markup of $5.16, which is 35% based on selling price. Find the cost of the weight gloves.

SOLUTION

Set up the problem.

$$
\begin{array}{lll}
\% & C & \$? \\
35\% & M & \$5.16 \\
\hline
100\% & S & \$ \\
\end{array}
$$

Identify the components.

$$
\begin{array}{llll}
\textbf{rate} & 65\% & C & \$? \\
\textbf{rate} & 35\% & M & \$5.16 & \textbf{part} \\
\hline
& 100\% & S & \$? & \textbf{base} \\
\end{array}
$$

Now solve for base (selling price), and subtract the markup from selling price to find the cost. Solve for base using the *corresponding* rate and part.

$$
\text{Base} = \frac{\textbf{Part}}{\textbf{Rate}} = \frac{5.16}{.35} = \$14.74 \text{ selling price } \left(\text{rounded}\right)
$$

Solve for cost.

$$
\begin{array}{ccc}
\text{Selling price} & - & \text{Markup} & = & \text{Cost} \\
\$14.74 & - & \$5.16 & = & \$9.58 \\
\end{array}
$$

The cost is $9.58.

$$
\begin{array}{lll}
65\% & C & \$\ 9.58 \\
35\% & M & \$\ 5.16 \\
\hline
100\% & S & \$14.74 \\
\end{array}
$$

EXAMPLE 3

Finding Markup When Selling Price Is Given

Nancy Gee, an employee at Olympic Sports, must calculate the markup on a three-pack of tennis balls. The selling price of the tennis balls is $3.95 and the markup is 20% of selling price. Find the markup.

SOLUTION

Set up the problem.

$$
\begin{array}{lll}
\% & C & \$ \\
20\% & M & \$? \\
\hline
\mathbf{100\%} & S & \$3.95 \\
\end{array}
$$

Identify the components.

	80%	C	$	
rate	20%	M	$?	part
	100%	S	$3.95	base

Solve for part

$$\text{Part} = \textbf{Base} \times \textbf{Rate} = \$3.95 \times .2 = \$.79 \text{ markup}$$

The markup is $.79.

> **QUICK TIP**   If the rate for cost, 80%, had been used in the formula, the result would have been the cost.

EXAMPLE 4

Finding Markup When Cost Is Given

Find the markup on a dartboard made in England if the cost is $27.45 and the markup is 25% of selling price.

SOLUTION
Set up the problem.

%	C	$27.45
25%	M	$?
100%	S	$

Identify the components.

rate	75%	C	$27.45	part
rate	25%	M	$?	part
	100%	S	$	base

Solve for base, using the rate and part that go together. In this example, use the rate and part for cost.

$$\text{Base} = \frac{\text{Part}}{\text{Rate}} = \frac{\$27.45}{.75} = \$36.60 \text{ selling price}$$

$$\text{Selling price} - \text{Cost} = \text{Markup}$$
$$\$36.60 - \$27.45 = \$9.15$$

75%	C	$27.45
25%	M	$ 9.15
100%	S	$36.60

> **QUICK TIP**   Remember, when calculating markup on selling price, *selling price* is *always the base* and 100% always goes next to selling price.

Objective 4 Determine percent markup on cost and the equivalent percent markup on selling price. Sometimes a markup based on cost must be compared with a markup based on selling price. For example, a salesperson who sells to both manufacturers who use markup on cost and to retailers who use markup on selling price might have to make quick conversions from one markup method to the other. Such a conversion might also be necessary for a manufacturer who thinks in terms of cost and who wants to understand a wholesaler or retail customer. Or perhaps a retailer or wholesaler might convert markup on selling price to markup on cost to better understand the manufacturer.

Make these comparisons by computing first the markup on cost and then the markup on selling price.

EXAMPLE 5

*Determining
Equivalent Markups*

Awanata Jackson sells fishing lures to both fishing-equipment wholesalers and sporting-goods stores. If the lure costs her $4.20 and she sells it for $5.25, what is the percent of markup on cost? What is the percent of markup on selling price?

SOLUTION

First, compute the rate of markup on cost. Set up the problem.

$$
\begin{array}{ll}
100\% & C \quad \$4.20 \\
\;?\% & M \quad \$ \\
\hline
\% & S \quad \$5.25
\end{array}
$$

Identify the components.

$$
\begin{array}{lll}
& 100\% & C \quad \$4.20 \quad \text{base} \\
\text{rate} & \;?\% & M \quad \$1.05 \quad \text{part} \\
\text{rate} & \% & S \quad \$5.25 \quad \text{part}
\end{array}
$$

Solve for rate.

$$
\text{Rate} = \frac{\text{Part}}{\text{Base}} = \frac{\$1.05}{\$4.20} = .25 = 25\% \text{ markup on cost}
$$

The markup on cost is 25%.

Next, compute the rate of markup on selling price. Set up the problem.

$$
\begin{array}{ll}
\% & C \quad \$4.20 \\
\;?\% & M \quad \$ \\
\hline
100\% & S \quad \$5.25
\end{array}
$$

Identify the components.

$$
\begin{array}{lll}
\text{rate} & \% & C \quad \$4.20 \quad \text{part} \\
\text{rate} & \;?\% & M \quad \$1.05 \quad \text{part} \\
& 100\% & S \quad \$5.25 \quad \text{base}
\end{array}
$$

Solve for rate.

$$
\text{Rate} = \frac{\text{Part}}{\text{Base}} = \frac{\$1.05}{\$5.25} = .20 = 20\% \text{ markup on selling price}
$$

The markup on selling price is 20%.

The results in this example show that a 25% markup on cost is equivalent to a 20% markup on selling price.

QUICK TIP   In Example 5, the markup on cost was determined first (25%). The problem was then reworked with the same dollar amounts but with the selling price as base. The result was 20%.

Objective 5   Convert markup percent on cost to selling price.   Another method for markup comparisons is to use **conversion formulas**. Convert markup percent on cost to markup percent on selling price with the following formula.

$$
\frac{\% \text{ markup on cost}}{100\% + \% \text{ markup on cost}} = \% \text{ markup on selling price}
$$

EXAMPLE 6

Converting Markup on Cost to Markup on Selling Price

Convert a markup of 25% on cost to its equivalent markup on selling price.

SOLUTION

Use the formula for converting markup on cost to markup on selling price.

$$\frac{\% \text{ markup on cost}}{100\% + \% \text{ markup on cost}} = \% \text{ markup on selling price}$$

$$\frac{25\%}{100\% + 25\%} = \frac{25\%}{125\%} = \frac{.25}{1.25} = .20 = 20\%$$

As shown, a markup of 25% on cost is equivalent to a markup of 20% on selling price.

The markup on cost (25%) is divided by 100% plus the markup on cost. The parentheses keys are used here.

$$25 \boxed{\%} \boxed{\div} \boxed{(} 100 \boxed{\%} \boxed{+} 25 \boxed{\%} \boxed{)} \boxed{=} 0.2$$

Note: Refer to Appendix C for calculator basics.

Objective ⬚6 **Convert markup percent on selling price to cost.** Convert markup percent on selling price to markup percent on cost with the following formula.

$$\frac{\% \text{ markup on selling price}}{100\% - \% \text{ markup on selling price}} = \% \text{ markup on cost}$$

EXAMPLE 7

Converting Markup on Selling Price to Markup on Cost

Convert a markup of 20% on selling price to its equivalent markup on cost.

SOLUTION

Use the formula for converting markup on selling price to markup on cost.

$$\frac{\% \text{ markup on selling price}}{100\% - \% \text{ markup on selling price}} = \% \text{ markup on cost}$$

$$\frac{20\%}{100\% - 20\%} = \frac{20\%}{80\%} = \frac{.2}{.8} = .25 = 25\%$$

A markup of 20% on selling price is equivalent to a markup of 25% on cost.

The following table shows common markups expressed as percent on cost and also on selling price. A table like this would be helpful to anyone using markup equivalents on a regular basis.

Markup Equivalents

Markup on Cost	Markup on Selling Price
20%	$16\frac{2}{3}\%$
25%	20%
$33\frac{1}{3}\%$	25%
50%	$33\frac{1}{3}\%$
$66\frac{2}{3}\%$	40%
75%	$42\frac{6}{7}\%$
100%	50%

Objective ⬚7 **Find the selling price for perishables.** When a business sells items that are perishable (such as baked goods, fruits, or vegetables), the fact that some items will spoil and become unsellable must be considered when determining the selling price of each item that is sold.

The bar graph below shows the top 10 retailers in the United States. Which of these companies do you think considers perishables as part of their business operations. It appears that half of them sell groceries as part of the product line.

Numbers in the News

What's in Store for You
The top ten retailers in thousands of dollars.

❶ **Wal-Mart** $246,525,000

❷ **Home Depot** $58,247,000

❸ **Kroger** $51,759,500

❹ **Target** $42,722,000

❺ **Sears** $41,366,000

❻ **Costco** $37,993,093

❼ **Albertsons** $35,626,000

❽ **Safeway** $32,399,200

❾ **J. C. Penney** $32,347,000

❿ **Kmart** $30,762,000

**DATA: Retail Forward Inc.,
National Retail Federation**

EXAMPLE 8

*Finding Selling Price
for Perishables*

The Bagel Boy bakes 60 dozen bagels at a cost of $2.16 per dozen. If a markup of 50% on selling price is needed and 5% of the bagels remain unsold and will be donated to a shelter, find the selling price per dozen bagels.

SOLUTION

Step 1 First find the cost of the bagels.

$$\text{Cost} = 60 \text{ dozen} \times \$2.16 = \$129.60$$

Step 2 Next, find the selling price, using a markup of 50% of selling price.

rate				
50%	C	$129.60	**part**
50%	M	$	
100%	S	$?	**base**

$$\text{Base} = \frac{\text{Part}}{\text{Rate}} = \frac{\$129.60}{.5} = \$259.20$$

The total selling price is $259.20.

Step 3 Now, find the number of dozen bagels that will be sold. Since 5% will not be sold, 95% (100% − 5%) will be sold.

$$95\% \times 60 \text{ dozen} = \mathbf{57} \text{ dozen bagels sold}$$

The selling price of $259.20 must be received from the sale of **57** dozen bagels.

Step 4 Find the selling price per dozen bagels by dividing the total selling price by the number of bagels sold.

$$\frac{\$259.20}{57} = \$4.55 \text{ selling price per dozen} \left(\text{rounded}\right)$$

A selling price of $4.55 per dozen gives the desired markup of 50% on selling price while allowing for 5% of the bagels to be unsold.

7.2 | Exercises

FOR EXTRA HELP

MyMathLab

InterActMath.com

MathXL

MathXL
Tutorials on CD

Addison-Wesley
Math Tutor Center

DVT/Videotape

The **Quick Start** exercises in each section contain solutions to help you get started.

Solve for the missing numbers. Markup is based on selling price. Round dollar amounts to the nearest cent. (See Examples 1–4.)

Quick Start

1.	75%	C	$21.00	2.	60%	C	$18.60
	25%	M	$ 7.00		40%	M	$12.40
	100%	S	$28.00		100%	S	$31.00

3.		C	$145.00	4.		C	
		M				M	$ 89.00
	100%	S	$250.00			S	

5.	50%	C	$2025	6.	65%	C	
		M				M	$ 527.80
		S				S	

Find the missing quantities by first computing the markup on one base and then computing the markup on the other. Round rates to the nearest tenth of a percent and dollar amounts to the nearest cent. (See Example 5.)

Quick Start

	Cost	Markup	Selling Price	% Markup on Cost	% Markup on Selling Price
7.	$1920	$480.00	$2400.00	25%	20%
8.	$357.52	$78.48	$436.00	22%	18%
9.	$13.80	_____	_____	_____	38%
10.	$33.75	_____	$67.50	_____	_____
11.	_____	$300.00	_____	40%	_____
12.	$5.15	_____	$15.45	_____	_____

Find the equivalent markups on either cost or selling price using the appropriate formula. Round to the nearest tenth of a percent. (See Examples 6 and 7.)

Quick Start

	Markup on Cost	Markup on Selling Price		Markup on Cost	Markup on Selling Price
13.	100%	50%	14.	_____	20%
15.	18%	_____	16.	50%	_____

For 13: $\frac{100\%}{100\% + 100\%} = \frac{1}{2} = .5 = 50\%$

17. Use the table on page 273 to find the three types of retail stores with the lowest markups. Why do markups differ so much from one type of retail store to another?

18. To have a markup of 100% or greater, the markup must be calculated on cost. Show why this is always true. (See Objective 6.)

Solve the following application problems. Round rates to the nearest tenth of a percent and dollar amounts to the nearest cent.

19. HOME-WORKOUT EQUIPMENT Olympics Sports has a markup of $280 on a Schwinn Bowflex. If this is a 35% markup on selling price, find **(a)** the selling price, **(b)** the cost, and **(c)** the cost as a percent of selling price.

    (a) _____
    (b) _____
    (c) _____

20. SWIMMING POOL PUMP Leslie's Pool Supply pays $187.19 for a Hydramax 11 pump. If the markup is 28% on the selling price, find **(a)** the cost as a percent of selling price, **(b)** the selling price, and **(c)** the markup.

    (a) _____
    (b) _____
    (c) _____

21. ITALIAN SILK TIES Dress for Success purchased 240 Italian silk ties for $2280. They sold 162 of the ties at $25 each, 45 of the ties at $15 each, 20 ties at $10 each, and the remainder at $5 each. Find **(a)** the total amount received for the ties, **(b)** the total markup, **(c)** the markup percent on selling price, and **(d)** the equivalent markup percent on cost.

    (a) _____
    (b) _____
    (c) _____
    (d) _____

22. RIVER-RAFT SALES Olympic Sports purchased a job lot of 380 river rafts for $7600. If they sold 158 of the rafts at $45 each, 74 at $35 each, 56 at $30 each, and the remainder at $25 each, find **(a)** the total amount received for the rafts, **(b)** the total markup, **(c)** the markup percent on selling price, and **(d)** the equivalent markup percent on cost.

    (a) _____
    (b) _____
    (c) _____
    (d) _____

23. SELLING BANANAS The produce manager at Tom Thumb Market knows that 10% of the bananas purchased will become unsaleable and will have to be thrown out. If she buys 500 pounds of bananas for $.28 per pound and wants a markup of 45% on the selling price, find the selling price per pound of bananas.

    23. _____

24. POTTERY-SHOP SALES The Aztec Pottery Shop finds that 15% of their production cannot be sold. If they produce 100 items at a cost of $2.15 each and desire a markup of 40% on selling price, find the selling price per item.

    24. _____

Supplementary Application Exercises on Markup

Solve each of the following application problems. Round rates to the nearest tenth of a percent and dollar amounts to the nearest cent.

 1. **TELESCOPES**   Olympic Sports pays $74.50 for a Bushnell telescope and the markup is 30% of cost. Find the markup.

1. _____

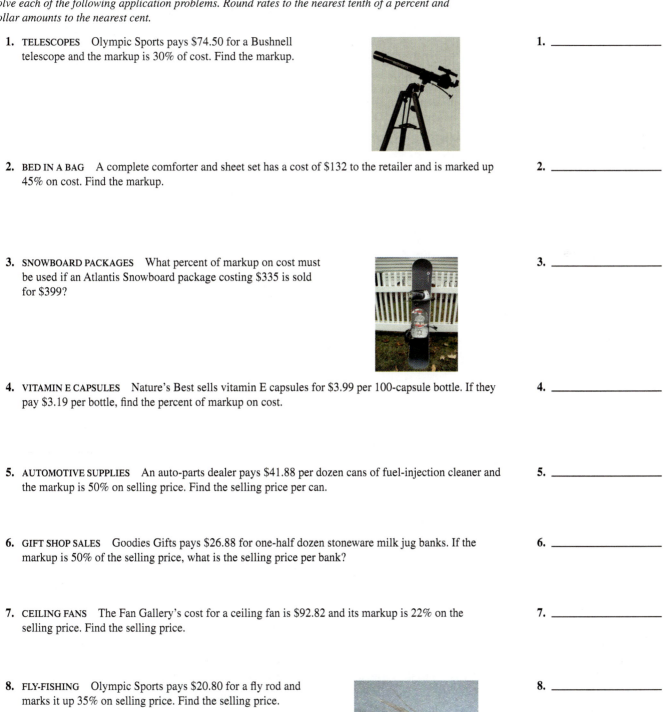

2. **BED IN A BAG**   A complete comforter and sheet set has a cost of $132 to the retailer and is marked up 45% on cost. Find the markup.

2. _____

3. **SNOWBOARD PACKAGES**   What percent of markup on cost must be used if an Atlantis Snowboard package costing $335 is sold for $399?

3. _____

4. **VITAMIN E CAPSULES**   Nature's Best sells vitamin E capsules for $3.99 per 100-capsule bottle. If they pay $3.19 per bottle, find the percent of markup on cost.

4. _____

5. **AUTOMOTIVE SUPPLIES**   An auto-parts dealer pays $41.88 per dozen cans of fuel-injection cleaner and the markup is 50% on selling price. Find the selling price per can.

5. _____

6. **GIFT SHOP SALES**   Goodies Gifts pays $26.88 for one-half dozen stoneware milk jug banks. If the markup is 50% of the selling price, what is the selling price per bank?

6. _____

7. **CEILING FANS**   The Fan Gallery's cost for a ceiling fan is $92.82 and its markup is 22% on the selling price. Find the selling price.

7. _____

 8. **FLY-FISHING**   Olympic Sports pays $20.80 for a fly rod and marks it up 35% on selling price. Find the selling price.

8. _____

 indicates an exercise that is related to the **Case in Point** feature within the section.

▽ C **9.** PING-PONG TABLE   Olympic Sports pays $136.79 for a ping-pong table. The markup is 24% on selling price. Find **(a)** the cost as a percent of selling price, **(b)** the selling price, and **(c)** the markup.

(a) _____
(b) _____
(c) _____

10. DOUBLE-PANE WINDOWS   Eastern Building Supply pays $3808 for all the double-pane windows needed for a 3-bedroom, 2-bath home. If the markup on the windows is 15% on selling price, what is **(a)** the cost as a percent of selling price, **(b)** the selling price, and **(c)** the markup?

(a) _____
(b) _____
(c) _____

11. COMMUNICATION EQUIPMENT   A discount store purchased touch-tone wall phones at a cost of $288 per dozen. If the store needs 20% of cost to cover operating expenses and 15% of cost for the net profit, what is **(a)** the selling price per phone and **(b)** the percent of markup on selling price?

(a) _____
(b) _____

12. BOWLING EQUIPMENT   The Bowlers Pro-Shop determines that operating expenses are 23% of selling price and desires a net profit of 12% of selling price. If the cost of a team shirt is $29.25, what is **(a)** the selling price and **(b)** the percent of markup on cost?

(a) _____
(b) _____

▽ C **13.** MOUNTAIN BIKE SALES   Olympic Sports advertises mountain bikes for $199.90. If the store's cost is $2100 per dozen, what is **(a)** the markup per bicycle, **(b)** the percent of markup on selling price, and **(c)** the percent of markup on cost?

(a) _____
(b) _____
(c) _____

14. HAND TOOL SALES   The Tool Shed advertises standard/metric socket sets for $39. Their cost is $351 per dozen sets. Find **(a)** the markup per set, **(b)** the percent of markup on selling price, and **(c)** the percent of markup on cost.

(a) _____
(b) _____
(c) _____

15. LONG-STEMMED ROSES   Farmers Flowers purchased 12 gross of long-stemmed roses at a cost of $945. If 25% of the roses cannot be sold and a markup of 100% on cost is needed, find the regular selling price per dozen roses.

15. _____

▽ C **16.** SPORTSWEAR   Olympic Sports buys 2000 baseball caps at $2.50 per hat. If a markup of 50% on selling price is needed and 5% of the caps are unsaleable, what is the selling price of each cap?

16. _____

7.3 | Markdown

Objectives

1. Define the term *markdown* when applied to selling.
2. Calculate markdown, reduced price, and percent of markdown.
3. Define the terms associated with loss.
4. Determine the break-even point and operating loss.
5. Determine the amount of a gross or absolute loss.

CASE *in* **POINT**

Maureen O'Connor, the manager of Olympic Sports, keeps a close eye on inventory. This January, some winter ski parkas are still on the shelves and she has decided to mark them down in order to sell them. Monitoring inventory is an important management function. Slow-selling and outdated merchandise must be moved out of the store to make room for new, more profitable merchandise.

Markdowns are used to stimulate sales volume. The following newspaper clipping shows how slashed prices have stimulated the sales of electronics items for holiday gift giving. Customer fears that retailers will run out of electronics items has caused the buying frenzy.

Shoppers Binge on Electronics

Holiday sales hot as prices fall on TVs to DVD players

By Lorrie Grant
USA TODAY

Strong sales of thin televisions, digital cameras and DVD players are pulling consumer electronics sellers through the holiday shopping season.

Two weekends of snowstorms in the East and consumers' tendency to shop later and later so far have given retailers more of a ho-hum than ho-ho-ho shopping season, overall.

But fear that the hottest digital items will sell out has boosted consumer electronics sales since Thanksgiving. Adding to the buzz: prices slashed as low as $29 on DVD players and below $500 on desktop PCs.

Objective 1 **Define the term *markdown* when applied to selling.** When merchandise does not sell at its marked price, the price is often reduced. The difference between the original selling price and the reduced selling price is called the **markdown**, with the selling price after the markdown called the **reduced price, sale price**, or **actual selling price**. The basic **formula for markdown** is as follows.

Reduced price = Original price − Markdown

EXAMPLE 1

Finding the Reduced Price

Olympic Sports has marked down an Atlas Home Fitness Center. Find the reduced price if the original price was $2879 and the markdown is 30%.

SOLUTION

The markdown is 30% of $2879, or .3 × $2879 = $863.70. Find the reduced price as follows.

Objective 2 Calculate markdown, reduced price, and percent of markdown.

$$
\begin{array}{ll}
\$2879.00 & \text{original price} \\
-\quad\ 863.70 & \text{markdown} \left(.30 \times \$2879\right) \\
\hline
\$2015.30 & \text{reduced price} \left(70\% \text{ of original price}\right)
\end{array}
$$

The calculator solution to this example uses the complement, with respect to one, of the discount.

$2879 [×] [(] 1 [−] .3 [)] [=] 2015.3

Note: Refer to Appendix C for calculator basics.

The next example shows how to find a **percent of markdown**.

> **QUICK TIP**   The original selling price is always the base or 100% and the percent of markdown is always calculated on the original selling price.

EXAMPLE 2

Calculating the Percent of Markdown

The total inventory of holiday cards at a gift shop has a retail value of $785. If the cards were sold at reduced prices that totaled $530, what is the percent of markdown on the original price?

SOLUTION

First find the amount of the markdown.

$$
\begin{array}{ll}
\$785 & \text{original price} \\
-\ 530 & \text{reduced price} \\
\hline
\$255 & \text{markdown}
\end{array}
$$

Finding the percent of the original price that is the markdown is a rate problem. (See **Chapter 3**.)

$$
\text{Rate} = \frac{\textbf{Part}}{\textbf{Base}} = \frac{255}{785} = .3248 = 32\% \text{ markdown rounded to the nearest whole percent}
$$

The cards were sold at a markdown of 32%.

EXAMPLE 3

Finding the Original Price

Bouza's Baby News offers a child's car seat at a reduced price of $63 after a 25% markdown from the original price. Find the original price.

SOLUTION

After the 25% markdown, the reduced price of $63 represents 75% of the original price. The original price, or base, must be found.

$$
\text{Base} = \frac{\textbf{Part}}{\textbf{Rate}} = \frac{63}{.75} = \$84 \text{ original price}
$$

The original price of the car seat was $84.

Check the answer by subtracting 25% of $84 from $84 $\left(\$84 - \left(.25 \times \$84\right) = \$63\right)$.

> **QUICK TIP**   In Example 3, notice that 75% is used in the formula rather than 25%. The reduced price, $63, is represented by 75%.

Objective **3** **Define the terms associated with loss.** The amount of a markdown must be large enough to sell the merchandise while providing as much profit as possible. Merchandise that is marked down will result in either a **reduced net profit**, **breaking even**, an **operating loss**, or a **gross** or **absolute loss**.

The following diagram illustrates the meaning of these terms.

Reduced net profit results when the reduced price is still within the net profit range—is greater than the cost plus operating expenses.

Objective **4** **Determine the break-even point and operating loss.** The **break-even point** is the point at which the reduced price just covers cost plus overhead (operating expenses).

An **operating loss** occurs when the reduced price is less than the break-even point. The operating loss is the difference between the break-even point and the reduced selling price.

Objective **5** **Determine the amount of a gross or absolute loss.** An **absolute loss** or **gross loss** is the result of a reduced price that is below the cost of the merchandise alone. The absolute or gross loss is the difference between the cost and reduced selling price.

The following formulas are helpful when working with markdowns.

> Break-even point = Cost + Operating expenses
> Operating loss = Break-even point − Reduced selling price
> Absolute loss = Cost − Reduced selling price

EXAMPLE 4

Determining a Profit or a Loss

Appliance Giant paid $400 for an energy-efficient refrigerator. If operating expenses are 30% of cost and the refrigerator is sold for $500, find the amount of profit or loss.

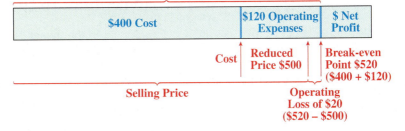

SOLUTION

Operating expenses are 30% of cost, or

$$\text{Operating expenses} = .30 \times \$400 = \$120$$

The break-even point for the refrigerator is

$$\text{Cost} + \text{Operating expenses} = \text{Break-even point}$$

$$\$400 + \left(.3 \times \$400\right) = \$400 + \$120 = \$520 \text{ Break-even point}$$

Since the break-even point is $520, and the selling price is $500, there is a loss of

$$\$520 - \$500 = \$20$$

The $20 loss is an operating loss, since the selling price is less than the break-even point but greater than the cost.

The calculator solution to this example follows.

$$400 \boxed{+} \boxed{(} .3 \boxed{\times} 400 \boxed{)} \boxed{-} 500 \boxed{=} 20$$

Note: Refer to Appendix C for calculator basics.

EXAMPLE 5

Determining the Operating Loss and the Absolute Loss

A game table normally selling for $360 at Olympic Sports is marked down 30%. If the cost of the game table is $260 and the operating expenses are 20% of cost, find **(a)** the operating loss and **(b)** the absolute loss.

Original Selling Price

| $260 | $312 | $360 |

| $260 Cost | $52 Operating Expenses | $48 Net Profit |

Reduced Selling Price $252   Cost $260   Break-even Point $312

Operating Loss $60

Absolute Loss $8

SOLUTION

(a) The break-even point (cost + operating expenses) is $312 **($260 + (.2 × $260) = $260 + $52**

$$\text{Reduced price} = \$360 - (.3 \times \$360) = \$360 - \$108 = \$252$$

$$\text{Operating loss} = \$312 \text{ break-even point} - \$252 \text{ reduced price} = \$60$$

(b) The absolute or gross loss is the difference between the cost and the reduced price.

$$\$260 \text{ cost} - \$252 \text{ reduced price} = \$8 \text{ absolute loss}$$

The following bar graph shows the percent of adults who get an emotional high as a result of making certain purchases. Customers love buying things—especially when they are on sale. This is valuable information to manufacturers, retailers, and all merchandisers.

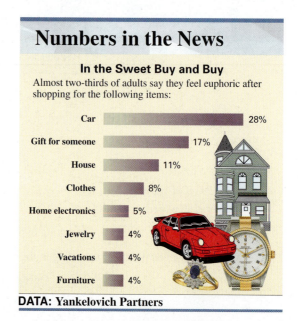

Numbers in the News

In the Sweet Buy and Buy

Almost two-thirds of adults say they feel euphoric after shopping for the following items:

Car 28%
Gift for someone 17%
House 11%
Clothes 8%
Home electronics 5%
Jewelry 4%
Vacations 4%
Furniture 4%

DATA: Yankelovich Partners

7.3 | Exercises

FOR EXTRA HELP

MyMathLab

InterActMath.com

MathXP MathXL

MathXL
Tutorials on CD

Tutor   Addison-Wesley
Center   Math Tutor Center

DVT/Videotape

The **Quick Start** *exercises in each section contain solutions to help you get started.*

Find the missing quantities. Round rates to the nearest whole percent and dollar amounts to the nearest cent. (See Examples 1–3.)

Quick Start

	Original Price	% Markdown	$ Markdown	Reduced Price
1.	$860	**25%**	$215	**$645**

$R = \frac{P}{B} = \frac{215}{860} = .25 = 25\%; \ \$860 - \$215 = \$645$

2.	**$240**	40%	**$96**	$144

$B = \frac{P}{R} = \frac{144}{.6} = \$240; \ \$240 - \$144 = \$96$

3.	$61.60	_____	_____	$43.12
4.	_____	$66\frac{2}{3}\%$	_____	$3.10
5.	$6.50	_____	$1.30	_____
6.	_____	50%	$65.25	_____

Complete the following. If there is no operating loss or absolute loss, write "none." (See Examples 4 and 5.)

Quick Start

	Cost	Operating Expense	Break-even Point	Reduced Price	Operating Loss	Absolute Loss
7.	$96	$24	**$120**	$100	**$20**	*none*

$\$96 + \$24 = \$120; \ \$120 - \$100 = \$20$

8.	$25	$8	**$33**	$22	**$11**	**$3**

$\$25 + \$8 = \$33; \ \$33 - \$22 = \$11; \ \$25 - \$22 = \$3$

9.	$50	_____	$66	$44	_____	_____
10.	$12.50	_____	$16.50	$11	_____	_____
11.	$310	$75	_____	_____	$135	_____
12.	$156	$44	_____	_____	$60	_____

 indicates an exercise that is related to the Case in Point feature within the section.

13. Describe five reasons why a store will markdown the price of merchandise to get it sold.

14. As a result of a markdown, there are three possible results: reduced net profit, operating loss, and absolute loss. As a business owner, which would concern you the most? Explain. (See Objectives 4 and 5.)

Solve the following application problems. Round rates to the nearest whole percent and dollar amounts to the nearest cent.

Quick Start

15. **AUTOMOBILE SOUND SYSTEMS**   Fry's Electronics prices their entire inventory of 200 watt AM/FM MP3/CD players with removable face at $133,509. If the original price of the inventory was $226,284, find the percent of markdown on the original price.

    $226,284 − $133,509 = $92,775$; $R = \frac{P}{B} = \frac{92,775}{226,284} = .409 = 41\%$

    15. <u>41%</u>

16. **OAK DESK**   An oak desk originally priced at $837.50 is reduced to $686.75. Find the percent of markdown on the original price.

    16. _____

 17. **TREADMILL**   Olympic Sports paid $720 for an Image 10.6 Qi treadmill. Their operating expenses are $33\frac{1}{3}\%$ of cost. If they sell the treadmill at a clearance price of $899.99, find the amount of profit or loss.

    17. _____

 18. **EXERCISE BICYCLES**   Olympic Sports has an end-of-season sale during which it sells a Proform Upright Bike for $265. If the cost was $198 and the operating expenses were 25% of cost, find the amount of profit or loss.

    18. _____

19. **TRUCK ACCESSORIES**   Pep Boys Automotive paid $208.50 for a pickup truck bedliner. The original selling price was $291.90, but this was marked down 35%. If operating expenses are 28% of the cost, find (a) the operating loss and (b) the absolute loss.

    (a) _____
    (b) _____

20. **ANTIQUES**   American Antiques paid $153.49 for a fern stand. The original selling price was $208.78, but this was marked down 46% in order to make room for incoming merchandise. If operating expenses are 14.9% of cost, find (a) the operating loss and (b) the absolute loss.

    (a) _____
    (b) _____

7.4 | Turnover and Valuation of Inventory

Objectives

1. Determine average inventory.
2. Calculate stock turnover.
3. Use uniform product codes.
4. Use the specific identification method to value inventory.
5. Determine inventory value using the weighted-average method.
6. Use the FIFO method to value inventory.
7. Use the LIFO method to value inventory.
8. Estimate inventory value using the retail method.

CASE *in* POINT

Many of the items stocked and sold by Olympic Sports are ordered year round. One example is the Explorer internal frame backpack. Maureen O'Connor, the manager, wants to make sure that her products are turning over (selling) so that the store does not have too much cash tied up in stock that is not selling. She also wants to be sure that the store has enough of the most popular products.

Objective 1 Determine average inventory. The average time for merchandise to sell is a common measure of a business's efficiency. The number of times that the merchandise sells during a certain period of time is called the **inventory turnover** or the **stock turnover**. A business such as a florist shop or produce stand has a very fast turnover of merchandise, perhaps just a few days. On the other hand, a furniture store normally has a much slower turnover, perhaps several months.

Find stock turnover by first calculating **average inventory**. The average inventory for a certain period is found by adding the inventories taken during the time period and then dividing the total by the number of times that the inventory was taken.

EXAMPLE 1

Determining Average Inventory

Inventory at Olympic Sports was $285,672 on April 1 and $198,560 on April 30. What was the average inventory?

SOLUTION

First add the inventory values.

$285,672   **April 1**
+ 198,560   **April 30**
$484,232

Then divide by the number of times inventory was taken.

$$\frac{\$484,232}{2} = \$242,116$$

The average inventory was $242,116.

QUICK TIP To find the average inventory for a period of time, an inventory must always be taken at the beginning of the period and at the end of the period. For example, to find average inventory for a full year, businesses commonly find inventory on the first day of each month and on the last day of the last month. They then find the average inventory by adding 13 inventory amounts and dividing by 13, the number of inventories taken.

Keeping a close watch on inventory is an ongoing concern of management. The following newspaper advertisement is promoting a year-end sale with no money to be paid until June of the following year. This sale should have a major impact in reducing the store's inventory.

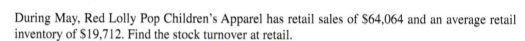

MONDAY–FRIDAY 10AM–9PM SATURDAY 10AM–8PM & SUNDAY 11AM–6PM. PRICES GOOD 'TIL TUESDAY!
GET HUGE STOREWIDE SAVINGS!
YEAR-END SALE!
No Money Down, No Interest & No Payment 'til June
...On Every Item ...On Every Room!
Same As Cash Option. On Approved Credit With No Down Payment, Interest Accrues From Delivery Date if not Paid in Full by June
NO DOWN PAYMENT, NO INTEREST & NO PAYMENTS 'TIL JUNE   ON 8-WAY, HAND-TIED LEATHER

Turnover is the number of times that the value of the merchandise or inventory in the store has sold during a period of time.

Objective **2** **Calculate stock turnover.** Businesses value inventory either at retail or at cost. For this reason, **stock turnover** is found by using either of these formulas.

$$\text{Turnover at retail} = \frac{\text{Retail sales}}{\text{Average inventory at retail}}$$

$$\text{Turnover at cost} = \frac{\text{Cost of goods sold}}{\text{Average inventory at cost}}$$

The turnover ratio may be identical by using either method. The variation that often exists is caused by stolen merchandise (called *inventory shrinkage*) or merchandise that has been marked down or has become unsellable. Normally, turnover at retail is slightly lower than turnover at cost. For this reason, many businesses prefer this more conservative figure.

EXAMPLE 2

Finding Stock Turnover at Retail

During May, Red Lolly Pop Children's Apparel has retail sales of $64,064 and an average retail inventory of $19,712. Find the stock turnover at retail.

SOLUTION

$$\text{Turnover at retail} = \frac{\text{Retail sales}}{\text{Average inventory at retail}} = \frac{\$64,064}{\$19,712} = \textbf{3.25 at retail}$$

On average, the store turned over or sold the value of its entire inventory 3.25 times during the month.

EXAMPLE 3

Finding Stock Turnover at Cost

If the average inventory value at cost for Red Lolly Pop Children's Apparel in Example 2 was $11,826, and the cost of goods sold was $38,792, find the stock turnover at cost.

SOLUTION

$$\text{Turnover at cost} = \frac{\text{Cost of goods sold}}{\text{Average inventory at cost}} = \frac{\$38,792}{\$11,826} = \textbf{3.28 at cost}\ (\textbf{rounded})$$

The stock turnover ratio is useful for comparison purposes only. Many trade organizations publish such operating statistics to permit businesses to compare their operation with the industry as a whole. In addition to this, management uses these rates to compare turnover from period to period and from department to department.

It is not always easy to place a value on each of the items in inventory. Many large companies keep a **perpetual inventory** by using a computer. As new items are received, the quantity, size, and cost of each are entered in the computer. Sales clerks enter product codes into the cash register (or uniform product codes are entered automatically with an optical scanner).

Objective 3   Use uniform product codes. Uniform product codes (**UPC**) are the black stripes that appear on the packaging for most items sold in stores. Each product and product size is assigned its own code number. These UPCs are a great help in keeping track of inventory.

A Cracker Jack box is shown at the left. The UPC number on the package is 4125723276. The checkout clerk in a retail store passes the coded lines over an optical scanner. The numbers are picked up by a computer, which recognizes the product by its code. The computer then forwards the price of the item to the cash register. At the same time the price is being recorded, the computer is subtracting the item automatically from inventory. After all the items being purchased have passed over the scanner, the customer receives a detailed cash-register receipt that gives a description of each item, the price of each item, and the total amount of the purchase.

Since the computer keeps track of stock on hand and is programmed to respond when inventory gets low, it provides more accurate inventory control and lower labor costs for the store.

Most businesses take a **physical inventory** which is an actual count of each item in stock at a given time at regular intervals. For example, inventory may be taken monthly, quarterly, semiannually, or just once a year. An inventory taken at regular intervals is called a **periodic inventory**.

There are four major methods used for inventory valuation: the specific identification method, the weighted-average method, the first-in first-out method, and the last-in, first-out method.

Objective 4   Use the specific identification method to value inventory. The **specific identification method** is useful if items are easily identified and costs do not fluctuate. Each item is cost coded with either numerals or letters. These costs are then added to find ending inventory.

Since the cost of many items changes with time, there may be several of the same item in stock that were purchased at different costs. For this reason, many businesses prefer taking inventory at retail. The retail value of all identical items is the same.

Objective 5   Determine inventory value using the weighted-average method. The **weighted average (average cost)** of inventory involves finding the average cost of an item and then multiplying the number of items remaining by the average cost per item.

EXAMPLE 4

Using Weighted Average (Average Cost) Inventory Valuation

Suppose Olympic Sports made the following purchases of the Explorer internal frame backpack during the year.

Beginning inventory	20 backpacks at $70
January	50 backpacks at $80
March	100 backpacks at $90
July	60 backpacks at $85
October	40 backpacks at $75

At the end of the year, there are 75 backpacks in inventory. Use the weighted-average method to find the inventory value.

SOLUTION

Find the total cost of all the backpacks.

Beginning inventory	**20**	×	$70	=	$1400
January	**50**	×	$80	=	$4000
March	**100**	×	$90	=	$9000
July	**60**	×	$85	=	$5100
October	**40**	×	$75	=	$3000
Total	**270**				**$22,500**

Find the average cost per backpack by dividing this total cost by the number purchased. The average cost per backpack is

$$\frac{\$22,500}{270} = \$83.33 \text{ (rounded)}.$$

Since the average cost is \$83.33 and 75 backpacks remain in inventory, the weighted-average method gives the inventory value of the remaining backpacks as \$83.33 × 75 = \$6249.75.

The calculator solution to this example has several steps. First, find the total number of backpacks purchased and place the total in memory.

20 + 50 + 100 + 60 + 40 = 270 **STO**

Next, find the total cost of all the backpacks purchased and divide by the number stored in memory. This gives the average cost per backpack.

20 × 70 + 50 × 80 + 100 × 90 + 60 × 85 +
40 × 75 = ÷ **RCL** = 83.3333

Finally, round the average cost to the nearest cent and multiply by the number of backpacks in inventory to get the weighted average inventory value.

83.33 × 75 = 6249.75

Note: Refer to Appendix C for calculator basics.

The cost of items purchased by the retailer is one of the greatest influences on the final retail price. The factors affecting cost include the quality of the product, the quantity purchased, and the geographic location of the purchaser. The following graphic shows the average price paid for a tennis racket in various countries around the world.

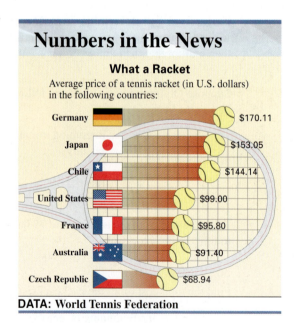

Numbers in the News

What a Racket

Average price of a tennis racket (in U.S. dollars) in the following countries:

Germany — \$170.11
Japan — \$153.05
Chile — \$144.14
United States — \$99.00
France — \$95.80
Australia — \$91.40
Czech Republic — \$68.94

DATA: World Tennis Federation

Objective **6** **Use the FIFO method to value inventory.** The **first-in, first-out (FIFO) method** of inventory valuation assumes a natural flow of goods through the inventory. The first goods to arrive are the first goods to be sold, so the last items purchased are the items remaining in inventory.

EXAMPLE 5

Using FIFO to Determine Inventory Valuation

Use the FIFO method to find the inventory value of the 75 backpacks from Olympics Sports in Example 4.

SOLUTION

With the FIFO method, the 75 remaining backpacks are assumed to consist of the 40 backpacks bought in October and 35 $(75 - 40 = 35)$ backpacks from the previous purchase in July. The value of the inventory is:

October	**40** backpacks at $75 =	$3000	value of last 40
July	**35** backpacks at $85 =	$2975	value of previous 35
	75 valued at	$5975	

The value of the backpack inventory is $5975 using the FIFO method.

Objective 7 Use the LIFO method to value inventory. The **last in, first-out (LIFO) method** of inventory valuation assumes a flow of goods through the inventory that is just the opposite of the FIFO flow. With LIFO, the goods remaining in inventory are those goods that were first purchased.

EXAMPLE 6

Using LIFO to Determine Inventory Valuation

Use the LIFO method to value the 75 backpacks in inventory at Olympic Sports. (See Example 4.)

SOLUTION

The calculation starts with the beginning inventory and moves through the year's purchases, resulting in 75 backpacks still in stock. The beginning inventory and January purchases come to 70 backpacks, so the cost of 5 more $(75 - 70 = 5)$ backpacks from the March purchase is needed.

Beginning inventory	**20** backpacks at $70 =	$1400	value of first 20
January	**50** backpacks at $80 =	$4000	value of next 50
March	**5** backpacks at $90 =	$ 450	value of last 5
Total	**75** valued at	$5850	

The value of the backpack inventory is $5850 using the LIFO method.

Depending on the method of valuing inventories that is used, Olympic Sports may show the inventory value of the 75 backpacks as follows.

Average cost method	$6249.75
FIFO	$5975
LIFO	$5850

The preferred inventory valuation method would be determined by Olympic Sports, perhaps on the advice of an accountant.

QUICK TIP While the FIFO method of inventory evaluation is the most commonly used method, accepted accounting practice insists that the method used to evaluate inventory be stated on the company's financial statements.

Objective 8 Estimate inventory value using the retail method. An estimate of the value of inventory may be found using the **retail method of estimating inventory**. With this method, the cost of goods available for sale is found as a percent of the retail value of the goods available for sale during the same period. This percent is then multiplied by the retail value of inventory at the end of the period. The result is an estimate of the inventory at cost.

EXAMPLE 7

Estimating Inventory Value Using the Retail Method

The inventory on December 31 at Olympic Sports was $129,200 at cost and $171,000 at retail. Purchases during the next three months were $165,400 at cost, $221,800 at retail, and net sales were $168,800. Use the retail method to estimate the value of inventory at cost on March 31.

SOLUTION

Step 1 Find the value of goods available for sale (inventory) at cost and at retail.

	At cost	At retail	
	$129,200	$171,000	beginning inventory
	+ 165,400	+ 221,800	purchases
	$294,600	$392,800	goods available for sale

Step 2 Find the retail value of current inventory.

− 168,800	net sales
$224,000	March 31 inventory at retail

Step 3 Now find the percent of the value of goods available for sale at cost to goods available for sale at retail (cost ratio).

$$\frac{\$294,600}{\$392,800} \quad \frac{\text{goods available for sale at cost}}{\text{goods available for sale at retail}} = .75 = 75\% \ (\text{cost ratio})$$

Step 4 Finally, the estimated inventory value at cost on March 31 is found by multiplying inventory at retail on March 31 by 75% (cost ratio)

Ending inventory at retail × % (cost ratio) = Inventory at cost

$224,000 × .75 = $168,000 March 31 inventory at cost.

Companies selling on the Internet often will be able to have products shipped directly from a manufacturer or wholesaler to their customers. This could result in decreased inventory requirements in the supply chain. The next visual shows projected sales growth in the fastest-growing online sales categories.

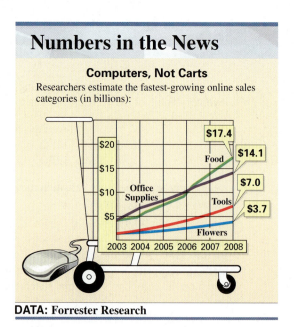

Numbers in the News

Computers, Not Carts

Researchers estimate the fastest-growing online sales categories (in billions):

$17.4 — Food
$14.1
$7.0 — Tools
$3.7 — Flowers

Office Supplies

$20
$15
$10
$5

2003 2004 2005 2006 2007 2008

DATA: Forrester Research

7.4 | Exercises

The **Quick Start** exercises in each section contain solutions to help you get started.

Find the average inventory in each of the following. (See Example 1.)

Quick Start

Date	Inventory Amount at Retail	Average Inventory	Date	Inventory Amount at Retail	Average Inventory
1. July 1	$18,300		**2.** January 1	$42,312	
October 1	$26,580		July 1	$38,514	
December 31	$23,139	**$22,673**	December 31	$30,219	**$37,015**
$68,019 total of inv. ÷ 3 = $22,673			$111,045 total of inv. ÷ 3 = $37,015		
3. January 1	$65,430		**4.** January 31	$69,480	
April 1	$58,710		April 30	$55,860	
July 1	$53,410		July 31	$80,715	
October 1	$78,950		October 31	$88,050	
December 31	$46,340	_____	January 31	$63,975	_____

Find the stock turnover at cost and at retail in each of the following. Round to the nearest hundredth. (See Examples 2 and 3.)

Quick Start

	Average Inventory at Cost	Average Inventory at Retail	Cost of Goods	Retail Sales	Turnover at Cost	Turnover at Retail
5.	$17,830	$35,390	$50,394	$99,450	**2.83**	**2.81**
	$50,394 ÷ $17,830 = 2.83 at cost; $99,450 ÷ $35,390 = 2.81 at retail					
6.	$15,140	$24,080	$67,408	$106,193	**4.45**	**4.41**
	$67,408 ÷ $15,140 = 4.45 at cost; $106,193 ÷ $24,080 = 4.41 at retail					
7.	$72,120	$138,460	$259,123	$487,379	_____	_____
8.	$38,074	$48,550	$260,420	$330,060	_____	_____
9.	$180,600	$256,700	$846,336	$1,196,222	_____	_____
10.	$411,580	$780,600	$1,905,668	$3,559,536	_____	_____

 indicates an exercise that is related to the Case in Point feature within the section.

Find the inventory values using (a) the weighted average method, (b) the FIFO method, and (c) the LIFO method for each of the following. Round to the nearest cent if necessary. (See Examples 4–6.)

Quick Start

Purchases	Now in Inventory	Weighted Average Method	FIFO Method	LIFO Method
11. Beginning inventory: 10 units at $8 June: 25 units at $9 August: 15 units at $10	20 units	$182	$195	$170

$(\$455 \text{ purchases} \div 50) \times 20 = \$9.10 \times 20 = \$182 \text{ average cost method}$

$$
\begin{array}{ll}
15 \times \$10 = \$150 & 10 \times \$8 = \$\ 80 \\
+\ 5 \times \$\ 9 = \$\ 45 & +\ 10 \times \$9 = \$\ 90 \\
\hline
20 \qquad \$195 \text{ FIFO} & 20 \qquad \$170 \text{ LIFO}
\end{array}
$$

Purchases	Now in Inventory	Weighted Average Method	FIFO Method	LIFO Method
12. Beginning inventory: 80 units at $14.50 July: 50 units at $15.80 October: 70 units at $13.90	90 units	_____	_____	_____
13. Beginning inventory: 50 units at $30.50 March: 70 units at $31.50 June: 30 units at $33.25 August: 40 units at $30.75	75 units	_____	_____	_____
14. Beginning inventory: 700 units at $1.25 May: 400 units at $1.75 August: 500 units at $2.25 October: 600 units at $3.00	720 units	_____	_____	_____

15. Identify three types of businesses that you think would have a high turnover. Identify three types of businesses that you think would have a low turnover.

16. Which departments in a grocery store do you think have the highest turnover? Which ones have the lowest turnover? Why do you think this is true?

Solve the following application problems. Round stock turnover to the nearest hundredth.

17. STOCK TURNOVER AT COST   The Glass Works has an average inventory at cost of $15,730, and cost of goods sold for the same period is $85,412. Find the stock turnover at cost.

$\frac{\$85,412}{\$15,730} = 5.43$ turnover at cost

17. <u>5.43 turnover at cost</u>

18. STOCK TURNOVER AT RETAIL   Jumbo Market has an average canned-fruit inventory of $2320 at retail. Retail sales of canned fruit for the year were $98,669. Find the stock turnover at retail.

18. _____

19. SPRAY-PAINT INVENTORY   The Graphic Hobby House made purchases of assorted colors of spray paint during the year as follows.

(a) _____
(b) _____
(c) _____

Beginning inventory   200 cans at $1.10
March   400 cans at $1.20
May   700 cans at $1.00
August   500 cans at $1.15
November   300 cans at $1.30

At the end of the year, they had 450 cans of spray paint in stock.
(a) Find the inventory value using the weighted average method.
(b) Find the inventory value using the FIFO method.
(c) Find the inventory value using the LIFO method.

20. SPORT T-SHIRTS   Olympic Sports made the following purchases of sport T shirts made in Taiwan: Beginning inventory was 650 shirts at $3.80 each; June, 500 shirts at $4.20 each; September, 450 shirts at $3.95 each; and December, 600 shirts at $4.05 each. An inventory at the end of the year shows that 775 T shirts remain. **(a)** Find the inventory value using the weighted average method. **(b)** Find the inventory value using the FIFO method. **(c)** Find the inventory value using the LIFO method.

(a) _____
(b) _____
(c) _____

21. ATHLETIC SOCKS   Olympic Sports made the following purchases of 3-pair packages of athletic socks.

(a) _____
(b) _____
(c) _____

Beginning inventory	200 packages at $3.10
May	250 packages at $3.50
August	300 packages at $4.25
October	280 packages at $4.50

An inventory at the end of October shows that 320 packages remain. **(a)** Find the inventory value using the weighted average method. **(b)** Find the inventory value using the FIFO method. **(c)** Find the inventory value using the LIFO method.

22. **AUTOMOBILE MUFFLERS**   Marco Muffler Wholesalers made purchases of automobile mufflers through the year as follows.

Beginning inventory	300 units at $21.60
March	400 units at $24.00
August	450 units at $24.30
November	350 units at $22.50

An inventory at the end of December shows that 530 mufflers remain. **(a)** Find the inventory value using the weighted average method. **(b)** Find the inventory value, using the FIFO method. **(c)** Find the inventory value using the LIFO method.

(a) _____
(b) _____
(c) _____

23. **PIANO REPAIR**   The September 30 inventory at Liverpool Piano Repair was $43,750 at cost and $62,500 at retail. Purchases during the next three months were $51,600 at cost, $73,800 at retail, and net sales were $92,500. Use the retail method to estimate the value of the inventory at cost on December 31.

23. _____

24. **EVALUATING INVENTORY**   Cell Phones Plus had an inventory of $27,000 at cost and $45,000 at retail on March 31. During the next three months, they made purchases of $108,000 at cost and $180,000 at retail and had net sales of $162,000. Use the retail method to estimate the value of inventory at cost on June 30.

24. _____

25. In your opinion, what are the benefits to a merchant who is using uniform product codes (UPCs)?

26. Which of the three inventory valuation methods discussed in this section was most interesting to you? Explain how this method determines inventory value.

Chapter 7 | Quick Review

CHAPTER TERMS *Review the following terms to test your understanding of the chapter. For each term you do not know, refer to the page number found next to that term.*

absolute loss [**p. 283**]	gross profit [**p. 260**]	net earnings [**p. 260**]	retail method of estimating
actual selling price [**p. 282**]	inventory turnover [**p. 287**]	net profit [**p. 260**]	inventory [**p. 291**]
average inventory [**p. 287**]	last-in, first-out (LIFO) method	operating expenses [**p. 260**]	sale price [**p. 282**]
break-even point [**p. 283**]	[**p. 291**]	operating loss [**p. 283**]	selling price [**p. 260**]
breaking even [**p. 283**]	margin [**p. 260**]	overhead [**p. 260**]	specific identification method
conversion formulas [**p. 273**]	markdown [**p. 281**]	percent of markdown [**p. 282**]	[**p. 289**]
cost [**p. 260**]	markup [**p. 260**]	periodic inventory [**p. 289**]	stock turnover [**p. 287**]
first-in, first-out (FIFO) method	markup based on cost [**p. 262**]	perpetual inventory [**p. 289**]	uniform product code (UPC)
[**p. 290**]	markup formula [**p. 261**]	physical inventory [**p. 289**]	[**p. 289**]
formula for markdown [**p. 282**]	markup on cost [**p. 260**]	reduced net profit [**p. 283**]	weighted-average (average cost)
gross loss [**p. 283**]	markup on selling price [**p. 260**]	reduced price [**p. 282**]	method [**p. 289**]

CONCEPTS

7.1 Markup on cost

$$
\begin{array}{ll}
100\% & \text{Cost} \qquad \text{(base)} \\
+ & \textbf{Markup? (part)} \\
\hline
& \text{Selling Price}
\end{array}
$$

Cost is base. Use the basic percent formula.

$$P = B \times R$$

7.1 Calculating the percent of markup

$$
(\text{rate})\begin{array}{lll}
100\% & C & \$ \ \text{(base)} \\
?\% & M & \$ \\
\hline
\% & S & \$
\end{array}
$$

Solve for rate.

7.1 Finding the cost and the selling price

$$
\begin{array}{lll}
100\% & C & \$? \ \textbf{(base)} \\
& M & \$ \\
\hline
S & \$? & \textbf{(part)}
\end{array}
$$

Solve for base.

7.2 Markup on selling price

$$
\begin{array}{lll}
\% & C & \$ \\
\% & M & \$? \ \textbf{(part)} \\
\hline
100\% & S & \$ \ \text{(base)}
\end{array}
$$

Solve for part.

EXAMPLES

$$
(\text{rate})\begin{array}{lll}
100\% & C & \$160 \ \text{(base)} \\
25\% & M & \textbf{\$?} \ \textbf{(part)} \\
\hline
& S & \$
\end{array}
$$

$$P = B \times R$$
$$P = \$160 \times \textbf{.25}$$
$$P = \$40 \ \text{markup}$$

$$
(\textbf{rate})\begin{array}{lll}
100\% & C & \$420 \ \text{(base)} \\
\textbf{?\%} & M & \$ \qquad \text{(part)} \\
\hline
\% & S & \$546 \ \text{(part)}
\end{array}
$$

$$\$546 - \$420 = \$126 \ \text{markup}$$

$$R = \frac{P}{B} = \frac{126}{420}$$
$$R = 30\%$$

$$
(\text{rate})\begin{array}{lll}
100\% & C & \$? \ \textbf{(base)} \\
50\% & M & \$56 \\
\hline
\% & S & \$? \ \textbf{(part)}
\end{array}
$$

$$B = \frac{P}{R} = \frac{56}{.5}$$
$$B = \$112 \ \text{cost}$$

$$\$112 + \$56 = \$168 \ \text{selling price}$$

$$
(\text{rate})\begin{array}{lll}
\% & C & \$ \\
25\% & M & \$? \ \textbf{(part)} \\
\hline
100\% & S & \$6.00 \ \text{(base)}
\end{array}
$$

$$P = B \times R$$
$$P = \$6.00 \times \textbf{.25}$$
$$P = \$1.50$$

CONCEPTS	EXAMPLES

7.2   Finding the cost

$$\begin{array}{lll} \% & C & \$\text{?} \ \textbf{(part)} \\ \% & M & \$ \\ \hline 100\% & S & \$ \end{array}$$

$$\left(\text{rate}\right)\begin{array}{lll} \% & C & \$\text{?} & \textbf{(part)} \\ 35\% & M & \$87.50 & \text{(part)} \\ \hline 100\% & S & \$ & \text{(base)} \end{array}$$

$$B = \frac{P}{R} = \frac{87.5}{.35} = \$250 \text{ selling price}$$

$$\$250 - \$87.50 = \$162.50 \text{ markup}$$

7.2   Calculating the selling price and the markup

$$\begin{array}{lll} \% & C & \$ \\ \% & M & \$\text{?} \ \textbf{(part)} \\ \hline 100\% & S & \$\text{?} \ \textbf{(base)} \end{array}$$

$$\left(\text{rate}\right)\begin{array}{lll} 75\% & C & \$150 & \left(\text{part}\right) \\ 25\% & M & \$\text{?} & \textbf{(part)} \\ \hline 100\% & S & \$\text{?} & \textbf{(base)} \end{array}$$

$$100\% - 25\% = 75\% \text{ cost}$$

$$B = \frac{P}{R} = \frac{150}{.75} = \$200 \text{ selling price}$$

$$\$200 - \$150 = \$50 \quad \text{markup}$$

7.2   Converting markup on cost to markup on selling price

Use the formula

$$\% \text{ markup on selling price} = \frac{\% \text{ markup on cost}}{100\% - \% \text{ markup on cost}}$$

Convert 25% markup on cost to markup on selling price.

$$\frac{\% \text{ markup on}}{\text{selling price}} = \frac{25\%}{\textbf{100\% + 25\%}}$$

$$= \frac{.25}{\textbf{1.25}}$$

$$= .2 = 20\%$$

7.2   Converting markup on selling price to markup on cost

Use the formula

$$\% \text{ markup on cost} = \frac{\% \text{ markup on selling price}}{100\% - \% \text{ markup on selling price}}$$

Convert 20% markup on selling price to markup on cost.

$$\frac{\% \text{ markup on}}{\text{cost}} = \frac{20\%}{\textbf{100\% - 20\%}}$$

$$= \frac{.2}{\textbf{.8}}$$

$$= .25 = 25\%$$

7.2   Finding selling price for perishables

1. Find total cost and selling price.
2. Subract the quantity not sold to find the number that are sold.
3. Divide the remaining sales by the number of saleable units to get selling price per unit.

60 doughnuts cost 15¢ each; 10 are not sold; 50% markup on selling price. Find selling price per doughnut.

$$\text{Cost} = 60 \times \$.15 = \$9$$

$$\left(\text{rate}\right)\begin{array}{lll} 50\% & C & \$9 \ \text{(part)} \\ 50\% & M & \\ \hline 100\% & S & \text{?} \ \text{(base)} \end{array}$$

$$B = \frac{P}{R} = \frac{9}{.5} = \$18$$

$$60 - 10 = 50 \text{ doughnuts sold}$$

$$\$18 \div 50 = \$.36 \text{ per doughnut}$$

7.3   Percent of markdown

Markdown is always a percent of the original price. Use the formula

$$R = \frac{P}{B}$$

$$\text{Markdown percent} = \frac{\text{Markdown amount}}{\text{Original price}}$$

Original price, $76; markdown, $19; find the percent of markdown.

$$R = \frac{P}{B} = \frac{19}{76} = .25$$

$$R = 25\% \text{ markdown}$$

CONCEPTS	EXAMPLES
7.3 Break-even point The cost plus operating expenses equals the break-even point.	Cost, $54; operating expenses, $16; find the break-even point. $54 cost + **$16 operating expenses** = $70 break-even point
7.3 Operating loss The difference between the break-even point and the reduced price (when below the break-even point) is the operating loss.	Break-even point, $70; reduced price, $58; find the operating loss. $$$70 break-even point $-$ **$58 reduced price** $$$12 operating loss
7.3 Absolute loss (gross loss) When the reduced price is below cost, the difference between the cost and reduced price is the absolute loss.	Cost, $54; reduced price, $48; find the absolute loss. $54 cost − **$48 reduced price** = $6 absolute loss
7.4 Average inventory Inventory is taken two or more times. Totals are added together, then divided by the number of inventories taken to get the average.	Inventories, $22,635, $24,692, and $18,796; find the average inventory. $$\frac{\$22{,}635 + \$24{,}692 + \$18{,}796}{3}$$ $$= \frac{\$66{,}123}{3}$$ $$= \$22{,}041 \text{ average inventory}$$
7.4 Turnover at retail Use the formula $$\text{Turnover} = \frac{\text{Retail sales}}{\text{Average inventory at retail}}$$	Retail sales, $78,496; average inventory at retail, $18,076; find turnover at retail. $$\frac{\$78{,}496}{\mathbf{\$18{,}076}} = 4.34 \text{ at retail} \quad \text{rounded}$$
7.4 Turnover at cost Use the formula $$\text{Turnover} = \frac{\text{Cost of goods sold}}{\text{Average inventory at cost}}$$	Cost of goods sold, $26,542; average inventory at cost, $6592; find turnover at cost. $$\frac{\$26{,}542}{\mathbf{\$6592}} = 4.03 \text{ at cost} \quad \text{rounded}$$
7.4 Specific identification to value inventory Each item is cost coded and the cost of each of the items is added to find total inventory.	Individual cost of each item in inventory is: item 1, $593; item 2, $614; item 3, $498; find total value of inventory. **$593** + **$614** + **$498** = $1705 total value of inventory
7.4 Weighted-average (average cost) method of inventory valuation This method values items in an inventory at the average cost of buying them.	Beginning inventory of 20 at $75; purchases of 15 at $80; 25 at $65; 18 at $70; 22 remain in inventory. Find the inventory value. **20** × $75 = $1500 **15** × $80 = $1200 **25** × $65 = $1625 **18** × $70 = $1260 Total **78**　　　　$5585 $$\frac{\$5585}{78} = \$71.60 \text{ average cost} \quad \text{rounded}$$ $71.60 × 22 = $1575.20 weighted-average method inventory value

CONCEPTS	EXAMPLES

7.4 First-in, first-out (FIFO) method of inventory valuation

First items in are first sold. Inventory is based on cost of last items purchased.

Beginning inventory of 25 items at $40; purchased on Aug. 7, 30 items at $35; 35 remain in inventory. Find the inventory

$$30 \times \$35 = \$1050 \quad \text{value of last 30}$$
$$\underline{5 \times \$40 = \$\ 200} \quad \text{value of previous 5}$$
$$35 \qquad\quad \$1250 \quad \text{value of inventory FIFO method}$$

7.4 Last-in, first-out (LIFO) method of inventory valuation

The items remaining in inventory are those items that were first purchased.

Beginning inventory of 48 items at $20 each; purchase on May 9, 40 items at $25 each; 55 remain in inventory. Find the inventory value.

$$48 \times \$20 = \$\ 960 \quad \text{value of first 48}$$
$$\underline{7 \times \$25 = \$\ 175} \quad \text{value of last 7}$$
$$55 \qquad\quad \$1135 \quad \text{value of inventory LIFO method}$$

7.4 Estimating inventory value using the retail method

$$\frac{\text{Goods available for sale at cost}}{\text{Goods available for sale at retail}} = \% \left(\text{cost ratio}\right)$$

$$\begin{array}{c}\text{Ending inventory}\\ \text{at retail}\end{array} \times \% \left(\text{cost ratio}\right) = \begin{array}{c}\text{Inventory}\\ \text{at cost}\end{array}$$

	Cost	Retail
beginning inventory	$9,000	$15,000
purchases	+ 36,000	+ 60,000
goods available for sale	$45,000	$75,000
net sales		− 54,000
ending inventory		$21,000

$$\frac{\$45,000}{\$75,000} \quad \begin{array}{l}\text{goods available for sale at cost}\\ \text{goods available for sale at retail}\end{array}$$

$$= .6 = 60\%$$

$$\$21,000 \times .6 = \$12,600 \text{ inventory value at cost}$$

Chapter 7 | Summary Exercise

Markdown: Reducing Prices to Move Merchandise

Olympic Sports purchased two dozen pairs of Nike Air Max in-line skates at a cost of $1950. Operating expenses for the store are 25% of cost while total markup on this type of product is 35% of selling price. Only 6 pairs of the skates sell at the original price and the manager decides to mark down the remaining skates. The price is reduced 25% and 6 more pairs sell. The remaining 12 pairs of skates are marked down 50% of the original selling price and are finally sold.

(a) Find the original selling price of each pair of skates.

(a) _____

(b) Find the total of the selling prices of all the skates.

(b) _____

(c) Find the operating loss.

(c) _____

(d) Find the absolute loss.

(d) _____

INVESTIGATE

Talk with the manager of a retail store. Does the store calculate markup based on cost or retail? Does the store use markdowns to promote the sale or liquidation of merchandise? How does the management decide how much to mark down merchandise? Ask the manager for an example of a product that had to be marked down so much that a gross loss resulted.

'NET ASSETS

Business on the Internet

REI

- 1938: Established by a group of 23 mountaineers.

- 2004: 2 million active members.

- 70 stores nationwide

- 7000 employees nationwide

- Named in "100 Best Companies to Work For" by Fortune magazine for 6 consecutive years.

REI was formed in 1938 by a group of 23 mountain climbers from Seattle, Washington. The company wanted the finest-quality climbing equipment and formed a buying cooperative (membership group) in order to find the best prices for their equipment. Today, anyone may shop at REI, but members—those who pay a one-time $15 fee to join—share in the company's profits through an annual patronage refund. In 2004, REI declared a total patronage refund of 10.2 percent to 1.8 million active members. REI is a privately held company and does not sell stock.

REI's easy-to-navigate Internet store provides access to more than 40,000 outdoor products and offers a variety of interactive education opportunities for outdoor enthusiasts. In addition to selecting from thousands of products and securely placing on-line orders, customers can use gear checklists, interact with gear experts, and learn basic outdoor skills by accessing educational clinics.

1. REI purchased one dozen Jansport backpacks at a cost of $504. If the company uses a markup of 25% on selling price, find the selling price of each backpack.

2. A two-person dome tent with a full rain fly has a wholesale price of $78. If the store has operating expenses of 24.5% of cost and desired a net profit of 10.5% of cost, find the selling price.

3. Do you have a special activity, sport, or hobby for which it is sometimes difficult to get the right kind and quality of equipment or supplies? Would you consider buying what you need by mail-order catalog or over the Internet? Why or why not?

4. List five possible advantages and five possible disadvantages of buying through the Internet. How could the disadvantages be eliminated or reduced? Do you see a time when you will be making half or more of your purchases through the Internet?

Chapter 7 | Test

To help you review, the numbers in brackets show the section in which the topic was discussed.

Solve for (a), (b), and (c). **[7.1 and 7.2]**

1. 100% C $64.00
    (a)% M $12.80
    (b)% S $(c)

2. 100% C $(b)
    38% M $(c)
    (a)% S $504.39

3. (a)% C $134.40
    (b)% M $(c)
    100% S $168.00

4. (a)% C $(c)
    (b)% M $ 6.15
    100% S $24.60

Find the equivalent markup on either cost or selling price, using the appropriate formula. Round to the nearest tenth of a percent. **[7.2]**

Markup on Cost	Markup on Selling Price	Markup on Cost	Markup on Selling Price
5. 25%	_____	6. 100%	_____

Complete the following. If there is no operating loss or absolute loss, write "none." **[7.3]**

Cost	Operating Expense	Break-even Point	Reduced Price	Operating Loss	Absolute Loss
7. $160	$40	_____	$186	_____	_____
8. $225	_____	$297	$198	_____	_____

Find the stock turnover at cost and at retail in the following. Round to the nearest hundredth. **[7.4]**

Average Inventory at Cost	Average Inventory at Retail	Cost of Goods Sold	Retail Sales	Turnover at Cost	Turnover at Retail
9. $14,120	$25,572	$81,312	$146,528	_____	_____

Solve the following application problems.

10. Olympic Sports buys jogging shorts manufactured in Indonesia for $97.50 per dozen pair. Find the selling price per pair if the retailer maintains a markup of 35% on selling price. **[7.2]**

10. _____

11. Best Buy sells a washer and dryer set for $790 while using a markup of 25% on cost. Find the cost. **[7.1]**

11. _____

12. The Computer Service Center sells a DeskJet print cartridge for $18.75. If the print cartridge costs the store $11.25, find the markup as a percent of selling price.   **[7.2]**

12. _____

13. Olympic Sports offers an inflatable boat for $199.95. If the boats cost $1943.52 per dozen, find **(a)** the markup, **(b)** the percent of markup on selling price, and **(c)** the percent of markup on cost. Round to the nearest tenth of a percent.   **[7.1 and 7.2]**

(a) _____
(b) _____
(c) _____

14. A motorcycle originally priced at $13,875 is marked down to $9990. Find the percent of markdown on the original price.   **[7.3]**

14. _____

15. Leslies' Pool Supply, a retailer, pays $285 for a diving board. The original selling price was $399, but it was marked down 40%. If operating expenses are 30% of cost, find **(a)** the operating loss and **(b)** the absolute loss.   **[7.3]**

(a) _____
(b) _____

16. Carpets Plus had an inventory of $117,328 on January 1, $147,630 on July 1, and $125,876 on December 31. Find the average inventory.   **[7.4]**

16. _____

Round to the nearest dollar amount.

17. Clutch Masters made the following purchases of universal joints during the year: Beginning inventory, 30 at $18.50 each; June, 25 at $21.80 each; September, 20 at $20.50 each; and November, 30 at $21.25 each. An inventory shows that 55 universal joints remain. Find the inventory value using the weighted average method.   **[7.4]**

17. _____

18. Find the inventory value listed in Exercise 17 using **(a)** the FIFO method and **(b)** the LIFO method.   **[7.4]**

(a) _____
(b) _____

Cumulative Review

Chapters 4–7

The following credit-card transactions were made at the Patio Store. Answer Exercises 1–5 using this information. Round to the nearest cent.   **[4.2]**

Sales			Credits
$428.80	$733.18	$22.51	$76.15
$316.25	$38.00	$162.15	$118.44
$68.95	$188.36		$13.86

1. Find the total amount of the sales slips.

1. _____

2. What is the total amount of the credit slips?

2. _____

3. Find the total amount of the deposit.

3. _____

4. Assuming that the bank charges the retailer a $3\frac{1}{4}\%$ discount charge, find the amount of the discount charge at the statement date.

4. _____

5. Find the amount of the credit given to the retailer after the fee is subtracted.

5. _____

Solve the following application problems.

6. Shaundra Brown worked 7 hours on Monday, 10 hours on Tuesday, 8 hours on Wednesday, 9 hours on Thursday, and 10 hours on Friday. Her regular hourly pay is $12.80. Find her gross earnings for the week if Brown is paid time-and-a-half for all hours over 8 worked in a day.   **[5.1]**

6. _____

7. The employees of Feather Farms paid a total of $968.50 in Social Security tax last month, $223.50 in Medicare tax, and $1975.38 in federal withholding tax. Find the total amount that the employer must send to the Internal Revenue Service.   **[5.4]**

7. _____

Find the net cost (invoice amount) for each of the following. Round to the nearest cent.   **[6.1]**

8. List price $475.50, less 20/20 _____

9. List price $375, less 25/10/5 _____

Find the single discount equivalent for each of the following series discounts.   **[6.2]**

10. 10/20 _____

11. 30/40/10 _____

Find the discount date and the net payment date for each of the following. The net payment date is 20 days after the final discount date.   **[6.4]**

		Date Goods	Final	Net
Invoice Date	Terms	Received	Discount Date	Payment Date
12. May 27	2/10 ROG	June 5	_____	_____
13. Oct. 9	3/15 EOM		_____	_____
14. June 24	4/10–30 ex.		_____	_____

Complete the following. If there is no operating loss or absolute loss, write "none."   **[7.3]**

Cost	Operating Expense	Break-even Point	Reduced Price	Operating Loss	Absolute Loss
15. $312	$88	_____	_____	$120	_____
16. _____	_____	_____	$220	$112	$32

Solve the following application problems.

17. The list price of a Great Basin Mountaineer 2-person tent at Olympic Sports is $149.99. Find the dealer's cost if given a 25/25 trade discount and a 3/20, n/30 cash discount. Assume that the dealer earns the maximum cash discount.   **[6.1 and 6.3]**

17. _____

18. Computer Towne purchases mouse pads for $43.20 per box of 3 dozen. If the store wants a markup of 52% on the selling price, find the selling price per mouse pad.   **[7.2]**

18. _____

19. The Retro-Fit Window Company has an average inventory of $18,784 at cost. If the cost of goods sold for the year was $241,938, find the stock turnover at cost. Round to hundredths.   **[7.4]**

19. _____

20. Inventory at a local store was taken at retail value four times and was found to be $53,820; $49,510; $60,820; and $56,380. Sales during the same period were $252,077. Find the stock turnover at retail. Round to hundredths.   **[7.4]**

20. _____

21. Thunder Manufacturing made the following purchases of rivet drums during the year: 25 at $135 each, 40 at $165 each, 15 at $108.50 each, and 30 at $142 each. An inventory shows that 45 rivet drums remain. Find the inventory value, using the weighted average method.   **[7.4]**

21. _____

22. Refer to Exercise 21. Find the inventory value, using **(a)** the FIFO method and **(b)** the LIFO method.   **[7.4]**

(a) _____

(b) _____

CHAPTER

Simple Interest

WITH THE SUPPORT OF AN ENCOURAGING BANKER, JANE BENSON was one of the first women to open an automobile dealership many years ago when she opened Benson Automotive. The dealership frequently borrows large

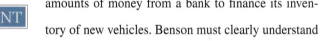

amounts of money from a bank to finance its inventory of new vehicles. Benson must clearly understand interest and notes, since these represent a significant cost to her business, literally the difference between profit and loss.

Interest is a fee charged to borrow money. Ancient clay tablets show that interest was being charged 5000 years ago. Banks, corporations, states, cities, countries, partnerships, and individuals borrow money. Large, financially strong corporations such as Microsoft and McDonald's borrow at the most favorable interest rate called the **prime rate**. The remainder of us must pay higher rates when we borrow to buy a car, a house, or to charge items to our charge cards. Interest rates vary greatly, as shown in the next figure. High interest rates make it more expensive to borrow for both individuals and corporations and can slow the entire economy.

There are two types of interest in common use today: simple interest and compound interest. **Simple interest** is commonly used for shorter loan periods, such as 9 months, and involves interest only on the **principal**, or debt. **Compound interest** requires interest to be paid on both principal and *also* on previously earned interest. This chapter discusses simple interest; compound interest is covered in Chapter 9.

8.1 | Basics of Simple Interest

Objectives

1. Solve for simple interest.
2. Calculate maturity value.
3. Use a table to find the number of days from one date to another.
4. Use the actual number of days in a month to find the number of days from one date to another.
5. Find exact and ordinary interest.
6. Define the basic terms used with notes.
7. Find the due date of a note.

CASE *in* POINT

Benson liked the new automobile models that just came out and thought she could sell a lot of them if her dealership had a good selection available. So Benson Automotive borrowed $1,350,000 from a bank to finance their inventory of new models. She hoped to borrow the money at 8%, but wasn't sure what rate the banker would charge her. She noted with relief that at least interest rates weren't as high as in the 1980s.

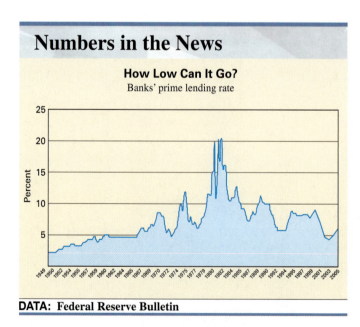

Numbers in the News

How Low Can It Go?
Banks' prime lending rate

DATA: Federal Reserve Bulletin

Objective **1** **Solve for simple interest.** Simple interest is interest charged on the entire principal for the entire length of the loan. It is found using the formula shown in the following box. **Principal** is the loan amount, **rate** is the interest rate, and **time** is the length of the loan *in years.*

$$\text{Simple Interest} = \text{Principal} \times \text{Rate} \times \text{Time}$$
$$I = P \times R \times T$$

QUICK TIP **Rate (R)** must be changed to a decimal or fraction before being substituted into $I = PRT$. For example, change 8% to .08 or $\frac{8}{100}$ before substituting into the formula in place of rate.

QUICK TIP **Time (T)** must be converted to years *before* being substituted into $I = PRT$. For example, convert 7 months to $\frac{7}{12}$ of a year before substituting in place of time.

EXAMPLE 1

Finding Simple Interest

Benson Automotive borrowed $1,350,000 at 8.5% to help purchase 75 new automobiles and SUV's. Benson borrowed the money for 9 months, anticipating that most of the vehicles would be sold by then and that she would have funds to pay back the loan with interest. One bank offered her 8.5%, and a second offered her 10%. Find the interest **(a)** at 8.5% and **(b)** at 10%. Then find **(c)** the interest she saved by going with the lower interest rate.

SOLUTION

(a) Substitute the values into $I = PRT$ and solve for simple interest.

$$I = PRT$$
$$I = \$1,350,000 \times .085 \times \frac{9}{12} \quad \text{9 months} = \frac{9}{12} \text{ of a year}$$
$$I = \$86,062.50 \text{ simple interest}$$

(b) Change the interest rate to 10% and find simple interest using $I = PRT$.

$$I = PRT$$
$$I = \$1,350,000 \times .10 \times \frac{9}{12}$$
$$I = \$101,250 \text{ simple interest}$$

(c) The difference in interest is $101,250 − $86,062.50 = $15,187.50. Benson noted that this is far more than she will make from selling five or even six vehicles.

The calculator solution for part **(a)** follows.

$$1350000 \boxed{\times} 8.5 \boxed{\%} \boxed{\times} 9 \boxed{\div} 12 \boxed{=} 86062.5$$

Note: Refer to Appendix C for calculator basics.

Objective **2** **Calculate maturity value.** The amount that must be repaid when the loan is due is the **maturity value** of the loan. Find this value by adding principal and interest.

$$\text{Maturity value} = \text{Principal} + \text{Interest}$$
$$M = P + I$$

EXAMPLE 2

Finding Maturity Value

Tom Swift needs to borrow $7200 to remodel his bookstore so that he can serve coffee to customers as they browse or sit and read. He borrows the funds from his uncle for 21 months at an interest rate of 9.25%. Find the interest due on the loan and the maturity value at the end of 21 months.

SOLUTION

Interest due is found using $I = PRT$, where T must be in years (21 months $= \frac{21}{12}$ years).

$$I = PRT$$
$$I = \$7200 \times .0925 \times \frac{21}{12} = \$1165.50$$

Find the maturity value using $M = P + I$, where $P = \$7200$ and $I = \$1165.50$.

$$M = P + I$$
$$M = \$7200 + \$1165.50 = \$8365.50$$

> **QUICK TIP**   A loan is an investment for a lender. The lender lends out the principal and later receives the maturity value, which includes interest.

Objective 3 **Use a table to find the number of days from one date to another.** The period of the loan was in months in Examples 1 and 2. However, it is common for loans to be for a certain number of days, such as 120 days. Or a loan may be due at a fixed date, such as April 17, and we may have to figure out the number of days until the loan must be paid off. One way to do this is to number the days of the year as in the table on the next page. Note that this table is also on the inside of the back cover of the book.

For example, June 11 is day 162 of the year and December 25 is day 359 of the year. Find the number of days from June 11 to December 25 by subtracting.

December 25 is day	359
June 11 is day	− 162
	197 days from June 11 to December 25

There are 197 days from June 11 to December 25.

> **QUICK TIP**   If it is leap year, June 11 is day 163 (add one day after the end of February during a leap year) and December 25 is day 360, so there are still 197 days from June 11 to December 25.

EXAMPLE 3

Finding the Number of Days from One Date to Another, Using a Table

Find the number of days from **(a)** March 24 to July 22, **(b)** April 4 to October 10, **(c)** November 8 to February 17 of the following year, and **(d)** December 2 to January 17 of the following year. Assume that it is not a leap year.

SOLUTION

(a)
July 22 is day	203
March 24 is day	− 83
	120 days from March 24 to July 22

(b)
October 10 is day	283
April 4 is day	− 94
	189 days from April 4 to October 10

(c) November 8 is day 312, so there are $365 - 312 = 53$ days from November 8 to the end of the year. Add days until the end of the year plus days into the next year to find the total.

November 8 to end of year	**53**
February 17 is day	+ **48**
	101 days from November 8 to February 17 of next year

The Number of Each of the Days of the Year*

Day of Month	Jan.	Feb.	Mar.	Apr.	May	June	July	Aug.	Sept.	Oct.	Nov.	Dec.	Day of Month
1	1	32	60	91	121	152	182	213	244	274	305	335	1
2	2	33	61	92	122	153	183	214	245	275	306	336	2
3	3	34	62	93	123	154	184	215	246	276	307	337	3
4	4	35	63	94	124	155	185	216	247	277	308	338	4
5	5	36	64	95	125	156	186	217	248	278	309	339	5
6	6	37	65	96	126	157	187	218	249	279	310	340	6
7	7	38	66	97	127	158	188	219	250	280	311	341	7
8	8	39	67	98	128	159	189	220	251	281	312	342	8
9	9	40	68	99	129	160	190	221	252	282	313	343	9
10	10	41	69	100	130	161	191	222	253	283	314	344	10
11	11	42	70	101	131	162	192	223	254	284	315	345	11
12	12	43	71	102	132	163	193	224	255	285	316	346	12
13	13	44	72	103	133	164	194	225	256	286	317	347	13
14	14	45	73	104	134	165	195	226	257	287	318	348	14
15	15	46	74	105	135	166	196	227	258	288	319	349	15
16	16	47	75	106	136	167	197	228	259	289	320	350	16
17	17	48	76	107	137	168	198	229	260	290	321	351	17
18	18	49	77	108	138	169	199	230	261	291	322	352	18
19	19	50	78	109	139	170	200	231	262	292	323	353	19
20	20	51	79	110	140	171	201	232	263	293	324	354	20
21	21	52	80	111	141	172	202	233	264	294	325	355	21
22	22	53	81	112	142	173	203	234	265	295	326	356	22
23	23	54	82	113	143	174	204	235	266	296	327	357	23
24	24	55	83	114	144	175	205	236	267	297	328	358	24
25	25	56	84	115	145	176	206	237	268	298	329	359	25
26	26	57	85	116	146	177	207	238	269	299	330	360	26
27	27	58	86	117	147	178	208	239	270	300	331	361	27
28	28	59	87	118	148	179	209	240	271	301	332	362	28
29	29		88	119	149	180	210	241	272	302	333	363	29
30	30		89	120	150	181	211	242	273	303	334	364	30
31	31		90		151		212	243		304		365	31

*Add 1 to each date after February 29 for a leap year

(d) December 2 is day 336, so there are $365 - 336 = 29$ days to the end of the year. Add days until the end of the year plus days into the next year to find the total.

December 2 to end of year   **29**
January 17 is day   **+ 17**
                     46 days from December 2 to January 17 of the next year

Objective **4** **Use the actual number of days in a month to find the number of days from one date to another.** The number of days between specific dates can be found using the number of days in each month of the year as shown.

Number of Days in Each Month

31 Days		30 Days	28 Days
January	August	April	February
March	October	June	(29 days in leap year)
May	December	September	
July		November	

Two other ways of remembering the number of days in each month are the rhyme method and the knuckle method, as seen below.

Rhyme Method:
30 days hath September
April, June, and November.
All the rest have 31, except February,
which has 28 and in a leap year 29.

Knuckle Method:

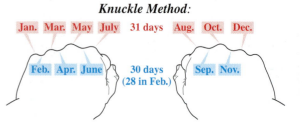

EXAMPLE 4

Finding the Number of Days from One Date to Another Using Actual Days

Find the number of days from **(a)** June 3 to August 14 and **(b)** November 4 to February 21.

SOLUTION

(a) June has 30 days so there are $30 - 3 = 27$ days from June 3 to the end of June.

June 3 to the end of June	27
31 days in July	31
14 days in August	$+ \ 14$
	72 days from June 3 to August 14

(b) November has 30 days so there are $30 - 4 = 26$ days from November 4 to the end of November.

November 4 to end of November	26
31 days in December	31
31 days in January	31
21 days in February	$+ \ 21$
	109 days from November 4 to February 21

QUICK TIP   To find the number of days from one date to another, do not count the day the loan was made, but do count the day that the loan is paid.

Objective 5   Find exact and ordinary interest. Simple interest rates are given as an annual rate, such as 7% per year. Since the rate is per year, time must also be given in years or fraction of a year when using $I = PRT$. If time is given in number of days, first change it to a fraction of a year.

$$T = \frac{\text{Number of days in the loan period}}{\text{Number of days in a year}}$$

Exact interest calculations require the use of the exact number of days in the year, 365 or 366 if a leap year. **Ordinary interest**, or **banker's interest**, calculations require the use of 360 days. Banks commonly used 360 days in a year for interest calculations before calculators and computers became widely available. Today, many institutions, the government, and the Federal Reserve Bank use the exact number of days in a year in interest calculations. However, some banks and financial institutions still use 360 days. You need to be able to use both.

For exact interest: Use 365 days (or 366 days if a leap year).

$$T = \frac{\text{Number of days in a loan period}}{365}$$

For ordinary, or banker's, interest: Use 360 days for the number of days.

$$T = \frac{\text{Number of days in a loan period}}{360}$$

Example 5 shows that **ordinary interest produces more interest** for the lending institution than does exact interest.

EXAMPLE 5

Finding Exact and Ordinary Interest

Radio station WERZ borrowed $28,300 on May 12 with interest due on August 27. Given 10% interest, find the interest on the loan using **(a)** exact interest and **(b)** ordinary interest.

SOLUTION

Either the table method or the method of the number of days in a month can be used to find that there are 107 days from May 12 to August 27.

(a) The **exact interest** is found from $I = PRT$ with $P = \$28,300$, $R = .1$, and $T = \frac{107}{365}$. (Remember to use 365 as the denominator with exact interest.)

$$I = PRT$$

$$I = \$28,300 \times .1 \times \frac{107}{365}$$

$$I = \$829.62 \left(\text{rounded}\right)$$

(b) Find **ordinary interest** with the same formula and values, except that $T = \frac{107}{360}$.

$$I = PRT$$

$$I = \$28,300 \times .1 \times \frac{107}{360}$$

$$I = \$841.14 \left(\text{rounded}\right)$$

In this example, the ordinary interest is $\$841.14 - \$829.62 = \$11.52$ more than the exact interest.

QUICK TIP   Orindary interest results in slightly more interest than exact interest. Thus ordinary interest favors those banks that use it when calculating loans.

QUICK TIP   *Use ordinary or banker's interest throughout the remainder of the book unless stated otherwise.*

Objective 6 **Define the basic terms used with notes.** A **promissory note** is a *legal document* in which one person or firm agrees to pay a certain amount of money, on a specific day in the future, to another person or firm. An example of a promissory note follows.

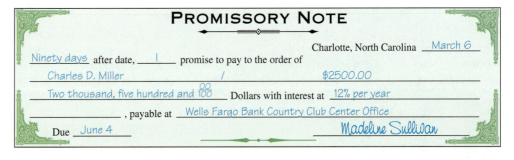

This type of promissory note is called a **simple interest note**, since simple interest calculations involving $I = PRT$ are used. Here are the terms of a simple interest promissory note.

Maker or **payer:** The person borrowing the money. (Madeline Sullivan in the sample note.)
Payee: The person who loaned the money and who will receive the payment. (Charles D. Miller in the sample note.)
Term: The length of time until the note is due. (90 days in the sample note.)
Face value or **principal:** The amount being borrowed. ($2500 in the sample note.)

Maturity value: The face value plus interest, also the amount due at maturity.
Maturity date or **due date:** The date the loan must be paid off with interest. (June 4 in the sample note.)

Find the interest on the loan in the sample note as follows.

$$\text{Interest} = \text{Face value} \times \text{Rate} \times \text{Time}$$

$$\text{Interest} = \$2500 \times .12 \times \frac{90}{360} = \$75$$

The maturity value is calculated as follows.

$$\text{Maturity value} = \text{Face value} + \text{Interest}$$

$$\text{Maturity value} = \$2500 + \$75 = \$2575$$

Madeline Sullivan must pay $2575 to Charles D. Miller on June 4th, the maturity date of the note.

Banks and financial institutions lend money only to individuals and firms they believe will repay the loan with interest. Even then, banks often require **collateral** or assets such as an automobile or stock in order to obtain a loan. If the loan is not repaid, the bank **forecloses** on the collateral, takes ownership, and sells or liquidates it. Funds from the sale of the collateral are first used to pay off the note and the expenses of the foreclosure. Any excess is returned to the maker of the note.

Objective 7 Find the due date of a note. Time is in months in some promissory notes. When this occurs, the loan is due after the given number of months has passed but on the same day of the month as the original loan was made. For example, a 4-month note made on May 25 would be due 4 months later on the 25th of September. Other examples follow.

Date Made	Length of Loan	Date Due
March 12	5 months	August 12
April 24	7 months	November 24
October 7	9 months	July 7
January 31	3 months	April 30

A loan made on January 31 for 3 months would normally be due on April 31. However, there are only 30 days in April, so the loan is due on April 30. Whenever a due date does not exist, such as February 30 or November 31, use the last day of the month (February 28 or November 30 in these examples).

EXAMPLE **6**

Finding Due Date, Interest, and Maturity Value

Find the due date, interest, and maturity value for a $600,000 loan made to Benson Automotive on July 31 for 7 months at 7.5% interest.

SOLUTION

Interest and principal is due seven months from July 31 or February 31 which *does not* exist. Since February only has 28 days in it (unless it is a leap year), interest and principal is due the last day of February, or February 28. If it were a leap year, the maturity value would be due February 29.

$$I = PRT = \$600,000 \times .075 \times \frac{7}{12} = \textbf{\$26,250}$$

$$M = P + I = \$600,000 + \textbf{\$26,250} = \$626,250$$

A total of $626,250 must be repaid on February 28.

QUICK TIP Do not convert the period of a loan from months to days to find the due date.

8.1 | Exercises

FOR EXTRA HELP

 MyMathLab

 InterActMath.com

 *Math*XP MathXL

 MathXL
Tutorials on CD

Tutor Center Addison-Wesley
Math Tutor Center

 DVT/Videotape

The **Quick Start** *exercises in each section contain solutions to help you get started.*

Find simple interest and maturity value to the nearest cent. (See Examples 1 and 2.)

Quick Start

		Interest	Maturity Value
1.	$3800 at 11% for 6 months	$209	$4009

$$I = \$3800 \times .11 \times \tfrac{6}{12} = \$209$$
$$M = \$3800 + \$209 = \$4009$$

2. $10,200 at 9.5% for 10 months   _____   _____

3. $5500 at 8% for 1 year   _____   _____

4. $800 at 6% for $2\frac{1}{2}$ years   _____   _____

Find the exact number of days from the first date to the second. In Exercises 7–8, assume that the second month given is in the following year, and assume no leap years. (See Examples 3 and 4.)

Quick Start

5. February 15 to April 24   **5.** 68 _____

From the table, February 15 is day 46; April 24 is day 114
Number of days = 114 − 46 = 68 days

6. May 22 to August 30   **6.** _____

7. December 1 to March 10   **7.** _____

8. October 12 to February 22   **8.** _____

C indicates an exercise that is related to the Case in Point feature within the section.

*Find (a) the exact interest and (b) the ordinary interest for each of the following to the nearest cent.
Then find (c) the amount by which the ordinary interest is larger. (See Example 5.)*

Quick Start

9. $52,000 at $8\frac{3}{4}\%$ for 200 days

    (a) Exact interest = $52,000 × .0875 × $\frac{200}{365}$ = $2493.15
    (b) Ordinary interest = $52,000 × .0875 × $\frac{200}{360}$ = $2527.78
    (c) Ordinary is larger by $2527.78 − $2493.15 = $34.63

    (a) $2493.15
    (b) $2527.78
    (c) $34.63

10. $185,000 at 7.5% for 180 days

    (a) _____
    (b) _____
    (c) _____

11. $29,500 at $11\frac{1}{4}\%$ for 120 days

    (a) _____
    (b) _____
    (c) _____

12. $52,610 at $8\frac{1}{2}\%$ for 82 days

    (a) _____
    (b) _____
    (c) _____

Identify each of the following from the promissory note shown. (See Objective 6.)

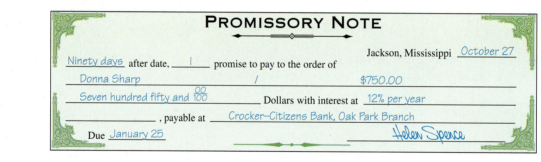

13. Maker   _____
14. Payer   _____
15. Payee   _____
16. Face value   _____
17. Term of loan   _____
18. Date loan was made   _____
19. Date loan is due   _____
20. Maturity value   _____

*Find the date due, the amount of interest (rounded to the nearest cent if necessary), and the
maturity value. (See Example 6.)*

Quick Start

	Date Loan Was Made	Face Value	Term of Loan	Rate	Date Loan Is Due	Maturity Value
21.	Mar. 12	$4800	220 days	9%	Oct. 18	$5064

    I = $4800 × .09 × $\frac{220}{360}$ = $264; *M* = $4800 + $264 = $5064

	Date Loan Was Made	Face Value	Term of Loan	Rate	Date Loan Is Due	Maturity Value
22.	Jan. 3	$12,000	100 days	9.8%	_____	_____

Date Loan Was Made	Face Value	Term of Loan	Rate	Date Loan Is Due	Maturity Value
23. Nov. 10	$6300	180 days	$9\frac{1}{4}\%$	_____	_____
24. July 14	$6800	90 days	$11\frac{3}{4}\%$	_____	_____

Solve the following application problems. Round dollar amounts to nearest cent.

Quick Start

25. **INVENTORY** Benson Automotive borrows $2,000,000 at $9\frac{1}{4}\%$ from a bank to buy land to build a building for a new dealership. Given that the loan is for 9 months, find **(a)** interest and **(b)** maturity value.

(a) $I = \$2,000,000 \times .0925 \times \frac{9}{12} = \$138,750$
(b) $M = \$2,000,000 + \$138,750 = \$2,138,750$

(a) $138,750
(b) $2,138,750

26. **LOANS BETWEEN BANKS** A bank in New York City borrows $25,000,000 at 9% for 90 days from a bank in Chicago. Find **(a)** the interest and **(b)** the maturity value.

(a) _____
(b) _____

27. **ROAD PAVING** Gilbert Construction Company needs to borrow $280,000 to build a short, paved road and install all utilities in a subdivision. The company decides to borrow the funds at 10% for 180 days. In 1980, the same note would have been at a rate of 22%. Find the difference in the interest charges based on the two rates.

27. _____

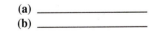

28. **INTERNATIONAL FINANCE** A small construction company in Mexico City borrows 300,000 pesos for 90 days at an interest rate of 18%. The same loan would have been at a rate of 35% several years ago. Find the difference in the interest charges based on the two rates.

28. _____

29. **CAPITAL IMPROVEMENT** Elizabeth Barton borrowed $6850 to install a small rock fountain and fish pond in front of her flower shop. She signed a 90-day note on July 5 at $9\frac{1}{4}\%$ interest. Find **(a)** the due date and **(b)** the maturity value of the note.

(a) _____
(b) _____

30. **COMPUTER STORE** ComputerTown signed a promissory note to a bank with a face value of $32,500 on September 10. The 90-day note is at 11% interest. Find **(a)** the due date and **(b)** the maturity value of the note.

(a) _____
(b) _____

31. **HEALTH FOOD** On March 10, the owner of The Granary borrowed $80,000 on a 180-day promissory note at 10.5% interest. Find **(a)** the due date and **(b)** the maturity value of the note.

(a) _____
(b) _____

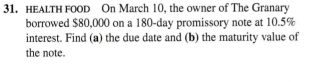

32. CORPORATE FINANCE   On October 15, IBM borrows $45,000,000 at 8% from a bank in San Francisco and agrees to repay the loan in 120 days using ordinary interest. Find **(a)** the due date and **(b)** the maturity value.

(a) _____
(b) _____

33. PENALTY ON UNPAID PROPERTY TAX   Joe Simpson's property tax is $683.21 and is due April 15. He does not pay until July 23. The county adds a penalty of 9.3% simple interest on his unpaid tax. Find the penalty using exact interest.

33. _____

34. PENALTY ON UNPAID INCOME TAX   On January 5, Helen Terry paid an income tax payment, which was due September 15. The penalty was 11% simple interest on the unpaid tax of $2100. Find the penalty using exact interest.

34. _____

35. EQUIPMENT SUPPLY   Fireman's Supply borrowed $48,000 to remodel the front of their store on January 31. The 8.75% simple interest note was due in 8 months. Find **(a)** the due date and **(b)** the maturity value of the note.

(a) _____
(b) _____

36. LOAN TO EMPLOYEE   On the last day in November, Terry Thompson loaned one of his employees $1600 for 3 months at 10% interest. Find **(a)** the due date and **(b)** the maturity value.

(a) _____
(b) _____

37. Explain the difference between exact interest and ordinary, or banker's, interest. (See Objective 5.)

38. List three companies that you have purchased products or services from in the past. List two reasons each of the companies may have needed to borrow money in the past.

8.2 Finding Principal, Rate, and Time

Objectives

1. Find the principal.
2. Find the rate.
3. Find the time.

Principal (P), rate (R), and time (T) were given for all problems in Section 8.1 and we calculated interest. In this section, interest is given, and we solve for principal, rate, or time.

QUICK TIP For simplicity, use banker's interest with 360 days for all problems in this section.

Objective 1 Find the principal. The principal (P) is found by dividing both sides of the simple interest equation $I = PRT$ by RT. See Appendix A for a review of algebra if needed.

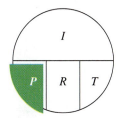

$$I = PRT$$

$$\frac{I}{RT} = \frac{P\cancel{RT}}{\cancel{RT}} \qquad \text{Divide both sides by } RT.$$

$$\frac{I}{RT} = P \qquad \text{or} \qquad P = \frac{I}{RT}$$

The various forms of the simple interest equation can be remembered using the circle sketch shown above. In the sketch, I (interest) is in the top half of the circle, with P (principal), R (rate), and T (time) in the bottom half of the circle. Find the formula for any one variable by covering that letter in the circle and then read the remaining letters noticing their position. For example, cover P and you are left with $\frac{I}{RT}$.

$$\text{Principal} = \frac{\text{Interest}}{\text{Rate} \times \text{Time}} \qquad \text{or} \qquad P = \frac{I}{RT}$$

QUICK TIP Remember that time must be in years or fraction of a year.

EXAMPLE 1

Finding Principal Given Interest in Days

Gilbert Construction Company borrows funds at 10% for 54 days to build a home. Find the principal that results in interest of $780.

SOLUTION

Write the rate as .10, the time as $\frac{54}{360}$, and then use the formula for principal.

$$P = \frac{I}{RT}$$

$$P = \frac{\$780}{.10 \times \dfrac{54}{360}}$$

$$.10 \times \frac{54}{360} = .015 \qquad \text{Simplify the denominator.}$$

$$P = \frac{\$780}{.015} = \$52,000 \qquad \text{Divide.}$$

The principal is $52,000.

Check the answer using $I = PRT$. The principal is \$52,000, the rate is 10%, and the time is $\frac{54}{360}$ year. The interest should be, and is, \$780.

$$I = \$52,000 \times .10 \times \frac{54}{360} = \textbf{\$780}$$

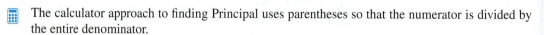

The calculator approach to finding Principal uses parentheses so that the numerator is divided by the entire denominator.

$$780 \boxed{\div} \boxed{(} .10 \boxed{\times} 54 \boxed{\div} 360 \boxed{)} \boxed{=} 52000$$

Note: Refer to Appendix C for calculator basics.

EXAMPLE 2

*Finding Principal
Given Length of Loan*

Frank Thomas took out a loan to pay his college tuition on February 2. The loan is due to be repaid on April 15 when Thomas expects to receive an income tax refund. The interest on the loan is \$37.80 at a rate of 10.5%. Find the principal.

SOLUTION
First find the number of days.

26	days remaining in February
31	March
+ 15	April
72	days from February 2 to April 15

$$T = \frac{72}{360}$$

Now use the following formula.

$$P = \frac{I}{RT}$$

$$P = \frac{\$37.80}{.105 \times \frac{72}{360}} \qquad \text{Substitute values into the formula.}$$

$$.105 \times \frac{72}{360} = .021 \qquad \text{Simplify the denominator.}$$

$$P = \frac{\$37.80}{.021} = \$1800 \qquad \text{Divide.}$$

The principal is \$1800. Check the answer using the formula for simple interest.

$$I = \$1800 \times .105 \times \frac{72}{360} = \textbf{\$37.80}$$

Objective 2 Find the rate. Solve the formula $I = PRT$ for rate (R) by dividing both sides of the equation by PT. The rate found in this manner will be the annual interest rate. See Appendix A for a review of algebra if needed.

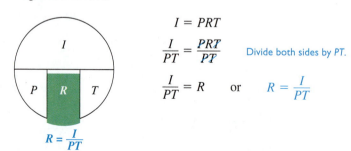

$$I = PRT$$

$$\frac{I}{PT} = \frac{PRT}{PT} \qquad \text{Divide both sides by } PT.$$

$$\frac{I}{PT} = R \qquad \text{or} \qquad R = \frac{I}{PT}$$

$$\text{Rate} = \frac{\text{Interest}}{\text{Principal} \times \text{Time (in years)}} \quad \text{or} \quad R = \frac{I}{PT}$$

EXAMPLE 3

Finding Rate Given the Length of Loan

An exchange student from the United States living in Brazil deposits $2500 in U.S. currency in a Brazilian bank for 45 days. Find the rate if the interest is $37.50 in U.S. currency.

SOLUTION

The rate of interest can be found with the following formula.

$$R = \frac{I}{PT}$$

$$R = \frac{\$37.50}{\$2500 \times \dfrac{45}{360}}$$

$$\$2500 \times \frac{45}{360} = \$312.50 \quad \text{Simplify the denominator.}$$

$$R = \frac{\$37.50}{312.50} = .12 \quad \text{Divide.}$$

Convert .12 to a percent to get 12%. Check the answer using the simple interest formula.

EXAMPLE 4

Finding Rate Given Length of Loan

Benson Automotive kept extra cash of $86,500 in an account from June 1 to August 16. Find the rate if the company earned $365.22 in interest during this period of time.

SOLUTION

Find the number of days using the table on page 000.

$$\begin{array}{r} \text{August 16 is day} \quad 228 \\ \text{June 1 is day} \quad -\ 152 \\ \hline \textbf{76 days} \end{array}$$

There are 76 days from June 1 to August 16.

$$T = \frac{76}{360}$$

Now use the formula.

$$R = \frac{I}{PT}$$

$$R = \frac{\$365.22}{\$86,500 \times \dfrac{76}{360}} = .02 \ \left(\text{rounded}\right)$$

The rate of interest is 2%.

Objective **3** **Find the time.** The time (T) is found by dividing both sides of the simple interest equation $I = PRT$ by PR. Note that time will be in years or fraction of a year. See Appendix A for a review of algebra if needed.

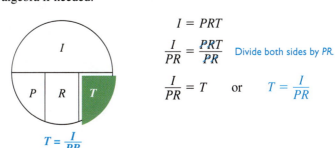

$$I = PRT$$

$$\frac{I}{PR} = \frac{PRT}{PR} \quad \text{Divide both sides by } PR.$$

$$\frac{I}{PR} = T \quad \text{or} \quad T = \frac{I}{PR}$$

$$T = \frac{I}{PR}$$

$$\text{Time (in years)} = \frac{\text{Interest}}{\text{Principal} \times \text{Rate}} \quad \text{or} \quad T = \frac{I}{PR}$$

This formula gives time in years, but we often need time in days. Convert time in years to time in days by multiplying the time in years by 360. For example, $\frac{1}{2}$ year is $\frac{1}{2} \times 360 = 180$ days.

$$\text{Time in days} = \frac{I}{PR} \times 360$$

Similarly, use the following if time is needed in months.

$$\text{Time in months} = \frac{I}{PR} \times 12$$

EXAMPLE 5

Finding Time in Days Given Principal and Rate

Roberta Sanchez deposited $6200 in an account paying 3% and she earned $72.33 in interest. Find the number of days that the deposit earned interest. Round to a whole number of days.

SOLUTION

$$T \text{ in days} = \frac{I}{PR} \times \mathbf{360}$$

$$T = \frac{\$72.33}{\$6200 \times .03} \times 360 = 140 \text{ days } (\text{rounded})$$

The money was on deposit for 140 days.

Use parentheses around the denominator of the fraction to make sure that the calculations are done in the correct order. Round to the nearest day.

72.33 $\boxed{\div}$ $\boxed{(}$ 6200 $\boxed{\times}$ .03 $\boxed{)}$ $\boxed{\times}$ 360 $\boxed{=}$ 140

Note: Refer to Appendix C for calculator basics.

Summary

Interest $I = PRT$

Principal $P = \dfrac{I}{RT}$

Rate $R = \dfrac{I}{PT}$

Time $T (\text{in years}) = \dfrac{I}{PR}$

              $T (\text{in months}) = \dfrac{I}{PR} \times 12$

              $T (\text{in days}) = \dfrac{I}{PR} \times 360$

All of these are modifications of the formula $I = PRT$.

8.2 | Exercises

FOR EXTRA HELP

MyMathLab

InterActMath.com

MathXL

MathXL
Tutorials on CD

Addison-Wesley
Math Tutor Center

DVT/Videotape

The **Quick Start** exercises in each section contain solutions to help you get started.

Find the principal in each of the following. Round to the nearest cent. (See Example 1.)

Quick Start

	Rate	Time (in days)	Interest	Principal
1.	$7\frac{3}{4}\%$	90	$271.25	**$14,000**

$$P = \frac{\$271.25}{.0775 \times \frac{90}{360}} = \$14,000$$

	Rate	Time (in days)	Interest	Principal
2.	9.5%	120	$63.79	**$2014.42**

$$P = \frac{\$63.79}{.095 \times \frac{120}{360}} = \$2014.42$$

3.	10%	80	$11.20	_____
4.	6%	24	$6.24	_____
5.	$8\frac{1}{2}\%$	120	$306	_____
6.	10.5%	140	$87.20	_____

Find the rate in each of the following. Round to the nearest tenth of a percent. (See Example 3.)

Quick Start

	Principal	Time	Interest	Rate
7.	$7600	200 days	$498.22	**11.8%**

$$R = \frac{\$498.22}{\$7600 \times \frac{200}{360}} = 11.8\%$$

	Principal	Time	Interest	Rate
8.	$15,600	90 days	$312	_____
9.	$42,800	60 days	$677.67	_____

C indicates an exercise that is related to the Case in Point feature within the section.

Principal	Time	Interest	Rate
10. $2000	90 days	$62.50	_____
11. $8000	4 months	$200	_____
12. $4800	5 months	$197.60	_____

Find the time in each of the following. In Exercises 13–16, round to the nearest day; in Exercises 17–18, round to the nearest month. (See Example 5.)

Quick Start

Principal	Rate	Interest	Time
13. $74,000	9.5%	$2343.33	**120 days**

$$T = \frac{\$2343.33}{\$74{,}000 \times .095} \times 360 = 119.9998 \text{ or } 120 \text{ days}$$

Principal	Rate	Interest	Time
14. $3600	9%	$58.50	_____
15. $2400	11%	$45.47	_____
16. $20,000	8%	$1200	_____
17. $3500	$10\frac{1}{4}\%$	$143.50	_____
18. $8400	$7\frac{1}{4}\%$	$357	_____

In each of the following application problems, find principal to the nearest cent, rate to the nearest tenth of a percent, or time to the nearest day.

Quick Start

19. **MONEY MARKET** Liz Nault earned $196.88 interest in 9 months from a money market account paying 3.5% interest. Find the amount initially invested.

19. $7500.19

$$P = \frac{\$196.88}{.035 \times \frac{9}{12}} = \$7500.19$$

20. **BANK LOAN** Citizens Bank earned $12,250 interest in 45 days from a short-term investment that paid 5.6% interest. Find the amount initially invested.

20. _____

21. **INVESTING IN BONDS**   Joan Gretz invested $3600 in a mutual fund containing bonds. Find the rate if she earned $237.50 in interest in 250 days.

21. _____

22. **LAW ENFORCEMENT**   The Smith County Police Department borrowed $120,000 for 135 days to purchase new radar-detection equipment to detect speeders. Find the rate if the interest was $4050.

22. _____

23. **INVENTORY**   Benson Automotive was offered a discount on all new models that were purchased within 15 days. Jane Benson worked out an agreement with her bank to borrow $180,000 with interest charges of $6300 in 140 days. Find the rate.

23. _____

24. **SAVING FOR RETIREMENT**   Mike Jordan deposited $2000 into a Roth Individual Retirement Account (IRA) investing in a mutual fund containing corporate bonds. Find the rate if he has $2192.50 in the account 15 months later.

24. _____

25. **RETIREMENT ACCOUNT**   Over a period of 300 days, Shawna Johnson earned $450 interest in a retirement account paying interest at a rate of 5%. Find **(a)** the principal at the beginning of the 300 days and **(b)** the amount in the account at the end of 300 days.

(a) _____
(b) _____

26. **INTEREST EARNINGS**   In 383 days, Troy McLain earned $75 interest in an account paying 4.5% interest. Find **(a)** the principal at the beginning of the 383 days and **(b)** the amount in the account at the end of 383 days.

(a) _____
(b) _____

27. **TIME OF DEPOSIT**   Benson automotive earned $69.46 interest on a $9400 deposit in an account paying 3.5%. Find the number of days that the funds were on deposit.

27. _____

28. **TIME OF DEPOSIT**   Find how long Quinlan Enterprises must deposit $7500 at 6% in order to earn $243.75 interest.

28. _____

29. **RATE OF INTEREST**   Ti Lee earns $223.03 in interest in 320 days after making a deposit of $6272.73. Find the interest rate.

29. _____

30. **PENALTY ON LATE PAYMENT**   Smithville Toyota lets an $1800 mortgage payment go 70 days overdue and is charged a penalty of $59.50. Find the rate of interest that was charged as a penalty. (*Note:* Penalty rates are frequently quite high.)

30. _____

31. COMPUTER PURCHASE   Ideal Computers purchased 10 computers from a Japanese computer manufacturer. Ideal paid their bill after 45 days with a finance charge of $150. If the Japanese company charges 10% interest, find **(a)** the cost of the 10 computers excluding the interest and **(b)** the cost per computer.

(a) _____
(b) _____

32. LAWN-MOWER PURCHASE   Yard Mowers, Inc. bought 15 self-propelled, 22-inch lawn mowers from Green Lawns, Ltd. The company paid after 90 days and was charged an annual finance charge of 12% or $126. Find **(a)** the cost of the 15 lawn mowers excluding the interest and **(b)** the cost per mower.

(a) _____
(b) _____

33. PROMISSORY NOTE   Jan Rice signed a promissory note for $640 at $11\frac{1}{2}\%$ interest with interest charges of $42.52. Find the term of the note to the nearest day.

33. _____

34. TIME OF DEPOSIT   The Frampton Chamber of Commerce earns $682.71 interest on a $16,385 investment at 5.5%. Find the length of time of the investment to the nearest day.

34. _____

35. HOME CONSTRUCTION   Gilbert Construction Company needs to borrow $220,000 for one year for materials needed to build three homes. They can borrow from either of two banks. Interest charges from Bank One would amount to $23,650, whereas interest charges from First National Bank would amount to $25,000. Find the interest rates associated with a loan from **(a)** Bank One versus **(b)** First National Bank.

(a) _____
(b) _____

36. INVENTORY PURCHASE   Forest Nursery needs to borrow $9500 on February 1 to buy additional inventory and will repay the loan on July 15. Interest charges for State Bank and First National Bank are $480 and $443.60, respectively. Find the rate for **(a)** the State Bank loan and **(b)** the First National Bank loan.

(a) _____
(b) _____

37. A retired couple receives $14,000 per year from Social Security and an additional $18,000 in interest from retirement plans and lifetime savings. They need all of their income to pay expenses including medical bills. What happens if the interest rate on their retirement plans and lifetime savings decreases significantly?

38. How would the formula for calculating time in days (given principal, interest, and rate) change if exact interest is used in contrast to ordinary interest?

8.3 | Simple Discount Notes

Objectives

1. Define the basic terms used with simple discount notes.
2. Find the bank discount and proceeds.
3. Find the face value.
4. Find the effective interest rate.
5. Understand U.S. Treasury bills.

CASE *in* **POINT**

In the past, Benson Automotive often borrowed funds to purchase new automobiles to place on their car lot. It wasn't that owner Jane Benson liked debt. In fact, she strongly disliked debt and was very careful with it. However, the cost of new cars was so high that she commonly had to borrow. Benson knows that interest rates change and that higher rates result in higher costs and lower profits. As a result, she often tries to understand the direction in which interest rates are going.

HERE & NOW

Objective 1 Define the basic terms used with simple discount notes. Face value (or principal) in the simple interest notes in Section 8.1 was the amount or the proceeds the borrower actually received from the lender. The borrower paid back the maturity value, which is the principal plus interest $(M = P + I)$.

In contrast, **simple discount notes** are notes in which the interest is deducted in advance. The borrower never receives the face value on these types of notes. Rather the borrower receives the **proceeds**, which is the face value on the note minus the interest. The **face value** (or **maturity value**) is the amount the borrower must pay when the note matures and it is the sum of the proceeds and the interest. Interest for these notes is called **bank discount**, or **discount**, so face value = proceeds + bank discount. These notes are also called **interest-in-advance notes**, since the interest is subtracted before funds are given to the borrower. Students sometimes borrow money from the government for college using Stafford loans, which are simple discount notes.

> **QUICK TIP** Simple discount notes are not necessarily better or worse than simple interest notes for a borrower. They simply represent a different way to determine the interest on a note and how the note is repaid.

Some banks work with simple interest notes, others work with simple discount notes. Either type of note can be used for any commercial or personal loan in which the repayment is made in one lump sum.

Simple Interest versus Simple Discount Notes

Type of Note	Loan Amount		Interest		Repayment Amount
Simple interest	Face value (Principal)	+	Interest	=	Maturity value
Simple discount	Proceeds	+	Discount (Interest)	=	Face value (Maturity value)

Here is an example of a *simple discount note*. A borrower signs a note for $2000, but receives $1850. The difference $\left(\$2000 - \$1850 = \$150\right)$ is interest, but it is *also called* the bank discount. The borrower must pay back $2000 on the maturity date of the loan.

Time

| Borrower receives proceeds of $1850 | → | Borrower pays back face value of $2000 |

QUICK TIP   Simple interest is calculated on the *principal* while simple discount is calculated on the *maturity value.*

Objective ② **Find the bank discount and proceeds.** The formula for finding the bank discount is a form of the basic percent equation of Chapter 3. The formula is similar to the one used to calculate simple interest, but different letters are used since the ideas differ slightly.

Calculating Bank Discount

$$\text{Bank discount} = \text{Face value} \times \text{Discount rate} \times \text{Time} \quad \text{or} \quad B = MDT,$$

where

$B = $ Bank discount
$M = $ Face value $\left(\text{Maturity value}\right)$
$D = $ Discount rate
$T = $ Time $\left(\text{in years}\right)$

Then, if P is the proceeds,

$$\text{Proceeds} \left(\text{Loan amount}\right) = \text{Face value} - \text{Bank discount} \quad \text{or} \quad P = M - B$$

Stated in another way,

Face Value = Proceeds $\left(\text{loan amount}\right)$ + Bank discount   **or** $M = P + B$

QUICK TIP   Simple interest rate applies *to the principal* of a simple interest note. Simple discount rate *applies to the face value or maturity value* of a simple discount note.

EXAMPLE 1

Finding Discount and Proceeds

Jim Peterson signs a simple discount note with a face value of $35,000 so that he can purchase a truck with plow for his snow removal business. The banker discounts the 10-month note at 9%. Find the amount of the discount and the proceeds.

SOLUTION

Find the discount with the formula $B = MDT$, where $M = \$35,000$, $D = 9\%$, and $T = \frac{10}{12}$, or $\frac{5}{6}$.

$$B = MDT$$

$$B = \$35,000 \times .09 \times \frac{5}{6} = \$2625$$

The discount of $2625 is the interest charge on the loan. The proceeds that Peterson actually receives when making the loan is found using $P = M - B$.

$$P = M - B$$

$$P = \$35,000 - \$2625 = \$32,375$$

Peterson signs the discount note with a face value of $35,000, but receives $32,375. Ten months later he must pay $35,000 to the bank.

EXAMPLE 2

Finding the Proceeds

Colleen Dee signs a 6-month simple discount note with a face value of $4500, which paid one-half of the cost to purchase and install a large electric billboard to put in front of her office building. She paid for the other one-half out of her checking account. Find the proceeds to Dee if the discount rate is 10.5%.

SOLUTION

The bank discount (B) is not known, but we do know that $B = MDT$. Therefore, we can substitute MDT in place of B.

$$P = M - B$$

$$P = M - MDT \quad \text{Substitute } MDT \text{ in place of } B.$$

$$P = \$4500 - \left(\$4500 \times .105 \times \frac{6}{12} \right) \quad \text{Substitute values.}$$

$$P = \$4263.75$$

Dee receives $4263.75, but must pay back $4500 in 6 months.

Objective 3 **Find the face value.** If the loan amount (proceeds) of a simple discount note are known, use the following formula to find the corresponding face value.

Calculating Face Value to Achieve Desired Proceeds

$$M = \frac{P}{1 - DT},$$

where

M = Face value of the simple discount note
P = Proceeds received by the borrower
D = Discount rate used by the bank   NOTE: The symbol D is the discount *rate*,
T = Time of the loan (in years)   not the bank discount.

The formula derivation follows.

$$P = M - B$$

$$P = M - MDT \quad \text{Substitute } MDT \text{ in place of } B.$$

$$P = M(1 - DT) \quad \text{Divide } M \text{ out of both terms.}$$

$$M = \frac{P}{(1 - DT)} \quad \text{Divide both sides by } (1 - DT).$$

EXAMPLE 3

Finding the Face Value

Bill Thompson needs $4000 for 180 days to rebuild his 1956 Oldsmobile. His banker agrees to lend him the money at a 10% discount rate. Find the face value of the simple discount note that would result in proceeds of $4000 to Thompson.

SOLUTION

Use the formula.

$$M = \frac{P}{1 - DT}$$

Replace P with $4000, D with .10, and T with $\frac{180}{360}$.

$$M = \frac{\$4000}{1 - \left(.10 \times \frac{180}{360}\right)} = \$4210.53 \ (\text{rounded})$$

The face value of the note is $4210.53. Thompson receives $4000 in proceeds on signing the note and 180 days later must repay $4210.53 to the bank.

The problem

$$\frac{\$4000}{1 - \left(.10 \times \frac{180}{360}\right)}$$

can be solved using a calculator by first thinking of the problem as shown below with brackets to set off the denominator:

$$\$4000 \div \left[1 - \left(.10 \times \frac{180}{360}\right) \right]$$

The parentheses inside the brackets are not really needed due to order of operations. The problem is then solved as follows.

4000 ÷ (1 − .10 × 180 ÷ 360) = 4210.53 (rounded)

Note: Refer to Appendix C for calculator basics.

QUICK TIP   A discount rate of 10% is not the same as an interest rate of 10% as shown in the next example.

EXAMPLE 4

Comparing Discount Notes and Simple Interest Notes

Jane Benson of Benson Automotive has been offered loans from two different banks. Each note has a face value of $75,000 and a time of 90 days. One note has a simple interest rate of 10% and the other a simple discount rate of 10%. She wants to know which is the better deal.

SOLUTION

Find the interest owed on each.

Simple Interest Note	Simple Discount Note
$I = PRT$	$B = MDT$
$I = \$75,000 \times .10 \times \frac{90}{360}$	$B = \$75,000 \times .10 \times \frac{90}{360}$
$I = \$1875$	$B = \$1875$

The amount of interest is the same in both notes. Now find the amount the borrower would receive.

Simple Interest Note	Simple Discount Note
Face Value = $75,000	Proceeds = $M - B$
	= $75,000 - $1875
	= $73,125

The borrower has use of $75,000 with the simple interest note, but only $73,125 with the simple discount note. Yet the amount of interest is identical. Therefore, the simple interest note is the better loan for Benson Automotive. However, it is not true that a simple interest note is necessarily better than a simple discount note. You must look at each note individually to understand the terms. Find the maturity value for each note.

Simple Interest Note	Simple Discount Note
$M = P + I$	Maturity = Face value
$= \$75{,}000 + \$1875$	$= \$75{,}000$
$= \$76{,}875$	

The differences between these two notes can be summarized as follows.

	Simple Interest Note	Simple Discount Note
Face value	$75,000	$75,000
Interest	$1875	$1875
Amount available to borrower	$75,000	$73,125
Maturity value	$76,875	$75,000

Objective 4 **Find the effective interest rate.** The different ways of finding interest can be confusing. As a result, the **Federal Truth-in-Lending Act** was passed in 1969. While this law does *not* regulate interest rates, it does require that all interest rates be given in a form so that one loan can easily be compared to another. While this law is discussed in more detail in Chapter 11, the next example shows how to get the simple interest rate corresponding to the discount rate given in Example 4.

The **effective rate**, also called the **Annual Percentage Rate**, **APR**, or **true rate**, of interest is the interest rate that is calculated based on the actual amount received by the borrower. The discount rate of 10% stated in Example 4 is called the **stated rate**, or **nominal rate**, since it is the rate written on the note. It is not the effective rate since the 10% applies to the maturity value of $75,000 and *not* to the proceeds of $73,125 actually received by the borrower. The next example shows how to find the effective rate for Example 4.

EXAMPLE 5

Finding the Effective Interest Rate

Find the effective rate of interest (APR) for the simple discount note of Example 4.

SOLUTION

Find the effective rate (APR) by using the formula for simple interest: $I = PRT$. In this case, $I = \$1875$ (the discount), $P = \$73{,}125$ (the proceeds), and $T = \frac{90}{360}$. Use the following formula from Section 8.2.

$$R = \frac{I}{PT}$$

$$R = \frac{\$1875}{\$73{,}125 \times \dfrac{90}{360}} = 10.26\% \quad \left(\text{rounded}\right)$$

The interest rate 10.26% is the **effective rate of interest** or **true rate of interest**. Federal regulations require that rates be rounded to the nearest quarter of a percent when communicated to a borrower. 10.26% is closer to 10.25% than to 10.50%. Therefore, an **annual percentage rate** (**APR**) of 10.25% must be reported to someone signing a discount note with a face value of $75,000 for 90 days at a discount rate of 10%.

QUICK TIP   The discount rate is not an interest rate to be applied to proceeds (loan amount)—rather it is applied to face value.

The table below shows two identical loans, one a simple interest loan and the other a simple discount loan. Proceeds, interest charges, terms of the loans and maturity values are all identical.

	Simple Interest Loan	Simple Discount Loan
Proceeds	$11,100	$11,100
Interest	$900	$900
Maturity Value	$12,000	$12,000
Face Value	$11,100	$12,000
Time	10 months	10 months
Interest Rate	9.73%	—
Discount Rate	—	9%

Since the loans are identical, a 9.73% simple interest loan for 10 months must be equivalent to a 9% simple discount note for 10 months. In both situations, the borrower receives $11,100 and must repay $12,000 ten months later. Therefore, the two loans have the *same effective rate*.

Objective 5 Understand U.S. Treasury bills. The U.S. government uses discount interest when it borrows money from the public or financial institutions through the sale of **U.S. Treasury bills** (also called **T-bills**). The government auctions T-bills with maturities of 13 weeks, 26 weeks, and 52 weeks. Large financial institutions sometimes buy large T-bills at an auction and then allow investors to buy a smaller portion of the T-bill through them. In this event, the maturity of an investment in a T-bill for an individual investor may be other than 13 weeks, 26 weeks, or 52 weeks. An individual can buy a T-bill from a broker or directly from the government.

EXAMPLE 6

Finding Facts About T-Bills

An Italian bank is worried about devaluation of their currency, so it purchases $1,000,000 in U.S. T-bills in order to place cash in a safe place for a short period of time. The T-bills are at a 4% discount rate for 26 weeks. Find **(a)** the total purchase price, **(b)** the total maturity value, **(c)** the interest earned, and **(d)** the effective rate of interest.

SOLUTION

$M = \$1,000,000; D = .04; T = \frac{26}{52}$

(a) Bank discount = Face value × Discount rate × Time
$$= \$1,000,000 \times .04 \times \frac{26}{52} = \$20,000$$

Purchase price = Face value − Bank discount
$$= \$1,000,000 - \$20,000 = \$980,000$$

(b) Maturity value = Face value
$$= \$1,000,000$$

(c) Interest = Bank discount
$$= \$20,000$$

(d) Effective rate = $\dfrac{\text{Interest earned}}{\text{Purchase price (proceeds)} \times \text{Time}}$
$$= \frac{\$20,000}{\$980,000 \times \frac{26}{52}} = .04081 = 4.08\%$$

8.3 Exercises

FOR EXTRA HELP

MyMathLab

InterActMath.com

MathXP MathXL

MathXL
Tutorials on CD

Addison-Wesley
Math Tutor Center

DVT/Videotape

The **Quick Start** *exercises in each section contain solutions to help you get started.*

Find the discount to the nearest cent, then find the proceeds. (See Example 1.)

Quick Start

Face Value	Discount Rate	Time (Days)	Discount	Proceeds or Loan Amount
1. $7800	9%	120	$234	$7566

$$B = \$7800 \times .09 \times \tfrac{120}{360} = \$234; \ P = \$7800 - \$234 = \$7566$$

Face Value	Discount Rate	Time (Days)	Discount	Proceeds or Loan Amount
2. $15,000	10.25%	90		
3. $19,000	10%	180		
4. $1250	11%	150		
5. $22,400	$8\frac{3}{4}\%$	75		
6. $18,050	8%	80		

Find the maturity date and proceeds for the following. Round to the nearest cent. (See Examples 1 and 2.)

Quick Start

Face Value	Discount Rate	Date Made	Time (Days)	Maturity Date	Proceeds or Loan Amount
7. $6400	9.5%	Mar. 22	90	Jun. 20	$6248

$$P = \$6400 - \left(\$6400 \times .095 \times \tfrac{90}{360}\right) = \$6248$$

Face Value	Discount Rate	Date Made	Time (Days)	Maturity Date	Proceeds or Loan Amount
8. $9500	12%	Oct. 12	100		
9. $1000	$10\frac{1}{4}\%$	July 12	150		
10. $18,500	$9\frac{1}{4}\%$	May 1	220		

 indicates an exercise that is related to the Case in Point feature within the section.

	Face Value	Discount Rate	Date Made	Time (Days)	Maturity Date	Proceeds or Loan Amount
11.	$24,000	10%	Dec. 10	60	_____	_____
12.	$8000	10.5%	Nov. 4	165	_____	_____

Solve each of the following application problems. Round rate to the nearest tenth of a percent, time to the nearest day, and money to the nearest cent.

Quick Start

13. BOAT PURCHASE   An ExxonMobil employee borrowed $6000 from First National Bank to purchase a boat. He plans to repay the loan with a Christmas bonus he is to receive in 120 days. If he borrowed the money at a discount rate of 11%, find **(a)** the discount and **(b)** the proceeds.

(a) $220
(b) $5780

(a) $B = \$6000 \times .11 \times \frac{120}{360} = \$220$; (b) $P = \$6000 - \$220 = \$5780$

14. INCOME-TAX PAYMENT   Managers at Benson Automotive sign a $48,000 simple discount note for six months for funds to pay corporate income taxes. If the discount rate is 8.5%, find **(a)** the discount and **(b)** the proceeds.

(a) _____
(b) _____

15. CHRISTMAS TREE FARM   To plant Christmas trees in another field, Tom Parsons signed a note with a face value of $25,000 at an 11% discount rate. Find the length of the loan in days if the discount is $1527.78.

15. _____

16. TIME OF NOTE   Dixie Schaitberger signed a $12,200 note, at a discount rate of 11%. She was told it would have $931.94 in interest. Find the length of the loan in days.

16. _____

17. WEB-PAGE DESIGN   Nance Sherman needed $8000 to start a Web-page design company and signed a 180-day note with a face value of $8398.95. Find the discount rate.

17. _____

18. CELL PHONES   Ben Delfs needed funds to open a cell phone store and his uncle agreed to put up collateral for the loan. Delfs signed a 130-day simple discount note with a face value of $12,000 and proceeds of $11,523.33. Find the discount rate.

18. _____

19. NEW ROOF   Roy Gerard needs $7260 to pay for a new roof on his house. His bank loans him money at a discount rate of 12%, and the loan is for 240 days. Find the face value of the loan so he will have $7260.

19. _____

20. AUTO REPAIR   Benson Automotive needs $120,000 to upgrade the various tools in their auto repair shop. The simple discount note has a 9.5% rate and matures in 80 days. Find the face value of the loan needed.

20. _____

21. POOR CREDIT   Cathy Cox has poor credit, but found a bank that would lend her $4200 when she uses some collateral. The bank charges a 12% discount rate. Find **(a)** the proceeds if the note is for 10 months and **(b)** the effective interest rate charged by the bank.

(a) _____
(b) _____

22. BAD CREDIT HISTORY   Tim Garcia has a bad credit history partly due to a divorce. The bank agrees to lend him funds based on a note with a face value of $9400, but they require him to use his truck as collateral. Even then, the bank charges him a high 16% discount rate. Find **(a)** the proceeds and **(b)** the effective rate if the note is for 7 months.

(a) _____
(b) _____

23. INTERNATIONAL FINANCE   A business owner in England signs a 10% discount note for 40,000 English pounds (£40,000) with a bank in London. If the proceeds are £38,833.33, find the time of the note in days.

23. _____

24. SIMPLE DISCOUNT RATE   A contractor receives proceeds of $4713.54 on a 12.5% simple discount note with a face value of $5000. Find the time of the note in days.

24. _____

25. EARTHQUAKE DAMAGE   A large bridge in Japan was damaged by an earthquake. The firm repairing the bridge needs proceeds of 165,000 Japanese yen (¥165,000) for 30 days to pay wages and buy supplies. A bank lends them the funds at an 8% discount rate. Find **(a)** the face value and **(b)** the effective rate.

(a) _____
(b) _____

26. INTERNATIONAL FINANCE   A Japanese electric company requires proceeds of $720,000 (local currency) and borrows from a bank in Thailand at 12% discount for 45 days. Find **(a)** the face value of the note and **(b)** the effective interest rate.

(a) _____
(b) _____

The following exercises apply to U.S. Treasury bills, discussed at the end of this section.
(Assume 52 weeks per year for each exercise, and round to the nearest hundredth of a percent.)
(See Example 6.)

27. **PURCHASE OF T-BILLS**   A large Japanese investment firm purchases $25,000,000 in U.S. T-bills at a 6%
discount for 13 weeks. Find **(a)** the purchase price of the T-bills, **(b)** the maturity value of the T-bills,
(c) the interest earned, and **(d)** the effective rate.

(a) _____

(b) _____

(c) _____

(d) _____

28. **T-BILLS**   Nina Horn buys a $50,000 T-bill at a 5.8% discount for 26 weeks. Find **(a)** the purchase price
of the T-bill, **(b)** the maturity value, **(c)** the interest earned, and **(d)** the effective rate of interest.

(a) _____

(b) _____

(c) _____

(d) _____

29. Explain the main differences between simple interest notes and simple discount notes. (See Objective 1.)

30. As a borrower, would you prefer a simple interest note with a rate of 11% or a simple discount note at a
rate of 11%? Explain using an example. (See Example 4.)

8.4 Discounting a Note before Maturity

Objectives

1. Understand the concept of discounting a note.
2. Find the proceeds when discounting simple interest notes.
3. Find the proceeds when discounting simple discount notes.

A note is a *legal responsibility* for one individual or firm to pay a specific amount on a specific date to another individual or firm. Notes can be bought and sold just as a tangible item, such as an automobile can be bought and sold. The clipping taken from a newspaper shows some firms that buy notes. This section shows how to find the value of a note that is sold before its maturity date.

Objective 1 Understand the concept of discounting a note. Businesses sometimes help their customers purchase products or services by accepting a promissory note rather than requiring an immediate cash payment. For example, a company that manufactures pleasure boats, a retailer that sells the boats, and a bank may do business as follows:

1. Boat manufacturer sells boats to a retailer and accepts a promissory note instead of cash.
2. Boat manufacturer needs cash and sells the note to a bank before it matures.
3. Retailer pays the maturity value of the note to the bank when due.

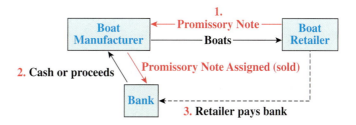

The bank deducts a fee from the maturity value of the note when it buys the note from the manufacturer. The fee is interest for the number of days, called the **discount period**, that the bank will hold the note until it is due. The fee charged by the bank is the **bank discount** or just **discount**. The **discount rate** is the percent used by the bank to find the discount. The process of finding the value of the note on a specific date before it matures is **discounting the note**.

> **QUICK TIP** Both simple interest notes and simple discount notes can be discounted before they mature.

Objective 2 Find the proceeds when discounting simple interest notes. The amount of cash actually received by the boat manufacturer on the sale of a promissory note is the **proceeds**. The bank then collects the maturity value from the maker of the note, the retailer, when it is due. These notes are usually sold with **recourse**. This means that the bank receives reimbursement from the manufacturer if the retailer does not pay the bank when the note matures. Thus the bank is protected against loss. Many banks refuse to buy these types of notes unless they have recourse.

As shown in the newspaper clipping on the next page, interest rates constantly change due to general economic conditions. Further, one firm may apply a different interest rate to a particular loan than another firm. As a result, it is common for the rate at which a note is discounted to differ from the original rate on the note.

Some economists see rates rising sooner than expected

WASHINGTON—Federal Reserve officials are likely to say once again after their meeting Tuesday that they plan to keep interest rates low for a long time.

But some private-sector economists question whether the Fed Chairman and his colleagues are at risk of getting behind the inflation curve.

Pointing to data suggesting percolating inflation pressures and stronger-than-expected growth, a handful of economists say the Fed will have to raise rates much earlier than most think, perhaps as early as March, or risk a surge in prices.

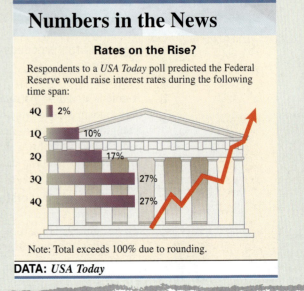

Numbers in the News

Rates on the Rise?

Respondents to a *USA Today* poll predicted the Federal Reserve would raise interest rates during the following time span:

4Q 2%
1Q 10%
2Q 17%
3Q 27%
4Q 27%

Note: Total exceeds 100% due to rounding.

DATA: *USA Today*

Source: USA Today.

Calculate the Proceeds when Discounting a Simple Interest Note

1. First, understand the simple interest note by finding
   (a) the **due date** of the original note and
   (b) the **maturity value** of the original note ($M = P + I$, where $I = PRT$).
2. Then discount the simple interest note.
   (a) Find the **discount period**, which is the time (e.g., number of days) from the sale of the note to the maturity date of the note.
   (b) Find the **discount**, using the formula

$$B = M \times D \times T$$
$$= \text{maturity value} \times \text{discount rate} \times \text{discount period}$$

   (c) Find the **proceeds** after discounting the original note using $P = M - B$.

EXAMPLE 1

Finding Proceeds

Blues Recording holds a 200-day simple interest note from a rock group that agreed to pay them to record an album and produce 1000 copies on CDs. The 12% simple interest note is dated March 24 and has a face value of $4800. Blues Recording wishes to convert the note to cash, so they sell it to a bank on August 15. If there is a discount rate of 12.5%, find the proceeds to the recording studio.

SOLUTION

Go through the four steps of discounting a note.

Step 1   *Find maturity value.* The note is dated March 24 and is due in 200 days. The due date is found as follows.

$$\text{day } 83 \,(\text{March 24}) + 200 \text{ days} = \text{day } 283 \,(\text{October 10})$$

Since this is a simple interest note, the proceeds are given but the maturity value must be found. First find the **interest** on the note if held until maturity.

$$I = \boldsymbol{PRT} = \$4800 \times .12 \times \frac{200}{360} = \$320$$

The **maturity value** is $4800 + $320 = $5120.

Step 2 Now discount this simple interest note.

(a) *Find the discount period.* The **discount period** is the number of days from August 15, which is the date the note is discounted (sold) to the bank, until the due date of the note (October 10).

Discount Period = 56 Days

Date Loan Was Made		Discount Date	Loan Due Date
March 24		August 15	October 10

Length of Loan: 200 Days

October 10 is day   283
August 15 is day − 227
discount period   **56 days**

Blues Recording holds the 200-day note for $200 - 56 = 144$ days before they sell it. The buyer of the note holds it for 56 days before the rock group must pay off the note.

(b) *Find the bank discount.* Find the **discount** by using the formula $B = MDT$, where $M = \$5120$, $D = 12.5\%$, and T is $\frac{56}{360}$.

$$B = MDT = \$5120 \times .125 \times \frac{56}{360} = \$99.56 \left(\text{rounded}\right)$$

The bank discount is $99.56.

(c) *Find the proceeds.* **Proceeds** are found by subtracting the bank discount from the maturity value.

$$P = M - B$$
$$P = \$5120 - \$99.56 = \$5020.44$$

Date	Transaction
March 24	Rock group signs 200-day simple interest note for $4800.
August 15	Blues Recording sells note to bank for $5020.44.
October 10	Bank receives $5120 from payer (rock group).

> **QUICK TIP** A company that sells a note before it matures receives less money than if they wait until the loan matures. However, the firm will receive funds earlier than if they wait.

EXAMPLE 2

Finding Proceeds

On March 27, Dayton Finance loans Jorge Rivera $9200 for 150 days at 11% simple interest. The finance company sells the note to a private investor on April 24. Find the proceeds to Dayton Finance if the note is sold at a discount rate of 12%.

SOLUTION

Step 1 *Find the maturity value.* The maturity date is March 27 (day 86) + 150 days = day 236 (August 24). The simple interest is found as follows.

$$I = PRT = \$9200 \times .11 \times \frac{150}{360} = \$421.67$$

The maturity value is $9200 **(face value)** + $421.67 **(interest)** = $9621.67.

Step 2 (a) *Find the discount period.* The discount period is the time from the day the loan is sold (April 24) to the maturity date of the loan (August 24).

Discount Period = 122 Days

Date Loan Discount Loan
Was Made Date Due Date

March 27 April 24 August 24

Length of Loan: 150 Days

$$\text{day } 236 \left(\text{August } 24\right) - \text{day } 114 \left(\text{April } 24\right) = \textbf{122 days}$$

(b) *Find the bank discount.* The discount is found using 122 days and the 12% discount rate.

$$B = MDT = \$9621.67 \times .12 \times \frac{122}{360} = \$391.28$$

(c) *Find the proceeds.* The proceeds are found as follows.

$$P = M - B = \$9621.67 - \$391.28 = \$9230.39$$

Dayton Finance receives $9230.39 from the sale of the note. Jorge Rivera must pay the full maturity value of the note to the private investor on the maturity date of August 24.

> **QUICK TIP** When finding the bank discount, be sure to use the maturity value of the original note.

It is also common for a business needing cash to sell part of its accounts receivable (money owed to the company) before it is due. The process is called **factoring**, and those who buy the accounts receivable are called **factors**. The calculations involved in factoring are the same as those for finding the discount discussed in this section.

Objective **3** **Find the proceeds when discounting simple discount notes.**

Calculate the Proceeds when Discounting a Simple Discount Note

1. First, understand the simple discount note by finding:
   (a) the **due date** of the original note,
   (b) the **discount** of the original note using $B = MDT$, and
   (c) the **proceeds** from the original note using $P = M - B$.
   The **maturity value (face value)** of the note is written on the note itself and is the value needed in step 2(b) below.
2. Then discount the simple discount note.
   (a) Find the **discount period**, which is the time (e.g., number of days) from the sale of the note to the maturity date of the note.
   (b) Find the **discount**, using the formula $B = MDT$.
   (c) Find the **proceeds** after discounting the original note using $P = M - B$.

> **QUICK TIP** There are two different discounts in problems of this type. The first occurs when the original note is signed. The second discount occurs when this note is sold before maturity.

The steps for finding the proceeds at the time of sale are shown for a U.S. Treasury bill in Example 3. The calculations to find the proceeds are the same for any simple discount note.

EXAMPLE **3**

Finding the Proceeds

Benson Automotive used excess cash to purchase a $100,000 Treasury bill with a term of 26 weeks at a 6.5% simple discount rate. However, the firm needs cash exactly 8 weeks later and sells the T-bill. During the 8 weeks, market interest rates moved up slightly so that the bill was sold at a 7% discount rate. Find **(a)** the initial purchase price of the T-bill, **(b)** the proceeds received by the firm at the subsequent sale of the T-bill, and **(c)** the effective interest rate received by the firm.

SOLUTION

(a) *Find the discount and proceeds.* The discount that Benson Automotive receives when buying the T-bill is found as follows.

$$B = MDT = \$100,000 \times .065 \times \frac{26}{52} = \$3250$$

The cost to the company is the maturity value minus the discount.

$$P = M - B = \$100,000 - \$3250 = \$96,750$$

Therefore, the U.S. government receives $96,750 from the sale of the T-bill.

(b) *Find the discount period, discount, and proceeds.* Now follow the steps in the preceding table to find the proceeds Benson Automotive receives for selling the T-bill 8 weeks later. The discount period is 18 weeks, since the T-bill is sold $26 - 8 = 18$ weeks before its due date.

The discount at the time of the sale is as follows.

$$B = MDT = \$100,000 \times .07 \times \frac{18}{52} = \$2423.08$$

Finally, the proceeds equal the maturity value of the T-bill ($100,000) less the discount at the time of the sale.

$$P = M - B = \$100,000 - \$2423.08 = \$97,576.92$$

(c) Benson Automotive paid $96,750 to buy the T-bill and received $97,576.92 for it 8 weeks later.

$$\text{Interest received} = \$97,576.92 - \$96,750 = \textbf{\$826.92}$$

$$R = \frac{\$826.92}{\$96,750 \times \frac{8}{52}} = 5.56\% \ \left(\text{rounded}\right)$$

The company would have earned 6.5% on the T-bill had it left the treasury bill invested until maturity. Instead, the company sold it after market interest rates rose, but before the T-bill matured. This caused the company to end up with an effective interest rate somewhat less than 6.5%.

General Motors (GM) has been in business for nearly 100 years. You may recognize some of the brand names that it produces: Buick, Cadillac, Chevrolet, GMC, Holden, HUMMER, Opel, Pontiac, Saab, Saturn, and Vauxhall. Additionally, it markets automobiles made by GM Daewoo, Isuzu, Subaru, and Suzuki.

GM is expanding globally with operations in 32 countries and vehicle sales in 190 countries. It sold more than 5.9 million cars and 2.7 million trucks in 2002 and had about 15% of the total worldwide market. The company also owns one of the world's largest financial services companies: GMC Financial Services. This subsidiary loans money to individuals to purchase cars or homes, but it also sometimes loans money to businesses. Total revenue for 2003 was over $190 billion in U.S. currency.

1. Assume that GM sells 2.7 million trucks and sport utility vehicles (SUV's) at an average price of $24,600 in one year. Find the total revenue.

2. Assume that GM borrows $850 million dollars for 216 days from Bank of America. Find the interest and maturity value if the rate is 6%.

3. Bank of America sells the above note after 96 days to the Bank of New York. Find the discount period and the proceeds to Bank of America if the discount rate is 6.3%.

4. Use the World Wide Web to find Total Revenue (from GM's Income Statement) and Total Current Liabilities (from its Balance Sheet) for the most recent year for GM.

8.4 | Exercises

FOR EXTRA HELP

MyMathLab

InterActMath.com

Math*XP* MathXL

MathXL
Tutorials on CD

Addison-Wesley
Math Tutor Center

DVT/Videotape

The **Quick Start** *exercises in each section contain solutions to help you get started.*

Find the discount period for each of the following. (See Examples 1 and 2, Step 2.)

Quick Start

Date Loan Was Made	Length of Loan	Date of Discount	Discount Period
1. Apr. 29	200 days	July 31	**107 days**
2. July 28	120 days	Sept. 20	_____
3. May 28	74 days	June 18	_____
4. Sept. 17	130 days	Jan. 13	_____

Find the proceeds to the nearest cent when each of the following is discounted. (Hint: The maturity value is given.) (See Examples 1 and 2.)

Quick Start

Maturity Value	Discount Rate	Discount Period	Proceeds
5. $10,400	8.5%	90 days	**$10,179**

$B = \$10,400 \times .085 \times \frac{90}{360} = \$221; \ P = \$10,400 - \$221 = \$10,179$

Maturity Value	Discount Rate	Discount Period	Proceeds
6. $4800	10.3%	200 days	_____
7. $2500	9%	30 days	_____
8. $3000	11%	60 days	_____

Find the maturity value of each of the following simple interest notes. Each note is then discounted at 12%. Find the discount period, the discount, and proceeds after discounting. (See Examples 1 and 2.)

Quick Start

Date Loan Was Made	Face Value	Length of Loan	Rate	Maturity Value	Date of Discount	Discount Period	Discount	Proceeds
9. Feb. 7	$6200	90 days	$10\frac{1}{2}\%$	**$6362.75**	Apr. 1	**37 days**	**$78.47**	**$6284.28**

$I = \$6200 \times .105 \times \frac{90}{360} = \$162.75; \ M = \$6200 + \$162.75 = \$6362.75$
Feb. 7 is day 38; Apr. 1 is day 91; $91 - 38 = 53$
Discount period $= 90 - 53 = 37$ days
$B = \$6362.75 \times .12 \times \frac{37}{360} = \$78.47$
Proceeds $= \$6362.75 - \$78.47 = \$6284.28$

 indicates an exercise that is related to the Case in Point feature within the section.

	Date Loan Was Made	Face Value	Length of Loan	Rate	Maturity Value	Date of Discount	Discount Period	Discount	Proceeds
10.	June 15	$9200	140 days	12%	_____	Oct. 22	_____	_____	_____
11.	July 10	$2000	72 days	11%	_____	Aug. 2	_____	_____	_____
12.	May 29	$5500	80 days	10%	_____	July 8	_____	_____	_____

First, find the initial proceeds of each of the following simple discount notes. Each note is then discounted at 11%. Find the discount period, the discount, and proceeds after discounting. (See Example 3.)

Quick Start

	Date Loan Was Made	Maturity Value	Length of Loan	Rate	Initial Proceeds	Date of Discount	Discount Period	Discount	Proceeds at Time of Sale
13.	Jan. 12	$17,800	90 days	10%	**$17,355**	Mar. 1	**42 days**	**$228.43**	**$17,571.57**

$B = MDT = \$17,800 \times .10 \times \frac{90}{360} = \$445$; $P = M - B = \$17,800 - \$445 = \$17,355$
Jan. 12 is day 12; Due date is $12 + 90 = 102$ or Apr. 12
Mar. 1 is day 60; Discount period is $102 - 60 = 42$ days
$B = MDT = \$17,800 \times .11 \times \frac{42}{360} = \$228.43$
$P = M - B = \$17,800 - \$228.43 = \$17,571.57$

	Date Loan Was Made	Maturity Value	Length of Loan	Rate	Initial Proceeds	Date of Discount	Discount Period	Discount	Proceeds at Time of Sale
14.	Aug. 4	$24,000	120 days	10.5%	_____	Oct. 8	_____	_____	_____
15.	May 4	$32,100	150 days	9.5%	_____	July 10	_____	_____	_____
16.	Apr. 30	$22,000	200 days	9%	_____	Jul. 12	_____	_____	_____

Solve the following application problems. Round interest and discount to the nearest cent.

17. ROCK CRUSHER First Bank loaned $360,000 for 180 days to a company purchasing a rock-crushing machine. The bank sold the 7% simple interest note 120 days later at an 8% discount rate. Find **(a)** the bank discount and **(b)** the proceeds.

(a) $4968
(b) $367,632

(a) $M = \$360,000 + \left(\$360,000 \times .07 \times \frac{180}{360}\right) = \$372,600$
discount period $= 180$ days $- 120$ days $= 60$ days
$B = \$372,600 \times .08 \times \frac{60}{360} = \$4968$
(b) $P = \$372,600 - \$4968 = \$367,632$

18. TRACTOR PURCHASE Cook and Daughters Farm Equipment accepts a $5800 simple interest note at 12% for 100 days, for a small used tractor. The note is dated May 12. On June 17, the firm discounts the note at the bank, at a 13% discount rate. Find **(a)** the bank discount and **(b)** the proceeds.

(a) _____
(b) _____

19. AUTOMOBILE DEALERSHIP Benson Automotive signed a 180-day simple discount note with a face value of $250,000 and a rate of 9% on March 19. The lender sells the note at an 8% discount on June 14. Find **(a)** the proceeds of the original note to the dealership, **(b)** the discount period, **(c)** the discount, and **(d)** the proceeds at the sale of the note on June 14.

(a) _____
(b) _____
(c) _____
(d) _____

20. SEWING CENTER Kathy Bates, owner of Sew What, agreed to a 10% simple discount note with a maturity value of $18,500 on July 30. She planned to add to her inventory of sewing machines with the funds. The 120-day note is sold by the lender at a 12% discount on September 2. Find **(a)** the proceeds of the original note to Bates, **(b)** the discount period, **(c)** the discount, and **(d)** the proceeds at the sale of the note on September 2.

(a) _____
(b) _____
(c) _____
(d) _____

21. FINANCING CONSTRUCTION To build a new building for its body shop, Benson Automotive signed a $300,000 simple interest note at 9% for 150 days with National Bank on November 20. On February 6, National Bank sold all of its notes to Bank One. Find **(a)** the maturity value of the note and **(b)** the proceeds to National Bank given a discount rate of 10.5%.

(a) _____
(b) _____

22. BATTERY STORE   An NTB outlet borrowed $48,500 on a
200-day simple interest note to expand the battery store.
The note was signed on December 28 and carried an interest
rate of 9.8%. The note was then sold on March 17 at a
discount rate of 10%. Find **(a)** the maturity value of the
note and **(b)** the proceeds to the seller of the note on May 17.

(a) _____

(b) _____

23. PURCHASE OF A T-BILL   Elizabeth Barton bought a $25,000, 26-week T-bill at a discount rate of 6.8%
on August 7. She sold it 10 weeks later at a discount rate of 7%. Find **(a)** Barton's purchase price,
(b) the discount 10 weeks later when she sold it, **(c)** the proceeds to Barton, and **(d)** the effective
interest rate rounded to the nearest hundredth of a percent for the time Barton held the note.

(a) _____

(b) _____

(c) _____

(d) _____

24. PURCHASE OF A T-BILL   Tina Klein bought a $10,000, 7.5%, 52-week T-bill on June 29 and sold it
26 weeks later at a discount rate of 8%. Find Klein's **(a)** purchase price for the T-bill, **(b)** the discount at
time of sale, **(c)** the proceeds to Klein, and **(d)** the effective interest rate rounded to the nearest
hundredth of a percent.

(a) _____

(b) _____

(c) _____

(d) _____

25. Explain the procedure used to determine the bank discount and the proceeds for a note. (See Objective 2.)

26. Explain the effect of a rise in general market interest rates on an investor who is holding notes. Give
an example. (See Example 3.)

Supplementary Application Exercises on Simple Interest and Simple Discount

The **Quick Start** *exercises in each section contain solutions to help you get started.*

There are similarities and differences between simple interest and simple discount calculations. This exercise set compares these two important concepts. First, the key similarities between the two are as follows.

1. Both types of notes involve lump sums repaid with a single payment at the end of a stated period of time.
2. The length of time is generally 1 year or less.

The following table compares simple interest and simple discount notes.

	Simple Interest Note	Simple Discount Note
Variables	I = Interest	B = Discount
	P = Principal (face value)	P = Proceeds
	R = Rate of interest	D = Discount rate
	T = Time, in years or fraction of a year	T = Time, in years or fraction of a year
	M = Maturity value	M = Maturity value
Face value	Stated on note	Same as maturity value
Interest charge	$I = PRT$	$B = MDT$
Maturity value	$M = P + I$	Same as face value
Amount received by borrower	Face value or principal	Proceeds: $P = M - B$
Identifying phrases	Interest at a certain rate	Discount at a certain rate
	Maturity value greater than face value	Proceeds
		Maturity value equal to face value
Effective interest rate	Same as stated rate, R	Greater than stated rate, D

QUICK TIP   The variable P is used for *principal or face value* in simple interest notes, but P is used for *proceeds* in simple discount notes. P represents the amount received by the borrower.

Solve the following application problems. Round rates to the nearest tenth of a percent, time to the nearest day, and money to the nearest cent.

Quick Start

1. The owner of Redwood Furniture, Inc., signed a 120-day note for $18,000 at 11% simple interest. Find **(a)** the interest and **(b)** the maturity value.

   (a) $I = PRT = \$18,000 \times .11 \times \frac{120}{360} = \$660$
   (b) $M = \$18,000 + \$660 = \$18,660$

   (a) $660
   (b) $18,660

2. Bill Travis signed a note for $18,500 with his uncle to start an auto repair shop on Commerce Street. The note is due in 300 days and has a discount rate of 14%. Travis hopes that a bank will refinance the note for him at a lower rate after he has been in business for 300 days. Find the proceeds.

   $B = \$18,500 \times .14 \times \frac{300}{360} = \$2158.33; \quad P = \$18,500 - \$2158.33 = \$16,341.67$

   2. $16,341.67

C indicates an exercise that is related to the Case in Point feature within the section.

3. James Watkins signed a note with a 10% simple interest rate, interest of $2400, and time of 180 days. Find the principal.

3. _____

4. Benson Automotive signed a $150,000 note at a simple discount rate of 10.5% and a discount of $8750. Find the length of the loan in days.

4. _____

5. A loan to a German bank was for $1,290,000 with a maturity value of $1,327,410 and a rate of 6%. Find the time.

5. _____

6. Jane Benson of Benson Automotive loaned her nephew $20,000 for 150 days at 9% simple interest. Find (a) the interest and (b) the maturity value.

(a) _____
(b) _____

7. John O'Neill borrowed $24,000 for 250 days at 7% simple interest. Find (a) the interest and (b) the maturity value.

(a) _____
(b) _____

8. Gilbert Construction Company signed a 5-month, $145,000 note at an 11.5% discount rate. Find the effective rate of interest.

8. _____

9. First Bank signed an 80 day, $82,000 note at a 12% discount rate. Find the effective rate of interest.

9. _____

10. On October 14, Citibank loaned $10,000,000 to Fleet Mortgage Company for 180 days at a 10.5% discount rate. Find (a) the due date and (b) the proceeds.

(a) _____
(b) _____

11. On December 24, Junella Martin signed a 100-day note for $80,000 for a new Jaguar. Given a discount rate of 11%, find (a) the due date and (b) the proceeds.

(a) _____
(b) _____

12. Lupe Galvez has a serious problem: two of her more energetic preschoolers keep getting out of the yard of her child-care center. She signs a note with interest charges of $670.83 to reinforce the fence around the entire yard. The simple interest note is for 140 days at 11.5%. Find the principal to the nearest dollar.

12. _____

13. Quality Furnishings accepted a 270-day, $8000 note on May 25. The interest rate on the note is 12% simple interest. The note was then discounted at 14% on August 7. Find the proceeds.

13. _____

14. On November 19, a firm accepts an $18,000, 150-day note with a simple interest rate of 9%. The firm discounts the note at 12% on February 2. Find the proceeds.

14. _____

15. Barton's Flowers accepted a $16,000, 150-day note from Wedded Bliss Catering. The note had a simple interest rate of 11% and was accepted on May 12. The note was then discounted at 13% on July 20. Find the proceeds to Barton's Flowers.

15. _____

16. Leon Herbert signed a 220-day, 10% simple interest note with a face value of $28,000. The bank he borrowed the money from turned around and sold the note 90 days later at an 11% discount rate. Find **(a)** the interest, **(b)** the maturity value, **(c)** the discount period, **(d)** the discount and **(e)** the proceeds to the bank.

(a) _____
(b) _____
(c) _____
(d) _____
(e) _____

17. Janice Dart signed a 140-day simple discount note at a rate of 9.9% with a maturity value of $82,000. The bank she borrowed the funds from sold the note 40 days later at a 10% discount rate. Find **(a)** the discount on the original note, **(b)** the proceeds of the original note, **(c)** the discount period, **(d)** the discount at the time of sale, and **(e)** the proceeds to the bank at the time of sale.

(a) _____
(b) _____
(c) _____
(d) _____
(e) _____

18. James and Tiffany Paterson need a 220-day loan for $68,000 to open Adventure Sports Unlimited. Bank One agrees to a simple interest note with a loan amount of $68,000 at $9\frac{1}{4}$% interest. Union Bank agrees to a simple discount note with proceeds of $68,000 and a 9.5% simple discount rate. Find **(a)** the interest for the simple interest note, **(b)** the maturity value of the discount note, **(c)** the interest for the discount note, and **(d)** the savings in interest charges of the simple interest note over the discount note.

(a) _____
(b) _____
(c) _____
(d) _____

19. Gilbert Construction Company needs to borrow $380,000 for $1\frac{1}{2}$ years to purchase some land to subdivide. One bank offers the firm a simple interest note with a principal of $380,000 and a rate of 12%. A second bank offers the company a discount note with proceeds of $380,000 and an 11% discount rate. **(a)** Which note produces the lower interest charges? **(b)** What is the difference in interest?

(a) _____

(b) _____

20. Show with an example that the effective interest rate is greater than the discount rate stated on a note.

21. Explain the difference in meaning of the variable P (principal) in a simple interest note and the variable P (proceeds) in a simple discount note.

Chapter 8 | Quick Review

CHAPTER TERMS *Review the following terms to test your understanding of the chapter. For each term you do not know, refer to the page number found next to that term.*

annual percentage rate [**p. 331**]
bank discount [**p. 327**]
collateral [**p. 314**]
compound interest [**p. 308**]
discount [**p. 327**]
discounting the note [**p. 337**]
discount period [**p. 337**]
discount rate [**p. 337**]
due date [**p. 314**]
effective rate of interest [**p. 331**]
exact interest [**p. 312**]

face value [**p. 327**]
factoring [**p. 340**]
factors [**p. 340**]
Federal Truth-in-Lending Act [**p. 331**]
foreclose [**p. 314**]
interest [**p. 308**]
interest-in-advance notes [**p. 327**]
maker of a note [**p. 313**]
maturity date [**p. 314**]

maturity value [**p. 309**]
nominal rate [**p. 331**]
ordinary interest [**p. 312**]
payee of a note [**p. 313**]
payer of a note [**p. 313**]
prime rate [**p. 308**]
principal [**p. 308**]
proceeds of a note [**p. 327**]
promissory note [**p. 313**]
rate [**p. 309**]

recourse [**p. 337**]
simple discount note [**p. 327**]
simple interest [**p. 308**]
simple interest note [**p. 313**]
stated rate [**p. 331**]
T-bills [**p. 332**]
term of a note [**p. 313**]
time [**p. 309**]
true rate of interest [**p. 331**]
U.S. Treasury bills [**p. 332**]

CONCEPTS | EXAMPLES

8.1 Finding simple interest when time is expressed in years
1. Use formula $I = PRT$.
2. Express R in decimal form.
3. Express time in years.
4. Substitute values for P, R, and T and multiply.

A loan of $5900 is made for $1\frac{3}{4}$ years at 10% per year; find the simple interest.
$$I = PRT$$
$$I = \$5900 \times .10 \times 1.75 = \$1032.50$$
The simple interest is $1032.50.

8.1 Finding simple interest when time is expressed in months
1. Use formula $I = PRT$.
2. Express R in decimal form.
3. Express time in years by dividing number of months by 12.
4. Substitute values for P, R, and T and multiply.

Find the simple interest on $24,000 for 8 months at 10%.
$$I = PRT$$
$$I = \$24,000 \times .10 \times \frac{8}{12} = \$1600$$
The simple interest is $1600.

8.1 Finding the maturity value of a loan
1. Find I, using the formula $I = PRT$.
2. Find the maturity value using the formula $M = P + I$.

A loan of $8500 is made for one year at 9%. Find the maturity value of the loan.
$$I = PRT$$
$$I = \$8500 \times .09 \times 1 = \$765$$
$$M = P + I$$
$$M = \$8500 + \$765 = \$9265$$
The maturity value is $9265.

8.1 Finding the number of days from one date to another using a table
1. Find the day corresponding to the final date using the table.
2. Find the day corresponding to the initial date.
3. Subtract the smaller number from the larger number.

Find the number of days from February 15 to July 28.
1. July 28 is day 209
2. Feb. 15 is day 46
3. Number of days is

$$\begin{array}{r} 209 \\ -\ 46 \\ \hline 163 \end{array}$$

There are 163 days from February 15 to July 28.

CONCEPTS	EXAMPLES
8.1 Finding the number of days from one date to another using actual number of days in a month Add actual number of days in each month or partial month from initial date to final date.	Find the number of days from April 20 to June 27. April 20 to April 30 10 days May 31 days June 27 days 68 days
8.1 Finding exact interest Use the formula $$I = PRT$$ with $T = \dfrac{\text{Number of days of loan}}{365}$	Find the exact interest on a \$9000 loan at 8% for 140 days. $$I = PRT$$ $$I = \$9000 \times .08 \times \frac{140}{365} = \$276.16$$ The exact interest is \$276.16.
8.1 Finding ordinary or banker's interest Use the formula $$I = PRT$$ with $T = \dfrac{\text{Number of days of loan}}{360}$	Find the ordinary interest on a loan of \$14,000 at 7% for 120 days. $$I = PRT$$ $$I = \$14,000 \times .07 \times \frac{120}{360} = \$326.67$$ The ordinary or banker's interest is \$326.67.
8.1 Finding the due date, interest, and maturity value of a simple interest promissory note when the term of loan is in months **1.** Add number of months in term of note to initial date of note. **2.** Use formula $I = PRT$ to find interest. **3.** Find maturity value as follows. Maturity value = Principal + Interest	Find the due date, the interest, and the maturity value of a loan made on February 15 for 7 months at 8% with a face value of \$9400. September 15 is 7 months from February 15, so note is due on September 15. $$I = PRT$$ $$I = \$9400 \times .08 \times \frac{7}{12} = \$438.67$$ **M = Principal + Interest** $$M = \$9400 + \$438.67 = \$9838.67$$
8.1 Finding due date of a promissory note when term of loan is expressed in days Use either a table or the actual number of days in each month.	A loan is made on August 14 and is due in 80 days. Find the due date. August 14 to August 31 17 days September 30 days October 31 days 78 days The loan is for 80 days, which is 2 days more than 78. Therefore, the loan is due on November 2.

CONCEPTS	EXAMPLES
8.2 Finding principal given interest, interest rate, and time Use the formula $$P = \frac{I}{RT}$$ 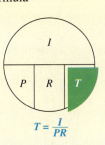 $P = \dfrac{I}{RT}$	Find the principal that produces an interest of \$240 at 9% for 60 days. $$P = \frac{I}{RT}$$ $$P = \frac{\$240}{.09 \times \frac{60}{360}} = \$16,000$$ The principal is \$16,000.
8.2 Finding the rate of interest given principal, interest, and time Use the formula $$R = \frac{I}{PT}$$ $R = \dfrac{I}{PT}$	A principal of \$8000 deposited for 45 days earns interest of \$75. Find the rate of interest. $$R = \frac{I}{PT}$$ $$R = \frac{\$75}{\$8000 \times \frac{45}{360}} = .075$$ Rate of interest = 7.5%.
8.2 Finding the time given principal, rate of interest, and interest To find the time in days, use the formula $$T \text{ (in days)} = \frac{I}{PR} \times 360$$ To find the time in months, use the formula $$T \text{ (in months)} = \frac{I}{PR} \times 12$$ $T = \dfrac{I}{PR}$	Tom Jones invested \$4000 at 8% and earned an interest of \$160. Find the number of days. $$T = \frac{I}{PR} \times 360$$ $$T = \frac{\$160}{\$4000 \times .08} \times 360 = 180 \text{ days}$$ The loan was for 180 days.
8.3 Finding the proceeds of a simple discount note Calculate bank discount using the formula $B = MDT$. Then calculate the proceeds or loan amount using the formula $P = M - B$.	Karen Pattern borrows \$6000 for 120 days at a discount rate of 9%. Find the proceeds. $$B = MDT$$ $$B = \$6000 \times .09 \times \frac{120}{360} = \$180$$ $$P = M - B$$ $$P = \$6000 - \$180 = \$5820$$
8.3 Finding the face value of a simple discount note Use the formula $$M = \frac{P}{1 - DT}$$	Sam Spade needs \$15,000 for new equipment for his restaurant. Find the face value of a note that will provide the \$15,000 in proceeds if he plans to repay the note in 180 days and the bank charges an 11% discount rate. $$M = \frac{P}{1 - DT}$$ $$M = \frac{\$15,000}{1 - \left(.11 \times \frac{180}{360}\right)} = \$15,873.02 \text{ (rounded)}$$

CONCEPTS	EXAMPLES
8.3 Finding the effective interest rate Find the interest (B) from the formula $$B = MDT.$$ Find proceeds from the formula $$P = M - B.$$ Then use the formula $$R = \frac{I}{PT}.$$	A 150-day, 11% simple discount note has a face value of $12,400. Find the effective rate to the nearest tenth of a percent. $$B = \$12,400 \times .11 \times \frac{150}{360} = \$568.33$$ $$P = \$12,400 - \$568.33 = \$11,831.67$$ $$R = \frac{\$568.33}{\$11,831.67 \times \dfrac{150}{360}} = 11.5\% \ (\text{rounded})$$

8.4 Finding the proceeds to an individual or firm that discounts a simple interest note

1. If necessary, find
   (a) the **due date** of the original note and
   (b) the **maturity value** of the original note $(M = P + I$ where $I = PRT)$.
2. (a) Find the **discount period**, which is the time (e.g., number of days) from the sale of the note to the maturity date of the note.
   (b) Find the **discount**, using the formula $B = MDT$.
   (c) Find the **proceeds** using $P = M - B$.

Moe's Ice Cream holds a 150-day note dated March 1 with a face value of $15,000 and a simple interest rate of 9%. Moe sells the note at a discount on June 1. Assume a discount rate of 11%. Find the proceeds.

1. Due date = day 60 (March 1) + 150 days = day 210 or July 29

$$I = PRT = \$15,000 \times .09 \times \frac{150}{360} = \$562.50$$

$$M = P + I = \$15,000 + \$562.50 = \$15,562.50$$

2. (a) The discount period is 58 days.

Discount Period = 58 Days

Date Loan Was Made   Discount Date   Loan Due Date

March 1   June 1   July 29

Length of Loan: 150 Days

(b) Bank Discount $= MDT =$

$$\$15,562.50 \times .11 \times \frac{58}{360} = \$275.80$$

(c) Proceeds $= M - B =$

$$\$15,562.50 - \$275.80 = \$15,286.70$$

CONCEPTS	EXAMPLES

8.4 **Finding the proceeds to an individual or firm that discounts a simple interest note**

1. If necessary find
   (a) the **due date** of the original note,
   (b) the **discount** of the original note using $B = MDT$, and
   (c) the **proceeds** from the original note using $P = M - B$.

   The **maturity value (face value)** of the note is written on the note itself.

2. (a) Find the **discount period**, which is the time (e.g., number of days) from the sale of the note to the maturity date of the note.
   (b) Find the **discount**, using the formula $B = MDT$.
   (c) Find the **proceeds** using $P = M - B$.

On May 10, Applecrest Farm Orchards signed a 120-day note for $22,000 at a simple discount rate of 10%. The note was sold on June 30 at a discount rate of 10.5%. Find **(a)** the proceeds from the original note and also **(b)** the proceeds at the time of sale.

(a) Due date is day 130 (May 10) + 120 days = day 250 or Sept. 7

$$B = MDT = \$22,000 \times .10 \times \frac{120}{360} = \$733.33$$

$$P = M - B = \$22,000 - \$733.33 = \$21,266.67$$

(b) June 30 is day 181
Discount period is 250 − 181 = 69 days

$$B = MDT = \$22,000 \times .105 \times \frac{69}{360} = \$442.75$$

(c) $P = M - B = \$22,000 - \$442.75 = \$21,557.25$

Chapter 8 | Summary Exercise

Banking in a Global World: How Do Large Banks Make Money?

Bank of America borrowed $80,000,000 at 5% interest for 180 days from a Japanese investment house. At the same time, the bank made the following loans, each for the exact same 180-day period:

1. A 7% simple interest note for $38,000,000 to a Canadian firm that extracts oil from Canadian tar sands;

2. An 8.2% simple discount note for $27,500,000 to a European contractor building a factory in South Africa; and

3. An 8% simple discount note for $14,500,000 to a Louisiana company building minesweepers in New Orleans for the British government.

(a) Find the difference between interest received and interest paid by the bank on these funds.

(a) _____

(b) The bank did not loan out all $80,000,000. Find the amount they actually loaned out.

(b) _____

INVESTIGATE

The very idea of interest is generally not acceptable in some third world countries. Here a bank takes partial ownership of a company when they lend to a company, at least until funds are repaid. Even in countries that do allow interest, interest rates vary considerable. Use financial newspapers, magazines, or the World Wide Web and find interest rates in three different countries and compare them to similar rates in the United States.

Chapter 8 | Test

To help you review, the numbers in brackets show the section in which the topic was discussed.

Find the simple interest for each of the following. Round to the nearest cent. **[8.1]**

1. $12,500 at $10\frac{1}{2}\%$ for 280 days

1. _____

2. $8250 at $9\frac{1}{4}\%$ for 8 months

2. _____

3. A loan of $6000 at 11% made on June 8 and due August 22

3. _____

4. A promissory note for $4500 at 10.3% made on November 13 and due March 8

4. _____

5. Joan Davies signed a 140-day simple interest note for $12,500 with a bank that uses *exact* interest. If the rate is 10.7%, find the maturity value. **[8.1]**

5. _____

6. Chez Bazan Bakery borrowed $24,300 for new ovens and other equipment. The simple interest loan was repaid in 6 months at $10\frac{1}{2}\%$. Find the amount of the repayment.   **[8.1]**

6. _____

7. Glenda Pierce plans to borrow $14,000 for a new hot tub and deck for her home. She has decided on a term of 200 days at 10.5% simple interest. However, she has a choice of two lenders. One calculates interest using a 360-day year and the other uses a 365-day year. Find the amount of interest Pierce will save by using the lender with the 365-day year.   **[8.1]**

7. _____

8. Lupe Gonzalez has $6500 in her retirement account. Find the interest rate required for the fund to grow to $7247.50 in 15 months.   **[8.2]**

8. _____

9. Hilda Heinz lends $1200 to her sister Olga at a rate of 9%. Find how long it will take for her investment to earn $100 in interest. (Round to the nearest day.)   **[8.2]**

9. _____

10. A woman invested money received from an insurance settlement for 7 months at 5% interest. If she received $1254.17 interest on her investment during this time, find the amount that she invested. (Round to the nearest dollar.)   **[8.2]**

10. _____

11. Mike Fagan needs $25,000 to expand his flower shop. Find the face value of a simple discount note that will provide the $25,000 in proceeds if he plans to repay the note in 240 days and the bank charges a 9% discount rate.   **[8.3]**

11. _____

Find the discount and the proceeds for the following simple discount notes.   **[8.3]**

Face Value	Discount Rate	Time (Days)	Discount	Proceeds
12. $9800	11%	120	_____	_____
13. $10,250	9.5%	60	_____	_____

14. Barbara Waters signed a simple discount note for $15,000 for 120 days at a rate of 9%. Find **(a)** the proceeds and **(b)** the effective interest rate based on the proceeds received by Waters.   **[8.3]**

(a) _____
(b) _____

15. Lizabeth Neault needed funds to open a law office. She borrowed $28,400 at 8.5% simple interest for 150 days on July 7. The bank she borrowed from sold the note at a 9% discount on August 20. Find the proceeds to the bank.   **[8.4]**

15. _____

16. A 90-day simple discount promissory note for $9200 with a simple discount rate of 11% was signed on January 25. It was discounted on March 2 at 12%. Find the proceeds at the time of the sale.   **[8.4]**

16. _____

17. A $20,000 T-bill is purchased at a 3.75% discount rate for 13 weeks. Find **(a)** the purchase price of the T-bill, **(b)** the maturity value, **(c)** the interest earned, and **(d)** the effective rate of interest to the nearest hundredth of a percent.   **[8.3]**

(a) _____
(b) _____
(c) _____
(d) _____

The following note was discounted at $12\frac{1}{2}\%$. Find the discount period, the discount, and the proceeds.   **[8.4]**

Date Loan Was Made	Face Value	Length of Loan	Rate	Date of Discount	Discount Period	Discount	Proceeds
18. Jan. 25	$9200	90 days	10%	Mar. 12	_____	_____	_____

19. Jan Guerra lends $9000 to her second cousin using a 180-day 10% simple interest note that was signed on October 30. Guerra subsequently has a car accident and desperately needs money so she sells the note at a discount of 15% on January 3 to an investor. Find **(a)** the discount, **(b)** the proceeds, and **(c)** the amount of money Guerra gains or loses.   **[8.4]**

(a) _____
(b) _____
(c) _____

CHAPTER

9

Compound Interest

BANK OF AMERICA IS ONE OF THE LARGEST BANKS IN THE WORLD with over 13,000 automatic-teller machines (ATMs). They also offer on-line

banking so that computer-savvy customers can pay bills, apply for loans and credit cards, balance check books, and transfer money at any time of the day using their computer.

Simple interest is interest charged on the entire principal, for the entire length of the loan. **Simple interest loans** are usually short-term loans that are repaid in one year or less. Banks usually use the simple interest calculations of Chapter 8 for short-term loans to businesses or individuals.

Compound interest is interest calculated on any interest previously credited to an account in addition to the original principal. Compound interest calculations are used in many long-term investments, such as savings accounts or certificates of deposit at a bank.

9.1 | Compound Interest

Objectives

1. Use the formula $I = PRT$ to calculate compound interest.
2. Identify interest rate per compounding period and number of compounding periods.
3. Use the formula $M = P(1 + i)^n$ to find compound amount.
4. Use the table to find compound amount.

CASE in POINT

Regina Foster worked overtime as a nurse and earned an extra $2000. She banks at Bank of America and wonders what the difference would be between a 6-year investment earning 5% offered by her bank and one earning 8% the bank once paid her father some years ago.

The amount or future value of an investment depends on three things: (1) whether interest is found using simple interest or compound interest, (2) the interest rate, and (3) the term or length of the investment. Look at the following options for a $10,000 investment.

Investment	Term	Annual Rate	Interest	Future value
1. Simple interest	6 years	5%	$3,000	$13,000
2. Compound interest	6 years	5%	$3,401	$13,401
3. Compound interest	6 years	8%	$5,869	$15,869
4. Compound interest	10 years	8%	$11,589	$21,589

Investments 1 and 2 show that compound interest generates more interest than simple interest. Investments 2 and 3 show the that a higher interest rate results in more interest. Finally, investments 3 and 4 show that the length of an investment can add significantly to the amount or future value.

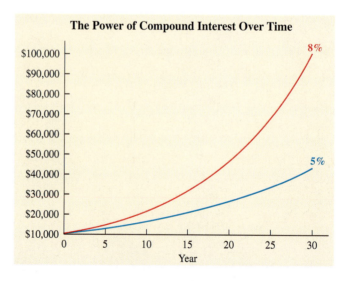

Objective **1** **Use the formula *I* = *PRT* to calculate compound interest.** **Compound interest** is interest calculated on previously credited interest in addition to the original principal. Compound interest calculations require that interest be calculated and credited to an account many times each year. The simple interest formulas $(I = PRT$ and $M = P + I)$ are used each time compound interest is calculated.

EXAMPLE **1**

*Comparing Simple
Interest to Compound
Interest*

Regina Foster wishes to compare simple interest to compound interest.

  (a) Find the interest for a $2000 investment earning 6% simple interest for 1 year.
  (b) Find the interest for a $2000 investment earning 6% interest compounded every 6 months for 1 year.
  (c) Find the additional interest at the end of 1 year when using compound interest.

SOLUTION

  (a) **Simple interest** on $2000 at 6% *for 1 year* is found as follows.

$$I = PRT = \$2000 \times .06 \times 1 = \$120$$

  (b) **Interest compounded every 6 months** means that interest must be calculated at the end of each 6-month interval. Interest is found using the simple interest formula $I = PRT$ for each 6-month period.

$$\text{Interest for the first 6 months} = PRT = \$2000 \times .06 \times \frac{1}{2} = \$60$$

The interest must be added to the principal before interest is calculated again. The new balance is found as follows.

$$\text{New balance} = \textbf{original principal} + \text{interest} = \textbf{\$2000} + \$60 = \textbf{\$2060}$$

This is the amount that earns interest in the second half of the year, **no longer just the $2000 originally deposited**.

new principal

$$\text{Interest for the second 6 months} = PRT = \textbf{\$2060} \times .06 \times \frac{1}{2} = \$61.80$$

Be sure to round interest to the nearest cent if needed.

$$\text{Total at end of year} = \underbrace{\text{principal at end of 6 months}} + \underbrace{\text{interest}}$$

$$= \qquad \textbf{\$2060} \qquad + \$61.80 = \textbf{\$2121.80}$$

This final amount on deposit at the end of the year is called the **compound amount**. The interest is the compound amount minus the original principal.

$$\text{Interest} = \text{Compound amount} - \text{Original principal}$$

$$\text{Interest} = \$2121.80 - \$2000 = \$121.80$$

Notice that interest earned in the second half of the year ($61.80) is more than that earned in the first half of the year ($60). This is because the interest earned at the end of the first six months also earns interest in the second half of the year. Interest earned during the second half of the year is based on original principal and interest earned in the first half of the year.

  (c) The difference in interest earned using simple and compound interest follows.

$$\text{Compound interest} - \text{Simple interest} = \$121.80 - \$120 = \textbf{\$1.80}$$

This difference may not seem like very much, but it amounts to *tremendous differences* when larger sums of money and longer time periods are used.

QUICK TIP Relatively small differences in interest rates can add up to large differences in compound amount over time.

> **QUICK TIP** Round interest amounts to the nearest cent each time interest is calculated.

EXAMPLE 2

Finding Compound Interest

The Peters hope to have $5000 in 4 years for a down payment on their first new car. They invest $3800 in an account that pays 6% interest at the end of each year, on previous interest in addition to principal. **(a)** Find the excess of compound interest over simple interest after 4 years. **(b)** Will they have enough money at the end of 4 years to meet their goal of a down payment?

SOLUTION

For each year, first calculate interest using $I = PRT$ and round to the nearest cent. Then find the new principal by adding the interest earned to the preceding principal.

(a)

Year	Interest	Compound Amount
1	$3800.00 × .06 × 1 = $228.00	$3800.00 + **$228.00** = $4028.00
2	$4028.00 × .06 × 1 = $241.68	$4028.00 + **$241.68** = $4269.68
3	$4269.68 × .06 × 1 = $256.18	$4269.68 + **$256.18** = $4525.86
4	$4525.86 × .06 × 1 = $271.55	$4525.86 + **$271.55** = $4797.41

$$\text{Compound interest} = \$4797.41 - \$3800 = \$997.41$$
$$\text{Simple interest} = \$3800 \times .06 \times 4 = \$912$$
$$\text{Difference} = \$997.41 - \$912 = \$85.41$$

(b) No, they will be short of their goal by $5000 − $4797.41 = **$202.59**.

Objective 2 **Identify interest rate per compounding period and number of compounding periods.** Compound interest often results in interest calculations more than once per year depending on the **compounding period**. The *interest rate per compounding period* is the annual rate divided by the number of compounding periods per year. The *number of compounding periods in 1 year* is the number of years multiplied by the number of compounding periods per year.

Interest Compounded	Compound at the End of Every	Number of Compounding Periods in 1 Year
semiannually	6 months	2
quarterly	3 months	4
monthly	1 month	12
daily	1 day	365*

*Leap year has 366 compounding periods.

EXAMPLE 3

Finding the Interest Rate per Compounding Period and the Number of Compounding Periods

Find the interest rate per compounding period and number of compounding periods over the life of each loan.

(a) 8% compounded semiannually, 3 years
(b) 12% per year, compounded monthly, $2\frac{1}{2}$ years
(c) 9% per year, compounded quarterly, 5 years

SOLUTION

(a) 8% compounded semiannually is $\frac{8\%}{2} = 4\%$ credited at the end of each 6 months. There are 3 years × 2 periods per year = 6 compounding periods in 3 years.
(b) 12% per year, compounded monthly, results in $\frac{12\%}{12} = 1\%$ credited at the end of each month. There are 2.5 years × 12 periods per year = 30 compounding periods in 2.5 years.
(c) 9% per year, compounded quarterly, results in $\frac{9\%}{4} = 2.25\%$ credited at the end of each quarter. There are 5 years × 4 periods per year = 20 compounding periods in 5 years.

> **QUICK TIP** The interest rate per compounding period is the annual rate times the fraction of a year over which compounding occurs. Assuming 10% compounded quarterly, the interest rate per compounding period = $10\% \times \frac{1}{4} = 2.5\%$ per quarter.

Objective ③ Use the formula $M = P(1 + i)^n$ to find compound amount. The **formula for compound interest** uses **exponents**, which is a short way of writing repeated products. For example,

Exponent: 3 tells how many times the base 2 is multiplied by itself.

$$2 \times 2 \times 2 = 2^3$$

base

Also, $4^2 = 4 \times 4 = 16$, and $5^4 = 5 \times 5 \times 5 \times 5 = 625$.

Assume that P dollars are deposited at a rate of interest i per compounding period for n periods. Then the compound amount and the interest are found as follows.

Formula for Compounding Interest

$$\text{Maturity value} = M = P(1 + i)^n$$
$$\text{Interest} = I = M - P$$

where

P = initial investment
n = total number of compounding periods
i = interest rate per compounding period

> **QUICK TIP** It is important to keep in mind that i is the interest rate *per compounding period*, not per year. Also that n is *the total number of compounding periods*.

EXAMPLE 4

Finding Compound Interest

An investment at Bank of America pays 7% interest per year compounded semiannually. Given an initial deposit of $2500, **(a)** use the formula to find the compound amount after 2 years, and **(b)** find the compound interest.

SOLUTION

(a) Interest is compounded at $\frac{7\%}{2} = 3.5\%$ every 6 months for 2 years × 2 periods per year = 4 periods. Therefore, 3.5% is the interest rate per compounding period (i) and 4 is the number of compounding periods (n).

$$M = P(1 + i)^n$$
$$= \$2500 \times (1 + .035)^4$$
$$= \$2500 \times (1.035)^4$$
$$= \$2500 \times 1.035 \times 1.035 \times 1.035 \times 1.035$$
$$= \$2868.81 \text{ rounded}$$

The compound amount is $2868.81.

(b)
$$I = M - P$$
$$= \$2868.81 - \$2500 = \$368.81$$
The compound interest is $368.81.

The calculator solution for part (a) is as follows.

2500 ⊠ ⦗ 1 ⊞ .035 ⦘ y^x 4 ⊜ 2868.81 (rounded)

Note: Refer to Appendix C for calculator basics.

Compound Interest Table
n = number of compounding periods; i = interest rate per compounding period

n	1%	1½%	2%	2½%	3%	4%	5%	6%	8%	10%	12%	n
1	1.01000	1.01500	1.02000	1.02500	1.03000	1.04000	1.05000	1.06000	1.08000	1.10000	1.12000	1
2	1.02010	1.03023	1.04040	1.05063	1.06090	1.08160	1.10250	1.12360	1.16640	1.21000	1.25440	2
3	1.03030	1.04568	1.06121	1.07689	1.09273	1.12486	1.15763	1.19102	1.25971	1.33100	1.40493	3
4	1.04060	1.06136	1.08243	1.10381	1.12551	1.16986	1.21551	1.26248	1.36049	1.46410	1.57352	4
5	1.05101	1.07728	1.10408	1.13141	1.15927	1.21665	1.27628	1.33823	1.46933	1.61051	1.76234	5
6	1.06152	1.09344	1.12616	1.15969	1.19405	1.26532	1.34010	1.41852	1.58687	1.77156	1.97382	6
7	1.07214	1.10984	1.14869	1.18869	1.22987	1.31593	1.40710	1.50363	1.71382	1.94872	2.21068	7
8	1.08286	1.12649	1.17166	1.21840	1.26677	1.36857	1.47746	1.59385	1.85093	2.14359	2.47596	8
9	1.09369	1.14339	1.19509	1.24886	1.30477	1.42331	1.55133	1.68948	1.99900	2.35795	2.77308	9
10	1.10462	1.16054	1.21899	1.28008	1.34392	1.48024	1.62889	1.79085	2.15892	2.59374	3.10585	10
11	1.11567	1.17795	1.24337	1.31209	1.38423	1.53945	1.71034	1.89830	2.33164	2.85312	3.47855	11
12	1.12683	1.19562	1.26824	1.34489	1.42576	1.60103	1.79586	2.01220	2.51817	3.13843	3.89598	12
13	1.13809	1.21355	1.29361	1.37851	1.46853	1.66507	1.88565	2.13293	2.71962	3.45227	4.36349	13
14	1.14947	1.23176	1.31948	1.41297	1.51259	1.73168	1.97993	2.26090	2.93719	3.79750	4.88711	14
15	1.16097	1.25023	1.34587	1.44830	1.55797	1.80094	2.07893	2.39656	3.17217	4.17725	5.47357	15
16	1.17258	1.26899	1.37279	1.48451	1.60471	1.87298	2.18287	2.54035	3.42594	4.59497	6.13039	16
17	1.18430	1.28802	1.40024	1.52162	1.65285	1.94790	2.29202	2.69277	3.70002	5.05447	6.86604	17
18	1.19615	1.30734	1.42825	1.55966	1.70243	2.02582	2.40662	2.85434	3.99602	5.55992	7.68997	18
19	1.20811	1.32695	1.45681	1.59865	1.75351	2.10685	2.52695	3.02560	4.31570	6.11591	8.61276	19
20	1.22019	1.34686	1.48595	1.63862	1.80611	2.19112	2.65330	3.20714	4.66096	6.72750	9.64629	20
21	1.23239	1.36706	1.51567	1.67958	1.86029	2.27877	2.78596	3.39956	5.03383	7.40025	10.80385	21
22	1.24472	1.38756	1.54598	1.72157	1.91610	2.36992	2.92526	3.60354	5.43654	8.14027	12.10031	22
23	1.25716	1.40838	1.57690	1.76461	1.97359	2.46472	3.07152	3.81975	5.87146	8.95430	13.55235	23
24	1.26973	1.42950	1.60844	1.80873	2.03279	2.56330	3.22510	4.04893	6.34118	9.84973	15.17863	24
25	1.28243	1.45095	1.64061	1.85394	2.09378	2.66584	3.38635	4.29187	6.84848	10.83471	17.00006	25
26	1.29526	1.47271	1.67342	1.90029	2.15659	2.77247	3.55567	4.54938	7.39635	11.91818	19.04007	26
27	1.30821	1.49480	1.70689	1.94780	2.22129	2.88337	3.73346	4.82235	7.98806	13.10999	21.32488	27
28	1.32129	1.51722	1.74102	1.99650	2.28793	2.99870	3.92013	5.11169	8.62711	14.42099	23.88387	28
29	1.33450	1.53998	1.77584	2.04641	2.35657	3.11865	4.11614	5.41839	9.31727	15.86309	26.74993	29
30	1.34785	1.56308	1.81136	2.09757	2.42726	3.24340	4.32194	5.74349	10.06266	17.44940	29.95992	30
31	1.36133	1.58653	1.84759	2.15001	2.50008	3.37313	4.53804	6.08810	10.86767	19.19434	33.55511	31
32	1.37494	1.61032	1.88454	2.20376	2.57508	3.50806	4.76494	6.45339	11.73708	21.11378	37.58173	32
33	1.38869	1.63448	1.92223	2.25885	2.65234	3.64838	5.00319	6.84059	12.67605	23.22515	42.09153	33
34	1.40258	1.65900	1.96068	2.31532	2.73191	3.79432	5.25335	7.25103	13.69013	25.54767	47.14252	34
35	1.41660	1.68388	1.99989	2.37321	2.81386	3.94609	5.51602	7.68609	14.78534	28.10244	52.79962	35
36	1.43077	1.70914	2.03989	2.43254	2.89828	4.10393	5.79182	8.14725	15.96817	30.91268	59.13557	36
37	1.44508	1.73478	2.08069	2.49335	2.98523	4.26809	6.08141	8.63609	17.24563	34.00395	66.23184	37
38	1.45953	1.76080	2.12230	2.55568	3.07478	4.43881	6.38548	9.15425	18.62528	37.40434	74.17966	38
39	1.47412	1.78721	2.16474	2.61957	3.16703	4.61637	6.70475	9.70351	20.11530	41.14478	83.08122	39
40	1.48886	1.81402	2.20804	2.68506	3.26204	4.80102	7.03999	10.28572	21.72452	45.25926	93.05097	40
41	1.50375	1.84123	2.25220	2.75219	3.35990	4.99306	7.39199	10.90286	23.46248	49.78518	104.21709	41
42	1.51879	1.86885	2.29724	2.82100	3.46070	5.19278	7.76159	11.55703	25.33948	54.76370	116.72314	42
43	1.53398	1.89688	2.34319	2.89152	3.56452	5.40050	8.14967	12.25045	27.36664	60.24007	130.72991	43
44	1.54932	1.92533	2.39005	2.96381	3.67145	5.61652	8.55715	12.98548	29.55597	66.26408	146.41750	44
45	1.56481	1.95421	2.43785	3.03790	3.78160	5.84118	8.98501	13.76461	31.92045	72.89048	163.98760	45
46	1.58046	1.98353	2.48661	3.11385	3.89504	6.07482	9.43426	14.59049	34.47409	80.17953	183.66612	46
47	1.59626	2.01328	2.53634	3.19170	4.01190	6.31782	9.90597	15.46592	37.23201	88.19749	205.70605	47
48	1.61223	2.04348	2.58707	3.27149	4.13225	6.57053	10.40127	16.39387	40.21057	97.01723	230.39078	48
49	1.62835	2.07413	2.63881	3.35328	4.25622	6.83335	10.92133	17.37750	43.42742	106.71896	258.03767	49
50	1.64463	2.10524	2.69159	3.43711	4.38391	7.10668	11.46740	18.42015	46.90161	117.39085	289.00219	50

Objective $\boxed{4}$ **Use the table to find compound amount.** Often, it is impractical to calculate the values of $(1 + i)^n$ in the compound amount formula. Instead, the value of $(1 + i)^n$ can be looked up in the compound interest table. The interest rate i at the top of the table is the interest rate *per compounding period*. The value of n down the far left (or far right) column of the table is *the number of compounding periods*. The value in the body of the table is the compound amount, or maturity value, for each $1 in principal.

> Compound amount = Principal × Number from compound interest table

Alternatively, financial calculators can be used to calculate compound amount directly without having to first find a value for $(1 + i)^n$. (**See Appendix C.**)

EXAMPLE 5

Finding Compound Interest

In each case, find the interest earned on a $2000 deposit

(a) for 3 years, compounded annually at 4%.
(b) for 5 years, compounded semiannually at 6%.
(c) for 6 years, compounded quarterly at 8%.
(d) for 4 years, compounded monthly at 12%.

SOLUTION

(a) In 3 years, there are $3 \times 1 = 3$ compounding periods. The interest rate per compounding period is $4\% \div 1 = 4\%$. Look across the top of the compound interest table, on page 000, for 4% and down the side for 3 periods to find **1.12486**.

$$\text{Compound amount} = M = \$2000 \times \mathbf{1.12486} = \mathbf{\$2249.72}$$
$$\text{Interest earned} = I = \mathbf{\$2249.72} - \$2000 = \$249.72$$

(b) In 5 years, there are $5 \times 2 = 10$ semiannual compounding periods. The interest rate per compounding period is $6\% \div 2 = 3\%$. In the compound interest table, find 3% across the top and 10 periods down the side to find **1.34392**.

$$\text{Compound amount} = M = \$2000 \times \mathbf{1.34392} = \mathbf{\$2687.84}$$
$$\text{Interest earned} = I = \mathbf{\$2687.84} - \$2000 = \$687.84$$

(c) Interest compounded quarterly is compounded 4 times a year. In 6 years, there are $4 \times 6 = 24$ quarters, or 24 periods. Interest of 8% per year is $\frac{8\%}{4} = 2\%$ per quarter. In the compound interest table, locate 2% across the top and 24 periods at the left, finding the number **1.60844**.

$$\text{Compound amount} = M = \$2000 \times \mathbf{1.60844} = \$3216.88$$
$$\text{Interest earned} = I = \$3216.88 - \$2000 = \$1216.88$$

(d) In 4 years, there are $4 \times 12 = 48$ monthly periods. Interest of 12% per year is $\frac{12\%}{12} = 1\%$ per month. Look in the compound interest table for 1% and 48 periods, finding the number **1.61223**.

$$\text{Compound amount} = M = \$2000 \times \mathbf{1.61223} = \$3224.46$$
$$\text{Interest earned} = I = \$3224.46 - \$2000 = \$1224.46$$

QUICK TIP Example 1 in Appendix D shows how a financial calculator can be used to solve this same type of problem.

The more often interest is compounded, the greater the amount of interest earned. Using a financial calculator, a compound interest table more complete than the compound interest table included on page 000 or the compound interest formula will give the results of interest on $1000 shown in the following table. (Leap years were ignored in finding daily interest.)

Interest on $1000 at 8% per Year for 10 Years	
Compounded	**Interest**
Not at all (simple interest)	$ 800.00
Annually	$1158.92
Semiannually	$1191.12
Quarterly	$1208.04
Monthly	$1219.64
Daily	$1225.35

EXAMPLE 6

Finding Compound Interest

John Smith inherited $15,000, which he deposited in a retirement account that pays interest compounded semiannually. How much will he have after 25 years if the funds grow at

(a) 6%   **(b)** 8%   **(c)** 10%?

SOLUTION

In 25 years, there are $2 \times 25 = 50$ semiannual periods. The semiannual interest rates are

(a) $\frac{6\%}{2} = 3\%$ **(b)** $\frac{8\%}{2} = 4\%$ **(c)** $\frac{10\%}{2} = 5\%$

Using factors from the table

(a) $15,000 × **4.38391** = $65,758.65
(b) $15,000 × **7.10668** = $106,600.20
(c) $15,000 × **11.46740** = $172,011

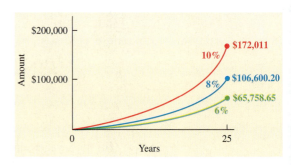

The graph shows the growth at the different interest rates over time.

> **QUICK TIP** Compound interest rate calculations are indicated by phrases such as compounded quarterly, 2% per quarter, or compounded daily.

Parents believe their children should study personal finance. The following pie chart indicates when parents believe this education should begin.

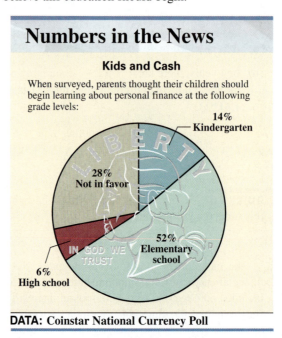

Numbers in the News

Kids and Cash

When surveyed, parents thought their children should begin learning about personal finance at the following grade levels:

14% Kindergarten
28% Not in favor
52% Elementary school
6% High school

DATA: Coinstar National Currency Poll

9.1 | Exercises

The **Quick Start** *exercises in each section contain solutions to help you get started.*

Use the formula for compound amount, not the table, to find the compound amount and interest. Round to the nearest cent. (See Examples 3 and 4.)

	Compound Amount	Interest
Quick Start		
1. $12,000 at 8% compounded annually for 4 years	**$16,325.87**	**$4325.87**

Compound interest is 8% per year for 4 years.
$M = \$12,000 \times (1 + .08)^4 = \$12,000 \times 1.08 \times 1.08 \times 1.08 \times 1.08 = \$16,325.87$
$I = \$16,325.87 - \$12,000 = \$4325.87$

2. $14,800 at 6% compounded semiannually for 4 years _____ _____

3. $28,000 at 10% compounded quarterly for 1 year _____ _____

4. $20,000 at 5% compounded quarterly for $\frac{3}{4}$ year _____ _____

Use values from the compound interest table on page 000 to find both the compound amount and the compound interest. Round compound amount to the nearest cent. (See Examples 3–6.)

	Compound Amount	Interest
Quick Start		
5. $32,350 at 6% compounded annually for 4 years	**$40,841.23**	**$8491.23**

Compound interest is 6% per year for 4 years.
$M = \$32,350 \times 1.26248 = \$40,841.23;\ I = \$40,841.23 - \$32,350 = \$8491.23$

6. $19,400 at 8% compounded quarterly for 3 years _____ _____

7. $14,500 at 10% compounded quarterly for 7 years _____ _____

C indicates an exercise that is related to the Case in Point feature within the section.

8. $12,500 at 8% compounded quarterly for 5 years   _____   _____

9. $45,000 at 6% compounded semiannually for 5 years   _____   _____

10. $82,000 at 8% compounded semiannually for 4 years   _____   _____

Find the simple interest for the period indicated. Then use table values to find the compound interest. Finally, find the difference between compound interest and simple interest. Round each to the nearest cent. (Interest is compounded annually.)

Quick Start

	Principal	Rate	Number of Years	Simple Interest	Compound Interest	Difference
11.	$5400	6%	4	$1296	$1417.39	$121.39
12.	$9200	5%	6	_____	_____	_____
13.	$1200	8%	15	_____	_____	_____
14.	$4625	4%	12	_____	_____	_____

Use the table to solve the following application problems. Round to the nearest cent.

Quick Start

15. **CREDIT UNION**   Bill Jensen deposits $8500 with Bank of America in an account paying 5% compounded semiannually. Find **(a)** the compound amount and **(b)** the interest in 6 years.

    Compound interest is $\frac{5\%}{2}$ = 2.5% and there are 2 × 6 = 12 compounding periods.
    (a) M = $8500 × 1.34489 = $11,431.57;   **(b)** I = $11,431.57 − $8500 = $2931.57

    (a) $11,431.57
    (b) $2931.57

16. **SAVINGS**   Vickie Ewing deposits her savings of $2800 in an account paying 6% compounded quarterly and she leaves it there for 9 years. Find **(a)** the compound amount and **(b)** the interest.

    (a) _____
    (b) _____

17. **INTERNATIONAL FINANCE**   Chi Tang, a businessperson from Taiwan, deposits 25,000 yen in a Japanese branch of Bank of America which pays 6% compounded semiannually. Find **(a)** the balance in the account after 4 years and **(b)** the interest.

    (a) _____
    (b) _____

18. **APPLIANCE STORE**   AAA Appliance places $42,000 in an account paying 6% compounded quarterly and leaves it there as collateral for a loan. Find **(a)** the balance in the account after one year and **(b)** the interest.

    (a) _____
    (b) _____

19. INVESTMENT DECISION   Bill Baxter has $25,000 to invest for a year. He can lend it to his sister who has agreed to pay 10% simple interest for the year. Or, he can invest it with a bank at 8% compounded quarterly for a year. How much additional interest would the simple interest loan to his sister generate?

19. _____

20. MAXIMIZING INTEREST   Bank of America has $850,000 to lend for 9 months. It can lend it to a local contractor at a simple interest rate of 12%, or it can lend it to a small business that will pay 12% compounded monthly. How much additional interest would the compound interest loan to the small business generate?

20. _____

21. INVESTING   Bob Williams has $25,000 to invest and believes that he will earn 6% compounded semiannually. Find the compound amount if he invests **(a)** for 2 years and **(b)** for 8 years. **(c)** Then find the additional amount earned due to the longer period.

(a) _____
(b) _____
(c) _____

22. WHICH INVESTMENT?   Jan Reus sold her home and has $18,000 to invest. She believes she can earn 8% compounded quarterly. Find the compound amount if she invests for **(a)** 3 years and **(b)** 9 years. **(c)** Then find the additional amount earned due to the longer time period.

(a) _____
(b) _____
(c) _____

23. Explain the difference between simple interest and compound interest. (See Objective 1.)

24. Explain the difference between 8% compounded monthly for 1 year and 8% simple interest for 1 year. If you were loaning money, which type of interest would you specify? Why?

25. Show the effect of both the interest rate and the period on an original investment of $2500. Decide on your own rates and time periods.

26. List three institutions that work with simple interest. List three others that use compound interest.

9.2 | Interest Bearing Bank Accounts and Inflation

Objectives

1. Define passbook, savings, and other interest bearing accounts.
2. Find interest compounded daily.
3. Define time deposit accounts.
4. Define inflation and the consumer price index.
5. Examine the effect of inflation on spendable income.
6. Understand the role of the government in regard to inflation.

CASE _in_ POINT

Individuals, businesses, and even countries have money on deposit at Bank of America. These deposits can be in many forms, including checking accounts, savings accounts, and time deposits, such as certificates of deposits.

Banks make money by charging _higher interest on funds they lend out_ to customers than on funds they have on deposit at the bank. They also make money from fees for services such as safe deposit boxes, transferring money from one place to another, use of the ATM, and checking accounts.

People sometimes think that banks have huge vaults of stored cash, but that is rarely the case. Banks usually only have enough cash to meet customers' needs for cash during the next few days. Most bank assets are in the loans to their many customers rather than in cash.

Objective 1 Define passbook, savings, and other interest bearing accounts. **Savings accounts** and **passbook accounts** are offered by banks and credit unions and can be a safe place to deposit money. These accounts are commonly insured by the Federal Deposit Insurance Corporation (FDIC) on deposits up to $100,000. Call your bank to see if it is federally insured. Savings accounts typically require a minimum balance. Interest rates paid on these accounts range from below 1% to over 5% and the interest is often compounded daily. The Truth in Savings Act of 1991 resulted in Regulation DD, which requires that interest on savings accounts be paid based on the _exact_ number of days.

Interest bearing checking accounts are also offered by many banks and credit unions. These accounts have a minimum balance, such as $1500, that must be maintained but they have the advantage that checks can be written on the account. They can be a good way to earn interest on your money as long as your balance does not fall too low, in which case the bank commonly charges a fee. However, interest rates are usually very low on these accounts. In order to earn as much interest as possible, many people today try to keep their money earning interest in one of these accounts and then pay their bills from this account using checks.

Objective 2 Find interest compounded daily. Interest on savings accounts, passbook accounts, and interest bearing checking accounts is found using compound interest. It is common for banks to pay interest **compounded daily** so that interest is credited for every day that the money is on deposit.

The formula for daily compounding is _exactly_ the same as that given in the last section. However, because the annual interest rate must be divided by 365 (for daily compounding), the arithmetic is very tedious. To avoid this, use the following special tables that give the necessary numbers for 1 to 90 days, as well as for 1 to 4 ninety day quarters, assuming $3\frac{1}{2}\%$ interest compounded daily. Even with daily compounding, interest is often credited to an account _only at the end of each quarter_ to make record keeping easier for the bank.

QUICK TIP   Interest rates vary widely. No one, including the authors of this textbook, know what interest rates will be in the future. Thus, the interest rate in this section is $3\frac{1}{2}\%$, or close to the historical average for accounts of this type.

Interest by Quarter for $3\frac{1}{2}$% Compounded Daily Assuming 90-Day Quarters

Number of Quarters	Value of $(1 + i)^n$
1	1.008667067
2	1.017409251
3	1.026227205
4	1.035121585

The four quarters in a year begin on January 1, April 1, July 1, and October 1. Although some quarters have 91 or 92 days in them, we assume 90-day quarters for convenience in calculation. Assuming daily compounding and a compounding period expressed in days or quarters, compound amount and interest are found as follows.

> **Compound amount = Principal × Number from table**
>
> Interest = Compound amount − Principal

Find the value from the table below if the number of days of the deposit is 90 days or less. Use the smaller table to the left if time is given in number of quarters.

> **QUICK TIP**   See Appendix C for financial calculator solutions that do not require the use of a table.

Values of $(1 + i)^n$ for $3\frac{1}{2}$% Compounded Daily

Number of Days n	Value of $(1 + i)^n$	n	Value of $(1 + i)^n$	n	Value of $(1 + i)^n$	n	Value of $(1 + i)^n$	n	Value of $(1 - i)^n$
1	1.000095890	19	1.001823491	37	1.003554076	55	1.005287650	73	1.007024219
2	1.000191790	20	1.001919556	38	1.003650307	56	1.005384048	74	1.007120783
3	1.000287699	21	1.002015631	39	1.003746548	57	1.005480454	75	1.007217357
4	1.000383617	22	1.002111714	40	1.003842797	58	1.005576870	76	1.007313939
5	1.000479544	23	1.002207807	41	1.003939056	59	1.005673296	77	1.007410531
6	1.000575480	24	1.002303909	42	1.004035324	60	1.005769730	78	1.007507132
7	1.000671426	25	1.002400021	43	1.004131602	61	1.005866174	79	1.007603742
8	1.000767381	26	1.002496141	44	1.004227888	62	1.005962627	80	1.007700362
9	1.000863345	27	1.002592271	45	1.004324184	63	1.006059089	81	1.007796990
10	1.000959318	28	1.002688410	46	1.004420489	64	1.006155560	82	1.007893628
11	1.001055300	29	1.002784558	47	1.004516803	65	1.006252041	83	1.007990276
12	1.001151292	30	1.002880716	48	1.004613127	66	1.006348531	84	1.008086932
13	1.001247293	31	1.002976882	49	1.004709460	67	1.006445030	85	1.008183598
14	1.001343303	32	1.003073058	50	1.004805802	68	1.006541538	86	1.008280273
15	1.001439322	33	1.003169243	51	1.004902153	69	1.006638056	87	1.008376958
16	1.001535350	34	1.003265438	52	1.004998513	70	1.006734583	88	1.008473651
17	1.001631388	35	1.003361641	53	1.005094883	71	1.006831119	89	1.008570354
18	1.001727435	36	1.003457854	54	1.005191262	72	1.006927665	90	1.008667067

Note: The value of $(1 + i)^n$ for $3\frac{1}{2}$% compounded daily for a quarter with 91 days is 1.008763788 and for a quarter with 92 days is 1.008860519.

EXAMPLE 1

Finding Daily Interest

Becky Gonzales received $12,500 from the sale of a piece of real estate. She wants to buy a new Toyota Camry, but decides to wait 60 days until the new models are out. She decides to place her money in a savings account earning $3\frac{1}{2}$% interest compounded daily for the 60 days. Find the amount of interest she will earn.

SOLUTION

The table value for 60 days is 1.005769730.

$$\text{Compound amount} = \$12{,}500 \times \mathbf{1.005769730} = \$12{,}572.12$$

$$\text{Interest} = \$12{,}572.12 - \$12{,}500 = \$72.12$$

The additional $72.12 isn't much money to Gonzales, but she is happy to earn some interest.

The next two examples show how interest is calculated when there are several deposits and/or withdrawals within a short period of time.

EXAMPLE 2

Finding Interest on Multiple Deposits

Tom Blackmore is a private investigator who keeps his extra cash in a savings account to earn interest. On January 10, he deposited $2463 in a savings account paying $3\frac{1}{2}$% compounded daily. He deposits an additional $1320 on February 18 and $840 on March 3. Find the interest earned through April 10.

SOLUTION

Treat each deposit separately. The $2463 was in the account for 90 days (21 days in January, 28 days in February, 31 days in March, and 10 days in April). The value for 90 days from the table is **1.008667067**.

Compound amount = $2463 × **1.008667067** = $2484.35   First deposit plus interest

The $1320 deposited on February 18 was in the account for 51 days (10 days in February, 31 days in March, and 10 days in April).

Compound amount = $1320 × **1.004902153** = **$1326.47**   Second deposit plus interest

The $840 was in the account for 38 days (28 days in March and 10 days in April).

Compound amount = $840 × **1.003650307** = $843.07   Final deposit plus interest

The total amount in the account on April 10 is found by adding the three compound amounts.

Total in account = **$2484.35** + **$1326.47** + **$843.07** = **$4653.89**

The interest earned is the total amount in the account less the deposits.

Interest earned = **$4653.89** − ($2463 + $1320 + $840) = $30.89

EXAMPLE 3

Finding Interest for the Quarter

Beth Gardner owns Blacktop Paving, Inc. She needs a place to keep extra cash, a place that will earn interest, but that will allow her to get funds when needed. She opened a savings account on July 20 with a $24,800 deposit. She then withdrew $3800 on August 29 for an unexpected truck repair and she made another withdrawal for $8200 on September 29 for payroll. Find the interest earned through October 1, given interest at $3\frac{1}{2}$% compounded daily.

SOLUTION

Of the original $24,800, a total of $24,800 − $3800 − $8200 = $12,800 earned interest from July 20 to October 1 or for 274 − 201 = 73 days. Find the factor 1.007024219 from the table.

Compound amount = $12,800 × 1.007024219 = $12,889.91

Interest = $12,889.91 − $12,800 = **$89.91**

The withdrawn $3800 earned interest from July 20 to August 29 or for 241 − 201 = 40 days.

Compound amount = $3800 × 1.003842797 = $3814.60

Interest = $3814.60 − $3800 = **$14.60**

Finally, the withdrawn $8200 earned interest from July 20 to September 29 or for 272 − 201 = 71 days.

Compound amount = $8200 × 1.006831119 = $8256.02

Interest = $8256.02 − $8200 = **$56.02**

The total interest earned is ($89.91 + $14.60 + $56.02) = **$160.53.** The total in the account on October 1 is found as follows.

deposits + interest −   withdrawals   = balance on October 1

$24,800 + **$160.53** − ($3800 + $8200) = $12,960.53

Objective 3 Define time deposit accounts. Banks pay higher interest rates on funds left on deposit for *longer time periods* in **time deposits.** A **certificate of deposit (CD)** requires a minimum amount of money, such as $1000, to be on deposit for a minimum period of time, such as 1 year. Find the compound amount of a time deposit as follows.

Compound amount = Principal × Number from the table

Interest = Compound amount − Principal

Compound Interest for Time Deposit Accounts Compounded Daily

Number of Years	3%	4%	5%	6%	7%	Number of Years
1	1.03045326	1.04080849	1.05126750	1.06183131	1.07250098	1
2	1.06183393	1.08328232	1.10516335	1.12748573	1.15025836	2
3	1.09417024	1.12748944	1.16182231	1.19719965	1.23365322	3
4	1.12749129	1.17350058	1.22138603	1.27122408	1.32309429	4
5	1.16182708	1.22138937	1.28400343	1.34982553	1.41901993	5
10	1.34984217	1.49179200	1.64866481	1.82202895	2.01361756	10

QUICK TIP The preceding compound interest table assumes daily compounding; the compound interest table on page 000 of **Section 9.1** *does not.*

EXAMPLE 4

Finding Interest and Compound Amount for Time Deposits

Tony Sanchez plans to purchase three machines for his auto-repair shop. Bank of America requires $20,000 in collateral before making the loan. Therefore, Tony deposits $20,000 with the bank in a 2-year certificate of deposit yielding 4% compounded daily. Find the compound amount and interest.

SOLUTION

Look at the table for 4% and 2 years, finding 1.08328232.

Compound amount = $20,000 × **1.08328232** = $21,665.65 rounded

Interest = $21,665.65 − $20,000 = $1665.65

Objective 4 Define inflation and the consumer price index. If you haven't done so, ask your parents or grandparents what salary they received when they were your age. Chances are it will be a lot less than you would be willing to work for today. Why? **Inflation** is the reason. Inflation results in the continuing rise in the general price level of goods and services. Look at the bar graph on the next page to see some possible effects of inflation during the next 20 years. Of course, wages also increase, so they will be higher in 20 years than they are today.

The **consumer price index (CPI)** is calculated by the government annually in the United States and is often referred to as the *cost of living index.* Other countries calculate similar indexes. The CPI can be used to track inflation—it measures the average change in prices from one year to the next for a common bundle of goods and services bought by the average consumer on a regular basis. The common bundle of goods and services is defined by the government to include food, housing, fuels, utilities, apparel, transportation, insurance, health care, recreation, and even pet care among other items.

Yearly inflation, as measured by the CPI, differs substantially from year to year as you can see from the following chart. Go to the web site for the Bureau of Labor Statistics at (http://www.bls.gov/) for information related to inflation. The Bureau has an inflation calculator that allows you to see the effects of inflation over the years.

Objective 5 Examine the effect of inflation on spendable income. Inflation can reduce the buying power of both wages and savings. If inflation goes up in a year in which you do not receive a raise, then you may have to purchase less or borrow money to buy the same items and

Numbers in the News

On the Up and Up

Inflation is likely to push the price of many common staples up considerably over the next 20 years. Such increases in the ordinary cost of living raise the pressure to save for retirement.

	Today	In 20 years*
New car	$25,000	$54,800
Movie tickets	$8.50	$18.50
Jeans	$35	$77
New house	$180,000	$394,400
Vacation	$1400	$3070

*Assumes an annual inflation rate of 4 percent.

DATA: Salomon Smith Barney; *Sacramento Bee*

Numbers in the News

Consumer Price Index

The annual inflation rate based on the consumer price index remains under control.

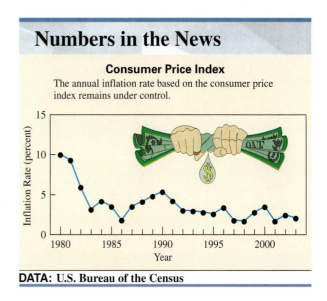

DATA: U.S. Bureau of the Census

services. Each of us would like to receive a raise that is larger than inflation for the year, but sometimes that does not happen. Example 5 illustrates the effect of inflation that increases more rapidly than wages.

EXAMPLE **5**

Estimating the Effects of Inflation

Inflation from one year to the next was 4.8% as measured by the CPI.

(a) Find the effect of the increase on a family with an annual income and budget of $19,800 (after taxes).

(b) What is the overall effect if the family members only receive a 2% (after tax) increase in pay for the year?

SOLUTION

(a) This is a percent problem. The cost of the goods and services that this family buys, if they buy the common bundle of goods and services, went up by 4.8% as measured by the CPI.

$$.048 \times \$19,800 = \$950.40$$

Therefore, these same goods and services will cost the family

$$\$19,800 + \$950.40 = \textbf{\$20,750.40 next year}.$$

(b) The family's income went up 2% after taxes or by

$$.02 \times \$19,800 = \textbf{\$396}.$$

Thus, their new income is $19,800 + $396 = $20,196. In effect, the family has lost $20,750.40 − $20,196 = **$554.40** in purchasing power without considering taxes.

QUICK TIP Example 5 can also be solved as follows.

Inflation rate	4.8%
− Raise	2.0%
Loss	2.8%

2.8% × $19,800 = **$554.40**

As shown in Example 5, inflation *slowly erodes* fixed income or incomes with a small annual increase built into them. Imagine the effect of losing purchasing power every year for 10 years in a row. Inflation can erode people's purchasing power *even as* their annual salaries are increasing.

Retired people are particularly concerned with inflation since they must live off Social Security and the assets they have accumulated during their lifetime. Some retired people do not have ways of increasing their income to keep pace with inflation and must *lower their standard of living* during their 10 to 30 or more years of retirement. Other retired people have investments such as stocks that help them stay ahead of inflation.

EXAMPLE 6

Estimating the Effects of Inflation

Joan Davies has $14,650 in a savings account paying $3\frac{1}{2}\%$ interest compounded daily. Ignoring taxes, what is her gain or loss in purchasing power in a year in which the CPI index increases by 4.2%?

SOLUTION

Use the interest-by-quarter table on page 000 to find that the compound amount factor for 4 quarters is 1.035121585.

$$\text{Compound amount at end of year} = \$14,650 \times 1.035121585 = \mathbf{\$15,164.53}$$

To keep up with inflation, Davies needs to earn 4.2% on her investment.

$$\text{Needed to keep up with inflation} = \$14,650 \times 1.042 = \mathbf{\$15,265.30}$$

The difference of $15,265.30 − $15,164.53 = $100.77 is the loss in purchasing power. The purchasing power of Davies's savings actually went *down*, even though she earned interest for the year. The problem worsens if Davies has to pay taxes on interest earned, since she will end up with even less interest.

Objective 6 Understand the role of the government in regard to inflation. The federal government generally tries to keep inflation at moderate levels, *since inflation can be so harmful*. When the economy becomes overheated (grows too quickly) and inflationary pressures increase, the Federal Reserve increases interest rates to reduce borrowing slightly and slow the economy. This action helps reduce inflationary pressures. Conversely, if inflation is low and the economy is growing very slowly or not at all, the Federal Reserve reduces interest rates to stimulate the economy and create more jobs. The Federal Reserve has been assigned the *very difficult task* of maintaining a growing and healthy economy with low levels of inflation.

The line graph shows the inflating costs of college education.

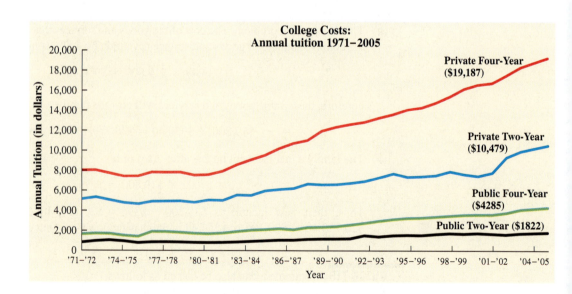

9.2 | Exercises

FOR EXTRA HELP

MyMathLab

InterActMath.com

MathXP MathXL

MathXL
Tutorials on CD

Tutor Center Addison-Wesley
Math Tutor Center

DVT/Videotape

The **Quick Start** *exercises in each section contain solutions to help you get started.*

Find the interest earned by the following. Assume $3\frac{1}{2}\%$ interest compounded daily. (See Examples 1–3.)

Quick Start

	Amount	Date Deposited	Date Withdrawn	Interest Earned
1.	$4800	July 6	September 30	**$39.75**

There are $(31 - 6) + 31 + 30 = 86$ days.
Interest is $4800 \times 1.008280273 - \$4800 = \$39.75$.

	Amount	Date Deposited	Date Withdrawn	Interest Earned
2.	$3850	January 5	February 9	**$12.94**

There are $(31 - 5) + 9 = 35$ days.
Interest is $3850 \times 1.003361641 - \$3850 = \$12.94$.

	Amount	Date Deposited	Date Withdrawn	Interest Earned
3.	$8200	October 4	December 7	_____
4.	$2830	May 4	June 23	_____
5.	$17,958	September 9	November 7	_____
6.	$12,000	December 3	February 20	_____

Find the compound amount for each of the following certificates of deposit. Assume daily compounding. (See Example 4.)

Quick Start

	Amount Deposited	Interest Rate	Time in Years	Compound Amount
7.	$3900	5%	4	**$4763.41**

$3900 \times 1.22138603 = \$4763.41$

	Amount Deposited	Interest Rate	Time in Years	Compound Amount
8.	$8000	4%	1	_____
9.	$12,900	3%	10	_____
10.	$3600.40	6%	10	_____

 indicates an exercise that is related to the Case in Point feature within the section.

11. Explain how you can use a time deposit to your advantage. Use an example. (See Objective 3.)

12. List five ways in which inflation affects your family. (See Objective 4.)

Solve the following application problems. If no interest rate is given, assume $3\frac{1}{2}\%$ interest compounded daily. Round to the nearest cent.

Quick Start

 13. SAVINGS ACCOUNT   Hilda Worth opened a savings account at Bank of America on April 1 with a $2530 deposit. She then deposited $150 on May 8 and $580 on May 24. Find **(a)** the balance on June 30 and **(b)** the interest earned through that date.

(a) $3284.75

(b) $24.75

(a) The $2530 (for 90 days) becomes $2530 × 1.008667067 = $2551.93
    The $150 (for 53 days) becomes $150 × 1.005094883 = $ 150.76
    The $580 (for 37 days) becomes $580 × 1.003554076 = $ 582.06
                                            Total is $3284.75

(b) Interest earned is $3284.75 − ($2530 + $150 + $580) = $24.75

14. PRINT SHOP   The manager of Quick Printing, Inc., is trying to get the most out of his assets including cash that has been sitting in a checking account that does not pay interest. He opened a savings account with a deposit of $8765 on January 4. On February 11, he deposited $936. Then, on March 21, he deposited a tax refund check for $650. Find **(a)** the balance on March 31 and **(b)** the interest earned through that date.

(a) _____

(b) _____

 15. SAVINGS ACCOUNT FOR EXTRA CASH   On April 1, MVP Sports opened a savings account at Bank of America with a deposit of $17,500. A withdrawal of $5000 was made 21 days later, and another withdrawal of $980 was made 12 days before July 1. Find **(a)** the balance on July 1 and **(b)** the interest earned through that date.

(a) _____

(b) _____

16. SAVINGS ACCOUNT   The owner of Rondo's Magic Shop opened a savings account at Bank of America for the extra cash in the firm. The initial deposit of $7800 was made on July 7. A withdrawal of $1500 was made 46 days later, and an additional withdrawal of $1000 was made 30 days before October 1. Find **(a)** the balance on October 1 and **(b)** the interest earned through that date.

(a) _____

(b) _____

17. TIME DEPOSIT   Wes Cockrell has $4000 to deposit in a certificate of deposit, but he is debating whether to leave it there for 2 years or for 3 years. Assume 5% compounded daily in both situations, and find the compound amount in each case.

17. _____

18. COMPOUNDING   Georgia Pastel Fabric has $15,000 to deposit in a certificate of deposit for either 3 or 5 years. Assume 6% compounded daily in both situations and find the compound amount in each case.

18. _____

19. LOAN COLLATERAL   An Italian firm deposited $800,000 in a 2-year time deposit earning 6% compounded daily with a New York bank as partial collateral for a loan. Find **(a)** the compound amount and **(b)** the interest earned.

(a) _____
(b) _____

20. PUTTING UP COLLATERAL   Joni Perez needs to borrow $20,000 to open a welding shop, but the bank will not lend her the money. Joni's uncle agrees to put up collateral for the loan with a $20,000, 4-year certificate of deposit paying 4% compounded daily. This means that the bank will take all or part of his deposit if Perez should fail to repay the loan. Find **(a)** the compound amount earned by her uncle and **(b)** the interest earned by her uncle.

(a) _____
(b) _____

21. RETIREMENT INCOME   The Walters accumulated $235,000 during more than 40 years of work. They originally deposited this money in a 5-year time deposit earning 6% and used the income for living expenses. On renewing the time deposit, they found that interest rates on a 5-year time deposit had fallen and that they were only going to receive 4%. Find the difference in their _annual income_ due to the decline in interest rates. (_Hint:_ Don't use the compound interest table.)

21. _____

22. INHERITANCE   Jessica Thompson inherited $80,000 and decided to put the money in one of two 4-year time deposits. The first time deposit yielded 5%, but the second only yielded 4%. Find the difference in the annual income. (_Hint:_ Don't use the compound interest table.)

22. _____

23. DETERMINING PURCHASING POWER   A family with a spending budget of $26,500 receives an increase in wages of 3% in a year in which inflation was 4.5%. Find the net gain or loss in their purchasing power.

23. _____

24. INFLATION AND RETIREMENT   Ben and Martha Wheeler are retired and after saving their entire life, they have $184,500 in a savings account at Bank of America paying $3\frac{1}{2}$% compounded daily. What is their gain or loss in purchasing power from interest in a year in which the CPI is 5%?

24. _____

Bank of America merged with Nations Bank to become one of the largest banks in the world. In keeping with the trend of globalization, Bank of America has offices in more than 30 countries, but serves customers in more than 150 countries around the world. The bank currently handles more than 150 transactions per second. More than 1 million bank customers pay 6.4 million bills regularly on-line, and the number is growing daily.

1. Assume that one month the bank has 1.1 million customers who pay a total of $258,500,000 in bills online. Find the average amount of per bill paid online for the month.

2. Assume that Bank of America borrows a total of $80,000,000 from a European bank at 4% interest for 1 year. They then lend out $65,000,000 of this amount at 8% compounded quarterly for 1 year. Find the difference between interest earned and interest paid out by the bank. (Use the compound interest table).

3. List three advantages and three disadvantages of banking with a very large bank.

4. Jason Clendenen has joined the Peace Corps and is in Cameroon, Africa. He runs out of money and his dad wants to wire him funds. Is there a Bank of America branch in Cameroon so that money can be wired through that bank?

9.3 | Present Value and Future Value

Objectives

1 Define the terms *future value* and *present value*.
2 Use tables to calculate present value.
3 Use future value and present value to estimate the value of a business.

Objective 1 **Define the terms *future value* and *present value*.** **Future value** is the amount available at a specific date in the future. It is the amount available after an investment has earned interest. All of the values we found in **Sections 9.1 and 9.2** were future values.

In contrast, **present value** is the amount needed today so that the desired future value will be available when needed. For example, an individual may need to know the present value that must be invested today in order to have a down payment for a new car in 3 years. Or a firm may need to know the present value that must be invested today in order to have enough money to purchase a new computer system in 20 months. The bar chart shows present value as the value today and future value as the value at a future date.

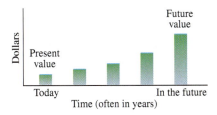

In this section, the future value, interest rate, and term are given and the present value that must be invested today to reach the future value is calculated.

Objective 2 **Use tables to calculate present value.** First, find the interest rate per compounding period (i) and the total number of compounding periods (n) of the investment. Then use these values to find the appropriate value in the present value of a dollar table. Finally, use the formula to find present value.

$$\text{Present value } (P) = \text{Future value} \times \text{Table value}$$

> **QUICK TIP** n is the number of compounding periods of the investment. The interest rate per compounding period (i) is found by dividing the annual interest rate by the number of compounding periods per year.

> **QUICK TIP**   See Appendix D for financial calculator solutions that do not require use of a table.

EXAMPLE 1

Finding Present Value

Betty Clark needs to purchase a walk-in freezer for her restaurant in 3 years at an estimated cost of $12,000. What lump sum deposited today at 5% compounded annually must she invest to have the needed funds? How much interest will she earn?

SOLUTION

Step 1   The interest rate is 5% per compounding period for 3 compounding periods (years in this case). Look across the top of the table for 5% and down the left column for 3 to find .86384.

$$\text{Present value} = \$12,000 \times .86384 = \mathbf{\$10,366.08}$$

Present Value of a Dollar Table
n = number of periods; *i* = interest rate per period

n	1%	1½%	2%	2½%	3%	4%	5%	6%	8%	10%	12%	n
1	.99010	.98522	.98039	.97561	.97087	.96154	.95238	.94340	.92593	.90909	.89286	1
2	.98030	.97066	.96117	.95181	.94260	.92456	.90703	.89000	.85734	.82645	.79719	2
3	.97059	.95632	.94232	.92860	.91514	.88900	.86384	.83962	.79383	.75131	.71178	3
4	.96098	.94218	.92385	.90595	.88849	.85480	.82270	.79209	.73503	.68301	.63552	4
5	.95147	.92826	.90573	.88385	.86261	.82193	.78353	.74726	.68058	.62092	.56743	5
6	.94205	.91454	.88797	.86230	.83748	.79031	.74622	.70496	.63017	.56447	.50663	6
7	.93272	.90103	.87056	.84127	.81309	.75992	.71068	.66506	.58349	.51316	.45235	7
8	.92348	.88771	.85349	.82075	.78941	.73069	.67684	.62741	.54027	.46651	.40388	8
9	.91434	.87459	.83676	.80073	.76642	.70259	.64461	.59190	.50025	.42410	.36061	9
10	.90529	.86167	.82035	.78120	.74409	.67556	.61391	.55839	.46319	.38554	.32197	10
11	.89632	.84893	.80426	.76214	.72242	.64958	.58468	.52679	.42888	.35049	.28748	11
12	.88745	.83639	.78849	.74356	.70138	.62460	.55684	.49697	.39711	.31863	.25668	12
13	.87866	.82403	.77303	.72542	.68095	.60057	.52032	.46884	.36770	.28966	.22917	13
14	.86996	.81185	.75788	.70773	.66112	.57748	.50507	.44230	.34036	.26333	.20462	14
15	.86135	.79985	.74301	.69047	.64186	.55526	.48102	.41727	.31524	.23939	.18270	15
16	.85282	.78803	.72845	.67362	.62317	.53391	.45811	.39365	.29189	.21763	.16312	16
17	.84438	.77639	.71416	.65720	.60502	.51337	.43630	.37136	.27027	.19784	.14564	17
18	.83602	.76491	.70016	.64117	.58739	.49363	.41552	.35034	.25025	.17986	.13004	18
19	.82774	.75361	.68643	.62553	.57029	.47464	.39573	.33051	.23171	.16351	.11611	19
20	.81954	.74247	.67297	.61027	.55368	.45639	.37689	.31180	.21455	.14864	.10367	20
21	.81143	.73150	.65978	.59539	.53755	.43883	.35894	.29416	.19866	.13513	.09256	21
22	.80340	.72069	.64684	.58086	.52189	.42196	.34185	.27751	.18394	.12285	.08264	22
23	.79544	.71004	.63416	.56670	.50669	.40573	.32557	.26180	.17032	.11168	.07379	23
24	.78757	.69954	.62172	.55288	.49193	.39012	.31007	.24698	.15770	.10153	.06588	24
25	.77977	.68921	.60953	.53939	.47761	.37512	.29530	.23300	.14602	.09230	.05882	25
26	.77205	.67902	.59758	.52623	.46369	.36069	.28124	.21981	.13520	.08391	.05252	26
27	.76440	.66899	.58586	.51340	.45019	.34682	.26785	.20737	.12519	.07628	.04689	27
28	.75684	.65910	.57437	.50088	.43708	.33348	.25509	.19563	.11591	.06934	.04187	28
29	.74934	.64936	.56311	.48866	.42435	.32065	.24295	.18456	.10733	.06304	.03738	29
30	.74192	.63976	.55207	.47674	.41199	.30832	.23138	.17411	.09938	.05731	.03338	30
31	.73458	.63031	.54125	.46511	.39999	.29646	.22036	.16425	.09202	.05210	.02980	31
32	.72730	.62099	.53063	.45377	.38834	.28506	.20987	.15496	.08520	.04736	.02661	32
33	.72010	.61182	.52023	.44270	.37703	.27409	.19987	.14619	.07889	.04306	.02376	33
34	.71297	.60277	.51003	.43191	.36604	.26355	.19035	.13791	.07305	.03914	.02121	34
35	.70591	.59387	.50003	.42137	.35538	.25342	.18129	.13011	.06763	.03558	.01894	35
36	.69892	.58509	.49022	.41109	.34503	.24367	.17266	.12274	.06262	.03235	.01691	36
37	.69200	.57644	.48061	.40107	.33498	.23430	.16444	.11579	.05799	.02941	.01510	37
38	.68515	.56792	.47119	.39128	.32523	.22529	.15661	.10924	.05369	.02673	.01348	38
39	.67837	.55953	.46195	.38174	.31575	.21662	.14915	.10306	.04971	.02430	.01204	39
40	.67165	.55126	.45289	.37243	.30656	.20829	.14205	.09722	.04603	.02209	.01075	40
41	.66500	.54312	.44401	.36335	.29763	.20028	.13528	.09172	.04262	.02009	.00960	41
42	.65842	.53509	.43530	.35448	.28896	.19257	.12884	.08653	.03946	.01826	.00857	42
43	.65190	.52718	.42677	.34584	.28054	.18517	.12270	.08163	.03654	.01660	.00765	43
44	.64545	.51939	.41840	.33740	.27237	.17805	.11686	.07701	.03383	.01509	.00683	44
45	.63905	.51171	.41020	.32917	.26444	.17120	.11130	.07265	.03133	.01372	.00610	45
46	.63273	.50415	.40215	.32115	.25674	.16461	.10600	.06854	.02901	.01247	.00544	46
47	.62646	.49670	.39427	.31331	.24926	.15828	.10095	.06466	.02686	.01134	.00486	47
48	.62026	.48936	.38654	.30567	.24200	.15219	.09614	.06100	.02487	.01031	.00434	48
49	.61412	.48213	.37896	.29822	.23495	.14634	.09156	.05755	.02303	.00937	.00388	49

Step 2 Interest earned = $12,000 − **$10,366.08** = $1633.92

Step 3 Check the answer by finding the future value of an investment of $10,366.08 in an account earning 5% compounded annually for 3 years. Use the table on page 000 to find **1.15763**.

$$\text{Future value} = \$10,366.08 \times \textbf{1.15763} = \$12,000.09$$

The reason it is not exactly $12,000 is due to rounding in the table value.

EXAMPLE 2

Finding Present Value

The local Harley-Davidson shop has seen business grow rapidly. The owners plan to double the size of their 6000-square-foot shop in one year at a cost of $280,000. How much should be invested in an account paying 6% compounded semiannually to have the funds needed?

SOLUTION

The interest rate per compounding period is

$$\frac{6\%}{2} = 3\%$$

and the number of compounding periods is

$$1 \text{ year} \times 2 \text{ periods per year} = 2.$$

Use the table to find **.94260**.

$$\text{Present value} = \$280,000 \times \textbf{.94260} = \$263,928$$

A deposit of $263,928 today at 6% compounded semiannually will provide $280,001.22 in one year. The difference is due to rounding.

EXAMPLE 3

Applying Present Value

Telco Telecommunications wishes to partner with a Korean company in the purchase of a satellite in 3 years. Telco plans to make a cash down payment of 40% of their anticipated $8,000,000 cost and borrow the remaining funds from a bank. Find the amount they should invest today in an account earning 6% compounded annually to have the down payment needed in 3 years.

SOLUTION

First find the down payment to be paid in 3 years.

$$\text{Down payment} = .40 \times \$8,000,000 = \$3,200,000$$

This is the future value needed exactly 3 years from now. Using the present value of a dollar table with 3 periods and 6% per period provides

$$\$3,200,000 \times \textbf{.83962} = \$2,686,784.$$

Telco must invest $2,686,784 today at 6% interest compounded annually to have the required down payment of $3,200,000 in 3 years.

Objective 3   Use future value and present value to estimate the value of a business.
Sometimes a business is growing fast. In this event, we need to take the growth rate into consideration when estimating the value of the business. To do this, first find the future value of the business a few years in the future. Then find the present value of this amount.

EXAMPLE 4

Evaluating a Business

Brianna Delfs and Tanya Zoban own Extreme Sports, Inc. whose value is $120,000 today assuming no further growth. However, the partners believe the value will grow at 12% per year compounded semiannually for the next four years. They want to take this rapid growth into consideration when valuing the business for a potential sale.

(a) Find the future value of the business in 4 years.
(b) Estimate the value of the retail store by finding the present value of the amount found in part (a) at 6% compounded quarterly.

SOLUTION

(a) The partners expect the business to grow at $\frac{12\%}{2} = 6\%$ per 6-month period for the next $4 \times 2 = 8$ compounding periods. Use the compound interest table from **Section 9.1** to find 1.59385.

$$\text{Future value} = \$120{,}000 \times \mathbf{1.59385} = \$191{,}262$$

This is an estimate of the value of the retail store in 4 years.

(b) Now find the present value of $191,262 assuming 6% compounded quarterly for 4 years or at $\frac{6\%}{4} = 1.5\%$ per quarter for $4 \times 4 = 16$ compounding periods. The value from the present value of a dollar table on page 000 is **.78803**.

$$\text{Present value} = \$191{,}262 \times \mathbf{.78803} = \$150{,}720.19$$

Thus, the partners should ask $150,720 for their business. They will only be able to get this if a buyer also believes that the store will continue to grow rapidly.

9.3 | Exercises

The **Quick Start** exercises in each section contain solutions to help you get started.

Find the present value of the following. Round to the nearest cent. Also, find the amount of interest earned. (See Examples 1 and 2.)

Quick Start

	Amount Needed	Time (Years)	Interest	Compounded	Present Value	Interest Earned
1.	$12,300	3	8%	annually	$9764.11	$2535.89

$P = \$12,300 \times .79383 = \$9764.11; \ I = \$12,300 - \$9764.11 = \$2535.89$

2.	$14,500	$2\frac{1}{2}$	8%	quarterly	$11,895.08	$2604.92

$P = \$14,500 \times .82035 = \$11,895.08; \ I = \$14,500 - \$11,895.08 = \$2604.92$

3.	$9350	4	5%	semiannually	_____	_____
4.	$850	10	8%	semiannually	_____	_____
5.	$18,853	11	6%	quarterly	_____	_____
6.	$20,984	9	10%	quarterly	_____	_____

Solve the following application problems.

Quick Start

7. **DIVORCE SETTLEMENT** The Prestons are getting a divorce, and part of the divorce settlement involves setting aside money for college tuition for their daughter who enters college in 7 years. They estimate the cost of four years tuition, food and lodging at the state university their daughter will attend at $40,000. Find **(a)** the lump sum that must be invested at 6% compounded semiannually and **(b)** the amount of interest earned.
(a) Lump sum $= P = \$40,000 \times .66112 = \$26,444.80$
(b) $I = \$40,000 - \$26,444.80 = \$13,555.20$

(a) $26,444.80
(b) $13,555.20

8. **SELF-EMPLOYMENT** Teresa Tabor wishes to start her own day-care business in 4 years and estimates that she will need $25,000 to do so. **(a)** What lump sum should be invested today at 6%, compounded semiannually, to produce the needed amount? **(b)** How much interest will be earned?

(a) _____
(b) _____

9. FINANCING COLLEGE EXPENSES   Mrs. Lorez wants all of her grandchildren to go to college and decides to help financially. How much must she give to each child at birth if they are to have $10,000 on entering college 18 years later, assuming 6% interest compounded annually?

9. _____

10. TIRE STORE   Felipe Bazan recently immigrated to the United States from Mexico. He has a little cash and hopes to open a small tire store in 3 years. How much must he deposit today if his credit union pays 6% compounded quarterly and if he needs $15,000 to open the store?

10. _____

11. EXPANDING MANUFACTURING OPERATIONS   Quantum Logic recently expanded their computer-chip assembly operations at a cost of $450,000. Management expects that the value of the investment will grow at a rate of 12% per year compounded annually for the next 5 years. **(a)** Find the future value of the investment. **(b)** Find the present value of the amount found in **(a)** at a rate of 6% compounded annually.

(a) _____
(b) _____

12. BUSINESS EXPANSION   Village Hardware expands its business at a cost of $20,000. They expect that the investment will grow at a rate of 10% per year compounded annually for the next 4 years. **(a)** Find the future value of the investment. **(b)** Find the present value of the amount found in **(a)** at a rate of 6% compounded annually.

(a) _____
(b) _____

13. VALUE OF A BUSINESS   Jessie Marquette believes her hair salon is worth $20,000 and estimates that its value will grow at 10% per year compounded annually for the next 3 years. If she sells the business, the funds will be invested at 8% compounded quarterly. **(a)** Find the future value if she holds onto the business. **(b)** What price should she insist on now if she sells the business?

(a) _____
(b) _____

14. VALUE OF A BUSINESS   John Fernandez figures his bike shop is worth $88,000 if sold today and that it will grow in value at 8% per year compounded annually for the next 6 years. If he sells the business, the funds will be invested at 5% compounded semiannually. **(a)** Find the future value of the shop. **(b)** What price should he insist on at this time if he sells the business?

(a) _____
(b) _____

15. Explain the difference between future value and present value. (See Objective 1.)

16. Explain how and when to use both the compound-interest table in **Section 9.1** and the present-value table in this section.

CHAPTER TERMS *Review the following terms to test your understanding of the chapter. For each term you do not know, refer to the page number found next to that term.*

CD [**p. 374**]

CPI [**p. 374**]

certificate of deposit [**p. 374**]

compound amount [**p. 361**]

compound interest [**p. 360**]

compounded daily [**p. 371**]

consumer price index [**p. 374**]

exponents [**p. 363**]

formula for compounding interest [**p. 363**]

future value [**p. 381**]

inflation [**p. 374**]

interest bearing checking accounts [**p. 371**]

passbook accounts [**p. 371**]

present value [**p. 381**]

savings accounts [**p. 371**]

simple interest [**p. 360**]

time deposit [**p. 374**]

CONCEPTS	EXAMPLES

9.1 Finding compound amount and compound interest

Find the number of compounding periods (n) and the interest rate per period (i).

Use the compound interest table to find the interest on $1.

Multiply the table value by the principal to obtain the compound amount.

Subtract principal from compound amount to obtain the interest.

EXAMPLE: Tom Jones invested $3000 at 6% compounded quarterly for 7 years.

There are $7 \times 4 = 28$ quarters or compounding periods in 7 years.

Interest of 6% per year $= \frac{6\%}{4} = 1\frac{1}{2}\%$ per period.

Find $1\frac{1}{2}\%$ across the top of the compound interest table and 28 down the left side to find **1.51722**.

Compound amount $= \$3000 \times 1.51722 = \$4551.66$

Interest $= \$4551.66 - \$3000 = \$1551.66$

9.2 Finding the interest earned when the interest is compounded daily

Find number of days that the deposit earns interest.

Use the 90-day or 1-quarter table to calculate interest on $1.

Find compound amount using the formula

Compound amount $=$ Principal $\times$ Table value.

Find interest earned using the formula

Interest $=$ Compound amount $-$ Principal.

EXAMPLE: Mary Carver deposits $1000 at $3\frac{1}{2}\%$ compounded daily on May 15. She withdraws the money on July 17. Find the compounded amount and interest earned.

May 15–May 31	16 days
June	30 days
July 1–July 17	17 days
	63 days

Table value $= 1.006059089$

Compound amount $= \$1000 \times 1.006059089$

Interest $= \$1006.06 - \$1000 = \$6.06$

9.2 Finding the interest on time deposits

Use the compound interest for time deposit accounts table to find the interest on $1 compounded daily.

Find the compound amount using the formula

Compound amount $=$ Principal $\times$ Table value.

Find interest using the formula

Interest $=$ Compound amount $-$ Principal.

EXAMPLE: Susan Barbee invests $50,000 in a certificate of deposit paying 5% compounded daily. Find the amount after 4 years.

Table value for 4 years at 5% $= 1.22138603$

Compound amount $= \$50,000 \times 1.22138603 = \$61,069.30$

Interest $= \$61,069.30 - \$50,000 = \$11,069.30$

9.2 Finding the effect of inflation on a pay raise

Find the new salary by multiplying the old salary by $(1 + \text{percent increase})$.

Find the salary needed to offset inflation by multiplying old salary by $(1 + \text{inflation rate})$.

Find the gain or loss by subtracting.

EXAMPLE: Leticia Jaramillo earns $45,000 per year as a computer programmer. She gets a raise of 3.5% in a year in which inflation is 5%. Ignoring taxes, find the effect on her purchasing power.

New salary $= \$45,000 \times 1.035 = \textbf{\$46,575}$

Salary needed to offset inflation $= \$45,000 \times 1.05 = \textbf{\$47,250}$

Loss in purchasing power $= \$47,250 - \$46,575 = \$675$

9.3 Finding the present value of a future amount

Determine the number of compounding periods (n).

Determine the interest per compounding period (i).

Use the values of n and i to determine the table value from the present value table.

Find present value from the following formula.

Present value $=$ Future value $\times$ Table value

EXAMPLE: Sue York must pay a lump sum of $4500 in 6 years. What lump sum deposited today at 6% compounded quarterly will amount to $4500 in 6 years?

Number of compounding periods $= 6 \times 4 = 24$

Interest per compounding period $= \dfrac{6\%}{4} = \mathbf{1\frac{1}{2}\%}$ **per period**

Table value $= \mathbf{.69954}$

Present value $= \$4500 \times .69954 = \$3147.93$

Chapter 9 | Summary Exercise

Valuing a Chain of McDonald's Restaurants

James and Mary Watson own a small chain of McDonald's restaurants that is valued at $2,300,000. They believe that the chain will grow in value at 12% per year compounded annually for the next 5 years. If they sell the chain, the funds will be invested at a rate of 6% compounded semiannually. They expect inflation to be 4% per year for the next 5 years. Ignore taxes, and answer the following, rounding answers to the nearest dollar.

(a) Find the future value of the chain after 5 years. Then find the price they should sell the chain for if they wish to have the same future value at the end of 5 years.

(a) _____

(b) Find the future value of the chain if it only grows at 2% per year for 5 years. Then find the price they should ask for the chain given a 2% growth rate per year.

(b) _____

(c) What future value would the chain be worth if it grew at their expected rate of inflation? Find the price they should ask for the chain if it grows at the rate of inflation.

(c) _____

(d) Complete the following table.

Growth Rate	Future Value	Market Value Today
2%	_____	_____
4% (inflation)	_____	_____
12%	_____	_____

QUICK TIP   The value of the chain varies by more than one million dollars, depending on the rate of growth assumed for the business for the next 5 years.

INVESTIGATE

The interest rates that a bank pays depend on whether the money is in a checking account, savings account or time deposit. Visit a local bank, and find the different interest rates that the bank will pay. Identify the conditions such as the minimum amount in an account, minimum deposit, and the length of time the money must be on deposit to earn each interest rate.

Chapter 9 | Test

To help you review, the numbers in brackets show the section in which the topic was discussed.

In each of these problems, round to the nearest cent. Find the compound amount and the interest earned for the following.   **[9.1]**

	Amount	Rate	Compounded	Time (Years)	Compound Amount	Interest Earned
1.	$8700	10%	annually	8	_____	_____
2.	$12,000	6%	semiannually	5	_____	_____
3.	$9800	6%	semiannually	5	_____	_____
4.	$12,500	10%	quarterly	4	_____	_____

Find the interest earned by the following. Assume $3\frac{1}{2}$% interest compounded daily.   **[9.2]**

5. $6400 deposited September 24 and withdrawn December 15

5. _____

6. $63,340 deposited December 5 and withdrawn March 2

6. _____

7. $37,650 deposited December 12 and withdrawn on February 29 (leap year)

7. _____

Find the present value of the following.   **[9.3]**

	Amount Needed	Time (Years)	Rate	Compounded	Present Value
8.	$35,000	20	8%	annually	_____
9.	$15,750	7	6%	quarterly	_____
10.	$56,900	10	4%	semiannually	_____

Solve the following application problems.

11. The Train Company deposited $12,500 in a savings account on July 3 and then deposited an additional $3450 in the account on August 5. Find the balance on October 1 assuming an interest rate of $3\frac{1}{2}$% compounded daily.   **[9.2]**

11. _____

12. Discount Auto Insurance deposited $1800 in a savings account paying $3\frac{1}{2}$% compounded daily on January 1 and deposited an additional $2300 in the account on March 12. Find the balance on April 1.   **[9.2]**

12. _____

13. Mike George deposits $4000 in a certificate of deposit for 5 years. Find the compound amount if the interest rate is 6% compounded daily.   **[9.2]**

13. _____

14. Benton Signs places $35,000 in a 2-year certificate of deposit yielding 5% compounded daily and uses it for collateral for a loan. Find the compound amount.   **[9.2]**

14. _____

15. Liz Mulig earns $52,000 per year as a chemistry professor. She receives a raise of 2.5% in a year in   15. _____
     which the CPI index increases by 3.8%. Ignoring taxes, find the effect of the two increases on her
     purchasing power.   **[9.2]**

16. James Arnosti makes $65,000 per year as an editor for a publisher. He was notified of a 1.5% raise in   16. _____
     a year in which the CPI index increased by 4%. Find the gain or loss in his purchasing power.   **[9.2]**

17. A note for $3500 was made at 8% per year compounded annually for 3 years. Find **(a)** the maturity value   **(a)** _____
     and **(b)** the present value of the note assuming 5% per year compounded semiannually.   **[9.3]**   **(b)** _____

18. Computers, Inc. accepted a 2-year note for $12,540 in lieu of immediate payment for computer   **(a)** _____
     equipment sold to a local firm. Find **(a)** the maturity value given a 10% rate compounded annually and   **(b)** _____
     (b) the present value of the note at 6% per year compounded semiannually.   **[9.3]**

19. A business worth $180,000 is expected to grow at 12% per year compounded annually for the next 4 years.   **(a)** _____
     (a) Find the expected future value. **(b)** If funds from the sale of the business today would be placed in an   **(b)** _____
     account yielding 8% compounded semiannually, what would be the minimum acceptable price for the
     business at this time?   **[9.3]**

20. A corporation worth 40 million dollars is expected to grow at 8% per year compounded annually for   **(a)** _____
     5 years. **(a)** Find the future value to the nearest million. **(b)** They then propose to sell the firm and   **(b)** _____
     invest the proceeds in a new venture that should grow at 12% compounded annually for 4 years.
     Beginning with the future value from part **(a)** rounded to the nearest million, find the expected future
     value to the nearest million at the end of 4 additional years.   **[9.1]**

Cumulative Review

Chapters 8–9

Round money amounts to the nearest cent, time to the nearest day, and rates to the nearest tenth of a percent.

Find the value of the unknown quantity using simple interest. Use banker's interest.   [8.1–8.2]

	Interest	Principal	Rate	Time		
1.	_____	$6800	8%	6 months	1.	_____
2.	_____	$6200	9.7%	250 days	2.	_____
3.	$46.67	_____	7%	100 days	3.	_____
4.	$475	_____ = $10,000	9.5%	180 days	4.	_____
5.	$50.93	$2100	_____	90 days	5.	_____
6.	$733.33	$12,000 = 11%	_____	200 days	6.	_____
7.	$202.22	$9100	10%	_____	7.	_____
8.	$915	$18,300	12%	_____	8.	_____

Find the discount and the proceeds.   [8.3]

	Face Value	Discount Rate	Time (Days)	Discount	Proceeds
9.	$9000	12%	90	_____	_____
10.	$875	$6\frac{1}{2}$%	210	_____	_____

Find the net proceeds when each of the following is discounted.   [8.4]

	Maturity Value	Discount Rate	Discount Period	Net Proceeds
11.	$5000	10%	90 days	_____
12.	$12,000	12%	150 days	_____

Find the compound amounts for the following.   **[9.1]**

13. $1000 at 4% compounded annually for 17 years

13. _____

14. $3520 at 8% compounded annually for 10 years

14. _____

Find the interest earned and compound amounts for each of the following. Assume $3\frac{1}{2}\%$ interest compounded daily.   **[9.2]**

Amount	Date Deposited	Date Withdrawn	Interest Earned	Compound Amount
15. $12,600	March 24	June 3	_____	_____
16. $7500	November 20	February 14	_____	_____

Find the present value and amount of interest earned for the following. Round to the nearest cent.   **[9.3]**

Amount Needed	Time (Years)	Interest	Compounded	Present Value	Interest
17. $1000	7	8%	annually	_____	_____
18. $19,000	9	5%	semiannually	_____	_____

Solve the following application problems. Use a 360-day year where applicable.

19. Cathy Cockrell signed a 180-day simple discount note with a rate of 10% and a face value of $25,000. Find **(a)** the interest and **(b)** the proceeds.   **[8.3]**

(a) _____
(b) _____

20. As a project manager, Regina Foster received a bonus of $18,000 for completing a difficult project on time. She invests it at 6% compounded quarterly for 5 years. Find the future value.   **[9.1]**

20. _____

21. A divorce settlement requires Samantha James to pay her ex-spouse $12,000 in 2 years. What lump sum can be invested today at 6% compounded semiannually so that enough will be available for the payment?   **[9.3]**

21. _____

22. Tom Davis owes $7850 to a relative. He has agreed to pay the money in 5 months, at an interest rate of 6%. One month before the loan is due, the relative discounts the loan at the bank. The bank charges a 7.92% discount rate. How much money does the relative receive?   **[8.4]**

22. _____

23. The owner of Jessica's Cookies has an extra $3200 that she puts into a savings account paying $3\frac{1}{2}\%$ per year compounded daily. Find the interest if the funds are left there for 65 days.   **[9.2]**

23. _____

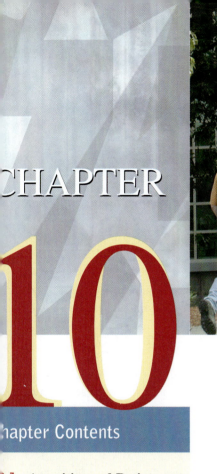

CHAPTER

10

Annuities, Stocks, and Bonds

ROMAN RODRIGUEZ RECEIVED HIS ASSOCIATE OF ARTS DEGREE IN computer information science and then received a bachelor's degree in the same

major. He had worked in the computer labs at the community college when a student there and happily accepted the permanent position they offered him. He was surprised at the choices related to retirement that he was given by the human resources department. Although he was only 25, Rodriguez knew that it was important to make good choices, since it would affect his future.

10.1 | Annuities and Retirement Accounts

Objectives

1. Define the basic terms involved with annuities.
2. Find the amount of an annuity.
3. Find the amount of an annuity due.
4. Understand different retirement accounts and find the amount of a retirement account.

CASE *in* POINT

The benefits coordinator asked Roman Rodriguez if he preferred an annuity paying a guaranteed interest or one invested in a mutual fund containing stocks. Rodriguez knew that he needed to be careful, since it was his future they were talking about.

Objective 1 Define the basic terms involved with annuities. Chapter 9 discussed lump sums of money such as present value (the value today) and future value (the value at some future date). However, business and individuals often use an **annuity** or a series of equal payments made at regular intervals. Examples of annuities include monthly mortgage payments on a building, a regular payment by a company into employees' retirement accounts, or monthly retirement checks made to an individual by an insurance company.

An **ordinary annuity** is an annuity in which payments are made *at the end of each period,* such as at the end of each month or quarter. The time between payments is the **payment period** and the time from the first payment through the last payment is the **term of the annuity**.

Compound interest is applied to each payment of an annuity. For example, consider regular monthly investments into a savings account. Each payment earns compound interest until the end of the term of the annuity. The total amount in the account at the end of the term is called the **amount of the annuity** or **future value of the annuity** since it is the amount available at a future date.

One use of an ordinary annuity is given by a small firm that is accumulating funds for a new vehicle. The firm makes deposits of $3000 at the end of each year for 6 years in an account earning 8% per year compounded annually. The first deposit of $3000 is made at the *end* of year 1 and earns interest *for only 5 years*. The compound interest table in Section 9.1 (page 000) for 5 years and 8% shows that the first payment grows to

$$\$3000 \times \mathbf{1.46933} = \$4407.99$$

in 5 years. The amount of the annuity is the sum of the compound amounts of all 6 payments.

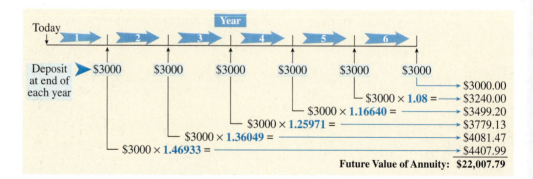

QUICK TIP   The annuity ends on the day of the last payment. Therefore, the last payment which is made at the end of year 6 earns no interest.

Objective 2 **Find the amount of an annuity.** The amount of an annuity can also be found using the amount of an annuity table on page 397. The number from the table is the amount or future value of an annuity with a payment of $1. The amount of an annuity with any payment is found as follows.

> Amount = Payment × Number from amount of an annuity table

As a check, go back to the annuity of $3000 at the end of each year for 6 years at 8% compounded annually. Locate 8% at the top of the table on page 397 and 6 periods in the far left (or far right) column to find **7.33593**.

$$\text{Amount} = \$3000 \times \mathbf{7.33593} = \$22{,}007.79$$

This amount is identical to the amount calculated on the previous page.

> **QUICK TIP** See Appendix D for financial calculator solutions that do not require the use of a table.

EXAMPLE 1

Finding the Value of an Annuity and Interest Earned

The community college will match Roman Rodriguez's contribution into his retirement plan, but only up to 5% of his salary. In other words, the community college will put one dollar into his retirement plan for every dollar that Rodriguez puts into it, but they will not contribute more than 5% of his $32,000 yearly salary. Rodriguez immediately decides to put 5% of his salary into the retirement plan. Using quarterly calculations, find the future value in 10 years **(a)** if the account earns 4% per year and **(b)** if the account earns 8% per year. **(c)** Then find the difference between the two future values.

SOLUTION

$$\text{Salary per quarter} = \$32{,}000 \div 4 = \$8000$$

Total contributions = Rodriguez's contributions + **matching contributions**
= .05 × $8000 + **.05 × $8000**
= $400 + **$400**
= $800 invested per quarter

(a) Interest of $\frac{4\%}{4} = 1\%$ is earned per quarter for $10 \times 4 = 40$ quarters. Look across the top of the table for 1% and down the side for 40 periods to find **48.88637**.

$$\text{Amount} = \$800 \times \mathbf{48.88637} = \mathbf{\$39{,}109.10}\text{ (rounded)}$$

(b) Interest of $\frac{8\%}{4} = 2\%$ is earned per quarter for $10 \times 4 = 40$ quarters. Look across the top of the table for 2% and down the side for 40 periods to find **60.40198**.

$$\text{Amount} = \$800 \times \mathbf{60.40198} = \mathbf{\$48{,}321.58}\text{ (rounded)}$$

(c) Difference = $48,321.58 − $39,109.10 = **$9212.48**

EXAMPLE 2

Finding the Amount of an Annuity and Interest Earned

At the birth of her grandson, Junella Smith commits to help pay for his college education. She decides to make deposits of $600 at the end of each 6 months into an account for 18 years. Find the amount of the annuity and the interest earned, assuming 6% compounded semiannually.

SOLUTION

Interest of $\frac{6\%}{2} = 3\%$ is earned each semiannual period. There are $2 \times 18 = 36$ semiannual periods in 18 years. Find 3% across the top and 36 periods down the side of the table to find **63.27594**.

$$\text{Amount} = \$600 \times \textbf{63.27594} = \$37,965.56$$
$$\text{Interest} = \$37,965.56 - (\textbf{36} \times \textbf{\$600}) = \$16,365.56$$

Although Smith knows that a college education will cost a lot more in 18 years than it does today, she feels that $37,965.56 will be of great help to her grandson when he enters college.

> **QUICK TIP**   Example 2 in Appendix D shows how a financial calculator can be used to solve these same types of problems.

Objective ③ **Find the amount of an annuity due.** Payments were made at the *end of each period* in the ordinary annuities discussed previously. In contrast, an annuity in which payments are made at the *beginning of each time period* is called **annuity due**. To find the amount of an annuity due, treat each payment as if it were made at *the end of the preceding period*, then

1. add 1 to the number of periods and find the amount using the table,
2. multiply the factor from the table by the periodic deposit to find the future value overstated by one payment, and
3. subtract 1 payment from this amount.

EXAMPLE 3

Find the Amount of an Annuity Due

Mr. and Mrs. Thompson set up an investment program using an *annuity due* with payments of $500 at the *beginning of each quarter*. Find **(a)** the amount of the annuity and **(b)** the interest if they make payments for 7 years into an investment account expected to pay 8% compounded quarterly.

SOLUTION

(a) **Step 1:** Interest of $\frac{8\%}{4} = 2\%$ is earned each quarter. There are $4 \times 7 = 28$ periods in 7 years. Since it is an annuity due, add 1 period to 28 making 29 periods.

Step 2: Look across the top of the table for 2% and down the side for 29 periods to find **38.79223**.

$$\$500 \times \textbf{38.79223} = \textbf{\$19,396.12} \text{ (rounded)}$$

Step 3: Now subtract one payment to find the amount of the annuity due.

$$\text{Amount of annuity due} = \textbf{\$19,396.12} - \$500 = \$18,896.12$$

(b) Subtract the 28 payments $(7 \text{ years} \times 4 \text{ payments per year})$ of $500 each to find the interest.

$$\text{Interest} = \$18,896.12 - (\textbf{28} \times \textbf{\$500}) = \$4896.12$$

The calculator solution to finding the interest in **(b)** follows.

$$18896.12 \; \boxed{-} \; 28 \; \boxed{\times} \; 500 \; \boxed{=} \; 4896.12$$

Note: Refer to Appendix C for calculator basics.

> **QUICK TIP**   For an annuity due, be sure to add 1 period to the number of compounding periods and subtract 1 payment from the amount calculated.

Amount of an Annuity Table
n = number of periods in annuity; i = interest per period

n	1%	1½%	2%	2½%	3%	4%	5%	6%	8%	10%	12%	n
1	1.00000	1.00000	1.00000	1.00000	1.00000	1.00000	1.00000	1.00000	1.00000	1.00000	1.00000	1
2	2.01000	2.01500	2.02000	2.02500	2.03000	2.04000	2.05000	2.06000	2.08000	2.10000	2.12000	2
3	3.03010	3.04522	3.06040	3.07562	3.09090	3.12160	3.15250	3.18360	3.24640	3.31000	3.37440	3
4	4.06040	4.09090	4.12161	4.15252	4.18363	4.24646	4.31013	4.37462	4.50611	4.64100	4.77933	4
5	5.10101	5.15227	5.20404	5.25633	5.30914	5.41632	5.52563	5.63709	5.86660	6.10510	6.35285	5
6	6.15202	6.22955	6.30812	6.38774	6.46841	6.63298	6.80191	6.97532	7.33593	7.71561	8.11519	6
7	7.21354	7.32299	7.43428	7.54743	7.66246	7.89829	8.14201	8.39384	8.92280	9.48717	10.08901	7
8	8.28567	8.43284	8.58297	8.73612	8.89234	9.21423	9.54911	9.89747	10.63663	11.43589	12.29969	8
9	9.36853	9.55933	9.75463	9.95452	10.15911	10.58280	11.02656	11.49132	12.48756	13.57948	14.77566	9
10	10.46221	10.70272	10.94972	11.20338	11.46388	12.00611	12.57789	13.18079	14.48656	15.93742	17.54874	10
11	11.56683	11.86326	12.16872	12.48347	12.80780	13.48635	14.20679	14.97164	16.64549	18.53117	20.65458	11
12	12.68250	13.04121	13.41209	13.79555	14.19203	15.02581	15.91713	16.86994	18.97713	21.38428	24.13313	12
13	13.80933	14.23683	14.68033	15.14044	15.61779	16.62684	17.71298	18.88214	21.49530	24.52271	28.02911	13
14	14.94742	15.45038	15.97394	16.51895	17.08632	18.29191	19.59863	21.01507	24.21492	27.97498	32.39260	14
15	16.09690	16.68214	17.29342	17.93193	18.59891	20.02359	21.57856	23.27597	27.15211	31.77248	37.27971	15
16	17.25786	17.93237	18.63929	19.38022	20.15688	21.82453	23.65749	25.67253	30.32428	35.94973	42.75328	16
17	18.43044	19.20136	20.01207	20.86473	21.76159	23.69751	25.84037	28.21288	33.75023	40.54470	48.88367	17
18	19.61475	20.48938	21.41231	22.38635	23.41444	25.64541	28.13238	30.90565	37.45024	45.59917	55.74971	18
19	20.81090	21.79672	22.84056	23.94601	25.11687	27.67123	30.53900	33.75999	41.44626	51.15909	63.43968	19
20	22.01900	23.12367	24.29737	25.54466	26.87037	29.77808	33.06595	36.78559	45.76196	57.27500	72.05244	20
21	23.23919	24.47052	25.78332	27.18327	28.67649	31.96920	35.71925	39.99273	50.42292	64.00250	81.69874	21
22	24.47159	25.83758	27.29898	28.86286	30.53678	34.24797	38.50521	43.39229	55.45676	71.40275	92.50258	22
23	25.71630	27.22514	28.84496	30.58443	32.45288	36.61789	41.43048	46.99583	60.89330	79.54302	104.60289	23
24	26.97346	28.63352	30.42186	32.34904	34.42647	39.08260	44.50200	50.81558	66.76476	88.49733	118.15524	24
25	28.24320	30.06302	32.03030	34.15776	36.45926	41.64591	47.72710	54.86451	73.10594	98.34706	133.33387	25
26	29.52563	31.51397	33.67091	36.01171	38.55304	44.31174	51.11345	59.15638	79.95442	109.18177	150.33393	26
27	30.82089	32.98668	35.34432	37.91200	40.70963	47.08421	54.66913	63.70577	87.35077	121.09994	169.37401	27
28	32.12910	34.48148	37.05121	39.85980	42.93092	49.96758	58.40258	68.52811	95.33883	134.20994	190.69889	28
29	33.45039	35.99870	38.79223	41.85630	45.21885	52.96629	62.32271	73.63980	103.96594	148.63093	214.58275	29
30	34.78489	37.53868	40.56808	43.90270	47.57542	56.08494	66.43885	79.05819	113.28321	164.49402	241.33268	30
31	36.13274	39.10176	42.37944	46.00027	50.00268	59.32834	70.76079	84.80168	123.34587	181.94342	271.29261	31
32	37.49407	40.68829	44.22703	48.15028	52.50276	62.70147	75.29883	90.88978	134.21354	201.13777	304.84772	32
33	38.86901	42.29861	46.11157	50.35403	55.07784	66.20953	80.06377	97.34316	145.95062	222.25154	342.42945	33
34	40.25770	43.93309	48.03380	52.61289	57.73018	69.85791	85.06696	104.18375	158.62667	245.47670	384.52098	34
35	41.66028	45.59209	49.99448	54.92821	60.46208	73.65222	90.32031	111.43478	172.31680	271.02437	431.66350	35
36	43.07688	47.27597	51.99437	57.30141	63.27594	77.59831	95.83632	119.12087	187.10215	299.12681	484.46312	36
37	44.50765	48.98511	54.03425	59.73395	66.17422	81.70225	101.62814	127.26812	203.07032	330.03949	543.59869	37
38	45.95272	50.71989	56.11494	62.22730	69.15945	85.97034	107.70955	135.90421	220.31595	364.04343	609.83053	38
39	47.41225	52.48068	58.23724	64.78298	72.23423	90.40915	114.09502	145.05846	238.94122	401.44778	684.01020	39
40	48.88637	54.26789	60.40198	67.40255	75.40126	95.02552	120.79977	154.76197	259.05652	442.59526	767.09142	40
41	50.37524	56.08191	62.61002	70.08762	78.66330	99.82654	127.83976	165.04768	280.78104	487.85181	860.14239	41
42	51.87899	57.92314	64.86222	72.83981	82.02320	104.81960	135.23175	175.95054	304.24352	537.63699	964.35948	42
43	53.39778	59.79199	67.15947	75.66080	85.48389	110.01238	142.99334	187.50758	329.58301	592.40069	1081.08262	43
44	54.93176	61.68887	69.50266	78.55232	89.04841	115.41288	151.14301	199.75803	356.94965	652.64076	1211.81253	44
45	56.48107	63.61420	71.89271	81.51613	92.71986	121.02939	159.70016	212.74351	386.50562	718.90484	1358.23003	45
46	58.04589	65.56841	74.33056	84.55403	96.50146	126.87057	168.68516	226.50812	418.42607	791.79532	1522.21764	46
47	59.62634	67.55194	76.81718	87.66789	100.39650	132.94539	178.11942	241.09861	452.90015	871.97485	1705.88375	47
48	61.22261	69.56522	79.35352	90.85958	104.40840	139.26321	188.02539	256.56453	490.13216	960.17234	1911.58980	48
49	62.83483	71.60870	81.94059	94.13107	108.54065	145.83373	198.42666	272.95840	530.34274	1057.18957	2141.98058	49
50	64.46318	73.68283	84.57940	97.48435	112.79687	152.66708	209.34800	290.33590	573.77016	1163.90853	2400.01825	50

Objective 4 **Understand different retirement accounts and find the amount of a retirement account.** The newspaper clipping shows that it can be difficult to know how much is needed to retire, but the consequences of having too little can be serious. The important thing is to begin saving as soon as possible and to save regularly. A great benefit of retirement plans is that an employed person can save for decades without paying taxes on gains or interest until the money is actually withdrawn. This makes it easier to accumulate large amounts of money. Retirement plans are for everyone including those who just graduated from college.

HERE & NOW

Retired without Enough Income

Tom Wheat retired in 2004 thinking that his social security check of $1040 per month and income from his savings would be enough to support his lifestyle. However, he subsequently developed diabetes, increasing his medical expenses to significantly more than he anticipated. Now, at 73 years of age, he can no longer work and must get by on a limited fixed income. He may have to sell his small home so that he can use the equity. But where would he live?

An **individual retirement account (IRA)** is a retirement account that an employed person establishes for him- or herself. There are two basic types of IRA accounts: regular and Roth. Deposits to a **regular IRA** are usually excluded from federal income taxes in the year they are made. For example, suppose you make $1800 per month and put $100 into a regular IRA. Then you may only have to pay taxes on $1800 − $100 = $1700 per month. Effectively, you deposit $100 into your personal retirement account, but because of taxes, your monthly paycheck may be reduced by perhaps only $80. Funds in a regular IRA grow tax-free, but income taxes must be paid when the money is taken out of the plan after age $59\frac{1}{2}$. You can take money out before $59\frac{1}{2}$, but there will be a penalty for doing so.

Another type of IRA is a **Roth IRA**. Deposits to a Roth IRA *are not excluded from federal taxes in the year paid*, so they do not reduce current income taxes. However, the deposits and interest grow tax-free. In addition, withdrawals from a Roth IRA at retirement **are not subject to income taxes** when withdrawn. This offers you a great opportunity to save money for retirement without having to pay taxes as you withdraw the funds. The following tables may help you.

Roth IRA

WHICH IRA IS BEST FOR YOU?		
	DEDUCTIBLE IRA	**ROTH IRA**
Tax deductible?	If you qualify	No
Taxable at withdrawal?	Yes	No
Penalty for early withdrawal?	Yes, prior to age 59.5	Yes*
Mandatory withdrawal age?	70.5	None
Penalty-free withdrawals?	$10,000 for first-time home buyers; unlimited for education	$10,000 for first-time home buyers, after five-year wait; unlimited for education

*Never any penalty for withdrawing your own contributions, but a penalty applies to withdrawal of any gains within five years of opening the account and/or before turning 59.5.

IRA Contribution Amount			
Year	**Contribution limit**	**Catch-up contribution (age 50 or older)**	**Maximum contribution (age 50 or older)**
2005	$4000	$500	$4500
2006	$4000	$1000	$5000
2007	$4000	$1000	$5000
2008	$5000	$1000	$6000

The following figure shows some of the many reasons companies offer retirement plans to full-time employees. There are several different types of company-sponsored retirement plans. Two common plans are the **401(k)** plan for individuals working for private-sector companies and the **403(b)** plan for employees of public schools and certain tax-exempt organizations. The names 401(k) and 403(b) refer to sections of the Internal Revenue code that define these plans. Each of these plans allows an employer to deduct a certain amount of money from your paycheck *before* taxes are calculated and to invest those funds in the plan. Companies also sometimes match all or part of the contribution of an employee.

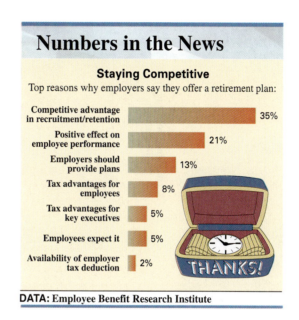

Numbers in the News

Staying Competitive

Top reasons why employers say they offer a retirement plan:

Competitive advantage in recruitment/retention	35%
Positive effect on employee performance	21%
Employers should provide plans	13%
Tax advantages for employees	8%
Tax advantages for key executives	5%
Employees expect it	5%
Availability of employer tax deduction	2%

DATA: Employee Benefit Research Institute

Regular contributions into either an IRA or a company-sponsored retirement plan are an annuity. The amount of the annuity is found using the same methods discussed earlier in this section. Payments are at the end of each period in an ordinary annuity and payments are at the beginning of each period in an annuity due.

> **QUICK TIP** Individuals who pay FICA (Social Security) taxes are eligible for monthly payments from the Social Security system at retirement. Social Security benefits are in addition to any income from IRAs or company retirement plans.

EXAMPLE 4

Finding the Value of an IRA

At 27, Joann Gretz sets up an IRA with Merrill Lynch where she plans to deposit $2000 at the end of each year until age 60. Find the amount of the annuity if she invests in **(a)** a bond fund that has historically yielded 6% compounded annually versus **(b)** a stock fund that has historically yielded 10% compounded annually. Assume that future yields equal historical yields.

SOLUTION

Age 60 is 60 − 27 = **33 years away**, so she will make deposits at the end of each year for 33 years.

(a) Bond Fund

Look down the left column of the amount of an annuity table on page 000 at 33 years and across the top for 6% to find **97.34316**.

$$\text{Amount} = \$2000 \times \textbf{97.34316} = \$194{,}686.32$$

(b) Stock Fund

Look down the left column of the table for 33 years and across the top for 10% to find **222.25154**.

$$\text{Amount} = \$2000 \times \textbf{222.25154} = \$444{,}503.08$$

Gretz can see the projected difference in the results of the treasury bill fund and the stock fund using the following graph. However, she is worried about the possibility of losing money in the stock fund. *See Exercise 20* at the end of this section to find her investment choice.

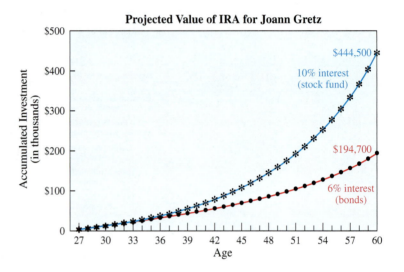

Currently, more than 45 million people receive regular payments from Social Security. Workers can retire as early as age 62 and get reduced benefits, or they can wait until full retirement age and receive full benefits. The full retirement age will increase gradually until it reaches age 67 for people born after 1959. The bar graph shows the importance of Social Security to Americans.

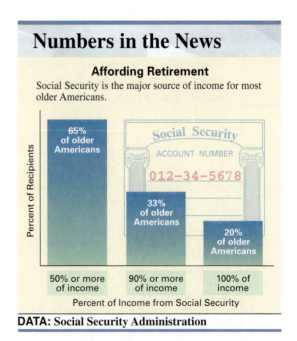

10.1 | Exercises

The **Quick Start** *exercises in each section contain solutions to help you get started.*

Find the amount of the following ordinary annuities rounded to the nearest cent. Find the total interest earned. (See Examples 1 and 2.)

Quick Start

Amount of Each Deposit	Deposited	Rate	Time (Years)	Amount of Annuity	Interest Earned
1. $900	annually	5%	18	$25,319.14	$9119.14
$900 × 28.13238 = $25,319.14; *I* = $25,319.14 − (18 × $900) = $9119.14					
2. $2900	annually	8%	5	$17,013.14	$2513.14
$2900 × 5.86660 = $17,013.14; *I* = $17,013.14 − (5 × $2900) = $2513.14					
3. $7500	semiannually	6%	10	_____	_____
4. $9200	semiannually	8%	5	_____	_____
5. $3500	quarterly	10%	7	_____	_____
6. $6900	quarterly	8%	4	_____	_____

Find the amount of the following annuities due rounded to the nearest cent. Find the total interest earned. (See Example 3)

Quick Start

Amount of Each Deposit	Deposited	Rate	Time (Years)	Amount of Annuity	Interest Earned
7. $1200	annually	8%	5	$7603.12	$1603.12
Look up 8%, 5 + 1 = 6 periods, finding 7.33593. $1200 × 7.33593 − $1200 = $7603.12 Interest = $7603.12 − 5 × $1200 = $1603.12					
8. $400	annually	6%	6	$2957.54	$557.54
Look up 6%, 6 + 1 = 7 periods, finding 8.39384. $400 × 8.39384 − $400 = $2957.54 Interest = $2957.54 − 6 × $400 = $557.54					
9. $9500	semiannually	4%	9	_____	_____
10. $1800	semiannually	5%	6	_____	_____

 indicates an exercise that is related to the Case in Point feature within the section.

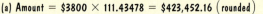

Amount of Each Deposit	Deposited	Rate	Time (Years)	Amount of Annuity	Interest Earned
11. $3800	quarterly	8%	3	_____	_____
12. $10,200	quarterly	10%	5	_____	_____

13. Explain the difference between an annuity and an annuity due. (See Objectives 2 and 3.)

14. Describe the differences between an IRA, a 401(k), and a 403(b). (See Objective 4.)

Solve the following application problems.

Quick Start

15. RETIREMENT PLANNING   Roman Rodriguez would like to know if he can retire in 35 years at age 60 when he plans to fish a lot. Assume the total deposit into his retirement account at the community college is $3800 at the end of each year and that the fund earns 6% per year. Find **(a)** the amount of the annuity and **(b)** the interest earned.

(a) $423,452.16
(b) $290,452.16

(a) Amount = $3800 × 111.43478 = $423,452.16 (rounded)
(b) Interest = $423,452.16 − 35 × $3800 = $290,452.16

16. SAVING FOR A HOME   Jim and Betty Collins want to save $6500 for a down payment on a home they hope to buy in 2 years. They invest $800 at the end of each quarter in an account earning 6% compounded quarterly. Find **(a)** the amount of the annuity and **(b)** the interest earned.

(a) _____
(b) _____

17. CHILD-CARE PAYMENTS   Monique Chaney places $250 of her quarterly child support check into an annuity for the education of her child. She does this at the beginning of each quarter for 9 years into an account paying 8% per year, compounded quarterly. Find **(a)** the amount of the annuity and **(b)** the interest earned.

(a) _____
(b) _____

_____

18. RETIREMENT CONTRIBUTION   Tim Johnson, who works at an oil refinery in Saudi Arabia, asked his employer to put his $450 quarterly retirement contribution into an annuity with National Insurance Company. Assume that the contribution is deposited at the beginning of each quarter for 10 years at 6% interest compounded quarterly. Find the future value of his retirement account.

18. _____

19. MUTUAL FUND INVESTING   Sandra Gonzales deposits $1000 into a mutual fund containing stocks at the end of each semiannual period for 12 years. Assume the fund earns 10% interest compounded semiannually and find the future value.

19. _____

20. T-BILL AND STOCK INVESTING   Joann Gretz (see Example 4, page 399) decides to place $\frac{1}{2}$ of her $2000 deposit at the end of each year into the bond fund and $\frac{1}{2}$ into the stock fund. Assume the bond fund earns 6% compounded annually and that the stock fund earns 10% compounded annually. Find the amount available in 33 years.

20. _____

10.2 | Present Value of an Ordinary Annuity

Objectives

1. Define the present value of an ordinary annuity.
2. Use the formula to find the present value of an ordinary annuity.
3. Find the equivalent cash price of an ordinary annuity.

Objective 1 Define the present value of an ordinary annuity. In Section 9.3, we saw how to find the amount that must be deposited today (the present value) to produce the desired value at a specific future date (the future value). In that section, we discussed only the lump sums of present value and future value since no periodic payments were made.

In Chapter 10, annuities are being discussed. Annuities involve regular periodic payments. Section 10.1 showed how to find the amount (or future value) of an annuity after a series of periodic payments. This section shows how to find the **present value of an annuity**, which is the present value of a series of periodic payments. There are two ways to think of the present value of an annuity. The first way described below involves an investment that grows to meet a need at some future date. The second way described below involves an investment that dwindles to zero as a series of payments are taken out of the account.

Present Value of an Ordinary Annuity

1. Suppose a firm needs $100,000 at a specific date in the future. They can achieve that goal either by
   (a) making periodic payments into an account for several years, or
   (b) depositing a lump sum into an account and letting the funds grow.
   The lump sum that can be deposited today is the present value of the annuity involving the periodic payments. (See Example 1.)
2. Suppose a retired couple needs $2000 per month for 10 years. The present value of this annuity is the lump sum that must be deposited today to generate the needed payments. (See Example 2.)

Objective 2 Use the formula to find the present value of an ordinary annuity. The present value of an annuity with periodic payments at the end of each period is found using values from the table.

$$\text{Present value of annuity} = \text{Payment} \times \frac{\text{Number from the present value}}{\text{of an annuity table}}$$

EXAMPLE 1

Finding the Present Value of an Annuity

Walter Bates wants to build up emergency funds for unexpected expenses in his small auto repair shop. He deposits $1200 at the end of each year for 15 years into a mutual fund that he believes will yield 8% compounded annually.

(a) Find the future value of this annuity using the concepts from Section 10.1.
(b) Find the lump sum that must be deposited today at 8% compounded annually to generate the same future value in 15 years. This lump sum is the present value of the annuity.

Payments

SOLUTION

(a) Use the amount of an annuity table in Section 10.1 along with 8% per compounding period and 15 compounding periods to find **27.15211**.

$$\text{Future value of annuity} = \$1200 \times \textbf{27.15211} = \$32{,}582.53 \text{ (rounded)}$$

(b) We do not need the future value of the annuity just calculated to find the present value of this annuity. Look in the present value of an annuity table on the next page for 8% across the top and 15 periods down the side to find **8.55948**.

$$\text{Present value of annuity} = \$1200 \times \textbf{8.55948} = \$10{,}271.38 \text{ (rounded)}$$

Therefore, $10,271.38 deposited today results in the same future value as $1200 deposited at the end of every year assuming 15 years and 8% compounded annually. Use the compound interest table in Section 9.1 with 8% per year and 15 years to check the results.

$$\text{Future value of lump sum} = \$10{,}271.38 \times \textbf{3.17217} = \$32{,}582.56$$

The two future values ($32,582.56 and $32,582.53) differ by 3 cents due to rounding.

> **QUICK TIP**   Either method produces $32,582.53 in 15 years at 8% compounded annually:
> 1. a single deposit of $10,271.38 today or
> 2. 15 end-of-year payments of $1200 each.

EXAMPLE 2

Finding the Present Value

James and Brandy Barrett recently divorced, and Brandy has custody of their 4-year-old son. As part of the divorce settlement, James must pay Brandy $900 at the end of each quarter until their son reaches age 16. James decides to put a lump sum into an account with a guaranteed yield of 6% compounded quarterly. Find the lump sum he must deposit today to make the required payments. Find the amount of interest that will be earned.

SOLUTION

Payments must be made for $16 - 4 = 12$ years, or for $12 \times 4 = 48$ quarters. The interest rate per quarter is $\frac{6\%}{4} = 1.5\%$ per quarter. Look across the top of the present value of an annuity table for 1.5% and down the side for 48 payments to find **34.04255**.

$$\text{Present value of annuity} = \$900 \times \textbf{34.04255} = \$30{,}638.30 \text{ (rounded)}$$

A deposit of $30,638.30 today will make 48 end-of-quarter payments of $900 each. Interest earned during the 12 years is the sum of all payments less the original lump sum.

$$\text{Interest} = \left(48 \times \$900\right) - \$30{,}638.30 = \textbf{\$12{,}561.70}$$

> **QUICK TIP**   Although the $900 withdrawals to Brandy are at the end of each quarter, the original lump sum must be deposited at the beginning of the first year.

EXAMPLE 3

Finding the Present Value

A French company hires a project manager to work in Saudi Arabia. The contract states that if the manager works there for 5 years he will receive an extra retirement benefit of $15,000 at the end of each semiannual period for the 8 years that follow. Find the lump sum that can be deposited today to satisfy the contract assuming 6% compounded semiannually.

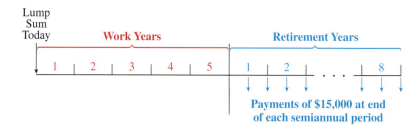

Present Value of an Annuity Table
n = number of periods; i = interest rate per period

n	1%	1½%	2%	2½%	3%	4%	5%	6%	8%	10%	12%	n
1	.99010	.98522	.98039	.97561	.97087	.96154	.95238	.94340	.92593	.90909	.89286	1
2	1.97040	1.95588	1.94156	1.92742	1.91347	1.88609	1.85941	1.83339	1.78326	1.73554	1.69005	2
3	2.94099	2.91220	2.88388	2.85602	2.82861	2.77509	2.72325	2.67301	2.57710	2.48685	2.40183	3
4	3.90197	3.85438	3.80773	3.76197	3.71710	3.62990	3.54595	3.46511	3.31213	3.16987	3.03735	4
5	4.85343	4.78264	4.71346	4.64583	4.57971	4.45182	4.32948	4.21236	3.99271	3.79079	3.60478	5
6	5.79548	5.69719	5.60143	5.50813	5.41719	5.24214	5.07569	4.91732	4.62288	4.35526	4.11141	6
7	6.72819	6.59821	6.47199	6.34939	6.23028	6.00205	5.78637	5.58238	5.20637	4.86842	4.56376	7
8	7.65168	7.48593	7.32548	7.17014	7.01969	6.73274	6.46321	6.20979	5.74664	5.33493	4.96764	8
9	8.56602	8.36052	8.16224	7.97087	7.78611	7.43533	7.10782	6.80169	6.24689	5.75902	5.32825	9
10	9.47130	9.22218	8.98259	8.75206	8.53020	8.11090	7.72173	7.36009	6.71008	6.14457	5.65022	10
11	10.36763	10.07112	9.78685	9.51421	9.25262	8.76048	8.30641	7.88687	7.13896	6.49506	5.93770	11
12	11.25508	10.90751	10.57534	10.25776	9.95400	9.38507	8.86325	8.38384	7.53608	6.81369	6.19437	12
13	12.13374	11.73153	11.34837	10.98318	10.63496	9.98565	9.39357	8.85268	7.90378	7.10336	6.42355	13
14	13.00370	12.54338	12.10625	11.69091	11.29607	10.56312	9.89864	9.29498	8.24424	7.36669	6.62817	14
15	13.86505	13.34323	12.84926	12.38138	11.93794	11.11839	10.37966	9.71225	8.55948	7.60608	6.81086	15
16	14.71787	14.13126	13.57771	13.05500	12.56110	11.65230	10.83777	10.10590	8.85137	7.82371	6.97399	16
17	15.56225	14.90765	14.29187	13.71220	13.16612	12.16567	11.27407	10.47726	9.12164	8.02155	7.11963	17
18	16.39827	15.67256	14.99203	14.35336	13.75351	12.65930	11.68959	10.82760	9.37189	8.20141	7.24967	18
19	17.22601	16.42617	15.67846	14.97889	14.32380	13.13394	12.08532	11.15812	9.60360	8.36492	7.36578	19
20	18.04555	17.16864	16.35143	15.58916	14.87747	13.59033	12.46221	11.46992	9.81815	8.51356	7.46944	20
21	18.85698	17.90014	17.01121	16.18455	15.41502	14.02916	12.82115	11.76408	10.01680	8.64869	7.56200	21
22	19.66038	18.62082	17.65805	16.76541	15.93692	14.45112	13.16300	12.04158	10.20074	8.77154	7.64465	22
23	20.45582	19.33086	18.29220	17.33211	16.44361	14.85684	13.48857	12.30338	10.37106	8.88322	7.71843	23
24	21.24339	20.03041	18.91393	17.88499	16.93554	15.24696	13.79864	12.55036	10.52876	8.98474	7.78432	24
25	22.02316	20.71961	19.52346	18.42438	17.41315	15.62208	14.09394	12.78336	10.67478	9.07704	7.84314	25
26	22.79520	21.39863	20.12104	18.95061	17.87684	15.98277	14.37519	13.00317	10.80998	9.16095	7.89566	26
27	23.55961	22.06762	20.70690	19.46401	18.32703	16.32959	14.64303	13.21053	10.93516	9.23722	7.94255	27
28	24.31644	22.72672	21.28127	19.96489	18.76411	16.66306	14.89813	13.40616	11.05108	9.30657	7.98442	28
29	25.06579	23.37608	21.84438	20.45355	19.18845	16.98371	15.14107	13.59072	11.15841	9.36961	8.02181	29
30	25.80771	24.01584	22.39646	20.93029	19.60044	17.29203	15.37245	13.76483	11.25778	9.42691	8.05518	30
31	26.54229	24.64615	22.93770	21.39541	20.00043	17.58849	15.59281	13.92909	11.34980	9.47901	8.08499	31
32	27.26959	25.26714	23.46833	21.84918	20.38877	17.87355	15.80268	14.08404	11.43500	9.52638	8.11159	32
33	27.98969	25.87895	23.98856	22.29188	20.76579	18.14765	16.00255	14.23023	11.51389	9.56943	8.13535	33
34	28.70267	26.48173	24.49859	22.72379	21.13184	18.41120	16.19290	14.36814	11.58693	9.60857	8.15656	34
35	29.40858	27.07559	24.99862	23.14516	21.48722	18.66461	16.37419	14.49825	11.65457	9.64416	8.17550	35
36	30.10751	27.66068	25.48884	23.55625	21.83225	18.90828	16.54685	14.62099	11.71719	9.67651	8.19241	36
37	30.79951	28.23713	25.96945	23.95732	22.16724	19.14258	16.71129	14.73678	11.77518	9.70592	8.20751	37
38	31.48466	28.80505	26.44064	24.34860	22.49246	19.36786	16.86789	14.84602	11.82887	9.73265	8.22099	38
39	32.16303	29.36458	26.90259	24.73034	22.80822	19.58448	17.01704	14.94907	11.87858	9.75696	8.23303	39
40	32.83469	29.91585	27.35548	25.10278	23.11477	19.79277	17.15909	15.04630	11.92461	9.77905	8.24378	40
41	33.49969	30.45896	27.79949	25.46612	23.41240	19.99305	17.29437	15.13802	11.96723	9.79914	8.25337	41
42	34.15811	30.99405	28.23479	25.82061	23.70136	20.18563	17.42321	15.22454	12.00670	9.81740	8.26194	42
43	34.81001	31.52123	28.66156	26.16645	23.98190	20.37079	17.54591	15.30617	12.04324	9.83400	8.26959	43
44	35.45545	32.04062	29.07996	26.50385	24.25427	20.54884	17.66277	15.38318	12.07707	9.84909	8.27642	44
45	36.09451	32.55234	29.49016	26.83302	24.51871	20.72004	17.77407	15.45583	12.10840	9.86281	8.28252	45
46	36.72724	33.05649	29.89231	27.15417	24.77545	20.88465	17.88007	15.52437	12.13741	9.87528	8.28796	46
47	37.35370	33.55319	30.28658	27.46748	25.02471	21.04294	17.98102	15.58903	12.16427	9.88662	8.29282	47
48	37.97396	34.04255	30.67312	27.77315	25.26671	21.19513	18.07716	15.65003	12.18914	9.89693	8.29716	48
49	38.58808	34.52468	31.05208	28.07137	25.50166	21.34147	18.16872	15.70757	12.21216	9.90630	8.30104	49
50	39.19612	34.99969	31.42361	28.36231	25.72976	21.48218	18.25593	15.76186	12.23348	9.91481	8.30450	50

SOLUTION

For clarity, number the years from 1 through 13. The project manager works years 1 through 5 then receives the annuity payments for years 6 through 13. First find the present value needed at the beginning of year 6 to fund the annuity using $\frac{6\%}{2} = 3\%$ per compounding period and $2 \times 8 = 16$ compounding periods in the present value of an annuity table.

Present value needed beginning of year 6 = $15,000 × **12.56110** = $188,416.50

Now find the present value needed at the beginning of year 1 to accumulate $188,416.50 by the end of year 5, which is the same as the beginning of year 6. Use the present value of a dollar table in Section 9.3 (page 382) with $\frac{6\%}{2} = 3\%$ per compounding period and $5 \times 2 = 10$ compounding periods.

Present value needed beginning of year 1 = $188,416.50 × **.74409** = $140,198.83

A lump sum of $140,198.83 deposited today will grow to $188,416.50 in 5 years, which will fund 16 semiannual payments of $15,000 each during the following 8 years.

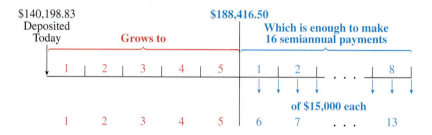

EXAMPLE 4

*Determining
Retirement Income*

Tish Baker plans to retire from nursing zat age 65 and hopes to withdraw $25,000 per year until she is 90. **(a)** If money earns 8% per year compounded annually, how much will she need at age 65? **(b)** If she deposits $2000 per year into her retirement plan beginning at age 32, and if the retirement plan earns 8% per year compounded annually, will her retirement account have enough for her to meet her goals?

SOLUTION

(a) The amount needed at age 65 is the present value of an annuity of $25,000 per year for $90 - 65 = 25$ years with interest of 8% compounded annually. The present value of an annuity table is used to find the following.

Present value = $25,000 × **10.67478** = $266,869.50

Baker will need $266,869.50 at age 65. This sum, at 8% compounded annually, will permit withdrawals of $25,000 per year until age 90.

(b) Baker makes payments of $2000 at the end of each year for $65 - 32 = 33$ years, at 8% compounded annually. These payments form a regular annuity. The amount of an annuity table in Section 10.1 is used to find the following.

Future value = $2000 × **145.95062** = $291,901.24

The value in the retirement account at 65 (**$291,901.24**) exceeds the amount needed to fund 25 yearly withdrawals of $25,000 each (**$266,869.50**). Therefore, Tish Baker will have more than enough money.

QUICK TIP   Example 7 in Appendix D shows how a financial calculator can be used to solve a problem similar to Example 4.

Objective $\boxed{3}$ **Find the equivalent cash price of an ordinary annuity.** Does 120 monthly payments of $1000 each have the same value as $120,000 in your hands today? No. An investment of $120,000 today will grow to substantially more than an investment of $1000 per month for the next 120 months. Therefore $120,000 today has more value than $1000 a month for 120 months.

In general, payments made at different times (for example, in different years) cannot be directly compared. To compare payments, first find the present value of each payment and then compare the present values to find the best alternative. The next example shows how this is done.

EXAMPLE **5**

Comparing Methods of Investment

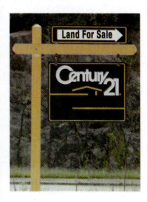

Jean Braddock is offering to sell a piece of property to two different real estate developers. Kapton Homes offers $200,000 in cash today for the land. RealProperty offers $80,000 now as a down payment and payments of $10,000 at the end of each quarter for 4 years. Assume that money can be invested at 8% per year compounded quarterly. Which offer should Braddock accept?

SOLUTION

Since payments from the two real estate developers occur over different time periods, the present value of each offer must be found to determine which is better.

The present value of Kapton Homes' offer is $200,000, since that payment is made now.

The present value of RealProperty's offer is the sum of the down payment plus the present value of the series of payments. Use $\frac{8\%}{4} = 2\%$ per compounding period for 4 years $\times$ 4 quarters per year $= 16$ compounding periods. Use the present value of an annuity table to find **13.57771**.

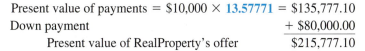

$$
\begin{array}{ll}
\text{Present value of payments} = \$10,000 \times \textbf{13.57771} = \$135,777.10 \\
\text{Down payment} \hspace{4.5cm} + \$80,000.00 \\
\hline
\text{Present value of RealProperty's offer} \hspace{1.8cm} \$215,777.10
\end{array}
$$

Therefore, $215,777.10 is the **equivalent cash price** to $80,000 down plus 16 quarterly payments of $10,000 each. RealProperty's offer is the better of the two by the following amount.

$$\$215,777.10 - \$200,000 = \$15,777.10$$

10.2 | Exercises

The **Quick Start** *exercises in each section contain solutions to help you get started.*

Find the present value of the following annuities. Round to the nearest cent. (See Examples 1–3.)

Quick Start

	Amount per Payment	Payment at End of Each	Time (Years)	Rate of Investment	Compounded	Present Value
1.	$1800	year	18	10%	annually	$14,762.54
	Present value = $1800 × 8.20141 = $14,762.54					
2.	$4100	year	7	6%	annually	$22,887.76
	Present value = $4100 × 5.58238 = $22,887.76					
3.	$2000	6 months	12	8%	semiannually	_____
4.	$1700	6 months	14	5%	semiannually	_____
5.	$894	quarter	6	4%	quarterly	_____
6.	$7500	quarter	5	10%	quarterly	_____

7. Explain the difference between the two ways to think of the present value of an annuity. (See Objective 1.)

8. Explain the meaning of equivalent cash price. (See Objective 3.)

Solve the following application problems. Round to the nearest cent.

Quick Start

9. **INJURY LAWSUIT**   The court ruled that Bakon Corporation was liable in the death of an employee. The settlement called for the company to pay the employee's widow $65,000 at the end of each year for 20 years. Find the amount the company must set aside today to satisfy this annuity assuming 5% compounded annually.

   Present value of annuity = $65,000 × 12.46221 = $810,043.65

   9. $810,043.65

10. **CARE OF THE DISABLED**   Mr. Roberts wishes to set up a 10-year annuity with quarterly payments of $8000 for the care of his disabled father. Find the amount he should deposit today at 6% interest compounded quarterly.

    10. _____

11. **TRACTOR REPLACEMENT**   Evergreen Farms, Inc. will need a new tractor in 3 years and has been told to deposit $3000 at the end of each quarter for 3 years to accumulate the needed funds. What lump sum deposited today will generate the same future value at 8% interest compounded quarterly?

    11. _____

12. The community college where Roman Rodriguez works sets aside an annual payment of $35,000 per year for 5 years so they will have funds to replace the personal computers, servers, and printers in their computer labs when needed. Assuming 5% compounded annually, what lump sum deposited today would result in the same future value?

    12. _____

 indicates an exercise that is related to the Case in Point feature within the section.

13. **COLLEGE EXPENSES**   Jason Clendenen needs $7000 every 6 months for living expenses and tuition at the University of Texas at Austin. As an engineering major, it will take him 5 years to earn his college degree. Find **(a)** the lump sum that must be deposited to meet this need and **(b)** the interest earned assuming 8% per year, compounded semiannually.

(a) _____
(b) _____

14. **DISASTER RELIEF**   After a terrible flood in Bangladesh, an international disaster relief organization agreed to help support families in a small city who lost everything with a payment of $25,000 every quarter for 5 years. Find **(a)** the lump sum that must be deposited to meet this need and **(b)** the interest earned assuming 6% per year, compounded quarterly.

(a) _____
(b) _____

15. **PAYING FOR COLLEGE**   Tom Potter estimates that his daughter's college needs, beginning in 8 years, will be $3600 at the end of each quarter for 4 years. **(a)** Find the total amount needed in 8 years assuming 8% compounded quarterly. **(b)** Will he have enough money available in 8 years if he invests $700 at the end of each quarter for the next 8 years at 8% compounded quarterly?

(a) _____
(b) _____

16. **VAN PURCHASE**   In 4 years, Jennifer Videtto will need a delivery van, for her office-supply store, that will require a down payment of $10,000 with payments of $1200 per month for 36 months. **(a)** Find the total amount needed in 4 years assuming 12% compounded monthly. **(b)** Will she have enough money available if she invests $1000 at the end of each month for the next 4 years at 12% compounded monthly?

(a) _____
(b) _____

17. **SELLING A BUSINESS**   Anna Stanley has two offers for her business. The first offer is a cash payment of $85,000 and the second is a down payment of $25,000 with payments of $3500 at the end of each quarter for 5 years. **(a)** Identify the better offer assuming 8% compounded quarterly. **(b)** Find the difference in the present values.

(a) _____
(b) _____

18. **GROCERY STORE**   Adolf Hegman has two offers for his Canadian grocery company. The first offer is a cash payment of $540,000, and the second is a down payment of $240,000 with payments of $65,000 at the end of each semiannual period for 4 years. **(a)** Identify the better offer assuming 10% compounded semiannually. **(b)** Find the difference in the present values.

(a) _____
(b) _____

19. **SEVERANCE PAY**   A company plans to lay off several workers in 3 years when a new automated assembly line is introduced. At that time, it anticipates a severance pay for the laid-off workers totaling $22,000 per month for 2 years. Find the shortage in 3 years if the company deposits $6000 per month for 36 months. Assume a rate of 12% compounded monthly.

19. _____

20. **RETIRING A MANAGER**   A commercial farmer plans to retire his manager in 5 years and will then pay him $10,000 at the end of each semiannual period for 15 years. Find the shortage in 5 years if the farmer deposits $12,000 at the end of each semiannual period for the next 5 years. Assume 8% compounded semiannually.

20. _____

10.3 | Sinking Funds (Finding Annuity Payments)

Objectives

1 Understand the basics of a sinking fund.
2 Set up a sinking fund table.

CASE *in* POINT

Roman Rodriguez works at a community college. The president of the college has decided to set up a sinking fund to accumulate funds needed in 5 years for a new building that will include a gymnasium and an indoor 50-meter swimming pool.

Objective 1 Understand the basics of a sinking fund. Individuals and businesses often need to raise a certain amount of money for use *at some fixed time in the future*. For example, Paul Pence needs $28,000 to purchase a truck in 3 years. Using 8% compounded quarterly and the amount of an annuity table in Section 10.1, one can guess the required payment at the end of each quarter needed to accumulate the $28,000:

Guess of quarterly payment	*From table*	*Future value*	*The guess is*
$1500	$1500 × **13.41209** = $20,118.14		too low
$2800	$2800 × **13.41209** = $37,553.85		too high

Clearly this method is awkward. The exact payment in this example can be found by dividing the future value of $28,000 by **13.41209**, or by using the table and methods provided in this section. In summary, this section shows how to find the periodic payment needed to achieve a specific future value at a specific date.

A fund set up to receive periodic payments is called a **sinking fund**. Sinking funds are used to provide money *to pay off a loan* in one lump sum *or to accumulate money* to build new factories, buy equipment, and so on. Large corporations and some government agencies use a form of debt called a **bond**, which is a promise to pay a fixed amount of money at some stated time in the future. Bonds are discussed in detail in **Section 10.5**. This section covers only the use of a sinking fund to pay off a bond when it is due.

The amount of the periodic payment needed, at the end of each period, to accumulate a fixed amount at a future date is found as follows.

> Payment = Future value × Number from sinking fund table

EXAMPLE 1

Finding Periodic Payments

The president of a community college wants to set up a sinking fund to accumulate funds needed in 5 years for a new sports complex. The cost that includes a gymnasium and an indoor 50-meter swimming pool is estimated to be $16,500,000. The school board decides to make end of each quarter deposits into a fund earning 6% compounded quarterly. Find **(a)** the amount of each quarterly payment and **(b)** the interest earned.

SOLUTION

(a) Use $\frac{6\%}{4}$ = 1.5% per compounding period for 4 × 5 years = 20 compounding periods in the sinking fund table on page 413 to find **.04325**.

$$\text{Quarterly payment} = \$16,500,000 \times \mathbf{.04325} = \mathbf{\$713,625}$$

Twenty end-of-quarter payments of $713,625 at 6% compounded quarterly will grow to $16,500,000.

(b) Interest is the future value minus the payments.

$$\text{Interest} = \$16,500,000 - \left(20 \times \mathbf{\$713,625}\right) = \$2,227,500$$

EXAMPLE 2

Finding Periodic Payments

Toys "R" Us Inc. sold $100,000 worth of bonds that must be paid off in 8 years. They now must set up a sinking fund to accumulate the necessary $100,000 to pay off their debt. Find the amount of each payment into a sinking fund if the payments are made at the end of each year and the fund earns 10% compounded annually. Find the amount of interest earned.

SOLUTION

Look along the top of the sinking fund table for 10% and down the side for 8 periods to find **.08744**.

$$\text{Payment} = \$100,000 \times .08744 = \$8744$$

Toys "R" Us must deposit $8744 at the end of each year for 8 years in an account paying 10% compounded annually to accumulate $100,000. The interest earned is the future value less all payments.

$$\text{Interest} = \$100,000 - (8 \times \$8744) = \$30,048$$

> **QUICK TIP** The interest rate a company earns on sinking fund investments frequently differs from the interest rate they must pay on debts such as bonds.

Objective 2 Set up a sinking fund table. A **sinking fund table** is used to show the interest earned and the accumulated amount of a sinking fund at the end of each period.

EXAMPLE 3

Setting up a Sinking Fund Table

Toys "R" Us in Example 2 deposited $8744 at the end of each year for 8 years in a sinking fund that earned 10% compounded annually. Set up a sinking fund table for these deposits.

SOLUTION

The sinking fund account contains no money until the end of the first year, when a single deposit of $8744 is made. Since the deposit is made at the end of the year, no interest is earned.

At the end of the second year, the account contains the original $8744, plus the interest earned by this money. This interest is found by the formula for simple interest.

$$I = \$8744 \times .10 \times 1 = \$874.40$$

An additional deposit is also made at the end of the second year, so that the sinking fund then contains the following total.

$$\$8744 + \$874.40 + \$8744 = \$18,362.40$$

Continue this work to get the following sinking fund table.

| | Beginning of Period | | End of Period | |
Period	Accumulated Amount	Periodic Deposit	Interest Earned	Accumulated Amount
1	$0	$8744.00	$0	$8744.00
2	$8744.00	$8744.00	$874.40	$18,362.40
3	$18,362.40	$8744.00	$1836.24	$28,942.64
4	$28,942.64	$8744.00	$2894.26	$40,580.90
5	$40,580.90	$8744.00	$4058.09	$53,382.99
6	$53,382.99	$8744.00	$5338.30	$67,465.29
7	$67,465.29	$8744.00	$6746.53	$82,955.82
8	$82,955.82	$8748.60	$8295.58	$100,000.00

Sinking Fund Table
n = number of periods; *i* = interest rate per period

i / n	1%	1½%	2%	2½%	3%	4%	5%	6%	8%	10%	12%	i / n
1	1.00000	1.00000	1.00000	1.00000	1.00000	1.00000	1.00000	1.00000	1.00000	1.00000	1.00000	1
2	.49751	.49628	.49505	.49383	.49261	.49020	.48780	.48544	.48077	.47619	.47170	2
3	.33002	.32838	.32675	.32514	.32353	.32035	.31721	.31411	.30803	.30211	.29635	3
4	.24628	.24444	.24262	.24082	.23903	.23549	.23201	.22859	.22192	.21547	.20923	4
5	.19604	.19409	.19216	.19025	.18835	.18463	.18097	.17740	.17046	.16380	.15741	5
6	.16255	.16053	.15853	.15655	.15460	.15076	.14702	.14336	.13632	.12961	.12323	6
7	.13863	.13656	.13451	.13250	.13051	.12661	.12282	.11914	.11207	.10541	.09912	7
8	.12069	.11858	.11651	.11447	.11246	.10853	.10472	.10104	.09401	.08744	.08130	8
9	.10674	.10461	.10252	.10046	.09843	.09449	.09069	.08702	.08008	.07364	.06768	9
10	.09558	.09343	.09133	.08926	.08723	.08329	.07950	.07587	.06903	.06275	.05698	10
11	.08645	.08429	.08218	.08011	.07808	.07415	.07039	.06679	.06608	.05396	.04842	11
12	.07885	.07668	.07456	.07249	.07046	.06655	.06283	.05928	.05270	.04676	.04144	12
13	.07241	.07024	.06812	.06605	.06403	.06014	.05646	.05296	.04652	.04078	.03568	13
14	.06690	.06472	.06260	.06054	.05853	.05467	.05102	.04758	.04130	.03575	.03087	14
15	.06212	.05994	.05783	.05577	.05377	.04994	.04634	.04296	.03683	.03147	.02682	15
16	.05794	.05577	.05365	.05160	.04961	.04582	.04227	.03895	.03298	.02782	.02339	16
17	.05426	.05208	.04997	.04793	.04595	.04220	.03870	.03544	.02963	.02466	.02046	17
18	.05098	.04881	.04670	.04467	.04271	.03899	.03555	.03236	.02670	.02193	.01794	18
19	.04805	.04588	.04378	.04176	.03981	.03614	.03275	.02962	.02413	.01955	.01576	19
20	.04542	.04325	.04116	.03915	.03722	.03358	.03024	.02718	.02185	.01746	.01388	20
21	.04303	.04087	.03878	.03679	.03487	.03128	.02800	.02500	.01983	.01562	.01224	21
22	.04086	.03870	.03663	.03465	.03275	.02920	.02597	.02305	.01803	.01401	.01081	22
23	.03889	.03673	.03467	.03270	.03081	.02731	.02414	.02128	.01642	.01257	.00956	23
24	.03707	.03492	.03287	.03091	.02905	.02559	.02247	.01968	.01498	.01130	.00846	24
25	.03541	.03326	.03122	.02928	.02743	.02401	.02095	.01823	.01368	.01017	.00750	25
26	.03387	.03173	.02970	.02777	.02594	.02257	.01956	.01690	.01251	.00916	.00665	26
27	.03245	.03032	.02829	.02638	.02456	.02124	.01829	.01570	.01145	.00826	.00590	27
28	.03112	.02900	.02699	.02509	.02329	.02001	.01712	.01459	.01049	.00745	.00524	28
29	.02990	.02778	.02578	.02389	.02211	.01888	.01605	.01358	.00962	.00673	.00466	29
30	.02875	.02664	.02465	.02278	.02102	.01783	.01505	.01265	.00883	.00608	.00414	30
31	.02768	.02557	.02360	.02174	.02000	.01686	.01413	.01179	.00811	.00550	.00369	31
32	.02667	.02458	.02261	.02077	.01905	.01595	.01328	.01100	.00745	.00497	.00328	32
33	.02573	.02364	.02169	.01986	.01816	.01510	.01249	.01027	.00685	.00450	.00292	33
34	.02484	.02276	.02082	.01901	.01732	.01431	.01176	.00960	.00630	.00407	.00260	34
35	.02400	.02193	.02000	.01821	.01654	.01358	.01107	.00897	.00580	.00369	.00232	35
36	.02321	.02115	.01923	.01745	.01580	.01289	.01043	.00839	.00534	.00334	.00206	36
37	.02247	.02041	.01851	.01674	.01511	.01224	.00984	.00786	.00492	.00303	.00184	37
38	.02176	.01972	.01782	.01607	.01446	.01163	.00928	.00736	.00454	.00275	.00164	38
39	.02109	.01905	.01717	.01544	.01384	.01106	.00876	.00689	.00419	.00249	.00146	39
40	.02046	.01843	.01656	.01484	.01326	.01052	.00828	.00646	.00386	.00226	.00130	40
41	.01985	.01783	.01597	.01427	.01271	.01002	.00782	.00606	.00356	.00205	.00116	41
42	.01928	.01726	.01542	.01373	.01219	.00954	.00739	.00568	.00329	.00186	.00104	42
43	.01873	.01672	.01489	.01322	.01170	.00909	.00699	.00533	.00303	.00169	.00092	43
44	.01820	.01621	.01439	.01273	.01123	.00866	.00662	.00501	.00280	.00153	.00083	44
45	.01771	.01572	.01391	.01227	.01079	.00826	.00626	.00470	.00259	.00139	.00074	45
46	.01723	.01525	.01345	.01183	.01036	.00788	.00593	.00441	.00239	.00126	.00066	46
47	.01677	.01480	.01302	.01141	.00996	.00752	.00561	.00415	.00221	.00115	.00059	47
48	.01633	.01437	.01260	.01101	.00958	.00718	.00532	.00390	.00204	.00104	.00052	48
49	.01591	.01396	.01220	.01062	.00921	.00686	.00504	.00366	.00189	.00095	.00047	49
50	.01551	.01357	.01182	.01026	.00887	.00655	.00478	.00344	.00174	.00086	.00042	50

> **QUICK TIP** The last payment differs by $4.60 due to rounding. The final accumulated amount must equal $100,000.

Frequently, an item costs more if its purchase is delayed a few years. The next example shows how to estimate the cost of a large purchase at a future date, and then how to find the payment needed to accumulate the necessary funds.

EXAMPLE 4

Finding Periodic Payments and Interest Earned

Lee Bareli manages a shopping mall that uses one large central air-conditioning unit for the entire mall. She estimates that the air-conditioning unit will need to be replaced in 4 years. Today, the unit would cost $850,000 and she has been told that its cost will increase at 5% per year. Assuming she can earn 8% per year compounded quarterly in a sinking fund, find the quarterly payments needed to accumulate the funds for the purchase.

SOLUTION

First, find the cost of the air-conditioning unit in 4 years. Use 5% per compounding period and 4 compounding periods in the compound interest table on page 000 to find **1.21551**.

$$\text{Cost in 4 years} = \$850{,}000 \times \textbf{1.21551} = \$1{,}033{,}183.50$$

This is the amount that must be accumulated in the sinking fund. The required quarterly payment is found using 4 years $\times$ 4 quarters per year = 16 compounding periods and $\frac{8\%}{4}$ = 2% per compounding period in the sinking fund table for a value of **.05365**.

$$\text{Quarterly payment} = \$1{,}033{,}183.50 \times \textbf{.05365} = \$55{,}430.29 \text{ (rounded)}$$

Payments of $55,430.29 at the end of each quarter for 4 years will result in the needed funds to purchase the air-conditioning unit.

Two different interest rates are involved in Example 4. The price is increasing at 5% per year compounded annually, but deposits in the sinking fund earn 8% compounded quarterly. **Interest rate spreads** such as this are common in business. For example, banks use an interest rate spread between what they pay for funds on deposit and what they charge on loans to customers.

10.3 | Exercises

FOR EXTRA HELP

 MyMathLab

 InterActMath.com

 MathXP MathXL

 MathXL Tutorials on CD

 Tutor Center Addison-Wesley Math Tutor Center

 DVT/Videotape

The **Quick Start** exercises in each section contain solutions to help you get started.

Find the amount of each payment needed to accumulate the indicated amount in a sinking fund. Round to the nearest cent. (See Examples 1–3.)

Quick Start

1. $12,000, money earns 5% compounded annually, 4 years
   Payment = $12,000 × .23201 = $2784.12

   1. **$2784.12**

2. $125,000, money earns 6% compounded annually, 25 years
   Payment = $125,000 × .01823 = $2278.75

   2. **$2278.75**

3. $8200, money earns 6% compounded semiannually, 5 years

   3. _____

4. $12,000, money earns 10% compounded semiannually, 3 years

   4. _____

5. $50,000, money earns 4% compounded quarterly, 5 years

   5. _____

6. $32,000, money earns 6% compounded quarterly, 3 years

   6. _____

7. $7894, money earns 12% compounded monthly, 3 years

   7. _____

8. $29,804, money earns 12% compounded monthly, 2 years

   8. _____

9. Explain the difference between a sinking fund (see Objective 1) and the present value of an annuity discussed in **Section 10.2**.

10. What is a sinking fund table? Who would use one? (See Objective 2.)

Solve each application problem. Round to the nearest cent.

Quick Start

11. A community college needs $920,000 in 3 years to remodel their student union. They decide to make payments into a sinking fund at the end of each semiannual period. **(a)** Find the amount of each payment assuming 5% per year compounded semiannually. **(b)** Find the total interest earned.
    (a) Payment = $920,000 × .15655 = $144,026
    (b) Interest = $920,000 − (6 × $144,026) = $55,844

    (a) **$144,026**
    (b) **$55,844**

12. SCUBA DIVING   The owner of Emerald Diving plans to buy all new scuba diving equipment to rent to divers in 5 years at a cost of $34,000. He believes that he can earn 8% compounded quarterly. Find **(a)** the amount of each of the quarterly payments needed and **(b)** the total interest earned.

    (a) _____
    (b) _____

indicates an exercise that is related to the Case in Point feature within the section.

13. ACCUMULATING $1 MILLION   Jessica Smith wants to know if she can accumulate $1,000,000 in her lifetime. She feels that she can earn 10% per year for 50 years if she invests in stocks. Find **(a)** the amount of each of the annual payments needed and **(b)** the total interest earned.

(a) _____
(b) _____

14. ALLIGATOR HUNTING   Cajun Jack needs $45,000 in 4 years for boats used to hunt alligators. **(a)** Find the amount of each payment if payments are made at the end of each quarter with interest at 6% compounded quarterly. **(b)** Find the total amount of interest earned.

(a) _____
(b) _____

15. NEW MACHINERY   Smith Dry Cleaning must buy a new cleaning machine in 7 years for $120,000. The firm sets up a sinking fund for this purpose. Find the payment into the fund at the end of each year if money in the fund earns 10% compounded annually.

15. _____

16. NEW AUDITORIUM   The membership of the Green Fields Baptist Church is large and growing rapidly. The leaders of the church are planning to build a new auditorium with special features for their televised broadcasts at a cost of $2,800,000 in 5 years. The membership has set up a sinking fund with the idea of making a payment at the end of each quarter. Find the payment needed if money earns 8% compounded quarterly.

16. _____

17. A NEW SHOWROOM   A Ford dealership wants to build a new showroom costing $2,300,000. It set up a sinking fund with end-of-the-month payments in an account earning 12% compounded monthly. Find the amount that should be deposited in this fund each month if the dealership wishes to build the showroom in **(a)** 3 years and **(b)** 4 years.

(a) _____
(b) _____

18. AIRPORT IMPROVEMENTS   A city near Chicago sold $9,000,000 in bonds to pay for improvements to an airport. It sets up a sinking fund with end-of-the-quarter payments in an account earning 8% compounded quarterly. Find the amount that should be deposited in this fund each quarter if the city wishes to pay off the bonds in **(a)** 7 years and **(b)** 12 years.

(a) _____
(b) _____

19. LAND SALE   Helen Spence sells some land in Nevada. She will be paid a lump sum of $60,000 in 4 years. Until then, the buyer pays 8% simple interest every quarter. **(a)** Find the amount of each quarterly interest payment. **(b)** The buyer sets up a sinking fund so that enough money will be present to pay off the $60,000. The buyer wants to make semiannual payments into the sinking fund. The account pays 8% compounded semiannually. Find the amount of each payment into the fund. **(c)** Prepare a table showing the amount in the sinking fund after each deposit.

(a) _____
(b) _____

Payment Number	Amount of Deposit	Interest Earned	Total in Account

20. RARE STAMPS Jeff Reschke bought a rare stamp for his collection. He agreed to pay a lump sum of $4000 after 5 years. Until then, he pays 6% simple interest every 6 months. **(a)** Find the amount of each semiannual interest payment. **(b)** Reschke sets up a sinking fund so that money will be present to pay off the $4000. He wants to make annual payments into the fund. The account pays 8% compounded annually. Find the amount of each payment into the fund. **(c)** Prepare a table showing the amount in the sinking fund after each deposit.

(a) _____
(b) _____

Payment Number	Amount of Deposit	Interest Earned	Total in Account

21. SPORTS COMPLEX Prepare a sinking fund table for the first 4 payments for the community college sports complex described in Example 1.

Payment Number	Amount of Deposit	Interest Earned	Total in Account

22. COMMERCIAL BUILDING Joan Miller plans to make a down payment of $70,000 on a commercial building for her plumbing company in 5 years. Construct a sinking fund table given semiannual payments of $6106.10 at the end of each period and an interest rate of 6% compounded semiannually.

Payment Number	Amount of Deposit	Interest Earned	Total in Account

Supplementary Application Exercises on Annuities and Sinking Funds

Solve the following application problems. Round to the nearest cent.

Quick Start

1. Bill Carrier deposits $500 at the end of each quarter for 6 years in a mutual fund that he believes will grow at 8% compounded quarterly. Find **(a)** the future value and **(b)** the interest.

   (a) Future value = $500 × 30.42186 = $15,210.93
   (b) Interest = $15,210.93 − (24 × $500) = $3210.93

   (a) $15,210.93
   (b) $3210.93

2. For 6 years, Jessica Savage deposits $1000 at the end of each quarter into an account earning 8% per year compounded quarterly. Find **(a)** the future value and **(b)** the interest.

   (a) _____
   (b) _____

3. Mr. and Mrs. Thompson deposit $2000 at the beginning of each year for 20 years into a retirement account earning 6% compounded annually. Find **(a)** the future value and **(b)** the interest.

   (a) _____
   (b) _____

4. Ben Chavez opens a Roth IRA, and he deposits $500 at the beginning of each quarter for 10 years. If the funds are invested at 10% interest per year compounded quarterly, find **(a)** the future value and **(b)** the interest.

   (a) _____
   (b) _____

5. Solectron needs to purchase new equipment for its production line in 3 years. The company has been advised to deposit $135,000 at the end of each quarter into an account which managers believe will yield 10% per year compounded quarterly. Find the lump sum that could be deposited today that will grow to the same future value.

   5. _____

6. Abel Plumbing saves $12,000 at the end of every semiannual period in an account earning 6% compounded semiannually to replace several of its trucks in 5 years. Find the lump sum that could be deposited today that will grow to the same future value.

   6. _____

7. Kitty Wysong was in a car accident and needs $3200 per month to live on for the next 3 years. Find the lump sum her insurance company must deposit in a fund earning 12% per year compounded monthly to make these payments.

   7. _____

8. Carl and Amy Glaser recently divorced. As part of the divorce settlement, Carl must pay Amy $1000 at the end of every quarter for 8 years. Find the lump sum he must deposit in an account earning 8% per year compounded quarterly to make the payments.

   8. _____

9. Ajax Coal sets up a sinking fund to purchase a new industrial bulldozer in 3 years at a price of $870,000. Find the annual payment the firm must make if funds are deposited in an account earning 8% compounded annually. Then set up a sinking-fund table.

Payment Number	Amount of Deposit	Interest Earned	Total in Account

10. Swift Petrochemicals wishes to purchase a new corporate jet costing $3,200,000 in 2 years. Find the semiannual payment the company must make into a sinking fund account earning 6% compounded semiannually. Then set up a sinking fund table.

Payment Number	Amount of Deposit	Interest Earned	Total in Account

10.4 | Stocks

Objectives

1. Define the types of stock.
2. Read stock tables.
3. Find the current yield on a stock.
4. Find the stock's PE ratio.
5. Define the Dow Jones Industrial Average and the Nasdaq Composite Index.
6. Define a mutual fund.

CASE *in* POINT

When Roman Rodriguez began his new job, he was given the choice of investing his retirement funds in a fixed interest fund or in a fund containing stocks and/or bonds. Which should he choose? Which would you choose? Why?

Almost all large businesses and also many smaller ones are set up as **corporations**. For example, companies that we refer to as Microsoft, Intel, Nike, McDonald's, and General Motors are actually corporations. **Publicly held** corporations are those owned by the public; their stocks are traded daily in markets called stock markets. **Privately held** corporations are owned by one or a few individuals, and their stock is not traded on a market. For example, your medical doctor or plumber may have organized a small business as a privately held corporation.

A corporation is a form of business that gives the owners (the stockholders) **limited liability**. You do not have to worry that lawsuits will be filed against you or your family just because you own stock in General Motors, Inc. The owners of corporations have protection through the limited-liability laws and can never lose more than they have invested in the corporation.

A corporation is set up with money, or **capital**, raised through the sale of shares of **stock**. A share of stock represents partial ownership of a corporation. If one million shares of stock are sold to establish a new firm, the owner of one share will own one-millionth of the corporation. The ownership of stock is shown by either paper **stock certificates**, like the one shown here, or by regular statements that are generated by the brokerage firm who bought the stock on your behalf.

In most states, corporations are required to have an **annual meeting**. At this meeting, open to all **stockholders** (owners of stock), the management of the firm is open to questions from stockholders. The stockholders also elect a **board of directors**—a group of people who represent the stockholders. The board of directors hires the **executive officers** of the corporation, such as the president, vice-presidents, and so on. The board of directors also distributes a portion of any profits in the form of **dividends** that are paid to the stockholders.

Objective 1 Define the types of stock. The two types of stock normally issued are **preferred stock** and **common stock**. As the name suggests, preferred stock *has certain rights* over common stock. For example, owners of preferred stock must be paid dividends *before* any dividends can be paid to owners of common stock. Also, corporate debt and preferred shareholders must be paid *before* common shareholders receive anything in the event that a corporation declares bankruptcy.

The shares of **publicly held corporations** are typically owned by many different individuals and institutions. Share prices of these firms are determined by supply and demand in public markets called **stock exchanges**. The New York Stock Exchange (NYSE) is the largest of the several exchanges in the United States. This exchange is located on Wall Street in New York City. Most foreign countries, including Japan, Taiwan, England, Canada, and Mexico, have their own stock exchanges. One of the most widely circulated financial newspapers in the country is the *Wall Street Journal*, which is published daily by Dow Jones & Company, Inc. It provides the reader with stock and bond quotes in addition to presenting general business, marketing, and financial news.

Objective 2 Read stock tables. Daily stock prices can easily be found on the World Wide Web, but many newspapers also print daily stock prices for at least some stocks. The format of the information given varies from one source of information to another, but we will be using the format used by the Wall Street Journal.

> **QUICK TIP** Historically, fractions were used for stock prices. That changed in 2001, and decimal numbers are now used.

EXAMPLE 1

Reading the Stock Table

After receiving his first paycheck from the community college, Roman Rodriguez went to his favorite store, Best Buy. At nearly 6 years old, his personal computer at home was ancient, and he needed to replace it. Analyze the information in the Wall Street Journal about Best Buy.

YTD % CHG	52-WEEK HI	LO	STOCK (SYM)	DIV	YLD %	PE	VOL 100s	CLOSE	NET CHG
116.1	62.70	22.48	BestBuy BBY	.40	.8	31	48286	52.19	1.63
(a)	(b)	(c)	(d)	(e)	(f)	(g)	(h)	(i)	(j)

SOLUTION

(a) The year-to-date increase in the price of Best Buy stock was 116.1%.
(b) The highest price the stock sold for during the year was $62.70.
(c) The lowest price the stock sold for during the year was $22.48.
(d) The stock symbol for Best Buy is BBY.
(e) Best Buy paid a dividend of $.40 per share during the year.
(f) The dividend of $.40 per share was .8% (rounded) of the closing price of $52.19.
(g) The ratio of the closing stock price for the day to the company earnings (PE) was 31.
(h) The number of Best Buy shares that traded that day was $100 \times 48{,}286 = 4{,}828{,}600$.
(i) The last price at which Best Buy shares traded that day was $52.19.
(j) Best Buy shares went up $1.63 on that day.

YTD % CHG	52-WEEK HI	LO	STOCK (SYM)	DIV	YLD %	PE	VOL 100s	CLOSE	NET CHG
41.6	23.63	11.37	AaronRent RNT s	.04f	.2	20	1436	20.65	−0.39
116.8	5.30	1.53	ABB ADS ABB s	1.23e	24.6	dd	1779	5.01	−0.05
▲17.9	47	33.75	Abbottlab ABT	.98	2.1	30	31676	47.15	0.31
20.8	33.65	19.36	Abercrombie A ANF	...		12	12219	24.71	0.54
0.8	8.54	6.10	Abitibi ABY	.10g	...		2224	7.77	0.06
69.1	12.68	7.40 ♣	AcadiaRity AKR x	.64f	5.1	21	283	12.55	0.14
46.2	26.35	13.45	Accenture ACN	...		25	11606	26.31	0.29
3.1	27.71	14.65	ActionPerf ATN	.20	1.0	15	1775	19.58	0.23
▲58.4	35.78	16.25	Actuant A ATU s	...		33	1863	36.80	1.25
▲93.1	25.39	12.24	Acuity Br AYI	.60	2.3	22	2411	26.14	1.29
64.8	16.55	6.35	Adecco ADO	.11e	.7	...	212	15.72	0.24
▲200.0	17.90	4.42	Admnstaff ASF	...		82	3035	18	0.71
▲18.4	26.70	21.55	AFP Prov ADS PVD	1.70e	6.2	...	348	27.33	0.86
63.7	83.65	36.99	AdvanceAuto AAP	...		29	1710	80.03	0.64
−26.5	15.18	9.35	AdvMktg MKT	.04	.4	20	997	10.81	0.02
67.7	20.67	11.30	AdvMedOp AVO	...		...	2764	20.07	0.14
136.5	18.50	4.73	AdvMicro AMD	...		dd	90636	15.28	0.58
84.2	5.51	2.23	AdSemEg ADS ASX s	stk		...	1302	4.89	−0.05
83.0	20.41	7.86	Advntst ADS ATE	.05e	.3	...	181	19.64	0.48
48.4	33.57	19.67	ADVO AD s	.11	.3	20	2148	32.49	0.44
▲18.3	14.29	6.44 ♣	Aegon AEG	.23e	1.6	12	4332	14.59	0.59
160.2	34.70	9.64	Aeropostale ARO	...		24	4344	27.50	0.32
62.1	70.25	39.90	Aetna AET	.04	.1	13	5783	66.67	0.91
109.5	25.29	11.51	AFCCapTrl Corts KRH	1.94	7.7	...	2	25.24	−0.01
3.9	56.56	40.01	AffiliCmptr A ACS	...		24	7227	54.68	0.60
38.8	73.11	36.52	AffilMangr AMG	...		27	1961	69.82	1.30
106.9	4.45	1.22	AgereSys A AGRA	...		dd	27052	2.98	0.01
100.0	3.75	1.19	AgereSys B AGRB	...		dd	23981	2.80	0.03
59.2	29.35	11.30	AgllentTch A	...		dd	14171	28.59	0.64
−18.2	16.47	9.72 ♣	AgnicoEgl AEM	.03g	.2	dd	9447	12.15	0.41
67.8	28.58	16.76	AgreeRfty ADC	1.94	6.8	14	188	28.35	0.21
45.1	16.51	9.40	Agrium AGU	.11g	.7	26	2257	16.41	0.08
−35.0	13.24	2.48	Ahold ADS AHO	.70e	9.3	...	6006	7.56	0.12
−25.4	5.40	2.05	AirNetSys ANS	...		22	513	3.67	−0.03
▲24.1	52.84	36.97	AirProduct APD	.92	1.7	30	5562	53.07	0.76
24.9	21.70	15.27	Airgas ARG	.12e	.6	22	3121	21.55	0.17
206.2	20.84	3.70	AirTranHldg AAI	...		11	11443	11.94	−0.07
25.5	15.81	11.25	AlamoGp ALG	.24	1.6	25	20	15.37	0.25
148.2	18.70	6	AlarisMed AMI	...		dd	1648	15.14	0.14
29.1	31.86	15.28	AlaskaAir ALK	...		dd	1228	27.94	0.29
▲64.2	33.65	19.95 ♣	AlbanyInt AIN	.28	.8	18	648	33.93	0.60
▲6.9	30.56	22.10	Albemarle ALB	.58f	1.9	15	1354	30.42	−0.02
25.4	64.40	47.08	AlbertoCl ACV	.42	.7	23	2012	63.22	1.40
1.3	24.31	17.76	Albertsons ABS	.76	3.4	13	16155	22.54	0.13
61.8	48.35	26.25	Alcan AL	.60	1.3	cc	12788	47.77	1.03
193.5	13.68	4.30	Aicatel ADS ALA	...		...	7716	13.03	0.16
▲70.8	38.18	18.45	Alcoa AA	.60	1.5	71	41620	38.91	1.42
▲53.4	59.50	34.70	Alcon ACL	.35e	.6	35	6521	60.51	1.28

YTD % CHG	52-WEEK HI	LO	STOCK (SYM)	DIV	YLD %	PE	VOL 100s	CLOSE	NET CHG
▲38.2	33.25	19.25	BankNY BK	.76	2.3	26	13584	33.12	0.10
49.9	50.83	32.15	BkNovaScotia BNS	2.00fg	4.0	11	119	50.23	−0.17
▲25.0	45.52	33.14	BkOne ONE	1.00	2.2	16	20794	45.70	0.57
101.4	19.75	8.76	BkAtlBcp A BBX	.13	.7	17	3515	19.03	0.63
43.7	33.57	20.60	Bknorth BNK	.76	2.3	15	2795	32.48	0.43
▲30.5	40.73	27	BantaCp BN	.68	1.7	30	1340	40.82	0.78
43.9	36.15	20.30	Barclays ADS BCS	1.30e	3.7	...	569	35.55	−0.08
39.5	81.41	54.03	Bard CR BCR	.92	1.1	22	2378	80.93	0.35
82.5	33.97	15.89	BarnesNoble BKS	...		18	3234	32.98	0.24
61.6	34.38	18.40 ♣	BarnesGp B	.80	2.4	23	858	32.89	0.17
79.8	85.36	43.12 ♣	BarrLabs BRL s	...		33	3684	78.04	0.58
▲48.8	22.82	14.10 ♣	BarckGld ABX	.22	1.0	62	22253	22.93	0.44
5.8	6.39	2.10	BarryRG RGB	...		dd	194	4.35	−0.24
44.3	52.66	29.35	BauschLomb BOL	.52	1.0	25	2442	51.96	0.59
8.0	31.32	18.18	BaxterInt BAX	.58	1.9	34	37804	30.25	0.26
9.4	56.10	37.80	BaxterInt un	3.50	6.4	...	184	54.80	0.40
7.0	6.18	5.25	BayVwCap BVC	4.61e	75.0	88	1398	6.15	0.01
▲35.3	28.95	10.80	Bayer ADS BAY	.97e	3.3	...	1166	29.30	0.40
32.5	83.12	57.58	BearSteam BSC	.80	1.0	9	10015	78.71	1.28
44.8	11.25	5.78	BearingPt BE	...		dd	3528	9.99	0.11
67.7	109.60	52.49	BeazerHm BZH	.10p		8	3138	101.60	−0.02
72.3	51.98	28.30	BeckmnCoultr BEC	.44	.9	19	2972	50.85	1.35
▲34.8	41.13	28.82	BectonDksn BDX	.60f	1.5	20	10659	41.38	1.13
13.0	29.55	24.32	BedfdPrpty BED x	2.04	7.0	16	876	29.04	0.05
▲44.1	20.91	10.50 ♣	Belden BWC	.20	.9	dd	812	21.93	1.02
2.5	28	25	BllSCpFd Corts KCO	1.78	6.7	...	2	26.69	0.12
8.9	30	19.79	BellSouth BLS	1.00f	3.5	15	40114	28.17	0.53
2.1	27.10	24.80	BllSoDeb7.00Corts KCH	1.75	6.7	...	15	25.99	−0.06
33.9	28.63	18.72	Belo BLC	.38	1.3	25	2262	28.55	0.20
−1.1	51.16	39.33 ♣	Bemis BMS	1.12	2.3	17	1716	49.10	0.34
84.2	38.38	16.58	BenchmkElec BHE s	...		25	3658	35.20	0.26
29.1	27.41	12.70	Benetton BNG	.81e	3.5	...	24	23.10	0.30
32.0	36.93	24.39	Berkley BER s	.28	.8	9	6052	34.85	0.53
15.7	84500	60600	BerkHathwy A BRKA	...		19	z450	84200	1399.99
15.6	2824	2015	BerkHathwy B BRKB	...			z8850	2800	33.10
▲21.5	20.57	14.36 ♣	BerryPete A BRY	.44	2.1	15	207	20.66	0.33
116.1	62.70	22.48	BestBuy BBY	.40	.8	31	48286	52.19	1.63
191.2	8.60	1.63	BeverlyEnt BEV	...		dd	4531	8.30	0.06
▲65.3	17.06	10.21	BHPBilton PLC BBL n	.16p		...	52	17.52	0.62
5.1	18.39	9.92	BigLots BLI	...		26	5256	13.90	0.39
−23.1	51.30	16.51	Biovail BVF	...		...	10222	20.30	0.61
−6.7	21	11.80	BisysGp BSG	...		18	4842	14.84	0.10
15.8	49.90	33.20	BlackDeck BDK	.84f	1.7	15	6361	49.65	0.44
11.2	33.54	21.85	BlackHills BKH	1.20	4.1	12	3283	29.50	0.64
▲35.1	53.10	38.70	BlkRk A BLK	.40e	.8	24	410	53.23	1.05
▲36.1	54.59	35.28	BlockHR HRB	.80	1.5	15	11056	54.70	0.65
44.0	23.07	11.80	Blkbstr A BBI	.08	.5	13	4895	17.64	0.29
103.4	8.20	3.66	BlountInt BLT	...		dd	70	7.75	0.09

Individuals must use **stockbrokers** to trade publicly held stocks. Regular stockbrokers charge more, but they offer financial advice. Some people trade stocks using **discount brokers** who offer less advice and a lower cost. Yet others trade stock over the Internet using E*Trade, Schwab, or Ameritrade, for example, where the costs of trading are very low. Some firms on the Internet will let you play a game of buying and selling stock to help you lean about trading stocks.

EXAMPLE 2

Finding the Cost of Stocks

Ignoring commissions, find Roman Rodriguez's cost of the following purchases.

(a) 100 shares of Best Buy at the close for the day.
(b) 200 shares of Alcoa at the close for the day.
(c) Then find the combined dividend these shares would have paid.

SOLUTION

(a) Cost of 100 Best Buy shares = 100 shares × $52.19 = $5219.
(b) Cost of 200 Alcoa shares = 200 shares × $38.91 = $7782.
(c) Dividend from 100 shares of Best Buy =   100 × $.40 = $ 40
     Dividend from 200 shares of Alcoa =   200 × $.60 = $120
                         Total dividend   = $160

Objective **3** **Find the current yield on a stock.** There is no certain way of choosing stocks that will go up in price. However, two **stock ratios** that people commonly look at before buying shares of a company are the **current yield** and the **price–earnings ratio**. Although current yield is shown in the stock tables as Yld %, we show how to find it here, since you may not always have the tables available. It is used to compare the dividends paid by stocks selling at different prices. The result is commonly rounded to the nearest tenth of a percent.

$$\text{Current yield} = \frac{\text{Annual dividend per share}}{\text{Closing price per share}}$$

EXAMPLE 3

Finding the Current Yield

Find the current yield for **(a)** Albertsons (Symbol ABS) and **(b)** Ahold ADS (Symbol AHO).

SOLUTION

(a) Current Yield for Albertsons $= \dfrac{\text{Annual Dividend}}{\text{Closing price}} = \dfrac{.76}{22.54} = 3.4\%$ (rounded)

(b) Current Yield for Ahold ADS $= \dfrac{\text{Annual Dividend}}{\text{Closing price}} = \dfrac{.70}{7.56} = 9.3\%$ (rounded)

QUICK TIP A company may not pay a dividend because
1. it is going through difficult financial times;
2. it needs the funds that would be paid out for research and development; or
3. it may be growing rapidly and needs the money to finance its growth.

Objective **4** **Find the stock's PE ratio.** One number that some people use to help decide which stock to buy is the **price–earnings ratio**, also called the **PE ratio**. It is often rounded to the nearest whole number.

$$\text{PE ratio} = \frac{\text{Closing price per share}}{\text{Annual net income per share}}$$

EXAMPLE 4

Finding the PE Ratio

Find the PE ratio for each of the following huge corporations. Round to the nearest whole number.

(a) Wal-Mart with a closing price of $52.77 and earnings of $1.972 and
(b) Microsoft with a closing price of $40.78 and earnings of $2.688.

SOLUTION

(a) PE ratio for Wal-Mart $= \dfrac{\$52.77}{\$1.972} = 27$ (rounded)

(b) PE ratio for Microsoft $= \dfrac{\$40.78}{\$2.688} = 15$ (rounded)

A company that is growing rapidly will probably have substantially more profits in a few years. As a result, investors are willing to pay more for rapidly growing companies, which tend to have higher PE ratios. A low PE ratio may suggest that a stock is a sleeper and is undervalued in the market. Alternatively, a low PE ratio may simply indicate that investors see a poor future for the company. It is best to compare the PE ratios of similar companies, for example, the PE ratios of Ford, General Motors, and Chrysler can reasonably be compared. It does not make much sense to compare the PE ratio of Wal-Mart to Microsoft, since they are in different industries.

Objective **5** **Define the Dow Jones Industrial Average and the Nasdaq Composite Index.** Both the **Dow Jones Industrial Average** and the **Nasdaq Composite Index** are used as indicators of trends in stock prices. The Dow Jones Industrial Average refers to an average of 30 very large, industrial companies. The Nasdaq Composite Index includes price information on over 5000 companies, many of which are involved with technology. Both of these indexes are commonly quoted by television, radio, in newsprint, and on the Internet.

The Dow Jones Industrial Average has been widely available for over 100 years, as you can see on the graph. The Great Depression from 1929–1939 devastated many investors, businesses, and families. Stock prices collapsed during the early years of the depression followed by many companies, including banks, going bankrupt. Many people were out of work and were unable to find a job. The entire world was mired in the depression until World War II when war-related activity brought us out of the depression. However, stock prices did not recover to pre-Depression levels until the 1950s.

In spite of the Great Depression and the many smaller recessions in the United States, you can clearly see that the trend of stock prices has been up over the past 100 years or so.

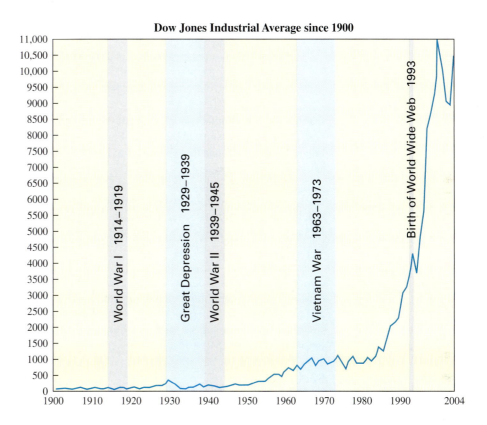

Dow Jones Industrial Average since 1900

Objective [6] **Define a mutual fund.** Ownership of shares in a single company *can be risky*—the company may suffer poor financial results causing the stock price to fall. This risk *can be reduced* by simultaneously investing in the stocks of several different companies, especially when they are in different industries.

One way to participate in the profits of successful corporations but to reduce risk is to purchase shares in a mutual fund that invests in stocks. A **mutual fund** receives money from many different investors and uses the money to purchase stocks or bonds in many different companies. For example, a $1000 investment in a typical mutual fund that owns stock, means that you own a very small piece of perhaps 100 different companies.

QUICK TIP Historically, stocks have consistently resulted in a greater return on investment than savings accounts, certificates of deposit, or bonds. Most financial planners agree that stocks should be a part of any long-range investment plan.

The table on the next page shows that some mutual funds *specialize* by investing in the stocks *of different types* of publicly held companies. For example, mutual funds may specialize in large-cap (large companies), small-cap (small companies), overseas (global), or specialty (real estate, oil, banking, etc.). Many financial planners say that *the first fund* you should invest in is an **index fund**

Numbers in the News

Going Global

By mixing just two or three of these low-priced exchange-traded funds, you can create a globally diversified portfolio

Exchange-traded fund (Ticker)	Invests in	Price	Expense ratio	One-year return
Vanguard Total Stock Market VIPERs (VTI)	U.S. stocks, all sizes	$102.00	21.0%	0.15%
iShares S&P 500 (IVV)	Large-cap U.S. stocks	106.03	17.4	0.09
iShares S&P Midcap 400 (IJH)	Mid-cap U.S. stocks	112.99	29.8	0.20
Midcap SPDRs (MDY)	Mid-cap U.S. stocks	103.25	29.6	0.25
iShares Russell 2000 (IWM)	Small-cap U.S. stocks	108.35	37.6	0.20
iShares MSCI Japan (EWJ)	Japanese stocks	8.84	27.2	0.84
iShares MSCI Pacific Ex-Japan (EPP)	Pacific Rim, but not Japan	68.74	32.3	0.50
iShares MSCI EAFE (EFA)	Europe and Pacific Rim	126.48	27.5	0.35
iShares MSCI Emerging Markets (EEM)	Emerging markets	148.50	48.9	0.75

DATA: Bloomberg

that tracks a broad index such as Standard and Poors 500, which includes 500 of the largest and best-managed companies in the world. Funds that specialize in companies in one industry, such as biotechnology, or funds that specialize in international stock are usually more volatile or risky.

QUICK TIP Many mutual-fund companies advertise on the World Wide Web, including Fidelity and Vanguard. Several of them will also let you track fund balances and make exchanges between funds using the World Wide Web.

EXAMPLE 5

Comparing Investment Alternatives

Cynthia Peck wants to know whether she should invest her retirement monies in certificates of deposit or in a mutual fund containing stock. Assume payments of $2000 per year for 30 years and **(a)** a certificate of deposit paying 4% compounded annually and **(b)** a mutual fund containing stock that has returned 8% per year. Find the future value for both and **(c)** compare the two investments.

SOLUTION

(a) Use 4% per year and 30 years in the table in Section 10.1 to find **56.08494**.

Future Value = $2000 × **56.08494** = $112,169.88

(b) Use 8% per year and 30 years in the table in Section 10.1 to find **113.28321**.

Future Value = $2000 × **113.28321** = $226,566.42

(c) Difference = $226,566.42 − $112,169.88 = $114,396.54

QUICK TIP Neither the 4% on the certificate of deposit nor the 8% on the mutual fund are guaranteed for 30 years. Stocks may do better or worse than bank deposits in any year, but tend to have higher returns over the long time periods required for retirement planning.

10.4 | Exercises

The **Quick Start** *exercises in each section contain solutions to help you get started.*

Find the following from the stock table on page 000 (not the preferred). (See Example 1.)

Quick Start

1. Change from the previous day for BlackDeck (BDK)   **1.** $\$.44$
2. Change from the previous day for BerkHathwy A (BRKA)   **2.** $\$1399.99$
3. Dividend for Blkbstr A (BBI)   **3.** $\$.08$

4. Dividend for Bard CR (BCR)   **4.** _____
5. Sales volume for the day for Airgas (ARG)   **5.** _____
6. Sales volume for the day for Belden (BWC)   **6.** _____
7. Current dividend yield for AbbottLab (ABT)   **7.** _____
8. Current dividend yield for Aetna (AET)   **8.** _____
9. PE ratio for BigLots (BLI)   **9.** _____
10. PE ratio for AaronRent (RNT)   **10.** _____
11. High for the year for BkOne (ONE)   **11.** _____
12. High for the year for Bayer ADS (BAY)   **12.** _____
13. Close price for BankNY (BK)   **13.** _____
14. Close price for AlamoGp (ALG)   **14.** _____
15. Low for the year for AgilentTch (A)   **15.** _____
16. Low for the year for AdvMicro (AMD)   **16.** _____
17. Percent change year-to-date for BauschLomb (BOL)   **17.** _____
18. Percent change year-to-date for BlockHR (HRB)   **18.** _____

Find the cost for the following stock purchases to the nearest cent. Ignore any broker's fee and assume that each purchase was at the close price for the day. Then find the annual dividend that would have been paid for the year on that number of shares. (See Example 2.)

Quick Start

Stock	Number of Shares	Cost	Dividend
19. Ahold (AHO)	300	$2268	$210
300 × 7.56 = $2268; 300 × .7 = $210			
20. BellSouth (BLS)	100	_____	_____
21. Aetna (AET)	80	_____	_____
22. Alcan (AL)	700	_____	_____

C indicates an exercise that is related to the Case in Point feature within the section.

23. Berkley (BER)   250   _____   _____

24. BarnesNoble (BKS)   200   _____   _____

25. Define and explain the following: **(a)** Current yield and **(b)** PE ratio. (See Objectives 3 and 4.)

26. Use the chart of the Dow Jones Industrial Average and estimate the years in which stocks fell by more than 10%. (See Objective 5.)

Find the current yield for each of the following stocks. Round to the nearest tenth of a percent.
(See Example 3.)

Quick Start

Stock	Current Price per Share	Annual Dividend	Current Yield
27. Coca-Cola	$50.75	$.88	1.7%
$.88 ÷ $50.75 = 1.7%			
28. Microsoft	$27.37	$.16	_____
29. McDonalds	$24.83	$.40	_____
30. General Motors	$53.40	$2.00	_____
31. Nike	$68.46	$.80	_____
32. Wal-Mart	$53.05	$.36	_____

Find the PE ratio for each of the following. Round all answers to the nearest whole number.
(See Example 4.)

Quick Start

Stock	Current Price per Share	Net Earnings per Share	PE Ratio
33. Pepsi	$46.38	$1.45	32
$46.38 ÷ $1.45 = 32			
34. General Motors	$53.01	$6.68	_____
35. General Electric	$46.18	$1.27	_____
36. Nike	$49	$2.19	_____

37. Dell $23.25 $.81   _____

38. Intel $30 $1.73   _____

Stock prices on consecutive days for a stock are shown next. Find the increase (decrease) in the price of each stock as a number and the percent increase (decrease) rounded to the nearest tenth of a percent.

39. 34.35, 35.20

40. 46.50, 45.90

Solve the following application problems.

Quick Start

41. Patsy Bonner buys 200 shares of General Electric at $56.30 and 100 shares of Safeway at $38.60. Find the total cost ignoring commissions.

    200 × $56.30 + 100 × $38.60 = $15,120

41. $15,120 _____

42. WRITING A WILL   In her will, Barbara Bains stated that the trustee should purchase 300 shares of AT&T and 200 shares of Wal-Mart and give the stock to her grandson on his 25th birthday. If the stocks are selling for $58.70 per share and $52.20 per share, respectively, find the total amount paid, ignoring broker's commissions.

42. _____

43. CDS OR GLOBAL STOCKS   Stan Walker is comparing certificates of deposit currently yielding 5% compounded semiannually to a mutual fund with international stocks that he believes will yield 8% compounded semiannually. Find the future value of an annuity with deposits of $600 every 6 months for 10 years for **(a)** the CDs and **(b)** the mutual fund. **(c)** Find the difference.

(a) _____
(b) _____
(c) _____

44. FIXED RATE OR STOCKS   Jesica Tate plans to contribute $2500 per year to a retirement plan and is debating the use of a fund that pays 4% per year versus a stock fund that she believes will yield 10% per year. Find the future value after 12 years for **(a)** the fixed rate and **(b)** the stock fund. **(c)** Find the difference.

(a) _____
(b) _____
(c) _____

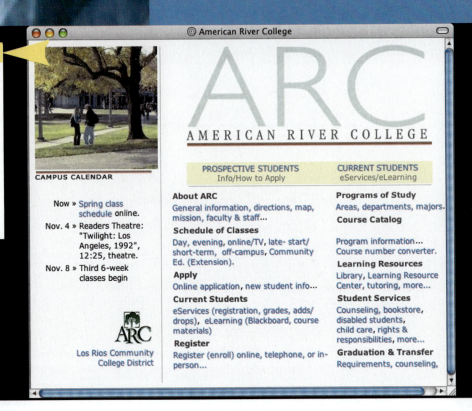

www.arc.losrios.edu

- Founded in 1955
- 26,800 students currently
- 45.7% of the students are under 25
- Offers on-line/TV and off-campus classes

ARC

AMERICAN RIVER COLLEGE

CAMPUS CALENDAR

Now » Spring class schedule online.
Nov. 4 » Readers Theatre: "Twilight: Los Angeles, 1992", 12:25, theatre.
Nov. 8 » Third 6-week classes begin

Los Rios Community College District

PROSPECTIVE STUDENTS Info/How to Apply	CURRENT STUDENTS eServices/eLearning
About ARC General information, directions, map, mission, faculty & staff...	**Programs of Study** Areas, departments, majors.
Schedule of Classes Day, evening, online/TV, late- start/ short-term, off-campus, Community Ed. (Extension).	**Course Catalog** Program information... Course number converter.
Apply Online application, new student info...	**Learning Resources** Library, Learning Resource Center, tutoring, more...
Current Students eServices (registration, grades, adds/ drops), eLearning (Blackboard, course materials)	**Student Services** Counseling, bookstore, disabled students, child care, rights & responsibilities, more...
Register Register (enroll) online, telephone, or in-person...	**Graduation & Transfer** Requirements, counseling,

Similar to many other community colleges across the country, American River College has continued to expand and diversify the programs of study that it offers. It now offers more than 60 different majors of study, including biology, engineering, hospitality management, mortuary science, collision repair, Japanese, and even fire technology. College personnel work very closely with students to help them find financial aid. Amazingly, nearly 48% of the students at American River College have some kind of financial aid.

1. American River College makes a contribution of $3200 per year to Roman Rodriguez's retirement plan. Assume that the college continues to make this contribution to his plan for 25 years and that funds are in stocks that average 8% per year. Find the future value.

2. Find the annual cost if American River College makes an average retirement plan contribution of $2400 per full-time employee, given that it has 734 full-time employees.

3. How many students at American River College have some kind of financial aid?

4. Assume that a wealthy donor has agreed to give American River College $250,000 per year for the next five years. Find the present value of these gifts, assuming 6% per year

5. What are the characteristics of a great instructor?

10.5 | Bonds

Objectives

1. Define the basics of bonds.
2. Read bond tables.
3. Find the commission charge on bonds and the cost of bonds.
4. Understand how mutual funds containing bonds are used for monthly income.

CASE *in* **POINT**

Roman Rodriguez has decided to include mutual funds holding stocks in his retirement plan, but he doesn't know about bonds. What are bonds? Should he invest in them?

Objective 1 **Define the basics of bonds.** Corporations can sell shares of stock to raise funds. Shares represent ownership in the corporation. However, managers sometimes prefer to borrow money rather than issuing stock. They borrow money for short-term needs from banks or insurance companies. Managers can also make long-term loans with banks, but they often prefer to borrow for the long term by issuing bonds. **Bonds** are legally binding promises to repay borrowed money at a specific date in the future. Corporations commonly pay interest on each bond each year. Unlike shareholders, bondholders do not own part of the corporation. Other entities, including countries, cities, and even churches, also borrow money by using bonds.

A company is said to go **bankrupt** if it can no longer meet its financial obligations to suppliers, banks, bondholders, etc. Bankruptcy is a complex process involving management, creditors, lawyers, and courts, but generally bankruptcy lawyers are paid first. Remaining assets are then used to pay off debt, including bonds. Shareholders *do not receive anything* unless there are assets remaining after all debts have been paid. Shareholders often receive very little or nothing from a bankruptcy, and bondholders often receive only a few cents on every dollar originally loaned.

Corporations frequently use substantial amounts of debt to build factories, expand operations, or buy other companies. The interest that must be paid on that debt is the *cost of having debt*. The larger the debt, the greater the amount of revenue the company must set aside to pay interest. As interest rates go higher, companies must set aside additional money to pay interest leaving less for other purposes including profits. On the other hand, interest costs go down when interest rates go

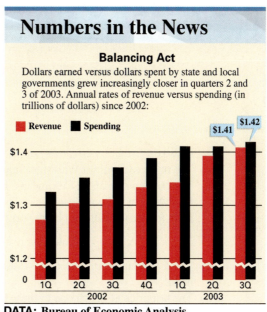

Numbers in the News

Balancing Act

Dollars earned versus dollars spent by state and local governments grew increasingly closer in quarters 2 and 3 of 2003. Annual rates of revenue versus spending (in trillions of dollars) since 2002:

DATA: Bureau of Economic Analysis

Argentina strikes a new deal to repay IMF

By James Cox
USA TODAY

Argentina moved toward ending two years of economic isolation Wednesday, entering a three-year aid deal with the International Monetary Fund just a day after defaulting on $201 million over the next three years. Kirchner pledged to set aside 3% of gross domestic product to pay Argentina's debt obligations but resisted IMF pressure for a bigger percentage.

The deal enhances Kirchner's

Source: USA Today, 9/11/03, Section 5B

down. The bar graph on the previous page and the clipping each show that both states and countries must deal with the consequences of the amount of debt they have accumulated.

Objective **2** **Read bond tables.** The **face value**, or **par value**, of a bond is *the original amount of money* borrowed by a company. Most corporations issue bonds with a par value of $1000. Suppose that a bond's owner needs money before the **maturity date** of the bond. In that event, the bond can be quickly sold through a bond dealer, such as Merrill Lynch. However, the price of the bond is determined, not by its initial price, but instead *by market conditions at the time of the sale*.

Market *interest rates fluctuate widely* from year to year, yet each bond pays exactly the same dollar amount of interest each year. If interest rates rise, investors will pay less for a bond because they want the new, higher interest yield. If interest rates fall, investors will pay more for a bond because they are satisfied with the lower yield. As a result, the price of a bond fluctuates in the opposite direction of interest rates. A bond may have a face value of $1000, but it often trades at a different value than $1000, as shown in Example 1.

EXAMPLE 1

Working with the Bond Table

Brandy Barrett was in an automobile accident that put her in the hospital for 3 weeks and required months of rehabilitation. The other driver was at fault, and his insurance company paid Barrett the liability limits on his policy of $50,000. Barrett needs monthly income and is thinking about investing the funds in Wal-Mart Stores bonds that mature in 2008. Data on this bond were taken from the *Wall Street Journal* bond table that follows. Analyze the information.

Company (Ticker)	Coupon	Maturity	Last Price	Last Yield	Est $ Vol (000's)
Wal-Mart Stores (**WMT**)	3.375	Oct 01, 2008	99.411	3.508	15,443

SOLUTION

(a) The ticker symbol for Wal-Mart Stores is WMT.

(b) The coupon rate of 3.375% means that each bond pays 3.375% simple interest on the $1000 face value of the bond each year. The face value of the bond of $1000 was the amount originally loaned to Wal-Mart by the original purchaser of the bond.

$$\text{Annual interest} = \$1000 \times .03375 = \$33.75$$

(c) This bond matures on October 1, 2008 when Wal-Mart must pay $1000 to the owner of the bond.

(d) The last price at which one of these bonds traded was **99.411%** of $1000.

$$\text{seller of 1 bond would receive} = .99411 \times \$1000 = \$994.11$$

The selling price on this day was $1000 - $994.11 = $5.89 less than the original $1000 price of the bond. Bond prices fluctuate significantly, depending on interest rates.

(e) Last Yield is the yield to the maturity date of the bond. It is found using a complicated formula that takes into consideration the current price of the bond, the maturity value of the bond, the time to maturity, and all interest payments that will be made between now and the maturity date.

(f) Since EST $ Volume is in thousands of dollars of bonds, sales for the day are as follows.

15,443 × $1000 or $15,443,000 in Wal-Mart bonds traded that day

Corporate Bonds

Monday, December 29, 2003

Forty most active fixed-coupon corporate bonds

COMPANY (TICKER)	COUPON	MATURITY	LAST PRICE	LAST YIELD	EST $ VOL (000's)
DaimlerChrysler North America Holding (DCX)	6.400	May 15, 2006	107.626	3.045	50,887
Tyson Foods (TSN)	8.250	Oct 01, 2011	118.483	5.307	34,350
BP Capital Markets PLC (BPLN)	2.625	Mar 15, 2007	99.178	2.895	30,985
Schering-Plough (SGP)	6.500	Dec 01, 2033	104.894	6.140	30,590
Comcast Holdings (CMCSA)	5.300	Jan 15, 2014	100.183	5.276	30,045
Sprint Capital (FON)	6.875	Nov 15, 2028	98.341	7.016	25,600
Altria Group (MO)	7.650	Jul 01, 2008	110.344	5.050	25,111
Time Warner (TWX)	7.700	May 01, 2032	117.540	6.356	23,830
Wells Fargo (WFC)	5.125	Feb 15, 2007	106.626	2.887	22,931
General Motors Acceptance (GMAC)	6.875	Aug 28, 2012	107.154	5.810	20,625
Target (TGT)	4.000	Jun 15, 2013	94.242	4.763	20,582
FirstEnergy (FE)	6.450	Nov 15, 2011	103.834	5.834	18,439
Bank of America (BAC)	5.125	Nov 15, 2014	99.969	5.128	15,770
Wal-Mart Stores (WMT)	3.375	Oct 01, 2008	99.411	3.508	15,443
Time Warner (TWX)	6.750	Apr 15, 2011	111.958	4.785	15,025
CIT Group (CIT)	4.750	Dec 15, 2010	101.176	4.550	14,490
Ford Motor Credit (F)	6.875	Feb 01, 2006	106.841	3.438	14,074
Abbott Laboratories (ABT)	5.125	Jul 01, 2004	101.951	1.200	13,865
Ford Motor Credit (F)	7.600	Aug 01, 2005	106.875	3.106	13,007
Wells Fargo Bank (WFC)	6.450	Feb 01, 2011	112.308	4.408	12,575
General Electric Capital (GE)	6.750	Mar 15, 2032	111.898	5.880	12,542
Morgan Stanley (MWD)	5.800	Apr 01, 2007	108.200	3.123	12,372
CIT Group (CIT)	7.625	Aug 16, 2005	108.829	2.061	12,354
Wells Fargo (WFC)	5.000	Nov 15, 2014	100.210	4.974	12,320
Bank of America (BAC)	7.875	May 16, 2005	108.258	1.755	12,100
General Motors (GM)	8.375	Jul 15, 2033	116.123	7.067	11,086
Wells Fargo (WFC)	4.800	Jul 29, 2005	104.495	1.889	11,015

Source: *Wall Street Journal*

EXAMPLE 2

Using the Bond Table

Find the volume sold and last sale price of the following bonds.

(a) Tyson Foods (TSN) maturing in 2011
(b) Target (TGT) maturing in 2013
(c) General Motors (GM) maturing in 2033

SOLUTION

Company	Volume Sold	Last Sale Price per Bond
(a) Tyson Foods	$34,350,000	$1000 × 1.18483 = $1184.83
(b) Target	$20,582,000	$1000 × .94242 = $942.42
(c) General Motors	$11,086,000	$1000 × 1.16123 = $1161.23

Objective **3** **Find the commission charge on bonds and the cost of bonds.** Commissions charged on bond sales vary among brokers. A common charge is $10 per bond, either to buy or to sell. However, commissions will be less for large volumes.

EXAMPLE 3

Find the Cost to Buy Bonds

Assume that the sales charge is $10 per bond, and find the following for Time Warner bonds maturing in 2032.

(a) The total cost of purchasing 20 bonds.
(b) The total annual interest paid on these bonds.
(c) The effective interest rate to the buyer including the cost of buying the bonds.

SOLUTION

(a) Total Cost = (Price per bond + Sales charge per bond) × Number of bonds

$$= (\mathbf{1.1754} \times \$1000 + \$10) \times 20 = \$23{,}708$$

(b) Annual interest for each bond is the coupon rate times the par value of the bond.

Annual interest = Interest per bond × Number of bonds

$$= (\mathbf{.077} \times \$1000) \times 20 = \$1540$$

(c) The effective interest rate is the total interest divided by the total cost of the bonds, including the cost of buying the bonds.

Effective interest rate = \$1540 ÷ \$23,708 = 6.5% (rounded)

QUICK TIP   The effective interest rate is not the same as the Last Yield. The Last Yield takes into consideration the cost of the bond, the maturity value of the bond, the time to maturity, and all interest payments.

EXAMPLE 4

Find the Net Amount from the Sale of Bonds

Find the amount received from the sale of 20 Bank of America bonds maturing in 2014.

SOLUTION

Amount received = (Sales price of a bond − Cost of selling bond) × Number of bonds

$$= \left(\mathbf{.99969} \times \$1000 - \$10\right) \times 20 = \$19{,}793.80$$

Objective 4   Understand how mutual funds containing bonds are used for monthly income. A mutual fund can invest everything in stocks, everything in bonds or part in stocks and part in bonds. Stock prices can be quite volatile, so financial planners recommend stock investments for people *who have a longer time horizon* over which to accumulate funds. Many planners recommend that *people invest in both stocks and bonds* during their lifetimes. Stocks may be a better investment *when investors are young*, since stocks have tended to have a higher return. Bonds may be a better investment *for investors close to retirement*, since there is less risk of losing principal in bonds and bonds pay regular interest.

EXAMPLE 5

Using a Bond Fund for Income

Brandy Barrett from Example 1 is undergoing rehabilitation and needs safety of principal. She also needs regular interest payments to help with medical expenses. She decides to place the \$50,000 received from the insurance company in a mutual fund containing bonds. **(a)** Find her annual income if the fund yields 6.5% per year. **(b)** How much would Barrett need to invest in the fund to earn \$10,000 per year?

SOLUTION

(a) Use the formula for simple interest: $I = PRT$.

Interest = \$50,000 × **.065** = \$3250

(b) Again use the formula for simple interest, but now the principal *(P)* is unknown. Divide both sides of $I = PRT$ by RT to find the following form of the equation.

$$\text{Principal} = P = \frac{I}{RT} = \frac{\$10{,}000}{.065 \times 1} = \$153{,}846.15$$

10.5 Exercises

*The **Quick Start** exercises in each section contain solutions to help you get started.*

Use the bond table in this section to find the following for General Electric Capital (GE) bonds maturing in 2032. (See Examples 1–2.)

Quick Start

1. Price per bond — **1.** $1118.98
2. Volume of bonds sold — **2.** _____
3. Date when bonds must be paid off by General Electric — **3.** _____
4. Annual interest paid — **4.** _____
5. Last yield or yield to maturity — **5.** _____
6. Price to buy 50 of these bonds including sales charge of $10 per bond — **6.** _____

Find the cost, including sales charges of $10 per bond, for each of the following transactions. (See Example 3.)

Quick Start

Bond	Maturity	Number Purchased	Cost
7. Sprint Capital (FON)	2028	50	$49,670.50
$(.98341 \times \$1000 + \$10) \times 50 = \$49,670.50$			
8. Tyson Foods (TSN)	2011	100	_____
9. CIT Group (CIT)	2010	40	_____
10. Altria Group (MO)	2008	200	_____
11. Schering-Plough	2033	150	_____
12. Comcast Holdings (CMCSA)	2014	80	_____

13. Explain the purpose of bonds. (See Objective 1.)

14. Explain how a bondholder can estimate the effective interest rate return on the total cost of the investment, including commissions. (See Example 3.)

C indicates an exercise that is related to the Case in Point feature within the section.

Solve each application problem. Assume a sales commission of $10 per bond, unless indicated otherwise, and use the table in the text. Round the rate to the nearest tenth of a percent.

Quick Start

15. **BOND PURCHASE**   Pete Fontaine purchased 20 Time Warner (TWX) bonds maturing in 2011. Find **(a)** the total cost of the purchase including commissions, **(b)** the annual interest payment, and **(c)** the effective interest rate to total cost including commissions. Round rate to the nearest tenth of a percent.

    (a) $22,591.60
    (b) $1350
    (c) 6.0%

    (a) Total cost = $\left(1.11958 \times \$1000 + \$10\right) \times 20 = \$22{,}591.60$
    (b) Annual interest = $\left(.0675 \times \$1000\right) \times 20 = \$1350$

    (c) Effective interest rate = $\dfrac{\$1350}{\$22{,}591.60} = 6.0\%$

16. New York City purchased 10,000 Wal-Mart Stores (WMT) bonds maturing in 2008. Find **(a)** the total cost of the purchase, including commissions (assume commissions of $1 per bond based on the large purchase). Then find **(b)** the annual interest payment and **(c)** the effective interest rate to total cost, including commissions.

    (a) _____
    (b) _____
    (c) _____

17. An investor bought 15 Sprint Capital (FON) bonds maturing in 2028. Find **(a)** the total cost of the purchase including commissions, **(b)** the annual interest payment, and **(c)** the effective interest rate to total cost including commissions.

    (a) _____
    (b) _____
    (c) _____

18. **RETIREMENT FUNDS**   The manager of a retirement account for United Pensions of America purchased 300 General Electric Capital (GE) bonds that mature in 2032 for the retirement account. Find **(a)** the total cost of the purchase including commissions, **(b)** the annual interest payment, and **(c)** the effective interest rate to total cost including commissions.

    (a) _____
    (b) _____
    (c) _____

19. Bernice Clarence places $45,000 in a bond fund that is currently yielding 8% compounded annually. **(a)** Find interest for the first year. She decides to let all interest payments remain in the account. **(b)** Find the amount in the account after 10 years if the fund continues to earn 8% compounded annually.

    (a) _____
    (b) _____

20. The community college where Roman Rodriguez works has an endowment funded by alumni and business owners in the community. The manager of the endowment invested $500,000 in a bond fund yielding 6% compounded semiannually. Find **(a)** the interest for the first year and **(b)** the future value of the account in 8 years.

    (a) _____
    (b) _____

Chapter 10 | Quick Review

CONCEPTS	EXAMPLES
10.1 Finding the amount of an ordinary annuity Determine the number of periods in the annuity *(n)* and the interest rate per annuity period *(i)*. Use *n* and *i* in the annuity table to find the value of $1 at the term of annuity. Find the value of an annuity using Amount = Payment × Number from table.	Ed Navarro deposits $800 at the end of each quarter for 10 years into an IRA. Given interest of 8% compounded quarterly, find the future value. $$n = 10 \times 4 = 40 \text{ periods}; \quad i = \frac{8\%}{4} = 2\% \text{ per period}$$ Number from table is **60.40198**. Amount = $800 × **60.40198** = $48,321.58
10.1 Finding the amount of an annuity due Determine the number of periods in the annuity. Add 1 to the value and use this as the value of *n*. Determine the interest rate per annuity period and use the table to find the value of $1 at term of annuity. The value of the annuity is Payment × Number from table − 1 payment.	Find the amount of an annuity due if payments of $700 are made at the beginning of each quarter for 3 years in an account paying 8% compounded quarterly. $$n = 3 \times 4 + 1 = 13; \quad i = \frac{8\%}{4} = 2\%$$ Number from table is **14.68033**. Amount = $700 × **14.68033** − $700 = $9576.23
10.2 Finding the present value of an annuity Determine the payment per period. Determine the number of periods in the annuity *(n)*. Determine the interest rate per period *(i)*. Use the values of *n* and *i* and find the number in the present value of an annuity table. The present value of an annuity is Present value = Payment × Number from table.	What lump sum deposited today at 8% compounded annually will yield the same total as payments of $600 at the end of each year for 10 years? Payment = $600; *n* = 10 Interest = 8% Number from table is **6.71008**. Present value = $600 × **6.71008** = **$4026.05**

CONCEPTS	EXAMPLES

10.2 Finding equivalent cash price

Determine the amount of the annuity payment.

Determine the number of periods in the annuity *(n)*.

Determine the interest rate per annuity period *(i)*.

Use *n* and *i* in the present value of an annuity table.

Add the present value of the annuity to the down payment to obtain today's equivalent cash price.

A buyer offers to purchase a business for $75,000 down and payments of $4000 at the end of each quarter for 5 years. Money is worth 8% compounded quarterly. How much is the buyer actually offering for the business?

Payment = $4000; $n = 20$

Interest = $\frac{8\%}{4}$ = 2%

Number from table is **16.35143**.

Present value = $4000 × **16.35143** = $65,405.72

Equivalent cash value = $75,000 + $65,405.72

$= \$140,405.72$

10.3 Determining the payment into a sinking fund

Determine the number of payments *(n)*. Determine the interest rate per period *(i)*. Find the value of the payment needed to accumulate $1 from the sinking fund table.

Calculate the payment using

Payment = Future value × Number from table.

No-Leak Plumbing plans to accumulate $500,000 in 4 years in a sinking fund for a new building. Find the amount of each semiannual payment if the fund earns 10% compounded semiannually.

$$n = 4 \times 2 = 8 \text{ periods}; \quad i = \frac{10\%}{2} = 5\% \text{ per period}$$

Number from table is **.10472**.

Payment = $500,000 × **.10472** = $52,360

10.3 Setting up a sinking fund table

Determine the required payment into the sinking fund.

Calculate the interest at the end of each period.

Add the previous total, next payment, and interest to determine total.

Repeat these steps for each period.

A company wants to set up a sinking fund to accumulate $10,000 in 4 years. It wishes to make semiannual payments into the account, which pays 8% compounded semiannually. Set up a sinking-fund table.

$n = 8$; $i = 4\%$

Number from table is **.10853**.

Payment = $10,000 × **.10853** = $1085.30

Payment	Amount of Deposit	Interest Earned	Total
1	$1085.30	$0	$1085.30
2	$1085.30	$43.41	$2214.01
3	$1085.30	$88.56	$3387.87
4	$1085.30	$135.51	$4608.68
5	$1085.30	$184.35	$5878.33
6	$1085.30	$235.13	$7198.76
7	$1085.30	$287.95	$8572.01
8	$1085.11	$342.88	$10,000.00

10.4 Reading the stock table

Locate the stock involved, and determine the various quantities required.

Use the stock table in Section 10.4 to find the following information for Advanced Marketing (MKT) and identify the values.

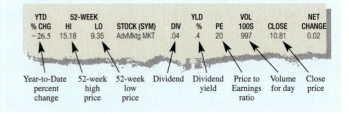

CONCEPTS	EXAMPLES
10.4 Finding the current yield on a stock To determine the current yield, use the formula $$\text{Current yield} = \frac{\text{Annual dividend}}{\text{Closing price}}$$	Find the current yield for a stock if the purchase price is \$35 and the annual dividend is \$.64. $$\text{Current yield} = \frac{\$.64}{\$35} = 1.8\% \text{ (rounded)}$$
10.4 Finding the price-earnings (PE) ratio To find the PE ratio, use the formula $$\text{PE ratio} = \frac{\text{Price per share}}{\text{Annual net income per share}}$$	Find the PE ratio for a stock priced at \$42.50 with earnings of \$2.11. $$\text{PE Ratio} = \frac{\$42.50}{\$2.11} = 20 \text{ (rounded)}$$
10.5 Determining the cost of purchasing bonds First locate the bond in the table. Then determine the price of the bond and multiply this value by \$1000 and the number of bonds purchased. Finally, add \$10 per bond to the total cost of bonds purchased.	Including sales charges of \$10 per bond, find the cost of 50 General Electric bonds that are selling at **111.898**. $$(1.11898 \times 1000 + \$10) \times 50 = \$56,449$$
10.5 Determining the amount received from the sale of bonds First locate the bond in the table. Then determine the price of the bond and multiply this value by 1000 and the number of bonds sold. Finally, subtract \$10 per bond from the total selling price.	Find the amount received after the sales charge from the sale of 20 Bank of America bonds selling at **99.969**. $$(.99969 \times \$1000 - \$10) \times 20 = \$19,793.80$$
10.5 Find the effective yield of a bond Find the cost of the bond after commission. Find the interest paid on the bond using the coupon rate. Finally, divide the interest by the cost of the bond.	Find the effective interest rate, to the nearest tenth, for a bond with a coupon rate of 3.8% and selling at 98.25. Price of bond = .9825 × \$1000 + \$10 = \$992.50 Interest = .038 × \$1000 = \$38 $$\text{Effective rate} = \frac{\$38}{\$992.50} = 3.8\% \text{ (rounded)}$$

Chapter 10 | Summary Exercise

Planning for Retirement

Roman Rodriguez wants to know how much he could have in his retirement fund if he saves until he turns 65. Rodriguez is 25 now, and is told that his employer will contribute $3200 at the end of each year for 40 years. He anticipates placing one-half of the contributions in a mutual-fund containing stock and the other one-half in a mutual fund containing bonds.

(a) Estimate his future value if the stock fund averages 10% per year compounded annually and the bond fund averages 6% per year compounded annually.

(a) _____

(b) Rodriguez is amazed that he could attain nearly one million dollars, but he knows that inflation will increase his cost of living significantly in 40 years. He assumes 3% inflation and wants to find the income needed at age 65 to have the same purchasing power as $30,000 today. (*Hint:* Look at the section on inflation in Section 9.2 and use the compound interest table in Section 9.1).

(b) _____

(c) Rodriguez wants to fund his retirement for 20 years, from age 65 to 85. Find the present value of the annual income that he needs at 65 as found in part **(b)**. Assume that the funds earn 8% per year compounded annually.

(c) _____

(d) Will his expected savings fund his anticipated payments?

(d) _____

(IN)VESTIGATE

There is some discussion about the ability of Social Security to pay retirement benefits. Do you think Social Security will be around to help you during your retirement? What percent of your retirement needs do you think Social Security will pay? Try to support your views with recent articles from newspapers, magazines, or information from the World Wide Web.

Chapter 10 Test

To help you review, the numbers in brackets show the section in which the topic was discussed.

Find the amounts of the following annuities. **[10.1]**

	Amount of Each Deposit	Deposited	Rate per Year	Number of Years	Type of Annuity	Amount of Annuity
1.	$1000	annually	6%	8	ordinary	_____
2.	$4500	semiannually	10%	9	ordinary	_____
3.	$30,000	quarterly	8%	6	due	_____
4.	$2600	semiannually	10%	12	due	_____

5. James Rivera earned his degree in drafting at a community college and recently began his new career. He was happy to learn that his new employer will deposit $2000 into his 401(k) retirement account at the end of each year. Find the amount he will have accumulated in 15 years if funds earn 8% per year. **[10.1]**

5. _____

6. James Rivera from Exercise 5 above has also decided to invest $500 at the end of each quarter in an IRA account that grows tax deferred. Find the amount he will have accumulated if he does this for 10 years and earns 8% compounded quarterly. **[10.1]**

6. _____

Find the present value of the following annuities. **[10.2]**

	Amount per Payment	Payment at End of Each	Number of Years	Interest Rate	Compounded	Present Value
7.	$1000	year	9	6%	annually	_____
8.	$4500	6 months	6	10%	semiannually	_____
9.	$708	month	3	12%	monthly	_____
10.	$14,000	quarter	6	8%	quarterly	_____

11. Betty Yowski borrows money for a new swimming pool and hot tub. She agrees to repay the note with a payment of $1200 per quarter for 6 years. Find the amount she must set aside today to satisfy this capital requirement in an account earning 10% compounded quarterly. **[10.2]**

11. _____

12. Dan and Mary Fisher just divorced. The divorce settlement included $650 a month payment to Dan for the 4 years until their son turns 18. Find the amount Mary must set aside today in an account earning 12% per year compounded monthly to satisfy this financial obligation. **[10.2]**

12. _____

Find the amount of each payment into a sinking fund for the following.   **[10.3]**

	Amount Needed	Years Until Needed	Interest Rate	Interest Compounded	Amount of Payment
13.	$100,000	9	6%	annually	_____
14.	$250,000	10	8%	semiannually	_____
15.	$360,000	11	6%	quarterly	_____
16.	$800,000	12	10%	semiannually	_____

Solve the following application problems.

17. The owner of Hickory Bar-B-Que plans to open a new restaurant in 4 years at a cost of $200,000. Find the required monthly payment into a sinking fund if funds are invested in an account earning 6% per year compounded semiannually.   **[10.3]**

17. _____

18. Lupe Martinez will owe her retired mother $45,000 for a piece of land. Find the required monthly payment into a sinking fund if Lupe pays it off in 4 years and the interest rate is 10% per year compounded quarterly.   **[10.3]**

18. _____

19. George Jones purchases 200 shares of Merck stock at $79.50 per share. Find **(a)** the total cost and **(b)** the annual dividend if the dividend per share is $1.36.   **[10.4]**

(a) _____
(b) _____

20. Belinda Deal purchases 25 IBM bonds that mature in 2010 at 95.1. They have a coupon rate of 4.2%. Find **(a)** the total cost if commissions are $10 per bond, **(b)** the annual interest and **(c)** the effective interest rate rounded to the nearest tenth.   **[10.5]**

(a) _____
(b) _____
(c) _____

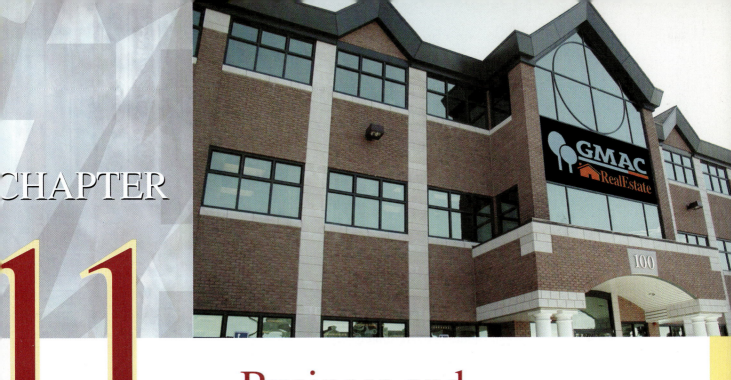

CHAPTER

11

Business and Consumer Loans

SUSAN BECKMAN GRADUATED FROM COLLEGE WITH A DEGREE IN

business and went to work for General Motors Acceptance Corporation (GMAC)

where she has been for eight years. GMAC is a sub-

sidiary of General Motors that now offers home and

personal loans to customers, as well as commercial loans to companies. Although

she has had several jobs at GMAC, Susan's current job is to evaluate potential

homeowners to see if they satisfy GMAC's requirements for a home loan.

It is almost impossible in our society to pay cash for everything. People borrow for purchases at department stores, gas stations, furniture stores, or automobile dealerships. Other examples of using credit that we don't commonly think of include turning on the lights, using the running water, or the much disliked roaming charges on our cell phones. Almost everyone borrows when they buy a home. This chapter looks at various methods of determining interest charges on purchases.

11.1 | Open-End Credit and Charge Cards

Objectives

1. Define open-end credit.
2. Define revolving charge accounts.
3. Use the unpaid balance method.
4. Use the average daily balance method.
5. Define loan consolidation.

CASE in POINT

Susan Beckman has worked with many families with serious debt problems. She often helps these families set up budgets and then tries to reduce their monthly debt payments by refinancing loans when possible.

Look at the median debt of students today. Students often borrow long term using student loans to help pay for tuition, fees and books. However, many students also have a lot of credit-card debt. How much debt do you have?

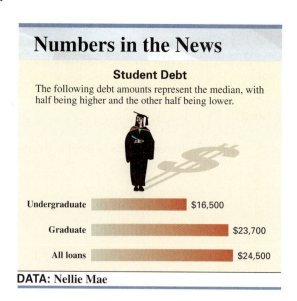

Numbers in the News

Student Debt

The following debt amounts represent the median, with half being higher and the other half being lower.

Undergraduate	$16,500
Graduate	$23,700
All loans	$24,500

DATA: Nellie Mae

Objective 1 Define open-end credit. A common way of buying on credit, called **open-end credit**, has no fixed payments. The customer continues making payments until no outstanding balance is owed. With open-end credit, additional credit is often extended before the initial amount is paid off. Examples of open-end credit include most department-store charge accounts and charge cards, including MasterCard and Visa. Individuals are given a **credit limit**, or a maximum amount that may be charged on these accounts. The lender determines the credit limit for each person based on his or her income, assets, other debts, and credit history.

A sale paid for with a **debit card** authorizes the retailer's bank to debit the purchaser's checking account immediately upon receipt. A Visa or MasterCard can be either a charge card or a debit card, depending on the bank and the card holder's preference.

QUICK TIP   **Debit cards** do not involve credit since their use results in an immediate debit.

Objective ② Define revolving charge accounts. With a typical department store account or bank card, a customer might make several small purchases during a month. Such accounts are often *never paid off*, although a minimum amount must be paid each month, and new purchases are continually being made. Since the account may never be paid off, it is called a **revolving charge account**.

Visa, MasterCard, Discover, and some oil-company cards use this method of extending credit. Sometimes there is an annual membership fee or a minimum monthly charge for the use of this service. A sample copy of a receipt signed by a customer using a credit card is shown on the next page. The clipping below shows the share of the U.S. credit card market by type of charge card. Notice that Visa has over 50% of the market and that Discover has only 5%.

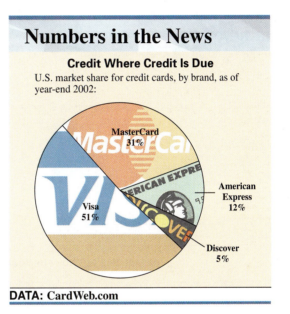

Numbers in the News

Credit Where Credit Is Due

U.S. market share for credit cards, by brand, as of year-end 2002:

MasterCard 31%

American Express 12%

Visa 51%

Discover 5%

DATA: CardWeb.com

At the end of a billing period, the customer receives a statement of payments and purchases made. This statement typically takes one of two forms. **Country club billing** provides a carbon copy of all original charge receipts. **Itemized billing**, more and more common because of its lower cost to credit card companies, provides an itemized listing of all charges, without copies of each individual charge. A typical itemized statement is also shown on the next page.

Finance charges are interest charges beyond the cash price of an item and may include interest, credit life insurance, a time-payment differential, and carrying charges. Interest charges can be avoided if the total balance is paid by the end of the **grace period**. Grace periods range between 15 and 30 days depending on the company. Often, there is no grace period on cash advances so that finance charges are assessed beginning immediately. Many lenders also charge **late fees** for payments that are received after the due date. **Over-the-limit fees** are charged by the lender when the borrower charges more than an approved maximum amount of debt.

QUICK TIP   Both late fees and over-the-limit fees are high. It is best to avoid them.

Objective ③ Use the unpaid balance method. Finance charges on open-end credit accounts may be calculated using the **unpaid balance method**. This method calculates finance charges based on the unpaid balance at *the end of the previous month*. Any purchases or returns during the current month are not used in calculating the finance charge, as you can see in the next example.

Society Bank Sales Slip (Merchant Copy)

7156 3120 1000 8581   **5928772**

DO NOT CIRCLE EXPIRATION DATE — USE BOX BELOW

02/9   01/9
Widjan Cadura

6388 312805392
Tower Sports
Phoenix MD

PURCHASER SIGN HERE

X *Widjan Cadura*

Cardholder acknowledges receipt of goods and/or services in the amount of the Total shown hereon and agrees to perform the obligations set forth in the cardholder's agreement with the issuer.

PRESS FIRMLY — USE BALLPOINT PEN

EXPIRATION DATE CHECKED ☒

QUAN.	CLASS	DESCRIPTION	PRICE	AMOUNT
1		Pr. Aerobic Shoes		39 95
2		Sweatshirts	20 00	40 00
1		Pr. Socks		2 95

DATE 10/3   AUTHORIZATION   SUB TOTAL 82 90
REFERENCE NO.   REG./DEPT.   TAX 4 15
FOLIO/CHECK NO.   SERVER/CLERK   TIPS/MISC.

Society BANK   **SALES SLIP**   **TOTAL** 87 05

IMPORTANT: RETAIN THIS COPY FOR YOUR RECORDS

MERCHANT COPY

AMERICA'S CHARGE CARD
★ ★ ★ ★ ★ ★ ★ ★

$ ___
WRITE IN THE AMOUNT OF YOUR PAYMENT

IMPORTANT! **1.** Return this portion of your statement with your check; please do not fold or bend **2.** Write your account number on the face of your check **3.** Make checks payable to America's Charge Card

ACCOUNT NUMBER	BILLING DATE	PAYMENT DUE DATE	NEW BALANCE	MINIMUM PAYMENT DUE
5211-1234-5678	10-18	11-12	609.26	30.00

JOHN Q. CARDHOLDER
1000 MAIN STREET
ANYTOWN, USA   00000

52112345678 0060926003000

PLEASE BE SURE OUR MAILING ADDRESS ON THE REVERSE SIDE APPEARS IN THE WINDOW OF THE RETURN ENVELOPE ▶

DO NOT INCLUDE INQUIRIES WITH PAYMENT. SEE REVERSE SIDE

PLEASE PRINT

FOR CHANGE OF ADDRESS	NEW STREET ADDRESS		APT. P.O. BOX NO.
	CITY	STATE	ZIP CODE

- -

RETAIN THIS PORTION FOR YOUR RECORDS

Account Number	Credit Line	Available Credit	Past Due Amount	BILLING DATE
5211-1234-5678	1,000.00	390.74	.00	10-18

TRAN DATE	POSTING DATE	REFERENCE NUMBER	TRANSACTION DESCRIPTION			TRANSACTION AMOUNT
09 06	09 09	12009230	TURNSTYLE 702	CHICAGO	IL	25.17CR
09 07	09 13	55190432	POLK BROS	CHICAGO	IL	150.00
09 13	09 21	86235671	MINNESOTA FABRICS 32	NORRIDGE	IL	39.54
09 14	09 19	21345678	SPIEGELS, INC.	CHICAGO	IL	60.10
09 20	09 20	93703523	PAYMENT THANK YOU			110.00CR
09 21	09 26	86245671	MINNESOTA FABRICS 32	NORRIDGE	IL	42.64
09 25	09 28	12001051	TURNSTYLE 702	CHICAGO	IL	14.49

PREVIOUS BALANCE	AVERAGE DAILY BALANCE	FINANCE CHARGE	NEW BALANCE	Minimum Payment
432.64	501.80	5.02	609.26	30.00

SEE ITEM NUMBER ON REVERSE SIDE FOR YOUR PERIODIC RATE(S). THE RANGE OF AVERAGE DAILY BALANCE(S) TO WHICH IT APPLIES AND THE CORRESPONDING

ANNUAL PERCENTAGE RATE

For Customer Service Call	Charge Transactions	PAYMENTS	CREDITS	PAYMENT DUE DATE
	306.77	110.00	25.17CR	11-12

Additional **FINANCE CHARGE** on purchases may be avoided if total NEW BALANCE is paid in full by PAYMENT DUE DATE

NOTICE: SEE REVERSE SIDE FOR IMPORTANT INFORMATION

YOU MAY AT ANY TIME PAY ALL OR PART OF THE BALANCE OWING ON YOUR ACCOUNT

EXAMPLE 1

*Finding Finance
Charge Using the
Unpaid-Balance
Method*

(a) Peter Brinkman's MasterCard account had an unpaid balance of $870.40 on November 1. During November, he made a payment of $100 and purchased a registered German shepherd for $150 using the card. Find the finance charge and the unpaid balance on December 1 if the bank charges 1.5% per month on the unpaid balance. A finance charge of 1.5% per month on the unpaid balance would be

$$\$870.40 \times .015 = \$13.06 \text{ for the month.}$$

Find the unpaid balance on December 1 as follows.

Previous balance	Finance charge	Purchases during month	Payment	New balance
↓	↓	↓	↓	↓

$$\$870.40 + \$13.06 + \$150 - \$100 = \$933.46$$

(b) During December, Brinkman made a payment of $50, charged $240.56 for Christmas presents, returned $35.45 worth of items, and took his family to dinner a few times with charges of $92.45. Find his unpaid balance on January 1.

The finance charge calculated on the unpaid balance is $933.46 × .015 = $14.00. The unpaid balance on January 1 follows.

$$\$933.46 + \$14.00 + \$240.56 + \$92.45 - \$35.45 - \$50 = \$1195.02$$

Month	Unpaid Balance at Beginning of Month	Finance Charge	Purchases during Month	Returns	Payment	Unpaid Balance at End of Month
November	$870.40	$13.06	$150.00	—	$100	$ 933.46
December	$933.46	$14.00	$333.01	$35.45	$ 50	$1195.02

The total finance charge during the 2-month period was $13.06 + $14.00 = $27.06.

(c) Brinkman knows that his debt is increasing. He moves the balance to another charge card that charges only .8% per month. Find his savings in finance charges for January.

$$\text{Savings} = \left(\underbrace{\$1195.02 \times .015}_{\text{old charge card}}\right) - \left(\underbrace{\$1195.02 \times .008}_{\text{new charge card}}\right) = \$8.37$$

Suppose you have an impulse and purchase a $1000 digital television set and charge it to a VISA card with finance charges of 1.5% per month on the unpaid balance. Further suppose that you make payments of $50 every month and don't charge anything else on the card. As shown next, it will take you 24 months to pay off the television set. The $1000 television set will cost you an extra $197.83 in finance charges for a total cost of $1197.83.

Month	Unpaid Balance at Beginning of Month	Finance Charge	Payment	Unpaid Balance at End of Month
1	$1,000.00	$15.00	$50.00	$965.00
2	$965.00	$14.48	$50.00	$929.48
3	$929.48	$13.94	$50.00	$893.42
⋮	⋮	⋮	⋮	⋮
22	$143.53	$ 2.15	$50.00	$95.68
23	$ 95.68	$ 1.44	$50.00	$47.12
24	$ 47.12	$ 0.71	$47.83	$ 0.00
Totals		$197.83	$1197.83	

QUICK TIP Notice that the unpaid balance decreases each month. Since the finance charge is based on the unpaid balance, it too decreases each month.

QUICK TIP   The cost of technology items often falls rapidly. If you had waited eight months and saved your money before buying, you might have been able to purchase the same television set for $700 cash rather than paying nearly $1200.

EXAMPLE 2

Finding Average Daily Balance

Objective 4   Use the average-daily-balance method. Most revolving charge plans now calculate finance charges using the **average-daily-balance method**. First, the balance owed on the account is found at the end of each day during a month or billing period. All of these amounts are added, and the total is divided by the number of days in the month or billing period. The result is the average daily balance of the account, which is then used to calculate the finance charge.

Beth Hogan's balance on a Visa card was $209.46 on March 3. Her activity for the next 30 days is shown in the table. **(a)** Find the average daily balance on April 3. Given finance charges based on $1\frac{1}{2}\%$ on the average daily balance, find **(b)** the finance charge for the month and **(c)** the balance owed on April 3.

Transaction Description	Transaction Amount
Previous balance $209.46	
March 3 Billing date	
March 12 Payment	**$50.00 CR***
March 17 Clothes	**$28.46**
March 20 Mail order	**$31.22**
April 1 Auto parts	**$59.10**

*CR represents *credit.*

SOLUTION

(a)

Date	Unpaid Balance	Number of Days Until Balance Changes
March 3	$209.46	9
March 12	$159.46 = $209.46 − **March 12 payment of $50**	5
March 17	$187.92 = $159.46 + **March 17 charge of $28.46**	3
March 20	$219.14 = $187.92 + **March 20 charge of $31.22**	12
April 1	$278.24 = $219.14 + **April 1 charge of $59.10**	2
April 3	end of billing cycle . . .	31 Total number of days in billing period

It is 9 days from March 3 to March 12, so the unpaid balance remains at $209.46 for 9 days.

There are 31 days in the billing period (March has 31 days). Find the average daily balance as follows:

Step 1 Multiply each unpaid balance by the number of days for that balance.

Step 2 Total these amounts.

Step 3 Divide by the number of days in that particular billing cycle (month).

Step 1

Unpaid Balance		Days		Total Balance
$209.46	×	9	=	$1885.14
$159.46	×	5	=	797.30
$187.92	×	3	=	563.76
$219.14	×	12	=	2629.68
$278.24	×	2	=	556.48
				$6432.36

← Step 2

Step 3

$$\frac{\$6432.36}{31} = \$207.50 \text{ average daily balance}$$

Hogan will pay a finance charge based on the average daily balance of $207.50.

(b) The finance charge is **.015 × $207.50** = $3.11 (rounded).

(c) The amount owed on April 3 is the beginning unpaid balance less any returns or payments, plus new charges and the finance charge.

$$\underset{\substack{\text{Previous}\\\text{balance}}}{\$209.46} - \underset{\text{Payment}}{\$50} + \underset{\text{New charges}}{\big(\$28.46 + \$31.22 + \$59.10\big)} + \underset{\substack{\text{Finance}\\\text{charge}}}{\$3.11} = \$281.35$$

> **QUICK TIP**   The billing period in Example 2 is 31 days. Some billing periods are 30 days (or 28 or 29 days in February). Be sure to use the correct number of days for the month of the billing period.

If the finance charges are expressed on a per-month basis, find the **annual percentage rate** by multiplying the monthly rate by 12, the number of months in a year. For example, $1\frac{1}{2}\%$ per month is found as follows:

$$1\tfrac{1}{2}\% \times 12 = 1.5\% \times 12 = 18\% \text{ per year}$$

Objective 5 **Define loan consolidation.**   Credit is *very easy to get* for individuals with a good credit history and a stable job. The clipping shows that too much spending and borrowing often creates problems. Below are some ways to help gain control of your finances.

Easy credit comes with big penalty

Young adults' free-spending ways set stage for bankruptcy.

By Margaret Webb Pressler
WASHINGTON POST

Christopher Siwy thinks he's pretty good with his finances, especially for a 23-year-old. Siwy just moved to Alexandria, Va., from Allentown, Pa., and with a starting job in information technology, he has no problem paying $160 on his student loan every month. He even saves a little bit out of each paycheck so that someday he can buy a condo.

But when Siwy wanted some wheels, he turned to the most popular financing plan for someone his age: a credit card.

Gain Control of your Finances

1.   Increase your income by investing in yourself. Choose a career you enjoy, and get training and education.
2.   Make a budget and stick to it.
3.   Spend less. Here are some suggestions.
     (a) Make sure you can afford your rent or mortgage payment.
     (b) Eat out less often.
     (c) Drive that old automobile one or two more years.
     (d) Purchase a less expensive automobile or reduce the number of automobiles in your family by one.
     (e) Be careful with the amount you spend on entertainment, hobbies, and travel.
     (f) Don't buy on impulse. If you want something, write it down on a piece of paper and stick it on your refrigerator for 30 days. After 30 days, ask yourself if you actually need the item.
4.   Try to pay cash for things the day you buy them rather than using credit.
5.   Save more by paying yourself first. Do this by saving a certain amount every month before you spend money on other things.
6.   Set some money aside for emergencies.
7.   Contribute to a long-range retirement plan.

Have you ever found yourself in the position that you cannot make all of your monthly payments? If so, you may be able to **consolidate your loans** into a single loan with one lower monthly payment. The new loan may have a lower interest rate and also a longer term, meaning that payments must be made for a longer period of time. This process can help you afford your monthly payments rather than defaulting on debt. **Defaulting on your debt**, or not making your payments, can mean repossession of your automobile or furniture, eviction from your apartment or house, and/or court appearances. Defaulting on your debt also ruins your credit history and can make it difficult to borrow money to buy a car or a home for years into the future.

QUICK TIP   Individuals who consolidate their loans and then borrow even more can get into very serious financial difficulties.

EXAMPLE **3**

Loan Consolidation

Bill and Jane Smith were married two years ago. Both were happy when they had their first child, but they needed to buy several things on credit. They now have the monthly payments shown below. The Smiths are having difficulties making the payments and they sometimes argue over money. They ask Susan Beckman at GMAC for help.

Revolving Accounts	Debt	Annual Percentage Rate	Minimum Monthly Payment
Sears	$3880.54	18%	$150
Dillards	$1620.13	16%	$ 60
MasterCard	$3140.65	14%	$100
Visa	$4920.98	18%	$200
Total	$13,562.30		Total $510

Other Payments	Monthly Payment
Rent	$800
Jane's car payment	$315
Bill's truck payment	$268
	Total $1383

SOLUTION

Susan Beckman:

1. Put the Smiths on a strict monthly budget.
2. Consolidated their revolving account debts into one longer-term, low-interest loan at GMAC (this required a loan guarantee from Bill's father).
3. Decreased one automobile payment by refinancing the loan over a longer term.

Here are their new monthly payments.

Account	Monthly Payment	New Status
Credit union loan for $13,562.30	$337.50	Revolving loans were consolidated
Rent	$800.00	Unchanged
Jane's car payment	$247.50	Refinanced using a longer term
Bill's truck payment	$268.00	Unchanged
	Total $1653.00	

$$\text{Reduction in Payments} = \left(\$510 + \$1383\right) - \$1653$$
$$= \textbf{\$240 per month}$$

The Smiths should be all right as long as they do the following.

1. Stay on their monthly budget.
2. Do not make additional credit purchases.
3. Continue to make all payments.

The Smiths may end up with severe debt problems if they borrow more before the existing loan balances are significantly reduced. Borrowing more could force them to declare bankruptcy.

11.1 | Exercises

FOR EXTRA HELP

MyMathLab

InterActMath.com

MathXL

MathXL
Tutorials on CD

Addison-Wesley
Math Tutor Center

DVT/Videotape

The **Quick Start** *exercises in each section contain solutions to help you get started.*

Find the finance charge on each of the following revolving charge accounts. Assume interest is calculated on the unpaid balance of the account. Round to the nearest cent. (See Example 1.)

Quick Start

	Unpaid Balance	Monthly Interest Rate	Finance Charge
1.	$6425.40	1.7%	**$109.23**
	$6425.40 × .017 = $109.23		
2.	$595.35	$1\frac{1}{2}\%$	_____
3.	$1201.43	$1\frac{1}{4}\%$	_____
4.	$2540.33	1.6%	_____

Complete the following tables, showing the unpaid balance at the end of each month. Assume an interest rate of 1.4% on the unpaid balance. (See Example 1.)

	Month	Unpaid Balance at Beginning of Month	Finance Charge	Purchases during Month	Returns	Payment	Unpaid Balance at End of Month
5.	October	$437.18	_____	$128.72	$27.85	$125	_____
	November	_____	_____	$291.64	—	$175	_____
	December	_____	_____	$147.11	$17.15	$150	_____
	January	_____	_____	$27.84	$127.76	$225	_____
6.	October	$255.40	_____	$27.50	—	$50	_____
	November	_____	_____	$59.60	$22.15	$45.50	_____
	December	_____	_____	$85.45	$32.00	$125	_____
	January	_____	_____	$325.68	—	$100	_____

7. Compare the unpaid balance method and the average daily balance method for calculating interest on open-end credit accounts. (See Objectives 3 and 4.)

8. Explain how consolidating loans may be of some advantage to the borrower. What disadvantages can you think of? (See Objective 5.)

C indicates an exercise that is related to the Case in Point feature within the section.

Find the finance charge for the following revolving charge accounts. Assume that interest is calculated on the average daily balance of the account. (See Example 2.)

Quick Start

Average Daily Balance	Monthly Interest Rate	Finance Charge
9. $1458.25	1.4%	**$20.42**

$1458.25 \times .014 = $20.42

10. $841.60	$1\frac{1}{2}\%$	**$12.62**

$841.60 \times .015 = $12.62

11. $389.95	$1\frac{1}{4}\%$	_____
12. $2235.46	1.6%	_____
13. $1235.68	1.4%	_____
14. $4235.47	$1\frac{3}{4}\%$	_____

Solve the following application problems.

15. HOT TUB PURCHASE   Betty Thomas borrowed $6500 on her Visa card to install a hot tub with landscaping around it. The interest charges are 1.6% per month on the unpaid balance. **(a)** Find the interest charges. **(b)** Find the interest charges if she moves the debt to a credit card charging 1% per month on the unpaid balance. **(c)** Find the savings.

(a) _____
(b) _____
(c) _____

16. CREDIT-CARD BALANCE   Alphy Jurarim used a credit card from GMAC to help pay for tuition expenses while in college and now owes $5232.25. The interest charges are 1.75% per month. **(a)** Find the interest charges. **(b)** Find the interest charges if he moves the debt to a credit card charging .8% per month on the unpaid balance. **(c)** Find the savings.

(a) _____
(b) _____
(c) _____

Find (a) the average daily balance for the following credit-card accounts. Assume one month between billing dates using the proper number of days in the month. (b) Then find the finance charge if interest is 1.5% per month on the average daily balance. (c) Finally, find the new balance. (See Example 2.)

Quick Start

17. Previous balance $139.56

September 12	Billing date	
September 20	Payment	$45
September 21	Athletic shoes	$37.25

(a) $132.64

(b) $1.99

(c) $133.80

Sep. 12 to Sep. 20 = 8 days at $139.56, gives $1116.48

Sep. 20 to Sep. 21 = 1 day at $139.56 − $45 = $94.56, gives $94.56

Sep. 21 to Oct. 12 = 21 days at $94.56 + $37.25 = $131.81, gives $2768.01

8 + 1 + 21 = 30 days

$1116.48 + $94.56 + $2768.01 = $3979.05

(a) Average daily balance = $\frac{\$3979.05}{30}$ = $132.64

(b) Finance charge = $132.64 × .015 = $1.99

(c) New balance = $139.56 + $1.99 + $37.25 − $45 = $133.80

18. Previous balance $228.95

January 27	Billing date	
February 9	Clothing	$11.08
February 13	Returns	$26.54
February 20	Payment	$29
February 25	Auto repairs	$71.19

(a) _____

(b) _____

(c) _____

19. Previous balance $312.78

June 11	Billing date	
June 15	Returns	$106.45
June 20	Jewelry	$115.73
June 24	Car rental	$74.19
July 3	Payment	$115

(a) _____

(b) _____

(c) _____

20. Previous balance $714.58

August 17	Billing date	
August 21	Mail order	$26.94
August 23	Returns	$25.41
August 27	Beverages	$31.82
August 31	Payment	$128.00
September 9	Returns	$71.14
September 11	Plane ticket	$110.00
September 14	Cash advance	$100.00

(a) _____
(b) _____
(c) _____

21. Previous balance $355.72

March 29	Billing date	
March 31	Returns	$209.53
April 2	Auto parts	$28.76
April 10	Gasoline	$14.80
April 12	Returns	$63.54
April 13	Returns	$11.71
April 20	Payment	$72.00
April 21	Flowers	$29.72

(a) _____
(b) _____
(c) _____

11.2 | Installment Loans

Objectives

1. Define installment loan.
2. Find the total installment cost and the finance charge.
3. Use the formula for approximate APR.
4. Use the table to find APR.

Objective 1 **Define installment loan.** A loan is **amortized** if both principal and interest are paid off by a sequence of equal periodic payments. An example is the paying of $250 per month for 48 months on a car loan. This type of loan is called an **installment loan**. Installment loans are used for cars, boats, home improvements, and even for consolidating several smaller loans into one affordable payment. The table shows the interest you might have to pay to finance a new automobile with an installment loan. Look on the World Wide Web to find competitive interest rates for car loans. Incidentally, you can also check the market value of an automobile on the Web.

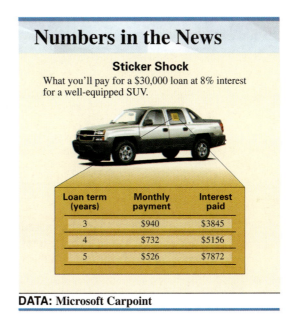

Numbers in the News

Sticker Shock

What you'll pay for a $30,000 loan at 8% interest for a well-equipped SUV.

Loan term (years)	Monthly payment	Interest paid
3	$940	$3845
4	$732	$5156
5	$526	$7872

DATA: Microsoft Carpoint

The **Federal Truth-in-Lending Act** (Regulation Z) of 1969 requires lenders to disclose their **finance charge** (the charge for credit) and **annual percentage rate (APR)** on installment loans. The federal government *does not* regulate rates. Each individual state sets the maximum allowable rates and charges.

The interest rate that is **stated** (in the newspaper, a marketing brochure, or a problem in a textbook) is also called the **nominal rate**. The nominal or stated rate can differ from the annual percentage rate or APR, which is based on the actual amount received by the borrower. The APR is the true effective annual interest rate for a loan. Information on two loans of $1000 each is shown below. An advertisement indicates a rate of 12% for each loan and the actual interest is $120 for each. However, the terms differ.

	Stated Rate	Interest	Term		APR
Loan 1	12%	$120	1 year	$R = \dfrac{I}{PT} = \dfrac{\$120}{\$1000 \times 1}$	= 12%
Loan 2	12%	$120	9 months	$R = \dfrac{I}{PT} = \dfrac{\$120}{\$1000 \times \frac{9}{12}}$	= 16%

Wow, look at the difference in the annual percentage rate between the two loans even though the stated rate, the principal, and the interest are the same. Why? Because *the term differs*. Which loan would you prefer?

Interest rate charges *vary significantly* from one loan source to another. The table below shows that the total finance charge for a $200 loan can be as high as $165 if the loan is rolled over a few times to extend it to six weeks. The finance charge also depends on the borrower's past credit history and income.

QUICK TIP   It pays to shop and make comparisons based on APR values before you borrow!

Numbers in the News

The Cost of Immediate Money

While a fee of 20% of a loan's face value may seem small, when compounded over six weeks it adds up when compared to a cash advance on a credit card. *Rollover* means extending the loan for *another* short period.

$200 Pawn Shop Loan for Six Weeks

Payday advance loan	
Interest rate (20% of loan's face value)	$40
Setup charge (weeks 1 and 2)	$15
First rollover (weeks 3 and 4)	$55
Second rollover (weeks 5 and 6)	$55
TOTAL COST	**$165**
Effective APR	**715%**
Credit-card cash advance	
TOTAL COST	**$5**
APR	**21%**

DATA: *Fortune* magazine

Objective **2** **Find the total installment cost and the finance charge.** Find the annual percentage rate by first finding the **total installment cost** (or the **deferred payment price**) and the **finance charge** on the loan. Do this with the following steps.

Step 1 Find the total installment cost.

Total installment cost = Down payment + Payment amount × Number of payments

Step 2 Find the finance charge (interest).

Finance charge = Total installment cost − Cash price

Step 3 Finally, find the amount financed (principal of loan).

Amount financed = Cash price − Down payment

QUICK TIP   The finance charge includes interest and any fees the lender charges. The amount financed is the loan amount, or the principal.

EXAMPLE 1

Finding Total Installment Cost

Frank Kimlicko recently received his master's degree and began work at a large community college as a music professor specializing in classical guitar. He purchased an exquisite sounding classical guitar costing $3800 with $500 down and 36 monthly payments of $109.61 each. Find **(a)** the total installment cost, **(b)** the finance charge, and **(c)** the amount financed.

SOLUTION

(a) The total installment cost is the down payment plus the total of all monthly payments.

$$\text{Total installment cost} = \$500 + (\$109.61 \times 36) = \mathbf{\$4445.96}$$

(b) The finance charge is the total installment cost less the cash price.

$$\text{Finance charge} = \mathbf{\$4445.96} - \$3800 = \$645.96$$

(c) The amount financed is $3800 - $500 = $3300.

QUICK TIP   In determining the total installment cost, the down payment is added to the total of the monthly payments.

Many students use an installment loan called a **Stafford loan** to help pay costs while in college. The government pays the interest on a *subsidized* Stafford loan while the student borrower is in school on at least a half-time basis. On the other hand, the student is responsible for interest on *unsubsidized* Stafford loans. Repayment of a loan begins six months after the borrower ceases at least half-time enrollment. You can find information about Stafford loans at the financial aid office at your college or at a bank.

Objective 3 Use the formula for approximate APR. The **approximate annual percentage rate (APR)** for a loan paid off in monthly payments can be found with the following formula.

$$\text{Approximate APR} = \frac{24 \times \text{Finance charge}}{\text{Amount financed} \times (1 + \text{Total number of payments})}$$

The formula is *only an estimate* of the APR. It is not accurate enough for the purposes of the Federal Truth-in-Lending Act, which requires the use of tables.

EXAMPLE 2

Finding Annual Percentage Rate

Ed Chamski decides to buy a used car for $6400. He makes a down payment of $1200 and monthly payments of $169 for 36 months. Find the approximate annual percentage rate.

SOLUTION

Use the steps outlined on page 456.

$$\begin{aligned}
\textbf{Total installment cost} &= \$1200 + (\$169 \times 36 \text{ months}) \\
&= \$7284 \text{ total installment cost} \\
\textbf{Finance charge} &= \$7284 - \$6400 = \$884 \\
\textbf{Amount financed} &= \$6400 - \$1200 = \$5200
\end{aligned}$$

Use the formula for approximate APR. Replace the finance charge with $884, the amount financed with $5200, and the number of payments with 36.

$$\begin{aligned}
\textbf{Approximate APR} &= \frac{\mathbf{24 \times \text{Finance charge}}}{\textbf{Amount financed} \times (\mathbf{1 + \text{Total number of payments}})} \\
&= \frac{24 \times \$884}{\$5200 \times (\mathbf{1 + 36})} \\
&= \frac{\$21,216}{\$192,400} \\
&= .110 \text{ or } 11\% \text{ approximate APR}
\end{aligned}$$

The approximate annual percentage rate on this loan is 11%. Example 3 shows how to find the actual APR for this loan.

> **QUICK TIP**   The precise APR can be found using a financial calculator as shown in examples in Appendix D.

Objective ☐4 **Use the table to find APR.** Special tables must be used to find annual percentage rates *accurate enough* to satisfy federal law. These tables are available from a Federal Reserve Bank or the Board of Governors of the Federal Reserve System, Washington, DC 20551. The table on page 459 shows a small portion of these tables. The APR is found from the APR table as follows:

Annual Percentage Rate (APR)

Step 1 Multiply the finance charge by $100, and divide by the amount financed.

$$\frac{\text{Finance charge} \times \$100}{\text{Amount financed}}$$

The result is the finance charge per $100 of the amount financed.

Step 2 Read down the left column of the annual percentage rate table to the proper number of payments. Go across to the number closest to the number found in step 1. Read the number at the top of that column to find the annual percentage rate.

> **QUICK TIP**   Federal law requires that a loan's annual percentage rate (APR) be stated to the nearest quarter of a percent.

EXAMPLE 3

Finding Annual Percentage rate

In Example 2, a used car costing $6400 was financed at $169 per month for 36 months after a down payment of $1200. The total finance charge was $884, and the amount financed was $5200. Find the annual percentage rate.

SOLUTION

Step 1 Multiply the finance charge by $100, and divide by the amount financed.

$$\frac{\$884 \times \$100}{\$5200} = \$17.00$$ Round to two decimal places for use in the table.

This gives the finance charge per $100 financed.

Step 2 Read down the left column of the annual percentage rate table to the line for 36 months (the actual number of monthly payments). Follow across to the right to find the number closest to $17.00. Here, find $17.01. Read across the top of this column of figures to find the annual percentage rate, 10.50%.

In this example, 10.50% is the annual percentage rate that must be disclosed to the buyer of the car. In Example 2, the formula for the approximate annual percentage rate gave an answer of 11%, which is not accurate enough to meet the requirements of the law.

> **QUICK TIP**   When using the annual percentage rate table, select the column with the table number that is closest to the finance charge per $100 of amount financed.

Annual Percentage Rate Table for Monthly Payment Plans

Annual Percentage Rate (Finance Charge per $100 of Amount Financed)

Number of Payments	10.00%	10.25%	10.50%	10.75%	11.00%	11.25%	11.50%	11.75%	12.00%	12.25%	12.50%	12.75%	13.00%	13.25%	13.50%	13.75%	Number of Payments
1	0.83	0.85	0.87	0.90	0.92	0.94	0.96	0.98	1.00	1.02	1.04	1.06	1.08	1.10	1.12	1.15	1
2	1.25	1.28	1.31	1.35	1.38	1.41	1.44	1.47	1.50	1.53	1.57	1.60	1.63	1.66	1.69	1.72	2
3	1.67	1.71	1.76	1.80	1.84	1.88	1.92	1.96	2.01	2.05	2.09	2.13	2.17	2.22	2.26	2.30	3
4	2.09	2.14	2.20	2.25	2.30	2.35	2.41	2.46	2.51	2.57	2.62	2.67	2.72	2.78	2.83	2.88	4
5	2.51	2.58	2.64	2.70	2.77	2.83	2.89	2.96	3.02	3.08	3.15	3.21	3.27	3.34	3.40	3.46	5
6	2.94	3.01	3.08	3.16	3.23	3.31	3.38	3.45	3.53	3.60	3.68	3.75	3.83	3.90	3.97	4.05	6
7	3.36	3.45	3.53	3.62	3.70	3.78	3.87	3.95	4.04	4.12	4.21	4.29	4.38	4.47	4.55	4.64	7
8	3.79	3.88	3.98	4.07	4.17	4.26	4.36	4.46	4.55	4.65	4.74	4.84	4.94	5.03	5.13	5.22	8
9	4.21	4.32	4.43	4.53	4.64	4.75	4.85	4.96	5.07	5.17	5.28	5.39	5.49	5.60	5.71	5.82	9
10	4.64	4.76	4.88	4.99	5.11	5.23	5.35	5.46	5.58	5.70	5.82	5.94	6.05	6.17	6.29	6.41	10
11	5.07	5.20	5.33	5.45	5.58	5.71	5.84	5.97	6.10	6.23	6.36	6.49	6.62	6.75	6.88	7.01	11
12	5.50	5.64	5.78	5.92	6.06	6.20	6.34	6.48	6.62	6.76	6.90	7.04	7.18	7.32	7.46	7.60	12
13	5.93	6.08	6.23	6.38	6.53	6.68	6.84	6.99	7.14	7.29	7.44	7.59	7.75	7.90	8.05	8.20	13
14	6.36	6.52	6.69	6.85	7.01	7.17	7.34	7.50	7.66	7.82	7.99	8.15	8.31	8.48	8.64	8.81	14
15	6.80	6.97	7.14	7.32	7.49	7.66	7.84	8.01	8.19	8.36	8.53	8.71	8.88	9.06	9.23	9.41	15
16	7.23	7.41	7.60	7.78	7.97	8.15	8.34	8.53	8.71	8.90	9.08	9.27	9.46	9.64	9.83	10.02	16
17	7.67	7.86	8.06	8.25	8.45	8.65	8.84	9.04	9.24	9.44	9.63	9.83	10.03	10.23	10.43	10.63	17
18	8.10	8.31	8.52	8.73	8.93	9.14	9.35	9.56	9.77	9.98	10.19	10.40	10.61	10.82	11.03	11.24	18
19	8.54	8.76	8.98	9.20	9.42	9.64	9.86	10.08	10.30	10.52	10.74	10.96	11.18	11.41	11.63	11.85	19
20	8.98	9.21	9.44	9.67	9.90	10.13	10.37	10.60	10.83	11.06	11.30	11.53	11.76	12.00	12.23	12.46	20
21	9.42	9.66	9.90	10.15	10.39	10.63	10.88	11.12	11.36	11.61	11.85	12.10	12.34	12.59	12.84	13.08	21
22	9.86	10.12	10.37	10.62	10.88	11.13	11.39	11.64	11.90	12.16	12.41	12.67	12.93	13.19	13.44	13.70	22
23	10.30	10.57	10.84	11.10	11.37	11.63	11.90	12.17	12.44	12.71	12.97	13.24	13.51	13.78	14.05	14.32	23
24	10.75	11.02	11.30	11.58	11.86	12.14	12.42	12.70	12.98	13.26	13.54	13.82	14.10	14.38	14.66	14.95	24
25	11.19	11.48	11.77	12.06	12.35	12.64	12.93	13.22	13.52	13.81	14.10	14.40	14.69	14.98	15.28	15.57	25
26	11.64	11.94	12.24	12.54	12.85	13.15	13.45	13.75	14.06	14.36	14.67	14.97	15.28	15.59	15.89	16.20	26
27	12.09	12.40	12.71	13.03	13.34	13.66	13.97	14.29	14.60	14.92	15.24	15.56	15.87	16.19	16.51	16.83	27
28	12.53	12.86	13.18	13.51	13.84	14.16	14.49	14.82	15.15	15.48	15.81	16.14	16.47	16.80	17.13	17.46	28
29	12.98	13.32	13.66	14.00	14.33	14.67	15.01	15.35	15.70	16.04	16.38	16.72	17.07	17.41	17.75	18.10	29
30	13.43	13.78	14.13	14.48	14.83	15.19	15.54	15.89	16.24	16.60	16.95	17.31	17.66	18.02	18.38	18.74	30
31	13.89	14.25	14.61	14.97	15.33	15.70	16.06	16.43	16.79	17.16	17.53	17.90	18.27	18.63	19.00	19.38	31
32	14.34	14.71	15.09	15.46	15.84	16.21	16.59	16.97	17.35	17.73	18.11	18.49	18.87	19.25	19.63	20.02	32
33	14.79	15.18	15.57	15.95	16.34	16.73	17.12	17.51	17.90	18.29	18.69	19.08	19.47	19.87	20.26	20.66	33
34	15.25	15.65	16.05	16.44	16.85	17.25	17.65	18.05	18.46	18.86	19.27	19.67	20.08	20.49	20.90	21.31	34
35	15.70	16.11	16.53	16.94	17.35	17.77	18.18	18.60	19.01	19.43	19.85	20.27	20.69	21.11	21.53	21.95	35
36	16.16	16.58	17.01	17.43	17.86	18.29	18.71	19.14	19.57	20.00	20.43	20.87	21.30	21.73	22.17	22.60	36
37	16.62	17.06	17.49	17.93	18.37	18.81	19.25	19.69	20.13	20.58	21.02	21.46	21.91	22.36	22.81	23.25	37
38	17.08	17.53	17.98	18.43	18.88	19.33	19.78	20.24	20.69	21.15	21.61	22.07	22.52	22.99	23.45	23.91	38
39	17.54	18.00	18.46	18.93	19.39	19.86	20.32	20.79	21.26	21.73	22.20	22.67	23.14	23.61	24.09	24.56	39
40	18.00	18.48	18.95	19.43	19.90	20.38	20.86	21.34	21.82	22.30	22.79	23.27	23.76	24.25	24.73	25.22	40
41	18.47	18.95	19.44	19.93	20.42	20.91	21.40	21.89	22.39	22.88	23.38	23.88	24.38	24.88	25.38	25.88	41
42	18.93	19.43	19.93	20.43	20.93	21.44	21.94	22.45	22.96	23.47	23.98	24.49	25.00	25.51	26.03	26.55	42
43	19.40	19.91	20.42	20.94	21.45	21.97	22.49	23.01	23.53	24.05	24.57	25.10	25.62	26.15	26.68	27.21	43
44	19.86	20.39	20.91	21.44	21.97	22.50	23.03	23.57	24.10	24.64	25.17	25.71	26.25	26.79	27.33	27.88	44
45	20.33	20.87	21.41	21.95	22.49	23.03	23.58	24.12	24.67	25.22	25.77	26.32	26.88	27.43	27.99	28.55	45
46	20.80	21.35	21.90	22.46	23.01	23.57	24.13	24.69	25.25	25.81	26.37	26.94	27.51	28.08	28.65	29.22	46
47	21.27	21.83	22.40	22.97	23.53	24.10	24.68	25.25	25.82	26.40	26.98	27.56	28.14	28.72	29.31	29.89	47
48	21.74	22.32	22.90	23.48	24.06	24.64	25.23	25.81	26.40	26.99	27.58	28.18	28.77	29.37	29.97	30.57	48
49	22.21	22.80	23.39	23.99	24.58	25.18	25.78	26.38	26.98	27.59	28.19	28.80	29.41	30.02	30.63	31.24	49
50	22.69	23.29	23.89	24.50	25.11	25.72	26.33	26.95	27.56	28.18	28.80	29.42	30.04	30.67	31.29	31.92	50
51	23.16	23.78	24.40	25.02	25.64	26.26	26.89	27.52	28.15	28.78	29.41	30.05	30.68	31.32	31.96	32.60	51
52	23.64	24.27	24.90	25.53	26.17	26.81	27.45	28.09	28.73	29.38	30.02	30.67	31.32	31.98	32.63	33.29	52
53	24.11	24.76	25.40	26.05	26.70	27.35	28.00	28.66	29.32	29.98	30.64	31.30	31.97	32.63	33.30	33.97	53
54	24.59	25.25	25.91	26.57	27.23	27.90	28.56	29.23	29.91	30.58	31.25	31.93	32.61	33.29	33.98	34.66	54
55	25.07	25.74	26.41	27.09	27.77	28.44	29.13	29.81	30.50	31.18	31.87	32.56	33.26	33.95	34.65	35.35	55
56	25.55	26.23	26.92	27.61	28.30	28.99	29.69	30.39	31.09	31.79	32.49	33.20	33.91	34.62	35.33	36.04	56
57	26.03	26.73	27.43	28.13	28.84	29.54	30.25	30.97	31.68	32.39	33.11	33.83	34.56	35.28	36.01	36.74	57
58	26.51	27.23	27.94	28.66	29.37	30.10	30.82	31.55	32.27	33.00	33.74	34.47	35.21	35.95	36.69	37.43	58
59	27.00	27.72	28.45	29.18	29.91	30.65	31.39	32.13	32.87	33.61	34.36	35.11	35.86	36.62	37.37	38.13	59
60	27.48	28.22	28.96	29.71	30.45	31.20	31.96	32.71	33.47	34.23	34.99	35.75	36.52	37.29	38.06	38.83	60

EXAMPLE **4**

Finding Annual
Percentage Rate

Thompson Plumbing borrowed $48,000 to purchase two vans. Management agreed to a note with payments of $1565.78 per month for 36 months with no down payment. Find the annual percentage rate.

SOLUTION

$$\text{Total installment cost} = \$0 \text{ down payment} + \$1565.78 \times 36$$
$$= \mathbf{\$56{,}368.08}$$

$$\text{Finance charge} = \mathbf{\$56{,}368.08} - \$48{,}000$$
$$= \$8368.08$$

$$\text{Amount financed} = \$48{,}000 - \$0 \text{ down payment}$$
$$= \mathbf{\$48{,}000}$$

Now use the formula for the APR.

$$\frac{\$8368.08 \times \$100}{\mathbf{\$48{,}000}} = 17.4335$$

Find the row associated with 36 payments in the annual percentage rate table. Look to the right across that row to find the number closest to 17.4335 which is 17.43. Look to the top of that column to find 10.75%. This is the APR to the nearest quarter of a percent.

11.2 | Exercises

The **Quick Start** *exercises in each section contain solutions to help you get started.*

Find the finance charge and the total installment cost for the following. (See Example 1.)

Quick Start

	Amount Financed	Down Payment	Cash Price	Number of Payments	Amount of Payment	Total Installment Cost	Finance Charge
1.	$1400	$400	$1800	24	$68.75	**$2050**	**$250**

TIC = $400 + (24 × $68.75) = $2050; FC = $2050 − $1800 = $250

2.	$650	$125	$775	24	$32	**$893**	**$118**

TIC = $125 + (24 × $32) = $893; FC = $893 − $775 = $118

3.	$150	none	$150	12	$15	_____	_____
4.	$1200	none	$1200	20	$70	_____	_____
5.	$2525	$375	$2900	18	$176	_____	_____
6.	$6388	$380	$6768	60	$136	_____	_____

Find the approximate annual percentage rate using the approximate annual percentage rate formula. Round to the nearest tenth of a percent. (See Example 2.)

Quick Start

	Amount Financed	Finance Charge	No. of Monthly Payments	Approximate APR
7.	$7600	$1200	30	**12.2%**

Approx. APR = $\dfrac{24 \times \$1200}{\$7600 \times (30 + 1)}$ = 12.2%

8.	$2200	$434	36	_____
9.	$7542	$1780	48	_____
10.	$4500	$650	36	_____

C indicates an exercise that is related to the Case in Point feature within the section.

Amount Financed	Finance Charge	No. of Monthly Payments	Approximate APR
11. $132	$11	12	_____
12. $8046	$973	24	_____

Find the annual percentage rate using the annual percentage rate table. (See Example 3.)

Quick Start

Amount Financed	Finance Charge	No. of Monthly Payments	APR
13. $1400	$185.68	24	12.25%

$$\frac{FC \times \$100}{AF} = \frac{\$185.68 \times \$100}{\$1400} = 13.26; \text{ from 24-payment row, APR} = 12.25\%$$

14. $345	$24.62	12	_____
15. $442	$28.68	14	_____
16. $4690	$1237.22	48	_____
17. $145	$13.25	18	_____
18. $650	$73.45	24	_____

19. Explain the difference between open-end credit and installment loans. (See Section 11.1 and Objective 1 of this section.)

20. Make a list of all of the items that you have bought on an installment loan. Make another list of things you plan to buy in the next 2 years on an installment loan. (See Objective 1.)

Solve the following application problems and use the table to find the annual percentage rate.

Quick Start

21. CHIP FABRICATION A Chinese computer chip manufacturer borrowed $84 million dollars worth of Chinese yuan to purchase some sophisticated equipment. The note required 24 monthly payments of $3.88 million each. Find the annual percentage rate.

21. <u>10.00%</u>

$$FC = 24 \times \$3.88 \text{ million} - \$84 \text{ million} = \$9.12 \text{ million}$$

$$\frac{FC \times \$100}{AF} = \frac{\$9.12 \times \$100}{\$84 \text{ million}} = 10.86; \text{ from 24-payment row, APR} = 10.00\%$$

22. **REFRIGERATOR PURCHASE** Sears offers a refrigerator for $1600 with no down payment, $294.06 in interest charges and 30 equal payments. Find the annual percentage rate.

22. _____

23. **AUTO PURCHASE** Pat Waller bought a new Toyota Civic for $20,800. She made a down payment of $2000 and agreed to make 60 payments to GMAC of $404 each. Find **(a)** the amount financed, **(b)** the total installment cost, **(c)** the total interest paid, and **(d)** the APR.

(a) _____
(b) _____
(c) _____
(d) _____

24. **SKI BOAT** James Berry purchased a ski boat costing $12,800 with $500 down and loan payments to GMAC of $399 per month for 36 months. Find **(a)** the amount financed, **(b)** the total installment cost, **(c)** the total interest paid, and **(d)** the APR.

(a) _____
(b) _____
(c) _____
(d) _____

25. **COMPUTER SYSTEM** A contractor in Mexico City purchased a computer system for 650,000 pesos. After making a down payment of 100,000 pesos, he agreed to make payments of 26,342.18 pesos per month for 24 months. Find **(a)** the total installment cost and **(b)** the annual percentage rate.

(a) _____
(b) _____

26. **TRACTOR PURCHASE** An electrical contractor in Hiroshima, Japan, purchased a tractor costing 2,700,000 yen. He made a down payment of 1,000,000 yen and agreed to monthly payments of 54,855 yen for 36 months. Find **(a)** the total installment cost and **(b)** the annual percentage rate.

(a) _____
(b) _____

11.3 | Early Payoffs of Loans

Objectives

- **1** Use the United States Rule for an early payment.
- **2** Find the amount due on the maturity date using the United States Rule.
- **3** Use the Rule of 78 when prepaying a loan.

Objective 1 **Use the United States Rule for an early payment.** It is common for a payment to be made on a loan *before it is due*. This may occur when a person receives extra money or when refinancing a debt at a lower interest rate somewhere else. Prepayments of loans are discussed in this section.

The first method for calculating early loan payment is the **United States Rule**, and it is used by the U.S. government, as well as most states and financial institutions. Under the United States Rule, any payment is first applied to any interest owed. The balance of the payment is then used to reduce the principal amount of the loan.

The United States Rule

Step 1 Find the simple interest due from the date the loan was made until the date the partial payment is made. Use the formula $I = PRT$.

Step 2 Subtract this interest from the amount of the payment.

Step 3 Any difference is used to reduce the principal.

Step 4 Treat additional partial payments in the same way, always finding interest on *only* the unpaid balance after the last partial payment.

Step 5 Remaining principal plus interest on this unpaid principal is then due on the due date of the loan.

Objective 2 **Find the amount due on the maturity date using the United States Rule.** If the partial payment is not large enough to pay the interest due, the payment is simply held until enough money is available to pay the interest due. This means that a partial payment smaller than the interest due offers no advantage to the borrower—the lender just holds the partial payment until enough money is available to pay the interest owed.

> **QUICK TIP** We will continue to use 360-day years in the calculations of this section.

EXAMPLE 1

Finding the Amount Due

On August 14, Dr. Jane Ficker signed a 180-day note for $28,500 for an X-ray machine for her dental office. The note has an interest rate of 10% compounded annually. On October 25, a payment of $8500 is made. **(a)** Find the balance owed on the principal. **(b)** If no additional payments are made, find the amount due at maturity of the loan.

SOLUTION

(a) **Step 1** Find interest from August 14 to October 25 $(298 - 226 = $ **72 days** $)$* using $I = PRT$.

$$\text{Interest} = \$28,500 \times .10 \times \frac{72}{360} = \textbf{\$570}$$

Step 2 Subtract interest from the October 25 payment to find amount of the payment to be applied to principal.

$$\text{Applied to principal} = \$8500 - \textbf{\$570} = \textbf{\$7930}$$

---

*The chart for the number of days of the year is inside the back cover of the book.

Step 3 Reduce the original principal by this amount.

$$\text{New principal} = \$28,500 - \$7930 = \$20,570$$

(b) Step 4 Since there are no additional partial payments, go onto Step 5.

Step 5 The note was originally for 180 days, but the partial payment was made after 72 days. Interest on the new principal of $20,570 will be charged for $180 - 72 = 108$ days.

$$\text{Interest} = \$20,570 \times .10 \times \frac{108}{360} = \$617.10$$

$$\text{Amount due at maturity} = \$20,570 + \$617.10 = \$21,187.10$$

EXAMPLE 2

Finding the Interest Paid and Amount Due

On March 1, Boston Dairy signs a promissory note for $38,500 to purchase milking equipment for their Holsteins. The note is for 180 days at a rate of 10%. The dairy makes the following partial payments: $6000 on June 9 and $3500 on July 11. Find the interest paid on the note and the amount due on the due date of the note.

SOLUTION

The first partial payment is on June 9, or using the number of days in each month, after $(30 + 30 + 31 + 9) = 100$ days.

$$\text{Interest for 100 days} = \$38,500 \times .10 \times \frac{100}{360} = \$1069.44 \ (\text{rounded})$$

First partial payment	$6000.00
Portion going to interest	−1069.44
Portion going to reduce debt	$4930.56

$$\text{Debt on June 9 } after\ 1st\ partial\ payment = \$38,500 - \$4930.56 = \$33,569.44$$

The second partial payment occurs $21 + 11 = 32$ days later.

$$\text{Interest for 32 days} = \$33,569.44 \times .10 \times \frac{32}{360} = \$298.40 \ (\text{rounded})$$

Second partial payment	$3500.00
Portion going to interest	−298.40
Portion going to reduce debt	$3201.60

$$\text{Debt on July 11 } after\ 2nd\ partial\ payment = \$33,569.44 - \$3201.60 = \$30,367.84$$

The first partial payment is made after 100 days, and the second partial payment is made after an additional 32 days. Thus, the due date of the note is $180 - 100 - 32 = 48$ days after the second partial payment.

$$\text{Interest for the last 48 days} = \$30,367.84 \times .10 \times \frac{48}{360} = \$404.90 \ (\text{rounded})$$

$$\text{Amount due } at\ maturity = \$30,367.84 + \$404.90 = \$30,772.74$$

Date Payment Made	Amount of Payment	Applied to Interest	Applied to Principal	Remaining Balance
June 9	$6,000.00	$1069.44	$4,930.56	$33,569.44
July 11	$3,500.00	$298.40	$3,201.60	$30,367.84
at maturity	$30,772.74	$404.90	$30,367.84	$0
Total	$40,272.74	$1772.74	$38,500.00	

Objective 3 Use the Rule of 78 when prepaying a loan. A variation of the United States Rule, called the **Rule of 78**, is still used by many lenders for installment loans. This rule allows a lender *to earn more of the finance charge during the early months* of the loan compared with the

United States Rule. Lenders typically use this rule to protect against early payoffs on small loans. Effectively, the lender will earn a higher rate of interest in the event of an early payoff under the Rule of 78 than under the United States Rule.

The Rule of 78 gets its name based on a loan of 12 months—the sum of the months $1 + 2 + 3 + \cdots + 12 = 78$. The finance charge for the first month is $\frac{12}{78}$ of the total charge, with $\frac{11}{78}$ in the second month, $\frac{10}{78}$ in the third month, and so on with $\frac{1}{78}$ in the final month. The Rule of 78 can be applied to loans *with terms other than 12 months*. For example, the sum of the months in a 6-month contract is $1 + 2 + 3 + 4 + 5 + 6 = 21$. The finance charge for the first month would be $\frac{6}{21}$, $\frac{5}{21}$ for the second month, and so on. Similarly, the sum of the months in a 15-month contract is $1 + 2 + \cdots + 15 = 120$. The finance charge for the first month of a 15-month contract is $\frac{15}{120}$ or $\frac{1}{8}$, and so on.

Monthly Finance Charges

Term of Loan	Month 1	Month 2	Month 3	Month 4	Month 5	Month 6	Month 7	Month 8	Month 9	Month 10	Month 11	Month 12
12 months	$\frac{12}{78}$	$\frac{11}{78}$	$\frac{10}{78}$	$\frac{9}{78}$	$\frac{8}{78}$	$\frac{7}{78}$	$\frac{6}{78}$	$\frac{5}{78}$	$\frac{4}{78}$	$\frac{3}{78}$	$\frac{2}{78}$	$\frac{1}{78}$
6 months	$\frac{6}{21}$	$\frac{5}{21}$	$\frac{4}{21}$	$\frac{3}{21}$	$\frac{2}{21}$	$\frac{1}{21}$						

The total finance charge on an installment loan is calculated when a loan is first made. Early payoff of a loan results in a lower finance charge. The portion of the finance charge that is *not earned* by the lender under the Rule of 78, called **unearned interest** or **refund**, is found using the following formula.

$$U = F\left(\frac{N}{P}\right)\left(\frac{1 + N}{1 + P}\right)$$

where

U = unearned interest F = finance charge
N = number of payments remaining P = original number of payments

QUICK TIP Unearned interest refers to interest that has not been earned by the lender.

EXAMPLE 3

Finding Unearned Interest and Balance Due

Adrian Ortega borrowed $600, which he is paying back in 24 monthly payments of $29.50 each. With 9 payments remaining, he decides to repay the loan in full. Find **(a)** the amount of unearned interest and **(b)** the amount necessary to repay the loan in full. Use the Rule of 78.

SOLUTION

(a) Total of all payments = 24 payments × $29.50 = **$708**

Finance charge = **$708** − $600 = **$108**

Find the amount of unearned interest as follows. The finance charge is $108, the scheduled number of payments is 24, and the loan is paid off with 9 payments left. Solve as follows.

$$\text{Unearned interest} = \mathbf{\$108} \times \frac{9}{24} \times \frac{(1 + 9)}{(1 + 24)} = \$16.20$$

(b) When Ortega decides to pay off the loan, he has 9 payments of $29.50 left.

Sum of remaining payments = 9 payments × $29.50 = **$265.50**

Ortega saves the unearned interest of $16.20 by paying off the loan early. Therefore, the amount needed to pay the loan in full is the sum of the remaining payments minus the unearned interest.

$$\text{Amount needed to repay the loan in full} = \text{Remaining payments} - \text{unearned interest}$$
$$= \$265.50 - \$16.20 = \$249.30$$

EXAMPLE 4

Finding Unearned Interest and Balance Due

Matt Thompson borrows $1200 for a new refrigerator and agrees to make 18 monthly payments of $74.30 each. After the 10th payment, he pays the loan in full. Find **(a)** the amount of unearned interest and **(b)** the amount needed to repay the loan in full using the Rule of 78.

SOLUTION

(a)
$$\text{Sum of payments} = 18 \text{ payments} \times \$74.30 = \$1337.40$$
$$\text{Finance charge} = \$1337.40 - \$1200 = \$137.40$$

Since there are $(18 - 10) = $ **8 payments remaining** when he pays the loan off, the unearned interest is found as follows.

$$\text{Unearned interest} = \$137.40 \times \frac{8}{18} \times \frac{(1 + 8)}{(1 + 18)} = \$28.93 \text{ (rounded)}$$

(b) The amount needed to pay the loan in full is the sum of the remaining payments less the unearned interest.

$$\text{Sum of remaining 8 payments} = 8 \text{ payments} \times \$74.30 = \$594.40$$
$$\text{Amount needed to repay the loan in full} = \$594.40 - \$28.93 = \$565.47$$

11.3 | Exercises

The **Quick Start** *exercises in each section contain solutions to help you get started.*

Find the balance due on the maturity date of the following notes. Find the total amount of interest paid on each note. Use the United States Rule. (See Examples 1 and 2.)

Quick Start

	Principal	Interest	Time (Days)	Partial Payments	Balance Due	Total Interest Paid
1.	$9800	$8\frac{1}{2}\%$	150	$1800 on day 50	**$8307.31**	**$307.31**
2.	$5800	12%	120	$2000 on day 45	**$3984.18**	**$184.18**
3.	$15,000	10.5%	200	$6500 on day 100	_____	_____
4.	$76,900	11%	180	$31,250 on day 75	_____	_____
5.	$18,457	12%	120	$5978 on day 34 $3124 on day 55	_____	_____
6.	$39,864	9%	105	$8458 on day 43 $11,354 on day 88	_____	_____

Each of the following loans is paid in full before the date of maturity. Find the amount of unearned interest. Use the Rule of 78. (See Example 3.)

Quick Start

	Finance Charge	Total Number of Payments	Remaining Number of Payments when Paid in Full	Unearned Interest
7.	$1050	24	11	**$231**

$$\$1050 \times \tfrac{11}{24} \times \frac{(1 + 11)}{(1 + 24)} = \$231$$

	Finance Charge	Total Number of Payments	Remaining Number of Payments when Paid in Full	Unearned Interest
8.	$422	30	16	_____
9.	$881	36	12	_____
10.	$325	24	22	_____
11.	$900	36	6	_____
12.	$1250	60	12	_____

C indicates an exercise that is related to the Case in Point feature within the section.

13. Explain why banks prefer the Rule of 78 to the United States Rule in the event of prepayment. (See Objective 3.)

14. Describe a situation in which a bank might prefer a loan to be prepaid.

Solve the following application problems using the United States Rule.

Quick Start

15. LANDSCAPING   Andrew Raring borrowed $8900 from GMAC to landscape his yard. The 240-day note had an interest rate of 12% compounded annually. He repaid the note in 140 days with his income tax refund. Find **(a)** the interest due and **(b)** the total amount due.

(a) $415.33
(b) $9315.33

(a) Loan is for 140 days
Interest = $8900 × .12 × $\frac{140}{360}$ = $415.33
(b) Amount due = $8900 + $415.33 = $9315.33

16. COMPUTER CONSULTANT   The computer system at Genome Therapy crashed several times last year. On January 10, the company borrowed $125,000 at 11% compounded annually for 250 days to pay a consultant to work on their Novell network. However, they decide to pay the loan in full on July 1. Find **(a)** the interest due and **(b)** the total amount due.

(a) _____
(b) _____

17. Thompson Packaging borrowed $92,000 on May 7, signing a note due in 90 days at 11.25% interest. On June 24, the company made a partial payment of $24,350. Find **(a)** the amount due on the maturity date of the note and **(b)** the interest paid on the note.

(a) _____
(b) _____

18. REMODELING   The Second Avenue Butcher Shop financed a remodeling program by giving the builder a note for $32,500. The note was made on September 14 and is due in 120 days. Interest on the note is 9.75%. On December 9, the firm makes a partial payment of $9000. Find **(a)** the amount due on the maturity date of the note and **(b)** the interest paid on the note.

(a) _____
(b) _____

19. **INVENTORY** Wholesale Paper orders large quantities of basic paper goods every 4 months to save on freight charges. For its last order, the firm signed a note on February 18, maturing on May 15. The face value of the note was $104,500, with interest of 11%. The firm made a partial payment of $38,000 on March 20 and a second partial payment of $27,200 on April 16. Find **(a)** the amount due on the maturity date of the note and **(b)** the amount of interest paid on the note.

(a) _____
(b) _____

20. **SURVEILLANCE CAMERAS** To help detect trespassers at night, a small security firm purchased some high-technology cameras using a note from GMAC for $32,000. The note was signed on July 26 and was due on November 20. The interest rate is 13%. The firm made a partial payment of $6000 on August 31 and a second partial payment of $11,700 on October 4. Find **(a)** the amount due on the maturity date of the note and **(b)** the interest paid on the note.

(a) _____
(b) _____

Solve the following application problems using the Rule of 78. (See Example 3.)

21. **ENGAGEMENT RING** Tom Stowe purchased a diamond engagement ring for $1150. He paid $100 down and agreed to 12 monthly payments of $95 each. After making 7 payments, he paid the loan in full. Find **(a)** the unearned interest and **(b)** the amount necessary to pay the loan in full.

(a) _____
(b) _____

22. **GARBAGE TRUCK** Haul-it-Away, Inc., purchased a garbage truck for $62,000. They paid a down payment of $22,000 and financed the remainder with 36 payments of $1328.57 each. They paid off the note with 12 payments remaining. Find **(a)** the amount of unearned interest and **(b)** the amount necessary to pay the loan in full.

(a) _____
(b) _____

23. PRINTING BlackTop Printing made a $5000 down payment on a special copy machine costing $23,800.
The loan agreement with GMAC called for 20 monthly payments of $1025 each. Find
(a) the finance charge, (b) the unearned interest, and (c) the amount necessary to pay the loan in full
after the 14th payment.

(a) _____

(b) _____

(c) _____

U

24. MOVIE PROJECTORS Movie 6, Inc., purchased several movie projectors
at a total cost of $12,200 with a down payment of $1500. The company
agreed to make 12 monthly payments of $945 each. Find (a) the finance
charge, (b) the unearned interest, and (c) the amount necessary to pay the
loan in full after the 8th payment.

(a) _____

(b) _____

(c) _____

U

11.4   Personal Property Loans

Objectives

1. Define personal property and real estate.
2. Use the formula for amortization to find payment.
3. Set up an amortization table.
4. Find monthly payments.

CASE *in* POINT

Before she was in mortgage lending at GMAC, Susan Beckman made personal property loans. Every day she would receive many loan applications showing income, assets, and debt, as well as employment information. She would then carefully check the credit worthiness of each applicant.

Automobile manufacturers sometimes use rebates and low cost loans as incentives to help persuade buyers to purchase their automobiles. **Rebates** are discounts that effectively reduce the purchase price and are discussed below.

Auto Rebates, Cheap Loans Coming Faster

By Earle Eldridge
USA TODAY

Automakers are offering cash rebates and cheap loans on new model-year cars much more quickly than they have in the past.

It's a sign that incentives are no longer being used just to reduce bloated inventory or clear out last year's models. Instead, the deals have become an important tool as automakers adjust prices throughout the year to meet swings in demand, both nationally and regionally.

Eleven non-luxury brands offered incentives on their 2004 models in September, according to the Power Information Network, an affiliate of J.D. Power and Associates. That's compared with eight brands offering incentives on new 2003 models last September. September is the traditional month for automakers to begin selling next year's models.

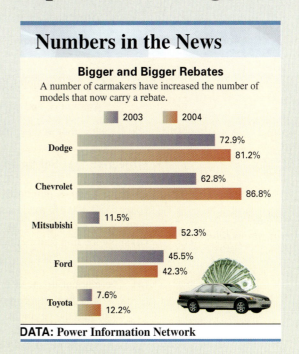

Numbers in the News

Bigger and Bigger Rebates
A number of carmakers have increased the number of models that now carry a rebate.

2003   2004

Dodge: 72.9% / 81.2%
Chevrolet: 62.8% / 86.8%
Mitsubishi: 11.5% / 52.3%
Ford: 45.5% / 42.3%
Toyota: 7.6% / 12.2%

DATA: Power Information Network

Objective 1   Define personal property and real estate. Items that can be moved from one location to another, such as an automobile, a boat, or a stereo, are called **personal property**. In contrast, land and homes cannot be moved and are called **real estate** or **real property**. Personal-property loans are discussed in this section, and real estate loans are discussed in the next section.

Banks, credit unions, finance companies and many other types of companies make money through personal property loans. Typically, these loans are repaid, or **amortized**, using monthly payments. Sometimes, a buyer is not able to make the payments as promised. In that event, the lender must **repossess** the personal property and sell it to someone else. Financial companies charge a slightly higher interest rate to everyone to make up for loans on which they never fully recover their money. Interest rates for personal property loans vary significantly from lender to lender and from year to year. Be sure to shop carefully before borrowing.

Objective **2** **Use the formula for amortization to find payment.** The periodic payment needed at the end of each period to amortize a loan with interest i per period, over n periods, is found by using the following formula.

$$\text{Payment} = \text{Loan amount} \times \text{Number from amortization table}$$

EXAMPLE 1

Amortizing a Loan

Sven Yarborough earned his degree at a community college and is now a mechanic at a Toyota dealership. He was so impressed with the quality of Toyotas that he purchased a Toyota SUV at a cost of $29,400, including tax, title and license, but after the rebate. He made a down payment of $3500 and was able to finance the balance at a special incentive rate of 6% per year for 4 years. Find **(a)** the monthly payment, **(b)** the portion of the first payment that is interest, **(c)** the balance due after one payment, **(d)** the interest owed for the second month, and **(e)** the balance after the second payment.

SOLUTION

(a) Amount financed = $29,400 − $3500 = $25,900

Use $\frac{6\%}{12}$ = .5% per month and 4 years × 12 = 48 months in the table to find **.02349**.

Monthly payment = $25,900 × **.02349** = $608.39 $\left(\text{rounded}\right)$

(b) Interest for month $I = PRT$ = $25,900 × .06 × $\frac{1}{12}$ = **$129.50**

Amount of 1st payment applied to principal = $608.39 − **$129.50** = **$478.89**

(c) Balance after 1st payment = $25,900 − **$478.89** = $25,421.11

(d) Interest for 2nd month = PRT = $25,421.11 × .06 × $\frac{1}{12}$ = **$127.11** $\left(\text{rounded}\right)$

Amount of 2nd payment applied to principal = $608.39 − **$127.11** = $481.28

(e) Balance after 2nd payment = $25,421.11 − $481.28 = **$24,939.83**

Objective **3** **Set up an amortization table.**

EXAMPLE 2

Creating an Amortization Table

Clarence Thomas purchased new commercial-grade washers and dryers for his laundromat at a cost of $22,300. He made a down payment of $5000 and agreed to pay the balance off in quarterly payments over 2 years at 12% compounded quarterly. Find **(a)** the quarterly payment and **(b)** show the first four payments in a table called an amortization schedule.

SOLUTION

(a) The amount financed is $22,300 − $5000 = $17,300. Find the factor from the table using 2 × 4 = 8 quarters and $\frac{12\%}{4}$ = 3% per quarter. Then multiply the amount financed by the factor to find the payment.

$$\text{Payment} = \$17,300 \times \textbf{.14246} = \$2464.56 \left(\text{rounded}\right)$$

(b)

Amortization Schedule

Payment Number	Amount of Payment	Interest for Period	Portion to Principal	Principal at End of Period
0	—	—	—	$17,300.00
1	$2464.56	$519.00	$1945.56	$15,354.44
2	$2464.56	$460.63	$2003.93	$13,350.51
3	$2464.56	$400.52	$2064.04	$11,286.47
4	$2464.56	$338.59	$2125.97	$ 9,160.50

n	½%	1%	1½%	2%	2½%	3%	4%	6%	8%	10%	12%	n
1	1.00500	1.01000	1.01500	1.02000	1.02500	1.03000	1.04000	1.06000	1.08000	1.10000	1.12000	1
2	.50375	.50751	.51128	.51505	.51883	.52261	.53020	.54544	.56077	.57619	.59170	2
3	.33667	.34002	.34338	.34675	.35014	.35353	.36035	.37411	.38803	.40211	.41535	3
4	.25313	.25628	.25944	.26262	.26582	.26903	.27549	.28859	.30192	.31547	.32923	4
5	.20301	.20604	.20909	.21216	.21525	.21835	.22463	.23740	.25046	.26380	.27741	5
6	.16960	.17255	.17553	.17853	.18155	.18460	.19076	.20336	.21632	.22961	.24323	6
7	.14573	.14863	.15156	.15451	.15750	.16051	.16661	.17914	.19207	.20541	.21912	7
8	.12783	.13069	.13358	.13651	.13947	.14246	.14853	.16104	.17401	.18744	.20130	8
9	.11391	.11674	.11961	.12252	.12546	.12843	.13449	.14702	.16008	.17364	.18768	9
10	.10277	.10558	.10843	.11133	.11426	.11723	.12329	.13587	.14903	.16275	.17698	10
11	.09366	.09645	.09929	.10218	.10511	.10808	.11415	.12679	.14008	.15396	.16842	11
12	.08607	.08885	.09168	.09456	.09749	.10046	.10655	.11928	.13270	.14676	.16144	12
13	.07964	.08241	.08524	.08812	.09105	.09403	.10014	.11296	.12652	.14078	.15568	13
14	.07414	.07690	.07972	.08260	.08554	.08853	.09467	.10758	.12130	.13575	.15087	14
15	.06936	.07212	.07494	.07783	.08077	.08377	.08994	.10296	.11683	.13147	.14682	15
16	.06519	.06794	.07077	.07365	.07660	.07961	.08582	.09895	.11298	.12782	.14339	16
17	.06151	.06426	.06708	.06997	.07293	.07595	.08220	.09544	.10963	.12466	.14046	17
18	.05823	.06098	.06381	.06670	.06967	.07271	.07899	.09236	.10670	.12193	.13794	18
19	.05530	.05805	.06088	.06378	.06676	.06981	.07614	.08962	.10413	.11955	.13576	19
20	.05267	.05542	.05825	.06116	.06415	.06722	.07358	.08718	.10185	.11746	.13388	20
21	.05028	.05303	.05587	.05878	.06179	.06487	.07128	.08500	.09983	.11562	.13224	21
22	.04811	.05086	.05370	.05663	.05965	.06275	.06920	.08305	.09803	.11401	.13081	22
23	.04613	.04889	.05173	.05467	.05770	.06081	.06731	.08128	.09642	.11257	.12956	23
24	.04432	.04707	.04992	.05287	.05591	.05905	.06559	.07968	.09498	.11130	.12846	24
25	.04265	.04541	.04826	.05122	.05428	.05743	.06401	.07823	.09368	.11017	.12750	25
26	.04111	.04387	.04673	.04970	.05277	.05594	.06257	.07690	.09251	.10916	.12665	26
27	.03969	.04245	.04532	.04829	.05138	.05456	.06124	.07570	.09145	.10826	.12590	27
28	.03836	.04112	.04400	.04699	.05009	.05329	.06001	.07459	.09049	.10745	.12524	28
29	.03713	.03990	.04278	.04578	.04889	.05211	.05888	.07358	.08962	.10673	.12466	29
30	.03598	.03875	.04164	.04465	.04778	.05102	.05783	.07265	.08883	.10608	.12414	30
31	.03490	.03768	.04057	.04360	.04674	.05000	.05686	.07179	.08811	.10550	.12369	31
32	.03389	.03667	.03958	.04261	.04577	.04905	.05595	.07100	.08745	.10497	.12328	32
33	.03295	.03573	.03864	.04169	.04486	.04816	.05510	.07027	.08685	.10450	.12292	33
34	.03206	.03484	.03776	.04082	.04401	.04732	.05431	.06960	.08630	.10407	.12260	34
35	.03122	.03400	.03693	.04000	.04321	.04654	.05358	.06897	.08580	.10369	.12232	35
36	.03042	.03321	.03615	.03923	.04245	.04580	.05289	.06839	.08534	.10334	.12206	36
37	.02967	.03247	.03541	.03851	.04174	.04511	.05224	.06786	.08492	.10303	.12184	37
38	.02896	.03176	.03472	.03782	.04107	.04446	.05163	.06736	.08454	.10275	.12164	38
39	.02829	.03109	.03405	.03717	.04044	.04384	.05106	.06689	.08419	.10249	.12146	39
40	.02765	.03046	.03343	.03656	.03984	.04326	.05052	.06646	.08386	.10226	.12130	40
41	.02704	.02985	.03283	.03597	.03927	.04271	.05002	.06606	.08356	.10205	.12116	41
42	.02646	.02928	.03226	.03542	.03873	.04219	.04954	.06568	.08329	.10186	.12104	42
43	.02590	.02873	.03172	.03489	.03822	.04170	.04909	.06533	.08303	.10169	.12092	43
44	.02538	.02820	.03121	.03439	.03773	.04123	.04866	.06501	.08280	.10153	.12083	44
45	.02487	.02771	.03072	.03391	.03727	.04079	.04826	.06470	.08259	.10139	.12074	45
46	.02439	.02723	.03025	.03345	.03683	.04036	.04788	.06441	.08239	.10126	.12066	46
47	.02393	.02677	.02980	.03302	.03641	.03996	.04752	.06415	.08221	.10115	.12059	47
48	.02349	.02633	.02938	.03260	.03601	.03958	.04718	.06390	.08204	.10104	.12052	48
49	.02306	.02591	.02896	.03220	.03562	.03921	.04686	.06366	.08189	.10095	.12047	49
50	.02265	.02551	.02857	.03182	.03526	.03887	.04655	.06344	.08174	.10086	.12042	50

> **QUICK TIP**   Notice that interest is large at first when the debt is high but that it decreases with every payment as the debt goes down. The last payment varies slightly from the regular monthly payments due to rounding.

Objective 4   Find monthly payments. The loan payoff table below can be used as an alternative to the amortization table on the previous page. The table on this page shows some higher interest rates and longer terms than the previous table. Personal property loans sometimes have higher interest rates since people are more likely to default on a loan for an expensive plasma television set than say on the loan on their home.

This table has a different format than the table on the previous page. The APR is down the left column, and the number of months is across the top of this table. You can see that values on the 12% APR row of this table (corresponding to 1% per month) match the values in the 1% column in the amortization table.

> Payment = Loan amount × Number from loan payoff table

Loan Payoff Table

Mos. APR	18	24	30	36	42	48	54	60	Mos. APR
8%	.05914	.04523	.03688	.03134	.02738	.02441	.02211	.02028	8%
9%	.05960	.04568	.03735	.03180	.02785	.02489	.02259	.02076	9%
10%	.06006	.04615	.03781	.03227	.02832	.02536	.02307	.02125	10%
11%	.06052	.04661	.03828	.03274	.02879	.02585	.02356	.02174	11%
12%	.06098	.04707	.03875	.03321	.02928	.02633	.02406	.02225	12%
13%	.06145	.04754	.03922	.03369	.02976	.02683	.02456	.02275	13%
14%	.06192	.04801	.03970	.03418	.03025	.02733	.02507	.02327	14%
15%	.06238	.04849	.04018	.03467	.03075	.02783	.02558	.02379	15%
16%	.06286	.04896	.04066	.03516	.03125	.02834	.02610	.02432	16%
17%	.06333	.04944	.04115	.03565	.03176	.02885	.02662	.02485	17%
18%	.06381	.04993	.04164	.03615	.03226	.02937	.02715	.02539	18%
19%	.06428	.05041	.04213	.03666	.03278	.02990	.02769	.02594	19%
20%	.06476	.05090	.04263	.03716	.03330	.03043	.02823	.02649	20%

EXAMPLE 3

Finding Amortization Payments

After a trade-in, Vickie Ewing owes $17,400 on a new Harley-Davidson motorcycle and wishes to pay the loan off in 60 months. She has found that she can finance the loan at 9% per year if she has a good credit history, but at 14% per year if she has a poor credit history.

(a) Find the monthly payment at both interest rates.
(b) Find the total finance charge at both interest rates.
(c) Find the extra cost of having poor credit.

SOLUTION

(a) Monthly payment at 9% = $17,400 × .02076 = $361.22 (rounded)
Monthly payment at 14% = $17,400 × .02327 = $404.90 (rounded)

(b) The finance charge is the sum of all of the payments minus the amount financed.

number of payments

Finance charge at 9% = 60 × $361.22 − $17,400 = $4273.20

Finance charge at 14% = 60 × $404.90 − $17,400 = $6894

(c) Extra cost of poor credit = $6894 − $4273.20 = $2620.80

11.4 | Exercises

FOR EXTRA HELP

 MyMathLab

 InterActMath.com

 MathXL

 MathXL Tutorials on CD

 Addison-Wesley Math Tutor Center

DVT/Videotape

The **Quick Start** *exercises in each section contain solutions to help you get started.*

Find the payment necessary to amortize the following loans using the amortization table. Round to the nearest cent if needed. (See Example 1.)

Quick Start

	Amount of Loan	Interest Rate	Payments Made	Number of Years	Payment
1.	$6800	8%	annually	5	$1703.13

Payment = $6800 × .25046 = $1703.13

2.	$2650	12%	annually	10	$469

Payment = $2650 × .17698 = $469

	Amount of Loan	Interest Rate	Payments Made	Number of Years	Payment
3.	$4500	8%	semiannually	$7\frac{1}{2}$	_____
4.	$12,000	6%	semiannually	8	_____
5.	$96,000	8%	quarterly	$7\frac{3}{4}$	_____
6.	$210,000	12%	quarterly	8	_____
7.	$4876	12%	monthly	3	_____
8.	$6800	6%	monthly	3	_____

Use the loan-payoff table to find the monthly payment and finance charge for each of the following loans. (See Example 2.)

Quick Start

	Amount Financed	Number of Months	APR	Monthly Payment	Finance Charge
9.	$5300	42	9%	$147.61	$899.62

MP = $5300 × .02785 = $147.61; FC = (42 × $147.61) − $5300 = $899.62

10.	$4800	24	12%	$225.94	$622.56

MP = $4800 × .04707 = $225.94; FC = (24 × $225.94) − $4800 = $622.56

 indicates an exercise that is related to the Case in Point feature within the section.

Amount Financed	Number of Months	APR	Monthly Payment	Finance Charge
11. $12,000	48	13%	_____	_____
12. $8102	48	8%	_____	_____
13. $11,750	60	11%	_____	_____
14. $16,000	60	10%	_____	_____

15. Explain the process of amortizing a loan. (See Objective 1.)

16. Explain why a loan officer at a bank might look at a credit report on someone before making a loan. Describe the steps you would take if your credit report was inaccurate.

Solve the following application problems using the amortization table.

Quick Start

17. **ROAD GRADER**   Revis Construction borrows $62,400 to purchase a grader to prepare roads before paving. The loan has a rate of 12% per year and requires 40 monthly payments. Find (a) the monthly payment and (b) the total interest paid.

    (a) Monthly payment = $62,400 × .03046 = $1900.70
    (b) Total interest = (40 × $1900.70) − $62,400 = $13,628

    (a) $1900.70 _____
    (b) $13,628 _____

18. **OPENING A RESTAURANT**   Chuck and Judy Nielson opened a restaurant at a cost of $340,000. They paid $40,000 of their own money and agreed to pay the remainder in quarterly payments over 7 years at 12%. Find (a) the quarterly payment and (b) the total amount of interest paid over 7 years.

    (a) _____
    (b) _____

19. An insurance firm pays $4000 for a new high-speed color printer for its computer. It amortizes the loan for the printer in 4 annual payments at 8%. Prepare an amortization schedule for this machine.

Payment Number	Amount of Payment	Interest for Period	Portion to Principal	Principal at End of Period

20. **TRACTOR PURCHASE** Long Haul Trucking purchases a used tractor for pulling large trailers on interstate highways at a cost of $72,000. It agrees to pay for it with a loan from GMAC that will be amortized over 9 annual payments at 8% interest. Prepare an amortization schedule for the truck.

Payment Number	Amount of Payment	Interest for Period	Portion to Principal	Principal at End of Period
	—	—	—	

Solve the following application problems. Use the loan payoff table. (See Objective 3.)

21. **ELECTRONIC EQUIPMENT** An engineering firm purchases 7 new workstations with laser printers for $3500 each. The firm makes a down payment of $10,000 and amortizes the balance with monthly loan payments to GMAC of 11% for 4 years. Prepare an amortization schedule showing the first 5 payments.

Payment Number	Amount of Payment	Interest for Period	Portion to Principal	Principal at End of Period
0	—	—	—	
1				
2				
3				
4				
5				

22. **AMORTIZING A LOAN** Rebecca Reed just graduated from dental school and borrows $120,000 from GMAC to purchase equipment for her own business. She agreed to amortize the loan with monthly payments at 10% for 4 years. Prepare an amortization schedule for the first 5 payments.

Payment Number	Amount of Payment	Interest for Period	Portion to Principal	Principal at End of Period
0	—	—	—	
1				
2				
3				
4				
5				

11.5 | Real Estate Loans

Objectives

1 Determine monthly payments on a home.
2 Prepare a repayment schedule.
3 Define escrow accounts.
4 Define fixed and variable rate loans.

> **CASE** *in* **POINT**
>
> Susan Beckman's current position at GMAC requires her to analyze the many home-loan applications she receives every week. Effectively, she makes the decision of whether or not GMAC will make a home loan to a particular applicant. Beckman then works with qualified applicants to make sure that GMAC does their part in making the loan.

Objective 1 Determine monthly payments on a home. A home is *one of the most expensive purchases* that a person makes in his or her lifetime. The monthly payment for a home mortgage depends on the amount borrowed, the interest rate, and the term of the loan.

The bar graph shows that home **mortgage** rates have fallen from a high of over 15% in 1981 to a low of about 6% in 2004. Interest rates are a very important factor in determining the monthly payment and therefore the affordability of a particular home. If interest rates fall, people **refinance** their home loans at lower rates to reduce their monthly payments. Sometimes, people are able to get cash out of the equity in their home when they refinance. The interest on a home loan may also be tax deductible, which can help people afford homes by reducing their income taxes.

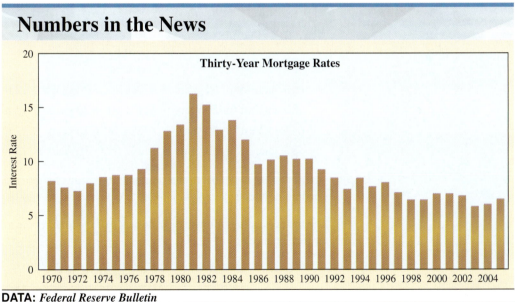

Numbers in the News

Thirty-Year Mortgage Rates

DATA: *Federal Reserve Bulletin*

The amount of the monthly payment is found by the methods given in the previous section, but special tables are used for real estate loans because of the long repayment periods. The real estate amortization table on the next page shows the monthly payment necessary to repay a $1000 loan for differing interest rates and lengths of repayment. To use the table, first find the amount to be financed in thousands by dividing the total amount to be borrowed by $1000. Multiply this value by the appropriate number from the table.

Real Estate Amortization Table (Principal and Interest per Thousand Dollars Borrowed)

Terms in Years	$6\frac{1}{2}\%$	$6\frac{3}{4}\%$	7%	$7\frac{1}{4}\%$	$7\frac{1}{2}\%$	$7\frac{3}{4}\%$	8%	$8\frac{1}{4}\%$	$8\frac{1}{2}\%$	$8\frac{3}{4}\%$	9%	$9\frac{1}{4}\%$	Terms in Years
10	11.35	11.48	11.62	11.75	11.88	12.01	12.14	12.27	12.40	12.54	12.67	12.81	10
15	8.71	8.85	8.99	9.13	9.28	9.42	9.56	9.71	9.85	10.00	10.15	10.30	15
20	7.46	7.60	7.76	7.91	8.06	8.21	8.37	8.53	8.68	8.84	9.00	9.16	20
25	6.75	6.91	7.07	7.23	7.39	7.56	7.72	7.89	8.06	8.23	8.40	8.57	25
30	6.32	6.49	6.66	6.83	7.00	7.17	7.34	7.52	7.69	7.87	8.05	8.23	30

EXAMPLE 1

Understanding the Effect of Rate and Term

The Stringers make a 10% down payment, but still need to borrow $140,000 to purchase a condominium. They want to know the effect of the interest rate and term of the loan on cost. Find **(a)** the monthly payment for both 20 and 30 years at $6\frac{1}{2}\%$ and at 8%. Then find **(b)** the total cost of the home with each loan and **(c)** the finance charge for each loan.

SOLUTION

(a) The amount to be financed in thousands = $140,000 ÷ $1000 = 140. Multiply this value by the appropriate factor from the real-estate amortization table.

	Monthly Payment
$6\frac{1}{2}\%$ interest for 20 years = 140 × **7.46** =	**$1044.40**
8% interest for 20 years = 140 × **8.37** =	**$1171.80**
$6\frac{1}{2}\%$ interest for 30 years = 140 × **6.32** =	**$ 884.80**
8% interest for 30 years = 140 × **7.34** =	**$1027.60**

Monthly payments range from $884.80 to $1171.80, depending on rate and term. The lower payment of $884.80 may look good to you at first, but look below.

(b) The total cost of financing the home is the sum of all payments.

	Monthly Payment		Total Cost
$6\frac{1}{2}\%$ interest for 20 years =	$1044.40 × 20 years × 12 month/yr =		**$250,656**
8% interest for 20 years =	$1171.80 × 20 years × 12 month/yr =		**$281,232**
$6\frac{1}{2}\%$ interest for 30 years =	$ 884.80 × 30 years × 12 month/yr =		**$318,528**
8% interest for 30 years =	$1027.60 × 30 years × 12 month/yr =		**$369,936**

Notice that the total cost of the condominium is much more than the original price of the condominium. Interest adds *a lot of cost* to the purchase as shown in **(c)**.

(c) The finance charge or interest cost of each of the loans is the total cost found in **(b)** minus the amount financed of $140,000.

Interest Rate

		$6\frac{1}{2}\%$	8%
Term of Loan	20 years	$110,656	$141,232
	30 years	$178,528	$229,936

Clearly, higher interest rates and longer terms add **HUGE** amounts to the interest that must be paid to purchase a property. Many people think it best to finance your home in a way to keep monthly payments low and then to pay cash for everything else. Home ownership offers tax advantages, and the value of the home often increases in value over time.

> **QUICK TIP** You can reduce your long-term cost of a mortgage by paying more than the required payment every month. Even an extra $60 per month can make a big difference over the long run.

> **QUICK TIP** Be sure to divide the loan amount by $1000 before calculating the monthly payment.

Mortgage payoffs of 25 or 30 years have been common in the past. However, **accelerated mortgages**, with payoffs of 15 or 20 years, are becoming more common for the reason pointed out in Example 1—lower total costs. Mortgages with shorter terms also tend to have slightly lower interest rates.

Objective 2 **Prepare a repayment schedule.** Many lenders use a computer to calculate an **amortization schedule**, also called a **repayment schedule**. This schedule separates each payment into the portion going to interest and the portion reducing the debt.

EXAMPLE 2

*Preparing a
Repayment Schedule*

The Jamisons have made payments on their home for more than 9 years. They have decided to refinance the $60,000 loan balance at 8% fixed for 30 years since it will lower their monthly payment to $440.40. Prepare a loan reduction schedule for this loan.

SOLUTION

Detailed loan payment calculations are shown only for the first two months. Interest for the first month is found using the simple interest formula.

$$\text{Interest} = P \times R \times T = \$60,000 \times .08 \times \frac{1}{12} = \mathbf{\$400.00}$$

monthly payment ⟶   ⟵ 1st month's interest

Amount of payment that reduces debt = $440.40 − **$400.00** = **$40.40**

Remaining debt = $60,000 − **$40.40** = $59,959.60

Repeat these steps for the second month only now the debt is $59,959.60.

$$\text{Interest} = P \times R \times T = \$59,959.60 \times .08 \times \frac{1}{12} = \mathbf{\$399.73}$$

Amount of payment that reduces debt = $440.40 − **$399.73** = **$40.67**

Remaining debt = $59,959.60 − **$40.67** = $59,918.93

The next table shows the first 12 payments, payments 262–269, and the last 3 payments. Notice that the interest charge is large and the principal payment is very small for the first payments. In fact, the principal is reduced by *only* $502.97 in the first 12 months, even though the payments total $5284.80. The remaining balance drops below one-half of the original loan of $60,000 only after 269 payments, or 22 years and 5 months out of a 30-year loan. It requires over 22 years to cut the loan balance in half and then close to 8 years to pay off the other half of the loan.

Loan Reduction Schedule
(Loan amount $60,000; Term 30 years; Rate 8%; Payment $440.40)

Payment Number	Interest Payment	Principal Payment	Remaining Balance	Payment Number	Interest Payment	Principal Payment	Remaining Balance
1	$400.00	$40.40	$59,959.60	262	$211.55	$228.85	$31,504.19
2	399.73	40.67	59,918.93	263	210.03	230.37	31,273.82
3	399.46	40.94	59,877.99	264	208.49	231.91	31,041.91
4	399.19	41.21	59,836.78	265	206.95	233.45	30,808.46
5	398.91	41.49	59,795.29	266	205.39	235.01	30,573.45
6	398.64	41.76	59,753.53	267	203.82	236.58	30,336.87
7	398.36	42.04	59,711.49	268	202.24	238.16	30,098.72
8	398.08	42.32	59,669.17	269	200.66	239.74	29,858.98
9	397.79	42.61	59,626.56	⋮	⋮	⋮	⋮
10	397.51	42.89	59,583.67	358	7.32	433.08	664.66
11	397.22	43.18	59,540.49	359	4.43	435.97	228.69
12	396.94	43.46	59,497.03	360	1.52	228.69	0.00
				Totals	$98,331.96	$60,000	

Objective 3   Define escrow accounts. To prevent losses from unpaid taxes or uninsured damages, many lenders require **escrow accounts** (also called **impound accounts**) for people taking out a mortgage. With an escrow account, buyers pay $\frac{1}{12}$ of the total estimated property tax and insurance each month. The lender holds these funds until the taxes and insurance fall due and then *pays the bills for the borrower*. Many consumer groups oppose this practice, since the lender earns interest on the money while waiting for payments to come due. In fact, a few states require that interest be paid on escrow accounts on any homes located in those states.

EXAMPLE 3

Finding Total Monthly Payment

Susan Beckman arranged for a client to receive a $75,000 loan for 25 years at $7\frac{1}{4}\%$ to purchase a summer cabin. Annual insurance and taxes on the property are $654 and $1329, respectively. Find the total monthly payment.

SOLUTION

Use the real-estate amortization table to find $7.23. Add monthly insurance and taxes to the payment amount.

Principal and interest   Insurance   Taxes

$$\text{Monthly Payment} = \left(75 \times \$7.23\right) + \left(\frac{\$654 + \$1329}{12}\right)$$
$$= \$542.25 + \$165.25$$
$$= \$707.50$$

Objective 4   Define fixed and variable rate loans. Home loans with fixed, stated interest rates are called **fixed-rate loans**. Many borrowers prefer fixed-rate loans, since they know that the principal and interest portion of their monthly payments will be fixed until they either sell the house or pay off the debt. In particular, it is good for a borrower (but bad for the lender) if he or she locks in a fixed-rate loan before interest rates go up. If interest rates fall after a person has borrowed money for a home, the person can usually refinance at a lower interest rate.

QUICK TIP   Managers of banks and mortgage companies seriously and continuously think about the effects of an increase or decrease in interest rates on company profits.

Another type of loan is the **adjustable rate mortgage (ARM)**, also called **variable interest rate loan**, as shown in the clipping. The interest rate on this type of loan is periodically adjusted either up or down depending on the movement of interest rates in general. Usually, there is a maximum limit in the increase in the rate from one year to the next.

With this type of mortgage, monthly payments increase if the interest rate is increased. Payments decrease if the rate is decreased. Thus, a borrower's monthly payment is not fixed but changes from time to time. The advantage of an adjustable rate mortgage is that it will originally have a lower interest rate than a fixed rate loan and thus a lower monthly payment. The potentially very serious disadvantage of an adjustable rate mortgage is that the monthly payment may eventually be increased to the point where the home is no longer affordable. Some people will not use adjustable rate mortgage because they want no risk the payment will increase significantly.

HERE & NOW

Adjustable-rate mortgages regain some popularity

By Thomas A. Fogarty
USA TODAY

Adjustable-rate mortgages are thriving these days even as interest rates remain low.

Interest rates on 1-year ARMs all this year have been running 2 to 3 percentage points lower than rates on 30-year, fixed-rate mortgages.

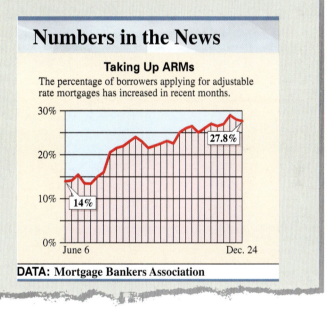

Numbers in the News

Taking Up ARMs

The percentage of borrowers applying for adjustable rate mortgages has increased in recent months.

27.8%

14%

June 6   Dec. 24

DATA: Mortgage Bankers Association

11.5 | Exercises

FOR EXTRA HELP

MyMathLab

InterActMath.com

*Math*XP MathXL

MathXL
Tutorials on CD

Tutor
Center Addison-Wesley
Math Tutor Center

DVT/Videotape

The **Quick Start** *exercises in each section contain solutions to help you get started.*

Use the real estate amortization table to find the monthly payment for the following loans. (See Example 1.)

Quick Start

Amount of Loan	Interest Rate	Term of Loan	Monthly Payment
1. $95,000	$8\frac{1}{2}\%$	30 years	**$730.55**
Payment = 95 × $7.69 = $730.55			
2. $149,000	$7\frac{3}{4}\%$	20 years	**$1223.29**
Payment = 149 × 8.21 = $1223.29			
3. $112,800	$8\frac{1}{2}\%$	15 years	_____
4. $132,000	$6\frac{1}{2}\%$	25 years	_____
5. $92,400	$6\frac{3}{4}\%$	30 years	_____
6. $88,200	$7\frac{1}{4}\%$	15 years	_____

7. Explain how different interest rates can make a large difference in interest charges over a number of years. (See Example 1.)

8. Explain how interest can result in a total cost that is over twice the original loan amount when a home is financed over 30 years. (See Example 1.)

Find the total monthly payment, including taxes and insurance, for the following loans. Round to the nearest cent.

Quick Start

Amount of Loan	Interest Rate	Term of Loan	Annual Taxes	Annual Insurance	Monthly Payment
9. $98,000	7%	30 years	$1250	$560	**$803.51**
98 × $6.66 = $652.68; $652.68 + $\frac{\$1250\ +\ \$560}{12}$ = $803.51					
10. $75,400	$8\frac{1}{2}\%$	20 years	$1177	$520	**$795.89**
75.4 × $8.68 = $654.47; $654.47 + $\frac{\$1177\ +\ \$520}{12}$ = $795.89					

 indicates an exercise that is related to the Case in Point feature within the section.

Amount of Loan	Interest Rate	Term of Loan	Annual Taxes	Annual Insurance	Monthly Payment
11. $58,600	8%	30 years	$745	$380	_____
12. $68,400	9%	30 years	$1256	$350	_____
13. $91,580	$8\frac{1}{4}\%$	25 years	$1326	$489	_____
14. $173,000	$6\frac{1}{2}\%$	30 years	$2800	$920	_____

Solve the following application problems.

15. **HOME PURCHASE**   The Potters want to buy a home costing $127,000 with annual insurance and taxes of $720 and $2300 respectively. They have saved $10,000 for a down payment, and they can get a $7\frac{1}{2}\%$, 30-year mortgage from GMAC. They are qualified for a home loan as long as the total monthly payment does not exceed $1200. Are they qualified?

16. **CONDOMINIUM PURCHASE**   Mr. and Mrs. Ariz wish to buy a condominium costing $95,000 with annual insurance and taxes of $680 and $2278 respectively. They have $6000 saved for a down payment and plan to amortize the balance at 9% for 25 years. They are qualified for a home loan as long as the total monthly payment does not exceed $850. Are they qualified for the loan?

16. _____

17. June and Bill Able borrow $122,500 at $7\frac{1}{2}\%$ for 15 years. Prepare a repayment schedule for the first two payments. (See Example 3.)

Payment Number	Total Payment	Interest Payment	Principal Payment	Balance of Principal

18. **ELDERLY HOUSING**   Tom Ajax purchases a home for his elderly mother. After a large down payment, he finances $44,300 at $7\frac{1}{4}\%$ for 10 years. Prepare a repayment schedule for the first two payments. (See Example 3.)

Payment Number	Total Payment	Interest Payment	Principal Payment	Balance of Principal

CHAPTER TERMS *Review the following terms to test your understanding of the chapter. For each term you do not know, refer to the page number found next to that term.*

accelerated mortgages **[p. 483]**

adjustable rate mortgages **[p. 485]**

amortization schedule **[p. 483]**

amortization table **[p. 474]**

amortize **[p. 455]**

annual percentage rate **[p. 455]**

approximate annual percentage rate **[p. 457]**

APR **[p. 455]**

average daily balance method **[p. 448]**

consolidate loans **[p. 450]**

country club billing **[p. 445]**

credit limit **[p. 444]**

debit cards **[p. 444]**

defaulting on debt **[p. 450]**

deferred payment price **[p. 456]**

escrow accounts **[p. 484]**

Federal Truth-in-Lending Act **[p. 455]**

finance charges **[p. 445]**

fixed rate loans **[p. 484]**

grace period **[p. 445]**

impound accounts **[p. 484]**

installment cost **[p. 456]**

installment loan **[p. 455]**

itemized billing **[p. 445]**

late fees **[p. 445]**

mortgages **[p. 481]**

nominal rate **[p. 455]**

open-end credit **[p. 444]**

over-the-limit fees **[p. 445]**

personal property **[p. 473]**

real estate **[p. 473]**

real property **[p. 473]**

refinance **[p. 481]**

refund of unearned interest **[p. 467]**

repayment schedule **[p. 483]**

repossess **[p. 473]**

revolving charge account **[p. 445]**

Rule of 78 **[p. 466]**

Stafford loan **[p. 457]**

stated rate **[p. 455]**

unearned interest **[p. 467]**

United States Rule **[p. 465]**

unpaid balance method **[p. 445]**

variable interest rate loans **[p. 485]**

CONCEPTS

11.1 Finding the finance charge on a revolving charge account, using the unpaid balance method

Start with the unpaid balance of the previous month. Then find the finance charge on the unpaid balance. Next, add the finance charge and any purchases. Finally, subtract any payments made.

EXAMPLES

Debbie Mahoney's MasterCard account had an unpaid balance of $385.65 on March 1. During March, she made a $100 payment and charged $68.92. Find the finance charge and the unpaid balance on April 1 if the bank charges 1.25% per month on the unpaid balance.

$$\text{Finance charge} = \$385.65 \times .0125 = \$4.82$$

Previous Balance	Finance Charge	Purchases	Payment	New Balance

$$\$385.65 + \$4.82 + \$68.92 - \$100 = \$359.39$$

11.1 Finding the finance charge on a revolving charge account, using the average daily balance method

First find the unpaid balance on each day of the month. Then add up the daily unpaid balances.

Next divide the total of the daily unpaid balances by the number of days in the billing period.

Finally, calculate the finance charge by multiplying the average daily balance by the finance charge.

The following is a summary of a credit-card account.

Previous balance $115.45
November 1 Billing date
November 15 Payment of $35.00
November 22 Charge of $45.00

Find the average daily balance and the finance charge if interest is 1% per month on the average daily balance.

Balance on November 1 = $115.45

$$\text{Nov. } 1 - 15 = 14 \text{ days at } \$115.45$$
$$14 \times \$115.45 = \$1616.30$$

Payment on Nov. 15 = $35.00
Balance on Nov. 15 = $115.45 - $35 = $80.45

$$\text{Nov. } 15 - \text{Nov. } 22 = 7 \text{ days at } \$80.45$$
$$7 \times \$80.45 = \$563.15$$

Charge on Nov. 22 of $45.00
Balance on Nov. 22 = $80.45 + $45 = $125.45

$$\text{Nov. } 22 - \text{Dec. } 1 = 9 \text{ days at } \$125.45$$
$$9 \times \$125.45 = \$1129.05$$

Daily Balances

$$\$1616.30 + \$563.15 + \$1129.05 = \mathbf{\$3308.50}$$

Average Daily Balance

$$= \frac{\$3308.50}{30} = \$110.28$$

Finance Charge

$$= \$110.28 \times .01 = \$1.10$$

CONCEPTS	EXAMPLES

11.2 Finding the total installment cost, finance charge, and amount financed

Total Installment Cost

= Down payment

+ (Amount of each payment × number of payments)

Finance Charge (interest)

= Total installment cost − Cash price

Amount Financed (principal of loan)

= Cash price − Down payment

Joan Taylor bought a leather coat for $1580. She put $350 down and then made 12 payments of $115 each. Find the total installment cost, the finance charge, and the amount financed.

Total Installment Cost

= $350 + (12 × $115) = $1730

Finance Charge

= $1730 − $1580 = $150

Amount Financed

= $1580 − $350 = $1230

11.2 Determining the approximate APR using a formula

First determine the finance charge. Then find the amount financed. Next calculate approximate APR using the formula.

Approximate APR

$$= \frac{24 \times \text{Finance charge}}{\text{Amt. fin.} \times \left(1 + \begin{array}{l}\text{Total no.}\\\text{of payments}\end{array}\right)}$$

Tom Jones buys a motorcycle for $8990. He makes a down payment of $1800 and then makes monthly payments of $230 for 36 months. Find the approximate APR.

Total Installment Cost

= $1800 + ($230 × 36) = $10,080

Finance Charge

= $10,080 − $8990 = **$1090**

Amount Financed

= $8990 − $1800 = $7190

Approximate APR

$$= \frac{24 \times \$1090}{\$7190 \times (1 + 36)} = 9.8\% \text{ (rounded)}$$

11.2 Finding APR using a table

First determine the finance charge per $100 of amount financed, using the formula

$$\frac{\text{Finance charge} \times \$100}{\text{Amount financed}}$$

Then read down the left column of the annual percentage rate table to the proper number of payments. Go across to the number closest to the number found above. Read across the top of the column to find the annual percentage rate.

Lupe Torres buys a used car for $6500. She makes a down payment of $1000 and agrees to make 24 monthly payments of $260.83. Use the table to find the APR.

Finance charge = $1000 + (24 × $260.83) − $6500

= $759.92

Amount financed = $5500

Finance charge per $100

$$\frac{\$759.92 \times \$100}{\$5500} = 13.817 \text{ (rounded)}$$

Use the 24-payment row in the table to find APR = 12.75%

CONCEPTS	EXAMPLES

11.3 Finding amount due on maturity date using the United States Rule

First determine the simple interest due from the date loan was made until the date of the partial payment. Then subtract this interest from the amount of the payment and reduce the principal by the difference. Next find the interest from the date of partial payment to the due date of the note, and add the unpaid balance and interest to find the amount due.

Sam Wiley signs a 90-day note on August 1 for $5000 at an interest rate of 12%. On September 15, he makes a payment of $1800. Find the balance owed on the principal. If no additional payments are made, find the amount due on the maturity date of the loan.

From August 1 to September 15, there are $30 + 15 = 45$ days.

$$I = \$5000 \times .12 \times \frac{45}{360} = \textbf{\$75} \text{ interest due}$$

$$
\begin{array}{rl}
\$1800 & \text{payment} \\
- \quad 75 & \text{interest due} \\
\hline
\$1725 & \text{applied to principal} \\
& \text{reduction}
\end{array}
$$

$$
\begin{array}{rl}
\$5000 & \text{amount owed} \\
- \ 1725 & \text{principal reduction} \\
\hline
\textbf{\$3275} & \textbf{balance owed}
\end{array}
$$

Note is for 90 days; a partial payment was made after 45 days. Interest on $3275 will be charged for $90 - 45 = 45$ days.

$$I = \$3275 \times .12 \times \frac{45}{360} = \textbf{\$49.13}$$

$$
\begin{array}{rl}
\$3275.00 & \text{principal owed} \\
+ \quad 49.13 & \text{interest} \\
\hline
\textbf{\$3324.13} & \textbf{amount due}
\end{array}
$$

11.3 Finding the unearned interest using the Rule of 78

First calculate the finance charge. Then find the unearned interest using the formula

$$U = F\left(\frac{N}{P}\right)\left(\frac{1+N}{1+P}\right)$$

where U = unearned interest,
$\quad F$ = finance charge,
$\quad N$ = number of payments remaining, and
$\quad P$ = total number of payments.

Next find the total of the remaining payments. Finally, subtract the unearned interest to find the balance remaining.

Tom Fish borrows $1500, which he is paying back in 36 monthly installments of $52.75 each. With 10 payments remaining, he decides to pay the loan in full. Find **(a)** the amount of unearned interest and **(b)** the amount necessary to pay the loan in full.

36 payments of $52.75 each for a total repayment of $36 \times \$52.75 = \$1899$.

Finance charge

$$= \$1899 - \$1500 = \textbf{\$399}$$

Unearned interest

$$= \textbf{\$399} \times \frac{10}{36} \times \frac{(1 + 10)}{(1 + 36)} = \textbf{\$32.95}$$

10 payments of $52.75 are left. These payments total

$$\$52.75 \times 10 = \$527.50$$

$$\$527.50 - \textbf{\$32.95} = \$494.55.$$

This is the amount needed to pay the loan in full.

CONCEPTS	EXAMPLES

11.4 **Finding the periodic payment for amortizing a loan**

First determine the number of periods for the loan and the interest rate per period.

The payment is found by multiplying the loan amount by the number from the amortization table.

Bob Smith agrees to pay $12,000 for a used car. The amount will be repaid in monthly payments over 3 years at an interest rate of 12%. Find the amount of each payment.

$$12 \times 3 = \textbf{36 periods} \left(\textbf{payments}\right)$$

$$\frac{12\%}{12} = \textbf{1\% per period}$$

Number from table is **.03321**.

Payment = $12,000 × **.03321** = $398.52

11.4 **Setting up an amortization schedule**

First find the periodic payment. Then calculate the interest owed in the first period using the formula $I = PRT$. Next subtract the value of I from the periodic payment.

This is the amount applied to the reduction of the principal. Then find the balance after the first periodic payment by subtracting the value of the debt reduction from the original amount. Now repeat the above steps until the original loan is amortized (paid off).

Terri Meyer borrows $1800. She will repay this amount in 2 years with semiannual payments at an interest rate of 8%. Set up an amortization schedule.

4 periods (payments); 4% per period

Number from table is **.27549**.

Payment = $1800 × **.27549** = $495.88

$I = PRT$

Interest owed = $1800 × .08 × $\frac{1}{2}$ = $72

Debt reduction = $495.88 − $72 = $423.88

Balance of loan = $1800 − $423.88 = $1376.12

Payment Number	Amount of Payment	Interest for Period	Portion to Principal	Principal at End of Period
0	—	—	—	$1800.00
1	$495.88	$72.00	$423.88	$1376.12
2	$495.88	$55.04	$440.84	$ 935.28
3	$495.88	$37.41	$458.47	$ 476.81
4	$495.88	$19.07	$476.81	$ 0.00

11.4 **Finding monthly payments, total amount paid, and finance charge**

First multiply the amount to be financed by the number from the amortization table or the loan payoff table to find the periodic payment. Then find the total amount repaid by multiplying the periodic payment by the number of payments. Finally, subtract the amount financed from the total amount repaid to obtain the finance charge.

Ben Apostolides purchased a new Toyota Camry and owes $16,400 after the trade-in. He decides on a term with 50 monthly payments. Assume an interest rate of 12% compounded monthly and find the amount of each payment and the finance charge.

Monthly payment = $16,400 × **.02551** = **$418.36**

Finance charge = 50 × **$418.36** − $16,400 = $4518

CONCEPTS	**EXAMPLES**
11.5 Finding the amount of monthly home loan payments and total interest charges over the life of a home loan Using the number of years and the interest rate, find the amortization value per thousand dollars from the real-estate amortization table. Next multiply the table value by the number of thousands in the principal to obtain the monthly payment. Then find the total amount of the payments and subtract the original amount owed from the total payments to obtain interest paid.	Lou and Rose Waters bought a house at the beach. After a down payment, they owe $75,000. Find the monthly payment at $8\frac{3}{4}\%$ and the total interest charges over the life of a 25-year loan. $$n = 25 \qquad i = 8\frac{3}{4}\%$$ **Table value** $= 8.23$ There are $\dfrac{\$75,000}{1000} = 75$ thousands in $75,000. **Monthly payment** $= 75 \times 8.23 = $ **$617.25** There are $25 \times 12 = 300$ payments. **Total payment** $= 300 \times $ **$617.25** $= \$185,175$ **Interest paid** $= \$185,175 - \$75,000 = \$110,175$

Summary Exercise

Consolidating Loans

John and Kathy MacGruder are struggling to make their monthly payments. Kathy works one job and takes care of their two small children. The interest rates on their numerous debts are high, since they have historically had a poor credit history. However, John has taken on a second job and they have somehow managed to make their payments regularly for a little over a year.

(a) Find the monthly payments on each of the following purchases and the total monthly payment.

Purchase	Original Loan Amount	Interest Rate	Term of Loan	Monthly Payment
Honda Accord	$18,800	12%	4 years	_____
Ford truck	$14,300	18%	4 years	_____
Home	$96,500	$8\frac{1}{2}\%$	15 years	_____
2nd mortgage on home	$ 4,500	12%	3 years	_____
			Total	_____

(b) These monthly expenses do not include car insurance ($215 per month), health insurance ($120 per month), or real-estate taxes on their home ($2530 per year), among other expenses. Find their total monthly outlay for all of these expenses.

Expense	Monthly Outlay
Payments on debt from **(a)**	_____
Car insurance	_____
Health insurance	_____
Real estate taxes on home	_____
Total	_____

(c) After discussing things with Susan Beckman at GMAC, the MacGruders have learned that they can (1) refinance the remaining $14,900 amount on the Honda Accord at 12% over 4 years, (2) refinance the remaining $8600 loan amount on the Ford truck at 12% over 3 years, (3) refinance the remaining $94,800 loan amount on their home at 8% over 30 years, and (4) reduce their car insurance payments by $28 per month. Complete the following table.

Item	Current Loan Amount	New Interest Rate	New Term of Loan	New Monthly Payment
Honda Accord				_____
Ford truck				_____
Home				_____
2nd mortgage on home				_____
Car insurance				_____
Health insurance				_____
Real estate taxes on home				_____
				_____

(d) Find the reduction in their monthly payments.

(d) _____

QUICK TIP   Part of the savings in the monthly payment came from reducing the interest rates. The remainder of the savings came from extending the loans further into the future.

INVESTIGATE

The interest rate that you are charged for borrowing money differs, depending on the bank you go to for a loan. Find current interest rates for financing a 2-year-old car from at least two banks in the area in which you live. Then go onto the World Wide Web and look for a lower interest rate.

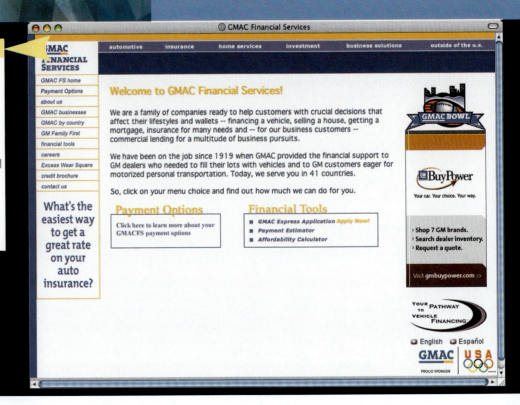

www.gmacfs.com/

- 1920 Expansion to Great Britain.
- 1958 Financed their 40 millionth vehicle.
- 1985 First-ever annual earnings of $1 billion.
- 2004 Operate in 41 countries.

GMAC is a wholly owned subsidiary of General Motors. GMAC was founded in 1919 to provide financial support to GM dealers who needed funds to fill their lots with vehicles and to customers purchasing GM cars. Today they offer many services for individuals including auto loans, home loans, other types of loans, charge cards, and insurance products. GMAC purchased the Bank of New York's commercial asset lending operations in 1999 and they now offer many different types of loans to business clients as well.

1. In one year, assume GMAC finances 3,000,000 automobiles with an average loan amount of $23,200. Find the total loan amount.

2. Assume their average interest rate earned last year was 7.18%. Find the annual interest on every $1 billion in loans.

3. Loan officer Susan Beckman made a $49,800 loan to Betsy Faber for a new Cadillac. Find the monthly payment if the interest rate is 6% and the loan is for 50 months.

4. Use GMAC's Web site to find 5 countries not in North America in which GMAC does business.

Chapter 11 Test

To help you review, the numbers in brackets show the section in which the topic was discussed.

Solve the following problems.

1. A cruise line needs to update the computer and sonar equipment on one of its luxury ships that sails the Caribbean. The cost of the equipment is $214,500. The company makes a down payment of $20,000 and agrees to 24 monthly payments of $8975 per month. Find the total finance charge.   **[11.1]**

1. _____

2. The balance on John Baker's MasterCard on November 1 is $680.45. In November, he charges an additional $337.32 of purchases, has returns of $45.42, and makes a payment of $50. If the finance charges are calculated at 1.5% per month on the unpaid balance, find his balance on December 1.   **[11.1]**

2. _____

Find the annual percentage rate, using the annual percentage rate table.   **[11.2]**

Amount Financed	Finance Charge	Number of Payments	APR
3. $5280	$1010.59	36	_____
4. $1130	$149.84	24	_____

Solve the following application problems.

5. Barton Springs Landscaping buys a used truck for $18,700 and agrees to make 36 payments of $612.25 each. Find the annual percentage rate on the loan.   **[11.2]**

5. _____

6. A note with a face value of $7000 is made on June 21. The note is for 90 days and carries interest of 13%. A partial payment of $2800 is made on July 17. Find the amount due on the maturity date of the note.   **[11.3]**

6. _____

7. Mockton Construction bought a truck and financed $7400 with 48 monthly payments of $228.14 each. Suppose the firm pays the loan off with 12 payments left. Use the Rule of 78 to find **(a)** the amount of unearned interest and **(b)** the amount necessary to pay off the loan.   **[11.3]**

(a) _____
(b) _____

Find the amount of each payment necessary to amortize the following loans.   **[11.4]**

8. Jenson SawLogs borrows $34,500 to buy a new electric generator. The company agrees to make quarterly payments for 2 years at 10% per year. Find the amount of the quarterly payment.

8. _____

9. Pizza for You has its building remodeled for $36,000. It pays $6000 down and pays off the balance in payments made at the end of each quarter for 5 years. Interest is 10% compounded quarterly. Find the amount of each payment so that the loan is fully amortized.

9. _____

Find the monthly payment necessary to amortize the following home mortgages.   **[11.5]**

10. $123,500, $7\frac{1}{2}$%, 30 years

10. _____

11. $134,560, 7%, 15 years

11. _____

Work the following application problems.   **[11.5]**

12. Mr. and Mrs. Zagorin plan to buy a $90,000 cabin, paying 20% down and financing the balance at 8% for 30 years. The taxes are $960 per year, with fire insurance costing $252 per year. Find the monthly payment (including taxes and insurance).

12. _____

13. General Business Forms purchases a $145,000 commercial building, pays 25% down and finances the balance at 9% for 15 years. **(a)** Find the total monthly payment given taxes of $2300 per year and insurance of $1350 per year. **(b)** Assume that insurance and taxes do not increase and find the total cost of owning the building for 15 years (including the down payment).

(a) _____
(b) _____

14. Jerome Watson, owner of Watson Welding, purchases a building for his business and makes a $25,000 down payment. He finances the balance of $122,500 for 20 years at 8%. **(a)** Find the total monthly payment given taxes of $3200 per year and insurance of $1275 per year. **(b)** Assume that insurance and taxes do not increase and find the total cost of owning the building for 20 years (including the down payment).

(a) _____
(b) _____

Cumulative Review

Chapters 10–11

Round money amounts to the nearest cent and rates to the nearest tenth of a percent.

Find the amount of each of the following ordinary annuities.   **[10.1]**

Amount of Each Deposit	Deposited	Rate	Time (Years)	Amount of Annuity	Interest Earned
1. $1000	annually	4%	8	_____	_____
2. $2000	quarterly	6%	5	_____	_____

Find the amount of each annuity due and the interest earned.   **[10.1]**

Amount of Each Deposit	Deposited	Rate	Time (Years)	Amount of Annuity	Interest Earned
3. $2500	annually	5%	6	_____	_____
4. $1800	semiannually	8%	5	_____	_____

Find the present value of the following annuities.   **[10.2]**

Amount per Payment	Payment at End of Each	Time (Years)	Rate of Investment	Compounded	Present Value
5. $925	6 months	11	8%	semiannually	_____
6. $27,235	quarter	8	8%	quarterly	_____

Find the required payment into a sinking fund.   **[10.3]**

Future Value	Interest Rate	Compounded	Time (Years)	Payment
7. $3600	8%	annually	7	_____
8. $4500	10%	quarterly	7	_____

Solve the following application problems using 360-day years where applicable.

9. At 58, Thomas Jones knows that he must start saving for his retirement. He decides to invest $300 per quarter in an account paying 10% compounded quarterly. Find the accumulated amount **(a)** at age 65 and **(b)** at age 70.   **[10.1]**

(a) _____

(b) _____

10. A public utility needs $60 million in 5 years for a major capital expansion. What annual payment must they place in a sinking fund earning 10% per year in order to accumulate the required funds?   **[10.3]**

10. _____

11. Jerry Walker purchased 100 shares of stock at $23.45 per share. The company had earnings of $1.56 and a yearly dividend of $.35. Find **(a)** the cost of the purchase ignoring commissions, **(b)** the price earnings ratio to the nearest whole number, and **(c)** the dividend yield.   **[10.4]**

(a) _____
(b) _____
(c) _____

12. Martin Wicker buys 9000 GM bonds due in 2020 at 104.38 for the pension fund he manages. The coupon rate is 6.4%. Find **(a)** the cost to purchase the bonds if the commission is $1 per bond, **(b)** the annual interest from all of the bonds, and **(c)** the effective interest rate.   **[10.5]**

(a) _____
(b) _____
(c) _____

13. James Thompson purchased a large riding lawnmower costing $2800 with $500 down and payments of $108.27 per month for 24 months. Find **(a)** the total installment cost, **(b)** the finance charge, and **(c)** the amount financed. Then **(d)** use the table to find the annual percentage rate to the nearest quarter of a percent.   **[11.2]**

(a) _____
(b) _____
(c) _____
(d) _____

14. Abbie Spring's unpaid balance on her Visa card on July 8 was $204.37. She made a payment of $100 on July 14 and had charges of $34.95 on July 16 and $95.12 on July 30. Assume an interest rate of 1.6% per month, and find the balance on August 8 using **(a)** the unpaid balance method and **(b)** the average-daily-balance method.   **[11.1]**

(a) _____
(b) _____

$\div$

15. Mayberry Pets borrows to purchase a van to transport animals and supplies. They agree to make quarterly payments on the $22,400 debt for 3 years at a rate of 8% compounded quarterly. Find **(a)** the quarterly payment and **(b)** the total amount of interest paid.   **[11.4]**

(a) _____
(b) _____

16. The Hodges purchase an older 4-bedroom home for $195,000 with 5% down. They finance the balance at $7\frac{1}{2}$% per year for 30 years. If insurance of $720 per year and taxes are $2940 per year, find the monthly payment.   **[11.5]**

16. _____

17. On January 10, Bob Jones signed a 200-day note for $24,000 to finance some work on a commercial building. The note was at 9% per year simple interest. Due to an unexpected income-tax refund, he was able to repay $10,000 on April 15. Use the United States Rule and find **(a)** the balance owed on the principal after the partial payment. Then find **(b)** the amount due at maturity of the loan.   **[11.3]**

(a) _____

(b) _____

18. Karoline Jacobs borrowed $2200 for new kitchen appliances. She agreed to pay the loan back with 8 payments of $290.69 each. After 3 payments, she decides to go ahead and pay off the loan in full. Use the Rule of 78 to find **(a)** the amount of unearned interest and **(b)** the amount needed to repay the loan in full.   **[11.3]**

(a) _____

(b) _____

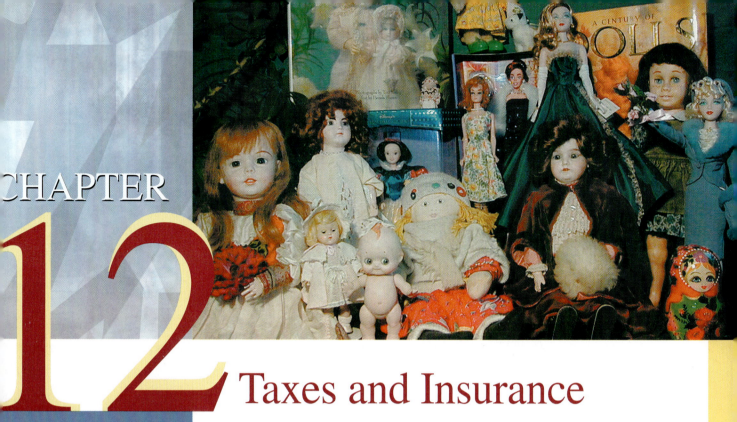

CHAPTER
12

Taxes and Insurance

MARTHA SPENCER OWNS THE DOLL HOUSE. SHE SELLS DOLLS FROM around the world and specializes in buying and selling Barbie dolls, both

CASE *in* POINT

collectible as well as newer ones. Her Web page has allowed her to do a lot of business over the Internet,

much of which is from other countries. Spencer owns the building in which her business is located and she has several employees. As a result, she must keep up with changes to the laws related to taxes and insurance.

Both **taxes** and **insurance** are facts of life. In one form or another, we pay taxes to many different entities including the federal government, states, counties, school districts, and cities. Supreme court Justice Oliver Wendell Holmes Jr. said, "taxes are the price we pay to live in a civilized society." In turn, individuals buy insurance to protect from catastrophic losses to their homes and automobiles and also to pay medical expenses or to pay any expenses at the time of death of a family member. Companies buy insurance to protect from potential losses to buildings and property, losses due to workers injured on the job, or even lawsuits from customers. This chapter first looks at two common types of taxes and then three common types of insurance.

12.1 | Property Tax

Objectives

1. Define *fair market value* and *assessed valuation.*
2. Find the tax rate.
3. Use the formula for property tax.
4. Express tax rate in percent, in dollars per $100, in dollars per $1000, and in mills.
5. Find taxes given the assessed valuation and the tax rate.

CASE *in* POINT

Martha Spencer was surprised by the amount of the property tax on her business this year. Her taxes went up significantly and she was determined to find out why. She also wanted to know where the money was going.

In virtually every area of the nation, owners of real property (such as buildings and land) must pay property tax. The money raised by this tax is used to provide services needed by the local community, such as police and fire protection, roads, schools, and other city and county services.

Objective 1 **Define *fair market value* and *assessed valuation.*** A local tax assessor estimates the market value of all the land and buildings in an area. The **fair market value** is the price at which a property can reasonably be expected to be sold. The **assessed valuation** of each property is found by multiplying the fair market value by a percent called the **assessment rate**. The assessment rate varies widely from one area to another, but normally remains constant within a state.

In some states, assessed valuation is 25% of fair market value. In other states, the assessed valuation is 40% to 60%, or even 100%, of fair market value. Using an assessed valuation rate that is a percent of fair market value has become an accepted practice over the years.

EXAMPLE 1

Finding the Assessed Value of Property

Find the assessed valuation for the following pieces of property owned by Martha Spencer.

(a) Home: Fair market value—$185,300; assessment rate 35%
(b) Business property: Fair market value—$328,500; assessment rate 35%
(c) House located in a different state: Fair market value—$123,800; assessment rate 60%

SOLUTION

Multiply the fair market value by the assessment rate.

(a) $185,300 × .35 = $64,855
(b) $328,500 × .35 = $114,975
(c) $123,800 × .60 = $74,280

QUICK TIP   Just because the assessment rate is higher in one area than another does not necessarily mean that the taxes are higher. The assessed value of the property must be multiplied by the tax rate to find the tax. (See Objectives 3–5.)

Objective 2 **Find the tax rate.** First, officials determine the amount of money needed by a taxing authority such as a local school district. Then the taxing authority finds the total fair market value of all real properties in the area. Finally, the **property tax rate** is the total tax amount needed, divided by the total of assessed values.

$$\text{Tax rate} = \frac{\text{Total tax amount needed}}{\text{Total assessed value}}$$

EXAMPLE 2

Finding the Tax Rate

Find the tax rate for the following park districts in River County.

(a) Total tax amount needed $368,400, total assessed value $7,368,000
(b) Total tax amount needed $633,750, total assessed value $28,800,000

SOLUTION

Divide the total tax amount needed by the total assessed value.

(a) $368,400 ÷ $7,368,000 = .05 = 5% tax rate
(b) $633,750 ÷ $28,800,000 = .022 = 2.2% tax rate   rounded to the nearest tenth of a percent

Taxes can be high. The pie chart shows where the average worker's money goes using an 8-hour workday for demonstration. Notice that Federal and State and Local taxes are a significant amount of every 8-hour workday.

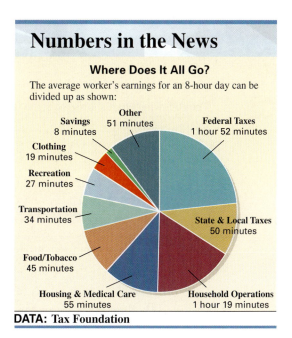

Numbers in the News

Where Does It All Go?

The average worker's earnings for an 8-hour day can be divided up as shown:

- Other 51 minutes
- Savings 8 minutes
- Federal Taxes 1 hour 52 minutes
- Clothing 19 minutes
- Recreation 27 minutes
- Transportation 34 minutes
- State & Local Taxes 50 minutes
- Food/Tobacco 45 minutes
- Housing & Medical Care 55 minutes
- Household Operations 1 hour 19 minutes

DATA: Tax Foundation

Objective 3 **Use the formula for property tax.** Property tax rates are expressed in different ways in different parts of the country. However, property tax is always found with this formula:

$$\text{Tax} = \text{Tax rate} \times \text{Assessed valuation}$$

QUICK TIP The market value of a piece of property usually differs considerably from the assessed value which is only used to calculate taxes.

Objective ④ **Express tax rate in percent, in dollars per $100, in dollars per $1000, and in mills. Percent**   Some areas express tax rates as a percent of assessed valuation. The tax on a piece of property with an assessed valuation of $74,000 at a tax rate of 9.42% follows.

$$\text{Tax} = .0942 \times \$74{,}000 = \$6970.80$$

Dollars per $100   In some areas, the tax rate is expressed as a number of dollars per $100 of assessed valuation. In this event, find the tax on a piece of land by first finding the number of hundreds in the assessed value and then multiplying the number of hundreds by the tax rate. For example, assume an assessed value of $56,300 and a tax rate of $11.42 per $100 of assessed value and find taxes as follows.

$$\$56{,}300 \div 100 = \textbf{563 hundreds}$$

Divide by 100 by moving the decimal point 2 places to the left.

$$\text{Tax} = 563 \times \$11.42 = \textbf{\$6429.46}$$

Multiply by the tax rate to find tax.

Dollars per $1000   In other areas, the tax rate is expressed as a number of dollars per $1000 of assessed valuation. If the tax rate is $98.12 per $1000, a piece of property having an assessed valuation of $197,000 would be taxed as follows.

$$\$197{,}000 = \textbf{197 thousands}$$

move the decimal point 3 places to the left to divide by 1000

$$\text{Tax} = \$98.12 \times \textbf{197} = \textbf{\$19,329.64}$$

Mills   Other taxing authorities express tax rates in mills or in one-thousandths of a dollar. For example, a tax rate might be expressed as 46 mills. Divide the 46 mills by 1000 to find $.046 per dollar of assessed value. Assuming a tax rate of 46 mills, the tax on a house assessed at $81,000 is found as follows.

46 mills = $.046

$$\text{Tax} = \textbf{.046} \times \$81{,}000$$
$$= \$3726$$

The following chart shows the same tax rates written in the four different systems.

Percent	Per $100	Per $1000	In Mills
12.52%	$12.52	$125.20	125.2
3.2%	$3.20	$32	32
9.87%	$9.87	$98.70	98.7

QUICK TIP   Although expressed differently, the rates in each row of this chart are equivalent tax rates.

Objective ⑤ **Find taxes given the assessed valuation and the tax rate.**   Property taxes are found by multiplying the tax rate and the assessed valuation as shown in the following example.

EXAMPLE 3

Finding the Property Tax

Find the taxes on each of the following pieces of property. Assessed valuations and tax rates are given.

(a) $58,975; 8.4%   **(b)** $875,400; $7.82 per $100
(c) $129,600; $64.21 per $1000   **(d)** $221,750; 94 mills

SOLUTION
Multiply the tax rate by the assessed valuation.

(a) 8.4% = **.084**

$$\text{Tax} = \text{Tax rate} \times \text{Assessed valuation}$$
$$\text{Tax} = \textbf{.084} \times \$58{,}975 = \$4953.90$$

(b) $875,400 = **8754 hundreds**

$$\text{Tax} = \$7.82 \times \textbf{8754} = \$68,456.28$$

(c) $129,600 = **129.6 thousands**

$$\text{Tax} = \$64.21 \times \textbf{129.6} = \$8321.62$$

(d) 94 mills = **.094**

$$\text{Tax} = \textbf{.094} \times \$221,750 = \$20,844.50$$

QUICK TIP Some states give a special tax break to homeowners called a homeowner's exemption. This feature allows homeowners to deduct a certain amount (varying from state to state) from the assessed value of their home before the taxes are calculated. This results in a lower tax.

The following figure shows the sales tax, state income tax, and the property tax rates in five locations across the United States. These towns were identified by the *Wall Street Journal* as "Five Hometowns for the Future." Notice the variations in the tax rates.

Numbers in the News

Someplace to Be

Location	Sales Tax	State Income Tax	Property Tax
San Juan Island, Washington	7.7%	None	$8.92 per $1000
Destin/South Walton Beach, Florida	7%	None	$1.55 per $1000 City of Destin; $7.59 per $1000 Walton County
Kailua-Kona, Hawaii	4%	2%–10% based on income	$8.50 per $1000
Petoskey/Harbor Springs, Michigan	6%	4%	$34.16 per $1000
Corolla, North Carolina	6%	6%–8.25%	$.64 per $100

DATA: *Wall Street Journal*

12.1 | Exercises

FOR EXTRA HELP

 MyMathLab

 InterActMath.com

 MathXL

 MathXL Tutorials on CD

 Addison-Wesley Math Tutor Center

DVT/Videotape

The **Quick Start** *exercises in each section contain solutions to help you get started.*

Find the assessed valuation for each of the following pieces of property. (See Example 1.)

	Fair Market Value	Rate of Assessment	Assessed Valuation		Fair Market Value	Rate of Assessment	Assessed Valuation
Quick Start							
1.	$85,000	40%	**$34,000**	**2.**	$68,000	60%	**$40,800**
	$85,000 × .4 = $34,000				$68,000 × .6 = $40,800		
3.	$142,300	50%	_____	**4.**	$98,200	42%	_____
5.	$1,300,500	25%	_____	**6.**	$2,450,000	80%	_____

Find the tax rate for the following. Write the tax rate as a percent rounded to the nearest tenth.
(See Example 2.)

	Total Tax Amount Needed	Total Assessed Value	Tax Rate		Total Tax Amount Needed	Total Assessed Value	Tax Rate
Quick Start							
7.	$18,300,000	$60,150,000	**30.4%**	**8.**	$7,600,000	$39,280,000	**19.3%**
	$18,300,000 ÷ $60,150,000 = .30424 = 30.4% (rounded)				$7,600,000 ÷ $39,280,000 = .19348 = 19.3% (rounded)		
9.	$1,580,000	$19,750,000	_____	**10.**	$2,175,000	$54,375,000	_____
11.	$1,224,000	$40,800,000	_____	**12.**	$2,941,500	$81,700,000	_____

Complete the following list comparing tax rates. (See Example 3.)

	Percent	Per $100	Per $1000	In Mills
Quick Start				
13.	4.84%	(a) **$4.84**	(b) **$48.40**	(c) **48.4**
14.	(a) _____	$6.75	(b) _____	(c) _____
15.	(a) _____	(b) _____	$70.80	(c) _____
16.	(a) _____	(b) _____	(c) _____	28

17. What is the difference between fair market value and assessed value? How is the assessment rate used when finding the assessed value? (See Objective 1.)

18. Select any tax rate and express it as a percent. Write this tax rate in three additional equivalent forms and explain what each form means. (See Objective 4.)

🔻 indicates an exercise that is related to the Case in Point feature within the section.

Find the property tax for the following. (See Example 3.)

Quick Start

	Assessed Valuation	Tax Rate	Tax		Assessed Valuation	Tax Rate	Tax
19.	$86,200	$6.80 per $100	$5861.60	**20.**	$41,300	$46.40 per $1000	_____
	862 × $6.80 = $5861.60						
21.	$128,200	42 mills	_____	**22.**	$37,250	3.4%	_____

Solve the following application problems.

Quick Start

23. REAL ESTATE TAXES   Martha Spencer owns the real estate used by The Doll House. The property has a fair market value of $328,500; the assessment rate is 35%, and the local tax rate is 5.2%. Find the tax.
$328,500 × .35 = $114,975; $114,975 × .052 = $5978.70

23. $5978.70

24. APARTMENT OWNER   Chad LeCompte owns a four-unit apartment building with a fair market value of $192,600. Property in the area is assessed at 40% of market value and the tax rate is 5.5%. Find the amount of the property tax.

24. _____

25. COMMERCIAL PROPERTY TAX   A new FM radio station broadcasts from a building having a fair market value of $334,400. The building is in an area where property is assessed at 25% of market value and the tax rate is $75.30 per $1000 of assessed value. Find the property tax.

25. _____

26. OFFICE COMPLEX   Huron Development just purchased a modern office complex for $12,380,000. The county assesses the property at 60% of market value and has a property tax of $18.40 per $1000 of assessed value. Find the property tax.

26. _____

27. The Savon Park office complex has a fair market value of $1,350,000. Property in the area is assessed at 40% of fair market value and the tax rate is $7.40 per $100 of assessed valuation. Find the property tax.

27. _____

28. Harley-Davidson of Lincoln has property with a fair market value of $518,600. The property is located in an area that is assessed at 35% of market value. The tax rate is $7.35 per $100. Find the property tax.

28. _____

29. COMPARING PROPERTY TAX RATES   In one parish (county), property is assessed at 40% of market value, with a tax rate of 32.1 mills. In a second parish, property is assessed at 24% of market value, with a tax rate of 50.2 mills. A telephone company is trying to decide where to place a small storage building with a fair market value of $95,000. **(a)** Which parish would charge the lower property tax? **(b)** Find the annual amount saved.

(a) _____
(b) _____

30. In one county, property is assessed at 30% of market value, with a tax rate of 45.6 mills. In a second county, property is assessed at 48% of market value, with a tax rate of 29.3 mills. If Henry Hernandez is trying to decide where to build a $140,000 house, **(a)** find which county would charge the lower property tax and **(b)** find the annual amount saved.

(a) _____
(b) _____

12.2 | Personal Income Tax

Objectives

1 List the four steps that determine tax liability.
2 Find adjusted gross income.
3 Know the standard deduction amounts.
4 Find the taxable income and income tax.
5 List possible deductions.
6 Determine a balance due or a refund from the Internal Revenue Service.
7 Prepare a 1040A and a Schedule 1 federal tax form.

CASE *in* POINT

As a business owner, Martha Spencer has her income taxes prepared by a qualified accountant. Still, she must keep track of all her financial information throughout the year for the time when preparing her taxes arrives. Even though she does not prepare her own taxes, Martha Spencer is still responsible for the accuracy of all the information contained on her personal income tax forms.

Income taxes are calculated based on income. They are a source of revenue for the federal government, most states, and many local governments. The pie charts below show the major categories of income and outlays for the U.S. government. Notice that personal income taxes and corporate income taxes combine to generate 50% of the government's income. Social Security, Medicare, unemployment and other retirement taxes are an additional 35% of the government's income. The government's largest outlay is Social Security, Medicare, and other retirement, which amounts to 38% of the total outlay.

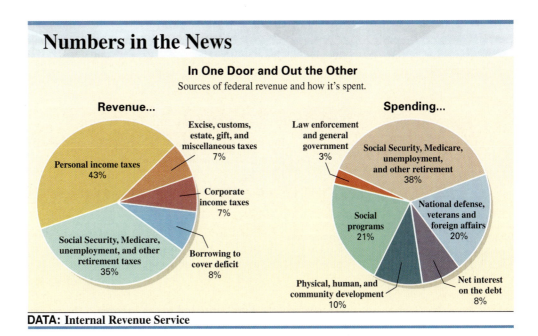

Numbers in the News

In One Door and Out the Other
Sources of federal revenue and how it's spent.

Revenue...

- Excise, customs, estate, gift, and miscellaneous taxes 7%
- Personal income taxes 43%
- Corporate income taxes 7%
- Social Security, Medicare, unemployment, and other retirement taxes 35%
- Borrowing to cover deficit 8%

Spending...

- Law enforcement and general government 3%
- Social Security, Medicare, unemployment, and other retirement 38%
- National defense, veterans and foreign affairs 20%
- Social programs 21%
- Physical, human, and community development 10%
- Net interest on the debt 8%

DATA: Internal Revenue Service

Some people believe that those with high incomes do not pay income taxes, but this is not true. As shown next, individuals making up the top 10% of earners pay 67.3% of all income taxes paid. The bottom 50% only pay 3.9% of all income taxes paid.

All Taxpayers	Group's Proportion of Total Income Taxes Paid
Top 1%	37.4%
Top 5%	56.5%
Top 10%	67.3%
Bottom 50%	3.9%

Source: Internal Revenue Service.

Most working adults are required to file an income tax return with the federal government every year. Married couples often submit one income tax return that shows income and deductions for both spouses. Forms for filing income taxes can be found at **Internal Revenue Service (IRS)** offices or at the IRS website, www.irs.gov. The IRS is the branch of the government responsible for collecting income taxes. Last year, more than 53 million people filed their income tax returns with the IRS electronically, using e-file options available at www.irs.gov. Filing electronically allows most filers access to free commercial online tax preparation software and can result in getting refunds in as few as 10 days.

There are thousands of rules relating to income tax preparation and many different forms each with its own instructions. However, many people have relatively simple income tax returns that do not require professionals to complete. No matter if you use a professional to file your income tax return or if you do it yourself, you should carefully save all records related to income and expenses. These records should be kept until the statue of limitations runs out which is 3 years from the date the tax return was filed or 2 years from the date the tax was paid, whichever is later. The IRS can audit you during this period and they require proof of income and deductions. Many accountants recommend that you keep records for 7 years, but you should keep records on property such as your home as long as they are needed, often longer than 3 years.

Objective 1 **List the four steps that determine tax liability.** There are four basic steps in finding total tax liability. They are listed here and described one at a time in the remainder of this section.

Steps for Preparing Your Income Tax Return

Step 1 Find the adjusted gross income (AGI) for the year.
Step 2 Find the taxable income.
Step 3 Find the tax.
Step 4 Check to see if a refund is due or if money is owed to the government.

QUICK TIP   It can be very complex to find the income tax for some people. This section introduces only the basic concepts.

Objective 2 **Find adjusted gross income.** The first step in finding personal income tax is to find **adjusted gross income**. Adjusted gross income is the total of all income less certain adjustments. Employers are required to send out **W-2 forms** showing wages paid, federal income taxes withheld, Social Security tax withheld, and Medicare tax withheld. Other types of income such as interest, dividends, and self-employment income are shown on **1099 forms**, which are also mailed to each individual. Sample W-2 and 1099-INT forms are shown on the next page.

a Control number	22222	Void ☐	**For Official Use Only**⊿ OMB No. 1545-0008		

b Employer identification number 94-1287319			**1** Wages, tips, other compensation **$24,738.41**	**2** Federal income tax withheld **$3275.60**

c Employer's name, address, and ZIP code	**3** Social security wages **$24,738.41**	**4** Social security tax withheld **$1533.78**

The Doll House
1568 Liberty Heights Ave.
Baltimore, MD 21230

5 Medicare wages and tips **$24,738.41**	**6** Medicare tax withheld **$358.71**
7 Social security tips	**8** Allocated tips

d Employee's social security number 123-45-6789	**9** Advance EIC payment	**10** Dependent care benefits

e Employee's first name and initial **Jennifer**	Last name **Crum**	**11** Nonqualified plans	**12a** See instructions for box 12

2136 Old Road
Towson, MD 21285

13 Statutory employee ☐	Retirement plan ☐	Third-party sick pay ☐	**12b**

14 Other	**12c**

	12d

f Employee's address and ZIP code

15 State **MD**	Employer's state ID number 600-5076	**16** State wages, tips, etc.	**17** State income tax	**18** Local wages, tips, etc.	**19** Local income tax	**20** Locality name

Form **W-2** **Wage and Tax Statement**

Department of the Treasury—Internal Revenue Service

9292   ☐ VOID   ☐ CORRECTED

PAYER'S name, street address, city, state, ZIP code, and telephone no.	Payer's RTN (optional)	OMB No. 1545-0112	
Employees Credit Union **2572 Brookhaven Drive** **Dundalk, MD 21222**		200⓪_ Form **1099-INT**	**Interest Income**

PAYER'S Federal identification number 94-1287319	RECIPIENT'S identification number 123-45-6789	**1** Interest income not included in box 3 $ **1624.01**	**Copy A** For **Internal Revenue Service Center** File with Form 1096.
RECIPIENT'S name **Jennifer Crum**		**2** Early withdrawal penalty $	**3** Interest on U.S. Savings Bonds and Treas. obligations $
Street address (including apt. no.) **2136 Old Road**		**4** Federal income tax withheld $	**5** Investment expenses $
City, state, and ZIP code **Towson, MD 21285**		**6** Foreign tax paid	**7** Foreign country or U.S. possession
Account number (optional)	2nd TIN not. ☐	$	For Privacy Act and Paperwork Reduction Act Notice, see the **2004 General Instructions for Forms 1099, 1098, 5498, and W-2G.**

Form **1099-INT**   Cat. No. 14410K   Department of the Treasury - Internal Revenue Service

---

Finding Adjusted Gross Income

1. Add amounts from all W-2 and 1099 forms along with dividends, capital gains, unemployment compensation, and tips or other employee compensation.
2. From this sum, subtract adjustments such as contributions to a *regular* **individual retirement account (IRA)** or alimony payments.

EXAMPLE **1**

Finding Adjusted Gross Income (AGI)

As an assistant manager at The Doll House, Jennifer Crum earned $24,738.41 last year and $1624.01 in interest from her credit union (see her W-2 and 1099 forms). She had $1500 in regular IRA contributions. Find her adjusted gross income.

SOLUTION

$$\text{Adjusted gross income} = \text{wages} + \text{interest} - \textbf{IRA contribution}$$
$$= \$24,738.41 + \$1624.01 - \textbf{\$1500}$$
$$= \$24,862.42$$

QUICK TIP   When filing your income tax, a copy of all W-2 forms is sent to the Internal Revenue Service along with the completed tax forms. However, the IRS does not require copies of 1099 forms to be sent to them unless income tax was withheld.

Objective ③   Know the standard deduction amounts.   Most people are almost finished at this point. A taxpayer next subtracts the larger of either the itemized deductions or the standard deduction from adjusted income. Itemized deductions are described in Objective 5 in this section and are associated with several limitations based on adjusted gross income. The **standard deduction** amount is based on the taxpayer's filing status as follows.

   $4750 for single taxpayers
   $9500 for married couples filing jointly or qualifying widow(er)
   $4750 for married taxpayers filing separately
   $7000 for head of household

Head of household refers to unmarried people who provide a home for other people such as a dependent child or even a dependent parent of the taxpayer.

Additional standard deductions are given for taxpayers and dependents who are blind or 65 years of age or older. The standard deduction amounts commonly change from one year to the next. The most current amounts can be obtained from the IRS.

The next step is to find the number of **personal exemptions**. One exemption is allowed for yourself, another for your spouse if filing a joint return, and yet another exemption for each child or other dependent. The number of exemptions does not depend on whether the filing status is single or married. The deduction for personal exemptions is $3050 times the number of exemptions. Here are some examples.

	Number of Exemptions	Personal Exemption(s)
Single individual	1	1 × $3050 = $3050
Married couple filing jointly, with 3 children	5	5 × $3050 = $15,250
Single woman, head of household with 2 dependent children	3	3 × $3050 = $9150

Taxable income is found by subtracting the standard deduction amount and personal exemptions from adjusted gross income. Taxes are then calculated based on taxable income as shown next.

Objective ④   Find the taxable income and income tax.   Taxable income is used along with data from the tax rate schedule shown next to find the income tax. As you can see from the tax rate schedules, individual income tax rates range from a low of 10% to a high of 35% depending on income and filing status. Assume a single filer has a taxable income of $37,300.

According to the tax rate schedule, the federal income tax would be $3910 + 25% of the excess over $28,400.

excess of taxable income over $28,400

$$\text{Federal Income Tax} = \$3910 + .25 \times \left(\$37,300 - \$28,400\right) = \$6135$$

The 25% tax rate is *only applied* to the amount above $28,400 and not to the entire taxable income of $37,300.

Tax Rate Schedules

Schedule X—Use if your filing status is **Single**

If the amount on Form 1040, line 40, is: Over—	But not over—	Enter on Form 1040, line 41	of the amount over—
$0	$7,000	.... 10%	$0
7,000	28,400	$700.00 + 15%	7,000
28,400	68,800	3,910.00 + 25%	28,400
68,800	143,500	14,010.00 + 28%	68,800
143,500	311,950	34,926.00 + 33%	143,500
311,950	....	90,514.50 + 35%	311,950

Schedule Y-2—Use if your filing status is **Married filing separately**

If the amount on Form 1040, line 40, is: Over—	But not over—	Enter on Form 1040, line 41	of the amount over—
$0	$7,000	.... 10%	$0
7,000	28,400	$700.00 + 15%	7,000
28,400	57,325	3,910.00 + 25%	28,400
57,325	87,350	11,141.25 + 28%	57,325
87,350	155,975	19,548.25 + 33%	87,350
155,975	....	42,194.50 + 35%	155,975

Schedule Y-1—Use if your filing status is **Married filing jointly** or **Qualifying widow(er)**

If the amount on Form 1040, line 40, is: Over—	But not over—	Enter on Form 1040, line 41	of the amount over—
$0	$14,000	.... 10%	$0
14,000	56,800	$1,400.00 + 15%	14,000
56,800	114,650	7,820.00 + 25%	56,800
114,650	174,700	22,282.50 + 28%	114,650
174,700	311,950	39,096.50 + 33%	174,700
311,950	....	84,389.00 + 35%	311,950

Schedule Z—Use if your filing status is **Head of household**

If the amount on Form 1040, line 40, is: Over—	But not over—	Enter on Form 1040, line 41	of the amount over—
$0	$10,000	.... 10%	$0
10,000	38,050	$1,000.00 + 15%	10,000
38,050	98,250	5,207.50 + 25%	38,050
98,250	159,100	20,257.50 + 28%	98,250
159,100	311,950	37,295.50 + 33%	159,100
311,950	....	87,736.00 + 35%	311,950

EXAMPLE 2

Finding Taxable Income and the Income Tax Amount

Find the taxable income and income tax for each of the following.

(a) Herbert White, married filing jointly, 5 daughters, adjusted gross income $42,300
(b) Onita Fields, single, no dependents, adjusted gross income $21,600
(c) Imogene Griffin, single, head of household, 2 children, adjusted gross income $64,300

SOLUTION

(a) Herbert White + spouse + 5 daughters = 7 exemptions

$$\text{Taxable income} = \$42,300 - \$9500 - \left(7 \times \$3050\right) = \$11,450$$

(Standard deduction, Deduction for exemptions, Taxable income)

The tax rate for married, filing jointly with less than $14,000 in taxable income is 10%.

$$\text{Income tax} = \$11,450 \times .10 = \$1145$$

(b) Onita Fields has no other dependents = 1 exemption

| | Standard deduction | Deduction for exemptions | Taxable income |

Taxable income = $21,600 − $4750 − $\left(1 \times \$3050\right)$ = $13,800

Income tax = $700 + **15% of excess over $7000**

= $700 + **.15** × $\left(\textbf{\$13,800} - \textbf{\$7000}\right)$ = $1720

(c) Imogene Griffin + 2 children = 3 exemptions

Taxable income = $64,300 − $7000 − $\left(3 \times \$3050\right)$ = $48,150

Income tax = $5207.50 + **25% of excess over $38,050**

= $5207.50 + **.25** × $\left(\textbf{\$48,150} - \textbf{\$38,050}\right)$ = $7732.50

Objective [5] **List possible deductions.** Actually, taxpayers may deduct *the larger of* **itemized deductions** *or* the standard deduction from their adjusted gross income *before* finding taxable income. This is particularly applicable to individuals that are paying interest on a home loan, but sometimes others can use this to their advantage. The most common **tax deductions** are listed below.

Medical and dental expenses Only medical and dental expenses exceeding 7.5% of adjusted gross income may be deducted. This deduction is effectively limited to catastrophic illnesses for most taxpayers. Expenses reimbursed by an insurance company are not deductible.

Taxes State and local income taxes, real estate taxes, and personal property taxes may be deducted (but not federal income or gasoline taxes).

Interest Home mortgage interest on the taxpayer's principal residence and a qualified second home is deductible. Other interest charges (including credit-card interest) may *not* be deducted.

Contributions Contributions to most charities may be deducted.

Miscellaneous deductions These expenses are only deductible to the extent that the total exceeds 2% of the taxpayer's adjusted gross income. They include unreimbursed employee expenses, tax preparation fees, appraisal fees for tax purposes, and legal fees for tax planning or tax litigation.

> **QUICK TIP**   The taxpayer may take the larger of the standard deduction or the itemized deductions. You must have documentation to show itemized expenses if you itemize.

EXAMPLE 3

Using Itemized Deductions to Find Taxable Income and Income Tax

Chris Kelly is single, has no dependents, and had an adjusted gross income of $26,735 last year, with deductions of $1352 for other taxes, $4118 for mortgage interest, and $317 for charity. Find his taxable income and his income tax.

SOLUTION

Itemized deductions = $1352 + $4118 + $317 = **$5787**

Kelly's itemized deductions of $5787 are larger than the $4750 standard deduction for a single person. Therefore, taxable income is adjusted gross income minus the itemized deductions minus the one personal exemption for being single with no dependents.

| | Itemized deductions | Personal exemption | |

Taxable income = $26,735 − $5787 − $\left(1 \times \$3050\right)$ = $17,898

Income tax = $700 + 15% of excess over $7000

= $700 + **.15** × $\left(\textbf{\$17,898} - \textbf{\$7000}\right)$ = $2334.70

> **QUICK TIP**   Every year, many of the rules related to the calculation of income taxes change. Always use the most current information and forms from the Internal Revenue Service when preparing your taxes.

Objective [6] **Determine a balance due or a refund from the Internal Revenue Service.** A taxpayer may have paid more to the IRS than is due. Add up the total amount of income tax paid using the W-2 forms. Usually, no taxes are withheld on 1099 forms. If the amount withheld is greater than the tax owed, the taxpayer is entitled to a refund. If the amount withheld is less than the tax owed, then the taxpayer must send the difference along with the tax return to the IRS.

EXAMPLE 4

Determining Tax Due or Refund

Tim Owen works as an attorney and his wife stays at home with their young daughter. Last year Tim had an adjusted gross income of $77,300 and $573.50 was withheld from his paycheck every month for federal income taxes. The Owens file a joint return and use the standard deduction. Find either the additional amount of income taxes they owe or the amount they overpaid.

SOLUTION

Adjusted gross income	$77,300	
standard deduction	− 9,500	married filing jointly
personal exemptions	− 9,150	3 exemptions × $3050
Taxable income	$58,650	

Income tax = $7820 + 25% of excess over $56,800

$$= \$7820 + .25 \times (\$58,650 - \$56,800) = \mathbf{\$8282.50}$$

Total amount withheld last year = $573.50 × 12 = $6882

Amount owed to IRS = **$8282.50** − $6882 = **$1400.50**

The Owens must send **$1400.50** with their tax return.

It is best to use the simplest form needed when filing with the IRS. The 1040EZ is the simplest form and is used by many students, but many people cannot use it. Here are IRS guidelines to help you determine the appropriate form.

1040EZ

1. Taxable income below $50,000
2. Single or Married Filing Jointly
3. Under age 65
4. No dependents
5. Interest income of $1500 or less

1040A

1. Taxable income below $50,000
2. Capital gain distributions, but no other capital gains or losses
3. Only tax credits for child tax, education, earned income, child and dependent care expenses, adoption and retirement savings contributions
4. Only deductions for IRA contributions, student loan interest, educator expenses or higher education tuition and fees
5. No itemized deductions

1040

1. Taxable income of $50,000 or more
2. Itemized deductions
3. Self-employment income
4. Income from sale of property

Objective 7 Prepare a 1040A and a Schedule 1 federal tax form. The next example shows how to complete an income tax return using **Form 1040A** and **Schedule 1 (Form 1040A)**.

QUICK TIP When completing income tax forms and calculations, notice that all amounts may be rounded to the nearest dollar.

EXAMPLE 5

Preparing a 1040A and a Schedule 1

Jennifer Crum is single and claims one exemption. Her income appears on the W-2 and 1099 forms on page 513. She contributes $1500 to a regular IRA. Crum satisfies the requirements to use Form 1040A, but must also fill out Schedule 1 (Form 1040A), since she has more than $1500 in interest.

Form
1040A

Department of the Treasury—Internal Revenue Service
U.S. Individual Income Tax Return   (99)   IRS Use Only—Do not write or staple in this space.

OMB No. 1545-0085

Label
(See page 19.)

Your first name and initial: **Jennifer**   Last name: **Crum**

Your social security number: **123 45 6789**

Use the IRS label.
Otherwise, please print or type.

If a joint return, spouse's first name and initial   Last name

Spouse's social security number

Home address (number and street). If you have a P.O. box, see page 20.   **2136 Old Road**   Apt. no.

▲ **Important!** ▲
You **must** enter your SSN(s) above.

City, town or post office, state, and ZIP code. If you have a foreign address, see page 20.   **Towson, MD 21285**

Presidential Election Campaign
(See page 20.)
Note. Checking "Yes" will not change your tax or reduce your refund.
Do you, or your spouse if filing a joint return, want $3 to go to this fund? . . . . ▶
You: ☒ Yes ☐ No   Spouse: ☐ Yes ☐ No

Filing status
Check only one box.

1 ☒ Single
2 ☐ Married filing jointly (even if only one had income)
3 ☐ Married filing separately. Enter spouse's SSN above and full name here. ▶
4 ☐ Head of household (with qualifying person). (See page 20.) If the qualifying person is a child but not your dependent, enter this child's name here. ▶
5 ☐ Qualifying widow(er) with dependent child (See page 21.)

Exemptions

6a ☒ **Yourself.** If your parent (or someone else) can claim you as a dependent on his or her tax return, **do not** check box 6a.

b ☐ **Spouse**

c **Dependents:**

If more than six dependents, see page 21.

(1) First name Last name	(2) Dependent's social security number	(3) Dependent's relationship to you	(4) ✓ if qualifying child for child tax credit (see page 23)
			☐
			☐
			☐
			☐
			☐
			☐

No. of boxes checked on 6a and 6b: **1**

No. of children on 6c who:
• lived with you
• did not live with you due to divorce or separation (see page 23)

Dependents on 6c not entered above

d Total number of exemptions claimed.

Add numbers on lines above: **1**

Income

Attach Form(s) W-2 here. Also attach Form(s) 1099-R if tax was withheld.

If you did not get a W-2, see page 24.

Enclose, but do not attach, any payment.

7 Wages, salaries, tips, etc. Attach Form(s) W-2.   7 **$24,738**

8a **Taxable** interest. Attach Schedule 1 if required.   8a **$1,624**
b **Tax-exempt** interest. **Do not** include on line 8a.   8b

9a Ordinary dividends. Attach Schedule 1 if required.   9a
b Qualified dividends (see page 25).   9b

10a Capital gain distributions (see page 25).   10a
b Post-May 5 capital gain distributions (see page 25).   10b

11a IRA distributions.   11a   11b Taxable amount (see page 25).   11b

12a Pensions and annuities.   12a   12b Taxable amount (see page 26).   12b

13 Unemployment compensation and Alaska Permanent Fund dividends.   13

14a Social security benefits.   14a   14b Taxable amount (see page 28).   14b

15 Add lines 7 through 14b (far right column). This is your **total income.** ▶   15 **$26,362**

Adjusted gross income

16 Educator expenses (see page 28).   16
17 IRA deduction (see page 28).   17 **$1,500**
18 Student loan interest deduction (see page 31).   18
19 Tuition and fees deduction (see page 31).   19
20 Add lines 16 through 19. These are your **total adjustments.**   20 **$1,500**

21 Subtract line 20 from line 15. This is your **adjusted gross income.** ▶   21 **$24,862**

For Disclosure, Privacy Act, and Paperwork Reduction Act Notice, see page 57.   Cat. No. 11327A   Form **1040A**

Form 1040A (2003)   Page **2**

Tax, credits, and payments	**22**	Enter the amount from line 21 (adjusted gross income).	22	$24,862

23a Check if:
- ☐ **You** were born before January 2, 1939, ☐ Blind
- ☐ **Spouse** was born before January 2, 1939, ☐ Blind

} **Total boxes checked** ▶ 23a [ ]

b If you are married filing separately and your spouse itemizes deductions, see page 32 and check here ▶ 23b ☐

Standard Deduction for—

- People who checked any box on line 23a or 23b **or** who can be claimed as a dependent, see page 32.
- All others:

Single or Married filing separately, $4,750

Married filing jointly or Qualifying widow(er), $9,500

Head of household, $7,000

24	Enter your **standard deduction** (see left margin).	24	$4,750
25	Subtract line 24 from line 22. If line 24 is more than line 22, enter -0-.	25	$20,112
26	Multiply $3,050 by the total number of exemptions claimed on line 6d.	26	$3,050
27	Subtract line 26 from line 25. If line 26 is more than line 25, enter -0-. This is your **taxable income.** ▶	27	$17,062
28	**Tax,** including any alternative minimum tax (see page 33).	28	$2,209

29	Credit for child and dependent care expenses. Attach Schedule 2.	29	
30	Credit for the elderly or the disabled. Attach Schedule 3.	30	
31	Education credits. Attach Form 8863.	31	
32	Retirement savings contributions credit. Attach Form 8880.	32	
33	Child tax credit (see page 37).	33	
34	Adoption credit. Attach Form 8839.	34	

35	Add lines 29 through 34. These are your **total credits.**	35	0
36	Subtract line 35 from line 28. If line 35 is more than line 28, enter -0-.	36	$2,209
37	Advance earned income credit payments from Form(s) W-2.	37	
38	Add lines 36 and 37. This is your **total tax.** ▶	38	$2,209

39	Federal income tax withheld from Forms W-2 and 1099.	39	$3,276
40	2003 estimated tax payments and amount applied from 2002 return.	40	

If you have a qualifying child, attach Schedule EIC.

41	**Earned income credit (EIC).**	41	
42	Additional child tax credit. Attach Form 8812.	42	
43	Add lines 39 through 42. These are your **total payments.** ▶	43	$3,276

Refund	**44**	If line 43 is more than line 38, subtract line 38 from line 43. This is the amount you **overpaid.**	44	$1,067

Direct deposit? See page 50 and fill in 45b, 45c, and 45d.

45a	Amount of line 44 you want **refunded to you.** ▶	45a	$1,067

b Routing number [ ][ ][ ][ ][ ][ ][ ][ ][ ]   **c** Type: ☐ Checking ☐ Savings

d Account number [ ][ ][ ][ ][ ][ ][ ][ ][ ][ ][ ][ ][ ][ ][ ][ ][ ]

46	Amount of line 44 you want **applied to your 2004 estimated tax.**	46	
Amount you owe	**47**	**Amount you owe.** Subtract line 43 from line 38. For details on how to pay, see page 51. ▶	47	
	48	Estimated tax penalty (see page 52).	48	

Third party designee

Do you want to allow another person to discuss this return with the IRS (see page 52)? ☐ **Yes.** Complete the following. ☒ **No**

Designee's name ▶ _____   Phone no. ▶ ( )   Personal identification number (PIN) ▶ [ ][ ][ ][ ][ ]

Sign here

Under penalties of perjury, I declare that I have examined this return and accompanying schedules and statements, and to the best of my knowledge and belief, they are true, correct, and accurately list all amounts and sources of income I received during the tax year. Declaration of preparer (other than the taxpayer) is based on all information of which the preparer has any knowledge.

Joint return? See page 20. Keep a copy for your records.

Your signature	Date	Your occupation	Daytime phone number
Jennifer Crum	4/14	Assistant Manager	(410) 286-2594
Spouse's signature. If a joint return, **both** must sign.	Date	Spouse's occupation	

Paid preparer's use only

Preparer's signature ▶	Date	Check if self-employed ☐	Preparer's SSN or PTIN
Firm's name (or yours if self-employed), address, and ZIP code ▶		EIN	
		Phone no. ( )	

Form **1040A**

Schedule 1
(Form 1040A)

Department of the Treasury—Internal Revenue Service

Interest and Ordinary Dividends
for Form 1040A Filers   (99)

OMB No. 1545-0085

Name(s) shown on Form 1040A

Jennifer Crum

Your social security number

123 45 6789

Part I

Interest

(See back
of schedule
and the
instructions
for Form
1040A,
line 8a.)

Note. If you received a Form 1099-INT, Form 1099-OID, or substitute statement from a brokerage firm, enter the firm's name and the total interest shown on that form.

			Amount	
1	List name of payer. If any interest is from a seller-financed mortgage and the buyer used the property as a personal residence, see back of schedule and list this interest first. Also, show that buyer's social security number and address.			
	Employees Credit Union	1	*$1,624*	
2	Add the amounts on line 1.	2	*$1,624*	
3	Excludable interest on series EE and I U.S. savings bonds issued after 1989. Attach Form 8815.	3		
4	Subtract line 3 from line 2. Enter the result here and on Form 1040A, line 8a.	4	*$1,624*	

Part II

Ordinary dividends

(See back
of schedule
and the
instructions
for Form
1040A,
line 9a.)

Note. If you received a Form 1099-DIV or substitute statement from a brokerage firm, enter the firm's name and the ordinary dividends shown on that form.

			Amount	
5	List name of payer.	5		
6	Add the amounts on line 5. Enter the total here and on Form 1040A, line 9a.	6		

For Paperwork Reduction Act Notice, see Form 1040A instructions.   Cat. No. 12075R   Schedule 1 (Form 1040A)

As you can see from the graph to the left, countries around the world collect income taxes. Notice how much lower taxes are in the United States compared to Germany and Denmark. The clipping to the right shows that 90% of taxpayers meet the April 15th deadline to file taxes.

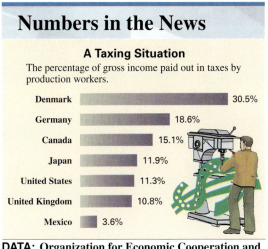

Numbers in the News

A Taxing Situation

The percentage of gross income paid out in taxes by production workers.

Country	Percentage
Denmark	30.5%
Germany	18.6%
Canada	15.1%
Japan	11.9%
United States	11.3%
United Kingdom	10.8%
Mexico	3.6%

DATA: Organization for Economic Cooperation and Development

Tax Facts

For 2003, 90% of taxpayers made the April 15 deadline while 6% request an extension. Other facts include:

Average taxable income
(after deductions) **$43,350**

Average tax owed **$9280**

Average refund **$1698**

Taxpayers who:

Itemized deductions **30.6%**

Received a refund **72%**

Used a tax preparer **54%**

DATA: Internal Revenue Service

QUICK TIP The October, 2003 Tax Act changed the way taxes on some dividends are calculated, at least until 2008 when this feature of the new law is rescinded. Since this law is complicated, we have chosen to include dividend income in adjusted gross income in the exercises that follow.

12.2 Exercises

FOR EXTRA HELP

 MyMathLab

 InterActMath.com

 MathXL

 MathXL Tutorials on CD

 Addison-Wesley Math Tutor Center

DVT/Videotape

The **Quick Start** *exercises in each section contain solutions to help you get started.*

Find the adjusted gross income for each of the following people. (See Example 1.)

	Name	Income from Jobs	Interest	Misc. Income	Dividend Income	Adjustments to Income	Adjusted Gross Income
Quick Start							
1.	R. Jacob	$22,840	$234	$1209	$48	$1200	__$23,131__
	$22,840 + $234 + $1209 + $48 − $1200 = $23,131						
2.	K. Chandler	$38,156	$285	$73	$542	$317	__$38,739__
	$38,156 + $285 + $73 + $542 − $317 = $38,739						
3.	The Hanks	$21,380	$625	$139	$184	$618	_____
4.	The Jazwinskis	$33,650	$722	$375	$218	$473	_____
5.	The Brashers	$38,643	$95	$188	$105	$0	_____
6.	The Ameens	$41,379	$1174	$536	$186	$2258	_____

Find the amount of taxable income and the tax owed for each of the following people. Use the tax rate schedule. The letter following the names indicates the marital status, and all married people are filing jointly. (See Examples 2 and 3.)

	Name	Number of Exemptions	Adjusted Gross Income	Total Deductions	Taxable Income	Tax Owed
Quick Start						
7.	R. Rodriguez, S	1	$32,400	$2398	__$24,600__	__$3340__
	$32,400 − $4750 − $3050 = $24,600; $700 + .15 × ($24,600 − $7000) = $3340					
8.	L. Pacos, S	1	$22,770	$898	__$14,970__	__$1895.50__
	$22,770 − $4750 − $3050 = $14,970; $700 + .15 × ($14,970 − $7000) = $1895.50					
9.	The Cooks, M	3	$38,751	$5968	_____	_____
10.	The Loveridges, M	7	$52,532	$6972	_____	_____
11.	The Jordans, M	5	$71,800	$9320	_____	_____

C indicates an exercise that is related to the Case in Point feature within the section.

Name	Number of Exemptions	Adjusted Gross Income	Total Deductions	Taxable Income	Tax Owed
12. G. Clarke, S	1	$32,322	$4318	_____	_____
13. D. Collins, S	2	$35,350	$6240	_____	_____
14. K. Tang, Head of Household	2	$93,240	$5480	_____	_____
15. B. Kammerer, S	1	$38,526	$5107	_____	_____
16. G. Nation, M	3	$143,420	$11,700	_____	_____
17. B. Albert, Head of Household	2	$62,613	$7681	_____	_____
18. B. Nelson, Head of Household	4	$58,630	$6290	_____	_____

Find the tax refund or tax due for the following people. The letter following the names indicates the marital status. Assume a 52-week year and that married people are filing jointly. (See Example 4.)

Quick Start

Name	Taxable Income	Federal Income Tax Withheld from Checks	Tax Refund or Tax Due
19. Karecki, S	$13,378	$243.10 monthly	$1260.50 tax refund
$2917.20 − $1656.70 = $1260.50			
20. Turner, K., S	$32,060	$347.80 monthly	_____
21. Hunziker, S	$23,552	$72.18 weekly	_____
22. The Fungs, M	$38,238	$119.27 weekly	_____
23. The Todds, M	$21,786	$208.52 monthly	_____
24. The Bensons, M	$46,850	$165.30 weekly	_____

25. List four sources of income for which an individual might receive W-2 and 1099 forms. (See Objective 3.)

26. List four possible tax deductions, and explain the effect that a tax deduction will have on taxable income and on income tax due. (See Objective 6.)

Quick Start

Find the tax in the following application problems.

27. **MARRIED—INCOME TAX**   The Tobins had an adjusted gross income of $45,378 last year. They had deductions of $482 for state income tax, $187 for city income tax, $472 for property tax, $3208 in mortgage interest, and $324 in contributions. They file a joint return and claim five exemptions.

27. $2394.20

$482 + $187 + $472 + $3208 + $324 = $4673 $\left(\text{below standard deduction} \right)$
$45,378 - $9500 - \left(5 \times $3050 \right) = $20,628$
Tax = $1400 + .15 \times \left($20,628 - $14,000 \right) = 2394.20

28. **SINGLE—INCOME TAX**   Diane Bolton works at The Doll House and had an adjusted gross income of $34,975 last year. She had deductions of $971 for state income tax, $564 for property tax, $3820 in mortgage interest, and $235 in contributions. Bolton claims one exemption and files as a single person.

28. _____

29. **SINGLE—INCOME TAX**   Martha Crutchfield has an adjusted gross income of $79,300 and files as a single person with only one exemption. Her deductions amounted to $4630.

29. _____

30. **MARRIED—INCOME TAX**   The Simpsons had an adjusted gross income of $48,260 last year. They had deductions of $1078 for state income tax, $253 for city income tax, $3240 for property tax, $5218 in mortgage interest, and $386 in contributions. They claim three exemptions and file a joint return.

30. _____

31. **HEAD OF HOUSEHOLD**   Martha Spencer, owner of The Doll House, had wages of $68,645, dividends of $385, interest of $672, and adjustments to income of $1058 last year. She had deductions of $877 for state income tax, $342 for city income tax, $986 for property tax, $5173 in mortgage interest, and $186 in contributions. She claims four exemptions and files as head of household.

31. _____

32. **HEAD OF HOUSEHOLD**   John Walker had wages of $30,364, other income of $2892, dividends of $240, interest of $315, and a regular IRA contribution of $750 last year. He had deductions of $1163 for state income tax, $1268 for property tax, $1294 in mortgage interest, and $540 in contributions. Walker claims two exemptions and files as head of household.

32. _____

33. MARRIED John and Vicki Karsten had combined wages and salaries of $64,280, other income of $5283, dividend income of $324, and interest income of $668. They have adjustments to income of $2484. Their itemized deductions are $7615 in mortgage interest, $2250 in state income tax, $1185 in real estate taxes, and $1219 in charitable contributions. The Karstens filed a joint return and claimed 6 exemptions.

33. _____

34. Eleanor Joyce is single and claims one exemption. As a television news reporter, her salary last year was $74,300, and she had other income of $2800 and interest income of $8400. She has an adjustment to income of $2200 for an Individual Retirement Account (IRA) contribution. Her itemized deductions are $5807 in mortgage interest, $2800 in state income tax, $1230 in real estate taxes, and $690 in charitable contributions.

34. _____

12.3 | Fire Insurance

Objectives

1. Define the terms *policy, coverage, face value,* and *premium.*
2. Find the annual premium for fire insurance.
3. Use the coinsurance formula.
4. Understand multiple carrier insurance.
5. List additional types of insurance coverage.

CASE *in* **POINT**

Martha Spencer owns the building in which The Doll House is located. A fire in that building could leave her in financial ruin. Although she hopes that there is never a fire, she carries fire insurance to protect her business in the event there is one.

Insurance protects against risk. For example, there is only a slight chance that a particular building will be damaged by fire in any year. However, the financial loss could be very large for the owner should his building be damaged by fire. Therefore, people or companies pay a small fee each year to an insurance company to protect them against catastrophic losses. The insurance company collects money from many different insureds and pays money to the few that suffer damages. The following chart shows that most fires in homes begin in the kitchen. Perhaps that is why many recommend that you keep a fire extinguisher in your kitchen. The clipping in the margin shows who is most at risk of fire death and also the common causes of fires in homes.

At Risk: Fire Deaths

- Seniors over age 70 and children under 5.
- Men twice as much as women.
- Blacks and Native Americans, whose fire death rates are higher than the national average.

What Causes Fires

- Cooking
- Careless smoking
- Heating systems
- Arson

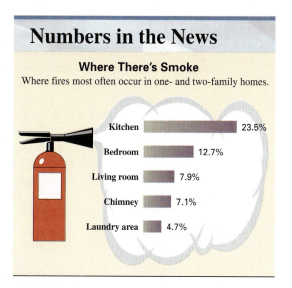

Individuals buy insurance to protect against losses due to fire, theft, illness or health problems, disability, car wrecks, lawsuits, and even death. Companies buy insurance to protect against losses due to fire, automobile accidents, employee illnesses, lawsuits, and worker accidents on the job.

> **QUICK TIP** Lenders almost always require fire insurance on any property that they use as collateral for a loan.

Objective 1 **Define the terms *policy, coverage, face value,* and *premium.*** The contract between the owner of a building and an insurance company is called a **policy** or an **insurance**

policy. A basic fire policy provides **coverage** or protection for both the owner of the building and also for the company that holds the mortgage on the building. The owner of a building can also purchase coverage on the contents of the building and liability insurance in the event someone is injured while on the property. Homeowners purchase a **homeowner's policy**, which includes all of these coverages. The line graph shows that homeowner's insurance continues to increase.

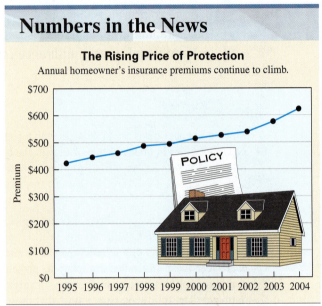

Numbers in the News

The Rising Price of Protection
Annual homeowner's insurance premiums continue to climb.

DATA: National Association of Insurance Commissioners; and Insurance Information Institute

The dollar value of the insurance coverage provided on a building itself is called the **face value** of the policy. The annual charge for the policy is called the **premium**. The premium is usually calculated based on factors such as age of the building, materials used in the construction of the building, crime rate of the neighborhood, presence of any safety features such as a sprinkler system, and any history of previous insurance claims on the property.

Objective 2 Find the annual premium for fire insurance. The amount of the premium charged by the insurance company depends on several factors. Among them are the type of construction of the building, the contents and use of the building, the location of the building, and the type and location of any fire protection that is available. Wood frame buildings generally are more likely to be damaged by fire than masonry buildings and thus require a larger premium. Building classifications are assigned to building types by insurance company **underwriters**. These categories are usually designated by letters such as *A*, *B*, and *C*. Underwriters also assign ratings called **territorial ratings** to each area that describe the quality of fire protection in the area. Although fire-insurance rates vary from state to state, the rates in the following table are typical.

Annual Rates for Each $100 of Insurance

Building Classification						
	A		B		C	
Territorial Rating	Building	Contents	Building	Contents	Building	Contents
1	$.25	$.32	$.36	$.49	$.45	$.60
2	$.30	$.44	$.45	$.55	$.54	$.75
3	$.37	$.46	$.54	$.60	$.63	$.80
4	$.50	$.52	$.75	$.77	$.84	$.90
5	$.62	$.58	$.92	$.99	$1.05	$1.14

EXAMPLE 1

Finding the Annual Fire Insurance Premium

The Doll House is in a building having a rating of class C. It is in territory 4. Find the annual premium if the replacement cost of the building is $640,000 and the contents are valued at $186,500.

SOLUTION

Building:

Replacement cost in hundreds = $640,000 ÷ 100 = **6400**

Insurance premium for the building = **6400** × $.84 = $5376

Contents:

Replacement cost in hundreds = $186,500 ÷ 100 = **1865**

Insurance premium for contents = **1865** × $.90 = $1678.50

Total Premium = $5376 + $1678.50 = **$7054.50**

Objective 3   Use the coinsurance formula. Most fires damage only a portion of a building and the contents. Since complete destruction of a building is rare, many owners save money by buying insurance for only a portion of the value of the building and contents. Realizing this, insurance companies place a **coinsurance clause** in almost all fire-insurance policies. Effectively, the business assumes part of the risk of a loss under coinsurance.

An 80% coinsurance clause in the policy requires that the amount of insurance carried by the owner of a building be at least 80% of the **replacement cost** of the building, where replacement cost is the cost to rebuild the entire building. Replacement cost is often higher than fair market value on older buildings. If the amount of insurance is less than 80% of the replacement cost, the insurance company pays only the portion of any loss as shown in the following formula.

$$\text{Amount insurance company will pay} \left(\text{assuming 80\% coinsurance}\right) = \text{Amount of loss} \times \frac{\text{Amount of policy}}{\text{80\% of replacement cost}}$$

> **QUICK TIP**   The insurance company never pays more than the face value of the policy, nor more than the amount of the loss.

EXAMPLE 2

Using the Coinsurance Formula

Buster Stetson owns a commercial building with a replacement cost of $760,000. His fire insurance policy has an 80% coinsurance clause and a face value of $570,000. The building suffers a fire loss of $144,000. Find the amount of the loss that the insurance company will pay.

SOLUTION

The policy must have a face value of at least 80% of $760,000 or $608,000 in order to receive the payment for the entire loss. Since the face value of $570,000 is less than 80% of the replacement cost, the company will pay only the following portion of the loss.

$$\text{Amount insurance company pays} = \$144,000 \times \frac{\textbf{\$570,000}}{\textbf{\$608,000}} = \$135,000$$

Amount not paid by insurance company = $144,000 − $135,000 = $9000

Stetson is responsible for the $9000.

The calculator solution to this example uses chain calculations and parentheses to set off the denominator. The result is then subtracted from the fire loss.

144,000 × 570,000 ÷ (80 % × 760,000) = 135,000

144,000 − 135,000 = 9000

Note: Refer to Appendix C for calculator basics.

EXAMPLE 3

Finding the Amount of Loss Paid by the Insurance Company

A Swedish investment group owns a warehouse with a replacement cost of $3,450,000. The company has a fire-insurance policy with a face value of $3,400,000. The policy has an 80% coinsurance feature. If the firm has a fire loss of $233,500, find the part of the loss paid by the insurance company.

SOLUTION

$$80\% \text{ of replacement cost} = .80 \times \$3,450,000 = \$2,760,000$$

The business has a fire-insurance policy with a face value of more than 80% of the value of the store. Therefore, the insurance company will pay the entire $233,500 loss.

Objective [4] **Understand multiple carrier insurance.** A business may have fire-insurance policies with several companies at the same time. Perhaps additional insurance coverage was purchased over a period of time, as new additions were made to a factory or building complex. Or perhaps the building is so large that one insurance company does not want to take the entire risk by itself, so several companies each agree to take a portion of the insurance coverage and thereby share the risk. In either event, the insurance coverage is divided among **multiple carriers**. When an insurance claim is made against multiple carriers, each insurance company pays its fractional portion of the total claim on the property.

EXAMPLE 4

Understanding Multiple Carrier Insurance

Youngblood Apartments has an insured loss of $1,800,000 while having insurance coverage beyond its coinsurance requirement. The insurance is divided among Company A with $5,900,000 coverage, Company B with $4,425,000 coverage, and Company C with $1,475,000 coverage. Find the amount of the loss paid by each of the insurance companies.

SOLUTION

Start by finding the total face value of all three policies.

$$\$5,900,000 + \$4,425,000 + \$1,475,000 = \$11,800,000 \text{ total face value}$$

$5,900,000	Company A pays $\dfrac{\$5,900,000}{\$11,800,000} = \dfrac{1}{2}$ of the loss
$4,425,000	Company B pays $\dfrac{\$4,425,000}{\$11,800,000} = \dfrac{3}{8}$ of the loss
+ $1,475,000	Company C pays $\dfrac{1,475,000}{\$11,800,000} = \dfrac{1}{8}$ of the loss
$11,800,000 total face value	

Since the insurance loss is $1,800,000, the amount paid by each of the multiple carriers is

Company A $\dfrac{1}{2} \times \$1,800,000 = \$900,000$

Company B $\dfrac{3}{8} \times \$1,800,000 = \$675,000$

Company C $\dfrac{1}{8} \times \$1,800,000 = \underline{\$225,000}$

Total loss = $1,800,000

QUICK TIP   If the coinsurance requirement is not met, the total amount of the loss paid by the insurance coverage is found, and then the amount that each of the carriers pays is found, as in Example 4.

The amount of money paid toward insurance premiums by businesses is often small when compared with other business expenses. Likewise, the average household pays only a small portion of its budget on insurance premiums. The following graph shows the percent of total household spending going to pay for insurance coverage.

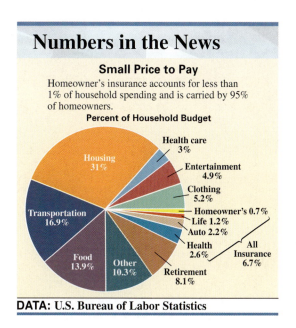

Numbers in the News

Small Price to Pay

Homeowner's insurance accounts for less than 1% of household spending and is carried by 95% of homeowners.

Percent of Household Budget

Housing 31%
Transportation 16.9%
Food 13.9%
Other 10.3%
Retirement 8.1%
Health 2.6%
Auto 2.2%
Life 1.2%
Homeowner's 0.7%
Clothing 5.2%
Entertainment 4.9%
Health care 3%
All Insurance 6.7%

DATA: U.S. Bureau of Labor Statistics

Objective 5 **List additional types of insurance coverage.** There are many other types of insurance coverages available as you can see.

Worker's compensation: pays employees for injuries while on the job
Liability: pays in the event of an injury to someone or to their property
Homeowner: protects a homeowner against fire and theft, but can also cover losses on stolen credit cards and medical costs for guests
Renter: protects a renter against loss of personal property
Medical: pays for medical expenses
Disability: pays monthly payments in the event of disability

QUICK TIP A renter without insurance on their contents will not be paid for damages to contents, in the event of a fire.

12.3 | Exercises

FOR EXTRA HELP

 MyMathLab

 InterActMath.com

 MathXL

 MathXL
Tutorials on CD

 Addison-Wesley
Math Tutor Center

 DVT/Videotape

The   **Quick Start**   *exercises in each section contain solutions to help you get started.*

Find the total annual premium for each of the following. Use the table on page 528. (See Example 1.)

Quick Start

	Territorial Rating	Building Classification	Building Value	Contents Value	Total Annual Premium
1.	2	B	$280,000	$80,000	**$1700**
2.	5	A	$220,500	$105,000	**$1976.10**
3.	1	C	$285,000	$152,000	_____
4.	2	B	$272,500	$111,500	_____
5.	5	B	$782,600	$212,000	_____
6.	3	A	$596,400	$206,700	_____

Find the amount to be paid by the insurance company in the following problems. Assume that each policy includes an 80% coinsurance clause. (See Examples 2 and 3.)

Quick Start

	Replacement Cost of Building	Face Value of Policy	Amount of Loss	Amount Paid
7.	$145,000	$85,000	$9600	**$7034.48**

$145,000 \times .8 = $116,000 \left(80\%\right); \frac{\$85,000}{\$116,000} \times \$9600 = \$7034.48$

8.	$187,400	$140,000	$10,850	**$10,132.07**

$187,400 \times .8 = $149,920 \left(80\%\right); \frac{\$140,000}{\$149,920} \times \$10,850 = \$10,132.07$

9.	$287,000	$232,500	$19,850	_____
10.	$780,000	$585,000	$10,400	_____

Replacement Cost of Building	Face Value of Policy	Amount of Loss	Amount Paid
11. $218,500	$195,000	$36,500	_____
12. $750,000	$500,000	$56,000	_____

Find the amount paid by each insurance company in the following problems involving multiple carriers. Assume that the coinsurance requirement is met. (See Example 4.)

Quick Start

Insurance Loss	Companies	Coverage	Amount Paid
13. $80,000	Company 1	$750,000	**$60,000**
	Company 2	$250,000	**$20,000**

$\frac{750,000}{1,000,000} \times \$80,000 = \$60,000; \frac{250,000}{1,000,000} \times \$80,000 = \$20,000$

Insurance Loss	Companies	Coverage	Amount Paid
14. $360,000	Company A	$1,200,000	_____
	Company B	$800,000	_____
15. $650,000	Company 1	$1,350,000	_____
	Company 2	$1,200,000	_____
	Company 3	$450,000	_____
16. $1,600,000	Company A	$4,800,000	_____
	Company B	$800,000	_____
	Company C	$2,400,000	_____

Find the annual fire-insurance premium in each of the following application problems. Use the table on page 528.

Quick Start

17. **FURNITURE STORE FIRE INSURANCE**   Billings Furniture owns a building with a replacement cost of $1,400,000 and with contents of $360,000. The building is class C with a territorial rating of 5.

17. $18,804

14,000 × 1.05 = $14,700; 3600 × $1.14 = $4104
$14,700 + $4104 = $18,804

18. **FIRE-INSURANCE PREMIUM**   Martha Spencer, owner of The Doll House, owns a class-B building with a replacement cost of $165,400. Contents are valued at $128,000. The territorial rating is 3.

18. _____

19. INDUSTRIAL FIRE INSURANCE   Valley Crop Dusting owns a class-B building with a
replacement cost of $107,500. Contents are worth $39,800. The territorial rating is 2.

19. _____

20. INDUSTRIAL BUILDING INSURANCE   London's Dredging Equipment is in a C-rated building with a
territorial rating of 4. The building has a replacement cost of $305,000 and the contents are worth
$682,000.

20. _____

21. Describe three factors that determine the premium charged for fire insurance. (See Objective 2.)

22. Explain the coinsurance clause and describe how coinsurance works. (See Objective 3.)

*In the following application problems, find the amount of the loss paid by (a) the insurance
company and (b) the insured. Assume an 80% coinsurance clause.*

23. FIRE LOSS   The Doll House is located in a building with a replacement cost of $328,500, but Martha Spencer insured it for only
$200,000 in order to save money on insurance premiums. An
electrical fire burns out of control and causes $180,000 in damage.

(a) _____
(b) _____

24. GIFT-SHOP FIRE LOSS   Indonesian Wonder gift shop has a replacement cost of $395,000. The shop is
insured for $280,000. Fire loss is $22,500.

(a) _____
(b) _____

25. SALVATION ARMY LOSS   The main office of the Salvation Army suffers a loss from fire of $45,000.
The building has a replacement cost of $550,000 and is insured for $300,000.

(a) _____
(b) _____

26. APARTMENT FIRE LOSS   Kathy Stephenson owns a triplex with a replacement cost of $185,000, and
it is insured for $111,000. Fire loss is $28,000.

(a) _____
(b) _____

27. Explain in your own words multiple-carrier insurance. Give two reasons for dividing insurance among multiple carriers. (See Objective 4.)

28. Several types of insurance coverage beyond basic fire coverage are included in a homeowner's policy. List and explain three losses that would be covered. (See Objective 5.)

In the following application problems, find the amount paid by each of the multiple carriers.
Assume that the coinsurance requirement has been met.

29. COINSURED FIRE LOSS   C. Wood Plumbing had an insured fire loss of $548,000. It had insurance coverage as follows: Company A, $600,000; Company B, $400,000; and Company C, $200,000.

A: _____
B: _____
C: _____

30. The Cycle Centre had an insured fire loss of $68,500. It has insurance as follows: Company 1, $60,000; Company 2, $40,000; and Company 3, $30,000.

1: _____
2: _____
3: _____

31. MAJOR FIRE LOSS   Gold's Gym had fire insurance coverage as follows: Company 1, $360,000; Company 2, $120,000; and Company 3, $240,000. The gym had an insured fire loss of $250,000.

1: _____
2: _____
3: _____

32. Global Manufacturing Company had an insured fire loss of $2,100,000. They had insurance as follows: Company A, $2,000,000; Company B, $1,750,000; Company C, $1,250,000.

A: _____
B: _____
C: _____

12.4 | Motor-Vehicle Insurance

Objectives

1. Describe the factors that affect the cost of motor-vehicle insurance.
2. Define liability insurance.
3. Define property damage insurance.
4. Describe comprehensive and collision insurance.
5. Define no-fault and uninsured motorist insurance.
6. Apply youthful operator factors.
7. Find the amounts paid by the insurance company and the insured.

> **CASE in POINT**
>
> Martha Spencer owns a van that is used by employees of The Doll House. She insures her van since an employee may be involved in an accident that hurts someone or damages property. Knowing that injuries can result in huge medical costs, Spencer carries more than the state-required minimum liability insurance.

Objective 1 **Describe the factors that affect the cost of motor-vehicle insurance.** Automobile insurance is one of the most common types of insurance. Banks or credit unions that finance automobiles require borrowers to buy automobile insurance. Many states require automobile insurance for people living and driving in those states. The average annual cost of automobile insurance continues to increase as shown in the bar graph. These costs continue to increase due to increasing costs associated with injuries, as well as repair costs.

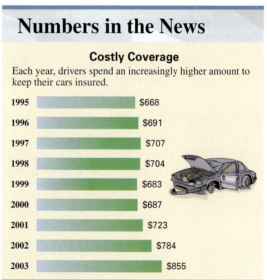

Numbers in the News

Costly Coverage

Each year, drivers spend an increasingly higher amount to keep their cars insured.

1995	$668
1996	$691
1997	$707
1998	$704
1999	$683
2000	$687
2001	$723
2002	$784
2003	$855

DATA: National Association of Insurance Commissioners; Insurance Information Institute

The **premium**, or cost, of an insurance policy is determined by **actuaries** who work for the insurance company. Actuaries look at the frequency and severity of accidents based on several factors including location of the insured vehicle, age and sex of the driver, and driving history of the driver. These factors help measure the risk of insuring a particular driver, which is used to determine the premium. For example, drivers between 16 and 25 years of age are more likely to be involved in accidents and are therefore charged a higher premium. Some automobiles are more likely to be stolen than others, resulting in higher premiums for their owners. Several automobile coverages are now discussed.

Objective ▣ **Define liability insurance.** **Liability** or **bodily injury insurance** protects the insured in case he or she injures someone with a car. Many states have minimum amounts of liability insurance coverage set by law. The amount of liability insurance is expressed as a fraction, such as 15/30. The fraction 15/30 means that the insurance company will pay up to $15,000 for injury to one person, and a total of $30,000 for injury to two or more persons in the same accident.

QUICK TIP   Claims following a car accident can be quite large. For example, imagine the loss of income of a surgeon who was disabled for life in a car accident.

The following table shows typical premium rates for various amounts of liability coverage. Included in the cost of the liability insurance is **medical insurance** for the driver and passengers in case of injury. For example, the table column 15/30 shows that the insured can also receive reimbursement for up to $1000 of his or her own medical expenses in an accident. Insurance companies divide the nation into territories based on past claims in all areas. Four territories are shown here. All tables in this section show annual premiums.

Liability (Bodily Injury) and Medical Insurance (Per Year)

| | \ \ \ \ \ \ \ \ \ Liability and Medical Expense Limits | | | | |
Territory	15/30 $1000	25/50 $2000	50/100 $3000	100/300 $5000	250/500 $10,000
1	$207	$222	$253	$282	$308
2	269	302	341	378	392
3	310	314	375	398	459
4	216	218	253	284	310

QUICK TIP   Drivers who have several speeding tickets or have been involved in accidents often have much higher premiums. This is especially true if the driver is young.

EXAMPLE 1

Finding the Liability and Medical Premium

Martha Spencer, owner of The Doll House, is in territory 2 and wants 100/300 liability coverage. Find the amount of the premium for this coverage and the amount of medical coverage included.

SOLUTION

Look up territory 2 and 100/300 coverage in the liability and medical insurance table to find an annual premium of $378. This cost includes $5000 medical coverage.

Objective ▣ **Define property damage insurance.** Liability coverage pays if you injure someone. **Property damage coverage** pays if you damage someone else's property such as their automobile or a building. The following table shows the annual cost for various **policy limits** on property damage. You are responsible for damages above the policy limit.

Property Damage Insurance (Per Year)

| | Property Damage Limits | | | |
Territory	$10,000	$25,000	$50,000	$100,000
1	$88	$93	$97	$103
2	168	192	223	251
3	129	134	145	158
4	86	101	112	124

QUICK TIP   In some states, drivers can get a discount on their insurance based on a good driving record, an antitheft system, good grades, or even by taking a defensive driving class. Contact your insurance agent to see if you are eligible for any discounts.

The following graph shows the number of states requiring various discounts on auto insurance premiums.

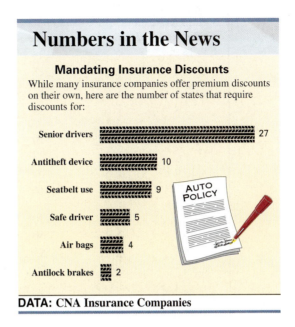

Numbers in the News

Mandating Insurance Discounts

While many insurance companies offer premium discounts on their own, here are the number of states that require discounts for:

Senior drivers	27
Antitheft device	10
Seatbelt use	9
Safe driver	5
Air bags	4
Antilock brakes	2

DATA: CNA Insurance Companies

EXAMPLE 2

Finding the Premium for Property Damage ■

Find the premium if Martha Spencer, in territory 2, wants property damage coverage of $50,000.

SOLUTION

Property damage coverage of $50,000 in territory 2 requires a premium of $223.

Objective 4 Describe comprehensive and collision insurance. Comprehensive insurance pays for damages to the insured's vehicle caused by a fire, by theft of the automobile, by vandalism (e.g., someone steals the CD player), by a tree falling onto the automobile, and other similar events. In general, comprehensive insurance pays for damages that are not covered by collision.

Collision coverage pays for repairs to the insured's vehicle when it is involved in a collision with another object. For example, collision insurance would pay for damages to your automobile should you lose control during an ice storm and run into a tree. Common collision insurance **deductibles** are $100, $250, $500, or $1000. You pay the deductible, and then the collision coverage pays for damages to the vehicle as shown below. The higher the deductible the lower the insurance premium, but of course you are responsible for the deductible in the event of an accident. Suppose your deductible is $250 but the damage to your automobile is $3280. The amount paid by the insurance company is found as follows.

Damages to your automobile	$3280
Deductible must be paid by you	− 250
Insurance company pays	$3030

QUICK TIP States require liability insurance but not comprehensive or collision. Lenders require comprehensive and collision coverages. If your car is paid for, it is your choice as to whether or not to purchase comprehensive and collision coverages. It may or may not be worthwhile on older vehicles.

The following table shows some typical rates for comprehensive and collision insurance. Rates are determined not only by territories, but also by age group and symbol. Here, age group refers to the age of the *vehicle*, not the driver. Age group 1 is a vehicle that is less than 2 years of age. Age group 2 is a vehicle that is at least 2, but less than 3, years of age. Age group 6 is a vehicle 6 years of age or older. Symbol is determined by the *cost* of the vehicle.

Comprehensive and Collision Insurance (Per Year)

Territory	Age Group	Comprehensive			Collision ($250 Deductible)		
		Symbol			Symbol		
		6	7	8	6	7	8
1	1	$58	$64	$90	$153	$165	$184
	2, 3	50	56	82	135	147	171
	4, 5	44	52	76	116	128	147
	6	34	44	64	92	110	128
2	1	$26	$28	$40	$89	$95	$104
	2, 3	22	24	36	80	86	98
	4, 5	20	24	34	71	77	86
	6	16	20	28	60	68	77
3	1	$70	$78	$108	$145	$157	$174
	2, 3	60	66	90	128	139	162
	4, 5	52	64	92	111	122	139
	6	20	22	32	66	74	81
4	1	$42	$46	$66	$97	$104	$124
	2, 3	36	40	58	87	94	107
	4, 5	32	38	54	77	84	94
	6	26	32	46	64	74	84

QUICK TIP Higher deductibles result in lower premiums. However, higher deductibles also mean that the insured may have to pay more in the event of an accident.

EXAMPLE 3

Finding the Comprehensive and Collision Premium

Martha Spencer, owner of The Doll House, is in territory 2 and has a 2-year-old minivan that has a symbol of 8. Use the comprehensive and collision-insurance table to find the cost for **(a)** comprehensive coverage and **(b)** collision coverage.

SOLUTION

(a) The cost of comprehensive coverage is $36.
(b) The cost of collision coverage is $98.

The following newspaper article shows the results of crash tests at 5 mph. The popular small pickup trucks racked up some large repair bills. The highest repair bill in this 5-mph test was Toyota Tacoma with $4361 in damage. This type of data helps insurance companies determine the premium for the vehicles they insure.

HERE & NOW

Hitting the Wall: Small Trucks Rack Up Big Repair Costs

An insurance industry study found that some popular small pickups racked up sizable repair bills in crash tests at just 5 mph. Leading the way was the Toyota Tacoma, which sustained $4361 in damage over four low-speed tests. Other trucks tested by the Insurance Institute for Highway Safety: the Chevrolet S-10 LS ($2246 in damage), the Ford Ranger XLT ($2952), the Dodge Dakota Sport ($3863) and the Nissan Frontier XE ($3867). Representatives of the automakers defended the trucks as crashworthy and said the insurance group released the study as part of a campaign to get the government to raise the bumper standard from 2.5 mph to 5 mph.

Source: Insurance Institute for Highway Safety

Objective ⑤ **Define no-fault and uninsured motorist insurance.** Some states have **no-fault** laws. Under no-fault insurance, all medical expenses and costs associated with an accident are paid to each individual by *his or her own insurance company*, no matter who is at fault. Legislators and insurance companies argue that no-fault insurance removes lawyers, courts, and juries from the process and results in quicker, less costly settlements. Others (including trial lawyers) argue that no-fault insurance leaves accident victims unable to recover all of their damages.

Most states do not have no-fault laws, thereby requiring the insurance company of the person at fault to pay for damages. A potential problem in these states is that the motorist who caused an accident either has no insurance at all, or has too little insurance for the damages that occurred. **Uninsured motorist insurance** protects a vehicle owner from financial liability when hit by a driver with no insurance. **Underinsured motorist insurance** provides protection to a vehicle owner when hit by a driver who has *too little* insurance. Typical costs for uninsured motorist insurance are shown in the table at the left.

Uninsured Motorist Insurance (Per Year)	
Territory	Basic Limit
1	$66
2	$44
3	$76
4	$70

EXAMPLE 4

Determining the Premium for Uninsured Motorist Coverage ■

Martha Spencer, in territory 2, wants uninsured motorist coverage. Find the premium in the uninsured motorist insurance table.

SOLUTION

The premium for the uninsured motorist coverage in territory 2 is $44.

Objective ⑥ **Apply youthful operator factors.** Sometimes young people feel that they are being charged *far too much* for automobile insurance. Insurance companies base their rates on probabilities calculated from statistical data. The graph below shows that drivers under 25 are more likely to be involved in a fatal automobile accident than drivers 25 or older. Therefore, insurance companies charge higher rates for **youthful** compared with **adult** drivers. The age at which a youth becomes an adult varies from company to company. Generally, drivers under 25 are considered youthful drivers, and drivers 25 or older are considered adults.

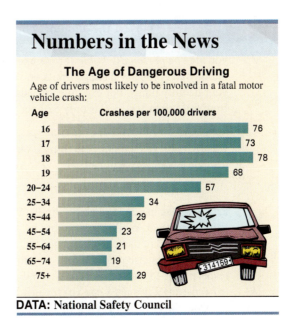

Numbers in the News

The Age of Dangerous Driving

Age of drivers most likely to be involved in a fatal motor vehicle crash:

Age	Crashes per 100,000 drivers
16	76
17	73
18	78
19	68
20–24	57
25–34	34
35–44	29
45–54	23
55–64	21
65–74	19
75+	29

DATA: National Safety Council

The following table shows a youthful-operator factor based on age and on whether or not the operator has taken driver's training. You can see that the factors are much higher for youthful operators who have not taken driver's training. The steps to apply these factors follow.

1. Determine the total premium for **all coverages** desired.
2. Multiply the premium by the youthful-operator factor from the table.

Youthful-Operator Factor

Age	With Driver's Training	Without Driver's Training
20 or less	1.55	1.75
21–25	1.15	1.40

EXAMPLE 5

Using the Youthful-Operator Factor

Janet Ito lives in territory 4, is 22 years old, has had driver's training, and drives a 5-year-old car with a symbol of 7. She wants a 25/50 liability policy, a $10,000 property damage policy, a comprehensive and collision policy, and uninsured motorist coverage. Find her annual insurance premium using the tables in this section.

SOLUTION

1. Determine the total premium for all coverages desired.

25/50 liability insurance	$218
$10,000 property damage	86
Comprehensive insurance	38
Collision	84
Uninsured motorist	+ 70
Subtotal	$496

2. Multiply the premium by the youthful-operator factor from the table.

$$\$496 \times 1.15 = \$570.40$$

The calculator solution to this example uses parentheses and chain calculations.

(218 + 86 + 38 + 84 + 70) × 1.15 = 570.4

Note: Refer to Appendix C for calculator basics.

Objective 7 **Find the amounts paid by the insurance company and the insured.** If you are at fault in an automobile accident and the damages *exceed the limits* on your insurance policy, you *may be personally liable* for the excess. You can help avoid this situation if you increase the liability and property damage limits on your policy. Sometimes, it does not cost much to increase your liability limits. For example, the additional cost of increasing liability coverage in territory 1 from 50/100 to 100/300 is only $29 per year ($282–$253).

QUICK TIP   Remember, the limits on the insurance policy are the maximum that the insurance company will pay.

EXAMPLE 6

Finding the Amounts Paid by the Insurance Company and the Insured

Eric Liwanag has 25/50 liability limits, $25,000 property damage limits, and $250 deductible collision insurance. While on vacation, he was at fault in an accident that caused $5800 damage to his car, $3380 in damage to another car, and resulted in severe injuries to the other driver and his passenger. A subsequent lawsuit for injuries resulted in a judgment of $45,000 and $35,000, respectively, to the other parties. Find the amounts that the insurance company will pay for (a) repairing Liwanag's car, (b) repairing the other car, and (c) paying the court judgment resulting from the lawsuit. (d) How much will Liwanag have to pay the injured parties?

SOLUTION

(a) The insurance company will pay $5550 ($5800 − $250 deductible) to repair Liwanag's car.
(b) Repairs on the other car will be paid to the property damage limits ($25,000). Here, the total repairs of $3380 are paid.
(c) Since more than one person was injured, the insurance company pays the limit of $50,000 ($25,000 to each of the two injured parties).
(d) Liwanag is liable for $30,000 ($80,000 − $50,000), the amount awarded over the insurance limits.

12.4 | Exercises

FOR EXTRA HELP

 MyMathLab

 InterActMath.com

 MathXL

 MathXL Tutorials on CD

Addison-Wesley Math Tutor Center

DVT/Videotape

The   Quick Start   *exercises in each section contain solutions to help you get started.*

Find the annual premium for the following. (See Examples 1–5.)

Quick Start

	Name	Territory	Age	Driver Training	Liability	Property Damage	Comprehensive Collision Age Group	Symbol	Uninsured Motorist	Annual Premium
1.	Smyth	3	42	—	50/100	$25,000	2	7	Yes	$790

$375 + $134 + $66 + $139 + $76 = $790

	Name	Territory	Age	Driver Training	Liability	Property Damage	Comprehensive Collision Age Group	Symbol	Uninsured Motorist	Annual Premium
2.	Morrissey	1	20	Yes	25/50	$25,000	4	7	No	$767.25

$222 + $93 + $52 + $128 + $0 = $495 × 1.55 = $767.25

	Name	Territory	Age	Driver Training	Liability	Property Damage	Comprehensive Collision Age Group	Symbol	Uninsured Motorist	Annual Premium
3.	Shraim	3	52	—	250/500	$50,000	2	8	Yes	_____
4.	Waldron	2	67	—	50/100	$100,000	1	6	Yes	_____

5. Describe four factors that determine the premium on an automobile insurance policy. (See Objective 1.)

6. Explain in your own words the difference between liability (bodily injury) and property damage. (See Objectives 2 and 3.)

Solve the following application problems.

Quick Start

7. **ADULT AUTO INSURANCE**   Bill Poole is 47 years old, lives in territory 4, and drives a 2-year-old car with a symbol of 7. He wants 250/500 liability limits, $100,000 property damage limits, comprehensive and collision insurance, and uninsured motorist coverage. Find his annual insurance premium.
7. $638
$310 + $124 + $40 + $94 + $70 = $638

8. Martha Spencer, owner of The Doll House, is thinking about moving from a house in territory 2 to a house in territory 3 and wants to know the effect on the insurance costs for her new car. She currently lives in territory 2 and the car has symbol 7. Her coverages are 250/500 for liability, $100,000 for property damage, comprehensive, collision, and uninsured motorist coverages. She is 53 years old. Find the change in annual cost.
8. _____

9. **YOUTHFUL-OPERATOR AUTO INSURANCE**   Brandy Barrett is 23 years old, took a driver's education course, lives in territory 1, and drives a 4-year-old car with a symbol of 6. She wants 50/100 liability limits, $25,000 property damage limits, comprehensive and collision insurance, and uninsured motorist coverage. Find her annual insurance premium.
9. _____

C indicates an exercise that is related to the Case in Point feature within the section.

10. YOUTHFUL OPERATOR—NO DRIVERS TRAINING   Karen Roberts' father gave her a new Honda
Accord to use at college under the condition she pay her own insurance. She is 17, has not had
driver's training, lives in territory 1, and her vehicle has a symbol of 6. She wants 50/100 liability
limits, $25,000 property damage limits, comprehensive and collision insurance, and uninsured
motorist coverage. Find her annual insurance premium.

10. _____

11. BODILY INJURY INSURANCE   Suppose your bodily injury policy has limits of 25/50, and you injure a
person on a bicycle. The judge awards damages of $36,500 to the cyclist. **(a)** How much will the
company pay? **(b)** How much will you pay?

(a) _____
(b) _____

12. BODILY INJURY INSURANCE   Three years ago Martha Spencer, owner of The Doll House, lost control of
her car while trying to find her cell phone and forced another driver off the road. The court awarded
$28,000 to the driver of the other car and $8,000 to a passenger of the other car. Spencer has limits of
15/30. **(a)** Find the amount the insurance company will pay. **(b)** Find the amount Spencer must pay.

(a) _____
(b) _____

13. MEDICAL EXPENSES AND PROPERTY DAMAGE   Wes Hanover accidentally backed into a parked car. He
caused $4300 in damage to the car and Hanover's passenger needed stitches in her forehead, which cost
$850. Hanover had 15/30 liability limits, $1000 medical expense, and property damage of $10,000. Find
the amount paid by the insurance company for **(a)** damages to the automobile and **(b)** for medical expenses.

(a) _____
(b) _____

14. MEDICAL EXPENSES AND PROPERTY DAMAGE   Jessica Wallace backed into a new Mercedes and caused
$12,800 in damage to the car. She also injured the vertebrae in her neck, requiring surgery costing $48,200.
She had 50/100 liability limits, $10,000 in property damage, and $3000 in medical expense coverage. Find the
amount paid by the insurance company for **(a)** damages to the automobile and **(b)** for medical expenses.

(a) _____
(b) _____

15. INSURANCE COMPANY PAYMENT   A reckless driver caused Leslie Silva to collide with a car in another
lane. Silva had 50/100 liability limits, $25,000 property damage limits, and collision coverage with a
$100 deductible. Silva's car had damage of $1878, while the other car suffered $6936 in damages. The
resulting lawsuit gave injury awards of $60,000 and $55,000, respectively, in damages for personal
injury to the two people in the other car. Find the amount that the insurance company will pay for
(a) repairing Silva's car, **(b)** repairing the other car, and **(c)** personal injury damages. **(d)** How much
must Silva pay beyond her insurance coverage, including the collision deductible?

(a) _____
(b) _____
(c) _____
(d) _____

16. INSURANCE PAYMENT   Driving a dangerous vehicle at an excessive speed caused the car driven by Bob
Armstrong to crash into another car. Armstrong had 15/30 liability limits, $10,000 property damage
limits, and collision coverage with a $100 deductible. Damage to Armstrong's car was $2980; the
other car, with a value of $22,800, was totaled. The results of a lawsuit awarded $75,000 and $45,000,
respectively, in damages for personal injury to the two people in the other car. Find the amount that
the insurance company will pay for **(a)** repairing Armstrong's car, **(b)** repairing the other car, and
(c) personal injury damages. **(d)** How much must Armstrong pay beyond his insurance coverage?

(a) _____
(b) _____
(c) _____
(d) _____

17. Explain why insurance companies charge a higher premium on auto insurance sold to a youthful
operator. Do you think that this higher premium is a good idea or not? (See Objective 6.)

18. Property damage pays for damage caused by you to the property of others. Since the average cost of a
new car today is over $20,000, what amount of property damage coverage would you recommend to a
friend who owns her own business?

12.5 | Life Insurance

Objectives

1. Understand life insurance that does not accumulate cash value (term and decreasing term).
2. Understand life insurance that accumulates cash value (whole life, universal life, variable life, limited payment, and endowment).
3. Find the annual premium for life insurance.
4. Use premium factors with different modes of premium payment.

CASE _in_ POINT

Martha Spencer owns and manages The Doll House and makes a living for herself and her two children. Recently divorced, Spencer worries about what would happen to her children if she became disabled or died. For this reason, she carries disability insurance in addition to $200,000 in life insurance on herself. She also has medical insurance on the entire family.

There is no doubt about it: Insurance is expensive. Yet, most of us need insurance (car, home, medical, disability, and life insurance). The graph on the left shows that one-fifth of adults have found themselves in a situation where they wished they had more insurance. The figure on the right illustrates that many adults are insured through either their employer or their spouse's employer.

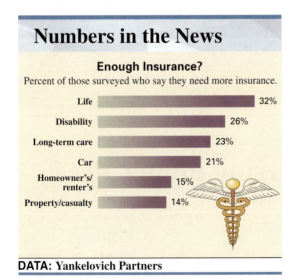

Numbers in the News

Enough Insurance?

Percent of those surveyed who say they need more insurance.

Life	32%
Disability	26%
Long-term care	23%
Car	21%
Homeowner's/renter's	15%
Property/casualty	14%

DATA: Yankelovich Partners

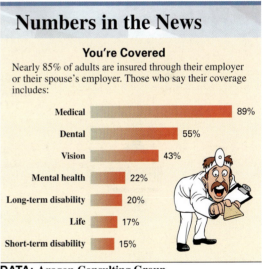

Numbers in the News

You're Covered

Nearly 85% of adults are insured through their employer or their spouse's employer. Those who say their coverage includes:

Medical	89%
Dental	55%
Vision	43%
Mental health	22%
Long-term disability	20%
Life	17%
Short-term disability	15%

DATA: Aragon Consulting Group

People buy **life insurance** to pay for their own burial expenses, to pay off a home mortgage or a car loan, to provide for a spouse and/or children, or to pay for their children's future college expenses. Some forms of life insurance build up a cash retirement value, other forms do not. Life insurance can also be important for the owner(s) of a business. Upon the death of the business owner, life insurance proceeds can provide a company with enough money to continue until it can be sold. Alternatively, life insurance proceeds can be used by one partner of a firm to buy out the ownership interest of a deceased partner.

QUICK TIP Proceeds from a life insurance policy are generally free of income taxes.

Objective **1** **Understand life insurance that does not accumulate cash value (term and decreasing term).** **Term insurance** Term insurance is the least expensive type of life insurance. It provides the most insurance per dollar spent, but it does not build up any cash values for retirement. This type of insurance coverage is usually renewable until some age, such as 70, when the insured is no longer allowed to renew it. As a result, most people discontinue term insurance before they die. However, term insurance is an excellent way to protect against an early death.

The premium on some term policies increases each year as the insured ages. Premiums on this type of policy become very expensive by the time a person is 60 years of age or so. As a result, many people prefer **level premium** term policies. Initially, these policies require a higher premium than policies with annually increasing premiums. However, the premium on a level premium policy is constant for a period of time such as 10 years or 20 years. Thereafter, the premiums would increase significantly. The graph shows that 20-year level term insurance costs have decreased significantly since 1930. The clipping below shows one reason why: We are living longer.

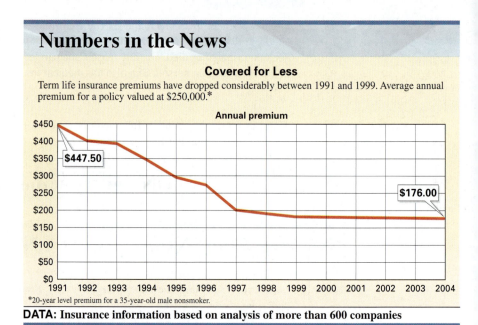

Numbers in the News

Covered for Less

Term life insurance premiums have dropped considerably between 1991 and 1999. Average annual premium for a policy valued at $250,000.*

Annual premium

$447.50

$176.00

*20-year level premium for a 35-year-old male nonsmoker.

DATA: Insurance information based on analysis of more than 600 companies

HERE & NOW

Insuring longer lives

We're living longer these days.

The evidence goes beyond the big jump in centenarians. (There are 50,000 centenarians nationally, and the numbers are expected to double each decade.)

The evidence now extends to insurance companies such as USAA Life Insurance of San Antonio, Texas.

The company has established lower premiums for its term life insurance products in 39 states, including California.

The new rates incorporate a decrease of 10 to 30 percent, on average.

Decreasing term insurance This is a type of term insurance with fixed premiums commonly to age 60 or 65, but the amount of life insurance decreases periodically. An example of this is a mortgage insurance policy on a home. The amount of life insurance on the owner decreases as the amount owed to the mortgage company decreases. Many large companies provide decreasing term insurance to employees as a benefit. The table provides an example of the benefits of one particular decreasing term policy.

Death Benefits for a Decreasing Term Policy
with a Premium of $11 per Month

Age	Amount of Life Insurance
Under 29	$40,000
30–34	$35,000
35–39	$30,000
40–44	$25,000
45–49	$18,000
50–54	$11,000
55–59	$7,000
60–66	$4,000
67 and over	$0

QUICK TIP Mortgage insurance is sometimes much more expensive than a regular term insurance policy. Compare prices before you buy.

The chart to the left shows the nation's leading causes of death, and the graph to the right shows the increase in life expectancy in the United States.

Nation's Top Killers

The 10 leading causes of death in the United States, ranked according to the number of lives lost in 2004:

1. Heart disease
2. Cancer
3. Stroke
4. Lung disease
5. Accidents
6. Diabetes
7. Influenza and pneumonia
8. Alzheimer's disease
9. Kidney disease
10. Bacterial infections

Source: National Center for Health Statistics

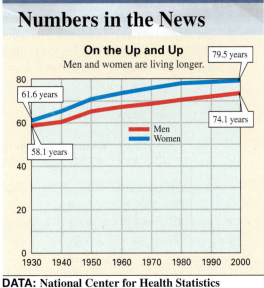

Numbers in the News

On the Up and Up
Men and women are living longer.

79.5 years
61.6 years
74.1 years
58.1 years

— Men
— Women

1930 1940 1950 1960 1970 1980 1990 2000

DATA: National Center for Health Statistics

QUICK TIP Scientists believe that diabetes will be much more common in North America during the coming decades. Therefore, this cause of death may eventually be ranked higher than the 6th leading cause of death.

Objective 2 Understand life insurance that accumulates cash value (whole life, universal life, variable life, limited payment, and endowment). Whole life (also called straight life, ordinary life, or permanent) This type of insurance provides a death benefit and a savings plan. The insured commonly pays a constant premium until death or retirement, whichever occurs first. If the policy is in force at the time of death, a death benefit is paid. Alternatively, the insured may choose to convert the accumulated **cash value** to a retirement benefit after paying the premium for many years.

Universal life   This type of insurance provides the life insurance protection of term insurance plus a tax-deferred way to accumulate assets. It sometimes allows people to establish a permanent policy at a lower premium than they would have to pay under a whole life policy and it gives the insured more flexibility. For example, universal life can help a family obtain more insurance when young children are at home and then help accumulate savings later after the children are grown. The portion of the premium going into retirement benefits receives money market interest rates and often has a guaranteed minimum rate of return regardless of what happens to market rates.

Variable life   This type of insurance allows the policyholder to make choices among a number of different investment options. It places the investment risk on the shoulders of the policyholder by allowing the insured to invest in any of the following: money market funds, bond funds, stock funds, or a combination of the three.

QUICK TIP   Variable life policies only accounted for 7% of all life insurance policies sold in 1990, but have since grown to nearly 40% of all life insurance policies sold.

Limited-payment life insurance   Limited-payment life is similar to whole life insurance, except that premiums are paid for only a fixed number of years, such as 20. This type of insurance is thus often called 20-pay life, representing payments for 20 years. The premium for limited-payment life is higher than that for whole life policies. Limited-payment life is most appropriate for athletes, actors, and others whose income is likely to be high for several years and then decline.

Endowment policies are the most expensive type of policy. These policies guarantee payment of a fixed amount of money to a given individual, whether or not the insured lives. Endowment policies might be taken out by parents to guarantee a sum of money for their children's college education. Because of the high premiums, this is one of the least popular types of policies today.

Objective 3   **Find the annual premium for life insurance.**   Calculation of life insurance rates by **actuaries** is based upon statistical data involving death rates, interest rates, and other factors. Women tend to live a few years longer than men, so a woman pays a lower life insurance premium than a man of the same age. Incidentally, women are more likely to be disabled than men and therefore have higher disability insurance rates than men. Use the actual age of a male to find the premium factor in the table below. However, subtract 5 from the age of a female before finding the premium factor in the table.

Annual Premium Rates* Per $1000 of Life Insurance

Age	10-Year Level Premium Term	Whole Life	Universal Life	20-Pay Life
20	1.60	4.07	3.48	12.30
21	1.65	4.26	3.85	12.95
22	1.69	4.37	4.10	13.72
23	1.73	4.45	4.56	14.28
24	1.78	4.68	4.80	15.95
25	1.82	5.06	5.11	16.60
30	1.89	5.66	6.08	18.78
35	2.01	7.68	7.45	21.60
40	2.56	12.67	10.62	24.26
45	3.45	19.86	15.24	28.16
50	5.63	26.23	21.46	32.59
55	8.12	31.75	28.38	38.63
60	14.08	38.42	36.72	45.74

*For women, subtract 5 years from the actual age. For example, rates for a 30-year old woman are shown for age 25 in the table.

QUICK TIP   Individuals with health problems must pay substantially more for life insurance than healthy individuals. Other factors that increase life insurance premiums include smoking or even some hobbies such as scuba diving or bungee jumping.

The premium for a life insurance policy is found with the following formula.

$$\text{Annual premium} = \text{Number of thousands} \times \text{Rate per } \$1000$$

EXAMPLE 1

Finding the Life Insurance Premium

Martha Spencer became the primary source of income for her family at age 35 after her divorce. At that time, she decided that she needed $200,000 in life insurance to pay off the mortgage on her home, some loans at her business, and to provide for her children. Find her annual premium for **(a)** a 10-year level premium term policy, **(b)** a whole life policy, **(c)** a universal life policy, and **(d)** a 20-pay life plan.

SOLUTION
First, divide the desired amount of life insurance by $1000 to find the number of thousands.

$$\$200,000 \div \$1000 = 200 \text{ thousands}$$

Since Spencer is a female, subtract 5 from her actual age before using the table $(35 - 5 = 30)$. Look in the table at age 30 for the rates for each type of insurance.

(a) 10-year level premium term $200 \times \mathbf{1.89} = \$378$
(b) Whole life $200 \times \mathbf{5.66} = \$1132$
(c) Universal life $200 \times \mathbf{6.08} = \$1216$
(d) 20-pay life $200 \times \mathbf{18.78} = \$3756$

Spencer wanted to buy universal life because of the savings feature, which would help her save for retirement. However, she purchased the level premium term instead, since her income was limited.

QUICK TIP   Use the actual age of a male when using the table of premiums. However, subtract 5 from the age of a female before using the table.

Objective 4   Use premium factors with different modes of premium payment. The annual life insurance premium is not always paid in a single payment. Many companies give the insured the option of paying the premium semiannually, quarterly, or monthly. For this convenience, the policyholder *pays an additional amount* that is determined by a **premium factor**. The following table shows typical premium factors.

Premium Factors

Mode of Payment	Premium Factor
Semiannually	.51
Quarterly	.26
Monthly	.0908

QUICK TIP   The premium factor for a semiannual payment is .51 rather than .50. This results in a little extra revenue to the insurance company.

EXAMPLE 2

Using a Premium Factor

The annual insurance premium on a $200,000 10-year level premium term life policy for Martha Spencer is $378. Use the premium factors table to find the amount of premium and the total annual cost if she pays **(a)** semiannually, **(b)** quarterly, or **(c)** monthly.

SOLUTION

		Premium	Annual Cost
(a) semiannually:	$378 × **.51**	= $192.78	$192.78 × **2 payments/year** = $385.56
(b) quarterly:	$378 × **.26**	= $98.28	$98.28 × **4 payments/year** = $393.12
(c) monthly:	$378 × **.0908**	= $34.32 (rounded)	$34.32 × **12 payments/year** = $411.84

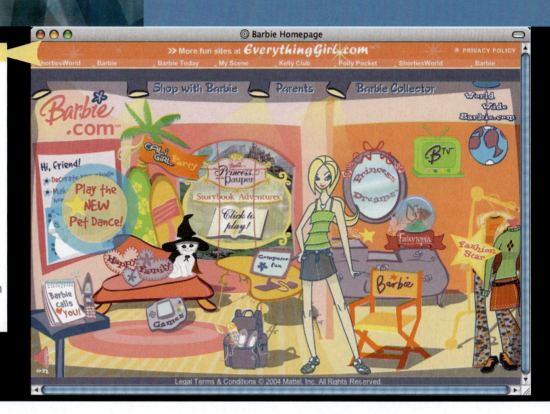

The Barbie doll is the most popular fashion doll ever created. If all the Barbie dolls that have been sold since 1959 were placed head-to-toe, the dolls would circle the earth more than seven times. The most popular Barbie ever sold was the Totally Hair Barbie, which was introduced in 1992. With hair from the top of her head to her toes, more than 10 million of these dolls were sold, resulting in revenue of $100 million. With current retail sales at an estimated $3.6 billion, Barbie is the #1 girls' brand worldwide.

Antique dolls representing adults from the seventeenth and eighteenth centuries have been found but are very rare. Individual craftsmen in England made most of these earliest dolls. The craftsmen carved the dolls of wood, painted their features, and also designed the costumes for the dolls. Some of these earliest dolls are valued at over $40,000.

1. In 2003, Mattel paid $203,222,000 in income taxes on an operating income of $785,710,000. Find the percent of operating income that went to income taxes to the nearest tenth of a percent.

2. Assume Mattel owns a building valued at $1,450,000 in an area with an assessment rate of 35% and a tax rate of 9%. Find the property tax on the building.

3. Mattel does some of their manufacturing in Mexico. Why do you think they do this?

4. Why do you think Barbie has been so popular?

12.5 | Exercises

The **Quick Start** *exercises in each section contain solutions to help you get started.*

Find the annual premium, the semiannual premium, the quarterly premium, and the monthly premium for each of the following. (Note: Subtract 5 years for women.) Round to the nearest cent.

Quick Start

	Face Value of Policy	Age of Insured	Sex of Insured	Type of Policy	Annual Premium	Semi-Annual Premium	Quarterly Premium	Monthly Premium
1.	$100,000	45	F	Term	$256	$130.56	$66.56	$23.24

100 × $2.56 = $256; $256 × .51 = $130.56; $256 × .26 = $66.56; $256 × .0908 = $23.24

2.	$60,000	30	M	Whole life	$339.60	$173.20	$88.30	$30.84

60 × $5.66 = $339.60; $339.60 × .51 = $173.20; $339.60 × .26 = $88.30; $339.60 × .0908 = $30.84

3.	$35,000	40	M	20-pay life				
4.	$60,000	50	F	20-pay life				
5.	$85,000	30	M	Universal life				
6.	$150,000	60	M	Term				
7.	$75,000	21	M	Whole life				
8.	$80,000	35	F	Term				
9.	$65,000	60	M	20-pay life				
10.	$50,000	45	F	Universal life				

11. Compare level premium term insurance to universal life insurance. Which would you prefer for yourself? Why? (See Objectives 1 and 2.)

12. Describe premium factors and how they are used. How often do you prefer paying an insurance premium: annually, semiannually, quarterly, or monthly? (See Objective 4.)

Solve the following application problems.

Quick Start

13. **LEVEL PREMIUM**   Tom Peters purchased a $200,000 10-year, level premium policy on his 40th birthday. Find the annual premium.

    200 × $2.56 = $512

    13. **$512**

14. **WHOLE LIFE INSURANCE**   Jessica Smith buys a whole life policy with a face value of $100,000 at age 35. Find the annual premium.

    14. _____

15. **KEY EMPLOYEE INSURANCE**   Martha Spencer owns the Doll House and has a 35-year old key male employee who she wants to insure for $50,000. Find the annual premium (a) for 10-year level term and (b) for whole life.

    (a) _____
    (b) _____

16. **20-PAY LIFE POLICY**   Luan Lee buys a $100,000, 20-pay life policy at age 45. Her son Bryan is the beneficiary, and will collect the face value of the policy. (a) Find the annual premium. (b) How much will Bryan get if his mother dies after making payments for 12 years?

    (a) _____
    (b) _____

17. **WHOLE LIFE INSURANCE**   Find the total premium paid over 30 years for a whole life policy with a face value of $20,000. Assume that the policy is taken out by a 25-year-old man.

    17. _____

18. **UNIVERSAL LIFE INSURANCE**   Richard Gonsalves takes out a universal life policy with a face value of $50,000. He is 40 years old. Find the monthly premium.

    18. _____

19. **PREMIUM FACTORS**   The annual premium for a whole life policy is $872. Using premium factors, find (a) the semiannual premium, (b) the quarterly premium, and (c) the monthly premium.

    (a) _____
    (b) _____
    (c) _____

20. **PREMIUM FACTORS**   A universal life policy has an annual premium of $2012. Use premium factors to find (a) the semiannual premium, (b) the quarterly premium, and (c) the monthly premium.

    (a) _____
    (b) _____
    (c) _____

Chapter 12 | Quick Review

CHAPTER TERMS   *Review the following terms to test your understanding of the chapter. For each term you do not know, refer to the page number found next to that term.*

CONCEPTS	EXAMPLES
12.1 Fair market value and assessed valuation The value of property is multiplied by a given percent to arrive at the assessed valuation. **Assessment Rate** × Market value = Assessed valuation	The assessment rate is 30%; fair market value is $115,000; find the assessed valuation. $$30\% \times \$115{,}000 = \$34{,}500$$
12.1 Tax rate The tax rate formula is $$\text{Tax rate} = \frac{\text{Total tax amount needed}}{\textbf{Total assessed value}}$$	Tax amount needed: $3,864,400; total assessed value: $107,345,000; find the tax rate. $$\frac{\$3{,}864{,}400}{\$107{,}345{,}000} = .036 = 3.6\% \ (\text{rounded})$$
12.1 Expressing tax rates in different forms and finding tax 1. **Percent**: multiply by assessed valuation. 2. **Dollars per $100**: move decimal 2 places to left in assessed valuation and multiply. 3. **Dollars per $1000**: move decimal 3 places to left in assessed valuation and multiply. 4. **Mills**: move decimal 3 places to the left in rate and multiply by assessed valuation. Use the formula Property tax = Assessed valuation × **Tax rate**	Assessed value, $90,000; tax rate, 2.5% $$\$90{,}000 \times \textbf{2.5\%} = \$2250$$ Tax rate, **$2.50** per $100 $$900 \times \textbf{\$2.50} = \$2250$$ Tax rate, **$25** per $1000 $$90 \times \textbf{\$25} = \$2250$$ Tax rate, **25 mills** $$\$90{,}000 \times \textbf{.025} = \$2250$$
12.2 Adjusted gross income Adjusted gross income includes wages, salaries, tips, dividends, and interest. Subtract IRA contributions and alimony.	Salary, $32,540; interest income, $875; dividends, $315; find adjusted gross income. $$\$32{,}540 + \textbf{\$875} + \textbf{\$315} = \$33{,}730$$

CONCEPTS	EXAMPLES
12.2 Standard deduction amounts The majority of taxpayers use the standard deduction allowed by the IRS.	$4750 for single taxpayers $9500 for married couples filing jointly or qualifying widow(er) $4750 for married taxpayers filing separately $7000 for head of a household
12.2 Taxable income The larger of either the total of itemized deductions or the standard deduction is subtracted from adjusted gross income along with **$3050** for each personal exemption.	Single taxpayer, adjusted gross income = $31,500, itemized deductions of $3850, find taxable income. Note that itemized deductions are less than standard deduction of $4750 and the personal exemption is $3050. $$\text{Taxable income} = \$31{,}500 - \$4750 - \$3050$$ $$= \$23{,}700$$
12.2 Tax rates There are six tax rates: 10%, 15%, 25%, 28%, 33%, and 35%.	Single: 10%; 15% over $7000; 25% over $28,400; 28% over $68,800; 33% over $143,500; 35% over $311,950 Married filing jointly or qualifying widow(er): 10%; 15% over $14,000; 25% over $56,800; 28% over $114,650; 33% over $174,700; 35% over $311,950 Married filing separately: 10%; 15% over $7000; 25% over $28,400; 28% over $57,325; 33% over $87,350; 35% over $155,975 Head of Household: 10%; 15% over $10,000; 25% over $38,050; 28% over $98,250; 33% over $159,100; 35% over $311,950
12.2 Balance due or a refund from the IRS If the total amount withheld by employers is greater than the tax owed, a refund results. If the tax owed is the greater amount, a balance is due.	Tax owed, $1253; tax withheld, $113 per month for 12 months. Find balance due or refund. $$\$113 \text{ withheld} \times \textbf{12} = \$1356 \text{ withheld}$$ $$\$1356 \text{ withheld} - \textbf{\$1253 owed} = \$103 \text{ refund}$$
12.3 Annual premium for fire insurance The building and territorial ratings are used to find the premiums per $100 for the building and contents. The two are added.	Building value, $80,000; contents, $35,000. Premiums are: building, $.75 per $100; contents, $.77 per $100. Find the annual premium. $$\text{Building: } 800 \text{ (hundreds)} \times \textbf{\$.75} = \$600$$ $$\text{Contents: } 350 \text{ (hundreds)} \times \textbf{\$.77} = \$269.50$$ $$\text{Total premium: } \textbf{\$600} + \textbf{\$269.50} = \$869.50$$
12.3 Coinsurance formula Part of the risk of fire is taken by the insured. An 80% coinsurance clause is common. $$\begin{array}{l} \text{Loss paid by} \\ \text{insurance} \\ \text{company} \end{array} = \begin{array}{l} \text{Amount} \\ \text{of loss} \end{array} \times \dfrac{\text{Policy amount}}{80\% \textbf{ of replacement cost}}$$	Replacement cost, $125,000; policy amount, $75,000; fire loss, $40,000; 80% coinsurance clause; find the amount of loss paid by insurance company. $$\$40{,}000 \times \frac{\textbf{\$75,000}}{\textbf{\$100,000}} = \$30{,}000 \begin{array}{l} \text{(amount insurance} \\ \text{company pays)} \end{array}$$

CONCEPTS	EXAMPLES
12.3 Multiple carriers Several companies insuring the same property, which limits the risk of the insurance company, with each paying its fractional portion of any claim.	Insured loss, $500,000 Insurance is Company A with $1,000,000; Company B with $750,000; Company C with $250,000; find the amount of loss paid by each company. Total insurance $= \$1,000,000 + \$750,000 + \$250,000 = \$2,000,000$ Company A $\dfrac{1,000,000}{2,000,000} \times \$500,000 = \$250,000$ Company B $\dfrac{750,000}{2,000,000} \times \$500,000 = \$187,500$ Company C $\dfrac{250,000}{2,000,000} \times \$500,000 = \$62,500$
12.4 Annual auto insurance premium Most drivers are legally required to purchase automobile insurance. The premium is determined by the types of coverage selected, the type of car, geographic territory, past driving record, and other factors.	Determine the premium: territory, 2; liability, 50/100; property damage, $50,000; comprehensive and collision, 3-year-old car with a symbol of 8; uninsured motorist coverage; driver is age 23 with driver's training. $\begin{array}{ll} \$341 & \text{liability} \\ \$223 & \text{property damage} \\ 36 & \text{comprehensive} \\ 98 & \text{collision} \\ \underline{44} & \text{uninsured motorist} \\ \mathbf{\$742 \times 1.15} & \text{youthful operator factor} \\ = \mathbf{\$853.30} & \end{array}$
12.5 Annual life insurance premium There are several types of life policies. Use the table and multiply by the number of $1000s of coverage. Subtract 5 years from the age of females. Premium = Number of thousands × **Rate per $1000**	Find the premiums on a $50,000 policy for a 30-year-old male. **(a)** 10-year level premium term $\qquad 50 \times \mathbf{\$1.89} = \$94.50$ **(b)** whole life $\qquad 50 \times \mathbf{\$5.66} = \$283$ **(c)** universal life $\qquad 50 \times \mathbf{\$6.08} = \$304$ **(d)** 20-pay life $\qquad 50 \times \mathbf{\$18.78} = \$939$
12.5 Premium factors Life insurance premiums may be paid semiannually, quarterly, or monthly. The annual premium is multiplied by the premium factor to determine the premium amount.	The annual life insurance premium is $740. Use the table to find the **(a)** semiannual, **(b)** quarterly, and **(c)** monthly premium. **(a)** Semiannual $\qquad \$740 \times \mathbf{.51} = \$377.40$ **(b)** Quarterly $\qquad \$740 \times \mathbf{.26} = \$192.40$ **(c)** Monthly $\qquad \$740 \times \mathbf{.0908} = \$67.19$

Chapter 12 | Summary Exercise

Financial Planning for Taxes and Insurance

Baker's Pottery manufactures and sells ceramic pots of all types, shapes, and styles. Planning ahead, the company set aside $53,500 to pay property taxes, fire insurance premiums, and life insurance premiums on the company president. All of these premiums happen to be due in the same month. Find each of the following.

(a) The company property has a fair market value of $1,990,000 and is assessed at 75% of this value. If the tax rate is $7.90 per $1000 of assessed value, find the annual property tax.

(a) _____

(b) The building occupied by the company is a class-B building worth $1,730,000. The contents are worth $3,502,000 and the territorial rating is 4. Find the annual fire insurance premium.

(b) _____

(c) The president of the company is a 50-year-old woman who lost the use of her legs in an automobile accident. She needs insurance and encourages the company to buy a $250,000, 10-year level-premium life insurance policy on her. Find the semiannual premium.

(c) _____

(d) Find the total amount needed to pay property taxes, the fire insurance premium, and the semiannual life insurance premium.

(d) _____

(e) How much more than the amount needed had the company set aside to pay these expenses?

(e) _____

INVESTIGATE

A recent article in *Consumer Reports* listed thirty Web sites that offer to help a person shop for term life insurance. The article also lists dial-up services that offer a similar service. Use the Web to find information and prices for term life insurance to meet your personal life insurance needs and look at several types of life insurance products.

Chapter 12 Test

To help you review, the numbers in brackets show the section in which the topic was discussed.

Complete the following chart comparing property tax rates. **[12.1]**

	Percent	Per $100	Per $1000
1.	5.76%	_____	_____
2.	_____	_____	$93.50

Find the taxable income and the tax for each of the following people. The letter following the names indicates the marital status. **[12.2]**

Name	Number of Exemptions	Adjusted Gross Income	Total Deductions	Taxable Income	Tax
3. J. Spalding, S	2	$38,295	$3648	_____	_____
4. The Bensons, M	4	$43,487	$8315	_____	_____

Find the tax owed in the following problems.

5. Bradkin's Toggery owns property with a fair market value of $104,600. Property in the area is assessed at 30% of fair market value with a tax rate of 3.65%. Find the annual tax. **[12.1]**

5. _____

6. The Blakely family has an adjusted gross income of $82,316. They are married and file jointly with five exemptions and have deductions of $6200. **[12.2]**

6. _____

7. Kari Heen had an adjusted gross income of $35,810 last year. She had deductions of $807 for state income tax, $729 for property tax, $1263 in mortgage interest, and $186 in contributions. Heen claims one exemption and files as a single person. **[12.2]**

7. _____

Find the annual fire insurance premium for the following. Use the table on page 528. **[12.3]**

8. Southside Plating owns a class-B building with a replacement cost of $147,000. Contents are valued at $83,500. The territorial rating is 5.

8. _____

9. Foxworthy's warehouse is valued at $220,000. The fire insurance policy (with an 80% coinsurance clause) has a face value of $150,000. If the building has a fire loss of $50,000, find the amount of the loss that the insurance company will pay.

9. _____

10. Dave's Body and Paint has an insurable loss of $72,000, while having insurance coverage beyond coinsurance requirements. The insurance is divided between Company *A* with $250,000 coverage, Company *B* with $150,000 coverage, and Company *C* with $100,000 coverage. Find the amount of loss paid by each of the insurance companies.

A: _____
B: _____
C: _____

Find the annual motor-vehicle insurance premium for the following people. **[12.4]**

Name	Territory	Age	Driver Training	Liability	Property Damage	Comprehensive Collision Age Group	Symbol	Uninsured Motorist	Annual Premium
11. Ramos	3	18	Yes	15/30	$10,000	5	7	Yes	_____
12. Larik	1	42	—	50/100	$100,000	1	8	Yes	_____

Find the annual premium, the semiannual premium, the quarterly premium, and the monthly premium for each of the following life-insurance policies. Use the tables in Section 12.5. **[12.5]**

	Annual	Semiannual	Quarterly	Monthly
13. Irene Chong, whole life, $28,000 face value, age 35	_____	_____	_____	_____
14. Gil Eckern, 20-pay life, $80,000 face value, age 40	_____	_____	_____	_____

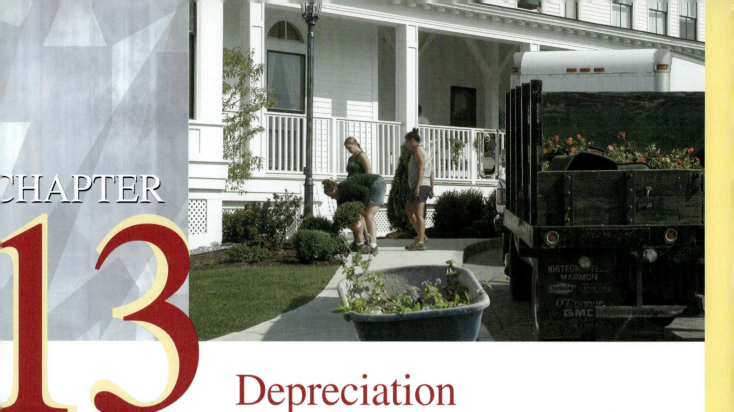

CHAPTER

13

Depreciation

CRUZ LANDSCAPING IS OWNED AND MANAGED BY JOSE AND RUDY

Cruz. The company does landscaping for the owners of new and existing residen-

tial and commercial properties. Its work includes the installation of sprinkler systems, walkways and con-

crete curbing, lawns, shrubberies, trees, and other plants. To offer this full serv-

ice, Cruz Landscaping has purchased many pieces of equipment, which include

trucks, trailers, tractors, trenchers, and rototillers.

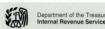

Department of the Treasury
Internal Revenue Service

Publication 946
Cat. No. 13081F

How To Depreciate Property

- **Section 179 Deduction**
- **Special Depreciation Allowance**
- **MACRS**
- **Listed Property**

For use in preparing
2005 Returns

Business expenses such as salaries, rent, and utilities must be subtracted from company revenues to determine net income. Other expenses including the cost of buildings, machinery, and fixtures are not subtracted all in one year. Since these items usually last several years, the cost of these purchases must be spread over the length of their useful life. This method is called **depreciation**.

Over the years, several methods of computing depreciation have been used, including **straight-line**, **declining-balance**, **sum-of-the-years'-digits**, and **units-of-production**. These methods are used in keeping company accounting records and, in many states, preparing state income tax returns. Items purchased after 1981 are depreciated for federal income tax returns with the accelerated cost recovery system or the modified accelerated cost recovery system, discussed later. The use of depreciation for federal income tax purposes is detailed in an Internal Revenue Service publication. The complete title of this publication is shown at the side.

A company need not use the same method of depreciation for all of its various assets. For example, the straight-line method of depreciation might be used on some assets and the declining-balance method on others. Furthermore, the depreciation method used in preparing a company's financial statement may be different from the method used in preparing income tax returns.

13.1 | Depreciation: Straight-Line Method

Objectives

1. Understand the terms used in depreciation.
2. Use the straight-line method of depreciation to find the amount of depreciation each year.
3. Use the straight-line method to find the book value of an asset.
4. Use the straight-line method to prepare a depreciation schedule.

CASE in POINT

The Cruz brothers and their accountant decide on which method of depreciation to use for the depreciable assets of Cruz Landscaping. The most commonly used method for both accounting and tax purposes is the straight-line method of depreciation.

Objective 1   Understand the terms used in depreciation. The physical assets of a company such as machinery, trucks, cars, or computers are **tangible assets**. Assets such as patents and copyrights, franchise fees, or customer lists are **intangible assets**. In general, either type of asset may be depreciated, as long as its useful life can be determined. The key terms in depreciation are summarized below.

Cost is the basis for determining depreciation. It is the total amount paid for the asset.

Useful life is the period of time during which the asset will be used. The Internal Revenue Service has guidelines for estimating the life of an asset used in a particular trade or business. However, useful life depends on the use of the asset, the repair policy, the replacement policy, obsolescence, and other factors.

Salvage value or **scrap value** (sometimes called **residual value**) is the estimated value of an asset when it is retired from service, traded in, disposed of, or exhausted. An asset may have a salvage value of zero, or **no salvage value**.

Accumulated depreciation is the amount of depreciation taken so far, a running balance of depreciation to date.

Book value is the cost of an asset minus the total depreciation to date. The book value at the end of an asset's life is equal to the salvage value. The book value can never be less than the salvage value.

Objective 2 **Use the straight-line method of depreciation to find the amount of depreciation each year.** The simplest method of depreciation, straight-line depreciation, assumes that assets lose an equal amount of value during each year of life. For example, suppose a heavy equipment trailer is purchased by Cruz Landscaping at a cost of $14,100. The trailer has a useful life of 8 years and a salvage value of $2100. Find the amount to be depreciated (**depreciable amount**) using the following formula.

$$\text{Amount to be depreciated} = \text{Cost} - \text{Salvage value}$$

Here, the amount to be depreciated over the 8-year period is:

$$
\begin{array}{rl}
\$14,100 & \text{cost} \\
-\quad 2,100 & \text{salvage value} \\
\hline
\$12,000 & \text{amount to be depreciated}
\end{array}
$$

With the straight-line method, an equal amount of depreciation is taken each year over the 8-year life of the trailer. The annual depreciation for this trailer is

$$\text{Depreciation} = \frac{\text{Depreciable amount}}{\text{Years of life}} = \frac{\$12,000}{8} = \$1500$$

Each year during the 8-year life of the trailer, the annual depreciation will be $1500, or $\frac{1}{8}$ of the depreciable amount. The annual rate of depreciation is $12\frac{1}{2}\%$ $\left(\frac{1}{8} = 12\frac{1}{2}\%\right)$.

Objective 3 **Use the straight-line method to find the book value of an asset.** The book value, or remaining value, of an asset at the end of a year is the original cost minus the depreciation up to and including that year (**accumulated depreciation**). With the trailer, the book value at the end of the first year is found as follows.

$$
\begin{array}{rl}
\$14,100 & \text{cost} \\
-\quad 1,500 & \text{first year's depreciation} \\
\hline
\$12,600 & \text{book value at end of the first year}
\end{array}
$$

Book value is found with the following formula.

$$\text{Book value} = \text{Cost} - \text{Accumulated depreciation}$$

Subtract the second year's depreciation from the book value at the end of year 1 to find the book value at the end of year 2 and so on.

EXAMPLE 1

Finding First Year Depreciation and Book Value

Krispy Kreme purchased a new doughnut fryer system at a cost of $26,500. The estimated life of the fryer is 5 years, with a salvage value of $3500. Find **(a)** the annual rate of depreciation, **(b)** the annual amount of depreciation, and **(c)** the book value at the end of the first year.

SOLUTION

(a) The annual rate of depreciation is 20% $\left(\text{5-year life} = \frac{1}{5} \text{ per year} = \mathbf{20\%}\right)$.

(b)
$$
\begin{array}{rl}
\$26,500 & \text{cost} \\
-\quad 3,500 & \text{salvage value} \\
\hline
\$23,000 & \text{depreciable amount}
\end{array}
$$

This $23,000 will be depreciated evenly over the 5-year life for an annual depreciation of $4600 $(\$23,000 \times 20\% = \$4600)$.

(c) Since the annual depreciation is $4600, the book value at the end of the first year will be

$26,500   cost
−   4,600   **depreciation in the first year**
$21,900   book value at the end of the first year

To solve Example 1 using a calculator, first use parentheses to find the depreciable amount. Next, divide to find depreciation. Finally, find the book value.

(26,500 − 3500) ÷ 5 = 4600
26,500 − 4600 = 21,900

Note: Refer to Appendix C for calculator basics.

If an asset is expected to have **no salvage value** at the end of its expected life, the entire cost will be depreciated over its life. In Example 1, if the doughnut fryer system had been expected to have no salvage value at the end of 5 years, the annual amount of depreciation would have been $5300 $\left(\$26{,}500 \times 20\% = \$5300\right)$.

> **QUICK TIP**   Find the book value at the end of any year by multiplying the annual amount of straight-line depreciation by the number of years and subtracting this result, the depreciation to date, from the cost.

EXAMPLE 2

Finding the Book Value at the End of Any Year

A lighted display case at Bead Works cost $3400, has an estimated life of 10 years, and a salvage value of $800. Find the book value at the end of 6 years.

SOLUTION

The annual rate of depreciation is 10%. (10-year life is $\frac{1}{10}$ or 10%.)

$3400   cost
−   800   **salvage value**
$2600   depreciable amount

Since $2600 is depreciated evenly over the 10-year life of the case, the annual depreciation is $260 $\left(\mathbf{\$2600 \times 10\% = \$260}\right)$.
The accumulated depreciation over the 6-year period is

$260 × **6 years** = $1560 accumulated depreciation $\left(6\text{ years}\right)$

Find the book value at the end of 6 years by subtracting the accumulated depreciation from the cost.

$3400   cost
−   1560   **accumulated depreciation $\left(\textbf{6 years}\right)$**
$1840   book value at the end of 6 years

After 6 years, this display case would be carried on the firm's books with a value of $1840.

> **QUICK TIP**   The book value helps the owner of a business estimate the value of the business, which is important when the owner is borrowing money or trying to sell the business.

Objective 4 **Use the straight-line method to prepare a depreciation schedule.** A **depreciation schedule** is often used to show the annual depreciation, accumulated depreciation, and book value over the useful life of an asset. As an aid in comparing three of the methods of depreciation discussed in the text, the depreciation schedule of Example 3 and the schedule shown in the double-declining-balance (see **Section 13.2**) and sum-of-the-years'-digits methods (see **Section 13.3**) use the same asset.

EXAMPLE 3

Preparing a Depreciation Schedule

Cruz Landscaping bought a new pickup truck for $21,500. The truck is estimated to have a useful life of 5 years, at which time it will have a salvage value (trade-in value) of $3500. Prepare a depreciation schedule using the straight-line method of depreciation.

SOLUTION

The annual rate of depreciation is 20% $\left(\text{5-year life} = \frac{1}{5} \textbf{ per year} = \textbf{20\%}\right)$. Find the depreciable amount as follows.

$$
\begin{array}{rl}
\$21,500 & \text{cost} \\
-\quad 3,500 & \text{salvage value} \\
\hline
\$18,000 & \text{depreciable amount}
\end{array}
$$

This $18,000 will be depreciated evenly over the 5-year life for an annual depreciation of $3600 $(\$18,000 \times \textbf{20\%} = \$3600)$.

This depreciation schedule includes a year zero that represents the initial purchase of the truck.

Year	Computation	Amount of Depreciation	Accumulated Depreciation	Book Value
0	—	—	—	$21,500
1	$(20\% \times \$18,000)$	$3600	$3,600	$17,900
2	$(20\% \times \$18,000)$	$3600	$7,200	$14,300
3	$(20\% \times \$18,000)$	$3600	$10,800	$10,700
4	$(20\% \times \$18,000)$	$3600	$14,400	$7,100
5	$(20\% \times \$18,000)$	$3600	$18,000	$3,500

The depreciation is $3600 each year, the accumulated depreciation at the end of 5 years is equal to the depreciable amount, and the book value at the end of 5 years is equal to the salvage value.

> **QUICK TIP** If the rate is a repeating decimal, use the fraction that is equivalent to the decimal. Instead of 33.3%, use the fraction $\frac{1}{3}$. Instead of 16.7%, use the fraction $\frac{1}{6}$.

Depreciation is used with assets having a useful life of *more than one year*. The asset to be depreciated must have a predictable life. A truck can be depreciated because its useful life can be estimated, but land cannot be depreciated because its life is considered to be indefinite. For example, the graph on the next page shows the remaining value of the pickup truck in Example 3 as it is depreciated over its useful life.

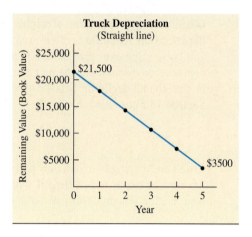

13.1 | Exercises

FOR EXTRA HELP

MyMathLab

InterActMath.com

Math XP   MathXL

MathXL
Tutorials on CD

Tutor
Center   Addison-Wesley
Math Tutor Center

DVT/Videotape

The **Quick Start** *exercises in each section contain solutions to help you get started.*

*Find the annual straight-line rate of depreciation, given the following estimated lives.
(See Example 1.)*

Quick Start

	Life	Annual Rate		Life	Annual Rate
1.	5 years	20%	**2.**	4 years	25%
	$\frac{1}{5} = 20\%$			$\frac{1}{4} = 25\%$	
3.	8 years	_____	**4.**	10 years	_____
5.	20 years	_____	**6.**	25 years	_____
7.	15 years	_____	**8.**	30 years	_____
9.	80 years	_____	**10.**	40 years	_____
11.	50 years	_____	**12.**	100 years	_____

*Find the annual amount of depreciation for the following, using the straight-line method. (See
Examples 1 and 2.)*

Quick Start

13.	Cost:	$9000	**14.**	Cost:	$3400
	Estimated life:	20 years		Estimated life:	4 years
	Estimated scrap value:	None		Estimated scrap value:	$800
	Annual depreciation:	**$450**		Annual depreciation:	**$650**
	$9000 × 5% = $450			$3400 − $800 = $2600	
				$2600 × 25% = $650	
15.	Cost:	$2700	**16.**	Cost:	$8100
	Estimated life:	3 years		Estimated life:	6 years
	Estimated scrap value:	$300		Estimated scrap value:	$750
	Annual depreciation:	_____		Annual depreciation:	_____
17.	Cost:	$4200	**18.**	Cost:	$12,200
	Estimated life:	5 years		Estimated life:	10 years
	Estimated scrap value:	None		Estimated scrap value:	$3200
	Annual depreciation:	_____		Annual depreciation:	_____

 indicates an exercise that is related to the Case in Point feature within the section.

Find the book value at the end of the first year for the following, using the straight-line method. (See Examples 1 and 2.)

Quick Start

19. Cost: $3200
    Estimated life: 8 years
    Estimated scrap value: $400
    Book value: **$2850**

    $3200 − $400 = $2800
    $2800 × 12.5% = $350
    $3200 − $350 = $2850

20. Cost: $35,000
    Estimated life: 10 years
    Estimated scrap value: $2500
    Book value: _____

21. Cost: $5400
    Estimated life: 12 years
    Estimated scrap value: $600
    Book value: _____

22. Cost: $4500
    Estimated life: 5 years
    Estimated scrap value: None
    Book value: _____

Find the book value at the end of 5 years for the following, using the straight-line method. (See Examples 1 and 2.)

Quick Start

23. Cost: $4800
    Estimated life: 10 years
    Estimated scrap value: $750
    Book Value: **$2775**

    $4800 − $750 = $4050
    $4050 × 10% = $405
    $405 × 5 = $2025
    $4800 − $2025 = $2775

24. Cost: $16,000
    Estimated life: 20 years
    Estimated scrap value: $2000
    Book value: _____

25. Cost: $80,000
    Estimated life: 50 years
    Estimated scrap value: $10,000
    Book value: _____

26. Cost: $660
    Estimated life: 8 years
    Estimated scrap value: $100
    Book value: _____

Solve the following application problems.

Quick Start

27. MACHINERY DEPRECIATION   Cruz Landscaping selects the straight-line method of depreciation for a Bob Cat tractor costing $12,000 with a 3-year life and an expected scrap value of $3000. Prepare a depreciation schedule.

Year	Computation	Amount of Depreciation	Accumulated Depreciation	Book Value
0	—	—	—	$12,000
1	$(33\frac{1}{3}\% \times \$9000)$	$3000	$3000	$9000
2	$(33\frac{1}{3}\% \times \$9000)$	$3000	$6000	$6000
3	$(33\frac{1}{3}\% \times \$9000)$	$3000	$9000	$3000

28. **LABORATORY EQUIPMENT**   Savannah Pipe has purchased a laboratory crusher costing $18,000, having an estimated life of 4 years and a salvage value of $1600. Prepare a depreciation schedule using the straight-line method of depreciation.

Year	Computation	Amount of Depreciation	Accumulated Depreciation	Book Value
0	—	—	—	$18,000
1				
2				
3				
4				

29. **VEHICLE DEPRECIATION**   Cruz Landscaping paid $25,600 for a $1\frac{1}{2}$-ton, dual-axle flatbed truck with an estimated life of 6 years and a salvage value of $7000. Prepare a depreciation schedule using the straight-line method of depreciation.

Year	Computation	Amount of Depreciation	Accumulated Depreciation	Book Value
0	—	—	—	$25,600
1				
2				
3				
4				
5				
6				

30. **BUSINESS FIXTURES**   Dorothy Sargent buys fixtures for her shop at a cost of $7800 and estimates the life of the fixtures as 10 years, after which they will have no salvage value. Prepare a depreciation schedule, calculating depreciation by the straight-line method.

Year	Computation	Amount of Depreciation	Accumulated Depreciation	Book Value
0	—	—	—	$7800
1				
2				
3				
4				
5				
6				
7				
8				
9				
10				

31. Develop a single formula that will show how to find annual depreciation using the straight-line method of depreciation. (See Objective 1.)

32. Explain the procedure used to calculate depreciation when there is no salvage value. Why will the book value always be zero at the end of the asset's life?

33. BARGE DEPRECIATION   A Dutch petroleum company purchased a barge for $1,300,000. The estimated life is 20 years, at which time it will have a salvage value of $200,000. Find **(a)** the annual amount of depreciation using the straight-line method and **(b)** the book value at the end of 5 years.

(a) _____
(b) _____

34. DEPRECIATING COMPUTER EQUIPMENT   The new computer equipment at Cruz Landscaping has a cost of $14,500, an estimated life of 8 years, and scrap value of $2100. Find **(a)** the annual depreciation and **(b)** the book value at the end of 4 years using the straight-line method of depreciation.

(a) _____
(b) _____

35. FURNITURE DEPRECIATION   A bookcase costs $880, has an estimated life of 8 years, and a scrap value of $160. Use the straight-line method of depreciation to find **(a)** the annual rate of depreciation, **(b)** the annual amount of depreciation, and **(c)** the book value at the end of the first year.

(a) _____
(b) _____
(c) _____

36. WAREHOUSE SHELVING   Levinson Supply purchased new warehouse shelving for $37,500. The estimated life is 10 years, with a salvage value of $7500. Use the straight-line method of depreciation to find **(a)** the annual rate of depreciation, **(b)** the annual amount of depreciation, and **(c)** the book value at the end of 5 years.

(a) _____
(b) _____
(c) _____

13.2 Depreciation: Declining-Balance Method

Objectives

1. Describe the declining-balance method of depreciation.
2. Find the double-declining-balance rate.
3. Use the double-declining-balance method to find the amount of depreciation and the book value for each year.
4. Use the double-declining-balance method to prepare a depreciation schedule.

CASE in POINT

Straight-line depreciation assumes that an asset loses an equal amount of value each year of its life. This is not realistic for most of the machinery and equipment owned by Cruz Landscaping. For example, a new tractor loses much more value during its first year of life than during its fifth year of life.

Objective 1   Describe the declining-balance method of depreciation.   Methods of **accelerated depreciation** are used to more accurately reflect the rate at which assets actually lose value. One of the more common accelerated methods of depreciation is the double-declining-balance method or **200% method**. With this method, the **double-declining-balance rate** is first established. This rate is multiplied by last year's book value to get this year's depreciation. Since the book value declines from year to year, the annual depreciation also declines, giving the origin of the name of this method.

Objective 2   Find the double-declining-balance rate.   Calculate depreciation using the double-declining-balance method, by first finding the straight-line rate of depreciation. Then adjust the straight-line rate to the desired declining-balance rate of 200% of the straight-line rate.

EXAMPLE 1

Finding the 200% Declining-Balance Rate

Find the straight-line rate and the double-declining-balance (200%) rate for each of the following years of life.

SOLUTION

Years of Life	Straight-Line Rate	Double-Declining-Balance Rate
3	33.33% $\left(\frac{1}{3}\right)$	$\times 2 = 66.67\%$ $\left(\frac{2}{3}\right)$
4	25%	$\times 2 = 50\%$
5	20%	$\times 2 = 40\%$
8	12.5%	$\times 2 = 25\%$
10	10%	$\times 2 = 20\%$
20	5%	$\times 2 = 10\%$
25	4%	$\times 2 = 8\%$
50	2%	$\times 2 = 4\%$

QUICK TIP   Throughout the remainder of this chapter, money amounts will be rounded to the nearest dollar, the common practice with depreciation.

Objective 3   Use the double-declining-balance method to find the amount of depreciation and the book value for each year.   Use the following formula and multiply the double-declining-balance rate by the declining balance. The declining balance is the total cost in the first year and the previous year's book value in following years. The answer is the amount of depreciation in that year.

Depreciation = Double-declining-balance rate $\times$ Declining balance

EXAMPLE 2

Finding Depreciation and Book Value Using Double-Declining-Balance

Cruz Landscaping purchased a portable storage building for $8100. It is expected to have a life of 10 years, at which time it will have no salvage value. Using the double-declining-balance method of depreciation, find the first and second years' depreciation and the book value at the end of the first and second year.

SOLUTION

The straight-line depreciation rate for a 10-year life is 10%. The double-declining-balance rate is 10% times 2, or 20%. The first year's depreciation is 20% of the declining balance or, in the first year, 20% of the cost.

$$20\% \times \$8100\,(\text{cost}) = \$1620 \text{ depreciation in the first year}$$

The book value at the end of the first year is

$$
\begin{array}{rl}
\$8100 & \text{cost} \\
-\ \ 1620 & \text{depreciation to date} \\
\hline
\$6480 & \text{book value at the end of the first year}
\end{array}
$$

The second year's depreciation rate is 20% of $6480 (last year's book value or declining balance) or

$$20\% \times \$6480\,(\text{declining balance}) = \$1296 \text{ depreciation in second year.}$$

The book value at the end of the second year is $8100 − $2916($1620 + $1296) = $5184.

QUICK TIP   *Never* subtract the salvage value from the cost when calculating depreciation using the double-declining-balance method.

Objective 4   Use the double-declining-balance method to prepare a depreciation schedule. The next example shows a depreciation schedule for the pickup truck discussed in Example 3 of **Section 13.1.** As this example shows, the same rate is used each year with the declining-balance method, and the rate is multiplied by the declining balance which is last year's book value. Also, the amount of depreciation in a given year may have to be adjusted so that **book value is never less than salvage value**.

EXAMPLE 3

Preparing a Depreciation Schedule

Cruz Landscaping bought a new pickup truck at a cost of $21,500. It is estimated the truck will have a useful life of 5 years, at which time it will have a salvage value (trade-in value) of $3500. Prepare a depreciation schedule using the double-declining-balance method of depreciation.

SOLUTION

The annual rate of depreciation is 40% (**20% straight-line × 2 = 40%**). Do not subtract salvage value from cost before calculating depreciation. In year 1, the full cost is used to calculate depreciation.

Year	Computation	Amount of Depreciation	Accumulated Depreciation	Book Value
0	—	—	—	$21,500
1	(40% × $21,500)	$8600	$8,600	$12,900
2	(40% × $12,900)	$5160	$13,760	$7,740
3	(40% × $7,740)	$3096	$16,856	$4,644
4		$1144*	$18,000	$3,500
5		$0	$18,000	$3,500

*QUICK TIP   In year 4 of the preceding table, 40% of $4644 is $1858. If this amount were subtracted from $4644, the book value would drop below the salvage value of $3500. Since book value may **never be less than salvage value**, depreciation of $1144 ($4644 − $3500) is taken in year 4, so that book value equals salvage value. No further depreciation remains for year 5 or subsequent years. The total amount of depreciation taken over the life of the asset is the same using either the straight-line or the double-declining-balance method of depreciation.

13.2 Exercises

The **Quick Start** *exercises in each section contain solutions to help you get started.*

Find the annual double-declining-balance (200% method) rate of depreciation, given the following estimated lives. (See Example 1.)

Quick Start

Life	Annual Rate	Life	Annual Rate
1. 5 years	**40%**	**2.** 20 years	**10%**
20% × 2 = 40%		5% × 2 = 10%	
3. 8 years	____	**4.** 25 years	____
5. 15 years	____	**6.** 4 years	____
7. 10 years	____	**8.** 30 years	____
9. 6 years	____	**10.** 40 years	____
11. 50 years	____	**12.** 100 years	____

Find the first year's depreciation for the following, using the double-declining-balance method of depreciation. (See Example 2.)

Quick Start

13. Cost:	$15,000		**14.** Cost:	$10,800	
Estimated life:	10 years		Estimated life:	20 years	
Estimated scrap value:	$3000		Estimated scrap value:	None	
Depreciation (year 1):	**$3000**		Depreciation (year 1):	**$1080**	
$15,000 × 20% = $3000			$10,800 × 10% = $1080		
15. Cost:	$22,500		**16.** Cost:	$38,000	
Estimated life:	5 years		Estimated life:	40 years	
Estimated scrap value:	$500		Estimated scrap value:	$5000	
Depreciation (year 1):	____		Depreciation (year 1):	____	
17. Cost:	$3800		**18.** Cost:	$1140	
Estimated life:	4 years		Estimated life:	6 years	
Estimated scrap value:	None		Estimated scrap value:	$350	
Depreciation (year 1):	____		Depreciation (year 1):	____	

C indicates an exercise that is related to the Case in Point feature within the section.

Find the book value at the end of the first year for the following, using the double-declining-balance method of depreciation. Round to the nearest dollar. (See Examples 1 and 2.)

Quick Start

19. Cost: $4200
    Estimated life: 10 years
    Estimated scrap value: $1000
    Book value: $3360

    $4200 × 20% = $840
    $4200 − $840 = $3360 book value

20. Cost: $2500
    Estimated life: 6 years
    Estimated scrap value: $400
    Book value: _____

21. Cost: $1620
    Estimated life: 8 years
    Estimated scrap value: None
    Book value: _____

22. Cost: $11,280
    Estimated life: 5 years
    Estimated scrap value: $1600
    Book value: _____

Find the book value at the end of 3 years for the following, using the double-declining-balance method of depreciation. Round to the nearest dollar. (See Examples 1 and 2.)

Quick Start

23. Cost: $16,200
    Estimated life: 8 years
    Estimated scrap value: $1500
    Book value $6834

    $16,200 × 25% = $4050 dep. year 1
    $16,200 − $4050 = $12,150
    $12,150 × 25% = $3038 dep. year 2
    $12,150 − $3038 = $9112
    $9112 × 25% = $2278 dep. year 3
    $9112 − $2278 = $6834 book value year 3

24. Cost: $8500
    Estimated life: 10 years
    Estimated scrap value: $1100
    Book value: _____

25. Cost: $6000
    Estimated life: 3 years
    Estimated scrap value: $750
    Book value: _____

26. Cost: $75,000
    Estimated life: 50 years
    Estimated scrap value: None
    Book value: _____

Solve the following application problems.

Quick Start

27. **WEIGHT-TRAINING EQUIPMENT**   Gold's Gym selects the double-declining-balance method of depreciation for some weight-training equipment costing $14,400. If the estimated life of the equipment is 4 years and the salvage value is zero, prepare a depreciation schedule.

Year	Computation	Amount of Depreciation	Accumulated Depreciation	Book Value
0	—	—	—	$14,400
1	(50% × $14,400)	$7200	$7,200	$7,200
2	(50% × $7,200)	$3600	$10,800	$3,600
3	(50% × $3,600)	$1800	$12,600	$1,800
4		$1800*	$14,400	$0

*To depreciate to $ scrap value

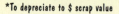

28. STUDIO SOUND SYSTEM   A studio sound system costing $11,760 has a 3-year life and a scrap value of $1400. Prepare a depreciation schedule using the double-declining-balance method of depreciation.

Year	Computation	Amount of Depreciation	Accumulated Depreciation	Book Value
0	—	—	—	$11,760
1				
2				
3				$1,400

29. CONVEYOR SYSTEM   Use the double-declining-balance method of depreciation to prepare a depreciation schedule for a new conveyor system installed at Camblin Steel. Cost = $14,000; estimated life = 5 years; estimated scrap value = $2500. (Round to the nearest dollar.)

Year	Computation	Amount of Depreciation	Accumulated Depreciation	Book Value
0	—	—	—	$14,000
1				
2				
3				
4				
5				$2,500

30. ELECTRONIC ANALYZER   Neilo Lincoln-Mercury decides to use the double-declining-balance method of depreciation on a Barnes Electronic Analyzer that was acquired at a cost of $25,500. If the estimated life of the analyzer is 8 years and the estimated scrap value is $3500, prepare a depreciation schedule. (Round to the nearest dollar.)

Year	Computation	Amount of Depreciation	Accumulated Depreciation	Book Value
0	—	—	—	$25,500
1				
2				
3				
4				
5				
6				
7				
8				$3,500

31. Another name for the double-declining-balance method of depreciation is the 200% method. Explain why the straight-line method of depreciation is often called the 100% method. (See Objective 2.)

32. Explain why the amount of depreciation taken in the last year of an asset's life may be zero when using the double-declining-balance method of depreciation. (See Objective 4.)

33. **CARPET-CLEANING EQUIPMENT**   John Walker, owner of The Carpet Solution, purchased some truck-mounted carpet-cleaning equipment at a cost of $8200. The estimated life of the equipment is 8 years, and the expected salvage value is $1250. Use the double-declining-balance method of depreciation to find the depreciation in the third year.

33. _____

34. **HARBOR BOATS**   A harbor boat costs $478,000 and has an estimated life of 10 years and a salvage value of $150,000. Find the depreciation in the second year using the double-declining-balance method of depreciation.

34. _____

35. **EXCAVATING MACHINERY**   Cruz Landscaping purchased a new backhoe. The cost of the backhoe was $39,240 and it has an estimated life of 5 years with no salvage value. Use the double-declining-balance method of depreciation to find the book value at the end of the third year.

35. _____

36. **COMMUNICATION EQUIPMENT**   Laura Rogers purchased some communication equipment for her public relations firm at a cost of $19,700. She estimates the life of the equipment to be 8 years, at which time the salvage value will be $1000. Use the double-declining-balance method of depreciation to find the book value at the end of 5 years.

36. _____

37. **CONSTRUCTION POWER TOOLS**   Cruz Landscaping purchased a portable cement mixer at a cost of $5800. The mixer has a life of 8 years and a scrap value of $1000. Use the double-declining-balance method of depreciation to find (a) the annual rate of depreciation, (b) the amount of depreciation in the first year, (c) the accumulated depreciation at the end of the fifth year, and (d) the book value at the end of the fifth year.

(a) _____
(b) _____
(c) _____
(d) _____

38. **PRIVATE SCHOOL EQUIPMENT**   Gale Klein bought some white boards to write on for her reading clinic at a cost of $3620. The estimated life of the white boards is 5 years, with a salvage value of $400. Use the double-declining-balance method of depreciation to find (a) the annual rate of depreciation, (b) the amount of depreciation in the first year, (c) the accumulated depreciation at the end of the third year, and (d) the book value at the end of the third year.

(a) _____
(b) _____
(c) _____
(d) _____

13.3 | Depreciation: Sum-of-the-Years'-Digits Method

Objectives

1. Understand the sum-of-the-years'-digits method of depreciation.
2. Find the depreciation fraction for the sum-of-the-years'-digits method.
3. Use the sum-of-the-years'-digits method to find the amount of depreciation for each year.
4. Prepare a depreciation schedule for the sum-of-the-years'-digits method.

Objective 1 Understand the sum-of-the-years'-digits method of depreciation. The sum-of-the-years'-digits method of depreciation is another accelerated depreciation method. The double-declining-balance method of depreciation produces more depreciation than the straight-line method in the early years of an asset's life and less depreciation in the later years. The sum-of-the-years'-digits method, however, produces results in between the straight-line and the double-declining-balance method—more than straight-line at the beginning and more than double-declining-balance at the end.

Objective 2 Find the depreciation fraction for the sum-of-the-years'-digits method. The use of the sum-of-the-years'-digits method requires a **depreciation fraction** instead of the depreciation rate used earlier. The annual depreciation is this depreciation fraction multiplied by the depreciable amount (cost minus salvage value). The depreciation fraction decreases annually, as does the depreciation.

To find the depreciation fraction, first find the denominator, which remains constant for every year of the life of the asset. The denominator is the sum of all the years of the estimated life of the asset (sum of the years' digits). For example, if the life is 6 years, the denominator is 21 **since $1 + 2 + 3 + 4 + 5 + 6 = 21$.** The numerator of the fraction, which decreases each year, gives the number of years of life remaining at the beginning of that year.

EXAMPLE 1

Finding the Depreciation Fraction

Find the depreciation fraction for each year if the sum-of-the-years'-digits method of depreciation is to be used for an asset with a useful life of 6 years.

SOLUTION

First determine the denominator of the depreciation fraction. The denominator is 21 $\left(1 + 2 + 3 + 4 + 5 + 6 = 21 \right)$. Next determine the numerator for each year. The number of years of life remaining at the beginning of any year is the numerator.

Year	Depreciation Fraction
1	$\frac{6}{21}$
2	$\frac{5}{21}$
3	$\frac{4}{21}$
4	$\frac{3}{21}$
5	$\frac{2}{21}$
6	$\frac{1}{21}$
21 ← sum of the years' digits	$\frac{21}{21}$

Under the sum-of-the-years'-digits method, an asset having a life of 6 years is assumed to lose $\frac{6}{21}$ of its value the first year, $\frac{5}{21}$ the second year, and so on. The sum of the six fractions in the table is $\frac{21}{21}$, or 1, so that the entire depreciable amount is used over the 6-year life.

> **QUICK TIP** It is common not to write these fractions in lowest terms, so that the year in question can be seen.

A quick method of finding the sum of the years' digits is by the formula

$$\frac{n(n + 1)}{2}$$

where n is the estimated life of the asset. For example, if the life is 6 years, **6** is multiplied by **6 + 1**, resulting in **6 × 7**, or 42. Then 42 is divided by 2, giving 21, the same denominator used in the table. This method eliminates adding digits and is especially useful when the life of an asset is long.

Objective 3 Use the sum-of-the-years'-digits method to find the amount of depreciation for each year. Use the following formula and multiply the depreciation fraction in any year by the depreciable amount (as in the straight-line method) to calculate the amount of depreciation in that year.

> Depreciation = Depreciation fraction × Depreciable amount

EXAMPLE 2

Finding Depreciation Using the Sum-of-the-Year's-Digits Method

Cruz Landscaping purchases a DitchMaster trencher at a cost of $8940. The trencher has a useful life of 8 years and an estimated salvage value of $1200. Find the first and second years' depreciation, using the sum-of-the-years'-digits method.

SOLUTION

The depreciation fraction has a denominator of 36 (or **1 + 2 + 3 + 4 + 5 + 6 + 7 + 8**). The numerator in the first year is 8. The first-year fraction, $\frac{8}{36}$, is then multiplied by the amount to be depreciated, $7740 ($8940 cost − $1200 salvage value).

$$\frac{8}{36} \times \$7740 = \$1720$$

The first year's depreciation is $1720.

The depreciation fraction for the second year, $\frac{7}{36}$, is multiplied by the original depreciable amount, $7740 ($8940 cost − $1200 salvage value). This gives

$$\frac{7}{36} \times \$7740 = \$1505$$

The second year's depreciation is $1505.

Objective 4 Prepare a depreciation schedule for the sum-of-the-years'-digits method. For comparison, the next example uses the same truck as in **Sections 13.1 and 13.2**.

EXAMPLE 3

Preparing a Depreciation Schedule

Cruz Landscaping bought a new pickup truck for $21,500. The truck is estimated to have a useful life of 5 years, at which time it will have a salvage value of $3500. Prepare a depreciation schedule using the sum-of-the-years'-digits method of depreciation.

SOLUTION

The depreciation fraction has a denominator of 15 (or **1 + 2 + 3 + 4 + 5**).

Year	Computation	Amount of Depreciation	Accumulated Depreciation	Book Value
0	—	—	—	$21,500
1	$\left(\frac{5}{15} \times \$18,000\right)$	$6000	$6,000	$15,500
2	$\left(\frac{4}{15} \times \$18,000\right)$	$4800	$10,800	$10,700
3	$\left(\frac{3}{15} \times \$18,000\right)$	$3600	$14,400	$7,100
4	$\left(\frac{2}{15} \times \$18,000\right)$	$2400	$16,800	$4,700
5	$\left(\frac{1}{15} \times \$18,000\right)$	$1200	$18,000	$3,500

QUICK TIP The sum-of-the-years'-digits method of depreciation allows rapid depreciation in the early years of the asset's life and yet also provides some depreciation during the last years.

13.3 | Exercises

FOR EXTRA HELP

MyMathLab

InterActMath.com

MathXL

MathXL
Tutorials on CD

Addison-Wesley
Math Tutor Center

DVT/Videotape

The **Quick Start** *exercises in each section contain solutions to help you get started.*

Find the sum-of-the-years'-digits depreciation fraction for the first year given the following estimated lives. (See Example 1.)

Quick Start

Life	First-Year Fraction		Life	First-Year Fraction
1. 4 years	$\frac{4}{10}$		**2.** 3 years	$\frac{3}{6}$
$4 + 3 + 2 + 1 = 10;\ \frac{4}{10}$			$3 + 2 + 1 = 6;\ \frac{3}{6}$	
3. 6 years	_____		**4.** 5 years	_____
5. 7 years	_____		**6.** 8 years	_____
7. 10 years	_____		**8.** 20 years	_____

Find the first year's depreciation for the following, using the sum-of-the-years'-digits method of depreciation. Round to the nearest dollar. (See Example 2.)

Quick Start

9. Cost:	$4800		**10.** Cost:	$5600
Estimated life:	4 years		Estimated life:	5 years
Estimated scrap value:	$700		Estimated scrap value:	$800
Depreciation (year 1):	**$1640**		Depreciation (year 1):	**$1600**
$4800 - \$700 = \$4100$			$5600 - \$800 = \$4800$	
$4100 \times \frac{4}{10} = \$1640$			$4800 \times \frac{5}{15} = \$1600$	

11. Cost:	$60,000		**12.** Cost:	$1440
Estimated life:	10 years		Estimated life:	8 years
Estimated scrap value:	$5000		Estimated scrap value:	None
Depreciation (year 1):	_____		Depreciation (year 1):	_____

13. Cost:	$18,500		**14.** Cost:	$97,400
Estimated life:	3 years		Estimated life:	8 years
Estimated scrap value:	$3500		Estimated scrap value:	$11,000
Depreciation (year 1):	_____		Depreciation (year 1):	_____

 indicates an exercise that is related to the Case in Point feature within the section.

Find the book value at the end of the first year for the following, using the sum-of-the-years'-digits method of depreciation. Round to the nearest dollar. (See Example 3.)

Quick Start

15.	Cost:	$9500		**16.**	Cost:	$14,800
	Estimated life:	8 years			Estimated life:	10 years
	Estimated scrap value:	$1400			Estimated scrap value:	None
	Book value:	**$7700**			Book value:	**$12,109**

$\$8100 \times \frac{8}{36} = \$1800$
$\$9500 - \$1800 = \$7700$

$\$14{,}800 \times \frac{10}{55} = \$2691$
$\$14{,}800 - \$2691 = \$12{,}109$

17.	Cost:	$3800		**18.**	Cost:	$15,650
	Estimated life:	5 years			Estimated life:	6 years
	Estimated scrap value:	$500			Estimated scrap value:	$2000
	Book value:	_____			Book value:	_____

Find the book value at the end of 3 years for the following using the sum-of-the-years'-digits method of depreciation. Round to the nearest dollar. (See Example 3.)

Quick Start

19.	Cost:	$2240		**20.**	Cost:	$27,500
	Estimated life:	6 years			Estimated life:	10 years
	Estimated scrap value:	$350			Estimated scrap value:	None
	Book value:	**$890**			Book value:	_____

$\frac{6}{21} + \frac{5}{21} + \frac{4}{21} = \frac{15}{21}$ dep. in 3 years
$\$2240 - \$350 = \$1890$
$\$1890 \times \frac{15}{21} = \$1350$
$\$2240 - \$1350 = \$890$

21.	Cost:	$4500		**22.**	Cost:	$6600
	Estimated life:	8 years			Estimated life:	5 years
	Estimated scrap value:	$900			Estimated scrap value:	$1500
	Book value:	_____			Book value:	_____

Solve the following application problems. Round to the nearest dollar.

Quick Start

23. **DEPRECIATING OFFICE FURNITURE**   Cruz Landscaping uses the sum-of-the-years'-digits method of depreciation to prepare a depreciation schedule for office furniture that costs $3900, has an expected life of 3 years, and has an estimated salvage value of $480.

Year	Computation	Amount of Depreciation	Accumulated Depreciation	Book Value
0	—	—	—	$3900
1	$\left(\frac{3}{6} \times \$3420\right)$	$1710	$1710	$2190
2	$\left(\frac{2}{6} \times \$3420\right)$	$1140	$2850	$1050
3	$\left(\frac{1}{6} \times \$3420\right)$	$570	$3420	$480

24. RESTAURANT EQUIPMENT   Old South Restaurant has remodeled its dining area at a cost of $14,400. The expected life of the remodeling is 4 years, at which time the salvage value is estimated to be $2400. Complete a depreciation schedule using the sum-of-the-years'-digits method of depreciation.

Year	Computation	Amount of Depreciation	Accumulated Depreciation	Book Value
0	—	—	—	$14,400
1				
2				
3				
4				

25. COMMERCIAL FREEZER   Big Town Market has purchased a new freezer case at a cost of $10,800. The estimated life of the freezer case is 6 years, at which time the salvage value is estimated to be $2400. Complete a depreciation schedule using the sum-of-the-years'-digits method of depreciation.

Year	Computation	Amount of Depreciation	Accumulated Depreciation	Book Value
0	—	—	—	$10,800
1				
2				
3				
4				
5				
6				

26. FORKLIFT DEPRECIATION   Cruz Landscaping purchased a new forklift. Prepare a depreciation schedule using the sum-of-the-years'-digits method of depreciation. Cost = $15,000; estimated life = 10 years; estimated scrap value = $4000.

Year	Computation	Amount of Depreciation	Accumulated Depreciation	Book Value
0	—	—	—	$15,000
1				
2				
3				
4				
5				
6				
7				
8				
9				
10				

27. Write a description of how the depreciation fraction is determined in any year of an asset's life when using the sum-of-the-years'-digits method of depreciation. (See Objective 2.)

28. If you were starting your own business, which of the three depreciation methods—straight-line, double-declining-balance, or sum-of-the-years'-digits—would you decide to use? Why?

29. SOLAR COLLECTOR   Find the depreciation in the third year for a solar collector, using the sum-of-the-years' digits method of depreciation. Cost is $23,000, estimated life is 8 years, and estimated scrap value is $5000.

29. _____

30. LANDSCAPE EQUIPMENT   Cruz Landscaping purchased a new tractor trencher at a cost of $32,000. The expected life of the unit is 8 years and the salvage value is expected to be $5000. Use the sum-of-the-years'-digits method of depreciation to determine the first year's depreciation.

30. _____

31. HOSPITAL EQUIPMENT   Orangevale Rental uses the sum-of-the-years'-digits method of depreciation on all hospital rental equipment. If it purchases new hospital beds at a cost of $12,800 and estimates the life of the beds to be 10 years with no scrap value, find the book value at the end of the fourth year.

31. _____

32. INDUSTRIAL TOOLING   Electro-car, a light-rail manufacturer, purchased a power-wheel and axle jig from a German manufacturer for $31,880. The jig has an expected life of 20 years and an estimated salvage value of $5000. Find the book value at the end of the third year.

32. _____

33. COMMERCIAL CARPET   Mercury Savings Bank has installed new floor covering at a cost of $25,200. It has a useful life of 6 years and no salvage value. Find **(a)** the first and **(b)** the second year's depreciation, using the sum-of-the-years'-digits method.

(a) _____
(b) _____

34. ASSET DEPRECIATION   Using the sum-of-the-years'-digits method of depreciation, find **(a)** the first and **(b)** the second year's depreciation for an asset that has a cost of $3375, an estimated life of 5 years and no salvage value.

(a) _____
(b) _____

35. FAST-FOOD RESTAURANTS   In-N-Out Burgers purchased a new deep-fry unit at a cost of $12,420. The expected life of the unit is 8 years with a scrap value of $1800. Use the sum-of-the-years'-digits method of depreciation to find **(a)** the first year's depreciation fraction, **(b)** the amount of depreciation in the first year, **(c)** the accumulated depreciation at the end of the eighth year, and **(d)** the book value at the end of the fourth year.

(a) _____
(b) _____
(c) _____
(d) _____

36. SEWER DRAIN SERVICE   Armour Drain bought a new sewer line root remover for $6725. The life of the machine is 10 years, and the scrap value is $1500. Use the sum-of-the-years'-digits method of depreciation to find **(a)** the first year's depreciation fraction, **(b)** the amount of depreciation in the first year, **(c)** the accumulated depreciation at the end of the tenth year, and **(d)** the book value at the end of the sixth year.

(a) _____
(b) _____
(c) _____
(d) _____

Supplementary Application Exercises on Depreciation

Round to the nearest dollar if necessary.

1. **WAREHOUSE CONSTRUCTION**   All Electronics Company built a new distribution center at a cost of $260,000. It has an estimated life of 40 years and an estimated scrap value of $40,000. Use the straight-line method of depreciation to find the book value of the distribution center at the end of 10 years.

1. _____

2. **COPY MACHINES**   The UPS store purchased a new copy machine at a cost of $9480. Use the straight-line-method of depreciation to find the amount of depreciation that should be charged off each year if the equipment has an estimated life of 4 years, and a scrap value of $1500.

2. _____

3. **WOODWORKING MACHINERY**   Custom Cabinets bought a lathe for turning table and chair legs at a total cost of $18,500. The estimated life of the lathe is 5 years and there is no scrap value. Find the depreciation in the first year using the double-declining-balance method of depreciation.

3. _____

4. **LANDSCAPING EQUIPMENT**   Cruz Landscaping purchased a scraper at a cost of $22,000. If the estimated life of the scraper is 10 years, find the book value at the end of 3 years using the double-declining-balance method of depreciation.

4. _____

5. **PERSONAL ORGANIZERS**   All-Pro Real Estate purchased some Sony personal organizers for its sales staff. The organizers have a total cost of $2850, an estimated life of 6 years, and a salvage value of $600. Use the sum-of-the-years'-digits method of depreciation to find the book value of all the organizers at the end of the third year.

5. _____

C indicates an exercise that is related to the Case in Point feature within the section.

6. BUSINESS SIGNAGE   The outdoor sign used by the World of Peace bookstore cost $7375, has an estimated life of 10 years, and has a salvage value of $500. Use the sum-of-the-years'-digits method of depreciation to find the book value at the end of the third year.

6. _____

7. JEWELRY DISPLAY CASES   The Diamond Center installed new shatterproof display cases at a cost of $45,600. Using the straight-line method of depreciation, find the amount of depreciation that should be charged off *each year* if the estimated life of the display cases is 10 years, and the scrap value is $8000.

7. _____

8. RECREATION EQUIPMENT   River Rentals purchased some canoes at a cost of $32,000. The estimated life of the canoes is 15 years, and the scrap value is $6500. Find the book value at the end of 10 years using the straight-line method of depreciation.

8. _____

9. BUSINESS SAFE   Schools Credit Union purchased a safe at a cost of $78,000. If the estimated life of the safe is 20 years, find the book value at the end of 2 years using the double-declining-balance method of depreciation.

9. _____

10. THEATER SEATING   The Music Circus has just installed new seating at a total cost of $228,000. The estimated life of the seating is 10 years, there is no salvage value, and the double-declining-balance method of depreciation is used. Find the depreciation in the first year.

10. _____

11. SHUTTLE SERVICE   Use the sum-of-the-years'-digits method of depreciation to find the amount of depreciation to be charged off *each year* on a $38,600 shuttle van that has an estimated life of 4 years and a scrap value of $4400.

11. _____
_____
_____
_____

12. **DIESEL TRACTOR**   Using the sum-of-the-years'-digits method of depreciation, find the amount of depreciation to be charged off each year on a diesel tractor purchased by Cruz Landscaping. The tractor has a cost of $85,000, an estimated life of 5 years, and a scrap value of $13,000.

12. _____
_____
_____
_____
_____
_____

13. **INDUSTRIAL FORKLIFT**   Lumber Plus buys 5 industrial lifts at a cost of $14,825 each. The life of the lifts is estimated to be 10 years and the scrap value is $3000 each. Use the sum-of-the-years'-digits method of depreciation to find the total book value of all the lifts at the end of the fourth year.

13. _____

14. **CAR-WASH MACHINERY**   Bubble Car Wash buys an automatic car-wash machine at a cost of $50,950. It has a scrap value of $10,000 and an estimated life of 6 years. Use the sum-of-the-years'-digits method of depreciation to find the book value at the end of the third year.

14. _____

15. **VIDEO EQUIPMENT**   Video Productions purchased some studio equipment manufactured in Australia. The total cost of the equipment was $21,600. It has an estimated life of 5 years and a salvage value of $2400. Use the sum-of-the-years'-digits method of depreciation to find the book value of the equipment at the end of the second year.

15. _____

16. RESTAURANT TABLES   King's Table bought new dining-room tables at a cost of $14,750. If the estimated life of the tables is 8 years, at which time they will be worthless, use the double-declining-balance method of depreciation to find the book value at the end of the third year.

16. _____

17. REFRIGERATED DISPLAY CASE   Goldi's Delicatessen purchased a new refrigerated display case for $10,800. It has an estimated life of 10 years and a salvage value of $1500. Use the straight-line method of depreciation to find **(a)** the accumulated depreciation and **(b)** the book value at the end of the sixth year.

(a) _____
(b) _____

18. SOFT DRINK BOTTLING   A small soft-drink bottler purchased two automatic filling and capping machines at a cost of $185,000 each. The machines have a scrap value of $30,000 each and an estimated life of 8 years. Use the double-declining balance method of depreciation to find **(a)** the accumulated depreciation and **(b)** the book value of the machines at the end of the third year.

(a) _____
(b) _____

13.4   Depreciation: Units-of-Production Method

Objectives

1. Describe the units-of-production method of depreciation.
2. Use the units-of-production method to find the depreciation per unit.
3. Calculate the annual depreciation using the units-of-production method.
4. Prepare a depreciation schedule using the units-of-production method.

CASE *in* **POINT**

Cruz Landscaping owns a stump chipper that is used to remove tree stumps in established landscapes. Since the machine is not used very often, the Cruz brothers' accountant suggests that the stump chipper be depreciated based on the number of hours it is used rather than by the number of years it is owned. Using this approach will be more realistic for depreciating the stump chipper and could result in faster or slower depreciation, depending on how often the chipper is used.

Objective 1 **Describe the units-of-production method of depreciation.** An asset often has a useful life given in terms of **units of production**, such as hours or miles of service. For example, an airplane or truck may have a useful life given as hours of air time or miles of travel. A steel press or stamping machine may have a useful life given as the total number of units that it can produce. For these assets, the units-of-production method is used. Just as with the straight-line method of depreciation, a constant amount of depreciation is taken with the units-of-production method. With the straight-line method a constant amount of depreciation is taken each year, while the units-of-production method depreciates a constant amount per unit of use or production.

Objective 2 **Use the units-of-production method to find the depreciation per unit.** Find the depreciation per unit with the following formula.

$$\text{Depreciation per unit} = \frac{\text{Depreciable amount}}{\text{Units of life}}$$

For example, suppose the stump chipper owned by Cruz Landscaping costs $15,000, has a salvage value of $3000, and is expected to operate 700 hours. Find the depreciation per hour by dividing the depreciable amount by the number of hours of life.

$$\begin{array}{rl} \$15,\!000 & \text{cost} \\ -\ \underline{3,\!000} & \text{salvage value} \\ \$12,\!000 & \text{depreciable amount} \end{array}$$

Objective 3 **Calculate the annual depreciation using the units-of-production method.**

$$\frac{\$12,\!000 \text{ depreciable amount}}{700 \text{ hours of life}} = \$17.14 \text{ (rounded) depreciation per hour}$$

Use the following formula and multiply the number of hours used during the year by the depreciation per unit to find the annual depreciation.

$$\text{Depreciation} = \text{Number of units (hours)} \times \text{Depreciation per unit (hour)}$$

Heavy-Truck Makers Find the Good Times Are Rolling

The newspaper headline above reports that sales of heavy-duty trucks, the ones that carry more than 33,000 pounds, are the highest they have been in almost 20 years. These behemoth trucks, also known as Class 8s or 18-wheelers, start at a base price of $70,000 and climb as high as $120,000 and more. To depreciate these trucks, the owners will often give them a useful life as high as 750,000 miles.

EXAMPLE 1

Using Units-of-Production Depreciation

North American Trucking purchased a new Kenworth truck for $95,000. The truck has a salvage value of $15,000 and an estimated life of 500,000 miles. Find the depreciation for a year in which the truck is driven 128,000 miles.

SOLUTION

First find the depreciable amount.

$$\begin{array}{rl} \$95,000 & \text{cost} \\ -\ \ 15,000 & \text{scrap value} \\ \hline \$80,000 & \text{depreciable amount} \end{array}$$

Next find the depreciation per unit.

$$\frac{\$80,000 \text{ depreciable amount}}{500,000 \text{ miles of life}} = \$.16 \text{ depreciation per mile}$$

Multiply to find the depreciation for the year.

$$128,000 \text{ miles} \times \$.16 = \$20,480 \text{ depreciation for the year}$$

EXAMPLE 2

Preparing a Depreciation Schedule

Bagel Boys purchased a packaging machine at a cost of $52,300. It has an estimated salvage value of $4000 and an expected life of 690,000 units. Prepare a depreciation schedule using the units-of-production method of depreciation. Use the following packaging schedule.

Year 1	240,000 units
Year 2	150,000 units
Year 3	90,000 units
Year 4	120,000 units
Year 5	90,000 units

SOLUTION

The depreciable amount is $48,300 $\left(\$52,300 - \$4000\right)$. The depreciation per unit is

$$\frac{\$48,300}{690,000 \text{ units}} = \$.07 \text{ per unit.}$$

The annual depreciation is found by multiplying the number of units packaged each year by the depreciation per unit.

Year 1	240,000 units × $.07 =	$16,800
Year 2	150,000 units × $.07 =	$10,500
Year 3	90,000 units × $.07 =	$6,300
Year 4	120,000 units × $.07 =	$8,400
Year 5	90,000 units × $.07 =	$6,300
Total	690,000 units	$48,300	depreciable amount

Objective **4** **Prepare a depreciation schedule using the units-of-production method.**
These results were used in the preparation of the following depreciation schedule.

Year	Computation	Depreciation	Accumulated Depreciation	Book Value
0	—	—	—	$52,300
1	$(240{,}000 \times \$.07)$	$16,800	$16,800	$35,500
2	$(150{,}000 \times \$.07)$	$10,500	$27,300	$25,000
3	$(90{,}000 \times \$.07)$	$6,300	$33,600	$18,700
4	$(120{,}000 \times \$.07)$	$8,400	$42,000	$10,300
5	$(90{,}000 \times \$.07)$	$6,300	$48,300	$4,000

QUICK TIP   In Example 2, the book value at the end of year 5 ($4000) is the amount of the salvage value. This is true because the total number of units of life (690,000) has been used up by the machine during the 5 years.

The newspaper clipping below shows the additional interest and funding being given to the study of entrepreneurship at universities. Many community college business programs are also offering courses in entrepreneurship. This knowledge will help prepare the student who plans to start and operate his or her own business. Being a business owner and not an employee, the student will find that there will be a greater need for understanding business operations including depreciation.

Venture Capital 101: Entrepreneur Courses Increase

University classes explore start-ups, business plans

By Jim Hopkins
USA TODAY

Entrepreneurship, a subject once snubbed by universities, is taking hold on U.S. campuses.

The latest sign: The Ewing Marion Kauffman Foundation recently awarded $25 million to eight U.S. universities pledging to spread entrepreneurship education beyond business classrooms.

Recipients include the University of Rochester, which will use part of its $3.5 million to teach music students to better manage orchestras and other professional music companies.

Historically, universities prepared students to manage *Fortune* 500 companies, education experts say. Schools taught big-company finance and organizational behavior but little about start-ups, such as developing business plans and seeking venture capital.

That changed, especially in the 1990s, after research showed most jobs and innovations are created by start-ups. Also driving academic interest in entrepreneurship:

• **Demand.** More students think self-employment is a safer haven than working for big corporations. "Many saw parents downsized out of work," says John Challenger, CEO of executive outplacement consultant Challenger Gray & Christmas.

Source: USA Today.

'NET ASSETS   Business on the Internet

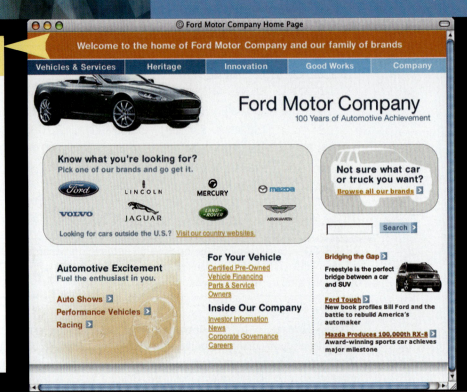

Ford Motor Company

- 1925: First Ford truck built
  — Price: $281

- 1936: Ford had already sold 3 million trucks

- 1950s: Moved to more car-like comfort and styling in their trucks

- 2003: Ford celebrates 100th anniversary

- 1983–2005: Ford F-150 best selling full-size pickup truck

In 1903, with $28,000 in cash, Henry Ford started Ford Motor Company which is now one of the worlds largest companies. Ford's greatest contribution to automobile manufacturing was the moving assembly that allowed individual workers to stay in one place and perform the same task on each vehicle as it passed by. Today, Ford Motor Company is a family of automotive brands consisting of Ford, Lincoln, Mercury, Mazda, Jaguar, Land Rover, Aston Martin, and Volvo.

Cruz Landscaping purchased another Ford F-150 Heritage pickup truck for $26,500. The useful life or the truck is 5 years and the estimated salvage value is $4500.

1. Using the information above, find the book value of the pickup truck after 3 years using the straight-line method of depreciation.

2. Find the book value of the truck at the end of 3 years using the double-declining-balance method of depreciation.

3. List four factors that affect how fast a pickup truck is used up or wears out. Does the book value of an asset necessarily reflect the true value (resale value) of that asset? Explain.

4. Talk to a business owner and list a few of the assets that they are depreciating. If the business had a choice of depreciating gradually over the life an asset (straight-line) or more rapidly using an accelerated depreciation method, what factors would the business owner consider when making their decision?

13.4 | Exercises

FOR EXTRA HELP

MyMathLab

InterActMath.com

Math XP MathXL

MathXL
Tutorials on CD

Tutor Center Addison-Wesley
Math Tutor Center

DVT/Videotape

The **Quick Start** *exercises in each section contain solutions to help you get started.*

Find the depreciation per unit in the following. Round to the nearest thousandth of a dollar. (See Example 1.)

Quick Start

Cost	Salvage Value	Estimated Life	Depreciation per Unit
1. $16,800	$1800	20,000 units	$.75
$16,800 − $1800 = $15,000; $15,000 ÷ 20,000 = $.75			
2. $22,500	$1500	60,000 units	_____
3. $3750	$250	120,000 units	_____
4. $7500	$500	15,000 miles	_____
5. $37,500	$7500	125,000 miles	_____
6. $300,000	$25,000	4000 hours	_____
7. $175,000	$25,000	5000 hours	_____
8. $125,000	$20,000	500,000 miles	_____

Find the amount of depreciation in each of the following. (See Example 1.)

Quick Start

Depreciation per Unit	Units Produced	Amount of Depreciation
9. $.46	55,000	$25,300
55,000 × $.46 = $25,300		
10. $.18	275,000	$49,500
275,000 × $.18 = $49,500		
11. $.54	32,000	_____
12. $.73	16,500	_____

 indicates an exercise that is related to the Case in Point feature within the section.

	Depreciation per Unit	Units Produced	Amount of Depreciation
13.	$.185	15,000	_____
14.	$.032	73,000	_____
15.	$.14	22,200	_____
16.	$.075	110,000	_____

17. In your own words, describe the conditions under which the units-of-production method of depreciation is most applicable.

18. Use an example of your own to demonstrate how the annual depreciation amount is found using the units-of-production method of depreciation. (See Objective 3.)

Solve the following application problems. Round to the nearest dollar.

Quick Start

19. DEEP FRYER   McDonald's purchased a new deep fryer at a cost of $6800. The expected life is 5000 hours of production, at which time it will have a salvage value of $500. Using the units-of-production method, prepare a depreciation schedule given the following production: year 1: 1350 hours; year 2: 1820 hours; year 3: 730 hours; year 4: 1100 hours.

Year	Computation	Amount of Depreciation	Accumulated Depreciation	Book Value
0	—	—	—	$6800
1	(1350 × $1.26)	$1701	$1701	$5099
2	(1820 × $1.26)	$2293	$3994	$2806
3	(730 × $1.26)	$920	$4914	$1886
4	(1100 × $1.26)	$1386	$6300	$500

 20. HEAVY-DUTY TRUCK   Cruz Landscaping purchased a Kenworth truck at a cost of $87,000. It estimates that the truck will have a life of 300,000 miles and will have a salvage value of $15,000. Use the units-of-production method to prepare a depreciation schedule given the following production: year 1: 108,000 miles; year 2: 75,000 miles; year 3: 117,000 miles.

Year	Computation	Amount of Depreciation	Accumulated Depreciation	Book Value
0	—	—	—	$87,000

13.5 | Depreciation: Modified Accelerated Cost Recovery System

Objectives

1. Understand the modified accelerated cost recovery system (MACRS).
2. Determine the recovery period of different types of property.
3. Find the depreciation rate given the recovery period and recovery year.
4. Use the MACRS to find the amount of depreciation.
5. Prepare a depreciation schedule using the MACRS.

CASE *in* POINT

No matter which methods of depreciation have been used by Cruz Landscaping for the company's accounting and state-tax purposes, the accountant tells the Cruz brothers that the modified accelerated cost recovery system (MACRS) must be used for federal income-tax purposes. This means that every depreciable asset owned by the company must have an MACRS depreciation schedule and that the Cruz brothers must also have an understanding of MACRS.

Objective 1 **Understand the modified accelerated cost-recovery system (MACRS).** A depreciation method known as the **accelerated cost recovery system (ACRS)** originated as part of the Economic Recovery Tax Act of 1981. It was later modified by the Tax Equity and Fiscal Responsibility Act of 1982 and again by the Tax Reform Act of 1984. The Tax Reform Act of 1986 brought the most recent and significant overhaul to the accelerated cost recovery system (ACRS) and applies to all property placed in service after 1986. The new method is known as the **modified accelerated cost recovery system (MACRS)**. The result is that there are now three systems for computing depreciation for *federal tax purposes*.

Federal Tax Depreciation Methods

1. The MACRS method of depreciation is used for all property placed in service after 1986.
2. The ACRS method of depreciation will continue to be used for all property placed in service from 1981 through 1986.
3. The straight-line, declining-balance, and sum-of-the-years'-digits methods continue to be used if the property was placed in service before 1981.

QUICK TIP The units-of-production method of depreciation is still allowed under the MACRS.

Keep two things in mind about the MACRS: First, the system is designed for tax purposes (it is sometimes called the **income-tax method**), and businesses often use some alternative method of depreciation (in addition to MACRS) for financial accounting purposes. Second, many states do not allow the modified accelerated cost recovery system of depreciation for finding state income tax liability. This means that businesses must use the *MACRS* on the *federal tax return* and one of the other methods on the *state tax return*.

Objective 2 **Determine the recovery period of different types of property.** Under the modified accelerated cost-recovery system, assets are placed in one of nine **recovery classes**, depending on whether the law assumes a 3-, 5-, 7-, 10-, 15-, 20-, 27.5-, 31.5-, or 39-year life for the asset. These lives, or **recovery periods**, are determined as follows.

MACRS Recovery Classes

3-year property	Tractor units for use over-the-road, any racehorse that is over 2 years old, any other horse that is over 12 years old, and qualified rent-to-own property
5-year property	Automobiles, taxis, trucks, buses, computers and peripheral equipment, office machinery (typewriters, calculators), copiers, and research equipment, breeding cattle, and dairy cattle
7-year property	Office furniture and fixtures (desks, files, safes), and any property not designated by law to be in any other class
10-year property	Vessels, barges, tugs, and similar water transportation equipment
15-year property	Improvements made directly to land, such as shrubbery, fences, roads, bridges, and any single-purpose agricultural or horticultural structure, and any tree or vine bearing fruits or nuts
20-year property	Certain farm buildings such as a storage shed
27.5-year property	Residential rental real estate such as rental houses, apartments, and mobile homes
31.5-year property	Nonresidential rental real estate such as office building, stores, and warehouses if placed in service before May 13, 1993
39-year property	Nonresidential property placed in service after May 12, 1993

EXAMPLE 1

Finding the Recovery Period for Property

Cruz Landscaping owns the following assets. Determine the recovery period for each of them.

(a) computer equipment   **(b)** an industrial warehouse (after May 12, 1993)
(c) a pickup truck   **(d)** office furniture   **(e)** a farm building (storage shed)

SOLUTION

Use the MACRS Recovery Classes list above.

(a) 5 years   **(b)** 39 years   **(c)** 5 years   **(d)** 7 years   **(e)** 20 years

Modified Accelerated Cost Recovery System (MACRS)

Useful Items
You may want to see:

Publication

☐ **225** Farmer's Tax Guide
☐ **463** Travel, Entertainment, and Gift Expenses
☐ **544** Sales and Other Dispositions of Assets
☐ **551** Basis of Assets
☐ **587** Business Use of Your Home

Objective 3   Find the depreciation rate, given the recovery period and recovery year. With MACRS, salvage value is ignored, so that *depreciation is based on the entire original cost of the asset*. The depreciation rates are determined by applying the double-declining-balance (200%) method to the 3-, 5-, 7-, and 10-year class properties, the 150% declining-balance method to the 15- and 20-year class properties, and the straight-line (100%) method to the 27.5-, 31.5-, and 39-year class properties. Since these calculations are repetitive and require additional knowledge, the Internal Revenue Service provides tables that show the depreciation rates. The rates are shown as **percents** in the table on the following page. To determine the rate of depreciation for any year of life, find the recovery year in the left-hand column, and then read across to the allowable recovery period.

Notice that the number of recovery years is one greater than the class life of the property. This is because only a half year of depreciation is allowed for the first year the property is placed in service, regardless of when the property is placed in service during the year. This is known as the **half-year convention** and is used by most taxpayers. A complete coverage of depreciation, including all depreciation tables, is included in the **Internal Revenue Service**, **Publication 946**, and may be obtained by contacting the IRS Forms Distribution Center. This publication (946) lists several items that the taxpayer or tax preparer might find useful. This list is shown at the side.

QUICK TIP   MACRS is the federal income-tax method of depreciation and several important points should be remembered:
1. No salvage value is used:
2. The life of the asset is determined by using the recovery periods assigned to different types of property; and
3. A depreciation rate is usually found for each year by referring to a MACRS Table of Depreciation Rates.

MACRS Depreciation Rates

Applicable Percent for the Class of Property

Recovery Year	3-Year	5-Year	7-Year	10-Year	15-Year	20-Year	27.5-Year	31.5-Year	39-Year
1	33.33	20.00	14.29	10.00	5.00	3.750	1.818	1.587	2.568
2	44.45	32.00	24.49	18.00	9.50	7.219	3.636	3.175	2.564
3	14.81	19.20	17.49	14.40	8.55	6.677	3.636	3.175	2.564
4	7.41	11.52	12.49	11.52	7.70	6.177	3.636	3.175	2.564
5		11.52	8.93	9.22	6.93	5.713	3.636	3.175	2.564
6		5.76	8.92	7.37	6.23	5.285	3.636	3.175	2.564
7			8.93	6.55	5.90	4.888	3.636	3.175	2.564
8			4.46	6.55	5.90	4.522	3.636	3.175	2.564
9				6.56	5.91	4.462	3.637	3.175	2.564
10				6.55	5.90	4.461	3.636	3.174	2.564
11				3.28	5.91	4.462	3.637	3.175	2.564
12					5.90	4.461	3.636	3.174	2.564
13					5.91	4.462	3.637	3.175	2.564
14					5.90	4.461	3.636	3.174	2.564
15					5.91	4.462	3.637	3.175	2.564
16					2.95	4.461	3.636	3.174	2.564
17						4.462	3.667	3.175	2.564
18						4.461	3.636	3.174	2.564
19						4.462	3.637	3.175	2.564
20						4.461	3.636	3.174	2.564
21						2.231	3.637	3.175	2.564
22							3.636	3.174	2.564
23							3.637	3.175	2.564
24							3.636	3.174	2.564
25							3.637	3.175	2.564
26							3.636	3.174	2.564
27							3.637	3.175	2.564
28							3.636	3.174	2.564
29								3.175	2.564
30								3.174	2.564
31								3.175	2.564
32								3.174	2.564
33–39									2.564

EXAMPLE 2

Finding the Rate of Depreciation with MACRS

Find the rate of depreciation given the following recovery year and recovery period.

	(a)	(b)	(c)	(d)
Recovery Year	3	4	2	12
Recovery Period	3 years	10 years	5 years	27.5 years

SOLUTION

(a) 14.81%   (b) 11.52%   (c) 32.00%   (d) 3.636%

Objective 4 **Use the MACRS to find the amount of depreciation.** No salvage value is subtracted from the cost of property under the MACRS method, and the depreciation rate multiplied by the original cost determines the depreciation amount.

Finding the Amount of Depreciation with MACRS

Cruz Landscaping purchased a pickup truck. Find the amount of depreciation in the fourth year for the pickup truck that had a cost of $21,500.

SOLUTION

A pickup truck has a recovery period of 5 years. From the table on page 595, the depreciation rate in the fourth year of recovery of 5-year property is **11.52%**. Multiply this rate by the full cost of the property to determine the amount of depreciation.

$$11.52\% \times \$21,500 = \$2476.80 = \$2477 \text{ (rounded)}$$

The amount of depreciation is $2477.

Preparing a Depreciation Schedule with MACRS

Omaha Insurance Company has purchased new office furniture at a cost of $24,160. Prepare a depreciation schedule, using the modified accelerated cost recovery system.

SOLUTION

No salvage value is used with MACRS. Office desks and chairs have a 7-year recovery period. The annual depreciation rates for 7-year properties are as follows.

Recovery Year	Recovery Percent (Rate)
1	14.29%
2	24.49%
3	17.49%
4	12.49%
5	8.93%
6	8.92%
7	8.93%
8	4.46%

Objective 5 Prepare a depreciation schedule using the MACRS. Multiply the appropriate percents by $24,160 to get results shown in the following depreciation schedule.

Year	Computation	Amount of Depreciation	Accumulated Depreciation	Book Value
0	—	—	—	$24,160
1	(14.29% × $24,160)	$3452	$3,452	$20,708
2	(24.49% × $24,160)	$5917	$9,369	$14,791
3	(17.49% × $24,160)	$4226	$13,595	$10,565
4	(12.49% × $24,160)	$3018	$16,613	$7,547
5	(8.93% × $24,160)	$2157	$18,770	$5,390
6	(8.92% × $24,160)	$2155	$20,925	$3,235
7	(8.93% × $24,160)	$2157	$23,082	$1,078
8	(4.46% × $24,160)	$1078	$24,160	$0

The MACRS method of depreciation allows a rapid rate of investment recovery and at the same time results in less complicated computations. By eliminating the necessity for estimating the life of an asset and the need for using a salvage value, the tables provide a more direct method of calculating depreciation.

13.5 | Exercises

FOR EXTRA HELP

MyMathLab

InterActMath.com

MathXL

MathXL
Tutorials on CD

Addison-Wesley
Math Tutor Center

DVT/Videotape

The **Quick Start** exercises in each section contain solutions to help you get started.

Use the MACRS depreciation rates table to find the recovery percent (rate), given the following recovery year and recovery period. (See Examples 1 and 2.)

Quick Start

	Recovery Year	Recovery Period	Recovery Percent (Rate)		Recovery Year	Recovery Period	Recovery Percent (Rate)
1.	3	5-year	19.2%	**2.**	5	7-year	8.93%
3.	9	10-year	_____	**4.**	1	3-year	_____
5.	1	5-year	_____	**6.**	5	20-year	_____
7.	14	27.5-year	_____	**8.**	10	31.5-year	_____
9.	6	5-year	_____	**10.**	4	27.5-year	_____
11.	14	39-year	_____	**12.**	4	31.5-year	_____

Find the first year's depreciation for each of the following using the MACRS method of depreciation and the MACRS depreciation rates table. Round to the nearest dollar. (See Example 3.)

Quick Start

13. Cost: $12,250
Recovery period: 7 years
Depreciation (year 1):   **$1751**

14.29% rate
$12,250 × .1429 = $1751 depreciation

14. Cost: $8790
Recovery period: 5 years
Depreciation (year 1):   **$1758**

20% rate
$8790 × .20 = $1758 depreciation

15. Cost: $430,500
Recovery period: 10 years
Depreciation (year 1):   _____

16. Cost: $72,300
Recovery period: 20 years
Depreciation (year 1):   _____

17. Cost: $48,000
Recovery period: 10 years
Depreciation (year 1):   _____

18. Cost: $12,340
Recovery period: 3 years
Depreciation (year 1):   _____

Find the book value at the end of the first year for each of the following using the MACRS method of depreciation and the MACRS depreciation rates table. Round to the nearest dollar. (See Example 4.)

Quick Start

19. Cost: $9380
Recovery period: 3 years
Book value: **$6254**

$9380 × .3333 = $3126
$9380 − $3126 = $6254 book value

20. Cost: $68,700
Recovery period: 5 years
Book value: _____

 indicates an exercise that is related to the Case in Point feature within the section.

21. Cost: $18,800
    Recovery period: 10 years
    Book value: _____

22. Cost: $137,000
    Recovery period: 27.5 years
    Book value: _____

Find the book value at the end of 3 years for each of the following using the MACRS method of depreciation and the MACRS depreciation rates table. Round to the nearest dollar. (See Example 4).

Quick Start

23. Cost: $9570
    Recovery period: 5 years
    Book value: **$2756**

    20% + 32% + 19.2% = 71.2% rate 3 years
    $9570 × .712 = $6813.84 = $6814 dep. 3 years
    $9570 − $6814 = $2756 book value year 3

24. Cost: $18,800
    Recovery period: 3 years
    Book value: _____

25. Cost: $87,300
    Recovery period: 27.5 years
    Book value: _____

26. Cost: $390,800
    Recovery period: 31.5 years
    Book value: _____

Solve the following application problems. Use the MACRS depreciation rates table. Round to the nearest dollar. (See Example 4).

Quick Start

27. STORAGE TANK   Blue Ribbon Septic purchased a storage tank for $10,980. Prepare a depreciation schedule using the MACRS method of depreciation (3-year property).

Year	Computation	Amount of Depreciation	Accumulated Depreciation	Book Value
0	—	—	—	$10,980
1	(33.33% × $10,980)	$3660	$3,660	$7,320
2	(44.45% × $10,980)	$4881	$8,541	$2,439
3	(14.81% × $10,980)	$1626	$10,167	$813
4	(7.41% × $10,980)	$813*	$10,980	$0

*due to rounding in prior years

28. COMPANY VEHICLES   Cruz Landscaping purchased a pickup truck at a cost of $21,500. Prepare a depreciation schedule using the MACRS method of depreciation (5-year property).

Year	Computation	Amount of Depreciation	Accumulated Depreciation	Book Value
0	—	—	—	$21,500
1				
2				
3				
4				
5				
6				

29. OFFSHORE DRILLING   Gulf Drilling purchased a tugboat for $122,700. Prepare a depreciation schedule, using the MACRS method of depreciation (10-year property).

Year	Computation	Amount of Depreciation	Accumulated Depreciation	Book Value
0	—	—	—	$122,700
1				
2				
3				
4				
5				
6				
7				
8				
9				
10				
11				

30. RESIDENTIAL RENTAL PROPERTY   George Kavooris purchased some residential rental real estate for $415,000. Find the book value at the end of the tenth year using the MACRS method of depreciation (27.5-year property).

30. _____

Year	Computation	Amount of Depreciation	Book Value
0	—	—	$415,000
1			
2			
3			
4			
5			
6			
7			
8			
9			
10			

31. The same business asset may be depreciated using two or more different methods. Explain why a business would do this. (See Objective 1.)

32. After learning about MACRS, what three features stand out to you as being unique to this method? (See Objective 3.)

Use the MACRS depreciation rates table in the following application problems. Round to the nearest dollar.

33. **COMMERCIAL FISHING BOAT**   Reef Fisheries purchased a new fishing boat (10-year property) for $74,125. Find the depreciation in year 8 using the MACRS method of depreciation.

33. _____

34. **SHOPPING CENTER**   A new parking lot was added to the Oak Shopping Center at a cost of $118,000. Find the depreciation in year 12 using the MACRS method of depreciation (15-year property).

34. _____

C 35. **LAPTOP COMPUTERS**   Cruz Landscaping purchased a laptop computer for the office for $1700. Find the book value at the end of the third year using the MACRS method of depreciation.

35. _____

36. Dr. Owens Family Dentistry purchased new furniture for its patient reception area at a cost of $27,400. Find the book value at the end of the fifth year using the MACRS method of depreciation.

36. _____

37. **BOOKKEEPING BUSINESS**   Jim Bralley, owner of Interlink Financial Services, purchased an office building at a cost of $480,000. Find the amount of depreciation for each of the first five years using the MACRS method of depreciation (39-year property).

Year 1: _____
Years 2-5: _____

38. **INDEPENDENT BOOKSTORE OWNERSHIP**   Maretha Roseborough, owner of the Barnstormer Bookstore, bought a building to use for her business. The cost of the building was $220,000. Find the amount of depreciation for each of the first five years using the MACRS method of depreciation (39-year property).

Year 1: _____
Years 2-5: _____

Chapter 13 | Quick Review

CHAPTER TERMS *Review the following terms to test your understanding of the chapter. For each term you do not know, refer to the page number found next to that term.*

200% method [**p. 571**]	depreciable amount [**p. 563**]	Internal Revenue Service [**p. 594**]	residual value [**p. 562**]
accelerated cost recovery system (ACRS) [**p. 593**]	depreciation [**p. 562**]	modified accelerated cost	salvage value [**p. 562**]
accelerated depreciation [**p. 571**]	depreciation fraction [**p. 577**]	recovery system (MACRS) [**p. 593**]	scrap value [**p. 562**]
accumulated depreciation [**p. 562**]	depreciation schedule [**p. 565**]		straight-line method [**p. 562**]
book value [**p. 562**]	double-declining-balance rate [**p. 571**]	no salvage value [**p. 562**]	sum-of-the-years'-digits [**p. 577**]
cost [**p. 562**]	half-year convention [**p. 594**]	Publication 946 [**p. 594**]	tangible assets [**p. 562**]
declining balance [**p. 571**]	income tax method [**p. 593**]	recovery classes [**p. 593**]	units of production [**p. 587**]
	intangible assets [**p. 562**]	recovery periods [**p. 593**]	useful life [**p. 562**]

CONCEPTS	EXAMPLES
13.1 Straight-line method of depreciation The depreciation is the same each year. $$\text{Depreciation} = \frac{\textbf{Depreciable amount}}{\textbf{Years of life}}$$	Cost, $500; scrap value, $100; life of 8 years; find the annual amount of depreciation. $\quad\quad$ $500 cost $\quad - $ **100 scrap** $\quad\quad$ $400 depreciable amount $$\frac{\$400}{8} = \$50 \text{ depreciation each year}$$
13.1 Book value Book value is the remaining value at the end of the year. **Book value = Cost − Accumulated depreciation**	Cost, $400; scrap value, $100; life of 3 years; find the book value at the end of the first year. $\quad\quad$ $400 cost $\quad - $ **100 scrap value** $\quad\quad$ $300 depreciable amount $$\frac{\$300}{3} = \$100 \text{ depreciation}$$ $\quad\quad$ $400 cost $\quad - $ **100 depreciation** $\quad\quad$ $300 book value year 1
13.2 Double-declining-balance rate First find the straight-line rate, and then adjust it. For the 200% method, **multiply by 2.**	The life of an asset is 10 years. Find the double-declining-balance (200%) rate. $$10 \text{ years} = \textbf{10\%} \left(\frac{1}{10}\right) \textbf{straight-line}$$ $$\textbf{2} \times 10\% = 20\% \text{ per year}$$
13.2 Double-declining-balance depreciation method First find the double-declining-balance rate and then multiply by the cost in year 1. The rate is then multiplied by the declining book value in the following years. **Depreciation =** **Double-declining-balance rate × Declining balance**	Cost, $1400; life of 5 years; find the depreciation in years 1 and 2. $$2 \times 20\% \text{ (straight-line rate)} = 40\%$$ year 1: **40%** × $1400 = $560 depreciation year 1 $1400 − **$560** = $840 book value year 1 year 2: **40%** × $840 = $336 depreciation year 2

CONCEPTS	EXAMPLES

13.3   Sum-of-the-years'-digits depreciation fraction

Add the year's digits together to get the denominator. The numerator is the number of years of life remaining.

The shortcut for finding the denominator follows:

$$\frac{n(n+1)}{2}$$

Useful life is 4 years. Find the depreciation fraction for each year.

$$1 + 2 + 3 + 4 = 10$$

Year	Depreciation Fraction
1	$\frac{4}{10}$
2	$\frac{3}{10}$
3	$\frac{2}{10}$
4	$\frac{1}{10}$

13.3   Sum-of-the-years'-digits depreciation method

First find the depreciation fraction, then multiply by the depreciable amount.

Depreciation =
Depreciation fraction × Depreciable amount

Cost, $2500; salvage value, $400; life of 6 years; find depreciation in year 1.

$$\text{depreciation fraction} = \frac{6}{21}$$

$$\text{depreciable amount} = \$2100 \, (\$2500 - \$400)$$

$$\text{depreciation} = \frac{6}{21} \times \$2100 = \$600$$

13.4   Units-of-production depreciation amount per unit

Use the following formula:

$$\textbf{Depreciation per unit} = \frac{\textbf{Depreciable amount}}{\textbf{Units of life}}$$

Cost $10,000; salvage value, $2500; useful life of 15,000 units; find depreciation per unit.

$$\$10,000 - \$2500 = \$7500 \text{ depreciable amount}$$

$$\text{Depreciation per unit} = \frac{\$7500 \text{ depreciable amount}}{\textbf{15,000 units of life}} = \$.50$$

13.4   Units-of-production depreciation method

Multiply the number of units (hours) of production by the depreciation per unit (per hour).

Depreciation =
Number of units (hours) × Depreciation per unit (hour)

Cost, $25,000; salvage value, $2000; useful life, 100,000 units; production in year 1, 22,300 units; find the first year's depreciation.

1. $25,000 − **$2000** = $23,000 depreciable amount

2. $\dfrac{\$23,000}{\textbf{100,000}} = \$.23$ depreciation per unit

3. 22,300 × **$.23** = $5129 depreciation year 1

13.5   Modified accelerated cost recovery system (MACRS)

Established in 1986 for federal income tax purposes. No salvage value. Recovery periods are:

3-year	5-year	7-year
10-year	15-year	20-year
27.5-year	31.5-year	39-year

Find the proper rate from the table and then multiply by the cost to find depreciation.

Use the table, finding the recovery period column at the top of the table and the recovery year in the left-hand column. Cost: $4850; recovery period 5 years; recovery year, 3; find the depreciation.

Rate is **19.20%** from table.

$$\$4850 \times \textbf{.192} = \$931 \text{ depreciation}$$

Chapter 13 | Summary Exercise

Comparing Depreciation Methods: A Business Application

Trader Joe's purchased freezer cases at a cost of $285,000. The estimated life of the freezer cases is 5 years, at which time they will have no salvage value. The company would like to compare allowable depreciation methods and decides to prepare depreciation schedules for the fixtures using the straight-line, double-declining-balance, and the sum-of-the-years'-digits methods of depreciation. Using depreciation schedules, find the answers to these questions for Trader Joe's.

(a) What is the book value at the end of 3 years using the straight-line depreciation method?

(a) _____

(b) Using the double-declining-balance method of depreciation, what is the book value at the end of the third year?

(b) _____

(c) With the sum-of-the-years'-digits method of depreciation, what is the accumulated depreciation at the end of 3 years?

(c) _____

(d) What amount of depreciation will be taken in year 4 with each of the methods?

(d) _____

_____

_____

(IN)VESTIGATE

Identify a store or business with which you are familiar and list six of their depreciable assets. Examples could be such items as buildings, computer equipment, vehicles, and fixtures. Using the information on the MACRS method of depreciation, give the recovery period and the first-year depreciation rate for each of the six depreciable assets you listed.

Chapter 13 | Test

To help you review, the numbers in brackets show the section in which the topic was discussed.

Find the annual straight-line and double-declining rates (percents) of depreciation and the sum-of-the-years'-digits fraction for the first year for each of the following estimated lives.   **[13.1–13.3]**

Life	Straight-Line Rate	Double-Declining Rate	Sum-of-the-Years'-Digits Fraction
1. 4 years	_____	_____	_____
2. 5 years	_____	_____	_____
3. 8 years	_____	_____	_____
4. 20 years	_____	_____	_____

Solve the following application problems. Round to the nearest dollar.

5. Cloverdale Creamery purchased a soft-serve ice cream maker at a cost of $12,400. The machine has an estimated life of 10 years and a scrap value of $3000. Use the straight-line method of depreciation to find the annual depreciation.   **[13.1]**

5. _____

6. Sunset Swimming Pools purchased a new dump truck for $38,000. If the estimated life of the dump truck is 8 years, find the book value at the end of 2 years using the double-declining-balance method of depreciation.   **[13.2]**

6. _____

7. The Feather River Youth Camp has purchased a diesel generator for $8250. Use the sum-of-the-years'-digits method of depreciation to determine the amount of depreciation to be taken during *each of the 4 years* on the diesel generator that has a 4-year life and scrap value of $1500.   **[13.3]**

Year 1: _____
Year 2: _____
Year 3: _____
Year 4: _____

8. A private road costs $56,000 and has a 15-year recovery period. Find the depreciation in the third year using the MACRS method of depreciation.   **[13.5]**

8. _____

9. The water filtration system at Micro Brew costs $74,000, has an estimated life of 20 years, and an estimated scrap value of $12,000. Use the straight-line method of depreciation to find the book value of the machinery at the end of 10 years.   **[13.1]**

9. _____

10. Karl Schmidt, owner of Toy Train Hobby Shop, has added paging and intercom features to the communication systems of his 4 stores at a cost of $2800 per store. The estimated life of the systems is 10 years, with no expected salvage value. Using the sum-of-the-years'-digits method of depreciation, find the total book value of all the systems at the end of the third year.   **[13.3]**

10. _____

11. Table Fresh Foods purchased a machine to package its presliced garden salads. The machine costs $20,100 and has an estimated life of 30,000 hours and a salvage value of $1500. Use the units-of-production method of depreciation to find **(a)** the annual amount of depreciation and **(b)** the book value at the end of each year, given the following use information: Year 1: 7800 hours; Year 2: 4300 hours; Year 3: 4850 hours; Year 4: 7600 hours.   **[13.4]**

(a) _____

(b)

12. The Rice Growers Cooperative paid $2,800,000 to build a new rice-drying plant in 2005. The recovery period is 39 years. Use the MACRS method of depreciation to find the book value of the rice-drying plant at the end of the fifth year.   **[13.5]**

12. _____

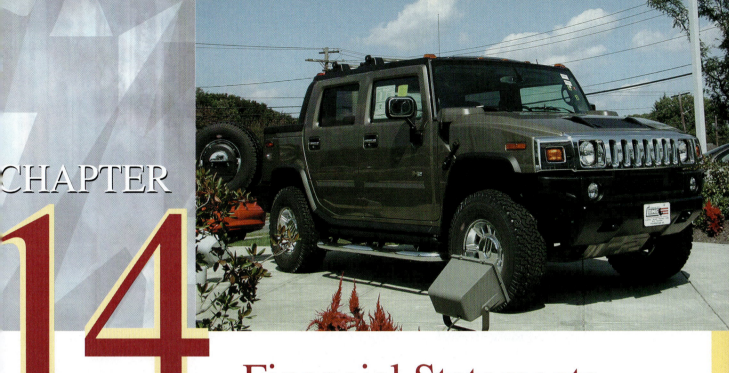

CHAPTER

14

Financial Statements and Ratios

GENERAL MOTORS CORPORATION (GM) IS THE WORLD'S LARGEST manufacturer of vehicles. They manufacture cars, pickups, SUVs, and fleet vehi-

cles for use in industry. Popular GM brands include Chevrolet, Saturn, Cadillac, and Saab. One unusual vehicle that they produce is the Hummer, which is a four-wheel-drive vehicle originally developed for the U.S. military.

Managers must keep careful records of expenses and income in order to operate the business, communicate with others about the business, provide information to lenders, and for tax purposes. **Accountants** track the income and expenses of the firm. They are concerned with issues such as meeting payroll, paying suppliers, estimating and paying taxes, and estimating and receiving revenues. This chapter looks at some tools used in business to gauge the financial health of a firm.

14.1 | The Income Statement

Objectives

1. Understand the terms on an income statement.
2. Prepare an income statement.

CASE *in* POINT

Every year, General Motors (GM) must prepare and distribute four quarterly reports and an annual report showing income, expenses, assets, and liabilities. In turn, managers and investors must be able to analyze these financial reports in order to understand the company's financial performance and their prospects for future growth.

Objective 1   **Understand the terms on an income statement.** An **income statement** is used to summarize all income and expenses for a given period of time, such as a month, a quarter, or a year. Here are some important definitions related to financial statements.

Gross sales, or **total revenue**, is the total amount of money received from customers.
Returns are returns by customers (usually used by smaller companies and not large ones).
Net sales is the value of goods and services bought and kept by customers.

$$\text{Net sales} = \text{Gross sales} - \text{Returns}$$

Cost of goods sold is the amount paid by the firm for items sold to customers.
Gross profit, or **gross profit on sales**, is the money left over after a firm pays for the cost of goods it sells.

$$\text{Gross profit} = \text{Net sales} - \text{Cost of goods sold}$$

Operating expenses, or **overhead**, is the firm's cost to run the business.
Net Income before taxes is the amount earned by the firm.
Net income, or **Net income after taxes**, is the income remaining after income taxes are paid.

$$\text{Net income before taxes} = \text{Gross profit} - \text{Operating expenses}$$
$$\text{Net income} = \text{Net income before taxes} - \text{Income taxes}$$

A portion of the 2002 income statement for General Motors is shown in Example 1. All data in the table are in millions of U.S. dollars. General Motors stock is traded on the New York Stock Exchange under the symbol GM. You can find information about the company on their website (www.gm.com), on many financial websites, or in some newspapers and magazines.

EXAMPLE **1**

Finding Net Income

In 2002, General Motors had gross sales and other income of $186,763,000,000 or one hundred eighty-six billion, seven hundred sixty-three million dollars. This number has been rounded to the nearest one million dollars. GM had no returns, but did have a large cost of goods sold that it paid to suppliers.

SOLUTION

The data in the following table are shown in millions of dollars for convenience.

General Motors Corporation Consolidated Statements of Income Year Ending December 31, 2002 (in millions of dollars)	
Gross Sales and Other Income	$186,763
Returns	– 0
Net Sales and Other Income	$186,763
Cost of Goods Sold	–153,344
Gross Profit	$ 33,419
Operating Expenses	– 31,150
Net Income before Taxes	$ 2,269
Income Taxes	– 533
Net Income	$ 1,736

In 2002, General Motors paid $533 million in income taxes and made *an after-tax profit* of $1.736 billion. General Motors is a huge company! Check the numbers as follows.

Cost of goods sold		Operating expenses		Taxes		Net income after taxes		Gross sales and other income in millions
$153,344	+	$31,150	+	$533	+	$1,736	=	$186,763

The income statement checks.

QUICK TIP   General Motors' research expenses for designing new vehicles falls under *operating expenses*.

The value of a company's stock is based on *financial results* in addition to *perceived opportunities*. As a publicly held corporation, General Motors must publish financial results. The firm's stock price generally rises when profits are rising and generally falls when company profits are falling. You can obtain General Motors' current financial statements mailed to you *free* by calling company headquarters in Detroit, Michigan or by using the World Wide Web.

Objective 2 Prepare an income statement. Example 1 gave the value for the cost of goods sold whereas this amount would normally need to be calculated. The cost of goods sold can be found using the following formula. **Initial inventory** is the value of all goods on hand for sale at the beginning of the period. **Ending inventory** is the value of all goods on hand for sale at the end of the period.

> Value of initial inventory
> + Cost of goods purchased during time period
> + Freight
> – Value of ending inventory
> Cost of goods sold

EXAMPLE 2

Preparing an Income Statement

In the first year of business, Lawn and Tractor, Inc. had gross sales of $159,000 with returns of $9000. Inventory on January 1 of last year was $47,000. A total of $104,000 worth of goods was purchased last year, with freight on the goods totaling $2000. Inventory on December 31 of last year was $56,000. Wages paid to employees totaled $18,000. Rent was $9000, advertising was

$1000, utilities totaled $2000, and taxes on inventory and payroll totaled $4000. Miscellaneous expenses totaled $6000 and income taxes were $500. Complete an income statement for the store using the following steps.

Preparing an Income Statement

Step 1 Enter gross sales and returns. Subtract returns from gross sales to find net sales. Net sales in this example were $150,000.

Step 2 Enter the cost of goods purchased and the freight. Add these two numbers.

Step 3 Add the inventory on January 1 and the total cost of goods purchased.

Step 4 Subtract the inventory on December 31 from the result of step 3. This gives the cost of goods sold.

Step 5 Subtract the cost of goods sold from net sales, which were found in step 1. The result is the gross profit.

Step 6 Enter all expenses and add them to get the total expenses.

Step 7 Subtract the total expenses from the gross profit to find the net income before taxes.

Step 8 Subtract income taxes from net income before taxes to find net income (net income after taxes).

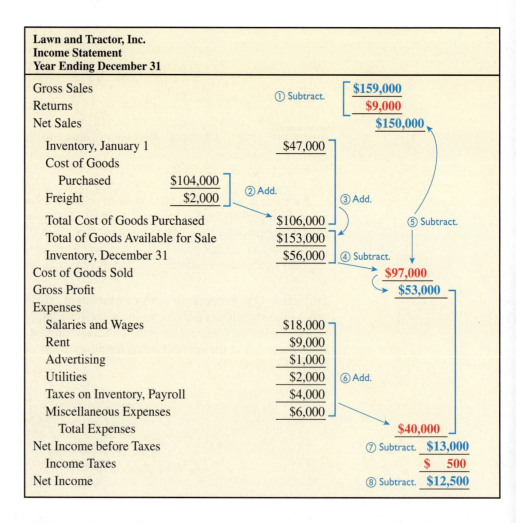

QUICK TIP   Be sure to check the results of your income statement by adding the cost of goods sold, expenses, net income, and income taxes. This total *should* equal net sales.

14.1 | Exercises

FOR EXTRA HELP

 MyMathLab

 InterActMath.com

 MathXL

 MathXL
Tutorials on CD

 Addison-Wesley
Math Tutor Center

DVT/Videotape

The **Quick Start** *exercises in each section contain solutions to help you get started.*

Find (a) the gross profit, (b) the net income before taxes, and (c) the net income after taxes for each firm. (See Example 1.)

Quick Start

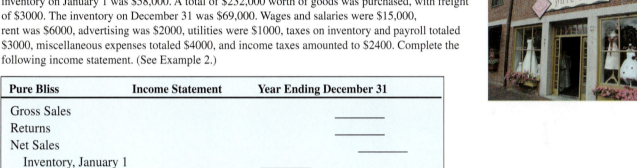

1. **BIKE STORE** Janis Jacobs opened a bike store in the mall three years ago. Last year, the cost of goods sold was $367,200, operating expenses were $228,300, income taxes were $22,700, gross sales were $685,900, and returns were $2350.

   (a) Gross profit = $685,900 − $2350 − $367,200 = $316,350
   (b) Net income before taxes = $316,350 − $228,300 = $88,050
   (c) Net income after taxes = $88,050 − $22,700 = $65,350

   (a) $316,350
   (b) $88,050
   (c) $65,350

2. **ICE CREAM SHOP** Ben's Ice Cream had net sales of $281,400, operating expenses of $119,380, a cost of goods sold of $103,800 and paid $7240 in taxes.

   (a) _____
   (b) _____
   (c) _____

3. **WOMEN'S CLOTHING** Pure Bliss had gross sales of $284,000 last year, with returns of $6000. The inventory on January 1 was $58,000. A total of $232,000 worth of goods was purchased, with freight of $3000. The inventory on December 31 was $69,000. Wages and salaries were $15,000, rent was $6000, advertising was $2000, utilities were $1000, taxes on inventory and payroll totaled $3000, miscellaneous expenses totaled $4000, and income taxes amounted to $2400. Complete the following income statement. (See Example 2.)

Pure Bliss	Income Statement	Year Ending December 31
Gross Sales		_____
Returns		_____
Net Sales		_____
Inventory, January 1	_____	
Cost of Goods		
Purchased	_____	
Freight	_____	
Total Cost of Goods Purchased	_____	
Total of Goods Available for Sale	_____	
Inventory, December 31	_____	
Cost of Goods Sold		_____
Gross Profit		_____
Expenses		
Salaries and Wages	_____	
Rent	_____	
Advertising	_____	
Utilities	_____	
Taxes on Inventory, Payroll	_____	
Miscellaneous Expenses	_____	
Total Expenses		_____
Net Income before Taxes		_____
Income Taxes		_____
Net Income		_____

4. Explain why a lender and an investor would want to look at an income statement before making a loan or an investment. (See Objective 1.)

5. Explain why a banker would look at your personal income statement before approving you for a loan. (See Objective 1.)

6. DENTAL-SUPPLY COMPANY   New England Dental Supply is a regional wholesaler that had gross sales last year of $2,215,000. Returns totaled $26,000. Inventory on January 1 was $215,000. Goods purchased during the year totaled $1,123,000. Freight was $4000. Inventory on December 31 was $265,000. Wages and salaries were $154,000, rent was $59,000, advertising was $11,000, utilities were $12,000, taxes on inventory and payroll totaled $10,000, and miscellaneous expenses were $9000. In addition, income taxes for the year amounted to $287,400. Complete the following income statement for this firm. (See Example 2.)

New England Dental Supply
Income Statement
Year Ending December 31

Gross Sales		_____
Returns		_____
Net Sales		_____
Inventory, January 1	_____	
Cost of Goods		
Purchased	_____	
Freight	_____	
Total Cost of Goods Purchased	_____	
Total of Goods Available for Sale	_____	
Inventory, December 31	_____	
Cost of Goods Sold		_____
Gross Profit		_____
Expenses		
Salaries and Wages	_____	
Rent	_____	
Advertising	_____	
Utilities	_____	
Taxes on Inventory, Payroll	_____	
Miscellaneous Expenses	_____	
Total Expenses		
Net Income before Taxes		_____
Income Taxes		_____
Net Income		_____

14.2 Analyzing the Income Statement

Objectives

1. Compare income statements using vertical analysis.
2. Compare income statements to published charts.
3. Compare income statements using horizontal analysis.

CASE *in* POINT

It is important for you to look carefully at the income statement of General Motors Corporation, or any other company, before investing in the stock of that company. Invest only if you are convinced the company will be profitable.

Objective 1 Compare income statements using vertical analysis. A firm can find its net income for a given period of time by going through the steps presented in the previous section. A question that might then be asked is "What happened to each part of the sales dollar?" The first step toward answering this question is to list each of the important items on the income statement as a percent of net sales. This process is called a **vertical analysis** of the income statement.

In a vertical analysis, each item on the income statement is found as a percent of the net sales.

$$R = \frac{P}{B} \quad \text{or} \quad R = \frac{\text{Particular item}}{\text{Net sales}}$$

For example, use data from the 2002 income statement for General Motors on page 609 to find the following.

$$\text{Percent cost of goods sold} = \frac{\$153,344}{\$186,763} = 82.1\%$$

In 2002, General Motors spent more than 82% of the revenue the company received for cost of goods sold, e.g., to purchase components from suppliers. Other companies may have little or no cost of goods sold. For example, Microsoft Corporation writes computer software and sells it to customers around the world. Microsoft has no cost of goods sold, since the cost of the components it buys, such as optical disks, is so small.

A **comparative income** statement is used to compare results from two or more years. It can be used to show how the company is doing over time.

EXAMPLE **1**

Performing a Vertical Analysis

First perform a vertical analysis of the 2001 and 2002 income statements shown next for General Motors. Then construct a comparative income statement by showing the results in a table.

General Motors Corporation
Consolidated Statements of Income

Year Ending December 31, 2002	(millions of dollars) 2001	2002
Gross Sales or Gross Revenue	$177,260	$186,763
Returns	− 0	− 0
Net Sales	$177,260	$186,763
Cost of Goods Sold	− 143,850	− 153,344
Gross Profit	$ 33,410	$ 33,419
Operating Expenses	− 32,041	− 31,150
Net Income before Taxes	$ 1,369	$ 2,269
Income Taxes	− 768	− 533
Net Income	$ 601	$ 1,736

SOLUTION

Calculate each value in the column labeled 2001 as a percent of 2001 net sales rounded to the nearest tenth of a percent. Then do the same for 2002.

<table>
<tr><th></th><th colspan="2">Comparative Income Statement</th></tr>
<tr><th></th><th>2001</th><th>2002</th></tr>
<tr><td>Percent Cost of Goods Sold</td><td>$\frac{\$143,850}{\$177,260} = 81.2\%$</td><td>$\frac{\$153,344}{\$186,763} = 82.1\%$</td></tr>
<tr><td>Percent of Gross Profit</td><td>$\frac{\$33,410}{\$177,260} = 18.8\%$</td><td>$\frac{\$33,419}{\$186,763} = 17.9\%$</td></tr>
<tr><td>Percent Operating Expenses</td><td>$\frac{\$32,041}{\$177,260} = 18.1\%$</td><td>$\frac{\$31,150}{\$186,763} = 16.7\%$</td></tr>
<tr><td>Percent Net Income before Taxes</td><td>$\frac{\$1,369}{\$177,260} = .8\%$</td><td>$\frac{\$2,269}{\$186,763} = 1.2\%$</td></tr>
<tr><td>Percent Net Income after Taxes</td><td>$\frac{\$601}{\$177,260} = .3\%$</td><td>$\frac{\$1,736}{\$186,763} = .9\%$</td></tr>
</table>

The cost of goods sold increased from 81.2% of net sales in 2001 to 82.1% of net sales in 2002. This resulted in a decrease in percent of gross profit from 2001 to 2002. However, this was offset by the reduction in operating expenses from 18.1% of net sales in 2001 to 16.7% of net sales in 2002. Profits increased from .3% of net sales in 2001 to .9% of net sales in 2002. Therefore, the average General Motors profit on a $28,000 automobile in 2002 was only about .9% × $28,000 or $252.

Objective 2 Compare income statements to published charts. If you own a business or are considering investing in a business, you would be wise to compare the financial figures for that business to industry averages. Published charts of industry averages can be obtained from the federal government.

Type of Business	Cost of Goods	Gross Profit	Total Expenses*	Net Income	Wages	Rent	Advertising
Supermarkets	82.7%	17.3%	13.9%	3.4%	6.5%	.8%	1.0%
Men's and women's apparel	67.0%	33.0%	21.2%	11.8%	8.0%	2.5%	1.9%
Women's apparel	64.8%	35.2%	23.4%	11.7%	7.9%	4.9%	1.8%
Shoes	60.3%	39.7%	24.5%	15.2%	10.3%	4.7%	1.6%
Furniture	68.9%	31.2%	21.7%	9.6%	9.5%	1.8%	2.5%
Appliances	66.9%	33.1%	26.0%	7.2%	11.9%	2.4%	2.5%
Drugs	67.9%	32.1%	23.5%	8.6%	12.3%	2.4%	1.4%
Restaurants	48.4%	51.6%	43.7%	7.9%	26.4%	2.8%	1.4%
Service stations	76.8%	23.2%	16.9%	6.3%	8.5%	2.3%	.5%

* Total Expenses represents the total of all expenses involved in running the firm. These expenses include, but are not limited to, wages, rent, and advertising.

EXAMPLE 2

Compare Business Ratios

Gina Burton wishes to compare the business ratios of her shoe store, Burton's Shoes, to industry averages. Figures from her store and industry averages for shoe stores are shown below.

	Cost of Goods	Gross Profit	Total Expenses	Net Income	Wages	Rent	Advertising
Burton's Shoes	**58.2%**	**41.8%**	**28.3%**	**13.5%**	**11.7%**	**5.6%**	**2.8%**
Shoes (from previous chart)	60.3%	39.7%	24.5%	15.2%	10.3%	4.7%	1.6%

SOLUTION

Burton's expenses are higher than the average for other shoe stores and her net income is lower. Wages are higher—perhaps because her store is located in an area with high wages or perhaps the store is not large enough to efficiently utilize its employees. Burton also spends a higher percent than average for advertising. Perhaps she can reduce her advertising expenses without lowering sales.

Objective ③ **Compare income statements using horizontal analysis.** Another way to analyze an income statement is to prepare a **horizontal analysis**. A horizontal analysis finds percent of change (either increases or decreases) between the current time period and a previous time period. This comparison can expose unusual changes such as a rapid increase in expenses or decline in net sales or profits.

Do a horizontal analysis by finding the amount of change from the previous year to the current year in dollars and as a percent. For example, the income statement of General Motors on page 000 shows that net sales increased from $177,260 in 2001 to $186,763 in 2002 (remember, all figures are in millions). Increase in sales = $186,763 − $177,260 = $9503.

$$\text{Percent increase in sales} = \frac{\$9503}{\$177,260} = \mathbf{5.4\%}$$

It is difficult for General Motors to grow rapidly, because the market is very competitive and includes a number of foreign automakers such as Toyota and because GM is already huge. However, a 5.4% growth rate was slightly higher than inflation during 2002. The following line graph shows that General Motors' stock price has not done as well as the Dow Jones Industrial Average during the past 5 years.

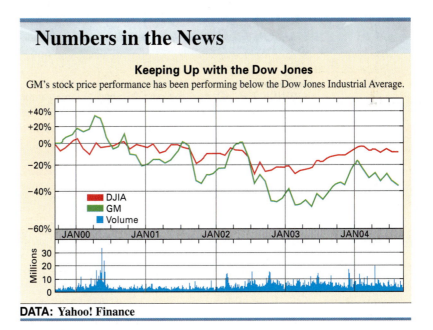

Numbers in the News

Keeping Up with the Dow Jones

GM's stock price performance has been performing below the Dow Jones Industrial Average.

DATA: Yahoo! Finance

To emphasize,

$$\% \text{ of change} = \frac{\text{Change}}{\text{Previous year's amount}}$$

Always use *last year* as the base.

EXAMPLE 3

Performing a Horizontal Analysis

Calculate a horizontal analysis for the 2001 and 2002 income statements for General Motors using the data given in Example 1.

SOLUTION

Find the increase by subtracting the 2001 figure from the 2002 figure. Then divide by the 2001 figure to find the percent increase (or decrease) to the nearest tenth of a percent.

General Motors Corporation
Consolidated Statements of Income
Year Ending December 31 (in millions of dollars)

	2001	2002	Increase	Percent
Net Sales	$177,260	$186,763	$9503	5.4%
Gross Profit	$ 33,410	$ 33,419	$ 9	.0%
Net Income before Taxes	$ 1,369	$ 2,269	$ 900	65.7%
Net Income after Taxes	$ 601	$ 1,736	$1135	188.9%

Although sales grew at 5.4%, gross profit did not grow. Net income before and after taxes increased nicely, but that was because net income in 2001 was so small. Earlier, we saw that the cost of goods sold had gone up from 2001 to 2002. An investor would want to carefully explore why costs of goods sold are going up and other issues before buying stock in the company.

Here are some of the things General Motors is doing to decrease costs and increase revenue:

1. reduce costs by building more components in foreign countries with cheap labor
2. reduce costs by engineering better quality materials and reducing waste in the manufacturing processes
3. increase revenue by increasing sales in foreign countries
4. increase sales by continuing to develop new styles and models of vehicles
5. increase revenue by forming alliances with other companies (such as Saab)

QUICK TIP   During the past 25 years, many companies have moved manufacturing jobs to 3rd world countries to decrease labor costs. Now there is evidence that some technology and professional jobs are also being moved to 3rd world countries for the same reasons. Education is one of the basic ways to help protect yourself and your family from the job displacement brought by this trend.

14.2 | Exercises

FOR EXTRA HELP

 MyMathLab

 InterActMath.com

 MathXL

 MathXL
Tutorials on CD

 Addison-Wesley
Math Tutor Center

 DVT/Videotape

The **Quick Start** *exercises in each section contain solutions to help you get started.*

Prepare a vertical analysis for each of the following firms. Round percents to the nearest tenth of a percent. (See Example 1.)

Quick Start

1. **SCUBA SHOPPE** Reef Scuba, Inc., had net sales of $439,000, operating expenses of $143,180, and a cost of goods sold of $198,400.

   Percent cost of goods sold $= \frac{\$198,400}{\$439,000} = $ **45.2%**

   Percent operating expenses $= \frac{\$143,180}{\$439,000} = $ **32.6%**

   1. <u>45.2%; 32.6%</u>

2. **COFFEE SHOP** Tatum's Coffee and Books had operating expenses of $198,400, a cost of goods sold of $287,104 and net sales of $589,250.

   2. _____

3. **GUITAR SHOP** In 2005, Classic Guitars had a cost of goods sold (guitars) of $243,570, operating expenses of $140,450, and net sales of $480,300.

   3. _____

4. **COIN SHOP** Traver's Coin Shop, Inc., had net sales of $294,380, operating expenses of $68,650, and a cost of goods sold of $163,890.

   4. _____

The following charts show some figures from the income statements of several companies. In each case, prepare a vertical analysis by expressing each item as a percent of net sales. Then write in the appropriate average profit from the table in the book. (See Objective 2.)

5.   **Gooden Drugs**

	Amount	Percent	Average Percent
Net Sales	$850,000	100%	100%
Cost of Goods Sold	$570,350	_____	_____
Gross Profit	$279,650	_____	_____
Wages	$106,250	_____	_____
Rent	$21,250	_____	_____
Advertising	$12,750	_____	_____
Total Expenses	$209,100	_____	_____
Net Income before Taxes	$70,550	_____	_____

6.   **Ellis Restaurant**

	Amount	Percent	Average Percent
Net Sales	$600,000	100%	100%
Cost of Goods Sold	$280,000	_____	_____
Gross Profit	$320,000	_____	_____
Wages	$160,600	_____	_____
Rent	$15,000	_____	_____
Advertising	$8,000	_____	_____
Total Expenses	$255,000	_____	_____
Net Income before Taxes	$65,000	_____	_____

7. Compare a vertical analysis to a horizontal analysis. (See Objectives 1 and 3.)

8. Why would a lender want to use both vertical and horizontal analyses before making a long-term loan to a firm?

Complete the following comparative income statement. Round to the nearest tenth of a percent.

9.   **Hernandez Nursery Comparative Income Statement**

	This Year		Last Year	
	Amount	Percent	Amount	Percent
Gross Sales	$1,856,000	_____	$1,692,000	_____
Returns	$6,000	_____	$12,000	_____
Net Sales	_____	100.0%	_____	100.0%
Cost of Goods Sold	$1,202,000	_____	$1,050,000	_____
Gross Profit	$648,000	_____	$630,000	_____
Wages	$152,000	_____	$148,000	_____
Rent	$82,000	_____	$78,000	_____
Advertising	$111,000	_____	$122,000	_____
Utilities	$32,000	_____	$17,000	_____
Taxes on Inv., Payroll	$17,000	_____	$18,000	_____
Miscellaneous Expenses	$62,000	_____	$58,000	_____
Total Expenses	$456,000	_____	$441,000	_____
Net Income before Taxes	_____	_____	_____	_____

Complete the following horizontal analysis for the Hernandez Nursery comparative income statement given above. Round to the nearest tenth of a percent.

10.   **Hernandez Nursery Horizontal Analysis**

			Increase or (Decrease)	
	This Year	Last Year	Amount	Percent
Gross Sales	$1,856,000	$1,692,000	_____	_____
Returns	$6,000	$12,000	_____	_____
Net Sales	$1,850,000	$1,680,000	_____	_____
Cost of Goods Sold	$1,202,000	$1,050,000	_____	_____
Gross Profit	$648,000	$630,000	_____	_____
Wages	$152,000	$148,000	_____	_____
Rent	$82,000	$78,000	_____	_____
Advertising	$111,000	$122,000	_____	_____
Utilities	$32,000	$17,000	_____	_____
Taxes on Inv., Payroll	$17,000	$18,000	_____	_____
Miscellaneous Expenses	$62,000	$58,000	_____	_____
Net Income before Taxes	$192,000	$189,000	_____	_____

The following table gives the percents for various items from the income statements of firms in various businesses. Complete these tables by including the appropriate percents from the table on page 614. Identify any areas that might require attention by management.

Type of Store	Cost of Goods	Gross Profit	Total Operating Expenses	Net Income	Wages	Rent	Advertising
11. Women's apparel	66.4%	33.6%	25.3%	8.3%	8.4%	6.5%	1.9%
	____	____	____	____	____	____	____
12. Drug store	71.2%	28.8%	26.5%	2.3%	12.9%	5.3%	2.0%
	____	____	____	____	____	____	____

13. Update the graph on page 615 using the Internet.

14. Use the Internet to find a graph of the stock-price history of a company you are familiar with.

14.3 | The Balance Sheet

Objectives

[1] Understand the terms on a balance sheet.
[2] Prepare a balance sheet.

CASE *in* **POINT**

> Before purchasing stock in General Motors, you should look at the company's balance sheet in addition to their income statement. The balance sheet shows information not shown in the income statement such as assets and liabilities on a specific day. It will tell you how much cash GM has and the amount of the firm's debts on that day.

Objective [1] **Understand the terms on a balance sheet.** An income statement summarizes the financial affairs of a business firm for a given period of time, such as a year. On the other hand, a **balance sheet** describes the financial condition of a firm *at one point in time*, such as the last day of a year. A balance sheet shows the **assets** of a firm, which is the total value of everything owned by a business at a particular time. Assets include property, equipment, money owed to the company, cash, and securities owned. A balance sheet also shows the amounts owed by the business to others, called **liabilities**.

Both assets and liabilities are divided into two categories, **long-term** and **current (short-term)**. Long-term generally applies to assets or liabilities with a life of more than a year. Short-term applies when the time involved is less than a year.

Assets

Current assets—cash or items that can be converted into cash within a short period of time such as a year
 Cash—in checking and savings accounts and money-market instruments
 Accounts receivable—funds owed by customers of the firm
 Notes receivable—value of all notes owed to the firm
 Inventory—cost of merchandise that the firm has for sale

Plant and equipment—assets that are expected to be used for more than one year (also called **fixed assets** or **plant assets**)
 Land—book value of any land owned by the firm
 Buildings—book value of any building owned by the firm
 Equipment—book value of equipment, store fixtures, furniture, and similar items owned by the firm

Liabilities

Current liabilities—items that must be paid by the firm within a short period of time, usually one year
 Accounts payable—amounts that must be paid to other firms
 Notes payable—value of all notes owed by the firm

Long-term liabilities—items that will be paid after one year
 Mortgages payable—total due on all mortgages
 Long-term notes payable—total of all other debts of the firm

The difference between the total of all assets and the total of all liabilities is called the **owners' equity**, which is also referred to as **net worth** or, for a corporation, **stockholders' equity**. The relationship between owners' equity, assets, and liabilities is shown in the fundamental formula below.

$$\text{Owners' equity} = \text{Assets} - \text{Liabilities}$$
$$\text{or}$$
$$\text{Assets} = \text{Liabilities} + \text{Owners' equity}$$

Objective 2 Prepare a balance sheet.

EXAMPLE 1

Preparing a Balance Sheet

General Motors' assets and liabilities, in millions of dollars, on December 31, 2002, were as follows: cash and short-term investment, $23,623; accounts receivable and other $33,495; property and equipment $226,001; other assets $88,752; accounts payable $69,517; other current liabilities $1516; long-term debt $200,424; and other liabilities $93,600. Complete a balance sheet.

General Motors Corporation
Consolidated Balance Sheet
Year Ending December 31, 2002 (millions of dollars)

Current Assets:		
Cash and Short-Term Investments	$ 23,623	
Accounts Receivable and Other	$ 33,495	
Total Current Assets	$ 57,118	sum of all current assets
Other Assets:		
Property and Equipment	$226,001	
Equity and Other Assets	$ 88,752	
Total Assets	$371,871	sum of current assets and all other assets
Current Liabilities:		
Accounts Payable	$ 69,517	
Other Current Liabilities	$ 1,516	
Total Current Liabilities	$ 71,033	sum of current liabilities
Other Liabilities:		
Long-Term Debt	$200,424	
Other Liabilities	$ 93,600	
Total Liabilities	$365,057	sum of current liabilities and all other liabilities
Stockholders' Equity:	$ 6,814	Total Assets − Total Liabilities
Total Liabilities and Equity	$371,871	

A balance sheet shows assets and liabilities at one point in time. For example, on December 31, 2002, General Motors had $23,623 million, or $23,623,000,000, in cash and short-term investments. This amount could be very different one month later. The balance sheet above does not give any information about cash and short-term investments held at any date other than on December 31, 2002.

QUICK TIP Stockholders' Equity = Total Assets − Liabilities. However, the stockholders' equity can be in different forms such as retained earnings and various types of stock.

14.3 Exercises

FOR EXTRA HELP

 MyMathLab

 InterActMath.com

 MathXL

 MathXL
Tutorials on CD

 Addison-Wesley
Math Tutor Center

DVT/Videotape

Complete the balance sheets for the following business firms. (See Example 1.)

1. **GROCERY CHAIN**   Brookshire's Grocery (all figures in millions): fixtures, $28; buildings, $290; land, $466; cash, $273; notes receivable, $312; accounts receivable, $264; inventory, $180; notes payable, $312; mortgages payable, $212; accounts payable, $63; long-term notes payable, $55.

Brookshire's Grocery Balance Sheet December 31 (in millions)

ASSETS

Current Assets
  Cash   _____
  Notes Receivable   _____
  Accounts Receivable   _____
  Inventory   _____

        Total Current Assets   _____

Plant Assets
  Land   _____
  Buildings   _____
  Fixtures   _____

        Total plant Assets   _____
Total Assets   =======

LIABILITIES

Current Liabilities
  Notes Payable   _____
  Accounts Payable   _____

        Total Current Liabilities   _____
Long-Term Liabilities
  Mortgages Payable   _____
  Long-Term Notes Payable   _____

        Total Long-Term Liabilities   _____

Total Liabilities   _____

OWNERS' EQUITY

Owners' Equity   _____
Total Liabilities and Owners' Equity   =======

2. LOPEZ MANUFACTURING (ALL FIGURES IN THOUSANDS) Land is $8750; accounts payable total $49,230; notes receivable are $2600; accounts receivable are $37,820; cash is $14,800; buildings are $21,930; notes payable are $3780; owners' equity is $54,320; long-term notes payable are $18,740; mortgages total $26,330; inventory is $49,680; fixtures are $16,820.

Lopez Manufacturing Balance Sheet—December 31

ASSETS

Current Assets
  Cash   _____
  Notes Receivable   _____
  Accounts Receivable   _____
  Inventory   _____

                Total Current Assets   _____
Plant Assets
  Land   _____
  Buildings   _____
  Fixtures   _____

                Total Plant Assets   _____
      Total Assets   ══════

LIABILITIES

Current Liabilities
  Notes Payable   _____
  Accounts Payable   _____

                Total Current Liabilities   _____
Long-Term Liabilities
  Mortgages Payable   _____
  Long-Term Notes Payable   _____

                Total Long-Term Liabilities   _____
      Total Liabilities   _____

OWNERS' EQUITY

Owners' Equity   _____
      Total Liabilities and Owners' Equity   ══════

3. Compare a balance sheet to an income statement.

4. Use the World Wide Web to find the amount of cash and equivalents at the end of the most recent fiscal year for GM and the Coca-Cola Company. Which company has more? Why do you suppose it has as much as it does?

14.4 | Analyzing the Balance Sheet

Objectives

1. Compare balance sheets using vertical analysis.
2. Compare balance sheets using horizontal analysis.
3. Find financial ratios.

Objective 1 **Compare balance sheets using vertical analysis.** A balance sheet can be analyzed in much the same way as an income statement. In a **vertical analysis**, each item on the balance sheet is expressed as a percent of total assets. A **comparative balance sheet** shows the vertical analysis for two different years.

EXAMPLE 1

Comparing Balance Sheets

First, do a vertical analysis for both the 2001 and 2002 balance sheets for General Motors by calculating each value as a percent of the total assets for the year. Round percents to the nearest tenth. Then compare the percents to identify changes from 2001 to 2002.

SOLUTION

General Motors Corporation
Consolidated Balance Sheet
December 31 (millions of dollars)

	2001		2002	
ASSETS:	**Amount**	**Percent**	**Amount**	**Percent**
Current Assets:				
Cash and Short-Term Investments	$ 19,345	6.0%*	$ 23,623	6.4%
Accounts Receivable and Other	$ 27,841	8.6%	$ 33,495	9.0%
Total Current Assets	$ 47,186	14.6%	$ 57,118	15.4%
Other Assets:				
Property and Equipment	$223,385	69.0%	$226,001	60.8%
Equity and Other Assets	$ 53,398	16.5%	$ 88,752	23.9%
Total Assets	$323,969	100.1%	$371,871	100.1%
LIABILITIES:	**Amount**	**Percent**	**Amount**	**Percent**
Current Liabilities:				
Accounts Payable	$ 53,513	16.5%	$ 69,517	18.7%
Other Current Liabilities	$ 10,733	3.3%	$ 1,516	0.4%
Total Current Liabilities	$ 64,246	19.8%	$ 71,033	19.1%
Other Liabilities:				
Long-Term Debt	$163,912	50.6%	$200,424	53.9%
Other Liabilities	$ 76,104	23.5%	$ 93,600	25.2%
Total Liabilities	$304,262	93.9%	$365,057	98.2%
Stockholders' Equity:	$ 19,707	6.1%	$ 6,814	1.8%
Total Liabilities and Equity	$323,969	100.0%	$371,871	100.0%

*$19,345 ÷ $323,969 = 6.0% (rounded)

Total amounts in the table may not add up to totals shown, due to rounding. Percents add up to total percent.

The dollar amount of current assets increased by $57,118 − $47,186 = $9932 in millions of dollars. Investors and bankers want sufficient current assets since it adds financial strength to the company in the event of unforeseen events. Current liabilities increased from $64,246 to $71,033 but declined from 19.8% to 19.1% of total liabilities and equity, a trend that investors and creditors like to see. However, long-term debt was up sharply from 50.6% to 53.9% of total liabilities and equity, so GM took on significant additional long-term debt. Finally, note that stockholders' equity fell from $19,707 to $6814 in millions. This is a significant fall in the equity of stockholders and reflects some difficult issues the General Motors managers faced during 2002. General Motors is facing relentless competition from both U.S. and foreign automakers and this can be seen in the sharply decreasing stockholders' equity.

Objective 2 **Compare balance sheets using horizontal analysis.** Perform a **horizontal analysis** by finding the change, both in dollars and in percent, for each item on the balance sheet from one year to the next. As before, always use the previous year as a base when finding the percents.

EXAMPLE 2

Using Horizontal Analysis

General Motors' cash and short-term investments increased from $19,345 at the end of 2001 to $23,623 at the end of 2002. The percent increase to the nearest tenth is found as follows.

$$\frac{\left(\$23{,}623 - \$19{,}345\right)}{\$19{,}345} = 22.1\%$$

Complete a horizontal analysis of the current assets portion of General Motors' balance sheet.

SOLUTION

Comparative Analysis of Consolidated Balance Sheets
General Motors Corporation (in millions of dollars)

Current Assets – December 31	2002	2001	Increase Amount	Percent
Cash and Short-Term Investments	$23,623	$19,345	$4278	22.1%
Accounts Receivable and Other	$33,495	$27,841	$5654	20.3%
Total Current Assets	$57,118	$47,186	$9932	21.0%

Total current assets grew 21.0% from 2001 to 2002. This growth rate is much larger than the 5.4% growth in revenue for General Motors during the same period. Thus, the managers made some structural changes to the corporation during 2002 including the substantial increase of long-term debt.

Objective 3 **Find financial ratios.** After preparing a balance sheet for a business firm, an accountant often calculates several different **financial ratios** for the firm. These ratios can be compared with financial ratios for other firms in the same industry or of the same size. A ratio *that is far out of line compared with other similar firms or with industry averages* might well indicate coming financial difficulties.

The first two ratios we discuss are designed to measure the **liquidity** of a firm. Liquidity refers to a firm's ability to raise cash quickly without being forced to sell assets at a big loss.

Find the **current ratio** by dividing current assets by current liabilities. This ratio is read as current assets to current liabilities.

$$\text{Current ratio} = \frac{\text{Current assets}}{\text{Current liabilities}}$$

QUICK TIP   The current ratio is also known as the **banker's ratio**.

EXAMPLE 3

Finding the Current Ratio

The December 31, 2002 balance sheet shows current assets of $57,118 and current liabilities of $71,033. Find the current ratio.

SOLUTION

$$\text{Current ratio} = \$57{,}118 \div \$71{,}033 = .8 \text{ (rounded)}$$

The ratio is often expressed as .8 to 1 or as .8 : 1.

Lending institutions look at the current ratio before making loans. A common rule of thumb, not necessarily applicable to all businesses, is that the current ratio should be at least 2:1. A firm with a current ratio much lower than 2 may have an increased risk of financial difficulties and may have difficulty borrowing. General Motors has a current ratio far below this value potentially signaling some problems with short term liquidity should unexpected events happen requiring cash quickly. However, the current ratio is only one indicator, so you must look beyond the current ratio.

One disadvantage of the current ratio is that inventory is included in current assets. In a period of financial difficulty, a firm might have trouble disposing of its inventory at a reasonable price. Some accountants feel that the **acid test** of a firm's financial health is to consider only **liquid assets**: assets that either are cash or can be converted to cash quickly, such as securities and accounts and notes receivable.

The **acid-test ratio**, also called the **quick ratio**, is defined as follows.

$$\text{Acid-test ratio} = \frac{\text{Liquid assets}}{\text{Current liabilities}}$$

QUICK TIP As a general rule based on past experience, the acid-test ratio should be at least 1 to 1, so that liquid assets are at least enough to cover current liabilities.

EXAMPLE 4

Finding the Acid-Test Ratio

Find the 2002 acid-test ratio for General Motors given that year-end inventory was $15,272.

SOLUTION

$$\text{Liquid assets} = \text{Current assets} - \text{Inventory}$$
$$= \$57{,}118 - \$15{,}272 = \$41{,}846$$
$$\text{Current liabilities} = \$71{,}033$$
$$\text{Acid-test ratio} = \$41{,}846 \div \$71{,}033 = .59$$

This ratio is well below 1:1 which indicates that General Motors was financially struggling during 2002.

A company with a large amount of capital invested should have a higher net income than a company with a small amount invested. To check on this, accountants often find the **ratio of net income after taxes to average owners' equity**. The **average owners' equity** is found by adding the owners' equity at the beginning and end of the year and dividing by 2.

$$\frac{\text{Average owners'}}{\text{equity}} = \frac{\text{Owners' equity at beginning} + \text{Owners' equity at end}}{2}$$

Then the ratio of net income after taxes to average owners' equity is found as follows.

$$\frac{\text{Ratio of net income after taxes}}{\text{to average owners' equity}} = \frac{\text{Net income after taxes}}{\text{Average owners' equity}}$$

EXAMPLE **5**

Finding the Return on
Average Equity

Find the 2002 ratio of net income after taxes to average owners' equity for General Motors.

SOLUTION

The stockholders' equity fell from $19,707 on December 31, 2001 to $6814 on December 31, 2002.

$$\text{Average owners'} \atop \text{equity} = \frac{\$19,707 + \$6814}{2} = \mathbf{\$13,260.5} \text{ (in millions)}$$

Use the net income after taxes in 2002 from the income statement in Section 14.1 on page 609 to find the following.

$$\text{Ratio of net income after taxes} \atop \text{to average owners' equity} = \frac{\$1736}{\mathbf{\$13,260.5}} = \mathbf{13.1\%} \text{ (rounded)}$$

The ratio of net income after taxes to average owners' equity should be significantly higher than the interest rate paid on savings accounts or even government bonds. Otherwise, the capital represented by these assets should be deposited in a bank account or in government bonds. After all, government bonds have *less risk* than an investment in a company. Generally, investments with a higher risk must give more reward to investors to encourage them to invest. Thus, increased risk should result in a higher return to the investor than that of a risk-free savings account. The ratio for GM exceeds the rates paid on government bonds, but it is not as high as the ratios of many growth companies, which often have returns higher than 15%.

QUICK TIP   The ratio of net income after taxes to average owners' equity is the only ratio of the three we have looked at that requires you to look at both the income statement and the balance sheet.

The financial ratios of this section can be summarized as follows.

Summary of Ratios

$$\text{Current ratio} = \frac{\text{Current assets}}{\text{Current liabilities}} \qquad \text{Acid-test ratio} = \frac{\text{Liquid assets}}{\text{Current liabilities}}$$

$$\text{Ratio of net income after taxes} \atop \text{to average owners' equity} = \frac{\text{Net income after taxes}}{\text{Average owners' equity}}$$

QUICK TIP   All of these ratios become more meaningful when a particular company is compared against other companies in the same industry (e.g., General Motors versus Ford).

14.4 | Exercises

1. **YACHT CONSTRUCTION**   Complete this balance sheet using vertical analysis. Round to the nearest tenth of a percent. (See Example 1.)

Comparative Balance Sheet for Pleasure Yacht, Inc. (in thousands of dollars)

	Amount This Year	Percent This Year	Amount Last Year	Percent Last Year
ASSETS				
Current Assets				
Cash	$52,000	___	$42,000	___
Notes Receivable	$8,000	___	$6,000	___
Accounts Receivable	$148,000	___	$120,000	___
Inventory	$153,000	___	$120,000	___
Total Current Assets	___	___	___	___
Plant Assets				
Land	$10,000	___	$8,000	___
Buildings	$14,000	___	$11,000	___
Fixtures	$15,000	___	$13,000	___
Total Plant Assets	___	___	___	___
TOTAL ASSETS	___	100%	___	100%
LIABILITIES				
Current Liabilities				
Accounts Payable	$3,000	___	$4,000	___
Notes Payable	$201,000	___	$152,000	___
Total Current Liabilities	___	___	___	___
Long-Term Liabilities				
Mortgages Payable	$20,000	___	$16,000	___
Long-Term Notes Payable	$58,000	___	$42,000	___
Total Long-Term Liabilities	___	___	___	___
TOTAL LIABILITIES	___	___	___	___
Owners' Equity	$118,000	___	$106,000	___
TOTAL LIABILITIES AND OWNERS' EQUITY	___	___	___	___

2. **POOLS AND SPAS**   Complete the following horizontal analysis for a portion of the balance sheet for Peerless Pools & Spas. Note that figures are shown in thousands of dollars. (Round to tenths of a percent). (See Example 1.)

Peerless Pools & Spas

	This Year	Last Year	Increase or (Decrease) Amount	Percent
ASSETS				
Current Assets				
Cash	$52,000	$42,000	_____	_____
Notes Receivable	$8,000	$6,000	_____	_____
Accounts Receivable	$148,000	$120,000	_____	_____
Inventory	$153,000	$120,000	_____	_____
Total Current Assets	$361,000	$288,000	_____	_____
Plant Assets				
Land	$10,000	$8,000	_____	_____
Buildings	$14,000	$11,000	_____	_____
Fixtures	$15,000	$13,000	_____	_____
Total Plant Assets	$39,000	$32,000	_____	_____
TOTAL ASSETS	$400,000	$320,000	_____	_____

In Exercises 3–6, find (a) the current ratio and (b) the acid-test ratio. Round each ratio to the nearest hundredth. (c) Do the ratios suggest that the company is financially healthy, according to the guidelines given in the text? (See Examples 3 and 4.)

3. Peerless Pools & Spas has the balance sheet given above and current liabilities of $204,000. Find the ratios for this year.

(a) _____
(b) _____
(c) _____

4. **PLUMBING COMPANY**   Tekla Plumbing has current liabilities of $356,800, cash of $32,800, notes and accounts receivable of $248,500, and an inventory valued at $82,400.

(a) _____
(b) _____
(c) _____

5. **CADILLAC DEALER**   Wagner Cadillac has current assets of $2,210,350, current liabilities of $1,232,500, total cash of $480,500, notes and accounts receivable of $279,050, and an inventory of $1,450,800.

(a) _____
(b) _____
(c) _____

_____

6. **OXYGEN SUPPLY**   BlueTex Oxygen Supply has current assets of $2,234,000, current liabilities of $840,000, total cash of $339,000, notes and accounts receivable of $1,215,000, and an inventory of $680,000.

(a) _____
(b) _____
(c) _____

A portion of a comparative balance sheet is shown next. First complete the chart, and then find the current ratio and the acid-test ratio for the indicated year. Round each ratio to the nearest hundredth.

	Amount This Year	Percent This Year	Amount Last Year	Percent Last Year
Current Assets				
Cash	$12,000	_____	$15,000	_____
Notes Receivable	$4,000	_____	$6,000	_____
Accounts Receivable	$22,000	_____	$18,000	_____
Inventory	$26,000	_____	$24,000	_____
Total Current Assets	$64,000	80%	$63,000	84%
Total Plant Assets	$16,000	_____	$12,000	_____
TOTAL ASSETS	_____	100.0%	_____	100.0%
Total Current Liabilities	$30,000	_____	$25,000	_____

7. This year

8. Last year

Find the ratio of net income after taxes to average owners' equity for the following. (See Example 5.) Round to the nearest tenth of a percent.

9. INTERNATIONAL AIRLINE   Stockholders' equity in TNA Airline, a small international airline that uses small airplanes to move freight to and from Mexico, is $845,000 at the beginning of the year and $928,500 at the end of the year. Net income after taxes for the year was $54,400.

9. _____

10. TRANSMISSION REPAIR   Owners' equity at Amgen Transmission was $372,600 at the beginning of the year and $402,100 at the end of the year. Net income after taxes for the year was $55,003.

10. _____

Calculate the current ratio and acid-test ratio for the following companies. Are the companies healthy, based on the guidelines given in the text?

11. Akron Hardware: Current assets: $268,700
                          Current liabilities: $294,200
                          Liquid assets: $109,900

11. _____

12. Lupe's Hair Styles: Current assets: $18,250
                              Current liabilities: $2,400
                              Liquid assets: $8,250

12. _____

13. Explain why the acid-test ratio is a better measure of the financial health of a firm than the current ratio. (See Objective 3.)

14. Explain why increased risk requires a higher return on investment. (See Objective 3.)

CONCEPTS

14.1 Finding the gross profit and net income

1. Find the net sales.
2. Determine the cost of goods sold.
3. Find gross profit from the formula.

Gross profit = Net sales − Cost of goods sold

4. Find the operating expenses.
5. Find the net income from the formula.

Net income = Gross profit − Operating expenses

14.2 Find the percent of net sales of individual items

1. Determine net sales using the formula.

Net sales = Gross sales − Returns

2. Use the formula

$$\text{Percent of net sales} = \frac{\text{Particular item}}{\text{Net sales}}$$

for each item.

EXAMPLES

Candy the Way You Like It had a cost of goods sold of $123,500, operating expenses of $48,950, and gross sales of $206,100. Find the gross profit and net income before taxes.

Gross profit = Gross sales − Returns − Cost of goods sold
$$= \$206,100 \quad - 0 \quad - \$123,500$$
$$= \$82,600$$

Net income before taxes = **Gross profit − Operating expenses**
$$= \$82,600 - \$48,950$$
$$= \$33,650$$

Bill's Appliances lists the following information.

Gross sales = $340,000	Salaries and wages = $19,000
Returns = $15,000	Rent = $8000
Cost of goods sold = $210,000	Advertising = $12,000

Express each item as a percent of net sales. (Round to the nearest tenth of a percent.)

Net sales = $340,000 − $15,000 = $325,000
Gross profit = $325,000 − $210,000 = $115,000
Total expenses = $19,000 + $8000 + $12,000 = $39,000
Net income = $115,000 − $39,000 = $76,000
Find all the desired percents to the nearest tenth of a percent by dividing each item by net sales.

$$\text{Percent gross sales} = \frac{\$340,000}{\$325,000} = 104.6\%$$

$$\text{Percent return} = \frac{\$15,000}{\$325,000} = 4.6\%$$

$$\text{Percent cost of goods sold} = \frac{\$210,000}{\$325,000} = 64.6\%$$

$$\text{Percent gross profit} = \frac{\$115,000}{\$325,000} = 35.4\%$$

$$\text{Percent expenses} = \frac{\$39,000}{\$325,000} = 12\%$$

$$\text{Percent salaries and wages} = \frac{\$19,000}{\$325,000} = 5.8\%$$

$$\text{Percent net income} = \frac{\$76,000}{\$325,000} = 23.4\%$$

$$\text{Percent rent} = \frac{\$8,000}{\$325,000} = 2.5\%$$

$$\text{Percent advertising} = \frac{\$12,000}{\$325,000} = 3.7\%$$

CONCEPTS	EXAMPLES

14.2 Comparing income statements with published charts

In one chart, list the percent of items from a published chart and the particular company.

In the previous example, prepare a vertical analysis of Bill's Appliances.

Bill's Appliances

	Cost of Goods	Gross Profit	Total Expenses	Net Income	Wages	Rent	Advertising
Bill's Appliances	64.6%	35.4%	12%	23.4%	5.8%	2.5%	3.7%
Published Data	66.9%	33.1%	26%	7.2%	11.9%	2.4%	2.5%

14.2 Preparing a horizontal analysis chart

1. List last year's and this year's values for each item.
2. Calculate the amount of the increase or decrease of each item.
3. Calculate the percent increase or decrease by dividing the change by last year's amount.

The results of a horizontal analysis of the portion of a business is given. Calculate the percent increases or decreases in each item.

Ocean Salvage

	This Year	Last Year	Increase or (Decrease) Amount	Percent
Gross Sales	$735,000	$700,000	$35,000	5%
Returns	$5,000	$10,000	($5,000)	(50%)
Net Sales	$730,000	$690,000	$40,000	5.8%
Cost of Goods Sold	$530,000	$540,000	($10,000)	(1.9%)
Gross Profit	$200,000	$150,000	$50,000	33.3%

14.3 Constructing a balance sheet

List all of the current assets, other assets, current liabilities, and other liabilities on one page.

Subtract total liabilities from total assets to find stockholders' equity.

Techno's Cell Phone and Pager has cash of $28,300, accounts receivable of $49,250, and inventories of $4900. Other assets consist of equipment and a truck with total value of $24,300. Accounts payable are $9300 and loans total $12,200. There is no long-term debt. Other liabilities amount to $12,400.

Techno's Cell Phone and Pager Balance Sheet

Current Assets:		Current Liabilities:	
Cash and Equivalents	$28,300	Accounts Payable	$9,300
Accounts Receivable	$49,250	Loans and Notes Payable	$12,200
Inventories	$4,900	*Total Current Liabilities*	$21,500
Total Current Assets	$82,450	Other Liabilities:	$12,400
Other Assets:		*Total Liabilities*	$33,900
Equipment and Truck	$24,300		
Total Assets	$106,750	*Stockholders' Equity*	$72,850
		Total Liabilities and Equity	$106,750

14.4 Determining the value of the current ratio

1. Determine the current assets.
2. Find the current liabilities.
3. Divide current assets by current liabilities.

The Circle Tour Agency has $250,000 in current assets and $110,000 in current liabilities. Find the current ratio.

$$\text{Current ratio} = \frac{\text{Current assets}}{\text{Current liabilities}}$$

$$= \frac{\$250,000}{\$110,000} = 2.27 \text{ (rounded)}$$

CONCEPTS	EXAMPLES

14.4   Finding the value of the acid-test ratio

1. Determine the liquid assets.
2. Find the current liabilities.
3. Divide liquid assets by liabilities.

If the Circle Tour Agency has $125,000 in liquid assets, find the acid-test ratio.

$$\text{Acid-test ratio} = \frac{\text{Liquid assets}}{\text{Current liabilities}}$$

$$= \frac{\$125,000}{\$110,000} = 1.14 \text{ (rounded)}$$

14.4   Determine the ratio of net income after taxes to the average owners' equity

1. Find the net income after taxes.
2. Determine the average owners' equity for the year using the formula

$$\text{Average owners' equity} = \frac{\left(\begin{array}{c}\text{Owners' equity} \\ \text{at beginning}\end{array} + \begin{array}{c}\text{Owners' equity} \\ \text{at end}\end{array}\right)}{2}$$

3. Divide the net income by the average owners' equity.

At the beginning of the year, the Circle Tour Agency had an owners' equity of $140,000. At the end of the year, the owners' equity was $180,000. The net income after taxes for the agency was $25,000. Find the ratio of net income after taxes to average owners' equity.

Net income after taxes = $25,000

Average owners' equity

$$= \frac{\$140,000 + \$180,000}{2} = \$160,000$$

Ratio of net income after taxes to average owners' equity

$$= \frac{\$25,000}{\$160,000} = 15.6\% \text{ (rounded)}$$

Chapter 14 | Summary Exercise

Owning Your Own Business

Tom Walker wants to expand his bicycle shop and has gone to a bank for a loan. The commercial loan officer asks Walker for his most recent income statement and balance sheets based on the following data.

Gross Sales	$212,000	Salaries and Wages	$37,000
Returns	$12,500	Rent	$12,000
Inventory on January 1	$44,000	Advertising	$2,000
Cost of Goods Purchased	$75,000	Utilities	$3,000
Freight	$8,000	Taxes on Inventory, Payroll	$7,000
Inventory on December 31	$26,000	Miscellaneous Expenses	$4,500
		Income taxes	$4,320

(a) Prepare an income statement.

Walker's Bicycle Shop
Income Statement
Year Ending December 31

Gross Sales		_____
Returns		_____
Net Sales		_____
Inventory, January 1	_____	
Cost of Goods Purchased	_____	
Freight	_____	
Total Cost of Goods Purchased		_____
Total of Goods Available for Sale		_____
Inventory, December 31		_____
Cost of Goods Sold		_____
Gross Profit		_____
Expenses		
Salaries and Wages	_____	
Rent	_____	
Advertising	_____	
Utilities	_____	
Taxes on Inventory, Payroll	_____	
Miscellaneous Expenses	_____	
Total Expenses		_____
Net Income Before Taxes		_____
Income Taxes		_____
NET INCOME AFTER TAXES		======

(b) Express the following items as a percent of net sales. Round to nearest tenth of a percent.

Gross Sales	_____	Salaries and Wages	_____
Returns	_____	Rent	_____
Cost of Goods Sold	_____	Utilities	_____

(c) After the year is completed, Walker has $62,000 in cash, $2500 in notes receivable, $8200 in accounts receivable, and $26,000 in inventory. He has land worth $7600, buildings valued at $28,000, and fixtures worth $13,500. He also has $4500 in notes payable and $27,000 in accounts payable, mortgages for $15,000, long-term notes payable of $8000, and owner's equity of $93,300. Prepare a balance sheet.

Walker's Bicycle Shop
Balance Sheet
December 31

ASSETS

Current Assets
  Cash   _____
  Notes Receivable   _____
  Accounts Receivable   _____
  Inventory   _____
      Total Current Assets   _____
Plant Assets
  Land   _____
  Buildings   _____
  Fixtures   _____
      Total Plant Assets   _____
Total Assets   _____

LIABILITIES

Current Liabilities
  Notes Payable   _____
  Accounts Payable   _____
      Total Current Liabilities   _____
Long-Term Liabilities
  Mortgages Payable   _____
  Long-Term Notes Payable   _____
      Total Long-Term Liabilities   _____
Total Liabilities   _____

OWNERS' EQUITY

Owners' Equity   _____
TOTAL LIABILITIES AND OWNERS' EQUITY   _____

(d) Find the current ratio and the acid-test ratio for Walker's business. (Round to nearest hundredths.) _____

(e) If you were the commercial loan officer, would you approve Walker's requested loan? Why or why not?

INVESTIGATE

Publicly held companies must publish their financial statements and make them available to anyone who wishes to look at them. Choose a publicly held company that you are familiar with and obtain their financial statements by either contacting the main offices of the company or by using the World Wide Web. Calculate the financial ratios introduced in the last section of this chapter for the company you choose (current ratio, acid-test ratio, average owners' equity, and ratio of net income after taxes to average owners' equity).

'NET ASSETS

General Motors

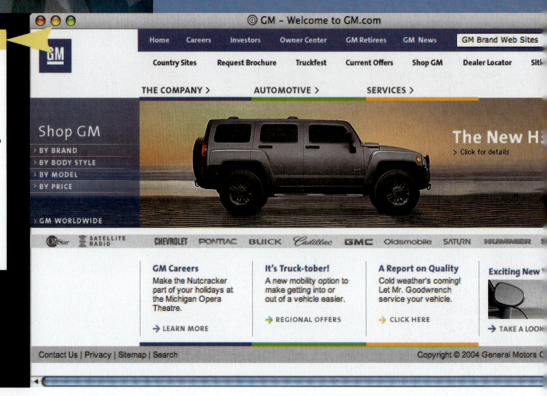

www.gm.com/

- 1908: Founded

- 1953: Introduced the Corvette

- 1989: Purchased 50% of Saab in Sweden

- 2001: GMC launched the Hummer H2 SUV

- 2004: 326,000 employees

General Motors has been building and selling automobiles for more than 100 years. The company has manufacturing operations in 32 countries and sells vehicles in more than 190 countries. GM has global partnerships with companies in Italy, Japan, South Korea, Germany, Russia, and France. In 2002, GM sold about 8.5 million cars and trucks worldwide. Wholly owned subsidiary GMAC Financial Services makes home and automobile loans to individuals and also lends money to companies.

GM set industry sales records in the United States in 2002 when it sold more than 2.7 million trucks and more than 1.2 million SUVs. Company managers must carefully watch the financial statements from the many different subsidiaries that GM owns.

1. General Motors' assets on December 31, 2002 were about $4.49 billion. Estimate the company's assets at the end of 2003 and 2004 assuming they grow at 5% per year. Round to the nearest hundredth of a billion.

2. General Motors recently entered into an e-business partnership with Sony. Here are some data on Sony (in thousands of dollars): current assets $29,180, total assets $69,475, current liabilities $20,211 and total liabilities $50,544. Find the current ratio for Sony.

3. Use the World Wide Web to find a graph showing GM's stock price for the past one year.

Chapter 14 | Test

To help you review, the numbers in brackets show the section in which the topic was discussed.

1. Benni's Catfish Distribution, Inc., had gross sales of $756,300 with returns of $285. The inventory on January 1 was $92,370, and the cost of goods purchased during the year was $465,920. Freight costs during the year were $1205. Total inventory on December 31 was $82,350. Salaries and wages totaled $84,900, advertising was $2800, rent was $42,500, utilities were $18,950, taxes on inventory and payroll were $4500, and miscellaneous expenses totaled $18,400. Income taxes were $25,450. Complete the following income statement.   **[14.1]**

Benni's Catfish Distribution, Inc.
Income Statement Year
Ending December 31

Gross Sales		_____
Returns		_____
Net Sales		_____
Inventory, January 1	_____	
Cost of Goods Purchased	_____	
Freight	_____	
Total Cost of Goods Purchased	_____	
Total of Goods Available for Sale	_____	
Inventory, December 31	_____	
Cost of Goods Sold		_____
Gross Profit		_____
Expenses		
Salaries and Wages	_____	
Rent	_____	
Advertising	_____	
Utilities	_____	
Taxes on Inventory and Payroll	_____	
Miscellaneous Expenses	_____	
Total Expenses		_____
Net Income Before Taxes		_____
Income Taxes		_____
Net Income After Taxes		══════

2. Complete a horizontal analysis for the following portion of an income statement. Round to the nearest tenth of a percent.   **[14.2]**

Marge's Television Shoppe
Comparative Income Statement (Portion)

	This Year	Last Year	Increase or (Decrease) Amount	Percent
Net Sales	$95,000	$60,000	_____	_____
Cost of Goods Sold	$63,000	$40,000	_____	_____
Gross Profit	$16,000	$12,000	_____	_____

3. Complete the following chart for Franklin's Service Station. Express each item as a percent of net sales, and then write in the appropriate average percent from the chart on page 614. Round to the nearest tenth of a percent. [14.2]

Franklin's Service Station

	Amount	Percent	Average Percent
Net Sales	$400,000	100%	100%
Cost of Goods Sold	$275,000	_____	_____
Gross Profit	$125,000	_____	_____
Net Income	$37,500	_____	_____
Wages	$37,500	_____	_____
Rent	$8,000	_____	_____
Total Expenses	$87,500	_____	_____

Find (a) the current ratio and (b) the acid-test ratio for each firm. (Round to the nearest hundredth.) [14.4]

4. Bonfry Bridge Construction:

Current assets:	$2,482,500
Current liabilities:	$1,800,200
Cash:	$850,000
Notes and Accounts Receivable:	$680,100
Inventory:	$952,400

(a) _____
(b) _____

5. Walter's Rifle Shop:

Current assets:	$154,000
Current liabilities:	$146,500
Cash:	$22,000
Note and accounts receivable:	$32,500
Inventory:	$99,500

(a) _____
(b) _____

Find the ratio of net income after taxes to average owners' equity for each of the following firms. (Round to the nearest tenth of a percent.) [14.4]

6. Talisman Imports:

Net income after taxes:	$148,200
Owners' equity	
beginning of year:	$472,600
end of year:	$514,980

6. _____

7. Baker Drilling Co.:

Net income after taxes:	$8,465,000
Owners' equity	
beginning of year:	$28,346,000
end of year:	$36,450,000

7. _____

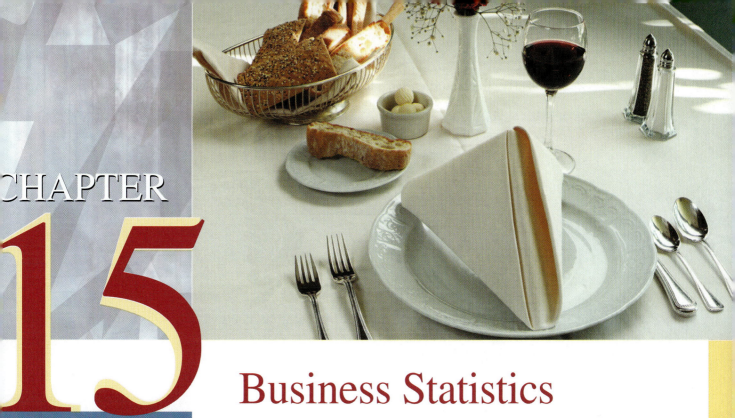

CHAPTER

15

Business Statistics

GINA HARDEN BEGAN WORKING IN A RESTAURANT AT AGE 16 AND

has managed one for several years. As an exchange student in France during

CASE *in* POINT college, she loved the sidewalk cafes in Paris and

decided that one day she would open her own. Using

money she had saved and additional funds borrowed from her brother, Gina

opened Gina's Bistro.

The word **statistics** comes from words that mean *state numbers*, or data gathered by the government, such as number of births and deaths. Today, the word *statistics* is used in a much broader sense to include data from business, economics, and many other fields. Statistics is a powerful and commonly used tool in business. A large company such as General Motors, for example, uses statistics for forecasting demand for their products.

> **QUICK TIP**   Statistics is commonly used in business and is a useful subject for individuals in most career paths. We encourage you to take a class in statistics at your college.

15.1 | Frequency Distributions and Graphs

Objectives

1. Construct and analyze a frequency distribution.
2. Make a bar graph.
3. Make a line graph.
4. Draw a circle graph.

CASE *in* POINT

Gina's Bistro uses only the freshest ingredients in its gourmet sandwiches and is already building a reputation for delicious soups and salads. The cafe has only been open for six months, and Harden has to watch sales very carefully since most restaurants fail within their first year.

Objective 1 Construct and analyze a frequency distribution. It can be difficult to interpret or find patterns in a large group of numbers. One way of analyzing the numbers is to organize them into a table that shows the frequency of occurrence of the various numbers. This type of table is called a **frequency distribution**.

EXAMPLE 1

Construction of a Frequency Distribution

Gina Harden is analyzing sales activity over the past 24 weeks at Gina's Bistro. The weekly sales data shown next are to the nearest thousand dollars. Read down the columns, beginning with the left column, for successive weeks of the year.

$3.9	$4.0	$4.3	$4.6	$5.1	$5.6
$3.2	$4.2	$4.8	$4.9	$4.8	$4.8
$3.3	$4.1	$4.1	$5.2	$5.0	$5.3
$3.5	$3.9	$4.8	$5.0	$5.3	$5.3

Construct a table that shows each value of sales. Then go through the data and place a tally mark (|) next to each corresponding value, thereby creating a frequency distribution table.

SOLUTION

Sales (thousands)	Tally	Frequency	Sales (thousands)	Tally	Frequency	Sales (thousands)	Tally	Frequency
$3.2			1	$4.2			1	$5.1			1
$3.3			1	$4.3			1	$5.2			1
$3.5			1	$4.6			1	$5.3	|||	3
$3.9	||	2	$4.8	||||	4	$5.6			1
$4.0			1	$4.9			1			
$4.1	||	2	$5.0	||	2			

This frequency distribution shows that the most common weekly sales amount was $4800, although there were three weeks with sales of $5300.

The frequency distribution given in the previous example contains a great deal of information, perhaps more than is needed. It can be simplified by combining weekly sales into groups, forming the following grouped data.

Grouped Data

Sales (thousands)	Frequency (number of weeks)
$3.1–$3.5	3
$3.6–$4.0	3
$4.1–$4.5	4
$4.6–$5.0	8
$5.1–$5.5	5
$5.6–$6.0	1

QUICK TIP   The number of groups in the left column of the preceding table is arbitrary and usually varies between 5 and 15.

EXAMPLE 2

Analyzing a Frequency Distribution

Based on the data from Gina's Bistro, answer the following questions.

(a) Harden can take no salary, and the business still loses money when sales are less than or equal to $4000 per week. During how many weeks did this occur?

(b) Harden can take a small salary out of the company once sales go above $5000 per week. During how many weeks did this occur?

SOLUTION

(a) The first two classes in the grouped data represent weeks in which sales were equal to or less than $4000. Thus, Harden took no salary and the restaurant lost money for 6 weeks.

(b) The last two classes in the grouped data table are the number of weeks during which sales were above $5000, or 6 weeks. Therefore, Harden took a small salary for 6 weeks.

Objective 2   Make a bar graph.   The next step in analyzing these data is to use them to make a **graph**. A graph is a visual presentation of numerical data. One of the most common graphs is a **bar graph**, where the height of a bar represents the frequency of a particular value. A bar graph for the sales data follows.

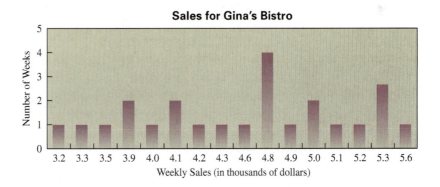

The information from the grouped data is shown in the following bar graph. This graph shows that weekly sales of between $4600 and $5000 were the most common. Notice that this graph *does not* show any trend that may be occurring over time.

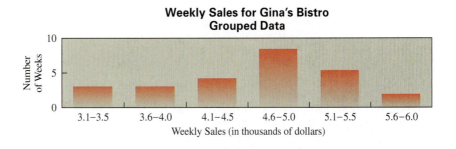

Objective **3** **Make a line graph.** Bar graphs show which numbers occurred and how many times, but do not necessarily show the order in which the numbers occurred. To discover any trends that may have developed, draw a **line graph**.

EXAMPLE **3**

Draw a Line Graph

Show the progression of sales at Gina's Bistro through the year using a line graph. Do this by totaling the first 4 weeks (the first column) of data in Example 1 for the first data point. Similarly, total the second 4 weeks (second column) of data for the next data point, and so on.

SOLUTION

The total for the first four weeks is $3.9 + $3.2 + $3.3 + $3.5 = $13.9 or $13,900 in sales for the first four weeks. The total for the second four weeks is $4.0 + $4.2 + $4.1 + $3.9 = $16.2, or $16,200 in sales. The six data points of the graph are $13.9, $16.2, $18, $19.7, $20.2, and $21 in thousands of dollars.

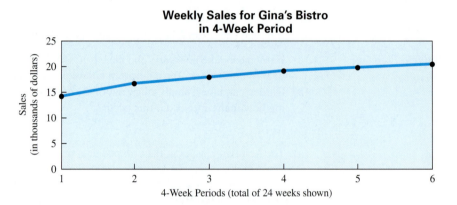

It is apparent that weekly sales are growing based on the line graph. Gina Harden is excited about this trend. She is determined to continue improving the business since her livelihood depends on the restaurant. She plans to work very hard over the next few months to further increase sales.

One advantage of line graphs is that two or more sets of data can be shown on the same graph. For example, suppose the manager of a local Radio Shack store wants to compare total sales, profits, and overhead using the following historical data.

Year	Total Sales	Overhead	Profit
2003	$740,000	$205,000	$83,000
2004	$860,000	$251,000	$102,000
2005	$810,000	$247,000	$21,000
2006	$1,040,000	$302,000	$146,000

Separate lines can be made on a line graph for each category so that necessary comparisons can be made. A graph such as this is called a **comparative line graph**.

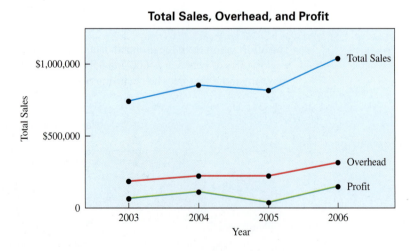

Objective 4 **Draw a circle graph.** Suppose a sales manager for Genome Pharmaceutical Company makes a record of the expenses involved in keeping a sales force on the road. After finding the total expenses, she could convert each expense into a percent of the total, with the following results. Notice that the percents add to 100%.

Sales Force Expenses

Item	Percent of Total
Travel	30%
Lodging	25%
Food	15%
Entertainment	10%
Sales meetings	10%
Other	10%

The sales manager can show these percents by using a **circle graph**, sometimes called a **pie chart**. A circle has 360 degrees $(360°)$. The 360° represents the total expenses, or 100%. Since entertainment is **10%** of the total expenses, she used

$$360° \times 10\% = 360° \times .10 = 36°$$

to represent her entertainment expense. Since lodging is **25%** of the total expenses, she used

$$360° \times 25\% = 360° \times .25 = 90°$$

to represent lodging. After she found the degrees that represent each of her expenses, she drew the circle graph shown here.

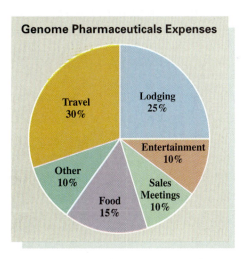

Circle graphs can be used to show comparisons *when one item is very small compared to another*. In the circle graph shown here, an item representing 1% of the total could be drawn as a very small but noticeable slice. Such a small item would hardly show up in a line graph.

EXAMPLE **4**	Based on the preceding circle graph of expenses, answer the following questions.

Interpreting a Circle Graph

(a) What percent of expenses was spent on travel and entertainment?
(b) What percent of expenses was spent on food and lodging?

SOLUTION

(a) Travel is   30%
      Entertainment is   + 10%
      Total spent   **40%**

(b) Food is   15%
      Lodging is   + 25%
      Total spent   **40%**

Graphs are great ways to communicate. The line graph on the left shows the rapid growth of grocery sales on the Internet. The bar chart on the right demonstrates why 74% of Americans with young children do not have a will. Courts are required to decide who raises the young child of a parent who has died without a will.

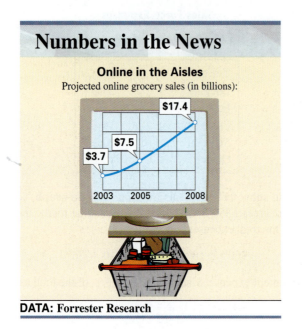

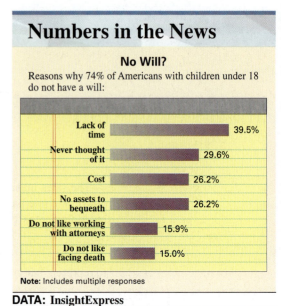

The damage due to computer viruses is shown below.

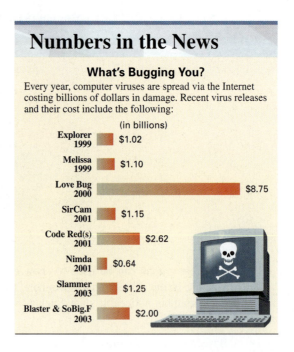

15.1 | Exercises

FOR EXTRA HELP

 MyMathLab

 InterActMath.com

 MathXL

 MathXL
Tutorials on CD

 Addison-Wesley
Math Tutor Center

DVT/Videotape

The **Quick Start** *exercises in each section contain solutions to help you get started.*

Answer Exercises 1–3 using the graphic on population growth and Exercises 4–6, using the line graph on people who live together.

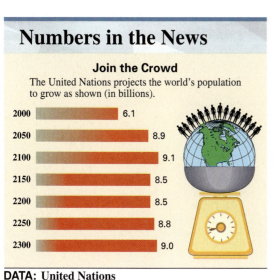

Numbers in the News

Join the Crowd

The United Nations projects the world's population to grow as shown (in billions).

Year	Billions
2000	6.1
2050	8.9
2100	9.1
2150	8.5
2200	8.5
2250	8.8
2300	9.0

DATA: United Nations

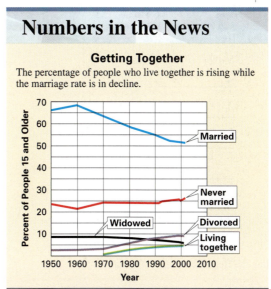

Numbers in the News

Getting Together

The percentage of people who live together is rising while the marriage rate is in decline.

DATA: U.S. Bureau of the Census

Quick Start

1. WORLD POPULATION   When is world population expected to reach its peak and what is the estimated population.

    1. 2100; 9.1 billion

2. Find the percent increase in world population from 2000 to 2050 to the nearest tenth of a percent.

    2. _____

3. In 2050, assume that 24% of the world's population are Chinese and 4% of the world's population are Americans. Estimate the number of Chinese and the number of Americans.

    3. _____

4. Approximately what percent of the U.S. population was married in 1950 and in 2000?

    4. _____

5. What has happened with the percent of people divorced during the past 50+ years?

    5. _____

6. Estimate the percent of widowed people in 2000.

    6. _____

 indicates an exercise that is related to the Case in Point feature within the section.

The following list shows the number of college credits completed by 30 employees of the Franklin Bank.

College Credits
Franklin Bank Employees

74	133	4	127	20	30
103	27	139	118	138	121
149	132	64	141	130	76
42	50	95	56	65	104
4	140	12	88	119	64

Use these numbers to complete the following table. (See Examples 1 and 2.)

Quick Start

Number of Credits	*Number of Employees*
7. 0–24	4
8. 25–49	
9. 50–74	
10. 75–99	
11. 100–124	
12. 125–149	

13. Make a line graph using the frequencies that you found.

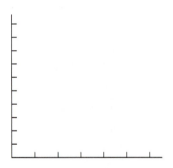

Quick Start

14. How many employees completed less than 25 credits?   **14.** 4

15. How many employees completed 50 or more credits?   **15.** _____

16. How many employees completed from 50 to 124 credits?   **16.** _____

17. How many employees completed from 0 to 49 credits?   **17.** _____

Six months' data on weekly sales (in thousands of dollars) for a chain of Wendy's restaurants follow. The numbers are in chronological order going down the columns. For example, sales for the first and fourth weeks, respectively, are $302,000 and $304,000.

302	304	318	301	330	337	335	348	339
265	275	279	283	322	349	330	325	334
315	288	299	326	325	342	328	347	

Use the numbers to complete the following table. (See Example 1.)

Quick Start

Sales (in thousands)	Frequency		Sales (in thousands)	Frequency
18. 260–269	1	**19.**	270–279	2
20. 280–289	____	**21.**	290–299	____
22. 300–309	____	**23.**	310–319	____
24. 320–329	____	**25.**	330–339	____
26. 340–349	____			

27. Make a bar graph using your answers to Exercises 18–26.

28. Make a line graph using the original numbers.

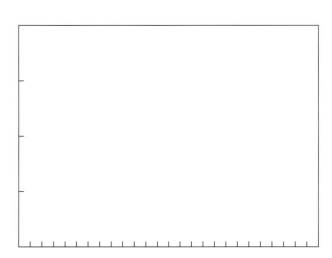

29. How many weeks did sales equal or exceed $300,000?

29. _____

30. How many weeks did sales fall below $270,000?

30. _____

The following numbers show the scores of 80 students on a marketing test.

Marketing Test Scores

79	60	74	59	55	98	61	67	83	71
71	46	63	66	69	42	75	62	71	77
78	65	87	57	78	91	82	73	94	48
87	65	62	81	63	66	65	49	45	51
69	56	84	93	63	60	68	51	73	54
50	88	76	93	48	70	39	76	95	57
63	94	82	54	89	64	77	94	72	69
51	56	67	88	81	70	81	54	66	87

Use these numbers to complete the following table. (See Example 1.)

Quick Start

Score	Frequency		Score	Frequency
31. 30–39	1	**32.**	40–49	6
33. 50–59	____	**34.**	60–69	____
35. 70–79	____	**36.**	80–89	____
37. 90–99	____			

38. Make a bar graph showing your answers to Exercises 31–37.

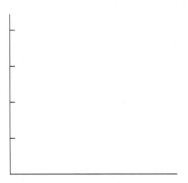

39. How many students passed the marketing test? (passing is 70)

39. _____

40. If a grade of B is achieved for a score of 80 or higher, how many students received a B or better?

40. _____

41. How many students failed the test? (scored below 70)

41. _____

42. How many students scored from 60 to 79?

42. _____

During one recent period Angela Rueben, a student, had $1400 in expenses, as shown in the following table. Find all numbers missing from the table in Exercises 43–48. (See Objective 4.)

Quick Start

Item	Dollar Amount	Percent of Total	Degrees of a Circle	Item	Dollar Amount	Percent of Total	Degrees of a Circle
43. Rent $\frac{72°}{360°} = .20 = 20\%$	$280	**20%**	72°	**44.** Clothing $\frac{\$210}{\$1400} = .15 = 15\%; .15 \times 360° = 54°$	$210	**15%**	**54°**
45. Books	$140	10%	_____	**46.** Entertainment	$210	_____	54°
47. Savings	$70	_____	_____	**48.** Other	_____	_____	36°

49. Draw a circle graph using this information. (See Objective 4.)

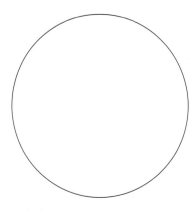

50. What percent did Rueben spend on food and rent?

50. _____

51. What percent did Rueben spend on savings and entertainment?

51. _____

Quick Start

52. FURNITURE Annual sales at Antique Furnishings are divided into five categories as follows.

Item	Annual Sales
Finishing materials	$25,000
Desks	$80,000
Bedroom sets	$120,000
Dining room sets	$100,000
Other	$75,000

Make a circle graph showing this distribution. (See Example 4.)

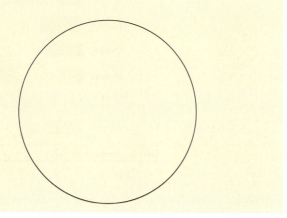

53. BOOK PUBLISHING Armstrong Publishing Company had 25% of its sales in mysteries, 10% in biographies, 15% in cookbooks, 15% in romance novels, 20% in science, and the rest in business books. Draw a circle graph with this information. (See Example 4.)

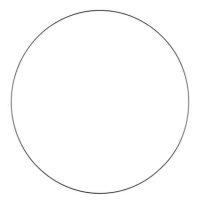

54. **FAST-FOOD** The market share of the different types of fast-food chains is shown in the figure at the right. Show this information in a circle graph. After making calculations, round degrees to the nearest whole degree and percents to the nearest whole percent.

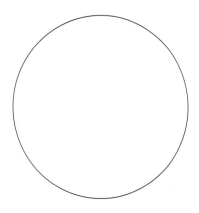

55. **MUSIC DOWNLOADS** The following graphic shows the amount spent to download individual songs from the Web along with forecasts. Show these data using a bar chart.

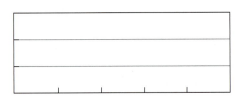

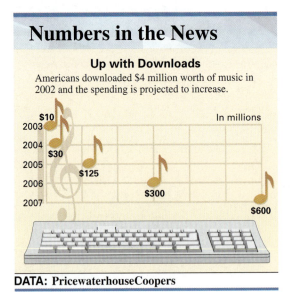

56. List the advantages of using a graph over a table when looking for trends.

57. Cut out three graphs from newspapers or magazines and tape them to your homework assignment. Be sure to explain the data in each case.

15.2 | Mean, Median, and Mode

Objectives

1. Find the mean of a list of numbers.
2. Find the weighted mean.
3. Find the median.
4. Find the mode.

Objective 1 Find the mean of a list of numbers. Businesses are often faced with the problem of analyzing a mass of raw data. Reports come in from many different branches of a company, or salespeople send in a large number of expense claims, for example. In analyzing all these data, one of the first things to look for is a **measure of central tendency**—a single number that is designed to represent the entire list of numbers. One such measure of central tendency is the **mean**, which is just the **average** of a collection of numbers or data.

$$\text{Mean} = \frac{\text{Sum of all values}}{\text{Number of values}}$$

For example, suppose milk sales at a local 7-Eleven for each of the days last week were $86, $103, $118, $117, $126, $158, and $149. The mean sales of milk (rounded to the nearest cent) follows.

$$\text{Mean} = \frac{86 + 103 + 118 + 117 + 126 + 158 + 149}{7} = \textbf{\$122.43}$$

One criticism of the mean is that its value *can be distorted* by one very large (or very small) value as shown in the next example. A better measure of central tendency in cases with one abnormally large (or small) value is shown later in this section (Objective 3).

EXAMPLE 1

Finding the Mean

Gina Harden has promised seven of her employees at Gina's Bistro that they will all work about the same number of hours. One employee commented that she worked considerably more hours than the other employees last month. The number of hours worked by each of the seven employees during the past month are given. Find the mean to the nearest hour.

75, 63, 76, 82, 70, 81, and 149

SOLUTION

Add the numbers and divide by 7, since there are 7 numbers. Check that the sum of the numbers is **596**.

distorted by 1 large number

$$\text{Mean} = \frac{\textbf{596}}{7} = 85 \left(\text{rounded}\right)$$

The mean of 85 seems a bit large since one employee worked a lot more hours than the other six employees. The mean without this value of 149 is the sum of the remaining 6 numbers divided by 6.

$$\text{Mean} = \frac{\textbf{447}}{6} = 75 \left(\text{rounded}\right)$$

This value seems more in line with the average number of hours worked. Perhaps there was an unusual reason the one employee worked 149 hours (e.g., someone else was sick).

Averages are commonly used. The graph on the next page shows that people today are waiting longer to get married than in the 1960s. The data is based on the average age at which people marry. The bar chart on the right shows average medical costs for individuals living in several different countries. Notice that the United States has, by far, the highest medical costs. Also notice that costs in neighboring Canada and Mexico are much lower than in the United States.

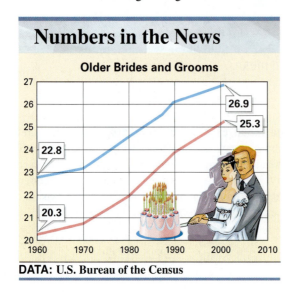

Numbers in the News

Older Brides and Grooms

DATA: U.S. Bureau of the Census

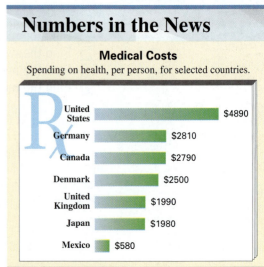

Numbers in the News

Medical Costs

Spending on health, per person, for selected countries.

United States	$4890
Germany	$2810
Canada	$2790
Denmark	$2500
United Kingdom	$1990
Japan	$1980
Mexico	$580

DATA: Organization for Economic Cooperation and Development

Objective **2** **Find the weighted mean.** Some of the items in a list might appear more than once. In this case, it is necessary to find the **weighted mean** or **weighted average**, where each value is the product of the number itself and the number of times it occurs.

EXAMPLE 2

Finding the Weighted Mean

Find the weighted mean of the numbers given in the following table.

Value	Frequency
3	4
5	2
7	1
8	5
9	3
10	2
12	1
13	2

SOLUTION

According to this table, the value 5 occurred 2 times, 8 occurred 5 times, 12 occurred 1 time, and so on. To find the mean, multiply each value by the frequency for that value. Then add the products. Finally, add the Frequency column to find the total number of values.

Value	Frequency	Product
3	4	12
5	2	10
7	1	7
8	5	40
9	3	27
10	2	20
12	1	12
13	2	26
Totals	**20**	**154**

$$\text{Weighted mean} = \frac{154}{20} = 7.7$$

A weighted average is used to find a student's grade point average, as shown by the next example.

EXAMPLE 3

*Finding the Grade
Point Average*

Mark Benson earned the following grades last semester. Find his grade point average to the nearest tenth. Assume A = 4, B = 3, C = 2, D = 1 and F = 0.

Course	Credits	Grade	Grade × Credits
Business Mathematics	3	A (=4)	4 × 3 = 12
Retailing	4	C (=2)	2 × 4 = 8
English	3	B (=3)	3 × 3 = 9
Computer Science	2	A (=4)	4 × 2 = 8
Computer Science Lab	2	D (=1)	1 × 2 = 2
Totals	14		39

SOLUTION

The grade point average for Benson is $\dfrac{39}{14} = 2.79 = 2.8$

This problem is solved using a scientific calculator as follows.

$$(\; 4 \times 3 + 2 \times 4 + 3 \times 3 + 4 \times 2 + 1 \times 2 \;) \div (\; 3 + 4 + 3 + 2 + 2 \;) = 2.8$$

(rounded)

Note: Refer to Appendix C for calculator basics.

QUICK TIP It is common to round grade point averages to the nearest tenth as we have done in the previous example.

Objective 3 Find the median. As we saw in Example 1, the mean is a poor indicator of central tendency in the presence of one very large or one very small number. This effect can be avoided by using another measure of central tendency called the **median**. The median divides a group of numbers in half—half the numbers lie at or above the median, and half lie at or below the median.

The process to find the median depends on whether there are an odd number of numbers in the list or an even number of numbers in the list. Find the median as follows.

1. List the numbers from smallest to largest as an **ordered array**.
2. Find the median.
   (a) If there are an odd number of numbers, then the median is the number in the middle. Find it by dividing the total number of numbers by 2. Use the next larger whole number in the array.
   (b) If there are an even number of numbers, then the median is the average of the two numbers in the middle. Find it by dividing the total number of numbers by 2. The median is the average of this number and the next larger number.

EXAMPLE 4

Finding the Median

Find the median of the following weights (in pounds):

(a) 30 lb, 25 lb, 28 lb, 23 lb, 24 lb

(b) 14 lb, 18 lb, 17 lb, 10 lb, 15 lb, 19 lb, 18 lb, 20 lb

SOLUTION

(a) First list the numbers from smallest to largest.

$$23, 24, 25, 28, 30 \qquad \text{ordered array}$$

There are 5 numbers, so divide 5 by 2 to get 2.5. The next larger whole number is 3, so the median is the third number, or 25. The numbers 23 and 24 are less than 25 and the numbers 28 and 30 are greater than 25.

(b) First list the numbers from smallest to largest.

$$10, 14, 15, 17, 18, 18, 19, 20 \qquad \text{ordered array}$$

There are 8 numbers, so divide 8 by 2 to get 4. The median is the mean of the numbers in the 4th and 5th positions.

$$\text{Median} = \frac{17 + 18}{2} = 17.5$$

The following graph refers to a median of 15 hours worked by teenage workers.

Numbers in the News

"Kids" on the Clock

Hours per week spent at either full- or part-time jobs by 12- to 17-year-olds. (Median is 15 hours.)

30 or more 10%
1–5 20%
21–30 14%
6–10 23%
11–20 33%

DATA: ICR's TeenEXCEL survey for Merril Lynch

Objective 4 **Find the mode.** The last important statistical measure is called the **mode**. The mode is the number that occurs most often. For example, if 10 students earned scores of

$$74, 81, 38, 74, 82, 80, 100, 92, 74, 85$$

on a business law examination, then the mode is 74. This is because more students obtained this score than any other score.

A data set in which every number occurs the same number of times is said to have *no mode*. A data set in which two different numbers occur the same number of times with each occurring more often than any other number in the data set, is said to be bimodal. The prefix "bi" means two and the root "modal" refers to modes. So **bimodal** means *two modes*.

EXAMPLE **5**

Finding the Mode

Professor Miller gave the same test to both his day and evening sections of Business Math at American River College. Find the mode of the tests given in each class. Which class has the lower mode?

(a) Day Class: 85, 92, 81, 73, 78, 80, 83, 80, 74, 69, 80, 65, 71,
65, 80, 93, 54, 78, 80, 45, 70, 76, 73, 80, 71, 68

(b) Evening Class: 68, 73, 59, 76, 79, 73, 85, 90, 73, 69, 73,
75, 93, 73, 76, 70, 73, 68, 82, 84, 77

SOLUTION

(a) The number 80 is the mode for the day class because it occurs more often than any other number.

(b) The number 73 is the mode for the evening class because it occurs more often than any other number.

The evening class has the lower mode.

QUICK TIP   It is not necessary to place the numbers in numerical order when looking for the mode, but it helps with a large set of numbers.

15.2 | Exercises

The **Quick Start** *exercises in each section contain solutions to help you get started.*

Find the mean for the following lists of numbers. Round to the nearest tenth. (See Example 1.)

Quick Start

1. Gallons of spoiled milk: 3.5, 1.1, 2.8, .8, 4.1

$$\frac{3.5 + 1.1 + 2.8 + .8 + 4.1}{5} = 2.5$$

1. **2.5** _____

2. Weeks premature: 2, 3, 1, 4, 5, 2

$$\frac{2 + 3 + 1 + 4 + 5 + 2}{6} = 2.8$$

2. **2.8** _____

3. Guests at board meetings: 40, 51, 59, 62, 68, 73, 49, 80

3. _____

4. Algebra quiz scores: 32, 26, 30, 19, 51, 46, 38, 39

4. _____

5. Number attending games: 21,900, 22,850, 24,930, 29,710, 28,340, 40,000

5. _____

6. Annual salaries: $38,500, $39,720, $42,183, $21,982, $43,250

6. _____

7. Average number of defects per batch: 10.6, 12.5, 11.7, 9.6, 10.3, 9.6, 10.9, 6.4, 2.3, 4.1

7. _____

8. Weight of dogs: 30.1, 42.8, 91.6, 51.2, 88.3, 21.9, 43.7, 51.2

8. _____

9. When is it better to use the median, rather than the mean, for a measure of central tendency? Give an example. (See Objective 3.)

10. List some situations where the mode is the best measure of central tendency to use to describe the data. (See Objective 4.)

 indicates an exercise that is related to the Case in Point feature within the section.

Find the weighted mean for the following. Round to the nearest tenth. (See Example 2.)

Quick Start

11.

Value	Frequency
9	3
12	4
18	2

12.3

$9 \times 3 = 27$
$12 \times 4 = 48$
$18 \times 2 = \underline{36}$
$\quad 9 \quad 111$

$\frac{111}{9} = 12.3$

12.

Value	Frequency
9	3
12	5
15	1
18	1

12

$9 \times 3 = 27$
$12 \times 5 = 60$
$15 \times 1 = 15$
$18 \times 1 = \underline{18}$
$\quad 10 \quad 120$

$\frac{120}{10} = 12$

13.

Value	Frequency
12	4
13	2
15	5
19	3
22	1
23	5

14.

Value	Frequency
25	1
26	2
29	5
30	4
32	3
33	5

15.

Value	Frequency
104	6
112	14
115	21
119	13
123	22
127	6
132	9

16.

Value	Frequency
243	1
247	3
251	5
255	7
263	4
271	2
279	2

Find the grade point average for the following students. Assume that A = 4, B = 3, C = 2, D = 1, and F = 0. Round to the nearest tenth. (See Example 3.)

17. Credits	Grade		18. Credits	Grade
4	B		3	A
2	A		3	B
5	C		4	B
1	F		2	C
3	B	_____	4	D	_____

Find the median for the following list of numbers. (See Example 4.)

Quick Start

19. Number of virus attacks on a network: 140, 85, 122, 114, 98

85, 98, 114, 122, 140

↑

Median

19. 114

20. Cost of new computers: $1400, $850, $975, $1045, $1190

20. _____

21. Number of books loaned: 125, 100, 114, 150, 135, 172

21. _____

22. Calories in menu items: 346, 521, 412, 515, 501, 528, 298, 621

22. _____

23. Number of students taking business math: 37, 63, 92, 26, 44, 32, 75, 50, 41

23. _____

24. Number of orders: 1072, 1068, 1093, 1042, 1056, 1005, 1009

24. _____

Find the mode or modes for each of the following lists of numbers. (See Example 5.)

Quick Start

25. Porosity of soil samples: 21%, 18%, 21%, 28%, 22%, 21%, 25%   **25.** <u>21%</u>

     If the data are listed according to how many times each number appears,
   18%, 21%, 22%, 25%, 28%
       21%
       21%
       ↑
       └──21% is the mode since 21% is listed more than any other number.

26. Number of students graduating with honors: 85, 69, 72, 69, 103, 81, 98   **26.** _____

27. Age of retirees: 80, 72, 64, 64, 72, 53, 64   **27.** _____

28. Number of pages read: 86, 84, 83, 84, 83, 86, 86   **28.** _____

29. Number of 5th-grade students: 32, 38, 32, 36, 38, 34, 35, 30, 39   **29.** _____

30. Number of people on flights from Chicago to Denver: 178, 104, 178, 150, 165, 165, 82   **30.** _____

A quality-control inspector in a plant that manufactures electric motors measured the following shaft diameters (in thousandths of an inch).

      35,   33,   32,   34,   35,   34,   35,   35,   34

Using these numbers, find each of the following. (Round to the nearest hundredth.)

31. The mean _____   **32.** The median _____

The quality-control inspector subsequently determined that he had made a mistake when he wrote 32 thousandths of an inch. Eliminate this number from the list above and find each of the following.

33. The mean _____   **34.** The median _____

35. If you want to avoid a single extreme value having a large effect on the average, would you use the mean or the median? Explain your answer.

36. Does an employer look at the mean, median, or mode grade on a college transcript when considering hiring a new employee? Which do you think the employer should look at? Explain.

Chapter 15 | Quick Review

CHAPTER TERMS *Review the following terms to test your understanding of the chapter. For each term you do not know, refer to the page number found next to that term.*

average **[p. 653]**
bar graph **[p. 643]**
bimodal **[p. 656]**
circle graph **[p. 645]**
comparative line graph **[p. 644]**

frequency distribution **[p. 642]**
graph **[p. 643]**
line graph **[p. 644]**
mean **[p. 653]**

measure of central tendency **[p. 653]**
median **[p. 655]**
mode **[p. 656]**
ordered array **[p. 655]**

pie chart **[p. 645]**
statistics **[p. 642]**
weighted average **[p. 654]**
weighted mean **[p. 654]**

CONCEPTS	EXAMPLES

15.1 Constructing a frequency distribution from raw data

1. Construct a table listing each value and the number of times this value occurs.
2. Combine the pieces of data into groups.

Construct a frequency distribution for weekly sales, in thousands, at a small concrete plant.

$22, $20, $22, $25, $18, $19, $22, $24, $24, $29, $19

Data	Tally	Frequency
$18	|	1
$19	||	2
$20	|	1
$22	|||	3
$24	||	2
$25	|	1
$29	|	1

Classes	Frequency
$18–$20	4
$21–$23	3
$24–$26	3
$27–$29	1

15.1 Constructing a bar graph from a frequency distribution

Draw a bar for each class using the frequency of the class as the height of the bar.

Construct a bar graph from the frequency distribution of the previous example.

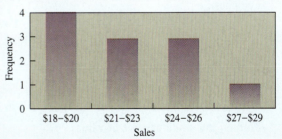

15.1 Constructing a line graph

1. Plot each year on the horizontal axis.
2. For each year, find the value of sales for that year and plot a point at that value.
3. Connect all points with straight lines.

Construct a line graph for the following sales data.

Year	Total Sales
2002	$850,000
2003	$920,000
2004	$875,000
2005	$975,000

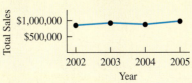

CONCEPTS	EXAMPLES

15.1 Constructing a circle graph

1. Determine the percent of the total for each item.
2. Find the number of degrees of a circle that each percent represents.
3. Draw the circle.

Construct a circle graph for the following table, which lists expenses for a business trip.

Item	Amount
Car	$200
Lodging	$300
Food	$250
Entertainment	$150
Other	$100
	$1000

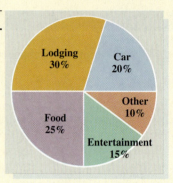

Item	Amount	Percent of Total
Car	$200	$\frac{\$200}{\$1000} = \frac{1}{5} = 20\%$; $360° \times 20\% = 360 \times .20 = 72°$
Lodging	$300	$\frac{\$300}{\$1000} = \frac{3}{10} = 30\%$; $360° \times 30\% = 360 \times .30 = 108°$
Food	$250	$\frac{\$250}{\$1000} = \frac{1}{4} = 25\%$; $360° \times 25\% = 360 \times .25 = 90°$
Entertainment	$150	$\frac{\$150}{\$1000} = \frac{3}{20} = 15\%$; $360° \times 15\% = 360 \times .15 = 54°$
Other	$100	$\frac{\$100}{\$1000} = \frac{1}{10} = 10\%$; $360° \times 10\% = 360 \times .10 = 36°$

15.2 Finding the mean of a set of numbers

1. Add all numbers to obtain the total.
2. Divide the total by the number of pieces of data.

The quiz scores for Pat Phelan in her business math course were as follows

$$85 \quad 79 \quad 93 \quad 91$$
$$78 \quad 82 \quad 87 \quad 85$$

Find Pat's quiz average.

$$\text{Mean} = \frac{85 + 79 + 93 + 91 + 78 + 82 + 87 + 85}{8} = 85$$

15.2 Finding the median of a set of numbers

1. Arrange the data in numerical order from lowest to highest.
2. Select the middle value or the average of the two middle values.

Find the median for Pat Phelan's grades from the previous example. The data arranged from the lowest to highest are

$$78 \quad 79 \quad 82 \quad 85 \quad 85 \quad 87 \quad 91 \quad 93$$

Middle two values are 85 and 85. The average of these two values is

$$\frac{85 + 85}{2} = 85$$

15.2 Finding the mode of a set of values

Determine the most frequently occurring value.

Find the mode for Phelan's grades in the previous example. The most frequently occurring score is 85 (it occurs twice), so the mode is 85.

Chapter 15 | Summary Exercise

Watching the Growth of a Small Business

Pat Gutierrez expanded her company, Christian Books Unlimited, this year by opening stores in different cities. One store was opened in February, and the second was opened in June. Sales, in thousands of dollars, for the two stores are as follows.

	Feb.	Mar.	Apr.	May	June	July	Aug.	Sep.	Oct.
Store 1	6.5	6.8	7.0	6.9	7.5	7.8	8.0	7.6	8.2
Store 2	—	—	—	—	8.2	6.2	8.2	8.7	9.6

(a) Find the median, mean, and mode sales for each store to the nearest tenth.

(b) Plot sales for both stores on the same line graph with month on the horizontal axis and sales on the vertical axis.

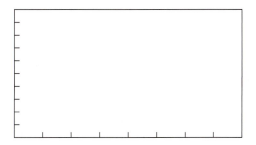

(c) What trends are apparent from the preceding line graph?

INVESTIGATE

Cut at least three graphs and charts out of recent newspapers or magazines. Then explain each of them in writing. Graphs and charts can be great tools for communicating with customers, fellow workers, or even your boss. How can you make sure that a graph or chart clearly communicates the message that you wish it to convey?

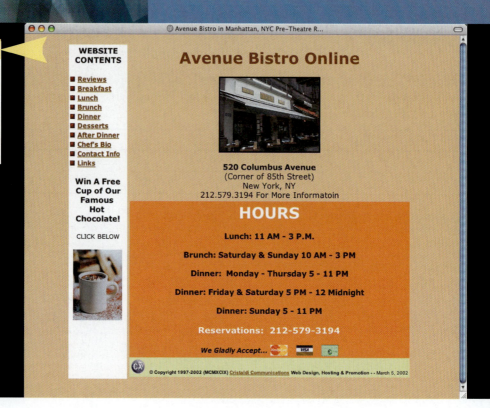

Christina Kelly is the Executive Chef at Avenue Bistro. As a child, Kelly was so devoted to baking that she sometimes woke up at three in the morning to experiment with meringue or other foods her family enjoyed. In fact, she loved cooking so much that she decided against going to Harvard in 1980 to take a position in the kitchen at The Commissary in Philadelphia. Subsequently, she has worked in London, Turkey, Morocco, and New York, extending her knowledge of cooking every step along the way.

Visit Avenue Bistro in New York and you will find delicious meals: scrambled eggs and croissants for breakfast, wonderful salads for lunch, and more ambitious meals for dinner such as Grilled Salmon with Black Truffle or Maine Cod with Lobster. Desserts include mouth-watering items such as Apple Galette with Milk Confit and Tapioca or Roasted Bananas Caramel with Ginger Ice Cream.

1. Assume that weekly sales for the six weeks before Christmas at Avenue Bistro were as follows (in thousands): $22.5, $21.7, $23.8, $27.2, $28.6, and $32.9. Find the mean, median, and mode. Round to the nearest tenth of a thousand if needed.

2. Show the sales data above in a line graph—be sure to label appropriately.

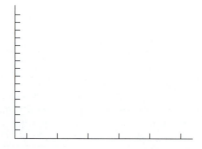

Chapter 15 | Test

To help you review, the numbers in brackets show the section in which the topic was discussed.

1. The following are the number of cases of motor oil sold per week, by a regional distributor on the east coast, for each of the past 20 weeks.

12,450	11,300	12,800	10,850	14,100
14,900	12,300	11,600	12,400	12,900
13,300	12,500	13,390	12,800	12,500
15,100	13,700	12,200	11,800	12,600

   Use these numbers to complete the following table.   **[15.1]**

Cases of Motor Oil	Number of Weeks
10,000–10,999	___
11,000–11,999	___
12,000–12,999	___
13,000–13,999	___
14,000–14,999	___
15,000–15,999	___

2. How many weeks had sales of 13,000 cases or more?   **[15.1]**

   2. _____

3. Use the numbers in the table above to draw a bar graph. Be sure to put a heading and labels on the graph.   **[15.1]**

4. During a 1-year period, the campus newspaper at Dallas Community College had the following expenses. Find all numbers missing from the table.   **[15.1]**

Item	Dollar Amount	Percent of Total	Degrees of a Circle
Newsprint	$12,000	20%	___
Ink	$6,000	___	36°
Wire Service	$18,000	30%	___
Salaries	$18,000	30%	___
Other	$6,000	10%	___

5. Draw a circle graph using the information in test question 4.   **[15.1]**

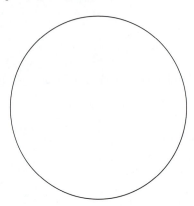

6. What percent of the expenses were for newsprint, ink, and wire service?   **[15.1]**

6. _____

Find the mean for the following. Round to the nearest tenth if necessary.   **[15.2]**

7. Number attending classical music recitals: 220, 275, 198, 212, 233, 246

7. _____

8. Length of boards (centimeters): 12, 18, 14, 17, 19, 22, 23, 25

8. _____

9. Weekly commission ($): 458, 432, 496, 491, 500, 508, 512, 396, 492, 504

9. _____

10. Volume			**11. Sales**	
(Quarts)	*Frequency*		*($)*	*Frequency*
6	7		150	15
10	3		160	17
11	4		170	21
14	2		180	28
19	3		190	19
24	1	_____	200	7

Find the median for the following lists of numbers.   **[15.2]**

12. Numbers of actors trying for a part: 22, 18, 15, 25, 20, 19, 7

12. _____

13. Number of telephone calls received per hour: 41, 39, 45, 47, 38, 42, 51, 38

13. _____

14. Hours worked per day: 7.6, 9.3, 21.8, 10.4, 4.2, 5.3, 7.1, 9.0, 8.3   **14.** _____

15. Trees planted: 58, 76, 91, 83, 29, 34, 51, 92, 38, 41   **15.** _____

Find the mode or modes for the following lists of numbers.   **[15.2]**

16. Contestants' ages: 51, 47, 48, 32, 47, 71, 82, 47   **16.** _____

17. Customers served: 32, 51, 74, 19, 25, 43, 75, 82, 98, 100   **17.** _____

18. Defectives per batch: 96, 104, 103, 104, 103, 104, 91, 74, 103   **18.** _____

Solve the following application problems.

19. Ragland's Fine Furnishings had the following sales (in thousand of dollars).

Year	Sales
2002	$754
2003	$782
2004	$853
2005	$592
2006	$680

The geographic area in which Ragland's operates had a business recession (slow down) in 2005. Do you think the recession may have affected their business? Support your view by drawing a line graph.

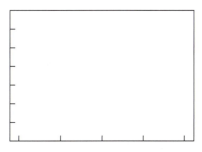

20. Ted Smith sells stocks and bonds at Merrill Lynch. His wife developed a serious illness at the beginning of 2005 and her condition slowly improved through the balance of the year. Do you think his personal problems may have influenced his work performance?

| | Quarterly Commissions | | | |
Year	Quarter 1	Quarter 2	Quarter 3	Quarter 4
2003	$14,250	$12,375	$15,750	$13,682
2004	$13,435	$14,230	$11,540	$15,782
2005	$8,207	$7,350	$10,366	$11,470

Support your view by drawing a line graph. Be sure to label the quarter in which Mrs. Smith became ill.

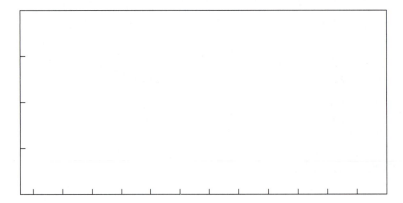

Appendix A

Equations and Formula Review

A.1 | Equations

Objectives

1. Learn the basic terminology of equations.
2. Use basic rules to solve equations.
3. Solve equations requiring more than one operation.
4. Combine similar terms in equations.
5. Use the distributive property to simplify equations.

Objective 1 **Learn the basic terminology of equations.** An equation is a statement that says two expressions are equal. For example, in the equation

$$x + 5 = 9$$

the expression $x + 5$ and the number 9 are equal to one another. The letter x is just a name for a value that is not yet known. Clearly, x must be 4 in order for $x + 5$ to be equal to 9. So $x = 4$ is the **solution** to the equation. Some other terminology used when working with equations follows.

Equation Terminology

The letter x is called a **variable**; it is a letter that represents a number. Any letter can be used for a variable.

A **term** is a single letter, a single number, or the product of a number and a letter. In $x + 5 = 9$, the variable x, as well as the numbers 5 and 9 are terms. In the expression $3y - 14.5$, both $3y$ and 14.5 are terms.

In $x + 5 = 9$, the expression $x + 5$ is the **left side** of the equation, and 9 is the **right side**.

A **solution** to the equation is any number that can replace the variable and result in a true statement. The solution for this equation is the number 4, since the replacement of the variable x with the number 4 results in a true statement.

$$x + 5 = 9$$
$$4 + 5 = 9 \quad \text{Let x = 4.}$$
$$9 = 9 \quad \text{True}$$

The check shows an example of **substitution**; the number 4 was substituted for the variable x.

QUICK TIP In the expression $x + 5$, x is the same thing as $+ 1x$, or 1 times x.

Objective 2 **Use basic rules to solve equations.** In solving equations, the object is to find a number that can be used to replace the variable so that the equation is a true statement. This is done by changing the equation so that all the terms containing a variable are on one side of the equation and all numbers are on the other side. An equation states that two expressions are equal to one another. As long as both sides of an equation are changed in exactly the same way, then the expressions are still equal to one another. Look at the following rules, which can be used when solving equations.

Rules for Solving Equations

Addition Rule The same number may be added or subtracted on both sides of an equation.
Multiplication Rule Both sides of an equation may be multiplied or divided by the same nonzero number.

QUICK TIP   Remember these two things when solving equations:

1.   What you do to one side of an equation, you must also do to the other side.
2.   Solve equations using the opposite math operation from those used in the equation.

EXAMPLE 1

Solving a Linear Equation Using Addition

Solve $x - 9 = 15$.

SOLUTION

To solve this equation, x must be alone on one side of the equal sign and all numbers collected on the other side. To change the $x - 9$ to x, perform the opposite operation to "undo" what was done. The opposite of subtraction is addition, so add 9 to both sides.

$$x - 9 = 15$$
$$x - 9 + 9 = 15 + 9 \qquad \text{Add 9 to both sides.}$$
$$x + 0 = 24$$
$$x = 24$$

To check this answer, substitute 24 for x in the original equation.

$$x - 9 = 15 \qquad \text{Original equation.}$$
$$24 - 9 = 15 \qquad \text{Let x = 24.}$$
$$15 = 15 \qquad \text{True.}$$

The answer of $x = 24$ checks.

EXAMPLE 2

Solving a Linear Equation Using Subtraction

Solve $k + 7 = 18$.

SOLUTION

To isolate k on the left side, do the opposite of adding 7, which is *subtracting* 7.

$$k + 7 = 18$$
$$k + 7 - 7 = 18 - 7 \qquad \text{Subtract 7.}$$
$$k = 11$$

EXAMPLE 3

Solving a Linear Equation Using Division

Solve $5p = 60$.

SOLUTION

The term $5p$ indicates the multiplication of 5 and p. Since the opposite of multiplication is division, solve the equation by *dividing* both sides by 5.

$$5p = 60$$
$$\frac{5p}{5} = \frac{60}{5} \qquad \text{Divide by 5.}$$
$$p = 12$$

Check by substituting 12 for p in the original equation.

Sometimes we put slash or cancel marks through the numbers used to divide both sides; slash marks would be used in Example 3, as follows.

$$5p = 60$$

$$\frac{\cancel{5}p}{\cancel{5}} = \frac{60}{5}$$

$$p = 12$$

EXAMPLE 4	Solve $\dfrac{y}{3} = 9$.

Solving a Linear Equation Using Multiplication

SOLUTION

The bar in $\frac{y}{3}$ means to divide, so solve the equation by multiplying both sides by 3. (The opposite of division is multiplication.) As in the following solution, it is common to use a dot to indicate multiplication.

$$\frac{y}{3} = 9$$

$$\frac{y}{3} \cdot \mathbf{3} = 9 \cdot \mathbf{3} \qquad \text{Multiply by 3.}$$

$$y = 27$$

Example 5 shows how to solve an equation using a reciprocal. To get the **reciprocal** of a nonzero fraction, exchange the numerator and the denominator. For example, the reciprocal of $\frac{7}{9}$ is $\frac{9}{7}$. The product of two reciprocals is 1:

$$\frac{\overset{1}{\cancel{7}}}{\underset{1}{\cancel{9}}} \cdot \frac{\overset{1}{\cancel{9}}}{\underset{1}{\cancel{7}}} = 1$$

EXAMPLE 5	Solve $\dfrac{3}{4}z = 9$.

Solving a Linear Equation Using Reciprocals

SOLUTION

Solve this equation by multiplying both sides by $\frac{4}{3}$, the reciprocal of $\frac{3}{4}$. This process will give $1z$, or just z, on the left.

$$\frac{3}{4}z = 9$$

$$\frac{3}{4}z \cdot \frac{\mathbf{4}}{\mathbf{3}} = 9 \cdot \frac{\mathbf{4}}{\mathbf{3}} \qquad \text{Multiply both sides by } \tfrac{4}{3}.$$

$$z = 12$$

Objective 3 **Solve equations requiring more than one operation.** The equation in Example 6 requires two steps to solve.

EXAMPLE 6	Solve $2m + 5 = 17$.

Solving a Linear Equation Using Several Steps

SOLUTION

To solve equations that require more than one step, first isolate the terms involving the unknown (or variable) on one side of the equation and constants (or numbers) on the other side by using addition and subtraction.

$$2m + 5 = 17$$

$$2m + 5 - \mathbf{5} = 17 - \mathbf{5} \qquad \text{Subtract 5 from both sides.}$$

$$2m = 12$$

Now divide both sides by 2.

$$\frac{2m}{2} = \frac{12}{2} \qquad \text{Divide by 2.}$$
$$m = 6$$

As before, check by substituting 6 for m in the original equation.

QUICK TIP The unknown can be on either side of the equal sign. $6 = m$ is the same as $m = 6$. The number is the solution when the equation has the variable by itself on either the left *or* the right side of the equation.

EXAMPLE 7

Solving a Linear Equation Using Several Steps

Solve $3y - 12 = 52$.

SOLUTION

$$3y - 12 = 52$$
$$3y - 12 + 12 = 52 + 12 \qquad \text{Add 12 to both sides.}$$
$$3y = 64$$
$$\frac{3y}{3} = \frac{64}{3} \qquad \text{Divide both sides by 3.}$$
$$y = 21\frac{1}{3}$$

Objective 4 Combine similar terms in equations. Some equations have more than one term with the same variable. Terms with the same variables can be *combined* by adding or subtracting the coefficients, as shown.

$$5y + 2y = (5 + 2)y = 7y$$
$$11k - 8k = (11 - 8)k = 3k$$
$$12p - 5p + 2p = (12 - 5 + 2)p = 9p$$
$$2z + z = 2z + 1z = (2 + 1)z = 3z$$

Terms with different variables in them *cannot* be combined into a single term. For example, $12y + 5x$ cannot be combined to make one term since y and x are different variables and may have different values.

EXAMPLE 8

Solving a Linear Equation Using Several Steps

Solve $8y - 6y + 4y = 24$.

SOLUTION
Start by combining terms on the left: $8y - 6y + 4y = 2y + 4y = 6y$. This gives the simplified equation $6y = 24$.

$$8y - 6y + 4y = 24$$
$$(8 - 6 + 4)y = 24 \qquad \text{Combine like terms.}$$
$$6y = 24$$
$$\frac{6y}{6} = \frac{24}{6} \qquad \text{Divide by 6.}$$
$$y = 4$$

EXAMPLE **9**

Solving a Linear Equation

Solve $7z - 3z + .5z = 15$.

SOLUTION

$$7z - 3z + .5z = 15$$
$$(7 - 3 + .5)z = 15 \qquad \text{Combine like terms.}$$
$$4.5z = 15$$
$$\frac{4.5z}{\mathbf{4.5}} = \frac{15}{\mathbf{4.5}} \qquad \text{Divide both sides by 4.5.}$$
$$z = 3\frac{1}{3}$$

Objective 5 **Use the distributive property to simplify equations.** Some of the more advanced formulas used in this book involve a number in front of terms in parentheses. These formulas often require use of the *distributive property*. According to the **distributive property**, a number on the outside of the parentheses can be multiplied by each term inside the parentheses, as shown here.

$$a(b + c) = ab + ac$$

The following diagram may help in remembering the distributive property.

Multiply a by b and c
$$a(b + c) = ab + ac$$

The a is *distributed* over the b and the c, as in the following examples.

Multiply 2 by m and 7
$$2(m + 7) = 2m + 2 \cdot 7 = 2m + 14$$
$$8(k - 5) = 8k - 8 \cdot 5 = 8k - 40$$

EXAMPLE **10**

Solving a Linear Equation Using the Distributive Property

Solve $8(t - 5) = 16$.

SOLUTION

First use the distributive property on the left to remove the parentheses.

$$8(t - 5) = 16$$
$$8t - 40 = 16$$
$$8t - 40 + \mathbf{40} = 16 + \mathbf{40} \qquad \text{Add 40 to both sides.}$$
$$8t = 56$$
$$\frac{8t}{8} = \frac{56}{8} \qquad \text{Divide by 8.}$$
$$t = 7$$

Use the following steps to solve an equation.

> ## Solving an Equation
>
> **Step 1** Remove all parentheses on both sides of the equation using the distributive property.
> **Step 2** Combine all similar terms on both sides of the equation.
> **Step 3** Place all terms containing a variable on the same side of the equation and all terms not containing a variable on the opposite side of the equation. Do this by adding or subtracting terms from both sides of the equation as needed.
> **Step 4** Multiply or divide the variable term by numbers as needed to produce a term with a coefficient of 1 in front of the variable. Do the same to the other side of the equation by multiplying or dividing by the same value.

EXAMPLE 11

Solving a Linear Equation Using the Distributive Property

Solve $5r - 2 = 2(r + 5)$.

SOLUTION

$$5r - 2 = 2(r + 5)$$ Use the distributive property on the right side.
$$5r - 2 = 2r + 10$$
$$5r - 2 + 2 = 2r + 10 + 2$$ Add 2 to both sides to get all numbers on the right side.
$$5r = 2r + 12$$
$$5r - 2r = 2r + 12 - 2r$$ Subtract 2r from both sides to get all variables on the left side.

$$5r - 2r = 12$$ Combine similar terms on the left side.
$$3r = 12$$

$$\frac{3r}{3} = \frac{12}{3}$$ Divide both sides by 3.

$$r = 4$$

Check by substituting 4 for r in the original equation.

EXAMPLE 12

Solving a Linear Equation Using the Distributive Property

Solve $3(t - .8) = 14 + t$.

SOLUTION

$$3(t - .8) = 14 + t$$
$$3t - 2.4 = 14 + t$$ Use the distributive property.
$$3t - 2.4 - t = 14 + t - t$$ Subtract t from both sides.
$$2t - 2.4 = 14$$ Combine like terms.
$$2t - 2.4 + 2.4 = 14 + 2.4$$ Add 2.4 to both sides.
$$2t = 16.4$$

$$\frac{2t}{2} = \frac{16.4}{2}$$ Divide both sides by 2.

$$t = 8.2$$

| **QUICK TIP**   Be sure to check the answer in the *original* equation and not in any other step. |

| **QUICK TIP**   You may wish to solve Exercises 1–24 at the end of this appendix before proceeding to the next section. |

A.2 | Business Applications of Equations

Objectives

1. Translate phrases into mathematical expressions.
2. Write equations from given information.
3. Solve application problems.

Objective 1 **Translate phrases into mathematical expressions.** Most problems in business are expressed in words. Before these problems can be solved, they must be converted into mathematical language.

Word problems tend to have certain phrases that occur again and again. The key to solving word problems is to correctly translate these expressions into mathematical expressions. The next few examples illustrate this process.

EXAMPLE 1

Translating Verbal Expressions Involving Addition

Write the following verbal expressions as mathematical expressions. Use x to represent the unknown. Note that other letters can also be used to represent this unknown quantity.

SOLUTION

Verbal Expression	Mathematical Expression	Comments
(a) 5 plus a number	$5 + x$	x represents the number, and *plus* indicates **addition**
(b) Add 20 to a number	$x + 20$	x represents the number, and *add* indicates **addition**
(c) The sum of a number and 12	$x + 12$	x represents the number, and *sum* indicates **addition**
(d) 6 more than a number	$x + 6$	x represents the number, and *more than* indicates **addition**

EXAMPLE 2

Translating Verbal Expressions Involving Subtraction

Write each of the following verbal expressions as a mathematical expression. Use p as the variable.

SOLUTION

Verbal Expression	Mathematical Expression	Comments
(a) 3 less than a number	$p - 3$	p represents the number, and *less than* indicates **subtraction**
(b) A number decreased by 14	$p - 14$	p represents the number, and *decreased by* indicates **subtraction**
(c) 10 fewer than p	$p - 10$	p represents the number, and *fewer than* indicates **subtraction**

EXAMPLE **3**

*Translating Verbal
Expressions Involving
Multiplication and
Division*

Write the following verbal expressions as mathematical expressions. Use y as the variable.

SOLUTION

Verbal Expression	Mathematical Expression	Comments
(a) The product of a number and 3	$3y$	y represents the number, and *product* indicates multiplication
(b) Four times a number	$4y$	y represents the number, and *times* indicates multiplication
(c) Two thirds of a number	$\frac{2}{3}y$	y represents the number, and *of* indicates multiplication
(d) The quotient of a number and 2	$\frac{y}{2}$	y represents the number, and *quotient* indicates division
(e) The sum of 3 and a number is multiplied by 5	$5(3+y)$ or $5(y+3)$	This requires parentheses.
(f) 7 is multiplied by the difference of an unknown number minus 14	$7(y-14)$	This requires parentheses.

QUICK TIP   When adding or multiplying, the order of the variable and the number doesn't matter. For example,

$$3 + x = x + 3 \quad \text{and} \quad 5 \cdot y = y \cdot 5$$

The order *cannot* be reversed when subtracting (or dividing).

For example, $8 - y$ is *not* the same thing as $y - 8$, and

$$6 \div z \text{ is } not \text{ the same thing as } z \div 6.$$

Objective **2** **Write equations from given information.** Since equal mathematical expressions represent the same number, any words that mean *equals* or *same* translate into an $=$. The $=$ sign produces an equation which can be solved.

EXAMPLE **4**

*Writing an Equation
from Words*

Translate "the product of 5 and a number decreased by 8 is 100" into an equation. Use y as the variable. Solve the equation.

SOLUTION
Translate as follows.

the product of 5 · (y − 8) = 100

Simplify and complete the solution of the equation.

$$5 \cdot (y - 8) = 100$$
$$5y - 40 = 100 \quad \text{Apply the distributive property.}$$
$$5y = 140 \quad \text{Add 40 to both sides.}$$
$$y = 28 \quad \text{Divide by 5.}$$

EXAMPLE 5

Writing an Equation from Words

Write "The sum of an unknown and 6, when divided by 15, is equal to 7" as an equation using r as the variable.

SOLUTION

The sum of an unknown and 6	when divided by 15	is equal to	7
↓	↓	↓	↓
$(r + 6)$	$\div\ 15$	$=$	7

Simplify and solve the equation.

$$(r + 6) \div 15 = 7$$
$$(r + 6) \div 15 \cdot \mathbf{15} = 7 \cdot \mathbf{15} \qquad \text{Multiply both sides by 15.}$$
$$r + 6 = 105$$
$$r + 6 - \mathbf{6} = 105 - \mathbf{6} \qquad \text{Subtract 6 from both sides.}$$
$$r = 99$$

Objective ③ **Solve application problems.** Now that statements have been translated into mathematical expressions, you can use this knowledge to solve problems. The following steps represent a systematic approach to solving application problems.

Solving Application Problems: A Systematic Approach

Step 1 Read the problem very carefully. Reread the problem to make sure that its meaning is clear.

Step 2 Decide on the unknown. Choose a variable to represent the unknown number.

Step 3 Identify the knowns. Use the given information to write an equation describing the relationship given in the problem.

Step 4 Solve the equation.

Step 5 Answer the question asked in the problem.

Step 6 Check the solution by using the original words of the problem.

Step 3 is often the hardest. To write an equation from the information given in the problem, convert the facts stated in words into mathematical expressions. This converted mathematical expression, or equation, is called the *mathematical model* of the situation described in the original words.

QUICK TIP Always check to see whether your answer to a word problem needs to be expressed in units such as dollars, gallons, ounces, or whatever. Your answer is incomplete unless you have the correct units.

EXAMPLE 6

Solving a Business Problem

A restaurant manager found that she has 18 more females than males scheduled to work next week. The total number scheduled to work next week is 64. Find the number of females.

SOLUTION

Let m represent the number of males, then $(m + 18)$ is the number of females. The number of males plus the number of females is equal to 64, or the total number scheduled to work. Write this using an equation.

number of males	plus	number of females	equals	total scheduled
↓	↓	↓	↓	↓
m	$+$	$(m + 18)$	$=$	64

Solve the equation for the number of males.

$$2m + 18 = 64$$

$$2m + 18 - \mathbf{18} = 64 - \mathbf{18} \qquad \text{Subtract 18 from each side.}$$

$$2m = 46$$

$$\frac{2m}{\mathbf{2}} = \frac{46}{\mathbf{2}} \qquad \text{Divide by 2.}$$

$$m = 23 \qquad \text{There are 23 males scheduled.}$$

Now find the number of females using $(m + 18) = (23 + 18) = 41$ females.

EXAMPLE 7

Applying Equation Solving

A mattress is on sale for $200, which is $\frac{4}{5}$ of its original price. Find the original price.

SOLUTION

Let p represent the original price, $200 is the sale price, and the sale price is $\frac{4}{5}$ of the original price. Use all this information to write the equation.

sale price	is	$\frac{4}{5}$	of	original price
↓	↓	↓	↓	↓
$200	=	$\frac{4}{5}$	×	p

Solve the equation.

$$200 = \frac{4}{5} \cdot p$$

$$\frac{5}{4} \cdot 200 = \frac{5}{4} \cdot \frac{4}{5} \cdot p \qquad \text{Multiply by reciprocal.}$$

$$\frac{1000}{4} = \mathbf{1} \cdot p$$

$$250 = p$$

The original price was $250.

EXAMPLE 8

Solving a Business Problem

The Eastside Nursery ordered 27 trees. Some of the trees were elms, costing $17 each. The remainder of the trees were maples at $11 each. The total cost of the trees was $375. Find the number of elms and the number of maples.

SOLUTION

Let x represent the number of elm trees in the shipment. Since the shipment contained 27 trees, the number of maples is found by subtracting the number of elms from 27.

$$27 - x = \text{number of maples}$$

If each elm tree costs $17, then x elm trees will cost $17x$ dollars. Also, the cost of $27 - x$ maple trees at $11 each is $11(27 - x)$. The total cost of the shipment was $375.

A table can be very helpful in identifying the knowns and unknowns.

	Number of Trees	Cost per Tree	Total Cost
Elms	x	$17	$17x$
Maples	$(27 - x)$	$11	$11(27 - x)$
Totals	27		375

The information in the table is used to develop the following equation.

$$\textbf{cost of elms} + \textbf{cost of maples} = \textbf{total cost}$$

$$17x + 11(27 - x) = 375$$

Now solve this equation. First use the distributive property.

$$17x + 297 - 11x = 375 \qquad \text{Combine terms.}$$
$$6x + 297 = 375 \qquad \text{Subtract 297 from each side.}$$
$$6x = 78 \qquad \text{Divide each side by 6.}$$
$$x = 13$$

There were $x = 13$ elm trees and $27 - 13 = 14$ maple trees.

EXAMPLE 9

Solving Investment Problems

Laurie Zimmerman has $15,000 to invest. She places a portion of the funds in a passbook account and $3000 more than twice this amount in a retirement account. How much is put into the passbook account? How much is placed in the retirement account?

SOLUTION

Let z represent the amount invested in the passbook account. To find the amount invested in the retirement account, translate as follows.

3000	**more than**	**2 times the amount**
↓	↓	↓
3000	+	$2z$

Since the sum of the two investments must be $15,000, an equation can be formed as follows.

Amount invested in passbook		**Amount invested in retirement account**		**Total amount invested**
z	+	$(3000 + 2z)$	=	$15,000

Now solve the equation.

$$z + (3000 + 2z) = 15{,}000$$
$$3z + 3000 = 15{,}000$$
$$3z = 12{,}000 \qquad \text{Subtract 3000.}$$
$$z = 4000 \qquad \text{Divide by 3.}$$

The amount invested in the passbook account is z, or $4000. The amount invested in the retirement account is $3000 + 2z$ or $3000 + 2(4000) = \$11{,}000$.

QUICK TIP You may wish to work Exercises 61–70 at the end of this appendix before proceeding to the next section.

A.3 Business Formulas

Objectives

1. Evaluate formulas for given values of the variables.
2. Solve formulas for a specific variable.
3. Use standard business formulas to solve word problems.
4. Evaluate formulas containing exponents.

Objective 1 Evaluate formulas for given values of the variables. Many of the most useful rules and procedures in business are given as **formulas**: equations showing how one number is found from other numbers. One of the single most useful formulas in business is the one for simple interest.

$$\text{Interest} = \text{Principal} \times \text{Rate} \times \text{Time}$$

When written out in words, as shown, a formula can take up too much space and be hard to remember. For this reason, it is common to *use letters as variables for the words* in a formula. Many times the first letter in each word of a formula is used, to make it easier to remember the formula. By this method, the formula for simple interest is written as follows.

$$\text{Interest} = \text{Principal} \times \text{Rate} \times \text{Time}$$
$$I = PRT$$

By using letters to express the relationship between interest, principal, rate, and time we have generalized the relationship so that any value can be substituted into the formula. Once three values are substituted into the formula, we can then find the value of the remaining variable.

EXAMPLE 1

Evaluating a Formula

Use the formula $I = PRT$ and find I if $P = 7000$, $R = .09$, and $T = 2$.

SOLUTION

Substitute 7000 for P, .09 for R, and 2 for T in the formula $I = PRT$. Remember that writing P, R, and T together as PRT indicates the product of the three letters.

$$I = PRT$$
$$I = 7000(.09)(2)$$

Multiply on the right to get the solution.

$$I = 1260$$

EXAMPLE 2

Evaluating a Formula

Use the formula $I = PRT$ and find P if $I = 5760$, $R = .16$, and $T = 3$.

SOLUTION

Substitute the given numbers for the letters of the formula.

$$I = PRT$$
$$5760 = P(.16)(3)$$
$$5760 = .48P \qquad \text{\small $P(.16)(3) = .48P$}$$
$$\frac{5760}{.48} = \frac{.48P}{.48} \qquad \text{\small Divide both sides by .48.}$$
$$12{,}000 = P$$

EXAMPLE 3

Evaluating a Formula

Solve for rate (R) given $M = \$12{,}540$, $P = \$12{,}000$, and $T = .5$ in the equation $M = P(1 + RT)$.

SOLUTION

$$M = P(1 + RT)$$
$$\$12{,}540 = \$12{,}000(1 + R \cdot .5)$$
$$12{,}540 = 12{,}000 + 12{,}000 \cdot R \cdot .5 \qquad \text{\small Use the distributive property.}$$
$$12{,}540 = 12{,}000 + 6000 \cdot R$$
$$12{,}540 - \mathbf{12{,}000} = 12{,}000 + 6000 \cdot R - \mathbf{12{,}000} \qquad \text{\small Subtract 12,000 from each side.}$$
$$540 = 6000 \cdot R$$
$$\frac{540}{\mathbf{6000}} = \frac{6000 \cdot R}{\mathbf{6000}} \qquad \text{\small Divide by 6000.}$$
$$.09 = R$$

Rate is .09 or 9%.

Objective 2 **Solve formulas for a specific variable.** In Example 2, we found the value of P when given the values of I, R, and T. If several problems of this type must be solved, it may be better to rewrite the formula $I = PRT$ so that P is alone on one side of the equation. Do this with the rules of equations given earlier. Since P is multiplied by RT, get P alone by dividing both sides of the equation by RT.

$$I = PRT$$

$$\frac{I}{RT} = \frac{PRT}{RT} \quad \text{Divide by RT.}$$

$$\frac{I}{RT} = P$$

QUICK TIP The process of rearranging a formula is sometimes called solving a formula for a specific variable.

EXAMPLE 4

Solving a Formula for a Specific Variable

Solve for T in the formula $M = P(1 + RT)$. This formula gives the maturity value (M) of an initial amount of money (P) invested at a specific rate (R) for a certain period of time (T).

SOLUTION
Start by using the distributive property on the right side.

$$M = P(1 + RT)$$
$$M = P + PRT$$

Now subtract P from both sides.

$$M - P = P + PRT - P$$
$$M - P = PRT$$

Divide each side by PR.

$$\frac{M - P}{PR} = \frac{PRT}{PR}$$

$$\frac{M - P}{PR} = T$$

The original formula is now solved for T.

EXAMPLE 5

Solving a Formula for a Specific Variable

Solve for T in the formula $D = \frac{B}{MT}$.

SOLUTION
This formula gives the discount rate (D) of a note in terms of the face value (B), the time of a note (T), and the maturity value (M). Solve for T.

$$D = \frac{B}{MT}$$

$$DMT = \frac{B}{MT}MT \quad \text{Multiply by MT.}$$

$$DMT = B$$

$$\frac{DMT}{DM} = \frac{B}{DM} \quad \text{Divide by DM.}$$

$$T = \frac{B}{DM}$$

Objective 3 **Use standard business formulas to solve word problems.** In the following examples, application problems that use some common business formulas are solved.

EXAMPLE 6

Finding Gross Sales

Find the gross sales amount from selling 481 fishing lures at $2.65 each.

SOLUTION

The formula for gross sales is $G = NP$. N is the number of items sold and P is the price per item. To find the gross sales from selling 481 fishing lures at $2.65 each, use the formula as shown.

$$G = NP$$
$$G = 481(\$2.65)$$
$$G = \mathbf{\$1274.65}$$

The gross sales will be $1274.65.

EXAMPLE 7

Finding Selling Price

A retailer purchased a personal computer with a microphone to digitize voice at a cost of $1265. He then adds a markup of $150 before placing it on the shelf to sell. Find the selling price.

SOLUTION

The selling price is found by adding the cost of the item and the markup.

$$S = C + M$$

The variable C is the cost and M is the markup, which is the amount added to the cost to cover expenses and profit. The selling price is found as shown.

$$S = \$1265 + \$150$$
$$S = \mathbf{\$1415}$$

The selling price is $1415.

Objective 4 **Evaluate formulas containing exponents.** Exponents are used to show repeated multiplication of a quantity called the *base*. For example,

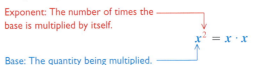

Exponent: The number of times the base is multiplied by itself.

$$x^2 = x \cdot x$$

Base: The quantity being multiplied.

Similarly,

$$z^3 = z \cdot z \cdot z \qquad \text{and} \qquad 5^4 = 5 \cdot 5 \cdot 5 \cdot 5$$

which is 625.

EXAMPLE 8

Finding Monthly Sales

Trinity Sporting Goods has found that monthly sales can be approximated using

$$\text{Sales} = 40 + 1.6 \times (\text{advertising})^2$$

as long as advertising is less than $4000. All of the figures in the equation above are in thousands. Estimate sales for a month with $3500 in advertising.

SOLUTION

Place 3.5 in the equation for the number of thousands of dollars of advertising and find sales.

$$\text{Sales} = 40 + 1.6(\mathbf{3.5})^2$$
$$\text{Sales} = 40 + 1.6(\mathbf{12.25})$$
$$\text{Sales} = 40 + 19.6$$
$$\text{Sales} = 59.6$$

Sales are projected to be $59,600 for the month.

QUICK TIP   Try solving Exercises 25–40 on page A-19 before starting Section A.4.

A.4 | Ratio and Proportion

Objectives

1. Define a ratio.
2. Set up a proportion.
3. Solve a proportion for unknown values.
4. Use proportions to solve problems.

Objective **1** **Define a ratio.** A **ratio** is a quotient of two quantities that can be used to *compare* the quantities. The ratio of the number a to the number b is written in any of the following ways.

$$a \text{ to } b, \qquad a{:}b, \qquad \text{or} \qquad \frac{a}{b}$$

This last way of writing a ratio is most common in mathematics, while $a{:}b$ is perhaps most common in business.

EXAMPLE 1

Writing Ratios

Write a ratio in the form $\frac{a}{b}$ for each word phrase. Notice in each example that the number mentioned first always gives the numerator.

SOLUTION

(a) The ratio of 5 hours to 3 hours is $\frac{5}{3}$.

(b) To find the ratio of 5 hours to 3 days, *first convert* 3 *days* to *hours*. Since there are 24 hours in 1 day, 3 days $= 3 \cdot 24 = 72$ hours. Then the ratio of 5 hours to 3 days is the quotient of 5 and 72.

$$\frac{5}{72}$$

(c) The ratio of \$700,000 in sales to \$950,000 in sales is written this way.

$$\frac{\$700{,}000}{\$950{,}000}$$

Write this ratio in lowest terms.

$$\frac{\$700{,}000}{\$950{,}000} = \frac{14}{19}$$

EXAMPLE 2

Writing Ratios

Burger King sold the following items in a one-hour period last Friday afternoon.

70 bacon cheeseburgers
15 plain hamburgers
30 salad combos
45 chicken sandwiches
40 fish sandwiches

Write ratios for the following items sold:

(a) bacon cheeseburgers to fish sandwiches
(b) salad combos to chicken sandwiches
(c) plain hamburgers to salad combos
(d) fish sandwiches to total items sold

SOLUTION

(a) $\dfrac{\text{bacon cheeseburgers}}{\text{fish sandwiches}} = \dfrac{70}{40} = \dfrac{7}{4}$

(b) $\dfrac{\text{salad combos}}{\text{chicken sandwiches}} = \dfrac{30}{45} = \dfrac{2}{3}$

(c) $\dfrac{\text{plain hamburgers}}{\text{salad combos}} = \dfrac{15}{30} = \dfrac{1}{2}$

(d) $\dfrac{\text{fish sandwiches}}{\text{total items sold}} = \dfrac{40}{200} = \dfrac{1}{5}$

Objective 2 **Set up a proportion.** A ratio is used to compare two numbers or amounts. A **proportion** says that two ratios are equal, as in the following example.

$$\frac{3}{4} = \frac{15}{20}$$

This proportion says that the ratios $\frac{3}{4}$ and $\frac{15}{20}$ are equal.

To see whether a proportion is true, use the method of **cross-products**.

Method of Cross-Products

The proportion

$$\frac{a}{b} = \frac{c}{d}$$

is true if the cross-products $a \cdot d$ and $b \cdot c$ are equal. (That is, if $ad = bc$.)

EXAMPLE 3

Determining if a Proportion Is True

Decide whether the following proportions are true.

(a) $\frac{3}{5} = \frac{12}{20}$ **(b)** $\frac{2}{3} = \frac{9}{16}$

SOLUTION

(a) Find each cross-product.

$$\frac{3}{5} = \frac{12}{20}$$
$$3 \times 20 = 5 \times 12$$
$$60 = 60$$

Since the cross-products are equal, the proportion is true.

(b) Find the cross-products.

$$\frac{2}{3} = \frac{9}{16}$$
$$2 \times 16 = 3 \times 9$$
$$32 \neq 27$$

This proportion is false, so $\frac{2}{3} \neq \frac{9}{16}$.

Objective 3 **Solve a proportion for unknown values.** The method of cross-products is just a shortcut version of solving an equation. To see how, start with the proportion

$$\frac{a}{b} = \frac{c}{d}$$

and multiply both sides by the product of the two denominators, bd.

$$bd \cdot \frac{a}{b} = bd \cdot \frac{c}{d}$$

or

$$ad = bc$$

The expressions *ad* and *bc* are the cross-products, and this solution shows that they are equal.

Four numbers are used in a proportion. If any three of these numbers are known, the fourth can be found.

EXAMPLE 4
Solving a Proportion

(a) Find *x* in this proportion.

$$\frac{3}{5} = \frac{x}{40}$$

SOLUTION
In a proportion, the cross-products are equal. The cross-products in this proportion are $3 \cdot 40$ and $5 \cdot x$. Setting these equal gives the following equation.

$$3 \cdot 40 = 5 \cdot x$$
$$120 = 5x$$

Divide both sides by 5 to find the solution.

$$24 = x$$

(b) Solve this proportion to find *k*.

$$\frac{3}{10} = \frac{5}{k}$$

SOLUTION
Find the two cross-products and set them equal.

$$3k = 10 \cdot 5$$
$$3k = 50$$
$$k = \frac{50}{3}$$

Write the answer as the mixed number $16\frac{2}{3}$ if desired.

EXAMPLE 5
Solving Proportions

A food wholesaler charges a restaurant chain $83 for 3 crates of fresh produce. How much should it charge for 5 crates of produce?

SOLUTION
Let *x* be the cost of 5 crates of produce. Set up a proportion with one ratio the number of crates and the other ratio the costs. Use this pattern.

$$\frac{\text{crates}}{\text{crates}} = \frac{\text{cost}}{\text{cost}}$$

Now substitute the given information.

3 crates cost $83

$$\frac{3}{5} = \frac{83}{x}$$

Use the cross-products to solve the proportion.

$$3x = 5(83)$$
$$3x = \$415$$
$$x = \$138.33 \quad \text{(rounded to the nearest cent)}$$

The 5 crates cost $138.33.

Objective **4** **Use proportions to solve problems.** Proportions are used in many practical applications as shown in the next two examples.

EXAMPLE 6

Solving Applications

A firm in Hong Kong and one in Thailand agree to jointly develop an engine-control microchip to be sold to North American auto manufacturers. They agree to split the development costs in a ratio of $8:3$ (Hong Kong firm to Thailand firm), resulting in a cost of \$9,400,000 to the Hong Kong firm. Find the cost to the Thailand firm.

SOLUTION

Let x represent the cost to the Thailand firm, then

$$\frac{8}{3} = \frac{9,400,000}{x}$$

$$8x = 3 \cdot 9,400,000 \qquad \text{Cross multiply.}$$

$$8x = 28,200,000$$

$$x = 3,525,000 \qquad \text{Divide by 8.}$$

The Thailand firm's share of the costs is \$3,525,000.

EXAMPLE 7

Solving Applications

Bill Thomas wishes to estimate the amount of timber on some forested land that he owns. One value he needs to estimate is the average height of the trees. One morning, Thomas notices that his own 6-foot body casts an 8-foot shadow at the same time that a typical tree casts a 34-foot shadow. Find the height of the tree.

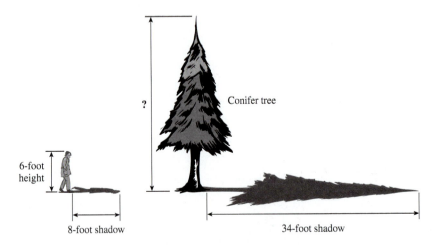

SOLUTION

Set up a proportion in which the height of the tree is given the variable name x.

$$\frac{6}{8} = \frac{x}{34}$$

$$6 \cdot 34 = 8 \cdot x \qquad \text{Cross multiply.}$$

$$\frac{204}{8} = \frac{8 \cdot x}{8} \qquad \text{Divide by 8.}$$

$$x = 25.5 \text{ feet}$$

The height of the tree is 25.5 feet.

Appendix A   Exercises

In Exercises 1–24, solve each equation for the variable.   **[A.1]**

1. $s + 12 = 15$ _____

2. $k + 15 = 22$ _____

3. $b - 7 = 24$ _____

4. $P - 13 = 52$ _____

5. $12 = b + 9$ _____

6. $7 = m - 3$ _____

7. $8k = 56$ _____

8. $3q = 120$ _____

9. $60 = 30m$ _____

10. $94 = 2z$ _____

11. $\dfrac{m}{5} = 6$ _____

12. $\dfrac{r}{7} = 1$ _____

13. $\dfrac{2}{3}a = 5$ _____

14. $\dfrac{3}{4}m = 18$ _____

15. $\dfrac{9}{5}r = 18$ _____

16. $2x = \dfrac{5}{3}$ _____

17. $3m + 5 = 17$ _____

18. $2y - 5 = 39$ _____

19. $4r + 3 = 9$ _____

20. $2p + \dfrac{1}{2} = \dfrac{3}{2}$ _____

21. $11r - 5r + 6r = 84$ _____

22. $5m + 6m - 2m = 72$ _____

23. $3x + 12 = 3(2x + 3)$ _____

24. $4z + 2 = 2(z + 2)$ _____

In Exercises 25–34, a formula is given, along with the values of all but one of the variables in the formula. Find the value of the variable that is not given.   **[A.1]**

25. $I = PRT$; $P = 2800, R = .09, T = 2$ _____

26. $S = C + M$; $C = 275, M = 49$ _____

27. $G = NP$; $N = 840,\ P = 3.79$ _____

28. $M = P(1 + RT)$; $P = 420,\ R = .07, T = 2\dfrac{1}{2}$ _____

29. $R = \dfrac{D}{1 - DT}$; $D = .04,\ T = 5$ _____

30. $A = \dfrac{S}{1 + RT}$; $S = 12{,}600,\ R = .12,$

$T = \dfrac{5}{12}$ _____

31. $T = \dfrac{D}{S}$; $T = 100, S = 2$ _____

32. $\dfrac{I}{PR} = T$; $P = 100, R = .02,\ T = 500$ _____

33. $d = rt$; $r = .07,\ t = 12$ _____

34. $I = PRT$; $P = 500,\ R = .08,\ T = 3$ _____

In Exercises 35–40, solve for the indicated variables.   **[A.3]**

35. $A = LW$; for W _____

36. $d = rt$; for r _____

37. $I = PRT$; for T _____

38. $P = 1 + RT$; for R _____

39. $A = P + PRT$; for T _____

40. $R(1 - DT) = D$; for R _____

In Exercises 41–46, write the ratio in lowest terms.   **[A.4]**

41. 250 pesos to 1250 pesos _____

42. 45 women to 110 men _____

43. $1.20 to 75¢ _____

44. 20 hours to 5 days _____

45. 35 dimes to 6 dollars _____

46. 30 inches to five yards _____

In Exercises 47–52, decide whether the proportions are true or false.   **[A.4]**

47. $\dfrac{2}{3} = \dfrac{42}{63}$ _____

48. $\dfrac{6}{9} = \dfrac{36}{52}$ _____

49. $\dfrac{18}{20} = \dfrac{56}{60}$ _____

50. $\dfrac{12}{18} = \dfrac{8}{12}$ _____

51. $\dfrac{420}{600} = \dfrac{14}{20}$ _____

52. $\dfrac{7.6}{10} = \dfrac{76}{100}$ _____

In Exercises 53–60, solve the proportions.   **[A.4]**

53. $\dfrac{y}{35} = \dfrac{25}{5}$ _____

54. $\dfrac{15}{s} = \dfrac{45}{117}$ _____

55. $\dfrac{a}{25} = \dfrac{4}{20}$ _____

56. $\dfrac{6}{x} = \dfrac{4}{18}$ _____

57. $\dfrac{z}{20} = \dfrac{80}{200}$ _____

58. $\dfrac{25}{100} = \dfrac{8}{m}$ _____

59. $\dfrac{1}{2} = \dfrac{r}{7}$ _____

60. $\dfrac{2}{3} = \dfrac{5}{s}$ _____

Solve the following application problems.

61. The sum of an unknown with eight is twenty. Find the unknown.   **[A.2]**

61. _____

62. The sum of four plus an unknown equals fifty-one. Find the unknown.   **[A.2]**

62. _____

63. An unknown times thirty equals one thousand eight hundred. Find the unknown.   **[A.2]**

63. _____

64. Twenty-four equals three times an unknown. Find the unknown.   **[A.2]**

64. _____

65. Three times an unknown plus five is equal to fifty. Find the unknown.   **[A.2]**

65. _____

66. Four plus seven times an unknown is eighteen. Find the unknown.   **[A.2]**

66. _____

67. The sum of two consecutive whole numbers is equal to ninety-one. Find the numbers.   **[A.2]**

67. _____

68. The sum of two consecutive odd whole numbers is equal to two hundred forty. Find both numbers.   **[A.2]**   68. _____

69. Tom throws some coins onto a table. His twin brother Joe throws coins worth twice as much onto the table. The total value of the coins on the table is $2.61. How much money did Joe place on the table?   **[A.2]**   69. _____

70. A business math class has 47 students, with 9 more females than males. Find the number of males and females in the class.   **[A.2]**   70. _____

71. Cajun Boatin' Inc. bought 5 small boats and 3 large boats for $14,878. A small boat costs $1742. Find the cost of a large boat.   **[A.2]**   71. _____

72. Mike paid $172,000 for an old 5-unit apartment house. Find the cost of a 12-unit apartment house.   **[A.4]**   72. _____

73. The tax on a $40 item is $3. Find the tax on a $160 item.   **[A.4]**   73. _____

74. Sam bought 17 table-model television sets for $1942.25. Find the cost of one set.   **[A.4]**   74. _____

75. The bookstore at Steeltrap Community College has a markup that is $\frac{1}{4}$ its cost on a book. Find the cost to the bookstore of a short paperback book selling for $20.   **[A.3]**   75. _____

76. An unknown principal (P) invested at 8% for $1\frac{3}{4}$ years yields a maturity value (M) of $1368. Use $M = P(1 + RT)$ to find the principal.   **[A.3]**   76. _____

77. Explain why all terms with a variable should be placed on one side of the equation, and all terms without a variable should be placed on the opposite side, when solving an equation.   **[A.1]**

78. In your own words, explain the terms formula, ratio, and proportion.   **[A.3–A.4]**

Appendix B

The Metric System

Today, the **metric system** is used just about everywhere in the world. In the United States, many industries are switching over to this improved system. The metric system is being taught in elementary schools, and it may eventually replace the current system.

A table on the English system is included here to refresh your memory. Notice that the time relationships are the same in the English and metric systems.

1 meter

1 meter is 39.37 inches.

1 yard

1 yard is 36 inches.

Length		Weight	
1 foot	= 12 inches (in.)	1 pound (lb)	= 16 ounces (oz)
1 yard (yd)	= 3 feet (ft)	1 ton (T)	= 2000 pounds (lb)
1 mile (mi)	= 5280 feet (ft)		

Capacity		Time	
1 cup (c)	= 8 fluid ounces	1 week (wk)	= 7 days
1 pint (pt)	= 2 cups	1 day	= 24 hours (hr)
1 quart (qt)	= 2 pints (pt)	1 hour (hr)	= 60 minutes (min)
1 gallon (gal)	= 4 quarts (qt)	1 minute (min)	= 60 seconds (sec)

Recall the basic metric units for length, volume, weight, and temperature:

Prefixes in the Metric System

deca- = 10 times
kilo- = 1000 times
deci- = $\frac{1}{10}$ times
centi- = $\frac{1}{100}$ times
milli- = $\frac{1}{1000}$ times

Objectives

1 Learn the metric system.
2 Learn how to convert from one system to the other.

Objective 1 Learn the Metric System. The basic unit of length in the metric system is the **meter**. A meter is a little longer than a yard. For shorter lengths, the units **centimeter** and **millimeter** are used. The prefix "centi" means hundredth, so 1 centimeter is one-hundredth of a meter. Thus

$$100 \text{ centimeters} = 1 \text{ meter}$$

The prefix "milli" means thousandth, so 1 millimeter means one-thousandth of a meter. Thus

$$1000 \text{ millimeters} = 1 \text{ meter}$$

"Meter" is abbreviated m, "centimeter" is cm, and "millimeter" is mm.

Convert from centimeters to millimeters to meters by moving the decimal point, as shown in the following example.

EXAMPLE 1

Converting Length Measurements

Convert the following measurements:

(a) 6.4 m to cm
(b) .98 m to mm
(c) 34 cm to m

SOLUTION

(a) A centimeter is a small unit of measure (a centimeter is about $\frac{1}{2}$ of the diameter of a penny) and a meter is a large unit (a little over 3 feet), so many centimeters make a meter. For this reason, *multiply* by 100 to convert meters to centimeters.

$$6.4 \text{ m} = 6.4 \times 100 = 640 \text{ cm}$$

(b) Multiply by 1000 to convert meters to millimeters.

$$.98 \text{ m} = .98 \times 1000 = 980 \text{ mm}$$

(c) A meter is a large unit of measure, and a centimeter is a smaller unit, so 34 cm is equivalent to a smaller number of meters. Thus, *divide* by 100 to convert centimeters to meters.

$$34 \text{ cm} = \frac{34}{100} = .34 \text{ m}$$

Long distances are measured in **kilometer** (km) units. The prefix "kilo" means one thousand. Thus,

$$1 \text{ kilometer} = 1000 \text{ meters}$$

Since a meter is about a yard, 1000 meters is about 1000 yards, or 3000 feet. Therefore, 1 kilometer is about 3000 feet. One mile is 5280 feet, so 1 kilometer is about 3000/5280 of a mile. Divide 3000 by 5280 to find that 1 kilometer is about .6 miles.

The basic unit of volume in the metric system is the **liter** (L) which is a little more than a quart. You may have noticed that Coca Cola is sometimes sold in 2-liter plastic bottles. Again the prefixes "milli" and "centi" are used. Thus,

$$1 \text{ liter} = 100 \text{ centiliters}$$

$$1 \text{ liter} = 1000 \text{ milliliters}$$

Milliliter (mL) and **centiliter** (cL) are such small volumes that they find their main uses in science. In particular, drug dosages are often expressed in milliliters.

Weight is measured in **grams** (g). A nickel weighs almost exactly 5 grams. **Milligrams** (mg; one-thousandth of a gram) and **centigrams** (cg; one-hundredth of a gram) are so small that they are used mainly in science. A more common measure is the **kilogram** (kg), which is 1000 grams. A kilogram weighs about 2.2 pounds.

$$1000 \text{ grams} = 1 \text{ kilogram}$$

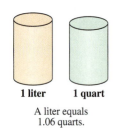

1 liter   1 quart

A liter equals
1.06 quarts.

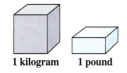

1 kilogram   1 pound

A kilogram equals
2.2 pounds.

EXAMPLE 2

Converting Weight Measurements

Convert the following measurements:

(a) 650 g to kg
(b) 9.4 L to cL
(c) 4350 mg to g

SOLUTION

(a) A gram is a small unit, and a kilogram is a larger unit. Thus, *divide* by 1000 to convert grams to kilograms.

$$650 \text{ g} = \frac{650}{1000} = .65 \text{ kg}$$

(b) *Multiply* by 100 to convert liters to centiliters.

$$9.4 \text{ L} = 9.4 \times 100 = 940 \text{ cL}$$

(c) *Divide* by 1000 to convert milligrams to grams.

$$4350 \text{ mg} = \frac{4350}{1000} = 4.35 \text{ g}$$

Objective 2   Learn How to Convert From One System to the Other. Most Americans do not think in the metric system as easily as in the **English system** of feet, quarts, pounds, and so on. So, they find it necessary to convert from one system to the other. Approximate conversion can be made with the aid of the next table.

English-Metric Conversion Table

From Metric	To English	Multiply By	From English	To Metric	Multiply By
Meters	Yards	1.09	Yards	Meters	0.914
Meters	Feet	3.28	Feet	Meters	0.305
Meters	Inches	39.37	Inches	Meters	0.0254
Kilometers	Miles	0.62	Miles	Kilometers	1.609
Grams	Pounds	0.00220	Pounds	Grams	454
Kilograms	Pounds	2.20	Pounds	Kilograms	0.454
Liters	Quarts	1.06	Quarts	Liters	0.946
Liters	Gallons	0.264	Gallons	Liters	3.785

EXAMPLE **3**

Converting Metric to English

Convert the following measurements:

(a) 15 meters to yards
(b) 39 yards to meters
(c) 47 meters to inches
(d) 87 kilometers to miles
(e) 598 miles to kilometers
(f) 12 quarts to liters

SOLUTION

(a) Look at the table for converting meters to yards, and find the number **1.09**. Multiply 15 meters by **1.09**.

$$15 \times 1.09 = 16.35 \text{ yards}$$

(b) Read the yards-to-meters row of the table. The number **.914** appears. Multiply 39 yards by **.914**.

$$39 \times .914 = 35.646 \text{ meters}$$

(c) 47 meters = $47 \times 39.37 = 1850.39$ inches
(d) 87 kilometers = $87 \times .62 = 53.94$ miles
(e) 598 miles = $598 \times 1.609 = 962.182$ kilometers
(f) 12 quarts = $12 \times .946 = 11.352$ liters

Temperature in the metric system is measured in degrees **Celsius** (abbreviated C). In the celsius scale, water freezes at 0°C and boils at 100°C. This is more sensible than degrees **fahrenheit** (abbreviated F) in use now, in which a mixture of salt and water freezes at 0°F, and 100°F represents the temperature inside the individual Gabriel Fahrenheit's mouth.

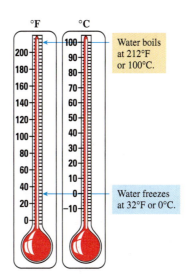

Water boils at 212°F or 100°C.

Water freezes at 32°F or 0°C.

Converting From Fahrenheit to Celsius

Step 1 Subtract 32.
Step 2 Multiply by 5.
Step 3 Divide by 9.

These steps can be expressed by the following formula.

$$C = \frac{5(F - 32)}{9}$$

EXAMPLE 4

Converting Fahrenheit to Celsius

Convert 68°F to Celsius:

SOLUTION

Use the steps above.

Step 1 Subtract 32. $68 - 32 = 36$
Step 2 Multiply by 5. $36 \times 5 = 180$
Step 3 Divide by 9. $\dfrac{180}{9} = 20$

Thus, 68°F = 20°C.

Converting From Celsius to Fahrenheit

Step 1 Multiply by 9.
Step 2 Divide by 5.
Step 3 Add 32.

These steps can be expressed by the following formula.

$$F = \frac{9 \times C}{5} + 32$$

EXAMPLE 5

Converting Celsius to Fahrenheit

Convert 11°C to Fahrenheit

SOLUTION

Use the steps above.

Step 1 Multiply by 9. $9 \times 11 = 99$
Step 2 Divide by 5. $99 \div 5 = 19.8$
Step 3 Add 32. $19.8 + 32 = 51.8°F$

Thus, 11°C = 51.8°F.

Appendix B | Exercises

Convert the following measurements.

1. 68 cm to m _____

2. 934 mm to m _____

3. 4.7 m to mm _____

4. 7.43 m to cm _____

5. 8.9 kg to g _____

6. 4.32 kg to g _____

7. 39 cL to L _____

8. 469 cL to L _____

9. 46,000 g to kg _____

10. 35,800 g to kg _____

11. .976 kg to g _____

12. .137 kg to g _____

Convert the following measurements. Round to the nearest tenth.

13. 36 m to yards _____

14. 76.2 m to yards _____

15. 55 yards to m _____

16. 89.3 yards to m _____

17. 4.7 m to feet _____

18. 1.92 m to feet _____

19. 3.6 feet to m _____

20. 12.8 feet to m _____

21. 496 km to miles _____

22. 138 km to miles _____

23. 768 miles to km _____

24. 1042 miles to km _____

25. 683 g to pounds _____

26. 1792 g to pounds _____

27. 4.1 pounds to g _____

28. 12.9 pounds to g _____

29. 38.9 kg to pounds _____

30. 40.3 kg to pounds _____

31. One nickel weighs 5 grams. How many nickels are in 1 kilogram of nickels?

31. _____

32. Seawater contains about 3.5 grams of salt per 1000 milliliters of water. How many grams of salt would 5 liters of seawater contain?

32. _____

33. Helium weighs about .0002 gram per milliliter. A balloon contains 3 liters of helium. How much would the helium weigh?

33. _____

34. About 1500 grams of sugar can be dissolved in a liter of warm water. How much sugar could be dissolved in 1 milliliter of warm water?

34. _____

35. Find your height in centimeters.

35. _____

36. Find your height in meters.

36. _____

Convert the following Fahrenheit temperatures to Celsius. Round to the nearest degree.

37. 104°F _____

38. 86°F _____

39. 536°F _____

40. 464°F _____

41. 98°F _____

42. 114°F _____

Convert each of the following Celsius temperatures to Fahrenheit.

43. 35°C _____

44. 100°C _____

45. 10°C _____

46. 25°C _____

47. 135°C _____

48. 215°C _____

In most cases today, medical measurements are given in the metric system. In each of the following problems, a doctor's prescription is given. Decide whether the dosage is or is not reasonable.

49. 1940 grams of Kaopectate after each meal

49. _____

50. 76.8 centiliters of cough syrup every 2 hours

50. _____

51. 943 milliliters of antibiotic every 6 hours

51. _____

52. 1.4 kilograms of vitamins every 3 hours

52. _____

Appendix C

Basic Calculators

Objectives

1. Learn the basic calculator keys.
2. Understand the $\boxed{\text{C}}$, $\boxed{\text{CE}}$, and $\boxed{\text{ON/C}}$ keys.
3. Understand the floating decimal point.
4. Use the $\boxed{\%}$ and $\boxed{\text{1/x}}$ keys.
5. Use the $\boxed{x^2}$, $\boxed{y^x}$, and $\boxed{\sqrt{\ \ }}$ keys.
6. Use the $\boxed{a^{b/c}}$ key.
7. Solve problems with negative numbers.
8. Use the calculator memory function.
9. Solve chain calculations using order of operations.
10. Use the parentheses keys.
11. Use the calculator for problem solution.

Calculators are among the more popular inventions of the last four decades. Each year, better calculators are developed and their cost drops. The first all-transistor desktop calculator was introduced to the market in 1966. It weighed 55 pounds, cost $2500, and was slow. Today, these same calculations are performed quite well on a calculator costing less than $10, and today's $200 pocket calculators have more ability to solve problems than some of the early computers.

Many instructors allow their students to use calculators in business mathematics courses. Some require calculator use. Many types of calculators are available, from the inexpensive basic calculator to the more complex **scientific**, **financial**, and **graphing** calculators.

In this Appendix, we discuss the basic calculator, which has a percent key, reciprocal key, exponent keys, square-root key, memory function, order of operations, and parentheses keys. In Appendix D, the financial calculator with its associated financial keys is discussed.

> **QUICK TIP**   The various calculator models differ significantly—*use the instruction booklet that came with your calculator* for specifics about that calculator if your answers differ from those in this section.

Objective 1 **Learn the basic calculator keys.** Most calculators use **algebraic logic**. Some problems can be solved by entering number and function keys in the same order as you would solve a problem by hand. Other problems require a knowledge of the order of operations when entering the problem.

EXAMPLE 1

Using the Basic Keys

(a) 12 + 25   (b) 456 ÷ 24

SOLUTION

(a) The problem 12 + 25 would be entered as

$$\boxed{1}\ \boxed{2}\ \boxed{+}\ \boxed{2}\ \boxed{5}\ \boxed{=}$$

and 37 would appear as the answer.

(b) Enter $456 \div 24$ as

$$\boxed{4}\ \boxed{5}\ \boxed{6}\ \boxed{\div}\ \boxed{2}\ \boxed{4}\ \boxed{=}$$

and 19 appears as the answer.

Objective $\boxed{2}$ **Understand the** $\boxed{C}$ $\boxed{CE}$**, and** $\boxed{ON/C}$ **keys.** All calculators have a $\boxed{C}$ key. Pressing this key erases everything in most calculators and prepares them for a new problem. Some calculators have a $\boxed{CE}$ key. Pressing this key erases only the number displayed, thus allowing for correction of a mistake without having to start the problem over. Many calculators combine the $\boxed{C}$ key and $\boxed{CE}$ key and use an $\boxed{ON/C}$ key. This key turns the calculator on and is also used to erase the calculator display. If the $\boxed{ON/C}$ is pressed after the $\boxed{=}$, or after one of the operations keys ($\boxed{+}$, $\boxed{-}$, $\boxed{\times}$, $\boxed{\div}$), everything in the calculator is erased. If the wrong operation key is pressed, simply press the correct key and the error is corrected. For example, in $7\ \boxed{+}\ \boxed{-}\ 3\ \boxed{=}$ 4, pressing the $\boxed{-}$ key cancels out the previous $\boxed{+}$ key entry.

Objective $\boxed{3}$ **Understand the floating decimal point.** Most calculators have a **floating decimal** that locates the decimal point in the final result.

EXAMPLE 2

Calculating with Decimal Numbers

Jennifer Videtto purchased 55.75 square yards of vinyl floor covering, at $18.99 per square yard. Find her total cost.

SOLUTION
Proceed as follows.

$$55.75\ \boxed{\times}\ 18.99\ \boxed{=}\ 1058.6925$$

The decimal point is automatically placed in the answer. Since money answers are usually rounded to the nearest cent, the answer is $1058.69.

In using a machine with a floating decimal, enter the decimal point as needed. For example, enter $47 as

$$\boxed{4}\ \boxed{7}$$

with no decimal point, but enter $.95 as follows.

$$\boxed{.}\ \boxed{9}\ \boxed{5}$$

One problem utilizing a floating decimal is shown by the following example.

EXAMPLE 3

Placing the Decimal Point in Money Answers

Add $21.38 and $1.22.

SOLUTION

$$21.38\ \boxed{+}\ 1.22\ \boxed{=}\ 22.6$$

The final 0 is left off. Remember that the problem deals with dollars and cents, and write the answer as $22.60.

Objective $\boxed{4}$ **Use the** $\boxed{\%}$ **and** $\boxed{1/x}$ **keys.** The $\boxed{\%}$ key moves the decimal point two places to the left when used following multiplication or division.

EXAMPLE 4

Using the $\boxed{\%}$ *Key*

Find 8% of $4205.

SOLUTION

$$4205\ \boxed{\times}\ 8\ \boxed{\%}\ \boxed{=}\ 336.4 = \$336.40$$

The $\boxed{1/x}$ key replaces a number with the reciprocal of that number.

EXAMPLE 5

Using the [**1/x**] *Key*

Find the inverse or reciprocal of 40.

SOLUTION

$$40 \boxed{1/x} .025$$

Objective 5   Use the [x^2], [y^x], **and** [$\sqrt{}$] **keys.** The product of 3×3 can be written as follows.

The exponent (2 in this case) shows how many times the base is multiplied by itself (multiply 3 by itself or 3×3). The [x^2] key can be used to quickly find the square of a number.

EXAMPLE 6

Using the [x^2] *Key*

Find 5^2 and 8.5^2.

SOLUTION

$$5 \boxed{x^2} 25 \quad \text{and} \quad 8.5 \boxed{x^2} 72.25 \left\{ \begin{array}{l} \text{Pushing } \boxed{=} \text{ is usually not necessary} \\ \text{when using the } \boxed{x^2} \text{ key.} \end{array} \right.$$

The [y^x] key raises any base number y to a power x. Use as follows.

1. Enter the base number first.
2. [y^x]
3. Enter the exponent.
4. [$=$]

EXAMPLE 7

Using the [y^x] *Key*

Find 5^3.

SOLUTION

$$5 \boxed{y^x} 3 \boxed{=} 125$$

Since $3^2 = 9$, the number 3 is called the **square root** of 9. Square roots of numbers are written with the symbol $\sqrt{}$.

$$\sqrt{9} = 3$$

EXAMPLE 8

Using the [$\sqrt{}$] *Key*

Find each square root.

(a) $\sqrt{144}$ **(b)** $\sqrt{20}$

SOLUTION

(a) Using the calculator, enter

$$144 \boxed{\sqrt{x}}$$

and 12 appears in the display. The square root of 144 is 12.

(b) The square root of 20 is

$$20 \boxed{\sqrt{x}} 4.472136$$

which may be rounded to the desired position.

Objective [6] **Use the** $\boxed{a^{b/c}}$ **key.** Many calculators have an $\boxed{a^{b/c}}$ key that can be used for problems containing fractions and mixed numbers. A mixed number is a number with both a whole number and a fraction such as $7\frac{3}{4}$, which equals $7 + \frac{3}{4}$. The rules for adding, subtracting, multiplying, and dividing both fractions and mixed numbers are given in Chapter 2. Here, we simply show how these operations are done on a calculator.

EXAMPLE 9

Using the $\boxed{a^{b/c}}$ *Key with Fractions*

Solve the following.

(a) $\dfrac{6}{11} + \dfrac{3}{4}$

(b) $\dfrac{3}{8} \div \dfrac{5}{6}$

SOLUTION

(a) $6 \; \boxed{a^{b/c}} \; 11 \; \boxed{+} \; 3 \; \boxed{a^{b/c}} \; 4 \; \boxed{=} \; 1\dfrac{13}{44}$

(b) $3 \; \boxed{a^{b/c}} \; 8 \; \boxed{\div} \; 5 \; \boxed{a^{b/c}} \; 6 \; \boxed{=} \; \dfrac{9}{20}$

QUICK TIP   The calculator automatically shows fractions in lowest terms and as mixed numbers when possible.

EXAMPLE 10

Using the $\boxed{a^{b/c}}$ *Key*

Solve the following.

(a) $4\dfrac{7}{8} \div 3\dfrac{4}{7}$ (b) $\dfrac{5}{3} \div 27.5$ (c) $65.3 \times 6\dfrac{3}{4}$

SOLUTION

(a) $4 \; \boxed{a^{b/c}} \; 7 \; \boxed{a^{b/c}} \; 8 \; \boxed{\div} \; 3 \; \boxed{a^{b/c}} \; 4 \; \boxed{a^{b/c}} \; 7 \; \boxed{=} \; 1\dfrac{73}{200}$

(b) $5 \; \boxed{a^{b/c}} \; 3 \; \boxed{\div} \; 27.5 \; \boxed{=} \; 0.060606061$

(c) $65.3 \; \boxed{\times} \; 6 \; \boxed{a^{b/c}} \; 3 \; \boxed{a^{b/c}} \; 4 \; \boxed{=} \; 440.775$

Objective [7] **Solve problems with negative numbers.** There are several calculations in business that result in a **negative number**, or **deficit amount**.

EXAMPLE 11

Working with Negative Numbers

The amount in the advertising account last month was $4800, while $5200 was actually spent. Find the balance remaining in the advertising account.

SOLUTION

Enter the numbers in the calculator.

$$4800 \; \boxed{-} \; 5200 \; \boxed{=} \; -400$$

The minus sign in front of the 400 indicates that there is a deficit or negative amount. This value can be written as −$400 or sometimes as ($400), which indicates a negative amount. Some calculators place the minus after the number, or as 400−.

Negative numbers may be entered into the calculator by using the $\boxed{-}$ before entering the number. For example, if $3000 is now added to the advertising account in Example 11, the new balance is calculated as follows.

$$\boxed{-} \; 400 \; \boxed{+} \; 3000 \; \boxed{=} \; 2600$$

The new account balance is $2600.

The $\boxed{+/-}$ key can be used to change the sign of a number that has already been entered. For example, $520 \; \boxed{+/-}$ changes $+520$ to -520.

Objective 8 **Use the calculator memory function.** Many calculators feature memory keys, which are a sort of electronic scratch paper. These **memory keys** are used to store intermediate steps in a calculation. On some calculators, a key labeled M or STO is used to store the numbers in the display, with MR or RCL used to recall the numbers from memory.

Other calculators have M+ and M− keys. The M+ key adds the number displayed to the number already in memory. For example, if the memory contains the number 0 at the beginning of a problem, and the calculator display contains the number 29.4, then pushing M+ will cause 29.4 to be stored in the memory (the result of adding 0 and 29.4). If 57.8 is then entered into the display, pushing M+ will cause

$$29.4 + 57.8 = 87.2$$

to be stored. If 11.9 is then entered into the display, with M− pushed, the memory will contain

$$87.2 - 11.9 = 75.3$$

The MR key is used to recall the number in memory as needed, with MC used to clear the memory.

QUICK TIP   Always clear the memory before starting a problem; not doing so is a common error.

Scientific calculators typically have one or more memory registers in which to store numbers. These memory keys are usually labeled as STO for store and RCL for recall. For example, 32.5 can be stored in memory register 1 by

$$32.5 \; \boxed{STO} \; 1$$

or it can be stored in memory register 2 by 32.5 STO 2 and so forth. Values are retrieved from a particular memory register by using the RCL key followed by the number of the register. For example, RCL 2 recalls the contents of memory register 2.

With a scientific calculator, a number stays in memory until it is replaced by another number or until the memory is cleared. The contents of the memory are saved even when the calculator is turned off.

EXAMPLE 12

Using the Memory Registers

An elevator technician counted the number of people entering an elevator and also measured the weight of each group of people. Find the average weight per person.

Number of People	Total Weight
6	839 pounds
8	1184 pounds
4	640 pounds

SOLUTION

First find the weight of all three groups and store in memory register 1.

$$839 \; \boxed{+} \; 1184 \; \boxed{+} \; 640 \; \boxed{=} \; 2663 \; \boxed{STO} \; 1$$

Then find the total number of people.

$$6 \; \boxed{+} \; 8 \; \boxed{+} \; 4 \; \boxed{=} \; 18$$

Finally, divide the contents of memory register 1 by the 18 people.

$$\boxed{RCL} \; 1 \; \boxed{\div} \; 18 \; \boxed{=} \; 147.94444 \text{ pounds}$$

This value can be rounded as needed.

Objective 9 **Solve chain calculations using order of operations.** Long calculations involving several operations (adding, subtracting, multiplying, and dividing) must be done in a specific sequence called the **order of operations** and are called **chain calculations**. The logic of the following order of operations is built into most scientific calculators and can help us work problems without having to store a lot of intermediate values.

Order of Operations

Step 1 Do all operations inside parentheses first.
Step 2 Simplify any expressions with exponents (squares) and find any square roots.
Step 3 Multiply and divide from left to right.
Step 4 Add and subtract from left to right.

EXAMPLE 13

Using the Order of Operations

Solve the following.

(a) $3 + 7 \times 9\frac{3}{4}$ **(b)** $42.1 \times 5 - 90 \div 4$

SOLUTION

The calculator automatically keeps track of the order of operations for us.

(a) 3 $+$ 7 $\times$ 9 $\boxed{a^{b/c}}$ 3 $\boxed{a^{b/c}}$ 4 $=$ $71\frac{1}{4}$

(b) 42.1 $\times$ 5 $-$ 90 $\div$ 4 $=$ 188

Objective 10 **Use the parentheses keys.** The parentheses keys can be used to help establish the order of operations in a more complex chain calculation. For example, $\frac{4}{5 + 7}$ can be written as $\frac{4}{(5 + 7)}$, which can be solved as follows.

Left-parenthesis key

4 $\div$ $($ 5 $+$ 7 $)$ $=$ 0.3333333

Right-parenthesis key

EXAMPLE 14

Using Parentheses

Solve the following problem.

$$\frac{16 \div 2.5}{39.2 - 29.8 \times .6}$$

SOLUTION

Think of this problem as follows:

$$\frac{(16 \div 2.5)}{(39.2 - 29.8 \times .6)}$$

Using parentheses to set off the numerator and denominator will help you minimize errors.

$($ 16 $\div$ 2.5 $)$ $\div$ $($ 39.2 $-$ 29.8 $\times$ $.6$ $)$ $=$ 0.3001876

Objective 11 **Use the calculator for problem solution.** Scientific calculators are great tools to help you solve problems.

EXAMPLE 15

Finding Sale Price

A compact-disc player with an original price of $560 is on sale at 10% off. Find the sale price.

SOLUTION

If the discount from the original price is 10%, then the sale price is 100% − 10% of the original price.

560 $\times$ $($ 100 $-$ 10 $)$ $\%$ $=$ 504

On some calculators the following keystrokes will also work:

560 − 10 % = 504.

EXAMPLE **16**

Applying Calculator Use to Problem Solving

A home buyer borrows $86,400 at 10% for 30 years. The monthly payment on the loan is $8.78 per $1000 borrowed. Annual taxes are $780 and fire insurance is $453 a year. Find the total monthly payment including taxes and insurance.

SOLUTION

The monthly payment is the *sum* of the monthly payment on the loan *plus* monthly taxes plus monthly fire insurance costs. The monthly payment on the loan is the number of thousands in the loan (86.4) times the monthly payment per $1000 borrowed (8.78).

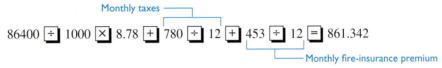

86400 ÷ 1000 × 8.78 + 780 ÷ 12 + 453 ÷ 12 = 861.342

To the nearest cent, this amount rounds to $861.34.

EXAMPLE **17**

Applying Calculator Use to Problem Solving

A Japanese company produces robot dogs for the retail market. The dogs talk, cock their heads, bark, and walk. The company sells the robotic dogs to distributors for $256.80 each.

(a) Find the revenue to the manufacturer if the company sells 26,340 of the robotic dogs.
(b) Find the final price to the customer if the robotic dogs are marked up an average of 87% on cost.

SOLUTION

(a) Revenue = Number sold × Price of each

26340 × 256.80 = 6764112   or $6,764,112

(b) Sales price = Cost × $(1 + \text{Markup percent})$

256.80 × 1.87 = 480.216   or $480.22

Appendix C | Exercises

Solve the following problems on a calculator. Round each answer to the nearest hundredth.

1.   384.92
       407.61
       351.14
   +   27.93

2.   85.76
       21.94
   +   39.89

3.   6850
          321
   +   4207

4.   781.42
       304.59
   +   261.35

5.   4270.41
   −   365.09

6.   3000.07
   −   48.12

7.   384.96
   −   129.72

8. $36.84 - 12.17$

9.   365
   ×   43

10.   27.51
   ×   1.18

11. 3.7×8.4

12. 62.5×81

13. $\dfrac{375.4}{10.6}$

14. $\dfrac{9625}{400}$

15. $96.7 \div 3.5$

16. $103.7 \div .35$

Solve the following chain calculations. Round each answer to the nearest hundredth.

17. $\dfrac{9 \times 9}{2 \times 5}$

18. $\dfrac{15 \times 8 \times 3}{11 \times 7 \times 4}$

19. $\dfrac{87 \times 24 \times 47.2}{13.6 \times 12.8}$

20. $\dfrac{2 \times (3 + 4)}{6 + 10}$

21. $\dfrac{2 \times 3 + 4}{6 + 10}$

22. $\dfrac{4200 \times .12 \times 90}{365}$

23. $\dfrac{640 - .6 \times 12}{17.5 + 3.2}$

24. $\dfrac{16 \times 18 \div .42}{95.4 \times 3 - .8}$

25. $\dfrac{14^2 - 3.6 \times 6}{95.2 \div .5}$

26. $\dfrac{9^2 + 3.8 \div 2}{14 + 7.5}$

Solve the following problems. Reduce any fractions to lowest terms or round to the nearest hundredth.

27. $7\frac{5}{8} \div \left(1 + \frac{3}{8}\right)$

28. $\left(5\frac{1}{4}\right)^2 \times 3.65$

29. $\left(\frac{3}{4} \div \frac{5}{8}\right)^3 \div 3\frac{1}{2}$

30. $\sqrt{6} \times \dfrac{3^2 + 2\frac{1}{2}}{7 \times \frac{5}{6}}$

31. Describe in your own words the order of operations to be used when solving chain calculations. (See Objective 9.)

32. Explain how the parentheses keys are used when solving chain calculations. (See Objective 9.)

Solve the following application problems on a calculator. Round each answer to the nearest cent.

33. Bucks County Community College Bookstore bought 397 used copies of a computer science book at a net cost of $46.40 each; 125 used copies of an accounting book at $38.40 each; and 740 used copies of a real estate text at $28.30 each. Find the total paid by the bookstore.

33. _____

34. Judy Martinez needs to file her expense account claims. She spent 5 nights at the Macon Holiday Inn at $104.19 per night and 4 nights at the Charlotte Sheraton at $86.80 per night. She then rented a car for 8 days at $36.40 per day. She drove the car 916 miles with a charge of $.28 per mile. Find her total expenses.

34. _____

35. In Virginia City, the sales tax is 6.5%. Find the tax on each of the following items: **(a)** a new car costing $17,908.43 and **(b)** a computer costing $1463.58.

(a) _____
(b) _____

36. Marja Strutz bought a two-year-old commercial fishing boat equipped for sardine fishing at a cost of $78,250. Additional safety equipment was needed at a cost of $4820, and sales tax of $7\frac{1}{4}\%$ was due on the boat and safety equipment. In addition she was charged a licensing fee of $1135 and a Coast Guard registration fee of $428. Strutz will pay $\frac{1}{3}$ of the total cost as a down payment and will borrow the balance. How much will she borrow?

36. _____

37. Ben Fick bought a small townhouse for $80,000. He paid $8000 down and agreed to make payments of $528.31 each month for 30 years. By how much does the down payment and the sum of the monthly payments exceed the purchase price?

37. _____

38. Linda Smelt purchased a 24-unit apartment house for $620,000. She made a down payment of
$150,000, which she had inherited from her parents, and agreed to make monthly payments of $5050
for 15 years. By how much does the sum of her down payment and all monthly payments exceed the
original purchase price?

38. _____

39. Ben Hurd wishes to open a small repair shop, but only has $32,400 in cash. He estimates that he will
need $15,000 for equipment, $2800 for the first month's rent on a building, and about $28,000
operating expenses until the business is profitable. How much additional funding does he need?

39. _____

40. Koplan Kitchens wishes to expand their retail store. In order to do so, they must first purchase the
$26,000 parcel of land next door to them. They then anticipate $120,000 in construction costs plus an
additional $28,500 for additional inventory. They have $50,000 in cash and must borrow the balance
from a bank. How much must they borrow?

40. _____

41. A college bookstore buys a used textbook for $24.50 at the end of a semester and sells it at the
beginning of the next semester for $60. Find the percent of markup on selling price to the nearest
percent.

41. _____

42. A homebuilder spent the following when building a home: $37,800 for a cleared parcel of land with
utility hookups, $59,600 for materials, and $24,300 for labor and other expenses. He then sold the
home for $136,500. Find the percent markup over cost to the nearest percent.

42. _____

Appendix D

Financial Calculators

Objectives

1. Learn the basic conventions used with cash flows.
2. Learn the basic financial keys.
3. Understand which keys to use for a particular problem.
4. Use the calculator to solve financial problems.

Calculators are among the more popular inventions of recent times. The power and capability of calculators has increased significantly even as their cost has continued to fall. Today, programmable calculators costing less than $100 have more ability to solve problems than some of the early computers.

Objective 1 Learn the basic conventions used with cash flows. There is a need to logically separate inflows of cash (cash received) from outflows of cash (cash paid out). The following convention is commonly used for this purpose, and will be used throughout this appendix.

1. Inflows of cash (cash received) are **positive**.
2. Outflows of cash (cash paid out) are **negative**.

For example, assume that you are making regular investments into an account. Your payments are *outflows* of cash and should be considered *negative* numbers. The future value of your savings will eventually be returned to you as an inflow of cash, thereby as a positive number.

Objective 2 Learn the basic financial keys. **Financial calculators** have special functions that allow the user to solve financial problems involving time, interest rates, and money. Many of the compound interest problems presented in this text can be solved using a financial calculator. Most financial calculators have financial keys similar to those shown below.

These keys represent the following functions (see Chapter 9 for a full definition of each term):

n —The number of compounding periods

i —The interest rate *per compounding period*

PV —Present value—the value in *today's* dollars

PMT —The amount of a level payment (for example, $625 per month); this is used for annuity type problems.

FV —Future value—the value at *some future date*

QUICK TIP Different financial calculators look and work somewhat differently from one another. You *must look at the instruction book* that came with your calculator to determine how the keys are used with that particular calculator.

QUICK TIP Different financial calculators sometimes give slightly different answers to the same problems due to rounding.

Objective ③ **Understand which keys to use for a particular problem.** Most simple financial problems require only four of the five financial keys described on the previous page. Both the number of compounding periods ⃞n⃞ and the interest rate per compounding period ⃞i⃞ *are needed for each financial problem*—these two keys will always be used. Which two of the remaining three financial keys (⃞**PV**⃞ , ⃞**PMT**⃞ , and ⃞**FV**⃞) are used depends on the particular problem. Using the convention described under Objective 1, one of these values will be negative and one will be positive. The process of solving a financial problem is to enter values for the three variables that are known, *then press the key for the unknown*, fourth variable.

For example, if you wish to know the future value of a series of known, equal payments, enter the specific values for ⃞n⃞ , ⃞i⃞ , and ⃞**PMT**⃞ . Then press ⃞**FV**⃞ for the result. Or, if you wish to know how long it will take for an investment to grow to some specific value at a given interest rate, enter values for ⃞**PV**⃞ , ⃞i⃞ , and ⃞**FV**⃞ . Then press ⃞n⃞ to find the required number of compounding periods.

QUICK TIP   Be sure to enter a cash inflow as a positive number or a cash outflow as a negative number. Also be sure to clear all values from the memory of your calculator before working a problem.

Objective ④ **Use the calculator to solve financial problems.**

EXAMPLE 1

Given n, i, and PV,
Find FV

Barbara and Ivan Cushing invest $2500 that they received from the sale of the old family car into a stock mutual fund that has recently paid 12% compounded quarterly. Find the future value in 5 years if the fund continues to do as well.

SOLUTION

The present value of $2500 (a cash outflow is entered as a negative number) is compounded at 3% per quarter $(12\% \div 4 = 3\%)$ for 20 quarters $(4 \times 5 = 20)$. Enter values for ⃞**PV**⃞ , ⃞i⃞ , and ⃞n⃞ .

$$-2500 \; \boxed{PV} \; 3 \; \boxed{i} \; 20 \; \boxed{n}$$

Then press ⃞**FV**⃞ to find the compound amount at the end of 5 years.

$$\boxed{FV} \; \$4515.28, \text{ which is the future value.}$$

EXAMPLE 2

Given n, i, and PMT,
Find FV

Joan Jones plans to invest $100 at the end of each month in a mutual fund that she believes will grow at 9% per year compounded monthly. Find the future value at her retirement in 20 years.

SOLUTION

Two hundred forty payments $(12 \times 20 = 240)$ of $100 each (cash outflows entered as a negative number) are made into an account earning .75% per month $(9\% \div 12 = .75\%)$. Enter values for ⃞n⃞ , ⃞**PMT**⃞ , and ⃞i⃞ .

$$240 \; \boxed{n} \; -100 \; \boxed{PMT} \; .75 \; \boxed{i}$$

Press ⃞**FV**⃞ for the result.

$$\boxed{FV} \; \$66,788.69, \text{ which is the future value.}$$

QUICK TIP   The order in which data are entered into the calculator does not matter—just remember to press the financial key for the unknown value last.

Any one of the four values used to solve a particular financial problem *can be unknown*. Look at the next three examples in which the number of compounding periods $\boxed{\textbf{n}}$, the payment amount $\boxed{\textbf{PMT}}$, and the interest rate per compounding period $\boxed{\textbf{i}}$, respectively, are unknown.

EXAMPLE 3

Given i, PMT, *and* FV,
Find n

Mr. Trebor needs $140,000 for a new tractor. He can invest $8000 at the end of each month in an account paying 6% per year compounded monthly. How many monthly payments are needed?

SOLUTION

The $8000 monthly payment (cash outflow) will grow at .5% per compounding period $\left(6\% \div 12 = .5\%\right)$ until a future value of $140,000 (cash inflow at a future date) is accumulated. Enter values for $\boxed{\textbf{PMT}}$, $\boxed{\textbf{i}}$, and $\boxed{\textbf{FV}}$.

$$-8000 \;\boxed{\textbf{PMT}}\; .5 \;\boxed{\textbf{i}}\; 140000 \;\boxed{\textbf{FV}}$$

Press $\boxed{\textbf{n}}$ to determine the number of payments.

$$\boxed{\textbf{n}} \;\; 17 \text{ monthly payments of \$8000 each are needed.}$$

Actually, 17 payments of $8000 each in an account earning .5% per month will grow to slightly more than $140,000:

$$-8000 \;\boxed{\textbf{PMT}}\; .5 \;\boxed{\textbf{i}}\; 17 \;\boxed{\textbf{n}}$$

Press $\boxed{\textbf{FV}}$ to determine the future value.

$$\boxed{\textbf{FV}} \;\; \textbf{\$141,578.41}, \text{ which is the future value.}$$

The 17th payment would only need to be

$$\$8000 - \left(\textbf{\$141,578.41} - \$140,000\right) = \$6421.59$$

to accumulate exactly $140,000.

EXAMPLE 4

Given n, i, *and* FV,
Find PMT

Jane Abel wishes to have $1,000,000 at her retirement in 40 years. Find the payment she must make at the end of each quarter into an account earning 10% compounded quarterly to attain her goal.

SOLUTION

One hundred sixty payments $\left(40 \times 4 = 160\right)$ are made into an account earning 2.5% per quarter $\left(10\% \div 4 = 2.5\%\right)$ until a future value of $1,000,000 (cash inflow at a future date) is accumulated. Enter values for $\boxed{\textbf{n}}$, $\boxed{\textbf{i}}$, and $\boxed{\textbf{FV}}$.

$$160 \;\boxed{\textbf{n}}\; 2.5 \;\boxed{\textbf{i}}\; 1000000 \;\boxed{\textbf{FV}}$$

Press $\boxed{\textbf{PMT}}$ for the quarterly payment.

$$\boxed{\textbf{PMT}} \;\; - \;\textbf{\$490.41}, \text{ which is the required quarterly payment of cash.}$$

One hundred sixty payments of $490.41 at the end of each quarter in an account earning 10% compounded quarterly will grow to $1,000,000.

EXAMPLE 5

Given n, PV, *and* FV,
Find i

Tom Fernandez bought 200 shares of stock in an oil company at $33.50 per share. Exactly three years later, he sold the stock at $41.25 per share. Find the annual interest rate, rounded to the nearest tenth of a percent, that Mr. Fernandez earned on this investment.

SOLUTION

In 3 years, the per share price increased from a present value of $33.50 to a future value of $41.25. The purchase of the stock is a cash outflow and the eventual sale of the stock is a cash inflow. It is not necessary to multiply the stock price by the number of shares—the interest rate indicating the return on the investment is the same whether 1 share or 200 shares are used. Enter values for n , PV , and FV .

$$3 \boxed{n} \; -33.50 \boxed{PV} \; 41.25 \boxed{FV}$$

Press i for the annual interest rate.

$$\boxed{i} \; \textbf{7.18\%}, \text{ or about 7.2\% per year.}$$

Mr. Fernandez's return on his original investment compounded at 7.2% per year.

Interest rates can have a great influence on both individuals and businesses. Individuals borrow for homes, cars, and other personal items whereas firms borrow to buy real estate, expand operations, or cover operating expenses. A small difference in interest rates can make *a large difference* in costs over time as shown in the next example.

EXAMPLE 6

Compare Monthly House Payments

John and Leticia Adams wish to borrow $62,000 on a 30-year home loan. Find the monthly payment at interest rates of **(a)** 8% and **(b)** 9%. Show **(c)** the monthly savings at the lower rate and **(d)** the total savings in monthly payments over the 30 years.

SOLUTION

(a) Enter a present value of $62,000 (cash inflow) with 360 compounding periods $(30 \times 12 = 360)$ and a rate of .666667% per month $(8\% \div 12 = .666667, \text{rounded})$ and press PMT to find the monthly payment.

$$62000 \boxed{PV} \; 360 \boxed{n} \; .666667 \boxed{i}$$

$\boxed{PMT}$ $-\$454.93$ is the monthly payment at 8% per year, rounded.

(b) Enter the values again using the new interest rate of .75% $(9\% \div 12 = .75\%)$.

$$62000 \boxed{PV} \; 360 \boxed{n} \; .75 \boxed{i}$$

$\boxed{PMT}$ $-\$498.87$ is the monthly payment at 9% per year, again, rounded.

(c) The difference in the monthly payment follows.

$$\$498.87 - \$454.93 = \textbf{\$43.94}$$

(d) The total difference saved over 30 years $(30 \times 12 = 360 \text{ payments})$ is

$$\textbf{\$43.94} \times 360 \text{ payments} = \$15,818.40.$$

The lower interest rate will reduce the Adams' mortgage payments by a total of $15,818.40 over 30 years.

EXAMPLE 7

Retirement Planning

Courtney and Nathan Wright plan to retire in 25 years and need $3500 per month for 20 years.

(a) Find the amount needed at retirement to fund the monthly retirement payments assuming the funds earn 9% compounded monthly while payments are being made.
(b) Find the amount of the quarterly payment they must make for the next 25 years to accumulate the necessary funds, assuming earnings of 12% compounded quarterly during the accumulation period.

SOLUTION

(a) The accumulated funds at the end of 25 years is, at their retirement, a present value that must generate a cash inflow to the Wrights of $3500 per month for 240 months $(20 \times 12 = 240)$ assuming earnings of .75% per month $(9\% \div 12 = .75\%)$. Enter values for $\boxed{\textbf{n}}$, $\boxed{\textbf{i}}$, and $\boxed{\textbf{PMT}}$.

240 $\boxed{\textbf{n}}$ .75 $\boxed{\textbf{i}}$ 3500 $\boxed{\textbf{PMT}}$

Press $\boxed{\textbf{PV}}$ to find the amount needed at the end of 25 years.

$\boxed{\textbf{PV}}$ **$389,007.34** is the amount they must accumulate.

(b) The Wrights have 25 years of quarterly payments (100 payments that are cash outflows) in an account earning 3% per quarter $(12\% \div 4 = 3\%)$ to accumulate a future value of $389,007.34. The question is one of what quarterly payment is required. Enter values for $\boxed{\textbf{n}}$, $\boxed{\textbf{i}}$, and $\boxed{\textbf{FV}}$.

100 $\boxed{\textbf{n}}$ 3 $\boxed{\textbf{i}}$ 389007.34 $\boxed{\textbf{FV}}$

Press $\boxed{\textbf{PMT}}$ to find the quarterly payment needed.

$\boxed{\textbf{PMT}}$ **−$640.57** is the required quarterly payment.

Thus, the Wrights must make 100 end-of-quarter deposits of $640.57 each into an account earning 3% per quarter in order to subsequently receive 20 years of payments of $3500 per month, assuming 9% per year during the time that payments are made.

Appendix D | Exercises

Using a financial calculator, solve the following problems for the missing quantity. Round dollar answers to the nearest hundredth, interest rates to the nearest hundredth of a percent, and number of compounding periods to the nearest whole number. Assume that any payments are made at the end of the period.

	n	i	PV	PMT	FV
1.	20	10%	$5800	—	_____
2.	7	8%	$8900	—	_____
3.	10	3%	_____	—	$12,000
4.	16	4%	_____	—	$8200
5.	7	8%	—	$300	_____
6.	25	2%	—	$1000	_____
7.	30	_____	—	$319.67	$12,000
8.	50	_____	—	$4718.99	$285,000
9.	360	1%	$83,500	_____	—
10.	180	.5%	$125,000	_____	—
11.	_____	4%	$85,383	$5600	—
12.	_____	2%	$3822	$100	—

Solve each of the following application problems.

13. Juanipa Manglimont inherited $23,500 from her father. She placed the money in a 5-year certificate of deposit earning 6% compounded quarterly. Find the future value at the end of 5 years.

13. _____

14. At the end of each month, Tina Ramirez has $50 per month taken out of her paycheck and invested in an account paying .5% per month. Find the future value at the end of 14 years.

14. _____

15. Mr. and Mrs. Thrash borrowed $86,500 on a 30-year home loan at 9% per year. Find the monthly payment.

15. _____

16. Terrance Walker wishes to have $20,000 in 10 years when his son begins college. What payment must he make at the end of each quarter in an investment earning 10% compounded quarterly?

16. _____

17. The *Daily Gazette* needs $340,000 for a new printing press. The *Gazette* can invest $12,000 per month in an account paying .8% per month. Find the number of payments that must be paid before reaching its goal. Round to the nearest whole number.

17. _____

18. Cathy Cockrell anticipates that she will need $70,000 when her son Sam enters college. She can save $500 per month and earn 7% per year compounded monthly. How long will it take her to save the needed funds?

18. _____

19. Mr. and Mrs. Peters wish to build their dream home and must borrow $110,000 on a 30-year mortgage to do so. Find the highest acceptable annual interest rate, to the nearest tenth of a percent, if they cannot afford a monthly payment above $845.

19. _____

20. Jim Blalock needs to borrow $28,000 for a new work truck, but cannot afford a payment of more than $700 per month. If a bank will finance the truck for 4 years, find the maximum interest rate Blalock can afford.

20. _____

Answers to Selected Exercises

Chapter 1

Section 1.1 Exercises (Page 11)

1. seven thousand, forty **3.** thirty-seven thousand, nine hundred one
5. seven hundred twenty-five thousand, nine **7.** 2070; 2100; 2000
9. 46,230; 46,200; 46,000 **11.** 106,050; 106,100; 106,000 **15.** 210
17. 2186 **19.** 1396 **21.** 983,493 **23.** 668 **25.** 2877 **27.** 21,546
29. 6,088,899 **31.** Totals vertically: $387,795; $426,869; $373,100;
$1,481,031; Totals horizontally: $269,761; $267,502; $206,932; $246,587;
$244,616; $245,633; $1,481,031 **33.** 9374 **35.** 117,552 **37.** 1,696,876
39. 8,107,899 **41.** Estimate: 12,760; Exact: 12,605 **43.** Estimate: 600;
Exact: 545 **45.** Estimate: 30,000; Exact: 29,986 **47.** $37 \times 18 = 666$;
66,600 **49.** $376 \times 6 = 2256$; 22,560,000 **51.** $1241\frac{1}{4}$ **53.** $458\frac{21}{43}$
57. $2385\frac{5}{18}$ **59.** $58\frac{4}{13}$ **61.** seven million, five hundred forty-three thousand,
five hundred **63.** three million, two hundred thousand **65.** 854,795 boxes
67. 353,000,000 **69.** 3000 balls **71.** 500 items per hour **73.** $2408
75. 293,387 acres **77.** 24,235 acres **79.** 4500 stores **81.** 6000 stores
83. 500 stores

Section 1.2 Exercises (Page 21)

1. 382 miles **3.** 467 passengers **5.** 401,500 veterans **7.** 2477 pounds
9. $26,898 **11.** 6,011,280 square feet **13.** $500 **15.** $20,961 **17.** $375
19. 20 seats

Section 1.3 Exercises (Page 25)

1. thirty-eight hundredths **3.** five and sixty-one hundredths **5.** seven and
four hundred eight thousandths **7.** thirty-seven and five hundred ninety-three
thousandths **9.** four and sixty-two ten-thousandths **13.** 438.4 **15.** 97.62
17. 1.0573 **19.** 3.5827 **21.** $.98 **23.** $.58 **25.** $1.17 **27.** 3.5; 3.52;
3.522 **29.** 2.5; 2.55; 2.548 **31.** 27.3; 27.32; 27.325 **33.** 36.5; 36.47;
36.472 **35.** .1; .06; .056 **37.** $5.06 **39.** $32.49 **41.** $382.01 **43.** $42.14
45. $.00 **47.** $1.50 **49.** $2.00 **51.** $752.80 **53.** $26 **55.** $0
57. $12,836 **59.** $395 **61.** $4700 **63.** $379 **65.** $722

Section 1.4 Exercises (Page 29)

1. $40 + 20 + 9 = 69$; 68.46 **3.** $6 + 4 + 5 + 7 + 2 = 24$; 23.82
5. $2000 + 5 + 3 + 7 = 2015$; 2171.414
7. $6000 + 500 + 20 + 8 = 6528$; 6666.061
9. $2000 + 70 + 500 + 600 + 400 = 3570$; 3451.446 **11.** 173.273
13. 59.3268 **17.** $5150.90 **19.** $5.73 per pound **21.** $20 - 7 = 13$; 13.16
23. $50 - 20 = 30$; 31.507 **25.** $300 - 90 = 210$; 240.034 **27.** $8 - 3 = 5$;
4.848 **29.** $5 - 2 = 3$; 3.0198 **31.** $43,815.81

Section 1.5 Exercises (Page 35)

1. $100 \times 4 = 400$; 406.56 **3.** $30 \times 7 = 210$; 231.88 **5.** $40 \times 2 = 80$;
89.352 **7.** 1.9152 **9.** 9.3527 **11.** .002448 **13.** $152.63 **15.** $418.10
17. 8.075 **19.** 27.442 **21.** 57.977 **25.** $14,790 **27.** 25.1 mpg **29.** 26
coins **31.** (a) .43 inch (b) 4.3 inches **33.** $129.25 **35.** (a) $70.05
(b) $25.80

Summary Exercise (Page 42)

(a) $24,168 (b) $3811 (c) 218 guests; $24 left over (d) $46.67
(e) 148 guests; $4 left over (f) 66 guests; $10 left over (g) $644.25

Chapter 1 Test (Page 43)

1. 840 **2.** 22,000 **3.** 672,000 **4.** 50,000 **5.** 900,000 **6.** $606 **7.** $8399
8. $21.06 **9.** $364.35 **10.** $7246 **11.** 181.535 **12.** 498.795 **13.** 133.6

14. 3.7947 **15.** 15.8256 **16.** 8.0882 **17.** 11.56 **18.** 23.8 **19.** 4.25
20. $505.31 **21.** $3942.90 (rounded) **22.** 14,454 gallons saved **23.** $3.93
24. $.79 per pound (rounded) **25.** 253 seedlings

Chapter 2

Section 2.1 Exercises (Page 51)

1. $\frac{29}{8}$ **3.** $\frac{17}{4}$ **5.** $\frac{38}{3}$ **7.** $\frac{183}{8}$ **9.** $\frac{55}{7}$ **11.** $\frac{364}{23}$ **13.** $3\frac{1}{4}$ **15.** $2\frac{2}{3}$ **17.** $3\frac{4}{5}$ **19.** $3\frac{7}{11}$
21. $1\frac{62}{63}$ **23.** $7\frac{8}{25}$ **27.** $\frac{1}{2}$ **29.** $\frac{2}{3}$ **31.** $\frac{6}{7}$ **33.** $\frac{5}{9}$ **35.** $\frac{7}{8}$ **37.** $\frac{1}{50}$ **41.** √ x √ x x √
x x **43.** √ √ √ √ √ x x √ **45.** √ √ x √ √ x √ √ **47.** √ x √ x x x x x

Section 2.2 Exercises (Page 57)

1. 16 **3.** 36 **5.** 48 **7.** 42 **9.** 24 **11.** 180 **13.** 480 **15.** 2100 **17.** 360
21. $\frac{2}{3}$ **23.** $\frac{1}{2}$ **25.** $\frac{17}{48}$ **27.** $1\frac{23}{36}$ **29.** $\frac{13}{14}$ **31.** $2\frac{1}{15}$ **33.** $2\frac{11}{36}$ **35.** $1\frac{13}{30}$ **37.** $\frac{9}{20}$
39. $\frac{7}{24}$ **43.** $\frac{23}{24}$ cubic yard **45.** $\frac{47}{60}$ inch **47.** $\frac{7}{24}$ of the contents **49.** $\frac{19}{24}$ of the
debt **51.** $\frac{3}{16}$ inch **53.** $\frac{7}{24}$ **55.** work and travel; 8 hours **57.** $\frac{1}{12}$ mile

Section 2.3 Exercises (Page 63)

1. $97\frac{4}{5}$ **3.** $80\frac{3}{4}$ **5.** $97\frac{7}{40}$ **7.** $53\frac{17}{24}$ **9.** $105\frac{107}{120}$ **11.** $7\frac{1}{8}$ **13.** $162\frac{1}{6}$ **15.** $9\frac{1}{24}$
17. $46\frac{23}{30}$ **19.** $\frac{7}{10}$ **23.** $116\frac{1}{2}$ inches **25.** 130 feet **27.** $1\frac{5}{8}$ cubic yards
29. $22\frac{7}{8}$ hours

Section 2.4 Exercises (Page 69)

1. $\frac{3}{10}$ **3.** $\frac{99}{160}$ **5.** $\frac{9}{32}$ **7.** $4\frac{3}{8}$ **9.** $9\frac{1}{3}$ **11.** $\frac{1}{3}$ **13.** $4\frac{7}{12}$ **15.** $1\frac{7}{9}$ **17.** $\frac{3}{5}$ **19.** $1\frac{1}{2}$
21. $\frac{2}{3}$ **23.** $2\frac{2}{3}$ **25.** $\frac{3}{20}$ **27.** $8\frac{2}{5}$ **31.** $12 **33.** $18.75 **37.** 6¢ **39.** 36 yards
41. 12 homes **43.** $2632\frac{1}{2}$ inches **45.** 2480 anchors **47.** 471 gallons
49. 88 dispensers **51.** 60 trips

Section 2.5 Exercises (Page 75)

1. $\frac{3}{4}$ **3.** $\frac{6}{25}$ **5.** $\frac{73}{100}$ **7.** $\frac{17}{20}$ **9.** $\frac{17}{50}$ **11.** $\frac{111}{250}$ **13.** $\frac{5}{8}$ **15.** $\frac{161}{200}$ **17.** $\frac{12}{125}$ **19.** $\frac{3}{80}$
21. $\frac{1}{16}$ **23.** $\frac{1}{625}$ **27.** .25 **29.** .375 **31.** .667 (rounded) **33.** .778 (rounded)
35. .636 (rounded) **37.** .88 **39.** .883 (rounded) **41.** .993 (rounded)

Summary Exercise (Page 79)

(a) $144,000 (b) $\frac{5}{12}$; $\frac{1}{4}$; $\frac{1}{12}$; $\frac{1}{16}$; $\frac{1}{16}$; $\frac{1}{8}$ (c) Miscellaneous: $\frac{1}{8}$; Insurance: $\frac{1}{16}$;
Advertising: $\frac{1}{16}$; Utilities: $\frac{1}{12}$; Rent: $\frac{1}{4}$; Salaries: $\frac{5}{12}$ (d) 150°; 90°; 30°; 22.5°;
22.5°; 45° (e) 360°; A full circle is 360°

Chapter 2 Test (Page 81)

1. $\frac{5}{6}$ **2.** $\frac{7}{8}$ **3.** $\frac{7}{11}$ **4.** $8\frac{1}{8}$ **5.** $4\frac{2}{3}$ **6.** $2\frac{2}{3}$ **7.** $\frac{31}{4}$ **8.** $\frac{94}{5}$ **9.** $\frac{147}{8}$ **10.** 30 **11.** 120
12. 72 **13.** $\frac{7}{8}$ **14.** $15\frac{1}{16}$ **15.** $36\frac{5}{16}$ **16.** 36 **17.** $1\frac{1}{2}$ **18.** $24\frac{1}{8}$ pounds
19. $510 **20.** $35\frac{7}{8}$ gallons **21.** 36 pull cords **22.** $\frac{5}{8}$ **23.** $\frac{41}{50}$ **24.** .25 inch
25. .875 inch

Chapter 3

Section 3.1 Exercises (Page 89)

1. 25% **3.** 72% **5.** 203.4% **7.** 362.5% **9.** 87.5% **11.** .05% **13.** 345%
15. 3.08% **17.** .625 **19.** .65 **21.** .125 **23.** .125 **25.** .0025 **27.** .8475
29. 1.75 **31.** .5; 50% **33.** $\frac{7}{8}$; 87.5% **35.** $\frac{1}{125}$; .008 **37.** 10.5; 1050%
39. $\frac{13}{20}$; 65% **41.** $\frac{1}{200}$; .5% **43.** .333$\overline{3}$; $33\frac{1}{3}$% **45.** $2\frac{1}{2}$; 250% **47.** $\frac{17}{400}$; .0425
49. .015; 1.5% **51.** $10\frac{3}{8}$; 10.375 **53.** $\frac{1}{400}$; .25% **55.** $\frac{3}{8}$; .375

Section 3.2 Exercises (Page 97)

1. 6 bicycles **3.** $604 **5.** 4.8 feet **7.** 10,185 miles **9.** 182 homes
11. 148.44 yards **13.** $5366.65 **17.** 264 adults **19.** $429.92 **21.** 44

females **23.** 8.95 ounces **25.** 4853 accidents **27. (a)** 28.6% female
(b) 313,099 female **29.** $249 **31.** 2156 products **33.** $135 million
35. $51,844.20 **37.** $6296.40 **39.** $199.89 **41.** $87.58

Section 3.3 Exercises (Page 105)

1. 2120 **3.** 325 **5.** 2000 **7.** 4800 **9.** 44,000 **11.** 20,000 **13.** $90,320
15. 1080 **17.** 312,500 **19.** 65,400 **21.** 40,000 **25.** 107.1 million households **27.** 7761 students **29.** $3800 **31.** 1055 people **33.** 180,000 jobs
35. $185,500

Supplementary Exercises (Page 107)

1. 16 ounces **3.** $288,150 **5.** 478,175 Mustangs **7.** 836 drivers
9. $39,000 **11.** 230 companies **13.** 55 companies

Section 3.4 Exercises (Page 113)

1. 10 **3.** 50 **5.** 28.3 **7.** 76 **9.** 4.1 **11.** 5.9 **13.** 1.25 **15.** 250 **17.** 27.8
21. 6.2% **23.** 2% **25.** 8.7% **27.** 9.1% **29.** 20%

Supplementary Exercises (Page 115)

1. 486 firms **3.** 40% **5.** $4.8 billion **7.** $396.05 **9.** $134 **11.** $284.40
13. 4% **15.** 4 million riders **17.** 24.4% **19.** 12.5% **21.** 5.5% **23.** 960
candy bars **25.** $8823 **27.** 2.1% **29.** 5742 deaths **31. (a)** 36% **(b)** 64%

Section 3.5 Exercises (Page 125)

1. $375 **3.** $27.91 **5.** $25 **7.** $854.50 **11.** $165,500 **13. (a)** $950
(b) $76 **15.** $306.7 million **17.** 22,000 people **19.** 747,126 people
21. $14.99 **23.** $3864 **25.** $145.24 million **27.** 51.2 million **29.** 25,000
students **31.** 695 deaths **33.** 6564 homes

Summary Exercise (Page 131)

Amazon.com, 172.7%; DaimlerChrysler, 39.45; Gateway Computer, 3.14;
Krispy Kreme, 18.5%; McDonald's, 52.1%; Merck, 55.30; Pepsi Bottling,
23.27; R. J. Reynolds, 42.11; Wal-Mart, 52.68; Yahoo, 159.9%

Chapter 3 Test (Page 133)

1. 300 home sales **2.** 12 open houses **3.** 1100 shippers **4.** 2.5% **5.** $3.15
6. $\frac{6}{25}$ **7.** 1920 purchase orders **8.** $\frac{7}{8}$ **9.** 8.5% **10.** $\frac{1}{200}$ **11.** $1.05
12. 224,000 units **13.** $5185; $25,315 **14.** 269.2 million people
15. (a) 11% **(b)** $4488 per year **16.** 75% **17.** $450 **18.** 1200 backpacks
19. 8.7% **20.** $1.47 billion

Cumulative Review Chapters 1–3 (Page 135)

1. 65,500 **3.** 78.4 **5.** 3609 **7.** 24,092 **9.** 85 **11.** 35.174 **13.** 12.218
15. $198 **17.** $31,658.27 **19.** $\frac{8}{9}$ **21.** $7\frac{2}{15}$ **23.** $13\frac{13}{24}$ **25.** 3 **27.** $42\frac{1}{2}$ acres
29. 130 feet **31.** $\frac{13}{20}$ **33.** 87.5% **35.** 2170 home loans **37.** 25%
39. 38.5% **41.** 97,757 copies **43.** 90,000% **45. (a)** 83%
(b) 1.909 billion pounds **47.** .69 billion pounds

Chapter 4

Section 4.1 Exercises (Page 147)

1. $14.20 **3.** $20.00 **5.** $17.10 **7.** $21.90 **9.** Mar. 8; $380.71; Nola
Akala; Tutoring; 3971.28; 79.26; 4050.54; 380.71; 3669.83 **11.** Dec. 4;
$37.52; Paul's Pools; Chemicals; 1126.73; 1126.73; 37.52; 1089.21
17. Oct. 10; $39.12; County Clerk; License; 5972.89; 752.18; 23.32; 6748.39;
39.12; 6709.27 **19.** 9412.64; 8838.86; 8726.71; 9479.99; 10,955.68;
10,529.13; 9891.20; 9825.58; 9577.41; 9913.26; 9462.76 **21.** 574.86; 384.36;
462.65; 620.07; 581.31; 405.43; 784.71; 587.51; 562.41; 487.41; 1209.76

Section 4.2 Exercises (Page 155)

1. $1595.36 **3.** $1387.67 **5.** $1332.16 **7.** $203.86 **9.** $66.48
11. $1064.72 **13.** $991.89 **15.** $962.13 **17.** $60.21 **19.** $59.25

Section 4.3 Exercises (Page 163)

1. $4870.24 **3.** $7690.62 **5.** $18,314.72 **11.** 421; $371.52; 424; $429.07;
427; $883.69; 429; $35.62; $1719.90; $6875.09; 701.56; 421.78; 689.35;

8687.78; 1719.90; $6967.88; $6965.92; 8.75; 6957.17; 10.71; $6967.88
13. 767, $63.24; 771, $135.76; $199.00; $5636.51; 220.16; 5856.67; 199.00;
$5657.67; $5858.85; 209.30; 5649.55; 8.12; $5657.67

Summary Exercise (Page 168)

(a) $8178.46 gross deposit **(b)** $7974 credit **(c)** $9810.36 total of checks
outstanding **(d)** $4882.58 deposits not recorded **(e)** $7274.56 balance

Chapter 4 Test (Page 169)

1. $19.90 **2.** $9.40 **3.** $17.40 **4.** Aug. 6; $6892.12; WBC Broadcasting;
Airtime; $16,409.82; 16,409.82; 6892.12; 9517.70 **5.** Aug. 8; $1258.36;
Lakeland Weekly; Space buy; 9517.70; 1572.00; 11,089.70; 1258.36; 9831.34
6. Aug. 14; $416.14; W. Wilson; Freelance Art; 9831.34; 10,000.00; 19,831.34;
416.14; 19,415.20 **7.** $1709.55 **8.** $81.99 **9.** $1627.56 **10.** $56.96
11. $1570.60 **12.** $5482.18

Chapter 5

Section 5.1 Exercises (Page 181)

1. 40; 0; $12.15 **3.** 38.75; 0; $16.05 **5.** 40; 5.25; $17.22 **7.** $329.60; $92.70;
$422.30 **9.** $380; $160.31; $540.31 **11.** $13.20; $347.60; $0; $347.60
13. $21.60; $576.00; $97.20; $673.20 **15.** $13.77; $367.20; $58.52; $425.72
17. 50.5; 10.5; $4.75; $479.75; $49.88; $529.63 **19.** 53.5; 13.5; $6.25;
$668.75; $84.38; $753.13 **21.** 35; 6; $14.10; $329.00; $84.60; $413.60
23. 39.5; 3.75; $16.20; $426.60; $60.75; $487.35 **25.** 39.75; 3.5; $15.30;
$405.45; $53.55; $459.00 **29.** $221.54; $443.08; $960; $11,520 **31.** $556.15;
$1112.31; $1205; $28,920 **33.** $830; $899.17; $1798.33; $21,580 **35.** $387
37. $467.25 **39.** $832 **41.** $788.80 **43.** $556.32 **45. (a)** $1260 biweekly
(b) $1365 semimonthly **(c)** $2730 monthly **(d)** $32,760 annually

Section 5.2 Exercises (Page 191)

1. $93.12 **3.** $153.60 **5.** $38.65 **7.** $52.68 **11.** $260.19 **13.** $284.65
15. $397.48 **17.** $471.50 **19.** $308.10 **21.** $421.65 **23.** $1405
25. $688.40 **27.** $478.10 **29.** $628.61

Section 5.3 Exercises (Page 199)

1. $20.13; $4.71 **3.** $28.72; $6.72 **5.** $52.99; $12.39 **7.** $189.39
9. $308.99 **11.** $41.56 **13.** $368.80; $76.07; $444.87; $27.58; $6.45; $4.45
15. $263.20; $49.35; $312.55; $19.38; $4.53; $3.13 **17.** $467.20; $122.64;
$589.84; $36.57; $8.55; $5.90 **19. (a)** $24.07 **(b)** $5.63 **21. (a)** $95.67
(b) $22.38 **(c)** $15.43 **23.** $7221.60; $1688.92 **25.** $3609; $844.04
27. $3328.61; $778.46

Section 5.4 Exercises (Page 211)

1. $195 **3.** $39 **5.** $52 **7.** $60 **9.** $37 **11.** $94 **13.** $6.87 **15.** $28.00
17. $69.11 **19.** $35.73; $8.36; $18.38; $513.81 **21.** $155.78; $36.43;
$275.63; $2044.69 **23.** $141.16; $33.01; $203.65; $1899.01 **25.** $122.21;
$28.58; $99.66; $1720.61 **27.** $44.05; $10.30; $25.09; $631.12
29. $110.76; $25.90; $370.35; $1279.43 **35.** $8748.05 **37.** $53,332.52
39. $44,323.10 **41.** $668.38 **43.** $544.76 **45.** $3745.64

Summary Exercise (Page 218)

(a) $818 **(b)** $368.10 **(c)** $1186.10 **(d)** $73.54 **(e)** $17.20 **(f)** $207.97
(g) $11.86 **(h)** $52.19 **(i)** $586.34

Chapter 5 Test (Page 219)

1. 40; 6.5; $537.30 **2.** 40; 7.5; $440.75 **3. (a)** $655 weekly **(b)** $1310
biweekly **(c)** $1419.17 semimonthly **(d)** $2838.33 monthly **4.** $134
5. $3525 **6. (a)** $499.10 **(b)** $116.73 **7. (a)** $89.90 **(b)** $116.73 **8.** $15
9. $47 **10.** $207 **11.** $62 **12.** $64 **13.** $1478.03 **14.** $421.15
15. $551.48 **16. (a)** $31.90 **(b)** $7.46 **(c)** $5.14 **17. (a)** $76.86
(b) $35.07 **18. (a)** $4552.55 **(b)** $1064.71 **19. (a)** $5255.20
(b) $1229.04 **20.** $2246.54

Chapter 6

Section 6.1 Exercises (Page 229)

1. $226.80 **3.** $126.36 **5.** $3610.36 **7.** $4887.73 **9.** $57.00
11. $28.40 **13.** $501.20 **15.** foot **17.** pair **19.** kilogram

21. case **23.** drum **25.** liter **27.** gallon **29.** cash on delivery
33. .9 × .8 = .72 **35.** .9 × .9 × .9 = .729 **37.** .75 × .95 = .7125
39. .6 × .7 × .8 = .336 **41.** .5 × .9 × .8 × .95 = .342 **43.** $267.52
45. $14.02 **47.** $722.93 **49.** $218.88 **51.** $16.83 **53.** $714.42 **55.** $972
57. $640 **63.** $242.99 **65.** (a) 20/15 (b) $4.08 **67.** $960.34
69. $326.40 undercharged

Section 6.2 Exercises (Page 235)

1. .72; 28% **3.** .68; 32% **5.** .504; 49.6% **7.** .5184; 48.16%
11. $720 **13.** $2280 **15.** (a) $25.89 (b) $28.76 (c) $2.87
17. (a) 20/20/20 is higher (b) .1% **19.** $740 **21.** 25.0%

Section 6.3 Exercises (Page 243)

1. May 14; June 3 **3.** July 25; Sep. 8 **5.** Oct. 1; Oct. 11 **7.** $1.70; $92.20
9. $0; $81.25 **11.** $21.60; $1120.55 **15.** $4542.69 **17.** $1798.92
19. (a) Jan. 28; Feb. 7; Feb. 17 (b) Mar. 9 **21.** (a) Apr. 25 (b) May 5

Section 6.4 Exercises (Page 249)

1. Mar. 10; Mar. 30 **3.** Dec. 22; Jan. 11 **5.** June 16; July 6
7. $20.47; $661.81 **9.** $0; $785.64 **11.** $229.60; $11,250.40
13. $.72; $23.23 **17.** (a) Dec. 13 (b) $2334.93 **19.** $4271.33
21. $1495.58 **23.** (a) $489.13 (b) $183.17 **25.** (a) June 10
(b) June 30 **27.** $1509.75 due **29.** (a) $3350.52 (b) $1052.06

Summary Exercise (page 256)

(a) $17,750.66 (b) October 15 (c) November 4 (d) $17,966.52
(e) $10,309.28; $8189.76

Chapter 6 Test (Page 257)

1. $225.65 **2.** $784.80 **3.** (a) .63 (b) 37% **4.** (a) .576 (b) 42.4%
5. Mar. 15 **6.** May 30 **7.** Jan. 15 **8.** Dec. 19 **9.** (a) $394.40 invoice total
(b) $386.51 after discount (c) $398.06 total amt due **10.** $91,300.78
11. (a) July 20 (b) $1013.76 **12.** $218.74 **13.** (a) Builders Supply
(b) $1.91 **14.** (a) $1762.20 (b) $1780 full amount **15.** $2438.58
16. (a) $1717.53 (b) $1198.47

Chapter 7

Section 7.1 Exercises (Page 267)

1. 140%; $4.96; $17.36 **3.** 100%; 20%; $27.17; $5.43 **5.** 100%; 130%;
$168.00; $218.40 **7.** $2.70; $11.70 **9.** 60%; $19.20 **11.** $61.44; 40%
13. $33.80; 25% **17.** $118.94 markup **19.** $12.95 selling price **21.** $221.40
selling price **23.** (a) $40 cost (b) 45% (c) 145% **25.** (a) 126% (b) $5.67
selling price (c) $1.17 markup

Section 7.2 Exercises (Page 277)

1. 75%; $21.00; $28.00 **3.** 58%; 42%; $105.00 **5.** 50%; 100%; $2025;
$4050 **7.** $1920; $2400.00 **9.** $8.46; $22.26; 61.3% **11.** $750; $1050;
28.6% **13.** 50% **15.** 15.3% **19.** (a) $800 selling price (b) $520 cost
(c) 65% **21.** (a) $4990 total received (b) $2710 markup (c) 54.3%
(d) 118.9% **23.** $.57

Supplementary Exercises (Page 279)

1. $22.35 markup **3.** 19.1% **5.** $6.98 **7.** $119 **9.** (a) 76% (b) $179.99
(c) $43.20 **11.** (a) $32.40 (b) 25.9% **13.** (a) $24.90 (b) 12.5%
(c) 14.2% **15.** $17.50

Section 7.3 Exercises (Page 285)

1. 25%; $645 **3.** 30%; $18.48 **5.** 20%; $5.20 **7.** $120; $20; none **9.** $16;
$22; $6 **11.** $385; $250; $60 **15.** 41% **17.** $60.01 operating loss
19. (a) $77.15 operating loss (b) $18.77 absolute loss

Section 7.4 Exercises (Page 293)

1. $22,673 **3.** $60,568 **5.** 2.83; 2.81 **7.** 3.59; 3.52 **9.** 4.69; 4.66
11. $182; $195; $170 **13.** $2352; $2385; $2312.50 **17.** 5.43 turnover at
cost **19.** (a) $508.50 weighted average method (b) $562.50 FIFO
(c) $520 LIFO **21.** (a) $1251.20 weighted average method (b) $1430 FIFO
(c) $1040 LIFO **23.** $30,660

Summary Exercise (Page 301)

(a) $125 original selling price (b) $2062.50 total selling price
(c) $375 operating loss (d) none

Chapter 7 Test (Page 303)

1. (a) 20 (b) 120 (c) 76.80 **2.** (a) 138 (b) 365.50 (c) 138.89
3. (a) 80 (b) 20 (c) 33.60 **4.** (a) 75 (b) 25 (c) 18.45 **5.** 20%
6. 50% **7.** $200; $14; none **8.** $72; $99; $27 **9.** 5.76; 5.73 **10.** $12.50
selling price per pair **11.** $632 **12.** 40% **13.** (a) $37.99 (b) 19.0%
(rounded) (c) 23.5% **14.** 28% **15.** (a) $131.10 operating loss (b) $45.60
absolute loss **16.** $130,278 average inventory **17.** $1124.75 weighted average method **18.** (a) $1157 FIFO (b) $1100 LIFO

Cumulative Review Chapters 4–7 (Page 305)

1. $1958.20 **3.** $1749.75 **5.** $1692.88 **7.** $4359.38 **9.** $240.47
11. 62.2% **13.** Nov. 15; Dec. 5 **15.** $400; $280; $32 **17.** $81.84 cost
(rounded) **19.** 12.88 turnover at cost **21.** $6489 weighted average
method

Chapter 8

Section 8.1 Exercises   (Page 315)

1. $209; $4009 **3.** $440; $5940 **5.** 68 **7.** 99 **9.** (a) $2493.15
(b) $2527.78 (c) $34.63 **11.** (a) $1091.10 (b) $1106.25 (c) $15.15
13. Helen Spence **15.** Donna Sharp **17.** 90 days **19.** Jan. 25 **21.** Oct. 18;
$5064 **23.** May 9; $6591.38 **25.** (a) $138,750 (b) $2,138,750
27. $16,800 **29.** (a) Oct. 3 (b) $7008.41 **31.** (a) Sep. 6 (b) $84,200
33. $17.23 **35.** (a) September 30 (b) $50,800

Section 8.2 Exercises   (Page 323)

1. $14,000 **3.** $504 **5.** $10,800 **7.** 11.8% **9.** 9.5% **11.** 7.5%
13. 120 days **15.** 62 days **17.** 5 months **19.** $7500.19 **21.** 9.5%
23. 9% **25.** (a) $10,800 (b) $11,250 **27.** 76 days **29.** 4%
31. (a) $12,000 (b) $1200 **33.** 208 days **35.** (a) 10.75% (b) 11.4%

Section 8.3 Exercises   (Page 333)

1. $234; $7566 **3.** $950; $18,050 **5.** $408.33; $21,991.67 **7.** Jun. 20; $6248
9. Dec. 9; $957.29 **11.** Feb. 8; $23,600 **13.** (a) $220 (b) $5780 **15.** 200
days **17.** 9.5% **19.** $7891.30 **21.** (a) $3780 (b) 13.3% **23.** 105 days
25. (a) 166,107.38 yen (b) 8.1% **27.** (a) $24,625,000 (b) $25,000,000
(c) $375,000 (d) 6.09%

Section 8.4 Exercises   (Page 343)

1. 107 days **3.** 53 days **5.** $10,179 **7.** $2481.25 **9.** $6362.75;
37 days; $78.47; $6284.28 **11.** $2044; 49 days; $33.39; $2010.61
13. $17,355; 42 days; $228.43; $17,571.57 **15.** $30,829.37; 83 days; $814.09;
$31,285.91 **17.** (a) $4968 (b) $367,632 **19.** (a) $238,750 (b) 93 days
(c) $5166.67 (d) $244,833.33 **21.** (a) $311,250 (b) $304,713.75
23. (a) $24,150 (b) $538.46 (c) $24,461.54 (d) 6.71%

Supplementary Exercises   (Page 347)

1. (a) $660 (b) $18,660 **3.** $48,000 **5.** 174 days **7.** (a) $1166.67
(b) 25,166.67 **9.** 12.3 % **11.** (a) Apr. 3 (b) $77,555.56 **13.** $8055.34
15. 16,243.88 **17.** (a) $3157 (b) $78,843 (c) 100 days (d) $2277.78
(d) $79,722.22 **19.** (a) simple interest note (b) $6689.82

Summary Exercise   (Page 356)

3. (a) $1,037,500 (b) $78,292,500

Chapter 8 Test   (Page 357)

1. $1020.83 **2.** $508.75 **3.** $137.50 **4.** $148.06 **5.** $13,013.01
6. $25,575.75 **7.** $11.19 **8.** 9.2% **9.** 333 days **10.** $43,000
11. $26,595.74 **12.** $359.33; $9440.67 **13.** $162.29; $10,087.71
14. (a) $14,550 (b) 9.3% **15.** $28,626.58 **16.** $9034.40
17. (a) $19,812.50 (b) $20,000 (c) $187.50 (d) 3.79% **18.** 44 days;
$144.07; $9285.93 **19.** (a) $452.81 (b) $8997.19 (c) loses $2.81

Chapter 9

Section 9.1 Exercises (Page 367)

1. $16,325.87; $4325.87   **3.** $30,906.76; $2906.76   **5.** $40,841.23; $8491.23
7. $28,949.25; $14,449.25   **9.** $60,476.40; $15,476.40   **11.** $1296; $1417.39;
$121.39   **13.** $1440; $2606.60; $1166.60   **15. (a)** $11,431.57   **(b)** $2931.57
17. (a) 31,669.25 yen   **(b)** 6669.25 yen   **19.** $439.25   **21. (a)** $28,137.75
(b) $40,117.75   **(c)** $11,980

Section 9.2 Exercises (Page 377)

1. $39.75   **3.** $50.48   **5.** $101.88   **7.** $4763.41   **9.** $17,412.96
13. (a) $3284.75   **(b)** $24.75   **15. (a)** $11,638.49   **(b)** $118.49
17. $4420.65; $4647.29   **19. (a)** $901,988.58   **(b)** $101,988.58
21. $4700   **23.** Loss of $397.50

Section 9.3 Exercises (Page 385)

1. $9764.11; $2535.89   **3.** $7674.01; $1675.99   **5.** $9792.06; $9060.94
7. (a) $26,444.80   **(b)** $13,555.20   **9.** $3503.40   **11. (a)** $793,053
(b) $592,616.78   **13. (a)** $26,620   **(b)** $20,989.60

Summary Exercise (Page 388)

(a) $4,053,382; $3,016,081   **(b)** $2,539,384; $1,889,530
(c) $2,798,295; $2,082,183;   **(d)** $2,539,384; $1,889,530; $2,798,295;
$2,082,183; $4,053,382; $3,016,081

Chapter 9 Test (Page 389)

1. $18,649.23; $9949.23   **2.** $16,127.04; $4127.04   **3.** $13,170.42; $3370.42
4. $18,556.38; $6056.38   **5.** $50.52   **6.** $530.60   **7.** $286.28   **8.** $7509.25
9. $10,380.83   **10.** $38,291.99   **11.** $16,077.25   **12.** $4120.01
13. $5399.30   **14.** $38,680.72   **15.** Loss of $676   **16.** Loss of $1625
17. (a) $4408.99   **(b)** $3801.87   **18. (a)** $15,173.40   **(b)** $13,481.41
19. (a) $283,233.60   **(b)** $206,955.96   **20. (a)** $59 million   **(b)** $93 million

Cumulative Review: Chapters 8–9 (Page 391)

1. $272   **3.** $2400.17   **5.** 9.7%   **7.** 80 days   **9.** $270; $8730   **11.** $4875
13. $1947.90   **15.** $86.07; $12,686.07   **17.** $583.49; $416.51   **19.** $1250;
$23,750   **21.** $10,661.88   **23.** $20.01

Chapter 10

Section 10.1 Exercises (Page 401)

1. $25,319.14; $9119.14   **3.** $201,527.78; $51,527.78   **5.** $139,509.30;
$41,509.30   **7.** $7603.12; $1603.12   **9.** $207,485.32; $36,485.32
11. $51,985.25; $6385.25   **15. (a)** $423,452.16   **(b)** $290,452.16
17. (a) $13,258.56   **(b)** $4258.56   **19.** $44,502

Section 10.2 Exercises (Page 409)

1. $14,762.54   **3.** $30,493.92   **5.** $18,991.59   **9.** $810,043.65
11. $31,726.02   **13. (a)** $56,776.30   **(b)** $13,223.70   **15. (a)** $48,879.76
(b) No   **17. (a)** First offer   **(b)** $2769.99   **19.** $208,893.30

Section 10.3 Exercises (Page 415)

1. $2784.12   **3.** $715.29   **5.** $2271   **7.** $183.22   **11. (a)** $144,026
(b) $55,844   **13. (a)** $860   **(b)** $957,000   **15.** $12,649.20   **17. (a)** $53,383
(b) $37,559   **19. (a)** $1200   **(b)** $6511.80

(c)	Payment Number	Amount of Deposit	Interest Earned	Total in Account
	1	$6511.80	$0	$6511.80
	2	$6511.80	$260.47	$13,284.07
	3	$6511.80	$531.36	$20,327.23
	4	$6511.80	$813.09	$27,652.12
	5	$6511.80	$1106.08	$35,270.00
	6	$6511.80	$1410.80	$43,192.60
	7	$6511.80	$1727.70	$51,432.10
	8	$6510.62	$2057.28	$60,000.00

21.	Payment Number	Amount of Deposit	Interest Earned	Total in Account
	1	$713,625	$0	$713,625.00
	2	$713,625	$10,704.38	$1,437,954.38
	3	$713,625	$21,569.32	$2,173,148.70
	4	$713,625	$32,597.23	$2,919,370.93

Supplementary Exercises (Page 419)

1. (a) $15,210.93   **(b)** $3210.93   **3. (a)** $77,985.46   **(b)** $37,985.46
5. $1,384,797.60   **7.** $96,344.03   **9.** $267,986.10

Payment Number	Amount of Deposit	Interest Earned	Total in Account
1	$267,986.10	$0	$267,986.10
2	$267,986.10	$21,438.89	$557,411.09
3	$267,996.02	$44,592.89	$870,000.00

Section 10.4 Exercises (Page 427)

1. $.44   **3.** $.08   **5.** 312,100   **7.** 2.1%   **9.** 26   **11.** $45.52   **13.** $33.12
15. $11.30   **17.** 44.3%   **19.** $2268; $210   **21.** $5333.60; $3.20
23. $8712.50; $70   **27.** 1.7%   **29.** 1.6%   **31.** 1.2%   **33.** 32   **35.** 36
37. 29   **39.** .85%, 2.5%   **41.** $15,120   **43. (a)** $15,326.80   **(b)** $17,866.85
(c) $2540.05

Section 10.5 Exercises (Page 435)

1. $1118.98   **3.** March 15, 2032   **5.** 5.88%   **7.** $49,670.50   **9.** $40,870.40
11. $158,841   **15. (a)** $22,591.60   **(b)** $1350   **(c)** 6.0%
17. (a) $14,901.15   **(b)** $1031.25   **(c)** 6.9%   **19. (a)** $3600
(b) $97,151.40

Summary Exercise (Page 440)

(a) $955,771.57   **(b)** $97,861.20   **(c)** $960,815.94   **(d)** It will be very close
to what is needed.

Chapter 10 Test (Page 441)

1. $9897.47   **2.** $126,595.71   **3.** $930,909   **4.** $121,490.46   **5.** $54,304.22
6. $30,200.99   **7.** $6801.69   **8.** $39,884.63   **9.** $21,316.12
10. $264,795.02   **11.** $21,461.99   **12.** $24,683.07   **13.** $8702   **14.** $8395
15. $5835.60   **16.** $17,976   **17.** $22,492   **18.** $2322   **19. (a)** $15,900
(b) $272   **20. (a)** $24,025   **(b)** $1050   **(c)** 4.4%

Chapter 11

Section 11.1 Exercises (Page 451)

1. $109.23   **3.** $15.02   **5.** October, $6.12; $419.17; November, $419.17,
$5.87, $541.68; December, $541.68, $7.58, $529.22;
January, $529.22, $7.41, $211.71   **9.** $20.42   **11.** $4.87   **13.** $17.30
15. (a) $104   **(b)** $65   **(c)** $39   **17. (a)** $132.64
(b) $1.99   **(c)** $133.80   **19. (a)** $312.91   **(b)** $4.69   **(c)** $285.94
21. (a) $139.71   **(b)** $2.10   **(c)** $74.32

Section 11.2 Exercises (Page 461)

1. $2050; $250   **3.** $180; $30   **5.** $3543; $643   **7.** 12.2%   **9.** 11.6%
11. 15.4%   **13.** 12.25%   **15.** 10.25%   **17.** 11.25%   **21.** 10.00%
23. (a) $18,800   **(b)** $26,240   **(c)** $5440   **(d)** 10.50%
25. (a) 732,212.32 pesos   **(b)** 13.75%

Section 11.3 Exercises (Page 469)

1. $8307.31; $307.31   **3.** $9198.18; $698.18   **5.** $9862.15; $507.15
7. $231   **9.** $103.18   **11.** $28.38   **15. (a)** $415.33   **(b)** $9315.33
17. (a) $69,936.02   **(b)** $2286.02   **19. (a)** $41,176.11   **(b)** $1876.11
21. (a) $17.31   **(b)** $457.69   **23. (a)** $1700   **(b)** $170   **(c)** $5980

Section 11.4 Exercises (Page 477)

1. $1703.13   **3.** $404.73   **5.** $4185.60   **7.** $161.93   **9.** $147.61; $899.62
11. $321.96; $3454.08   **13.** $255.45; $3577   **17. (a)** $1900.70
(b) $13,628

19.

Payment Number	Amount of Payment	Interest for Period	Portion to Principal	Principal at End of Period
0	—	—	—	$4000.00
1	$1207.68	$320.00	$887.68	$3112.32
2	$1207.68	$248.99	$958.69	$2153.63
3	$1207.68	$172.29	$1035.39	$1118.24
4	$1207.70	$89.46	$1118.24	$0

21.

Payment Number	Amount of Payment	Interest for Period	Portion to Principal	Principal at End of Period
0	—	—	—	$14,500.00
1	$374.83	$132.92	$241.91	$14,258.09
2	$374.83	$130.70	$244.13	$14,013.96
3	$374.83	$128.46	$246.37	$13,767.59
4	$374.83	$126.20	$248.63	$13,518.96
5	$374.83	$123.92	$250.91	$13,268.05

Section 11.5 Exercises (Page 487)

1. $730.55 **3.** $1111.08 **5.** $599.68 **9.** $803.51 **11.** $523.87
13. $873.82 **15.** Yes, they are qualified

17.

Payment Number	Total Payment	Interest Payment	Principal Payment	Balance of Principal
0	—	—	—	$122,500.00
1	$1136.80	$765.63	$371.17	$122,128.83
2	$1136.80	$763.31	$373.49	$121,755.34

Monthly payment = $122.5 \times \$9.28 = \$1136.80$

Summary Exercise (Page 494)

(a) $495.00; $420.13; $950.53; $149.45; $2015.11 **(b)** $2015.11; $215.00; $120.00; $210.83; $2560.94 **(c)** $392.32; $285.61; $695.83; $149.45; $187.00; $120.00; $210.83; $2041.04 **(d)** $519.90

Chapter 11 Test (Page 497)

1. $20,900 **2.** $932.56 **3.** 11.75% **4.** 12.25% **5.** 11% **6.** $4364.31
7. (a) $235.51 **(b)** $2502.17 **8.** $4811.72 **9.** $1924.50 **10.** $864.50
11. $1209.69 **12.** $629.48 **13. (a)** $1407.98 **(b)** $289,686.40
14. (a) $1398.25 **(b)** $360,580

Cumulative Review Chapters 10–11 (Page 499)

1. $9214.23; $1214.23 **3.** $17,855.03; $2855.03 **5.** $13,367.29 **7.** $403.45
9. (a) $11,957.94 **(b)** $27,257.87 **11. (a)** $2345 **(b)** 15 **(c)** 1.5%
13. (a) $3098.48 **(b)** $298.48 **(c)** $2300 **(d)** 12.00% **15. (a)** $2118.14
(b) $3017.68 **17. (a)** $14,570 **(b)** $14,952.46

Chapter 12

Section 12.1 Exercises (Page 509)

1. $34,000 **3.** $71,150 **5.** $325,125 **7.** 30.4% **9.** 8% **11.** 3%
13. (a) $4.84 **(b)** $48.40 **(c)** 48.4 **15. (a)** 7.08% **(b)** $7.08 **(c)** 70.8
19. $5861.60 **21.** $5384.40 **23.** $5978.70 **25.** $6295.08 **27.** $39,960
29. (a) The second parish **(b)** $75.24

Section 12.2 Exercises (Page 523)

1. $23,131 **3.** $21,710 **5.** $39,031 **7.** $24,600; $3340 **9.** $20,101;
$2315.15 **11.** $47,050; $6357.50 **13.** $23,010; $3101.50 **15.** $30,369;
$4402.25 **17.** $48,832; $7903 **19.** $1260.50 tax refund **21.** $570.56 tax
refund **23.** $65.66 tax due **27.** $2394.20 **29.** $14,766 **31.** $7915
33. $5840.30

Section 12.3 Exercises (Page 533)

1. $1700 **3.** $2194.50 **5.** $9298.72 **7.** $7034.48 **9.** $19,850
11. $36,500 **13.** $60,000; $20,000 **15.** $292,500; $260,000; $97,500
17. $18,804 **19.** $702.65 **23. (a)** $136,986.30 **(b)** $43,013.70
25. (a) $30,681.82 **(b)** $14,318.18 **29.** *A:* $274,000 *B:* $182,666.67
C: $91,333.33 **31. 1:** $125,000 **2:** $41,666.67 **3:** $83,333.33

Section 12.4 Exercises (Page 543)

1. $790 **3.** $932 **7.** $638 **9.** $657.80 **11. (a)** $25,000 **(b)** $11,500
13. (a) $4300 **(b)** $850 **15. (a)** $1778 **(b)** $6936 **(c)** $100,000
(d) $15,100

Section 12.5 Exercises (Page 553)

1. $256; $130.56; $66.56; $23.24 **3.** $849.10; $433.04; $220.77; $77.10
5. $516.80; $263.57; $134.37; $46.93 **7.** $319.50; $162.95; $83.07; $29.01
9. $2973.10; $1516.28; $773.01; $269.96 **13.** $512 **15. (a)** $100.50
(b) $384 **17.** $3036 **19. (a)** $444.72 **(b)** $226.72 **(c)** $79.18

Summary Exercise (Page 558)

(a) $11,790.75 **(b)** $39,940.40 **(c)** $439.88 **(d)** $52,171.03
(e) $1328.97

Chapter 12 Test (Page 559)

1. $5.76; $57.60 **2.** 9.35%; $9.35 **3.** $27,445; $3766.75 **4.** $21,787;
$2568.05 **5.** $1145.37 **6.** $8011.50 **7.** $3851.50 **8.** $2179.05
9. $42,613.64 **10.** *A:* $36,000 *B:* $21,600 *C:* $14,400 **11.** $1086.55
12. $696 **13.** $158.48; $80.82; $41.20; $14.39 **14.** $1940.80; $989.81;
$504.61; $176.22

Chapter 13

Section 13.1 Exercises (Page 567)

1. 20% **3.** 12.5% **5.** 5% **7.** $6\frac{2}{3}$% **9.** 1.25% **11.** 2% **13.** $450
15. $800 **17.** $840 **19.** $2850 **21.** $5000 **23.** $2775 **25.** $73,000
27.

Year	Computation	Amount of Depreciation	Accumulated Depreciation	Book Value
0	—	—	—	$12,000
1	$\left(33\frac{1}{3}\% \times \$9000\right)$	$3000	$3000	$9000
2	$\left(33\frac{1}{3}\% \times \$9000\right)$	$3000	$6000	$6000
3	$\left(33\frac{1}{3}\% \times \$9000\right)$	$3000	$9000	$3000

29.

Year	Computation	Amount of Depreciation	Accumulated Depreciation	Book Value
0	—	—	—	$25,600
1	$\left(16\frac{2}{3}\% \times \$18,600\right)$	$3100	$3100	$22,500
2	$\left(16\frac{2}{3}\% \times \$18,600\right)$	$3100	$6200	$19,400
3	$\left(16\frac{2}{3}\% \times \$18,600\right)$	$3100	$9300	$16,300
4	$\left(16\frac{2}{3}\% \times \$18,600\right)$	$3100	$12,400	$13,200
5	$\left(16\frac{2}{3}\% \times \$18,600\right)$	$3100	$15,500	$10,100
6	$\left(16\frac{2}{3}\% \times \$18,600\right)$	$3100	$18,600	$7000

33. (a) $55,000 depreciation **(b)** $1,025,000 book value **35. (a)** 12.5%
(b) $90 **(c)** $790

Section 13.2 Exercises (Page 573)

1. 40% **3.** 25% **5.** $13\frac{1}{3}$% **7.** 20% **9.** $33\frac{1}{3}$% **11.** 4% **13.** $3000
15. $9000 **17.** $1900 **19.** $3360 **21.** $1215 **23.** $6834 **25.** $750

27.

Year	Computation	Amount of Depreciation	Accumulated Depreciation	Book Value
0	—	—	—	$14,400
1	$(50\% \times \$14,400)$	$7200	$7,200	$7,200
2	$(50\% \times \$7,200)$	$3600	$10,800	$3,600
3	$(50\% \times \$3,600)$	$1800	$12,600	$1,800
4		$1800*	$14,400	$0

*To depreciate to 0 scrap value.

29.

Year	Computation	Amount of Depreciation	Accumulated Depreciation	Book Value
0	—	—	—	$14,000
1	$(40\% \times \$14,000)$	$5600	$5,600	$8,400
2	$(40\% \times \$8,400)$	$3360	$8,960	$5,040
3	$(40\% \times \$5,040)$	$2016	$10,976	$3,024
4		$524*	$11,500	$2,500
5		$0	$11,500	$2,500

*To depreciate to $2500 scrap value.

33. $1153   **35.** $8476   **37. (a)** 25%   **(b)** $1450   **(c)** $4425   **(d)** $1375

Section 13.3 Exercises (Page 579)

1. $\frac{4}{10}$ **3.** $\frac{6}{21}$ **5.** $\frac{7}{28}$ **7.** $\frac{10}{55}$ **9.** $1640   **11.** $10,000   **13.** $7500   **15.** $7700
17. $2700   **19.** $890   **21.** $2400

23.

Year	Computation	Amount of Depreciation	Accumulated Depreciation	Book Value
0	—	—	—	$3900
1	$(\frac{3}{6} \times \$3420)$	$1710	$1710	$2190
2	$(\frac{2}{6} \times \$3420)$	$1140	$2850	$1050
3	$(\frac{1}{6} \times \$3420)$	$570	$3420	$480

25.

Year	Computation	Amount of Depreciation	Accumulated Depreciation	Book Value
0	—	—	—	$10,800
1	$(\frac{6}{21} \times \$8400)$	$2400	$2400	$8,400
2	$(\frac{5}{21} \times \$8400)$	$2000	$4400	$6,400
3	$(\frac{4}{21} \times \$8400)$	$1600	$6000	$4,800
4	$(\frac{3}{21} \times \$8400)$	$1200	$7200	$3,600
5	$(\frac{2}{21} \times \$8400)$	$800	$8000	$2,800
6	$(\frac{1}{21} \times \$8400)$	$400	$8400	$2,400

29. $3000 depreciation   **31.** $4887   **33. (a)** $7200   **(b)** $6000   **35. (a)** $\frac{8}{36}$
(b) $2360   **(c)** $10,620   **(d)** $4750

Supplementary Exercise (Page 583)

1. $205,000   **3.** $7400   **5.** $1242   **7.** $3760   **9.** $63,180   **11.** $13,680;
$10,260; $6840; $3420   **13.** $37,575   **15.** $10,080   **17. (a)** $5580
(b) $5220

Section 13.4 Exercises (Page 591)

1. $.75   **3.** $.029   **5.** $.24   **7.** $30   **9.** $25,300   **11.** $17,280   **13.** $2775
15. $3108

19.

Year	Computation	Amount of Depreciation	Accumulated Depreciation	Book Value
0	—	—	—	$6800
1	$(1350 \times \$1.26)$	$1701	$1701	$5099
2	$(1820 \times \$1.26)$	$2293	$3994	$2806
3	$(730 \times \$1.26)$	$920	$4914	$1886
4	$(1100 \times \$1.26)$	$1386	$6300	$500

Section 13.5 Exercises (Page 597)

1. 19.2%   **3.** 6.56%   **5.** 20%   **7.** 3.636%   **9.** 5.76%   **11.** 2.564%
13. $1751   **15.** $43,050   **17.** $4800   **19.** $6254   **21.** $16,920   **23.** $2756
25. $79,364

27.

Year	Computation	Amount of Depreciation	Accumulated Depreciation	Book Value
0	—	—	—	$10,980
1	$(33.33\% \times \$10,980)$	$3660	$3,660	$7,320
2	$(44.45\% \times \$10,980)$	$4881	$8,541	$2,439
3	$(14.81\% \times \$10,980)$	$1626	$10,167	$813
4	$(7.41\% \times \$10,980)$	$813*	$10,980	$0

*due to rounding in prior years

29.

Year	Computation	Amount of Depreciation	Accumulated Depreciation	Book Value
0	—	—	—	$122,700
1	$(10\% \times \$122,700)$	$12,270	$12,270	$110,430
2	$(18\% \times \$122,700)$	$22,086	$34,356	$88,344
3	$(14.4\% \times \$122,700)$	$17,669	$52,025	$70,675
4	$(11.52\% \times \$122,700)$	$14,135	$66,160	$56,540
5	$(9.22\% \times \$122,700)$	$11,313	$77,473	$45,227
6	$(7.37\% \times \$122,700)$	$9,043	$86,516	$36,184
7	$(6.55\% \times \$122,700)$	$8,037	$94,553	$28,147
8	$(6.55\% \times \$122,700)$	$8,037	$102,590	$20,110
9	$(6.56\% \times \$122,700)$	$8,049	$110,639	$12,061
10	$(6.55\% \times \$122,700)$	$8,037	$118,676	$4,024
11	$(3.28\% \times \$122,700)$	$4,024*	$122,700	$0

*due to rounding in prior years

33. $4855   **35.** $490   **37. Year 1:** $12,326; **Years 2–5:** $12,307

Summary Exercise (Page 603)

(a) $114,000   **(b)** $61,560   **(c)** $228,000   **(d)** $57,000 straight-line;
$24,624 double-declining-balance; $38,000 sum-of-the-years'-digits

Chapter 13 Test (Page 604)

1. 25%; 50%; $\frac{4}{10}$ **2.** 20%; 40%; $\frac{5}{15}$ **3.** $12\frac{1}{2}$%; 25%; $\frac{8}{36}$ **4.** 5%; 10%; $\frac{20}{210}$
5. $940   **6.** $21,375   **7. Year 1:** $2700; **Year 2:** $2025; **Year 3:** $1350;
Year 4: $675   **8.** $4788 dep. year 3   **9.** $43,000   **10.** $5702 book value
3 years   **11. (a)** Year 1: $4836; Year 2: $2666; Year 3: $3007; Year 4: $4712
(b) Year 1: $15,264; Year 2: $12,598; Year 3: $9591; Year 4: $4879
12. $2,440,928

Chapter 14

Section 14.1 Exercises (Page 611)

1. (a) $316,350   **(b)** $88,050   **(c)** $65,350   **3.** Gross Sales, $284,000;
Returns, $6,000; Net Sales, $278,000; Inventory, January 1, $58,000; Cost of

Goods Purchased, $232,000; Freight, $3,000; Total Cost of Goods Purchased, $235,000; Total of Goods Available for Sale, $293,000; Inventory, December 31, $69,000; Cost of Goods Sold, $224,000; Gross Profit, $54,000; Salaries and Wages, $15,000; Rent, $6,000; Advertising, $2,000; Utilities, $1,000; Taxes on Inventory, Payroll, $3,000; Miscellaneous Expenses, $4,000; Total Expenses, $31,000; Net Income before Taxes, $23,000; Income Taxes, $2,400; Net Income, $20,600

Section 14.2 Exercises (Page 617)

1. 45.2%; 32.6% **3.** 50.7%; 29.2% **5.** Cost of Goods Sold, 67.1%, 67.9%; Gross Profit, 32.9%, 32.1%; Wages, 12.5%, 12.3%; Rent, 2.5%, 2.4%; Advertising, 1.5%, 1.4%; Total Expenses, 24.6%, 23.5%; Net Income before Taxes, 8.3%, 8.6% **9.** Gross Sales, 100.3%, 100.7%; Returns, .3%, .7%; Net Sales, $1,850,000, $1,680,000; Cost of Goods Sold, 65.0%, 62.5%; Gross Profit, 35.0%, 37.5%; Wages, 8.2%, 8.8%; Rent, 4.4%, 4.6%; Advertising, 6.0%, 7.3%; Utilities, 1.7%, 1.0%; Taxes on Inv., Payroll, .9%, 1.1%; Miscellaneous Expenses, 3.4%, 3.5%; Total Expenses, 24.6%, 26.3%; Net Income before Taxes, $192,000, 10.4%, $189,000, 11.3% **11.** 64.8%, 35.2%, 23.4%, 11.7%, 7.9%, 4.9%,1.8%

Section 14.3 Exercises (Page 623)

1. Cash, $273; Notes Receivable, $312; Accounts Receivable, $264; Inventory, $180; Total Current Assets, $1,029; Land, $466; Buildings, $290; Fixtures, $28; Total plant Assets, $784; Total Assets, $1,813; Notes Payable, $312; Accounts Payable, $63; Total Current Liabilities, $375; Mortgages Payable, $212; Long-Term Notes Payable, $55; Total Long-Term Liabilities, $267; Total Liabilities, $642; Owners' Equity, $1,171; Total Liabilities and Owners' Equity, $1,813

Section 14.4 Exercises (Page 629)

1. Cash, 13%, 13.1%; Notes Receivable, 2%, 1.9%; Accounts Receivable, 37%, 37.5%; Inventory, 38.3%, 37.5%; Total Current Assets, $361,000, 90.3%, $288,000, 90%; Land, 2.5%, 2.5%; Buildings, 3.5%, 3.4%; Fixtures, 3.8%, 4.1%; Total Plant Assets, $39,000, 9.8%, $32,000, 10%; Total Assets, $400,000, $320,000; Accounts Payable, .8%, 1.3%; Notes Payable, 50.3%, 47.5%; Total Current Liabilities, $204,000, 51%, $156,000, 48.8%; Mortgages Payable, 5%, 5%; Long-Term Notes Payable, 14.5%, 13.1%; Total Long-Term Liabilities, $78,000, 19.5%, $58,000, 18.1%; Total Liabilities, $282,000, 70.5%, $214,000, 66.9%; Owners' Equity, 29.5%, 33.1%; Total Liabilities and Owners' Equity, $400,000, 100%, $320,000, 100% **3. (a)** 1.77 **(b)** 1.02 **(c)** No, current ratio is low **5. (a)** 1.79 **(b)** .62 **(c)** No, acid-test ratio is low **7.** 2.13; 1.27 **9.** 6.1% **11.** .91; .37; Not healthy; very low liquidity

Summary Exercise (Page 636)

(a) Gross Sales, $212,000; Returns, $12,500; Net Sales, $199,500; Inventory, January 1, $44,000; Cost of Goods Purchased, $75,000; Freight, $8,000; Total Cost of Goods Purchased, $83,000; Total of Goods Available for Sale, $127,000; Inventory, December 31, $26,000; Cost of Goods Sold, $101,000; Gross Profit, $98,500; Salaries and Wages, $37,000; Rent, $12,000; Advertising, $2,000; Utilities, $3,000; Taxes on Inventory, Payroll, $7,000; Miscellaneous Expenses, $4,500; Total Expenses, $65,000; Net Income before Taxes, $33,000; Income Taxes, $4,320; Net Income after Taxes, $28,680 **(b)** Gross Sales, $106.3%; Returns, 6.3%; Cost of Goods Sold, 50.6%; Salaries and Wages, 18.5%; Rent, 6%; Utilities, 1.5% **(c)** Cash, $62,000; Notes Receivable, $2,500; Accounts Receivable, $8,200; Inventory, $26,000; Total Current Assets, $98,700; Land, $7,600; Buildings, $28,000; Fixtures, $13,500; Total Plant Assets, $49,100; Total Assets, $147,800; Notes Payable, $4,500; Accounts Payable $27,000; Total Current Liabilities, $31,500; Mortgages Payable, $15,000; Long-Term Notes Payable, $8,000; Total Long-Term Liabilities, $23,000; Total Liabilities, $54,500; Owners' Equity, $93,300; Total Liability and Owners' Equity, $147,800 **(d)** 3.13; 2.31

Chapter 14 Test (Page 639)

1. Gross Sales, $756,300; Returns, $285; Net Sales, $756,015; Inventory, January 1, $92,370; Cost of Goods Purchased, $465,920; Freight, $1,205; Total Cost of Goods Purchased, $467,125; Total of Goods Available for Sale, $559,495; Inventory, December 31, $82,350; Cost of Goods Sold, $477,145; Gross Profit, $278,870; Salaries and Wages, $84,900; Rent, $42,500; Advertis-

ing, $2,800; Utilities, $18,950; Taxes on Inventory and Payroll, $4,500; Miscellaneous Expenses, $18,400; Total Expenses, $172,050; Net Income Before Taxes, $106,820; Income Taxes, $25,450; Net Income After Taxes, $81,370 **2.** Net Sales, $35,000, 58.3%; Cost of Goods Sold, $23,000, 57.5%; Gross Profit, $4,000, 33.3% **3.** Cost of Goods Sold, 68.8%, 76.8%; Gross Profit, 31.3%, 23.2%; Net Income, 9.4%, 6.3%; Wages, 9.4%, 8.5%; Rent, 2%, 2.3%; Total Expenses, 21.9%, 16.9% **4. (a)** 1.38 **(b)** .85 **5. (a)** 1.05 **(b)** .37 **6.** 30.0% **7.** 26.1%

Chapter 15
Section 15.1 Exercises (Page 647)

1. 2100; 9.1 billion **3.** 2.136 billion Chinese; .356 billion Americans **5.** It is increasing **7.** 4 **9.** 6 **11.** 5

13.

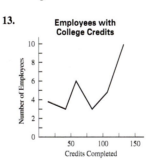

15. 23 **17.** 7 **19.** 2 **21.** 1 **23.** 2 **25.** 6

27.

29. 20 **31.** 1 **33.** 13 **35.** 17 **37.** 8 **39.** 38 **41.** 42 **43.** 20% **45.** 36° **47.** 5%, 18°

49.

51. 20%

53.

55.

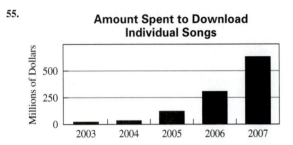

Amount Spent to Download
Individual Songs

Section 15.2 Exercises (Page 659)

1. 2.5 **3.** 60.3 **5.** 27,955 **7.** 8.8 **11.** 12.3 **13.** 17.2 **15.** 118.8 **17.** 2.6
19. 114 **21.** 130 **23.** 44 **25.** 21% **27.** 64 **29.** Bimodal with modes 32
and 38 **31.** 34.11 **33.** 34.38

Summary Exercise (Page 665)

(a) Store 1 mean = $7.4; Store 2 mean = $8.2; Store 1 median = $7.5; Store
2 median = $8.2; Store 1 has no mode; Store 2 mode = $8.2

(b)

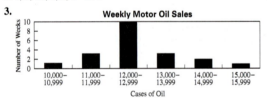

Sales at Christian Books Unlimited

(c) Sales at Store 2 seem to be growing faster than at Store 1.

Chapter 15 Test (Page 667)

1. 1, 3, 10, 3, 2, 1 **2.** 6

3.

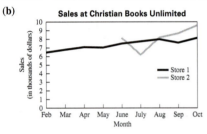

Weekly Motor Oil Sales

4. Newsprint, 72°; Ink, 10%; Wire Service, 108°; Salaries, 108°; Other, 36°

5.

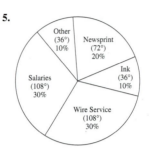

6. 60% **7.** 230.7 **8.** 18.8 centimeters **9.** $478.90 **10.** 11.3 **11.** 173.7
12. 19 **13.** 41.5 **14.** 8.3 **15.** 54.5 **16.** 47 **17.** No mode **18.** bimodal,
103 and 104

19.

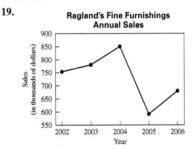

Ragland's Fine Furnishings
Annual Sales

It would appear as if the recession had an effect on their business.

20.

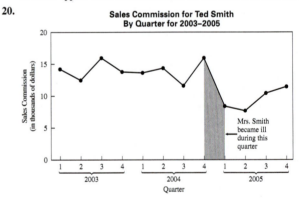

Sales Commission for Ted Smith
By Quarter for 2003–2005

It would appear that Mrs. Smith's illness affected Mr. Smith's work.

Glossary

401 (k):   A retirement plan for individuals working for private-sector companies.

403 (b):   A retirement plan for employees of public schools and certain tax-exempt organizations.

A

Absolute or gross, loss: The loss resulting when the selling price is less than the cost.

Accelerated depreciation: A technique to increase the depreciation taken during the early years of an asset's useful life.

Accelerated mortgages: Mortgages with payoffs of less than 30 years, such as 15, 20, or 25 years.

Accountant: A person who maintains financial data for a firm or individual and then prepares the income tax return.

Accounts payable: A business debt that must be paid.

Accumulated depreciation: A running balance or total of the depreciation to date on an asset.

Acid-test ratio: The sum of cash, notes receivable, and accounts receivable, divided by current liabilities.

ACRS (Accelerated cost recovery system):   A depreciation method introduced as part of the Economic Recovery Tax Act of 1981.

Actual rate of interest: The true annual percentage rate that can be used to compare loans.

Actuary: A person who determines insurance premiums.

Addends: The numbers added in an addition problem.

Addition rule: The same number may be added or subtracted on both sides of an equation.

Adjustable rate mortgage: A home loan where the interest rate is adjusted up or down depending on a benchmark interest rate.

Adjusted bank balance: The actual current balance of a checking account after reconciliation.

Adjusted gross income: An individual's or family's income for a year, including all sources of income, and after subtracting certain expenses, such as moving expenses and sick pay.

Algebraic logic: Rules used by most calculators for entering and evaluating arithmetic expressions.

Allowances: The number of allowances claimed by a taxpayer affects the amount withheld for income taxes.

American Express: A widely accepted credit card that requires an annual fee.

Amortization table: A table showing the level (unchanging) payment necessary to pay in full a loan for a specific amount of money including interest over a specific length of time.

Amortize: The process of paying off a loan with a sequence of periodic payments over a period of time.

Amount of an annuity: The future value of the annuity.

Amount of depreciation: The dollar amount of depreciation taken. This is usually an annual figure.

Annual meeting: Corporations have annual meetings for stockholders.

Annual percentage rate (APR):   The true annual percentage rate that can be used to compare loans. It is required by the federal Truth-in-Lending Act.

Annual percentage rate table: A table used to find the annual percentage rate (APR) on a loan or installment purchase.

Annual rate of depreciation: The percent or fraction of the depreciable amount or declining balance to be depreciated each individual year of an asset's useful life.

Annuity: Periodic payments of a given, fixed amount of money.

Annuity due: An annuity whose payments are made at the beginning of a time period.

APR (Annual percentage rate):   The true annual percentage rate that can be used to compare loans. It is required by the federal Truth-in-Lending Act.

"AS OF":   A later date that appears on an invoice. The given sales terms may start at this time.

Assessed value: The value of a piece of property. Set by the county assessor, assessed value is used in figuring property taxes.

Assessment rate: The assessed valuation of a property is found by multiplying the fair market value by the assessment rate.

Asset: An item of value owned by a firm.

ATM (Automated teller machine):   A machine that allows bank customers to make deposits, withdrawals, and fund transfers.

Automatic savings transfer account: A bank account that automatically transfers funds from one account to another.

Average: *See* mean.

Average cost method: An inventory valuation method whereby the cost of all purchases during a time period is divided by the number of units purchased.

Average daily balance method: A method used to calculate interest on open-end credit accounts.

Average inventory: The sum of all inventories taken divided by the number of times inventory was taken.

Average owner's equity: Sum of owner's equity at the beginning and end of the year divided by 2.

B

Bad checks: A check that is not honored because there are insufficient funds in the checking account.

Balance brought forward (Current balance): The amount left in a checking account after previous checks written have been subtracted.

Balanced: In agreement. When the bank statement amount and the depositor's checkbook balance agree, they are balanced.

Balance sheet: A summary of the financial condition of a firm at one point in time.

Bank discount: A bank fee charged on a note. It is subtracted from the face value to find the proceeds loaned.

Banker's ratio: *See* current ratio.

Bank statement: Usually sent out monthly by the bank, a list of all charges and deposits made against and to a checking account.

Banker's rule: A formula used to calculate interest by dividing exact days by 360.

Bankrupt: A company or individual whose liabilities exceed assets can declare bankruptcy, which is a legal process of working with debtors to pay off debts.

Bar graph: A graph using bars to compare various numbers.

Base: The starting point or reference point or that to which something is being compared.

bbl.: Abbreviation for *barrel*.

Bimodal: A set of data with two modes.

Blank endorsement: A signature on the back of a check by the person to whom the check is made.

Board of directors: A group of people who represent the stockholders of a corporation.

Bodily injury insurance: A type of automobile insurance that protects a driver in case he or she injures someone with a car.

Bond: A contractual promise by a corporation, government entity, or church to repay borrowed money at a specified rate and time.

Book value: The cost of an asset minus depreciation to date.

Break-even point: The cost of an item plus the operating expenses associated with the item. Above this amount a profit is made; below it, a loss is incurred.

Broker: A person who sells stocks, bonds, and other investments owned by others.

Business account: The type of checking account used by businesses.

bx.: Abbreviation for *box*.

C

C: Roman numeral for 100.

Canceled check: A check is canceled after the amount of the check has been transferred from the payer's bank account into the account of the receiver of the check.

Cancellation: A process used to simplify multiplication and division of fractions.

Capital: The amount of money originally invested in a firm. The difference between the total of all the assets and the total of all the liabilities is called the capital or net worth.

Capital gains: Profits made on investments such as stocks or real estate.

cart.: Abbreviation for *carton*.

Cash discount: A discount offered by the seller allowing the buyer to take a discount if payment is made within a specified period of time.

Cashier's check: A check written by a financial institution, such as a bank, that is guaranteed by the institution.

Cash value: Money that has built up in an ordinary life insurance policy.

Centi-: A prefix used in the metric system meaning hundredth. (For example, a centiliter is one one-hundredth of a liter.)

Centimeter: One one-hundredth of a meter. There are 2.54 centimeters to an inch.

Central tendency: The middle of a set of data.

Certificate of deposit (CD): A savings account in which a minimum amount of money must be deposited and left for a minimum period of time.

Chain calculations: Long calculations done on a calculator.

Chain discount: Two or more discounts that are combined into one discount.

Check register: A table usually found in a check book that is used by the check writer to list all checks written, deposits and withdrawals made, and ATM transactions.

Checks outstanding: Checks written that have not reached and cleared the bank as of the statement date.

Check stub: A stub attached to the check and retained as a record of checks written.

Circle graph: A circle divided into parts that are labeled and often colored or shaded to show data.

COD: Cash on delivery.

Coinsurance: The portion of a loss that must be paid by the insured.

Collateral: Assets foreclosed on by a lender should the borrower default on payments.

Collision insurance: A form of automobile insurance that pays for car repairs in case of an accident.

Commission: A fee charged by a broker for buying and selling stocks and bonds.

Commissions: Payments to an employee that represent a certain percent of the total sales produced by the employee's efforts.

Common denominator: Two or more fractions with the same denominator are said to have common denominators.

Common stock: Ownership of a corporation, held in portions called shares.

Comparative balance sheet: An analysis for two or more periods that compares asset categories such as cash.

Comparative income statement: A vertical analysis for two or more years that compares incomes or balance sheet items for each year analyzed.

Comparison graph (Comparative line graph):   One graph that shows how several things relate.

Compensatory time (Comp time):   Time off given to an employee to compensate for previously worked overtime.

Compound interest: Interest charged or received on both principal and interest.

Comprehensive insurance: A form of automobile insurance that pays for damage to a car caused by fire, theft, vandalism, and weather.

Consolidated statement: A financial statement showing the combined results of all subsidiaries of a firm.

Consumer price index (CPI):   A measure of the cost of living calculated by the government and used to estimate inflation.

Conventional loan: A loan made by a bank, savings and loan, or other lending agency that is not guaranteed or insured by the federal government.

Corporation: A form of business that gives the owners limited liability.

Cost: The total cost of an item, including shipping, insurance, and other charges. Most often, the cost is the basis for calculating depreciation of an asset.

Cost (Cost price):   The price paid to the manufacturer or supplier after trade and cash discounts have been taken. This price includes transportation and insurance charges.

Cost of goods sold: The amount paid by a firm for the goods it sold during the time period covered by an income statement.

Country club billing method: A billing method that provides copies of original charge receipts to the customer.

cpm.:   Abbreviation for *cost per thousand*.

Credit card (transactions):   The purchase or sale of goods or services using a credit card in place of cash or a check.

Credit union: A financial institution similar to a bank, except that it is owned by its member customers.

Credit union share draft account: A credit union account that may be used as a checking account.

Cross-products: The equal products obtained when each numerator of a proportion is multiplied by the opposite denominator.

cs.:   Abbreviation for *case*.

ct.:   Abbreviation for *crate*.

ctn.   Abbreviation for *carton*.

Current assets: Cash or items that can be converted into cash within a given period of time, such as a year.

Current liability: Debts that must be paid by a firm within a given period of time, such as a year.

Current ratio: The quotient of current assets and current liabilities.

Current yield: The annual dividend per share of stock divided by the current price per share.

cwt.:   Abbreviation for *per hundredweight* or *per one hundred pounds*.

D

Daily interest charge: The amount of interest charged per day on a loan.

Daily overtime: The amount of overtime worked in a day.

Debit card: A card that results in a debit to a bank account when the card is used for a purchase.

Decimal: A number written with a decimal point, such as 4.3 or 7.22.

Decimal equivalent: A decimal that has the same value as a fraction.

Decimal point: The starting point in the decimal system (.).

Decimal system: The numbering system based on powers of 10 and using the 10 one-place numbers 0, 1, 2, 3, 4, 5, 6, 7, 8, and 9, which are called *digits*.

Declining-balance depreciation: An accelerated depreciation method.

(200%) Declining-balance method: An accelerated method of depreciation using twice, or 200% of, the straight-line rate.

Decrease problem (Difference problem):   A percentage problem in which something is taken away from the base. Usually the base must be found.

Decreasing term insurance: A form of life insurance in which the insured pays a fixed premium until age 60 or 65, with the amount of life insurance decreasing periodically.

Deductible: An amount paid by the insured, with the balance of the loss paid by the insurance company.

Deductions: Amounts that are subtracted from the gross earnings of an employee to arrive at the amount of money the employee actually receives.

Defaulting on debt: Failure to pay back a debt.

Dependents: An extra deduction is allowed on income taxes for each dependent.

Deposit slip: A slip for listing all currency and checks that are part of a deposit into a bank account.

Denominator: The number below the line in a fraction. For example, in the fraction $\frac{7}{9}$, 9 is the denominator.

Depreciation: The decrease in value of an asset caused by normal use, aging, or obsolescence.

Depreciable amount: The amount of an asset's value that can be depreciated.

Depreciation schedule: A schedule or table showing the depreciation rate, amount of depreciation, book value, and accumulated depreciation for each year of an asset's life.

Difference (Remainder):   The answer in a subtraction problem.

Differential piece rate: A rate paid per item that depends on the number of items produced.

Digits: One-place numbers in the decimal system. They are 0, 1, 2, 3, 4, 5, 6, 7, 8, and 9.

Discount: (1) To reduce the price of an item. (2) The amount subtracted from the face value of a note to find the proceeds loaned.

Discount broker: A stockbroker who charges a reduced fee to customers (and, generally, reduced services).

Discount date: The last date on which a cash discount may be taken.

Discounting a note: Cashing or selling a note at a bank before the note is due from the maker.

Discount method of interest: A method of calculating interest on a loan by subtracting the interest from the amount of the loan. The borrower receives the amount borrowed less the discounted interest.

Discount period: The discount period is the period from the time of sale of a note to the note's due date.

Discount rate: The discount rate is a percent that is multiplied by the face value and time to find bank discount.

Discover: A credit card that sometimes pays the card holder back a percentage of the amount charged.

Distributive property: The property that states the product of the sum of two numbers equals the sum of the individual products; that is $a(b + c) = ab + ac$.

Dividend: (1) The number being divided by another number in a division problem. (2) A return on an investment; money paid by a company to the holders of stock.

Divisor: The number doing the dividing in a division problem.

Double-declining balance: A method of accelerated depreciation that doubles depreciation in the early years compared to straight-line depreciation.

Double time: Twice the regular hourly rate. A premium often paid for working holidays and Sunday.

Dow Jones Industrial Average: A frequently quoted average price of the stocks of 30 large industrial companies.

doz.:   Abbreviation for *dozen.*

Draw: A draw is an advance on future earnings.

Drawing account: An account from which a salesperson can receive payment against future commissions.

drm.:   Abbreviation for *drum.*

E

ea.:   Abbreviation for *each.*

Effective rate: The true rate of interest.

Effective rate of interest: The true annual percentage rate that can be used to compare loans. It is required by the federal Truth-in-Lending Act.

Electronic banking: Banking activities that take place over a network, such as the World Wide Web.

Electronic commerce: Purchases that take place over a network, such as the World Wide Web.

Electronic funds transfer: Moving money electronically over a network, such as the World Wide Web.

End-of-month dating (EOM):   A system of cash discounts in which the time period begins at the end of the month the invoice is dated. *Proximo* and *prox.* have the same meaning.

Endowment policy: A life insurance policy guaranteeing the payment of a fixed amount of money to a given individual whether or not the insured person lives.

Equation: Two algebraic expressions that are equal to one another.

Escrow account into which monies are paid: *See* impound account.

Exact interest: A method of calculating interest based on 365 days per year.

Executive officers: The top few officers in a corporation.

Expenses: The costs a firm must pay to operate and sell its goods or services.

Extension total: The number of items purchased times the price per unit.

Extra dating (ex., x):   Extra time allowed in determining the net payment date of a cash discount.

F

Face value: The amount shown on the face of a note.

Face value of a bond (Par value of a bond):   The amount the company has promised to repay.

Face value of a policy: The amount of insurance provided by the insurance company.

Factors: Companies that buy accounts receivable.

Factoring: The process of selling accounts receivable for cash.

Fair Labor Standards Act: A federal law that sets the minimum wage and also a 40-hour workweek.

Fair market value: The price for which a piece of property could reasonably be expected to be sold in the market.

FAS (Free alongside ship):   Free alongside the ship on the loading dock.

Federal Insurance Contributions Act (FICA):   An emergency measure passed by Congress in the 1930s that established the so-called social security tax. *See* FICA tax.

Federal Reserve Bank: Today, all banks are part of the Federal Reserve system. The Federal Reserve is our national bank.

Federal Truth-in-Lending Act: An act passed in 1969 that requires all interest rates to be given as comparable percents.

Federal Unemployment Tax Act (FUTA):   An unemployment insurance tax paid entirely by employers to the federal government for administrative costs of federal and state unemployment programs.

FHA loan: A real estate loan that is insured by the Federal Housing Administration, an agency of the federal government.

FICA tax (Social Security tax):   The amount of money deducted from the paychecks of almost all employees, used by the federal government to pay pensions to retired people, survivors' benefits, and disability.

FIFO: A method of inventory accounting in which the first items received are considered to be the first ones shipped.

Finance charge: The difference between the cost of something paid for in installments and the cash price.

Financial ratio: A number found using financial data that is used to compare different companies within the same industry.

Fixed assets: Items owned by a firm that will not be converted to cash within a year.

Fixed liabilities: Items that will not be paid off within a year.

Fixed-rate loan: A loan made at a fixed, stated rate of interest.

Flat-fee checking account: A checking account in which the bank supplies check printing, a bank charge card, and other services for a fixed charge per month.

Floating decimal: A feature on most calculators that positions the decimal point where it should be in the final answer.

FOB (Free on board): A notation sometimes used on an invoice. "Free on board shipping point" means the buyer pays for shipping. "Free on board destination" means the seller pays for shipping.

Foreclose: The process by which a lender takes back the property when payments are not made.

Form 941: The Employer's Quarterly Federal Tax Return form that must be filed by the employer with the Internal Revenue Service.

Form 1040A: The form used by most federal income tax payers.

Form 1040EZ: A simplified version of the 1040A federal income tax form.

Fraction: An indication of a part of a whole. (For example, $\frac{3}{4}$ means that the whole is divided into 4 parts, of which 3 are being considered.)

Frequency distribution table: A table showing the number of times one or more events occur.

Fringe benefits: Benefits offered by an employer, not including salary, that can include medical, dental, life insurance, and day care for employee's children.

Front-end rounding: Rounding so that all digits are changed to zero except the first digit.

Future value: The value, at some future date, of an investment.

G

GI (VA) loan: A loan guaranteed by the Veterans Administration and available only to qualified veterans.

Grace period: The period between the due date of a payment and the time the lending institution assesses a penalty for the payment being late, usually a few days after the payment is due.

Gram: The unit of weight in the metric system. (A nickel weighs about 5 grams.)

Graph: A visual presentation of numerical data.

Gr. gro. (Great gross): Abbreviation for 12 gross ($144 \times 12 = 1728$).

Gro.: Abbreviation for *gross*.

gross: A dozen dozen, or 144 items.

Gross earnings: The total amount of money earned by an employee before any deductions are taken.

Gross loss: *See* absolute loss.

Gross profit: The difference between the amount received from customers for goods and what the firm paid for the goods.

Gross profit on sales: *See* gross profit.

Gross sales: The total amount of money received from customers for the goods or services sold by the firm.

H

Half-year convention: Method of depreciation used for the first year the property is placed in service.

Head of household: An unmarried person who has dependents can use the head of household category when filing income taxes.

High: The highest price reached by a stock during the day.

Homeowner's policy: An insurance policy that covers a home against fire, theft, and liability.

Horizontal analysis: An analysis that shows the amount of any change from last year to the current year, both in dollars and as a percent.

I

Impound account (Escrow account): An account at a lending institution into which taxes and insurance are paid on a monthly basis by a borrower on real estate. The lender then pays the tax and insurance bills from this account when they become due.

Improper fraction: A fraction with a numerator larger than the denominator. (For example, $\frac{7}{5}$ is an improper fraction; $\frac{1}{9}$ is not.)

Incentive rate: A payment system based on the amount of work completed.

Income statement: A summary of all the income and expenses involved in running a business for a given period of time.

Income tax: The tax based on income that both individuals and corporations are required to pay to the federal government and sometimes to a state.

Income tax withholding: Federal income tax that the employer withholds from gross earnings.

Income-to-monthly-payment ratio: A ratio used to determine from an income standpoint whether a prospective borrower meets the lender's qualifications.

Increase problem (Amount problem): A percentage problem in which something has been added to the base. Usually the base must be found.

Index fund: A mutual fund that holds the stocks that are in a particular market index such as the Dow Jones Industrial Average.

Indicator words: Key words that help indicate whether to add, subtract, multiply, or divide.

Individual retirement account (IRA): An account designed to help people prepare for future retirement.

Inflation: Inflation results in a continuing rise in the cost of goods and services. *See* CPI.

Installment loan: A loan that is paid off with a sequence of periodic payments.

Insurance: Individuals and firms purchase insurance from insurance companies to protect them in the event of an unexpected loss.

Insured: A person or business that has purchased insurance.

Insurer: The insurance company.

Intangible assets: Assets such as patents, copyrights, or customer lists that have a value that cannot be immediately converted to cash, unlike jewelry or stocks.

Interest-bearing checking account: A checking account that earns interest.

Interest-in-advance notes: *See* simple discount notes.

Interest: A charge paid for borrowing money or a fee received for lending money.

Interest rate spread: The difference between the interest rate paid to depositors and the rates charged to borrowers by the same lender.

Internal Revenue Service: The branch of the U.S. federal government responsible for collecting taxes.

Inventory: The value of the merchandise that a firm has for sale on the date of balance sheet.

Inventory-to-net-working-capital ratio: Inventory divided by working capital, where working capital is current assets minus current liabilities.

Inventory turnover: The number of times during a certain time period that the average inventory is sold.

Invoice: A printed record of a purchase and sales transaction.

Invoice amount: List price minus trade discounts.

Invoice date: The date an invoice is printed.

Itemized billing method: A billing method that provides an itemization of the customer's charge purchases, but not copies of the original charge receipts.

Itemized deductions: Tax deductions, such as interest, taxes, and medical expenses, that are listed individually on a tax return in order to affect the total amount of taxes payable at the end of the year.

J

Joint return: An income tax return filed by both husband and wife.

K

Kilo-: A prefix used in the metric system to represent 1000.

Kilogram: A unit of weight in the metric system meaning 1000 grams. One kilogram is about 2.2 pounds.

Kilometer: One thousand meters. A kilometer is about .6 mile.

L

Late fees: Fees required because payments were made after a specific due date.

Least common denominator: The smallest whole number that all the denominators of two or more fractions evenly divide into. (For example, the least common denominator of $\frac{3}{4}$ and $\frac{5}{6}$ is 12.)

Level premium: A level premium insurance policy is one with a level premium throughout its life.

Liability: An expense that must be paid by a firm.

LIFO: A method of inventory accounting in which the most recent items received are considered to be the first ones shipped.

Like fractions: Fractions with the same denominator.

Limited liability: A form of protection that shields a company and its shareholders from having to pay large sums of money in the event that the company loses a lawsuit.

Limited-pay life insurance: Life insurance for which premiums are paid for only a fixed number of years.

Line graph: A graph that uses lines to compare numbers.

Liquidity: The ability of a firm or individual to raise cash quickly without being forced to sell assets at a loss.

Liquid assets: Cash or items that can be converted to cash quickly.

List price: The suggested retail price or final consumer price given by the manufacturer or supplier.

Liter: A measure of volume in the metric system. One liter is a little more than one quart.

Loan reduction schedule: *See* repayment schedule.

Long-term liabilities: Money owed by a firm that is not expected to be paid off within a year.

Long-term notes payable: The total of all debts of a firm, other than mortgages, that will not be paid within a year.

Low: The lowest price reached by a stock during the day.

Lowest terms: The form of a fraction if no number except the number 1 divides evenly into both the numerator and denominator.

M

M: Roman numeral for 1000.

MACRS (modified accelerated cost recovery system): A depreciation method introduced as part of the Tax Reform Act of 1986.

Maintenance charge per month: The charge to maintain a checking account (usually determined by the minimum balance in the account).

Maker of a note: A person borrowing money from another person.

Manufacturers: Businesses that buy raw materials and component parts and assemble them into products that can be sold.

Margin: The difference between cost and selling price.

Marital status: An individual can claim married, single or head of household when filing income taxes.

Markdown: A reduction from the original selling price. It may be expressed as a dollar amount or as a percent of the original selling price.

Marketing channels: The path of products and services beginning with the manufacturer and ending with the consumer.

Markup (Margin, Gross profit): The difference between the cost and the selling price.

Markup on cost: Markup that is calculated as a percent of cost.

Markup on selling price: Markup that is calculated as a percent of selling price.

Markup with spoilage: The calculation of markup including deduction for spoiled or unsaleable merchandise.

MasterCard: A credit-card plan (formerly known as Master-Charge).

Maturity value: The amount that a borrower must repay on the maturity date of a note.

Mean: The sum of all the numbers divided by the number of numbers.

Median: A number that represents the middle of a group of numbers.

Medical insurance: Insurance providing medical protection in the event of accident or injury.

Medicare tax: The amount of money deducted from the paychecks of almost all employees, used by the federal government to pay for Medicare.

Memory function: A feature on some calculators that stores results internally in the machine for retrieval and future use.

Merchant batch header ticket: The bank form used by businesses to deposit credit-card transactions.

Meter: A unit of length in the metric system that is slightly longer than 1 yard.

Metric system: A system of weights and measures based on decimals, used throughout most of the world. It is gradually being adopted in the United States.

Milli-: A prefix used in the metric system meaning thousandth. (For example, a milligram is one one-thousandth of a gram.)

Millimeter: One one-thousandth of a meter. There are 25.4 millimeters to an inch.

Mills: A way of expressing a real estate tax rate that is based on thousandths of a dollar.

Minuend: The number from which another number (the subtrahend) is subtracted.

Mixed number: A number written as a whole number and a fraction. (For example $1\frac{3}{4}$ and $2\frac{5}{9}$ are mixed. numbers.)

Mode: The number that occurs most often in a group of numbers.

Modified accelerated cost recovery system: *See* MACRS.

Money order: A document that looks similar to a check and issued by a bank, other financial institution, or a retail store that is often used in place of cash.

Mortgage: A loan on a home.

Mortgages payable: The balance due on all mortgages owed by a firm.

Multiple carrier insurance: The sharing of risk by several insurance companies.

Multiplicand: A number being multiplied.

Multiplication rule: The same nonzero number may be multiplied or divided on both sides of an equation.

Multiplier: A number doing the multiplying.

Mutual fund: A mutual fund accepts money from many different investors and uses it to purchase stocks or bonds of numerous companies.

N

NASDAQ composite index: A commonly quoted stock index composed of the stock prices of several technology companies.

Negative numbers: Numbers that are the opposite of positive numbers.

Net cost: The cost or price after allowable discounts have been taken. *See* net price.

Net cost equivalent: The decimal number derived from the complement of the single trade discount. This number multiplied by the list price gives the net cost.

Net earnings: The difference between gross margin and expenses. After the cost of goods and operating expenses are subtracted from total sales, the remainder is net profit.

Net income: The difference between gross margin and expenses.

Net pay: The amount of money actually received by an employee after deductions are taken from gross pay.

Net payment date: The date by which an invoice must be paid.

Net price: The list price less any discounts. *See* net cost.

Net proceeds: The amount received from the bank for a discounted note.

Net profit: *See* net earnings.

Net sales: The value of goods bought by customers after the value of goods returned is subtracted.

Net worth (Capital, Stockholder's equity, Owner's equity): The difference between assets and liabilities.

No-fault insurance: A guarantee of reimbursement (provided by the insured's own insurance company) for medical expenses and costs associated with an accident no matter who is at fault.

Nominal rate: The interest rate stated in connection with a loan. It may differ from the annual percentage rate.

No scrap value: The value of an item is assumed to be zero at the end of its useful life.

Nonsufficient funds (NSF): When a check is written on an account for which there is an insufficient balance, the check is returned to the depositor for nonsufficient funds.

Notes payable: The value of all notes owed by a firm.

Notes receivable: The value of all notes owed to a firm.

NOW account (Negotiable order or withdrawal): Technically a savings account with special withdrawal privileges. It looks the same and is used the same as a checking account.

Numerator: The number above the line in a fraction. (For example, in the fraction $\frac{5}{8}$, 5 is the numerator.)

O

Odd lot: Fewer than 100 shares of stock.

Open-end-credit: Credit with no fixed number of payments. The consumer continues making payments until no outstanding balance is owed.

Operating expenses (Overhead): Expenses of operating a business. Wages, salaries, rent, utilities, and advertising are examples.

Operating loss: The loss resulting when the selling price is less than the break-even point.

Ordered array: A list of numbers arranged from smallest to largest.

Order of operations: The rules that are used when evaluating long arithmetic expressions.

Ordinary annuity: An annuity whose payments are made at the end of a given period of time.

Ordinary dating: A method for calculating the discount date and the net payment date. Days are counted from the date of the invoice.

Ordinary interest: A method of calculating interest, assuming 360 days per year. *See* banker's rule.

Ordinary life insurance (Whole life insurance, Straight life insurance):   A form of life insurance whereby the insured pays a constant premium until death or retirement, whichever occurs sooner. Upon retirement, monthly payments are made by the company to the insured until the death of the insured.

Overhead: Expenses involved in running a firm. *See* operating expenses.

Over-the-limit fees: Fees charged when the balance on a credit-card account exceeds the account's credit limit.

Overdraft: An event that results when there is not enough money in a bank account to cover a check that is written from that account.

Overtime: The number of hours worked by an employee in excess of 40 hours per week.

Owner's equity: *See* net worth.

P

Part: The result of multiplying the base times the rate.

Partial payment: A payment made on an invoice that is less than the full amount of the invoice.

Partial product: Part of the process of getting the answer in a multiplication problem.

Par value of a bond: *See* face value of a bond.

Passbook account: A type of savings account for day-in and day-out savings.

Payee: The person who lends money and will receive repayment on a note.

Payer of a note: A person borrowing money from another person. *See* maker of a note.

Payroll: A record of the hours each employee of a firm worked and the amount of money due each employee for a given pay period.

Payroll card: A card maintained by employers showing the name of employee, dates of pay period, days, times, and hours worked.

Payroll ledger: A chart showing all payroll information.

Percent (Rate):   Some parts of a whole: hundredths, or parts of a hundred. (For example, a percent is one one-hundredth. Two percent means two parts of a hundred, or $\frac{2}{100}$.)

Percentage method: A method of calculating income tax withholding that is based on percentages.

Per debit charge: A charge per check (usually continues regardless of the number of checks written).

Periodic inventory: A physical inventory taken at regular intervals.

Perpetual inventory: A continuous inventory system normally involving a computer.

Personal account: The type of checking account used by individuals.

Personal exemption (Exemption):   A deduction allowed each taxpayer for each dependent and the taxpayer himself or herself.

Personal identification number (PIN):   A lettered or numbered code that allows a person with a credit or debit card to gain access to credit or cash.

Personal property: Property such as a boat, a car, or a stereo.

Piecework: A method of pay by which an employee receives so much money per item produced or completed.

Plant assets: *See* fixed assets.

Point-of-sale terminal: A machine that allows a customer to make purchases using a credit or debit card.

Policy: A contract outlining the insurance agreement between an insured and an insurance company.

Policy limits: The maximum amount that an insurance company will pay as defined in the policy.

Postdating: Dating in the future; on an invoice, "AS OF" dating.

pr.:   Abbreviation for *pair*.

Preferred stock: A type of stock that offers investors certain rights over holders of common stock.

Premium: The amount of money charged for insurance policy coverage.

Premium factor: A factor used to adjust an annual insurance premium to semiannually, quarterly, or monthly.

Premium payment: An additional payment for extra service.

Present value: The amount that must be deposited today to generate a specific amount at a specific date in the future.

Price-earnings (PE) ratio: The price per share divided by the annual net income per share of stock.

Prime interest rate: The interest rate banks charge their largest and most financially secure borrowers.

Prime number: A number that can be divided without remainder by exactly two distinct numbers: itself and 1.

Principal: The amount of money either borrowed or deposited.

Privately held corporation: A corporation that has relatively few owners, or perhaps a single owner. Its stock is not traded on a large exchange such as the New York Stock Exchange.

Proceeds: The amount of money a borrower receives after subtracting the discount from the face value of a note.

Product: The answer in a multiplication problem.

Promissory note: A business document in which one person agrees to repay money to another person within a specified amount of time and at a specified rate of interest in exchange for money borrowed.

Proper fraction: A fraction in which the numerator is smaller than the denominator. (For example, $\frac{2}{3}$ is a proper fraction; $\frac{9}{5}$ is not.)

Property damage insurance: A type of automobile insurance that pays for damages that the insured causes to the property of others.

Proportion: A mathematical statement that two ratios are equal.

Proprietorship: Stockholder's equity.

Proximo (Prox.):   *See* end-of-month dating.

Publicly held corporations: Corporations that are owned by the public and have stock that trades freely.

Purchase invoice: A list of items purchased, prices charged for the items, and payment terms.

Q

Qualifying for a loan: A person applying for a loan is said to qualify if his or her credit history, income, and financial statement satisfy the requirements of the lending institution.

Quick ratio: The quotient of liquid assets and current liabilities.

Quota: An expected level of production. A premium may be paid for surpassing quota.

Quotient: The answer in a division problem.

R

Rate: Parts of a hundred. *See* percent.

Rate of interest: The percent of interest charged on a loan for a certain time period.

Ratio: A comparison of two (or more) numbers, frequently indicated by a common fraction.

Real estate: Real property such as a home or a parcel of land.

Receipt-of-goods dating (ROG): A method of determining cash discounts in which time is counted from the date that goods are received.

Reciprocal: A fraction formed from a given fraction by interchanging the numerator and denominator.

Reconciliation: The process of checking a bank statement against the depositor's own personal records.

Recourse: Should the maker of a note not pay, the bank may have recourse to collect from the seller of the note.

Recovery classes: Classes used to determine depreciation under the modified accelerated cost recovery system.

Recovery period: The life of property depreciated under the accelerated cost recovery system.

Recovery year: The year of life of an asset when using the MACRS method of depreciation.

Reduced net profit: The situation that occurs when a markdown decreases the selling price to a point that is still above the break-even point.

Refinance: A borrower can go to a lender and refinance their existing loan with a different interest rate, period, and payment.

Regulation DD: A Federal Reserve System document that specifies how interest paid to savers is to be calculated.

Regulation Z: A Federal Reserve System document that implements the Truth-in-Lending Act.

Renter's coverage: Insurance that covers only the possessions of a renter and not the house or apartment in which the possessions are kept.

Repayment schedule: A schedule showing the amount of payment going toward interest and principal and the balance of principal remaining after each payment is made.

Repeating decimals: Decimal numbers that do not terminate, but that contain numbers that repeat themselves.

Replacement cost: The cost of replacing a property that is completely destroyed.

Repossess: The taking back of property by a lender when payments have not been made to the lender.

Residual value: *See* scrap value.

Restricted endorsement: A signature or imprint on the back of a check that limits the ability to cash the check.

Retailer: A business that buys from the wholesaler and sells to the consumer.

Retail method: A method used to estimate inventory value at cost that utilizes both cost and retail amounts.

Returned check: A check that was deposited and then returned due to lack of funds in the payer's account.

Return on average total assets: Net income divided by average total assets.

Returns: The total value of all goods returned by customers.

Revolving charge account: A charge account that never has to be paid off.

Roth IRA: Contributions to a Roth Individual Retirement Account (Roth IRA) are not deductible when made. However, funds in the account grow tax free and withdrawals are not taxed once the account holder reaches a certain age.

Round lot: A multiple of 100 shares of stock.

Rounding off: The reduction of a number with more decimals to a number with fewer decimals.

Rounding whole numbers: Reduction of the number of nonzero digits in a whole number.

Rule of 78: A method of calculating a partial refund of interest that has already been added to the amount of a loan. This calculation is done when the loan is paid off early.

S

Salary: A fixed amount of money per pay period.

Salary plus commission: Earnings based on a fixed salary plus a percent of all sales.

Sale price: The price of an item after markdown.

Sales invoice: *See* purchase invoice.

Sales tax: A tax placed on sales to the final consumer. The tax is collected by the state, county, or local government.

Salvage value: *See* scrap value.

Savings account: An interest paying account that allows day-to-day savings and withdrawals.

Schedule 1: The part of the 1040A federal tax form that is used to list all interest and dividends.

Scrap value (Salvage value): The value of an asset at the end of its useful life. For depreciation purposes, this is often an estimate.

SDI deduction: State disability insurance pays the employee in the event of disability and is paid for by the employee.

Self-employed people: People who work for themselves instead of for the government or for a private company.

Series discount: *See* chain discount.

Shift differential: A premium paid for working a less desirable shift, such as the swing shift or the graveyard shift.

Simple interest note: A note in which
interest = principal × interest rate × time in years.

Simple discount note: A note in which the interest is deducted from the face value in advance.

Simple interest: Interest received on only the principal.

Single discount equivalent: A series, or chain, discount expressed as a single discount.

Single return: An income tax return filed by a single person.

Sinking fund: A fund set up to receive periodic payments in order to pay off a debt at some time in the future.

sk.:   Abbreviation for *sack*.

Sliding scale: Commissions that are paid at increasing levels as sales increase.

Social Security tax: *See* FICA tax.

Special endorsement: A signature on the back of a check that passes the ownership of the check to someone else.

Specific identification method: An inventory valuation method that identifies the cost of each item.

Split-shift premium: A premium paid for working a split shift, for example, for an employee who is on 4 hours, off 4 hours, and then on 4 hours.

Square root: One of two equal positive factors of a number.

Stafford loan: A loan taken out by college students to help pay tuition.

Standard deduction: A tax preparer may use the higher of the itemized deductions or the standard deduction established by the government.

State income tax: An income tax that is paid to a state government on income earned in that state.

Stated rate: The interest rate stated in connection with a loan. It may differ from the annual percentage rate.

Statement: Usually sent out monthly by the bank, a list of all charges and deposits made against and to a checking account.

Statistics: Refers both to data and to the techniques used in analyzing data.

Stock: A form of ownership in a corporation that is measured in units called *shares*.

Stockbroker: A person who buys and sells stock at the stock exchange.

Stock exchange: An institution where stock shares are bought and sold.

Stockholders: Individuals who own stock in a particular company.

Stockholder's equity: *See* net worth.

Stock ratios: Ratios calculated from the financial statements of a company—used to determine the financial health of the firm.

Stock turnover: *See* inventory turnover.

Stop payment: A request from a depositor that the bank not honor a check that the depositor has written.

Straight commission: A salary that is a fixed percent of sales.

Straight life insurance: *See* ordinary life insurance.

Straight-line depreciation: A depreciation method in which depreciation is spread evenly over the life of the asset.

Substitution: Method for checking the solution to an equation.

Subtrahend: The number being subtracted or taken away in a subtraction problem.

Sum: The total amount; the answer in addition.

Sum-of-the-years'-digits method: An accelerated depreciation method that results in larger amounts of depreciation taken in earlier years of an asset's life.

T

T-bill: A short-term note issued by the federal government that pays interest to the note holder. Issuing T-bills allows the federal government to raise cash without having to borrow the money from a bank and pay interest.

Tangible assets: Assets such as a car, machinery, or computers.

Taxable income: Adjusted income subject to taxation.

Tax deduction: Any expense that the Internal Revenue Service allows the taxpayers to subtract from adjusted gross income.

Taxes: A sum of money that is paid by an individual based on the individual's yearly income. Many states also require payment of income taxes. Money raised by taxes helps pay for a variety of programs offered by the federal or state government.

Telephone transfer account: An interest-bearing checking account into which funds may be transferred by the customer over the telephone.

Term insurance: A form of life insurance providing protection for a fixed length of time.

Term of an annuity: The length of time that an annuity is in effect.

Term of a note: The length of time between the date a note is written and the date the note is due.

Terms: The area of an invoice where cash discounts are indicated if any are offered. The words "terms discount" are often used in place of "cash discount."

Territorial ratings: Ratings used by insurance companies that describe the quality of fire protection in a specific area.

Time-and-a-half rate: One and one-half times the normal rate of pay for any hours worked in excess of 40 per week.

Time card: A card filled out by an employee that shows the number of hours worked by that employee.

Time deposit account: A savings account in which the depositor agrees to leave money on deposit for a certain period of time.

Time rate: Earnings based on hours worked, not for work accomplished.

Total installment cost: Includes the down payment plus the sum of all payments.

Total revenue: The total of all revenue from all sources.

Trade discount: A discount offered to businesses. This discount is expressed either as a single discount (like 25%) or a series discount (like 20/10) and is subtracted from the list price.

True rate of interest: *See* effective rate of interest.

Turnover at cost: The cost of goods sold, divided by the average inventory at cost.

Turnover at retail: Sales, divided by the average inventory at retail.

U

Underinsured motorist: A motorist who does not carry enough insurance to cover the costs of an accident.

Underwriters: Term applied to any insurer. Usually associated with an insurance company.

Unearned interest: Interest that a company has received but has not yet earned so that it is not shown in revenues.

Uniform product code (UPC):   The series of black vertical stripes seen on products in stores that cashiers scan. Also called the *bar code.*

Uninsured motorist insurance: Insurance coverage that covers the insured when involved in an accident with a driver who is not insured.

United States Rule: The rule by which a loan payment is first applied to the interest owed, with the balance used to reduce the principal amount of the loan.

Unit price: The cost of one item.

Units-of-production: A depreciation method by which the number of units produced determines the depreciation allowance.

Universal life policy: A policy whose premiums flow into a general account from which the insurance company makes investments.

Unlike fractions: Fractions with different denominators.

Unpaid-balance method: A method used to calculate interest on open-end credit accounts.

Useful life: The estimated life of an asset. The Internal Revenue Service gives guidelines of useful life for depreciation purposes.

V

Valuation of inventory: Determining the value of merchandise in stock. Four common methods are specific-identification, average cost, FIFO, and LIFO.

Variable: A letter that stands for a number.

Variable commission: A commission whose rate depends on the total amount of the sales.

Variable interest rate loan: A loan on which the interest rate can go up or down.

Variable life policy: A policy most of whose premiums the insured may earmark for one or more separate investment funds.

Verbal form: Word form (the form of numbers expressed in words).

Vertical analysis: The listing of each important item on an income statement as a percent of total net sales or each item on a balance sheet as a percent of total assets.

Visa: A credit-card plan (formerly known as Bank Americard).

W

Wage: A rate of pay expressed as a certain amount of dollars per hour.

Wage bracket method: A method of calculating income tax withholding that is based on tables that list income ranges.

Weighted average method: A method for calculating the arithmetic mean for data where each value is weighted (or multiplied) according to its importance.

Whole life insurance: *See* ordinary life insurance.

Whole number: A number made up of digits to the left of the decimal point.

Wholesaler: A business that buys directly from the manufacturer or other wholesalers and sells to the retailer.

Withholding allowance: An allowance for the employee, spouse, and dependents that determines the amount of withholding tax taken from gross earnings.

Withholding tax: The money withheld from an employee's paycheck and deposited to the account of the employee with the federal or state government to cover the amount of income tax owed by the employee.

With recourse: An understanding that the seller of a note is responsible for payment of the note if the original maker of the note does not make payment. The note is sold with recourse.

Worker's compensation: Insurance purchased by companies to cover employees against work-related injuries.

W-2 form: The wage and tax statement given to the employee each year by the employer.

W-4 form: A form usually completed at the time of employment, on which an employee states the number of withholding allowances being claimed.

Y

Youthful operator: A driver of a motor vehicle who is under a certain age, usually 25.

Index

Acknowledgments

Page 14, Parachutists © PhotoDisc

Page 21, Iwo Jima flag raising © Joe Rosenthal/Corbis

Page 21, World War II veterans © Courtesy of Beth Anderson

Page 32, SuperBowl XXXVIII © Richard Carson/Reuters/Corbis

Page 50, http://www.homedepot.com/ © 2004 Homer TLC, Inc. and The Home Depot U.S.A., Inc.

Page 99, New England Patriots' quarterback, Tom Brady © Marc Serota/Reuters/Corbis

Page 114, Women in the military © AP Photo

Page 124, http://www.century21.com/ © 2004 Century 21 Real Estate Corporation

Page 153, http://www.jacksonandperkins.com/ © 2004 Harry and David

Page 197, Ask Marilyn column © 2002 Reprinted with permission from PARADE, September 22, 2002

Page 242, http://www.biggeorge.com/ © 2004 Big George Foreman's Place

Pages 638, http://www.gm.com/ © 2004 General Motors Corporation

Page 371, Bank vault © Digital Vision RF PP

Page 380, http://www.bankofamerica.com/ © 2004 Bank of America Corporation

Page 395, Roman Rodriquez monitoring his investments at his computer © PhotoDisc

Page 402, Roman Rodriquez retired and fishing with his grandson © PhotoDisc

Page 404, Saudi Arabian oil refinery © Langevin Jacques/Corbis Sygma

Page 406, Tish Baker working toward retirement © PhotoDisc

Page 424, Albertsons grocery store © David Young-Wolff/Photo Edit

Page 430, http://www.arc.losrios.edu/ © 2004 American River College. All Rights Reserved.

Page 436, New York City/Statue of Liberty skyline © PhotoDisc

Page 445, Credit cards © Digital Vision (PP)

Page 496, http://www.gmacfs.com/ © 2004 General Motors Acceptance Corporation

Page 529, Office building on fire © PhotoDisc

Page 535, Fire Department fighting a building fire © PhotoDisc (PP)

Page 539, Car accident caused by an ice storm © AP Photo/ J. Pat Carter

Page 551, http://www.barbie.everythinggirl.com/ BARBIE® is a trademark owned by and used with permission from Mattel, Inc. © 2005 Mattel, Inc. All Rights Reserved.

Percent Formula	Part = Base $\times$ Rate or $P = B \times R$ or $P = BR$
Markup Formula	Cost + Markup = Selling price or $C + M = S$
Converting Markup Percent on Cost to Selling Price	$\dfrac{\text{Markup on}}{\text{selling price}} = \dfrac{\text{Markup on cost}}{100\% \ + \ \text{Markup on cost}}$
Converting Markup Percent on Selling Price to Cost	$\dfrac{\text{Markup on}}{\text{cost}} = \dfrac{\text{Markup on selling price}}{100\% \ - \ \text{Markup on selling price}}$

Terms Associated with Loss

Stock Turnover	$\text{Turnover at retail} = \dfrac{\text{Retail sales}}{\text{Average inventory at retail}}$ $\text{Turnover at cost} = \dfrac{\text{Cost of goods sold}}{\text{Average inventory at cost}}$
Simple Interest	The *simple interest*, I, on a principal of P dollars at a rate of interest R per year for T years is given by $I = PRT$

Number of Days in Each Month

31 Days		30 Days	28 Days
January	August	April	February
March	October	June	(29 days in leap year)
May	December	September	
July		November	

Types of Interest	The method for forming the *time fraction* used for T in the formula $I = PRT$ is summarized as follows:

The *numerator* is the *exact number* of days in a loan period.
The *denominator* is one of the following:
- *exact interest* assumes 365 days in a year and uses 365 as denominator.
- *ordinary*, or *banker's interest* assumes 360 days in a year and uses 360 as denominator.

Interest, Principal, Rate, and Time Formulas	Interest: $I = PRT$ Principal: $P = \dfrac{I}{RT}$ Rate: $R = \dfrac{I}{PT}$
	Time: Time in days $= \dfrac{I}{PR} \times 360$ Time in years $= \dfrac{I}{PR}$
Maturity Value	The *maturity value* M of a loan having a principal P and interest I is given by $M = P + I$.

Simple Interest and Simple Discount

	Simple Interest Note	Simple Discount Note
	I = Interest	B = Discount
	P = Principal (face value)	P = Proceeds
	R = Rate of interest	D = Discount rate
	T = Time, in years or fraction of a year	T = Time, in years or fraction of a year
	M = Maturity value	M = Maturity value
Face value	Stated on note	Same as maturity value
Interest charge	$I = PRT$	$B = MDT$
Maturity value	$M = P + I$	Same as face value
Amount received by borrower	Face value or principal	Proceeds: $P = M - B$
Identifying phrases	Interest at a certain rate	Discounted at a certain rate
	Maturity value greater than face value	Proceeds
		Maturity value equal to face value
True annual interest rate	Same as stated rate R	Greater than stated rate D

Compound Interest	If P dollars are deposited at a rate of interest t per period for n periods, then the *compound amount* M, or the fir amount of deposit, is

$$M = P(1 + i)^n$$

The interest earned I is

$$I = M - P$$

(Use the compound interest table.)